# Lie Algebras, Lie Superalgebras, Vertex Algebras and Related Topics

Proceedings of Symposia in
PURE MATHEMATICS

Volume 92

# Lie Algebras, Lie Superalgebras, Vertex Algebras and Related Topics

2012–2014 Southeastern Lie Theory Workshop Series

Categorification of Quantum Groups and Representation Theory
April 21–22, 2012, North Carolina State University

Lie Algebras, Vertex Algebras, Integrable Systems and Applications
December 16–18, 2012, College of Charleston

Noncommutative Algebraic Geometry and Representation Theory
May 10–12, 2013, Louisiana State University

Representation Theory of Lie Algebras and Superalgebras
May 16–17, 2014, University of Georgia

Kailash C. Misra
Daniel K. Nakano
Brian J. Parshall
Editors

American Mathematical Society
Providence, Rhode Island

2010 *Mathematics Subject Classification.* Primary 17B37, 17B55, 17B56, 17B65, 17B67, 17B69, 20C08, 20C11, 20G05, 20G42.

---

**Library of Congress Cataloging-in-Publication Data**

Names: Misra, Kailash C., 1954- editor. — Nakano, Daniel K. (Daniel Ken), 1964- editor. — Parshall, Brian, 1945- editor. Title: Lie algebras, lie superalgebras, vertex algebras, and related topics : Southeastern Lie Theory Workshop Series 2012-2014 : Categorification of Quantum Groups and Representation Theory, April 21-22, 2012, North Carolina State University : Lie Algebras, Vertex Algebras, Integrable Systems and Applications, December 16-18, 2012, College of Charleston : Noncommutative Algebraic Geometry and Representation Theory, May 10-12, 2013, Louisiana State University : Representation Theory of Lie Algebras and Lie Superalgebras, May 16-17, 2014, University of Georgia / Kailash C. Misra, Daniel K. Nakano, Brian J. Parshall, editors. Description: Providence, Rhode Island : American Mathematical Society, [2016] — Series: Proceedings of symposia in pure mathematics ; volume 92 — Includes bibliographical references. Identifiers: LCCN 2015043322 — ISBN 9781470418441 (alk. paper) Subjects: LCSH: Lie algebras–Congresses. — Lie superalgebras–Congresses. — Vertex operator algebras–Congresses. — AMS: Nonassociative rings and algebras – Lie algebras and Lie superalgebras – Quantum groups (quantized enveloping algebras) and related deformations. msc — Nonassociative rings and algebras – Lie algebras and Lie superalgebras – Homological methods in Lie (super)algebras. msc — Nonassociative rings and algebras – Lie algebras and Lie superalgebras – Cohomology of Lie (super)algebras. msc — Nonassociative rings and algebras – Lie algebras and Lie superalgebras – Infinite-dimensional Lie (super)algebras. msc — Nonassociative rings and algebras – Lie algebras and Lie superalgebras – Kac-Moody (super)algebras; extended affine Lie algebras; toroidal Lie algebras. msc — Nonassociative rings and algebras – Lie algebras and Lie superalgebras – Vertex operators; vertex operator algebras and related structures. msc — Group theory and generalizations – Representation theory of groups – Hecke algebras and their representations. msc — Group theory and generalizations – Representation theory of groups – $p$-adic representations of finite groups. msc — Group theory and generalizations – Linear algebraic groups and related topics – Representation theory. msc — Group theory and generalizations – Linear algebraic groups and related topics – Quantum groups (quantized function algebras) and their representations. msc Classification: LCC QA252.3 .L5534 2016 — DDC 512/.482–dc23 LC record available at http://lccn.loc.gov/2015043322

---



∞ The paper used in this book is acid-free and falls within the guidelines
established to ensure permanence and durability.
Visit the AMS home page at `http://www.ams.org/`

10 9 8 7 6 5 4 3 2 1      21 20 18 19 17 16

# Contents

# Preface

Algebraic, analytic and geometric representations of Lie algebras, quantum groups and related algebraic structures have become a comprehensive and mainstream research area in mathematics with numerous applications in mathematics and theoretical physics. Research in representation theory includes quantized enveloping algebras and their connections with representations of algebraic and finite groups, quantum function algebras, Kac-Moody Lie algebras, Hecke algebras, vertex (operator) algebras, Hall algebras, A-infinity algebras, quivers, cluster algebras, Hopf algebras, and Khovanov-Lauda-Rouquier algebras. In particular, representation theory of quantized Kac-Moody Lie algebras and their categorical and geometric constructions have taken the lead not only within Lie theory but also in other areas of mathematics and physics such as combinatorics, group theory, number theory, integrable systems, partial differential equations, topology and conformal field theory.

The notion of categorification has been prevalent in the works of I. Frenkel, Ariki, Grojnowski, Brundan and Kleshchev since Chuang and Rouquier systematically developed the theory and showed that it could be a powerful tool in proving significant results. The ideas and methods soon brought about the discovery of Khovanov-Lauda-Rouquier algebras to categorify quantum groups and their representations. Moreover, this categorical approach was used in proving the Kazhdan-Lusztig positivity conjecture using Soergel bimodules. It is anticipated that these methods will lead to substantial progress in the field. The workshop at North Carolina State University in April 2012 brought together specialists in these areas to explore these topics in more depth.

In the late 1970s, Lepowsky-Wilson and I. Frenkel-Kac gave constructions of representations of affine Kac-Moody Lie algebras in terms of certain differential operators on the space of polynomials in infinitely many variables, that were recognized as the so-called vertex operators appearing in string theory. These constructions spurred the development of the field in various directions. One such application is related to combinatorial identities, including a representation-theoretic proof of the Rogers-Ramanujan type identities. Another is a beautiful connection to integrable systems such as the KP hierarchy, developed by the Kyoto school. The mysterious connections to the largest sporadic simple finite group (known as the "Monster") and to modular forms led R. Borcherds to introduce the notion of a vertex algebra. It turns out that this mathematical notion is essentially equivalent to the physical notion of a chiral algebra in conformal field theory. At the same time, vertex algebras provide a natural framework for the representation theory of infinite-dimensional Lie algebras and generalizations. The $q$-analog of the Frenkel-Kac construction given by Frenkel and Jing motivated the consideration of "quantum" vertex algebras, which are expected to play the same role with respect to quantum affine algebras. The workshop at the College of Charleston in December 2012 focused on these topics.

In the past few decades, the interactions between noncommutative geometry and representation theory have led to new and interesting developments in these areas. In his celebrated ICM talk Drinfeld laid the foundation for the interplay between the geometry of Poisson structure on Lie groups and their homogeneous spaces, and the representation theory of the related quantum function algebras. Representation theory also plays a fundamental role in related approaches to noncommutative algebraic geometry via Calabi–Yau algebras, superpotential algebras and quiver algebras. Calabi–Yau algebras and categories are nowadays an important ingredient in the categorification theory of cluster algebras. These subjects also have connections to geometric representation theory, whose main theme is the use of (derived) categories of sheaves on various varieties to describe categories of representations. An early success was the use of perverse sheaves in the proof of the Kazhdan–Lusztig conjectures. Since then, perverse sheaves have led to important results in the representation theory of finite reductive groups, Hecke algebras, affine Lie algebras, and quantum groups. More recently, derived categories of coherent sheaves on varieties such as the nilpotent cone and the Springer resolution have begun to play a role, especially in the context of the geometric Langlands program, and this has led to connections between noncommutative geometry and algebra. The workshop at Louisiana State University in May 2013 focused on research in this direction.

The workshop at the University of Georgia in May 2014 featured new developments in the representation theory of Lie superalgebras and their connections with geometry and Lie theory. In recent work, Boe, Kujawa and Nakano have utilized methods from tensor triangular geometry to reveal the Balmer spectrum for classical Lie superalgebras. These underlying spaces are prime candidates for applying methods in derived algebraic geometry. Progress in this direction, along with open conjectures, was presented by Kujawa in a series of lectures. Another major development has been the formulation of character formulas for the Lie superalgebra in type A by Brundan and the recent verification by Cheng, Lam and Wang via canonical basis and super duality. In another series of lectures, Wang outlined the general program and presented new work on this question for other classical Lie superalgebras.

Six years ago the three editors established a consortium called the "Southeastern Lie Theory Network" to enhance regional research collaboration and provide a stronger educational environment for graduate students and junior researchers. The editors also initiated an annual (three to four day) workshop series in Lie theory. The aim of these workshops was to bring together senior and junior researchers as well as graduate students to build and foster cohesive research groups in the region. Each workshop included expository talks by senior researchers and discussion sessions for junior researchers and graduate students with the purpose of communicating new ideas and results in the field. These workshops have been supported by the National Science Foundation and the host universities in the region. Since the inception of the consortium, eight workshops have been held. Each of these workshops have been well attended. The Proceedings of the first three workshops were published in this book series (volume 86) in 2012. The plenary speakers in the following four workshops during 2012–2014 were invited to contribute to this proceedings. Most of the articles presented in this book are self-contained. The survey articles, by Martin Schlichenmaier and Oleg K. Sheinman are accessible to a wide audience of readers.

The editors would like to take this opportunity to thank the conference participants, the contributors, and the editorial offices of the American Mathematical Society for making this volume possible.

Kailash C. Misra
Daniel K. Nakano
Brian J. Parshall

North Carolina State University, April 2012

College of Charleston, December 2012

Louisiana State University, May 2013

University of Georgia, May 2014

Proceedings of Symposia in Pure Mathematics
Volume **92**, 2016
http://dx.doi.org/10.1090/pspum/092/01583

# Modular affine vertex algebras and baby Wakimoto modules

Tomoyuki Arakawa and Weiqiang Wang

ABSTRACT. We develop some basic properties such as $p$-centers of affine vertex algebras and free field vertex algebras in prime characteristic. We show that the Wakimoto-Feigin-Frenkel homomorphism preserves the $p$-centers by providing explicit formulas. This allows us to formulate the notion of baby Wakimoto modules, which in particular provides an interpretation in the context of modular vertex algebras for Mathieu's irreducible character formula of modular affine Lie algebras at the critical level.

## 1. Introduction

Let $\mathbb{K}$ be an algebraically closed field of prime characteristic $p$. Denote by $U = \mathbb{K} \otimes_{\mathbb{Z}} U_{\mathbb{Z}}$, where $U_{\mathbb{Z}}$ is the Kostant-Garland $\mathbb{Z}$-form (including divided powers) of the universal enveloping algebra of $\mathfrak{g}$. Mathieu [**Ma**] established a character formula for the irreducible highest weight $U$-module $L(-\rho)$ at the critical level (see (5.3)), which can be rephrased as that the Wakimoto module of highest weight $-\rho$ over the complex field $\mathbb{C}$ remains irreducible over $U$ after reduction modulo $p$. Mathieu also gave a character formula for $\mathfrak{l}(-\rho)$ (and also for $L((p-1)\rho)$); see (5.1)-(5.2). Here $\mathfrak{l}(-\rho)$ denotes the irreducible quotient $\mathfrak{g}$-module of the Verma $\mathfrak{g}$-module of high weight $-\rho$, which can be regarded as an irreducible module over the restricted enveloping algebra $\mathfrak{u}_0(\mathfrak{g})$ (and $\mathfrak{u}_0(\mathfrak{g}) \subset U$). These two irreducible character formulas are equivalent by the Steinberg tensor product theorem and noting that $(p-1)\rho$ is a restricted weight.

Modular vertex algebras (i.e., vertex algebras in prime characteristic) were first considered in [**BR**] by Borcherds and Ryba in their study of modular moonshine. This paper is motivated by putting Mathieu's result in a proper context of modular Lie algebras and modular vertex algebras (where the algebra $U$ plays no role). We formulate the notion of $p$-centers for vertex algebras associated to Heisenberg algebras, affine algebras, and some other free fields, and this gives rise to corresponding $p$-restricted vertex algebras. We show that the $p$-centers and the state-field correspondence for these vertex algebras are compatible in a simple manner; cf. Proposition 2.6.

Wakimoto modules (over $\mathbb{C}$) were introduced by Wakimoto [**Wak**] for $\mathfrak{sl}_2$ and then by Feigin and E. Frenkel for general semisimple Lie algebras [**FF**]. Wakimoto modules have played a fundamental role in the affine vertex algebra setting and applications to the geometric Langlands program, cf. [**Fr1**, **Fr2**]. The construction of Wakimoto modules relies on the Wakimoto-Feigin-Frenkel homomorphism $\mathtt{w}$ from an affine vertex algebra to a bosonic free field vertex algebra. As a main result of this note we show that $\mathtt{w}$ (over the field $\mathbb{K}$) preserves the $p$-centers, and indeed we provide explicit formulas for the restriction of $\mathtt{w}$ on the $p$-center. This allows us to formulate a notion of baby Wakimoto modules, which is analogous to the more familiar notion of baby Verma modules for modular Lie algebras. Now Mathieu's result can be restated that the baby Wakimoto module of highest weight $-\rho$ is irreducible as module over $\mathfrak{g}$ or over $\mathfrak{u}_0(\mathfrak{g})$ (that is, it coincides with $\mathfrak{l}(-\rho)$ in the above notation).

This paper is organized as follows. In Section 2, we prove some basic properties of the modular affine vertex algebras including the $p$-centers. In Section 3, we describe the $p$-centers of the Heisenberg vertex algebra and of a symplectic bosonic vertex algebra. We formulate the main construction of the baby Wakimoto modules. In Section 4, we establish the formulas for the WFF homomorphism on the $p$-center of the affine vertex algebra. In Section 5, we give a reformulation of Mathieu's main result in terms of the irreducibility of the baby Wakimoto module of highest weight $-\rho$. We end with some conjectures and open problems on further development of modular representation theory of affine Lie algebras.

**Acknowledgments.** We have been working on this project on and off since 2007. The results were presented in the Taitung Workshop on "Group theory, VOA and algebraic combinatorics", Taiwan, in March 2013, organized by C.-H. Lam. There is some overlap of our work with a recent paper by Li and Mu [**LM**], where one can find more references on modular vertex algebras in recent years. The first author is partially supported by JSPS KAKENHI Grant Numbers 25287004, 26610006. The second author is partially supported by an NSF grant DMS-1405131.

## 2. Modular affine algebras and modular vertex algebras

**2.1. Affine Lie algebra in prime characteristic.** Let $\bar{\mathfrak{g}}$ be a finite-dimensional semisimple Lie algebra, which is a Lie algebra of a simply connected algebraic group $\bar{G}$ over an algebraically closed field $\mathbb{K}$ of characteristic $p > 0$. Then $\bar{\mathfrak{g}}$ is a restricted Lie algebra (also called a $p$-Lie algebra) with $p$-power map denoted by $-^{[p]}$; cf. [**Jan**] for a review of modular Lie algebras. Moreover, $\bar{\mathfrak{g}}$ affords a non-degenerate bilinear form $\langle \cdot, \cdot \rangle$, which induces a linear isomorphism $\bar{\mathfrak{g}} \to \bar{\mathfrak{g}}^*$. We fix a Chevalley basis $h_i (1 \leq i \leq \ell), e_\alpha, f_\alpha (\alpha \in \bar{\Delta}^+)$ of $\bar{\mathfrak{g}}$, where $\bar{\Delta}^+$ is a set of positive roots for $\bar{\mathfrak{g}}$ corresponding to a set of simple roots $\bar{\Pi} = \{\alpha_1, \ldots, \alpha_\ell\}$. We further write $e_i = e_{\alpha_i}, f_i = f_{\alpha_i}$. We denote by $\bar{B}$ (respectively, $\bar{B}_-$) the Borel subgroup of $\bar{G}$ whose Lie algebra $\bar{\mathfrak{b}}$ (respectively, $\bar{\mathfrak{b}}_-$) is spanned by root vectors from $\bar{\Delta}^+$ (respectively, $\bar{\Delta}^- = -\bar{\Delta}^+$).

We consider the affine Lie algebra

$$\mathfrak{g} \cong L\bar{\mathfrak{g}} \oplus \mathbb{K}c$$

where $L\bar{\mathfrak{g}} \cong \mathbb{K}[t, t^{-1}] \otimes \bar{\mathfrak{g}}$. We shall write $x_n = t^n \otimes x$ for $x \in \bar{\mathfrak{g}}$ and $n \in \mathbb{Z}$. Then $\bar{\mathfrak{g}}$ is naturally a Lie subalgebra of $\mathfrak{g}$ by the identification $1 \otimes \bar{\mathfrak{g}} \cong \bar{\mathfrak{g}}$. We denote by $h^\vee$ the dual Coxeter number for the affine Lie algebra $\mathfrak{g}$.

A Cartan subalgebra $\mathfrak{h}$ of the affine Lie algebra $\mathfrak{g}$ is

$$\mathfrak{h} = \bar{\mathfrak{h}} + \mathbb{K}c$$

and a Borel subalgebra of $\mathfrak{g}$ is $\mathfrak{b} = \mathbb{K}c + t\mathbb{K}[t] \otimes \bar{\mathfrak{g}} + \bar{\mathfrak{b}}$ with nilradical $\mathfrak{n} = t\mathbb{K}[t] \otimes \bar{\mathfrak{g}} + \bar{\mathfrak{n}}$, so that $\mathfrak{g} = \mathfrak{n}^- \oplus \mathfrak{h} \oplus \mathfrak{n}$. Denote by $\Delta_+$ the set of positive roots associated to $\mathfrak{n}$, and by $\Delta_+^{\mathrm{re}}$ the subset of real roots in $\Delta_+$. Let $\mathfrak{g}^*$ denote the restricted dual of $\mathfrak{g}$ associated to the root space decomposition of $\mathfrak{g}$.

Denote by $\bar{T} \subset \bar{G}$ the maximal torus with Lie algebra $\bar{\mathfrak{h}}$. Let $\mathbb{K}^* = \mathbb{K} - \{0\}$ be the torus corresponding to the derivation $d$ on $\mathfrak{g}$, where $[d, c] = 0$ and $[d, t^n \otimes x] = -nt^n \otimes x$ for $x \in \bar{\mathfrak{g}}$ and $n \in \mathbb{Z}$. Set $T = \bar{T} \times \mathbb{K}^*$.

LEMMA 2.1 (cf. [**Ma**], (1.4)). *There is a restricted Lie algebra structure on the affine Lie algebra $\mathfrak{g}$ as an extension of the one on $\bar{\mathfrak{g}}$, whose $p$-power map is given by*

$$c^{[p]} = c, \quad (t^n \otimes x)^{[p]} = t^{np} \otimes x^{[p]}, \quad \text{for } n \in \mathbb{Z}, x \in \bar{\mathfrak{g}}.$$

Then as usual one has the $p$-center $\mathcal{Z}_0(\mathfrak{g})$ in the enveloping algebra $U(\mathfrak{g})$ which is generated by $x^p - x^{[p]}$ for all $x \in \mathfrak{g}$. The subalgebra of $\mathcal{Z}_0(\mathfrak{g})$ generated by $x^p - x^{[p]}$ for all $x \in L\bar{\mathfrak{g}}$ will be denoted by $\mathcal{Z}_0'(\mathfrak{g})$ and referred to as the *proper p-center*.

Each $\chi \in (L\bar{\mathfrak{g}})^*$ defines a $p$-character and gives rise to the reduced enveloping algebra by

$$\mathfrak{u}_\chi(\mathfrak{g}) = U(\mathfrak{g})/I_\chi$$

where $I_\chi$ is the ideal generated by $a^p - a^{[p]} - \chi(a)^p$ for all $a \in L\bar{\mathfrak{g}}$. In particular, $\mathfrak{u}_0(\mathfrak{g})$ is called the restricted enveloping algebra of $\mathfrak{g}$. Note that according to our definition $c^p - c$ is not in the ideal $I_0$.

A distinguished restricted Lie subalgebra of $\mathfrak{g}$ is the Heisenberg algebra

$$\mathfrak{hs} = L\bar{\mathfrak{h}} \oplus \mathbb{K}c = \mathfrak{hs}^- \oplus \mathfrak{h} \oplus \mathfrak{hs}^+,$$

where $\mathfrak{hs}^\pm = \oplus_{n \in \pm\mathbb{N}} t^n \otimes \bar{\mathfrak{h}}$. The Lie algebra $\mathfrak{hs}$ has a large center spanned by $c, t^{pn} \otimes \bar{\mathfrak{h}}$ for $n \in \mathbb{Z}$.

The $p$-center $\mathcal{Z}_0(\mathfrak{hs})$ of $U(\mathfrak{hs})$ is generated by $x^p - x^{[p]}$ for all $x \in \mathfrak{hs}$ and the proper $p$-center $\mathcal{Z}_0'(\mathfrak{hs})$ of $U(\mathfrak{hs})$ is by definition the subalgebra generated by $x^p - x^{[p]}$ for all $x \in L\bar{\mathfrak{h}}$. The whole center of $U(\mathfrak{hs})$ is generated by $\mathcal{Z}_0(\mathfrak{hs})$ and $c, t^{pn} \otimes \bar{\mathfrak{h}}$ for $n \in \mathbb{Z}$, though this fact will not be needed below.

**2.2. Vertex algebras in prime characteristic.** The usual notion of vertex algebras can be readily made sense over the field $\mathbb{K}$ of characteristic $p > 0$ (cf. Borcherds-Ryba [**BR**]). All one needs is to use the divided power of the translation operator $T^{(i)} = T^i/i!, i \geq 1$ and noting that

$$Y(T^{(i)}a, z) = \partial^{(i)} Y(a, z),$$

where $\partial^{(i)}$ denotes the $i$th divided power of the derivative with respect to $z$.

Denote $L_+\bar{\mathfrak{g}} = \sum_{n \in \mathbb{Z}_+} t^n \otimes \bar{\mathfrak{g}}$. It is well known that the vacuum $\mathfrak{g}$-module of level $\kappa \in \mathbb{K}$

$$V^\kappa(\mathfrak{g}) = U(\mathfrak{g}) \bigotimes_{U(L_+\bar{\mathfrak{g}} + \mathbb{K}c)} \mathbb{K}_\kappa$$

carries a canonical structure of a vertex algebra (cf. e.g. [**Fr2**]), where $L\bar{\mathfrak{g}}^+$ acts on $\mathbb{K}_\kappa = \mathbb{K}$ trivially and $c$ as scalar $\kappa$. Denote by $|0\rangle = 1 \otimes 1$ the vacuum vector in $V^\kappa(\mathfrak{g})$.

**2.3. The $p$-centers of modular vertex algebras.** Let

$$x(z) = \sum_{n \in \mathbb{Z}} x_n z^{-n-1}, \qquad x \in \bar{\mathfrak{g}}.$$

The next lemma on vertex operators is standard (cf. [**Fr2**]), except the divided power notation.

LEMMA 2.2. *The following formulas hold in the vertex algebra* $V^\kappa(\mathfrak{g})$*:*

$$
\begin{aligned}
Y(x_{-r}|0\rangle, z) = \partial^{(r-1)}x(z) &= \sum_{n\in\mathbb{Z}} \binom{-n-1}{r-1} x_n z^{-n-r}, \qquad (2.1)\\
Y(x_{-r_1}y_{-r_2}\cdots|0\rangle, z) &= :\partial^{(r_1-1)}x(z)\,\partial^{(r_2-1)}y(z)\cdots:
\end{aligned}
$$

*for* $x, y, \ldots \in \bar{\mathfrak{g}}$, *and* $r, r_1, r_2, \ldots \in \mathbb{N}$.

We shall need some more formulas for vertex operators in characteristic $p$.

LEMMA 2.3. *The following identities hold for the vertex algebra* $V^\kappa(\mathfrak{g})$*: for* $x \in \bar{\mathfrak{g}}$ *and* $r \geq 1$, *we have*

$$
Y(x_{-rp}|0\rangle, z) = \partial^{(rp-1)}x(z) = \sum_{n\in\mathbb{Z}} \binom{-n-1}{r-1} x_{np}\, z^{-np-rp}, \tag{2.2}
$$

$$
Y(x_{-r}^p|0\rangle, z) = :(\partial^{(r-1)}x(z))^p: = \sum_{n\in\mathbb{Z}} \binom{-n-1}{r-1} x_n^p\, z^{-np-rp}. \tag{2.3}
$$

The special case of Lemma 2.3 for $r = 1$ reads:

$$
Y(x_{-p}|0\rangle, z) = \partial^{(p-1)}x(z) = \sum_{n\in\mathbb{Z}} x_{np}\, z^{-np-p}, \tag{2.4}
$$

$$
Y(x_{-1}^p|0\rangle, z) = :x(z)^p: = \sum_{n\in\mathbb{Z}} x_n^p\, z^{-np-p}. \tag{2.5}
$$

To prove Lemma 2.3, we shall need the following classical formula.

LEMMA 2.4. *For* $a = a_0 + pa' \in \mathbb{Z}_{\geq 0}, b = b_0 + pb'$ *with* $0 \leq a_0, b_0 \leq p-1$ *and* $a' \geq 0$, *we have*

$$
\binom{b}{a} \equiv \binom{b'}{a'}\binom{b_0}{a_0} \mod p.
$$

*(All the* $a$*'s and* $b$*'s involved are integers.)*

PROOF OF LEMMA 2.3. By Lemma 2.4, we obtain that $\binom{-m-1}{rp-1} \equiv 0 \mod p$ if $p \nmid m$, and $\binom{-np-1}{rp-1} \equiv \binom{-n-1}{r-1} \mod p$ for $n \in \mathbb{Z}$. Now (2.2) follows from (2.1).

We write $A(z) \equiv \partial^{(r-1)}x(z) = \sum_{n\in\mathbb{Z}} \binom{-n-1}{r-1} x_n z^{-n-r} = A_+(z) + A_-(z)$, where $A_\pm(z) = \sum_{n \lesseqgtr -r} \binom{-n-1}{r-1} x_n z^{-n-r}$. By the definition of normal ordered product and induction on $m \geq 1$, we have

$$
\begin{aligned}
:A(z)^m: &= A_+(z):A(z)^{m-1}: + :A(z)^{m-1}:A_-(z)\\
&= \sum_{i=0}^{m} \binom{m}{i} A_+(z)^i A_-(z)^{m-i}.
\end{aligned}
$$

Note that $A_+(z)^p = \sum_{n\leq -r} \binom{-n-1}{r-1} x_n^p z^{-np-p}$ since $x_n$ with $n < 0$ commute and $b^p = b$ for $b \in \mathbb{F}_p$. Similarly, $A_-(z)^p = \sum_{n\geq 0} \binom{-n-1}{r-1} x_n^p z^{-np-p}$. Hence, $:A(z)^p: = A_+(z)^p + A_-(z)^p$, whence (2.3). □

REMARK 2.5. Lemmas 2.2 and 2.3 are applicable to other modular vertex algebras, e.g. $\mathcal{F}$.

Denote

$$\iota(x_n) = x_n^p - x_{np}^{[p]}, \quad x \in \bar{\mathfrak{g}}.$$

We also denote $\iota(z) = z^p$ (the Frobenius morphism). In the next proposition, which follows directly from Lemmas 2.2 and 2.3, we formulate a basic property of modular affine vertex algebras.

PROPOSITION 2.6 (Commutativity of $\iota$ and $Y$). *For $x \in \bar{\mathfrak{g}}$ and $r \geq 1$, we have*

$$\begin{aligned} Y(\iota(x_{-r})|0\rangle, z) &= Y\left((x_{-r}^p - (x^{[p]})_{-rp})|0\rangle, z\right) \\ &= \left(\partial^{(r-1)} x(z)\right)^p - \partial^{(rp-1)} x^{[p]}(z) = \iota\left(\partial^{(r-1)} x(z)\right). \end{aligned} \tag{2.6}$$

When $r = 1$, we have

$$Y(\iota(x_{-1})|0\rangle, z) = \iota Y(x_{-1}|0\rangle, z) = \sum_{n \in \mathbb{Z}} (x_n^p - (x^{[p]})_{np})\, z^{-np-p}.$$

By definition, the center of a vertex algebra $V$ consists of all vectors $v \in V$ such that $Y(a, z)v \in V[[z]]$ for all $a \in V$. The center of a vertex algebra is a commutative vertex algebra (cf. [**Fr2**]).

DEFINITION 2.7. The *p-center* (or the *Frobenius center*) $\mathfrak{z}_0(V^\kappa(\mathfrak{g}))$ of the vertex algebra $V^\kappa(\mathfrak{g})$ is defined to be the subspace $\mathcal{Z}_0'(\mathfrak{g})|0\rangle \subset V^\kappa(\mathfrak{g})$.

Clearly these $p$-centers (and other $p$-centers below) are vertex subalgebras of the centers of the corresponding vertex algebras.

PROPOSITION 2.8. (1) *The p-center $\mathfrak{z}_0(V^\kappa(\mathfrak{g}))$ is a commutative vertex subalgebra of $V^\kappa(\mathfrak{g})$.*
(2) *$U(\mathfrak{g}) \cdot \mathfrak{z}_0(V^\kappa(\mathfrak{g}))$ is an ideal of the vertex algebra $V^\kappa(\mathfrak{g})$, and so the quotient*

$$V_0^\kappa(\mathfrak{g}) \stackrel{def}{=} V^\kappa(\mathfrak{g})/(U(\mathfrak{g}) \cdot \mathfrak{z}_0(V^\kappa(\mathfrak{g})))$$

*carries an induced vertex algebra structure.*

PROOF. Part (2) follows from (1) easily, and we shall prove (1).

Observe that the Fourier components of (2.6) are of the form $x_n^p - (x^{[p]})_{np}$ (up to a scalar multiple), and hence belong to the $p$-center $\mathfrak{z}_0(V^\kappa(\mathfrak{g}))$. By definition, the $p$-center $\mathfrak{z}_0(V^\kappa(\mathfrak{g}))$ is spanned by elements of the form $x = \iota(a_{-r_1} b_{-r_2} \cdots |0\rangle)$ with $a, b \ldots \in \bar{\mathfrak{g}}$ and $r_1, r_2, \ldots > 0$. By Proposition 2.6, the vertex operator

$$Y(x, z) = {:}Y(\iota(a_{-r_1})|0\rangle, z) Y(\iota(b_{-r_2})|0\rangle, z) \cdots{:}$$

is a linear combination of operators composed from those of the form $x_n^p - (x^{[p]})_{np}$, and hence clearly preserves $\mathfrak{z}_0(V^\kappa(\mathfrak{g}))$. □

Following the standard terminology in the theory of modular Lie algebras, we shall refer to the vertex algebras $V_0^\kappa(\mathfrak{g})$ as the *restricted* (or more precisely *p-restricted*) vertex algebras associated to $\mathfrak{g}$.

REMARK 2.9. Since $c^p - c \notin \mathcal{Z}_0'(\mathfrak{g})$ by definition, the central charges for the restricted vertex algebras $V_0^\kappa(\mathfrak{g})$ can be any scalar in $\mathbb{K}$.

A *baby Verma* $\mathfrak{g}$*-module* (associated to a weight $\lambda$ on $\mathfrak{h}$ of level $\kappa$) is a $\mathfrak{g}$-module of the form

$$V(\lambda) \equiv V^\kappa(\lambda) = \mathfrak{u}_0(\mathfrak{g}) \bigotimes_{\mathfrak{u}_0(\mathfrak{n}+\mathfrak{h})} \mathbb{K}_\lambda$$

where $\mathfrak{n}$ acts trivially on the one-dimensional space $\mathbb{K}_\lambda \cong \mathbb{K}$ and $\mathfrak{h}$ acts by the weight $\lambda \in \mathfrak{h}^*$. These baby Verma modules are modules of the restricted vertex algebra $V_0^\kappa(\mathfrak{g})$.

## 3. The baby Wakimoto modules

**3.1. A vertex algebra $M$.** Let $\mathcal{A}^\mathfrak{g}$ be the Weyl algebra over $\mathbb{K}$ with generators $a_{\alpha,n}, a^*_{\alpha,n}$ with $\alpha \in \overline{\Delta}_+, n \in \mathbb{Z}$, and relations

$$[a_{\alpha,n}, a^*_{\beta,m}] = \delta_{\alpha,\beta}\delta_{n,-m}, \quad [a_{\alpha,n}, a_{\beta,m}] = [a^*_{\alpha,n}, a^*_{\beta,m}] = 0.$$

A restricted Lie algebra structure on $\mathcal{A}^\mathfrak{g}$ is given as follows:

$$a_{\alpha,n}^{[p]} = (a^*_{\alpha,n})^{[p]} = 0, \quad n \in \mathbb{Z}.$$

Introduce the fields

$$a_\alpha(z) = \sum_{n\in\mathbb{Z}} a_{\alpha,n} z^{-n-1}, \quad a^*_\alpha(z) = \sum_{n\in\mathbb{Z}} a^*_{\alpha,n} z^{-n}, \qquad \alpha \in \overline{\Delta}_+.$$

Let $M$ be the Fock representation of $\mathcal{A}^\mathfrak{g}$ generated by $|0\rangle$ such that

$$a_{\alpha,n}|0\rangle = 0, \quad n \geq 0; \qquad a^*_{\alpha,n}|0\rangle = 0, \quad n > 0.$$

As a vector space, $M \cong \mathbb{K}[a_{\alpha,n-1}, a^*_{\alpha,n}]_{\alpha\in\bar{\Delta}_+, n\leq 0}$. It is well known that $M$ carries a vertex algebra structure with state-field correspondence

$$\begin{aligned} &Y(a_{\alpha_1,-r_1}\cdots a_{\alpha_k,-r_k} a^*_{\beta_1,-s_1}\cdots a^*_{\beta_m,-s_m}|0\rangle) \\ &= {:}\partial^{(r_1-1)}a_{\alpha_1}(z)\cdots\partial^{(r_k-1)}a_{\alpha_k}(z)\partial^{(s_1)}a^*_{\beta_1}(z)\cdots\partial^{(s_m)}a^*_{\beta_m}(z){:} \end{aligned}$$

and with the translation operator $T$ such that

$$T|0\rangle = 0, \ [T, a_{\alpha,n}] = -n a_{\alpha,n-1}, \ [T, a^*_{\alpha,n}] = -(n-1)a^*_{\alpha,n-1}.$$

PROPOSITION 3.1.

(1) *The $p$-center $\mathcal{Z}_0(\mathcal{A}^\mathfrak{g})$ is equal to $\mathbb{K}[a^p_{\alpha,n}, (a^*_{\alpha,n})^p]_{\alpha\in\bar{\Delta}_+, n\in\mathbb{Z}}$; and moreover, $\mathfrak{z}_0(M) \cong \mathbb{K}[a^p_{\alpha,n-1}, (a^*_{\alpha,n})^p]_{\alpha\in\bar{\Delta}_+, n\leq 0}$.*

(2) *The space $U(\mathcal{A}^\mathfrak{g})\cdot\mathfrak{z}_0(M)$ is an ideal of the vertex algebra $M$, so the quotient*

$$M_0 := M/(U(\mathcal{A}^\mathfrak{g})\cdot\mathfrak{z}_0(M))$$

*carries an induced vertex algebra structure.*

**3.2. Realization of the contragredient Verma modules.** We first recall (cf. e.g. [**Fr2**, pp.135]) that the contragredient Verma module of $\bar{\mathfrak{g}}$ can be realized via its identification with the space of regular functions $\mathcal{O}_{\bar{N}_+}$ on the unipotent subgroup $\bar{N}_+$ of the algebraic group $\bar{G}$; equivalently, this is described as follows: let $\bar{U} = \bar{N}_+\bar{B}_-/\bar{B}_-$ be the open cell of the flag variety $\bar{\mathcal{B}} := \bar{G}/\bar{B}_-$. Let

$$\phi : U(\bar{\mathfrak{g}}) \longrightarrow \mathcal{D}_{\bar{\mathcal{B}}}(\bar{U}) \tag{3.1}$$

denote the composition of the restriction to the open cell $\Gamma(\bar{\mathcal{B}}, \mathcal{D}_{\bar{\mathcal{B}}}) \to \mathcal{D}_{\bar{\mathcal{B}}}(\bar{U})$ with an algebra homomorphism $U(\bar{\mathfrak{g}}) \to \Gamma(\bar{\mathcal{B}}, \mathcal{D}_{\bar{\mathcal{B}}})$ where $\mathcal{D}_{\bar{\mathcal{B}}}$ denotes the sheaf of crystalline differential operators on the flag variety $\bar{\mathcal{B}}$ (i.e. no divided powers of differential operators, cf. e.g., [**BMR**]). Then the contragredient Verma module of $\bar{\mathfrak{g}}$ is the pullback of the $\mathcal{D}_{\bar{\mathcal{B}}}(\bar{U})$-module $\mathcal{O}_{\bar{U}}$ via the algebra homomorphism $\phi$. The restriction $\phi|_{\bar{\mathfrak{g}}} : \bar{\mathfrak{g}} \to \mathrm{Vect}_{\bar{U}}$ is a Lie algebra homomorphism, where $\mathrm{Vect}_{\bar{U}}$ the vector fields over $\bar{U}$.

Let us fix some coordinates $y_\gamma$ (and $\partial_\gamma := \frac{\partial}{\partial y_\gamma}$) for the open cell $\bar{U}$. The following lemma is standard.

LEMMA 3.2. *We have* $\mathcal{Z}_0(\mathcal{D}_{\bar{\mathcal{B}}}(U)) = \mathbb{K}[y_\gamma^p, \partial_\gamma^p]_{\gamma\in\bar{\Delta}_+}$.

The following is known (cf., e.g., [**BMR**, §1.3]).

LEMMA 3.3. *The restriction* $\phi : \bar{\mathfrak{g}} \longrightarrow \mathrm{Vect}_{\bar{U}}$ *is a homomorphism of restricted Lie algebras, and the homomorphism* $\phi : U(\bar{\mathfrak{g}}) \longrightarrow \mathcal{D}_{\bar{\mathcal{B}}}(\bar{U})$ *maps the p-center of* $U(\bar{\mathfrak{g}})$ *to the p-center of* $\mathcal{D}_{\bar{\mathcal{B}}}(\bar{U})$.

In some cases, we can make this fairly explicit as follows. For $x \in \bar{\mathfrak{n}} \oplus \bar{\mathfrak{h}}$, one can write

$$\phi(x) = \sum_{\beta\in\bar{\Delta}_+} c_\beta m_\beta(y_\gamma)\partial_\beta \tag{3.2}$$

where $c_\beta \in \mathbb{K}$ and $m_\beta(y_\gamma)$ denotes some monomials in the variables $y_\gamma$, for $\gamma \in \bar{\Delta}_+$.

LEMMA 3.4. *Let* $x \in \bar{\mathfrak{n}} \oplus \bar{\mathfrak{h}}$ *and retain the above notation* (3.2). *Then we have*

$$\phi(\iota(x)) = \sum_{\beta\in\bar{\Delta}_+} c_\beta^p m_\beta(y_\gamma^p)\partial_\beta^p.$$

PROOF. We know that

$$\phi(\iota(x)) = \phi(x)^p - \phi(x^{[p]}) = \big(\sum c_\beta m_\beta(y_\gamma)\partial_\beta\big)^p - \phi(x^{[p]})$$

lies in the $p$-center $\mathcal{Z}_0(\mathcal{D}_{\bar{\mathcal{B}}}(U))$ and $\phi(x^{[p]})$ is a sum of differential operators of order one (plus some possible constants). We expand this $p$th power and move the differential operator $\partial_\beta$ to the right by using commutators. Lemma 3.2 ensures all the commutators will cancel out with each other as they would produce differential operators of order between 1 and $p-1$. □

REMARK 3.5. Let $\mu : T^*\bar{\mathcal{B}} \to \bar{\mathcal{N}}$ be the Springer resolution, $\mu^{(1)} : T^*\bar{\mathcal{B}}^{(1)} \to \bar{\mathcal{N}}^{(1)}$ be the induced map between the corresponding Frobenius twists. Then the same argument as in proof of Lemma 3.4 shows that $\phi(\iota(x)) = (\mu^{(1)})^*(\bar{x})|_U$ ([**BMR**, 1.3.3]), where $\bar{x}$ is the image of $x \in \bar{\mathfrak{g}}$ by the projection $\mathbb{K}[\bar{\mathfrak{g}}^*] \to \mathbb{K}[\bar{\mathcal{N}}]$.

**3.3. Heisenberg vertex algebra $\pi^\kappa$.** Let $\mathcal{B}_\kappa^{\mathfrak{h}}$ be a copy of Heisenberg algebra (of the affine Lie algebra $\mathfrak{g}$), with generators $\mathbf{1}$ and $b_{i,n}$ $(i = 1, \ldots, \ell, n \in \mathbb{Z})$ and subject to the relations

$$[b_{i,n}, b_{j,m}] = n\kappa\langle h_i, h_j\rangle\delta_{n,-m}\mathbf{1}.$$

Denote by $\pi^\kappa$ the vertex algebra $\mathbb{K}[b_{i,n}]_{1\le i\le\ell; n<0}$ (where $\mathbf{1}$ acts as the identity map), with

$$Y(b_{i,-1}, z) \equiv b_i(z) = \sum_{n<0} b_{i,n} z^{-n-1},$$

and the translation $T$ given by

$$T \cdot b_{i_1,n_1} \cdots b_{i_m,n_m} = -\sum_{j=1}^m n_j b_{i_1,n_1} \cdots b_{i_j,n_j-1} \cdots b_{i_m,n_m}.$$

A restricted Lie algebra structure on $\mathcal{B}_\kappa^{\mathfrak{h}}$ is given by

$$b_{i,n}^{[p]} = b_{i,np}, \quad \mathbf{1}^{[p]} = \mathbf{1}, \qquad 1 \le i \le \ell, n \in \mathbb{Z}.$$

The proper $p$-center of the vertex algebra $\pi^\kappa$ (which excludes $\mathbf{1}^p - \mathbf{1}$) is then equal to

$$\mathfrak{z}'_0(\pi^\kappa) = \mathbb{K}[b^p_{i,n} - b_{i,np}]_{1\le i\le \ell; n<0}.$$

Denote by $\pi^\kappa_0$ the quotient vertex algebra of $\pi^\kappa$ by the ideal (in the sense of vertex algebras) generated by $\mathfrak{z}'_0(\pi^\kappa)$.

For $\kappa = 0$, $\pi^0$ is naturally a commutative vertex algebra.

**3.4. The Wakimoto-Feigin-Frenkel (WFF) homomorphism.** Let $\kappa_c$ denote the critical level for $\mathfrak{g}$. There exists a homomorphism of vertex algebras

$$\mathtt{w} = \mathtt{w}_\kappa : V^\kappa(\mathfrak{g}) \to M \otimes \pi^{\kappa-\kappa_c},$$

which is roughly speaking an affinization of $\phi$ defined in (3.1); see [**Fr2**, Theorem 6.1.6]. We shall call $\mathtt{w}$ the *WFF homomorphism*, since this was introduced by Wakimoto [**Wak**] in the $\mathfrak{sl}_2$ case and by Feigin-Frenkel [**FF**] for general semisimple Lie algebras $\bar{\mathfrak{g}}$. On generating fields, the formulas for $\mathtt{w}$ read as follows:

$$e_i(z) \mapsto a_{\alpha_i}(z) + \sum_{\alpha_i \neq \beta \in \bar{\Delta}_+} {:}P^i_\beta\big(a^*_\alpha(z)\big)a_\beta(z){:}, \tag{3.3}$$

$$h_i(z) \mapsto \sum_{\beta\in\bar{\Delta}_+} \beta(h_i){:}a^*_\beta(z)a_\beta(z){:} + b_i(z), \tag{3.4}$$

$$\begin{aligned} f_i(z) \mapsto &\sum_{\beta\in\bar{\Delta}_+} {:}Q^i_\beta\big(a^*_\alpha(z)\big)a_\beta(z){:} + \big(c_i + (\kappa-\kappa_c)\langle e_i, f_i\rangle\big)\partial_z a^*_{\alpha_i}(z) \\ &+ a^*_{\alpha_i}(z)b_i(z). \end{aligned} \tag{3.5}$$

We shall take for granted that $c_i$ are integers. The polynomial $Q^i_\beta\big(a^*_\alpha(z)\big)$ for $\beta = \alpha_i$ can be determined explicitly as follows; cf. [**Fr2**].

LEMMA 3.6. *We have* $Q^i_{\alpha_i}\big(a^*_\alpha(z)\big) = -{:}a^*_{\alpha_i}(z)^2{:}$.

EXAMPLE 3.7. For $\bar{\mathfrak{g}} = \mathfrak{sl}_2$, the formulas for $\mathtt{w}$ are greatly simplified (where we drop the indices of the Chevalley generators and of the free fields) as follows:

$$\begin{aligned} &e(z) \mapsto a(z), \qquad h(z) \mapsto -2{:}a^*(z)a(z){:} + b(z), \\ &f(z) \mapsto -{:}a^*(z)^2a(z){:} + \kappa\partial_z a^*(z) + a^*(z)b(z). \end{aligned} \tag{3.6}$$

The following is a main result of this paper, which will be proved in Section 4.

THEOREM 3.8. *The homomorphism* $\mathtt{w}: V^\kappa(\mathfrak{g}) \to M\otimes\pi^{\kappa-\kappa_c}$ *sends the $p$-center* $\mathfrak{z}_0(V^\kappa(\mathfrak{g}))$ *to the $p$-center* $\mathfrak{z}_0(M)\otimes\mathfrak{z}_0(\pi^{\kappa-\kappa_c})$.

We have the following immediate consequence.

COROLLARY 3.9. *The homomorphism $\mathtt{w}$ induces naturally a homomorphism of vertex algebras* $\mathtt{w}_0 : V^\kappa_0(\mathfrak{g}) \to M_0 \otimes \pi_0^{\kappa-\kappa_c}$.

**3.5. The baby Wakimoto modules.** We define baby Wakimoto modules (associated to $p$-characters) at the critical level $\kappa_c$ as follows.

Let $\xi$ be a $p$-character on the Lie algebra $\mathcal{A}^{\mathfrak{g}}$ such that

$$\xi((a^*_{\alpha,n})^p) = 0 = \xi(a^p_{\alpha,n-1}), \qquad \text{for } n>0, \alpha\in\bar{\Delta}_+. \tag{3.7}$$

Let $\xi^\pi$ be a $p$-character on the Lie algebra $\mathcal{B}^{\mathfrak{h}}$. Take a weight $\lambda(t) = \sum_{n\in\mathbb{Z}}\lambda_n t^{-n-1} \in \bar{\mathfrak{h}}^*((t))$ which is compatible with the $p$-character $\xi^\pi$ in the sense that

$$\lambda^p_{i,n} - \lambda_{i,np} = \xi^\pi(b_{i,n})^p, \qquad \text{for all } 1\le i\le\ell, n\in\mathbb{Z},$$

where $\lambda_{i,n} = \lambda_n(h_i)$. Such a weight gives rise to the one-dimensional $\pi^0$–module $\mathbb{K}_{\lambda(t)}$ on which $b_{i,n}$ acts by multiplication by $\lambda_{i,n}$. This defines a $\mathfrak{g}$-module at the critical level on $M$ (which is identified with $M \otimes \mathbb{K}_{\lambda(t)}$). Identifying $M$ as the polynomial algebra $\mathbb{K}[a_{\alpha,n-1}, a^*_{\alpha,n}]_{\alpha\in\bar{\Delta}_+, n\le 0}$, we let $I_\xi$ be the subalgebra of $M$ spanned by $(a^*_{\alpha,n})^p - \xi((a^*_{\alpha,n})^p)$, $a^p_{\alpha,n-1} - \xi(a^p_{\alpha,n-1})$, where $n \le 0$ and $\alpha \in \bar{\Delta}_+$. Since all the generators of $I_\xi$ are central in $U(\mathfrak{g})$, $I_\xi$ is clearly a $\mathfrak{g}$-submodule of $M$, and this gives rise to a quotient $\mathfrak{g}$-module

$$\mathfrak{w}_\xi(\lambda) := M/I_\xi.$$

We refer to the $\mathfrak{g}$-module $\mathfrak{w}_\xi(\lambda)$ as the *baby Wakimoto module* of high weight $\lambda$ (and $p$-character $\xi$).

Now assume $\kappa \neq \kappa_c$. Given $\lambda \in \mathfrak{h}^*$, let $\pi_0^{\kappa-\kappa_c}(\lambda)$ be the Fock representation of $\mathcal{A}^{\mathfrak{g}}$ generated by a vector $|\lambda\rangle$ such that

$$b_{i,n}|\lambda\rangle = 0 \quad (n > 0), \qquad b_{i,0}|\lambda\rangle = \lambda(h_i)|\lambda\rangle, \qquad \mathbf{1}|\lambda\rangle = |\lambda\rangle.$$

Any module over the vertex algebra $M \otimes \pi^{\kappa-\kappa_c}$ becomes a module over $V^\kappa(\mathfrak{g})$ via the pullback of homomorphism $\mathtt{w} : V^\kappa(\mathfrak{g}) \to M \otimes \pi^{\kappa-\kappa_c}$. In particular,

$$W(\lambda) := M \otimes \pi^{\kappa-\kappa_c}(\lambda)$$

becomes a module over $\mathfrak{g}$ of level $\kappa$. This is the *Wakimoto module* of high weight $\lambda$ and level $\kappa$ defined in [**FF**].

Let $\xi$ be a $p$-character on the Lie algebra $\mathcal{A}^{\mathfrak{g}}$ satisfying (3.7). Let $\xi^\pi$ be a $p$-character on the Lie algebra $\mathcal{B}^{\mathfrak{h}}$ such that

$$\xi^\pi(b^p_{i,n} - b_{i,np}) = 0, \qquad \text{for } n \ge 0.$$

Assume that $\lambda \in \mathfrak{h}^*$ is compatible with the $p$-character $\xi^\pi$ in the sense that

$$\lambda(h_i)^p - \lambda(h_i) = \xi^\pi(h_i)^p, \qquad \text{for all } 1 \le i \le \ell.$$

Identifying $\pi^{\kappa-\kappa_c}(\lambda)$ as the polynomial algebra $\mathbb{K}[b_{i,n}]_{n<0, 1\le i\le \ell}$, we let $I_{\xi^\pi}$ be the subalgebra of $\pi^{\kappa-\kappa_c}(\lambda)$ spanned by $b^p_{i,n} - b_{i,np} - \xi^\pi(b_{i,n})^p$, where $n < 0$ and $1 \le i \le \ell$. Since all the generators of $I_{\xi^\pi}$ are central in $U(\mathfrak{g})$, $I_{\xi^\pi}$ is clearly a $\mathfrak{g}$-submodule of $\pi^{\kappa-\kappa_c}(\lambda)$, and this gives rise to a quotient $\mathfrak{g}$-module

$$\mathfrak{w}^\kappa_{\xi,\xi^\pi}(\lambda) := (M \otimes \pi^{\kappa-\kappa_c}(\lambda))/(I_\xi \otimes I_{\xi^\pi}) \cong (M/I_\xi) \otimes (\pi^{\kappa-\kappa_c}(\lambda)/I_{\xi^\pi}).$$

We refer to the $\mathfrak{g}$-module $\mathfrak{w}^\kappa_{\xi,\xi^\pi}(\lambda)$ as the *baby Wakimoto module* of high weight $\lambda$ (and $p$-character $(\xi, \xi^\pi)$).

Assume for now that $\pi^{\kappa-\kappa_c}(\lambda)/I_{\xi^\pi}$ is a module over the restricted vertex algebra $\pi_0^{\kappa-\kappa_c}$. Since $\iota(b_{i,-1})|0\rangle = 0 \in \pi_0^{\kappa-\kappa_c}$, we have

$$0 = Y(\iota(b_{i,-1})|0\rangle, z) = \sum_{n\in\mathbb{Z}} (b^p_{i,n} - b_{i,np}) z^{-np-p}$$

when acting on $\pi^{\kappa-\kappa_c}(\lambda)/I_{\xi^\pi}$. Hence $\xi^\pi = 0$, and $\lambda(h_i) \in \mathbb{F}_p$ for $1 \le i \le \ell$. The converse is also true: if $\xi^\pi = 0$ and $\lambda(h_i) \in \mathbb{F}_p$ for $1 \le i \le \ell$, then $\pi^{\kappa-\kappa_c}(\lambda)/I_{\xi^\pi}$ is a module over the restricted vertex algebra $\pi_0^{\kappa-\kappa_c}$. Similarly, $M/I_\xi$ is a module over the restricted vertex algebra $M_0$ if and only if $\xi = 0$. Summarizing, we have proved the following.

PROPOSITION 3.10. *Let $\kappa \neq \kappa_c$. If $\lambda \in \mathfrak{h}^*$ satisfies $\lambda(h_i) \in \mathbb{F}_p$ for each $1 \le i \le \ell$ and $\lambda(c) = \kappa$, then $\mathfrak{w}^\kappa_{0,0}(\lambda)$ is a module over the vertex algebra $M_0 \otimes \pi_0^{\kappa-\kappa_c}$, and hence a module over the vertex algebra $V^\kappa(\mathfrak{g})$.*

## 4. Proof of Theorem 3.8

**4.1. Restriction of $\mathtt{w}$ to the $p$-center.** Theorem 3.8 follows readily from the explicit description of the restriction map of $\mathtt{w}$ to $\mathfrak{z}_0(V^\kappa(\mathfrak{g}))$ given in the following theorem. Recall that

$$\iota(e_i(z)) = e_i(z)^p, \qquad \iota(h_i(z)) = h_i(z)^p - h_i(z), \qquad \iota(f_i(z)) = f_i(z)^p. \tag{4.1}$$

THEOREM 4.1. *The restriction of $w$ to $\mathfrak{z}_0(V^\kappa(\mathfrak{g}))$ is given in terms of fields as follows:*

$$\iota(e_i(z))^w = a_{\alpha_i}(z)^p + \sum_{\alpha_i \neq \beta \in \bar{\Delta}_+} P^i_\beta\big(a^*_\alpha(z)\big)^p a_\beta(z)^p, \tag{4.2}$$

$$\iota(h_i(z))^w = -\sum_{\beta\in\bar{\Delta}_+} \beta(h_i) a^*_\beta(z)^p a_\beta(z)^p + b_i(z)^p - \partial^{(p-1)} b_i(z), \tag{4.3}$$

$$\begin{aligned}\iota(f_i(z))^w = &\sum_{\beta\in\bar{\Delta}_+} Q^i_\beta\big(a^*_\alpha(z)\big)^p a_\beta(z)^p + (\kappa^p - \kappa)\langle e_i, f_i\rangle \big(\partial_z a^*_{\alpha_i}(z)\big)^p \\ &+ a^*_{\alpha_i}(z)^p\big(b_i(z)^p - \partial^{(p-1)} b_i(z)\big).\end{aligned} \tag{4.4}$$

Actually $P^i_\beta\big(a^*_\alpha(z)\big)^p$ and $Q^i_\beta\big(a^*_\alpha(z)\big)^p$ are simply polynomials in the commuting vertex operators $a^*_\alpha(z)^p$, for $\alpha \in \bar{\Delta}_+$. Note that the normal orderings are no longer needed in the above formulas.

EXAMPLE 4.2. The formulas above read in the case of $\widehat{\mathfrak{sl}}_2$ as follows:

$$\begin{aligned}\iota(e(z))^{\mathtt{w}} &= a(z)^p,\\ \iota(h(z))^{\mathtt{w}} &= -2a^*(z)^p a(z)^p + b(z)^p - \partial^{(p-1)} b(z),\\ \iota(f(z))^{\mathtt{w}} &= -a^*(z)^{2p} a(z)^p + (\kappa^p - \kappa)\partial_z a^*(z) + a^*(z)^p\left(b(z)^p - \partial^{(p-1)} b(z)\right).\end{aligned}$$

The remainder of this section is devoted to the proof of Theorem 4.1.

**4.2. Proof of Theorem 4.1.** Let

$$A = \mathbb{K}[a^*_{\alpha,0}]_{\alpha\in\bar{\Delta}_+},$$

which can be identified with $\mathbb{K}[\bar{N}_+] = \mathbb{K}[\bar{U}]$. The space $V := \sum_{\alpha\in\bar{\Delta}_+} A a_{\alpha,-1}$ is identified with $\mathcal{T}_{\bar{\mathcal{B}}}(\bar{U})$, where $\mathcal{T}_{\bar{\mathcal{B}}}$ is the tangent sheaf of the flag variety $\bar{\mathcal{B}}$. Set $\Omega := TA$, where $T$ is the translation operator in the vertex algebra $M$. Note that $(M_0)_0 = A$ and $(M_0)_1 = V \oplus \Omega$. An element $D \in V$ acts on $A$ as a derivation by the correspondence $D \mapsto D_{(0)}$, where we denote $a_{(n)} = Y(a,z)_{(n)} = \mathrm{Res}_z z^n Y(a,z)$. One has

$$[D_{(m)}, f_{(n)}] = (Df)_{(m+n)} \quad \text{for } D\in V \text{ and } f \in A. \tag{4.5}$$

Recall $\phi : U(\bar{\mathfrak{g}}) \to \mathcal{D}_{\bar{\mathcal{B}}}(\bar{U})$, $x \mapsto D^x$. The algebra $\mathcal{D}_{\bar{\mathcal{B}}}(\bar{U})$ can be identified with the $A$-algebra generated by $V$ such that $D\cdot f = f\cdot D + D_{(0)}f$, and thus we have $D^x \in V$ for each $x \in \bar{\mathfrak{g}}$. It follows by (3.3)-(3.5) that the image of $x_{-1}|0\rangle$ under the vertex algebra homomorphism $\mathtt{w} : V^\kappa(\mathfrak{g}) \to M \otimes \pi^{\kappa-\kappa_c}$, denoted by $x^{\mathtt{w}}_{-1}|0\rangle$, is of the form

$$x^{\mathtt{w}}_{-1}|0\rangle = D^x_{(-1)} + f^x_{(-2)} + b^x_{(-1)} \tag{4.6}$$

for $x \in \bar{\mathfrak{g}}$, some $f^x \in A$ and some conformal weight one vector $b^x \in \pi^{\kappa-\kappa_c}$ depending on $x$.

LEMMA 4.3. *Let $x \in \bar{\mathfrak{g}}$ and $m \in \mathbb{Z}$. Then $\big(\iota(x_{-1})^{w}|0\rangle\big)_{(m)}$ commutes with $a^*_{\alpha,n}$ for all $\alpha \in \bar{\Delta}_+$ and $n \in \mathbb{Z}$.*

PROOF. Since the algebra homomorphism $\phi : U(\bar{\mathfrak{g}}) \to \mathcal{D}_{\bar{\mathcal{B}}}(U)$ is compatible with the restricted Lie algebra structures (see Lemma 3.3), we have, for $f \in A$, that

$$[D^x, D^y]f = D^{[x,y]}f, \quad (D^x)^p f = D^{x^{[p]}}f. \tag{4.7}$$

Recall from Proposition 2.6 that

$$Y(\iota(x_{-1})^{\mathtt{w}}|0\rangle, z) = \sum_{m\in\mathbb{Z}} \Big((x^{\mathtt{w}}_m)^p - (x^{[p]})^{\mathtt{w}}_{mp}\Big) z^{-mp-p}.$$

By (4.5), (4.6) and (4.7), one has, for $f \in A$,

$$\begin{aligned}
[(x^{\mathtt{w}}_m)^p, f_{(n)}] &= (\operatorname{ad} x^{\mathtt{w}}_m)^p(f_{(n)}) = (\operatorname{ad} D^x_{(m)})^p(f_{(n)})\\
&= ((D^x)^p f)_{(n+mp)} = (D^{x^{[p]}}f)_{(n+mp)}\\
&= [D^{x^{[p]}}_{(mp)}, f_{(n)}] = [(x^{[p]})^{\mathtt{w}}_{mp}, f_{(n)}].
\end{aligned}$$

Now the lemma follows by taking $f = a^*_{\alpha,0}$. □

LEMMA 4.4. *Let $x \in \bar{\mathfrak{g}}$ and $m \in \mathbb{Z}$. Then $(\iota(x_{-1})^{w}|0\rangle)_{(m)}$ commutes with $a_{\alpha,n}$ for all $\alpha \in \bar{\Delta}_+$ and $n \in \mathbb{Z}$.*

PROOF. Recall from [**Fr1**] that, for $\alpha \in \bar{\Delta}_+$, $e^{\mathtt{w}}_{\alpha,-1}$ is of the form

$$e^{\mathtt{w}}_{\alpha,-1} = a_{\alpha,-1} + \sum_{\substack{\beta\in\bar{\Delta}_+\\ \beta>\alpha}} P^{\alpha}_{\beta} a_{\beta,-1}, \tag{4.8}$$

where $P^{\alpha}_{\beta}$ is a polynomial in $A = \mathbb{K}[a^*_{\alpha,0}]_{\alpha\in\bar{\Delta}_+}$ of weight $\alpha-\beta$. Indeed, the weight of the right-hand-side of (4.8) equals $\alpha$ (see for example the first line of [**Fr1**, §1.3]), and hence the coefficient $P^{\alpha}_{\beta} \in A$ must be zero unless $\beta > \alpha$ in the standard dominance order of the root lattice of $\bar{\mathfrak{g}}$. It follows that each $a_{\alpha,-1}$ can be written as

$$a_{\alpha,-1} = e^{\mathtt{w}}_{\alpha,-1} + \sum_{\substack{\beta\in\bar{\Delta}_+\\ \beta>\alpha}} Q^{\alpha}_{\beta} e^{\mathtt{w}}_{\beta,-1} \tag{4.9}$$

for some polynomials $Q^{\alpha}_{\beta} \in A$.

Since $(\iota(x_{-1})|0\rangle)_{(m)} = x^p_m - (x^{[p]})_{mp}$ is central in $U(\mathfrak{g})$ and $\mathtt{w}$ is a $\mathfrak{g}$-homomorphism, $(\iota(x_{-1})^{\mathtt{w}}|0\rangle)_{(m)}$ commutes with $e^{\mathtt{w}}_{\gamma,n}$ for any $n$ and any $\gamma \in \bar{\Delta}_+$ (in particular for $\gamma \geq \alpha$). By Lemma 4.3, $(\iota(x_{-1})^{\mathtt{w}}|0\rangle)_{(m)}$ also commutes with $(Q^{\alpha}_{\beta})_{(n)}$ for any $n$. Now by applying $Y(-,z)_{(n)}$ to (4.9), $(\iota(x_{-1})^{\mathtt{w}}|0\rangle)_{(m)}$ commutes with $a_{\alpha,n}$. □

REMARK 4.5. It is elementary to show by induction on $k$ that for any $k, n, m, i, \beta$ one has

$$[(h^{\mathtt{w}}_{i,n})^k, a_{\beta,m}] = \sum_{d=1}^{k} \binom{k}{d} \beta(h_i) \cdot a_{\beta,nd+m}(h^{\mathtt{w}}_{i,n})^{k-d}.$$

It follows that $[(h^{\mathtt{w}}_{i,n})^p - h^{\mathtt{w}}_{i,np}, a_{\beta,m}] = 0$. Similarly, $[(h^{\mathtt{w}}_{i,n})^p - h^{\mathtt{w}}_{i,np}, a^*_{\beta,m}] = 0$.

PROPOSITION 4.6. *Formulas* (4.2)-(4.3) *hold.*

PROOF. Let us fix $i$. The strategy is similar to the proof of Lemma 3.4. We extend the notation to write $P^i_{\alpha_i}(a^*_\alpha(z)) = a_{\alpha_i}(z)$. Then by (4.1) and (3.3), we can write

$$(e_i(z)^p)^{\mathtt{w}} = \sum_{\beta\in\bar{\Delta}_+} P^i_\beta\big(a^*_\alpha(z)\big)^p a_\beta(z)^p + \sum_{t=1}^{p-1} {:}Y_t(a^*_\alpha(z))a_\beta(z)^t{:}, \tag{4.10}$$

for some differential polynomials $Y_t$, where the last summand arises from contractions in Wick's formula. Lemma 4.3 can be rephrased by saying that $(e_i(z)^p)^{\mathtt{w}}$ commutes with $a^*_\beta(z)$. The first summand on the right-hand side of (4.10) commute with $a^*_\beta(z)$, but $a_\beta(z)^t$, for $1 \le t \le p-1$, do not commute with $a^*_\beta(z)$. Hence we must have $Y_t = 0$ for all $t$ by a downward induction on $t$. This proves (4.2).

Similarly, noting $\beta(h_i)$ is integral and using (4.1) and (3.4), we have

$$(h_i(z)^p - h_i(z))^{\mathtt{w}} = -\sum_{\beta\in\bar{\Delta}_+} \beta(h_i)\big[({:}a^*_\beta(z)a_\beta(z){:})^p - {:}a^*_\beta(z)a_\beta(z){:}\big] + b_i(z)^p - \partial^{(p-1)}b_i(z). \tag{4.11}$$

Now write $({:}a^*_\beta(z)a_\beta(z){:})^p - {:}a^*_\beta(z)a_\beta(z){:} = a^*_\beta(z)^p a_\beta(z)^p + \sum_{t=1}^{p-1} {:}X_t(a^*_\beta(z))a_\beta(z)^t{:}$, for some differential polynomials $X_t$ (Here $\beta$ is fixed). But by considering the commutation of (4.11) with $a^*_\beta(z)$ and applying Lemma 4.3, we conclude that $X_t = 0$ for each $t$. This proves (4.3). □

To complete the proof of Theorem 4.1 it remains to prove (4.4). Denote by $\bar{\mathfrak{g}}_{\mathbb{Z}}$ the $\mathbb{Z}$-lattice generated by the Chevalley generators $e_\alpha, f_\alpha$ and $h_i$, for $\alpha \in \bar{\Delta}^+$ and $i = 1, \dots, \ell$. Denote by $V_{\mathbb{Z}}$ the $\mathbb{Z}$-lattice of $V^\kappa(\mathfrak{g})$ spanned by all possible $a_{-i_1}b_{-i_2}c_{-i_3}\dots|0\rangle$, where $a, b, c\dots \in \bar{\mathfrak{g}}_{\mathbb{Z}}$ and $i_1, i_2, i_3, \dots \ge 1$. Writing a general vertex operator $Y(a,z) = \sum_{n\in\mathbb{Z}} a_{(n)}z^{-n-1}$, we recall a general formula from the theory of vertex algebras (cf. [**Fr2**]):

$$[x_{(m)}, y_{(n)}] = \sum_{i\ge 0} \binom{m}{i} \big(x_{(i)}y\big)_{(m+n-i)}, \qquad i \ge 0. \tag{4.12}$$

From a similar consideration as in the proof of Proposition 4.6 above, we conclude that $\iota(f_i(z))^{\mathtt{w}}$ is of the form

$$\iota(f_i(z))^{\mathtt{w}} = \sum_{\beta\in\bar{\Delta}_+} Q^i_\beta\big(a^*_\alpha(z)\big)^p a_\beta(z)^p + \eta\big(\partial_z a^*_{\alpha_i}(z)\big)^p + a^*_{\alpha_i}(z)^p R(b_i(z)), \tag{4.13}$$

where $\eta \in \mathbb{K}$ and

$$R(b_i(z)) = Y(r_i, z) = b_i(z)^p + \dots \tag{4.14}$$

is a (normal ordered) polynomial in $b_i(z)$ and its derivatives and $r_i \in \pi^{\kappa-\kappa_c}$. Formula (4.4) now follows from the proposition below.

PROPOSITION 4.7. *We have*

(1) $R(b_i(z)) = b_i(z)^p - \partial^{(p-1)}b_i(z)$;
(2) $\eta = (\kappa^p - \kappa)\langle e_i, f_i\rangle$.

SKETCH OF A PROOF. It is possible to realize Wakimoto modules over $\mathbb{Z}$ as limits of twisting Verma modules, denoted by $W_{\mathbb{Z}}(\lambda)$, on which $e_{i,n}, h_{i,n}, f_{i,n}$ act. Then the formulas (3.3)-(3.5) are understood as a congruence equation modulo $pW_{\mathbb{Z}}(\lambda)$ when acting on any $v \in W_{\mathbb{Z}}(\lambda)$; moreover $(e^p_{-1}|0\rangle)^{\mathtt{w}}_{(n)}v \in pW_{\mathbb{Z}}(\lambda), (f^p_{-1}|0\rangle)^{\mathtt{w}}_{(n)}v \in$

$pW_{\mathbb{Z}}(\lambda)$, thanks to Lemmas 4.3 and 4.4. From weight consideration, we have $(e^p_{-1}|0\rangle)_{(p-1)}f^p_{-1}|0\rangle \equiv -p(h^p_{-1} - h_{-p}) \mod p^2V_{\mathbb{Z}}$.

On the other hand, for $n \geq 0$ and $v \in W_{\mathbb{Z}}(\lambda)$, we have

$$\Big((e^p_{-1}|0\rangle)_{(p-1)}f^p_{-1}|0\rangle\Big)^{\mathtt{W}}_{(n)} v = \sum_{i=0}^{p-1}(-1)^i\binom{p-1}{i}\Big((e^p_{-1}|0\rangle)^{\mathtt{W}}_{(p-1-i)}(f^p_{-1}|0\rangle)^{\mathtt{W}}_{(n+i)}v$$

$$- (-1)^{p-1}(f^p_{-1}|0\rangle)^{\mathtt{W}}_{(n+p-1-i)}(e^p_{-1}|0\rangle)^{\mathtt{W}}_{(i)}v\Big). \tag{4.15}$$

But if we compute (4.15) by applying (3.3)–(3.5), the only term involving $b_{i,n}$ is given by $-r_iv$, which by (4.14) must be equal to $-(b^p_{i,-1} - b_{i,-p})v$ modulo $pW_{\mathbb{Z}}(\lambda)$. Part (1) now follows from this together with (4.14).

Part (2) reduces to the $\mathfrak{sl}_2$ case by Lemma 3.6. □

## 5. Irreducible baby Wakimoto modules $\mathfrak{w}(-\rho)$

**5.1. Mathieu's character formula reformulated.** For an integral weight $\lambda \in \mathfrak{h}^*$, denote by $\mathfrak{l}(\lambda)$ the irreducible quotient $\mathfrak{g}$-module of the Verma $\mathfrak{g}$-module of high weight $\lambda$. Recall the torus $T$ from \$2.1. Then $\mathfrak{l}(\lambda)$ is naturally an $\mathfrak{g}$-$T$-module in the sense of Jantzen [**Jan**], and this allows one to makes sense its (formal) character $\mathrm{ch}\, \mathfrak{l}(\lambda)$ in the usual sense.

Mathieu [**Ma**] proved the following character formula

$$\mathrm{ch}\ \mathfrak{l}(-\rho) = e^{-\rho} \prod_{\alpha\in\Delta^{\mathrm{re}}_+} \frac{(1-e^{-p\alpha})}{(1-e^{-\alpha})}. \tag{5.1}$$

Note that $-\rho$ is a weight at the critical level $\kappa_c$. We have the following reformulation of a main result of Mathieu, which has the advantage that the irreducible $\mathfrak{g}$-module $\mathfrak{l}(-\rho)$ is realized explicitly as the baby Wakimoto module $\mathfrak{w}(-\rho)$ in terms of (restricted) free fields.

THEOREM 5.1. *The baby Wakimoto module $\mathfrak{w}(-\rho)$ is the irreducible high weight $\mathfrak{g}$-module of high weight $-\rho$.*

PROOF. By construction of the baby Wakimoto module, we have the following character formula:

$$\mathrm{ch}\ \mathfrak{w}(-\rho) = e^{-\rho} \prod_{\alpha\in\Delta^{\mathrm{re}}_+} \frac{(1-e^{-p\alpha})}{(1-e^{-\alpha})}.$$

The (obvious) surjective homomorphism $\mathfrak{w}(-\rho) \to \mathfrak{l}(-\rho)$ must be an isomorphism by a character comparison. □

As modules over $\mathfrak{g}$, we have $\mathfrak{l}(-\rho) = \mathfrak{l}((p-1)\rho)$. (More general $\mathfrak{l}(\lambda) = \mathfrak{l}(\mu)$ if $\lambda - \mu$ is a $p$-multiple of an integral weight of $\mathfrak{g}$.) Denote by $U = \mathbb{K} \otimes_{\mathbb{Z}} U_{\mathbb{Z}}$, where $U_{\mathbb{Z}}$ is the Kostant-Garland $\mathbb{Z}$-form of the universal enveloping algebra of $\mathfrak{g}$. Denote by $L(\lambda)$ (the notation $l(\lambda)$ was used in [**Ma**]) the irreducible highest weight $U$-module of highest weight $\lambda$ (which is assumed to be integral). Note that the restricted enveloping algebra is a subalgebra of $U$, i.e., $\mathfrak{u}_0(\mathfrak{g}) \subseteq U$. Since $(p-1)\rho$ is a restricted weight, it follows by Mathieu [**Ma**, Lemma 1.7] that $L((p-1)\rho)$ when restricted to $\mathfrak{u}_0(\mathfrak{g})$ remains to be irreducible, and hence $L((p-1)\rho) \cong \mathfrak{l}((p-1)\rho)$ as $\mathfrak{g}$-modules. Therefore Theorem 5.1 and (5.1) have the following implication.

COROLLARY 5.2 ([**Ma**]). *We have the following character formulas:*

$$\operatorname{ch} l((p-1)\rho) = e^{(p-1)\rho} \prod_{\alpha\in\Delta_+^{\mathrm{re}}} \frac{(1-e^{-p\alpha})}{(1-e^{-\alpha})}, \tag{5.2}$$

$$\operatorname{ch} l(-\rho) = e^{-\rho} \prod_{\alpha\in\Delta_+^{\mathrm{re}}} \frac{1}{1-e^{-\alpha}}. \tag{5.3}$$

The above two formulas are equivalent by Steinberg tensor product theorem.

**5.2. Conjectures and further problems.** Recall the vertex algebra $V^\kappa(\mathfrak{g}_\mathbb{C})$ over $\mathbb{C}$ has trivial center at a non-critical level $\kappa$; at the critical level, $V^{\kappa_c}(\mathfrak{g}_\mathbb{C})$ has a large center, which is explicitly described in [**Fr1**, **Fr2**]. This center continues to make sense for $V^{\kappa_c}(\mathfrak{g})$ over $\mathbb{K}$ in characteristic $p$; we shall refer to this as the Harish-Chandra center of $V^{\kappa_c}(\mathfrak{g})$ and denote it by $\mathfrak{z}_{\mathrm{HC}}(V^{\kappa_c}(\mathfrak{g}))$.

CONJECTURE 5.3. (1) For $\kappa \neq \kappa_c$, the center of the vertex algebra $V^\kappa(\mathfrak{g})$ coincides with the $p$-center $\mathfrak{z}_0(V^\kappa(\mathfrak{g}))$.
(2) The center of the vertex algebra $V^{\kappa_c}(\mathfrak{g})$ is generated by the Harish-Chandra center $\mathfrak{z}_{\mathrm{HC}}(V^{\kappa_c}(\mathfrak{g}))$ and the $p$-center $\mathfrak{z}_0(V^{\kappa_c}(\mathfrak{g}))$.

A $p$-character $\xi^M$ of $\mathcal{A}^\mathfrak{g}$ is called graded if $\xi^M(a_{\alpha,n}) = 0 = \xi^M(a^*_{\alpha,n})$ for all $n \neq 0$. A $p$-character $\xi^\pi$ of $\mathcal{B}^\mathfrak{h}_\kappa$ is graded if $\xi^\pi(b_{i,n}) = 0$ for all $n \neq 0$ and $1 \le i \le \ell$. Similarly, a graded $p$-character for $\mathfrak{g}$ can be defined.

The modular representation theory of (finite-dimensional) Lie algebras has been well developed; cf. the review of Jantzen [**Jan**]. It will be of great interest to develop modular representation theory for an affine Lie algebra $\mathfrak{g}$, say when the $p$-character is (graded) semisimple or nilpotent. In particular, one may ask if the baby Wakimoto modules are irreducible for *generic* (graded) semisimple $p$-characters. The modular representation theory of the algebra $U$ (or the corresponding algebraic group of $\mathfrak{g}$) has been very challenging; we refer to [**Lai**] and the references therein for results in this direction. The modular representation theory of $\mathfrak{g}$ should be somewhat more accessible and flexible by imposing various conditions on $p$-characters.

## References

[BMR] R. Bezrukavnikov, I. Mirković, and D. Rumynin, *Localization of modules for a semisimple Lie algebra in prime characteristic*, Ann. of Math. (2) **167** (2008), no. 3, 945–991, DOI 10.4007/annals.2008.167.945. With an appendix by Bezrukavnikov and Simon Riche. MR2415389 (2009e:17031)

[BR] R. E. Borcherds and A. J. E. Ryba, *Modular Moonshine. II*, Duke Math. J. **83** (1996), no. 2, 435–459, DOI 10.1215/S0012-7094-96-08315-5. MR1390654 (98b:17030)

[FF] B. L. Feĭgin and È. V. Frenkel′, *A family of representations of affine Lie algebras* (Russian), Uspekhi Mat. Nauk **43** (1988), no. 5(263), 227–228, DOI 10.1070/RM1988v043n05ABEH001935; English transl., Russian Math. Surveys **43** (1988), no. 5, 221–222. MR971497 (89k:17016)

[Fr1] E. Frenkel, *Wakimoto modules, opers and the center at the critical level*, Adv. Math. **195** (2005), no. 2, 297–404, DOI 10.1016/j.aim.2004.08.002. MR2146349 (2006d:17018)

[Fr2] E. Frenkel, *Langlands correspondence for loop groups*, Cambridge Studies in Advanced Mathematics, vol. 103, Cambridge University Press, Cambridge, 2007. MR2332156 (2008h:22017)

[Jan] J. C. Jantzen, *Representations of Lie algebras in positive characteristic*, Representation theory of algebraic groups and quantum groups, Adv. Stud. Pure Math., vol. 40, Math. Soc. Japan, Tokyo, 2004, pp. 175–218. MR2074594 (2005i:17025)

[Lai] C.-J. Lai, *On Weyl modules over affine Lie algebras in prime characteristic*, preprint, arXiv:1310.3696v4, Transformation Groups (to appear).

[LM] H. Li and Q. Mu, *Heisenberg VOAs over Fields of Prime Characteristic and Their Representations*, arXiv:1501.04314.

[Ma] O. Mathieu, *On some modular representations of affine Kac-Moody algebras at the critical level*, Compositio Math. **102** (1996), no. 3, 305–312. MR1401425 (97g:17026)

[Wak] M. Wakimoto, *Fock representations of the affine Lie algebra $A_1^{(1)}$*, Comm. Math. Phys. **104** (1986), no. 4, 605–609. MR841673 (87m:17011)

RIMS, Kyoto University, Kyoto 606-8502, Japan
*E-mail address*: arakawa@kurims.kyoto-u.ac.jp

Department of Mathematics, University of Virginia, Charlottesville, Virginia 22904
*E-mail address*: ww9c@virginia.edu

Proceedings of Symposia in Pure Mathematics
Volume **92**, 2016
http://dx.doi.org/10.1090/pspum/092/01580

# Howe correspondence and Springer correspondence for dual pairs over a finite field

A.-M. Aubert, W. Kraśkiewicz, and T. Przebinda

ABSTRACT. We study the Howe correspondence for the unipotent representations of the irreducible dual reductive pairs $(\mathrm{G}', \mathrm{G}) = (\mathrm{GL}_{n'}(\mathbb{F}_q), \mathrm{GL}_n(\mathbb{F}_q))$ with $n' \leq n$, and $(\mathrm{G}', \mathrm{G}) = (\mathrm{Sp}_4(\mathbb{F}_q), \mathrm{O}_{2n}^+)(\mathbb{F}_q)$, where $\mathbb{F}_q$ is a finite field with $q$ elements ($q$ odd), and $\mathrm{O}_{2n}^+$ is the $\mathbb{F}_q$-split orthogonal group. We show how to extract a "preferred" irreducible representation of $G$ from the image by the (conjectural in the second case) correspondence of a given irreducible representation of $G'$.

## Contents



## 1. Introduction

In case of a dual pair $(\mathrm{G}', \mathrm{G})$ defined over a finite field, the integral

$$\int_{\mathrm{G}'} \Theta(g'g)\, \Theta_{\Pi'^c}(g')\, dg' \qquad (g \in \mathrm{G}),$$

where $\Theta$ is the character of the Weil representation and $\Pi'^c$ is the representation contragredient to $\Pi'$, is a finite sum which obviously converges and defines a class function on G. This class function decomposes into a sum of several irreducible characters $\Theta_\Pi$. In other words Howe correspondence often does not associate a single irreducible representation of G to a given irreducible representation $\Pi'$ of $\mathrm{G}'$ and the situation is quite complex.

Then the following question arises naturally: is there a "preferred" representation among the irreducible representations of G which correspond to $\Pi'$? It is the aim of this article to propose a candidate for such a preferred irreducible representation, assuming that $\Pi'$ is unipotent.

---

2010 *Mathematics Subject Classification.* Primary 22E45; Secondary 20C33, 22E46.

The second author was partially support by the NSA grant H98230-13-1-0205.

Let $\mathbb{F}_q$ be a finite field of $q$ elements of characteristic $p$. As a consequence of our main result (assuming here for the simplicity of the exposition that $p$ of is large enough), we obtain that, in the following situations

(1) the dual pair is of type II, *i.e.*, $(\mathrm{G}', \mathrm{G}) = (\mathrm{GL}_{n'}(\mathbb{F}_q), \mathrm{GL}_n(\mathbb{F}_q))$, and $\Pi'$ is unipotent;
(2) the dual pair is $(\mathrm{G}', \mathrm{G}) = (\mathrm{Sp}_4(\mathbb{F}_q), \mathrm{O}^+_{2n}(\mathbb{F}_q))$, where $\mathrm{O}^+_{2n}(\mathbb{F}_q)$ denotes the split orthogonal group, and $\Pi'$ is unipotent and belongs to the principal series of $\mathrm{G}'$,

the preferred representation is an irreducible representation $\Pi_{\mathrm{pref}}$ of G that corresponds to $\Pi'$ by Howe correspondence and is the unique such representation the wave front set of which contains the wave front set of any irreducible representation of G which correspond to $\Pi'$ (see Corollary 14).

More generally, we consider an irreducible dual pair $(\mathrm{G}', \mathrm{G})$ over $\mathbb{F}_q$, with $p$ odd (without further assumption on it). As shown in [**AM93**], Howe correspondence for this pair induces a (non-bijective) correspondence between unipotent representations of $\mathrm{G}'$ and G. This correspondence between unipotent representations has been described in [**AMR96**, Théorème 5.5] in the case of $(\mathrm{G}', \mathrm{G}) = (\mathrm{GL}_{n'}(\mathbb{F}_q), \mathrm{GL}_n(\mathbb{F}_q))$. Recall that unipotent representations of $\mathrm{G}'$ are parametrized by partitions of $n'$. Assume that $n' \leq n$. We will prove that the unipotent representation of $G$, say $\Pi_{\mathrm{pref}}$, that is parametrized by the *joint partition* $\mu' \cup (n - n')$ (see Definition 1), occurs in the image by the correspondence of the representation of $\mathrm{G}'$, say $\Pi'$, that is parametrized by $\mu'$. Moreover, every representation $\Pi$ of G which occurs in the image of $\Pi'$ is parametrized by a partition of $n$ which is larger than $\mu' \cup (n-n')$ for the usual order on partitions. It follows that the closure of the unipotent support of each such representation $\Pi$ contains the unipotent support of $\Pi_{\mathrm{pref}}$.

In the case of ortho-symplectic dual pairs, the correspondence between unipotent representations has been described conjecturally in [**AMR96**, Conjecture 3.11] in terms of a (in general non-bijective) correspondence between irreducible representations of two Weyl groups.

In [**KS05**, Theorem 5.15], Kable and Sanat have proved the validity of the conjecture for the dual pair $(\mathrm{Sp}_4(\mathbb{F}_q), \mathrm{SO}^+_{2n}(\mathbb{F}_q))$ in the case of unipotent representations that belong to the principal series. Let $\mathrm{O}^-_{2n}(\mathbb{F}_q)$ denote the non-split orthogonal group, and let $\varepsilon = \pm$. The conjecture for the dual pair $(\mathrm{Sp}_{2n'}(\mathbb{F}_q), \mathrm{O}^\varepsilon_{2n}(\mathbb{F}_q))$ has been also confirmed computationally in [**AMR96**] for $n, n' \leq 11$ up-to a slight ambiguity in the principal series case $\mathrm{O}^\varepsilon_{2n}(\mathbb{F}_q)$ (in which case we have only the restriction of Weil Representation to $\mathrm{Sp}_{2n'}(\mathbb{F}_q) \cdot \mathrm{SO}^\varepsilon_{2n}(\mathbb{F}_q)$).

We will compute explicitly that correspondence between representations of Weyl groups in the case where one of them is $\mathsf{W}_2 = \mathsf{W}(\mathrm{B}_2)$ (see Proposition 8), give its translation into a correspondence between $u$-symbols and extract a bijective correspondence which behaves well with respect to unipotent classes (see Theorem 10).

For instance, for unipotent representations in the principal series of split groups, assuming the validity of the conjectural description of the correspondence in this case, and that we have $n \geq 3$ and $n' = 2$, we prove that the representation of $G$, say $\Pi_{\mathrm{pref}}$, that is parametrized by the pair of partitions $(\xi', \eta' \cup (n - n'))$ of $n$, occurs in the image by the correspondence of the representation of $\mathrm{G}'$, say $\Pi'$, that is parametrized by the pair of partitions $(\xi', \eta')$ of $n'$. Moreover, every representation of G which occurs in the image of $\Pi'$ is such that the closure of its unipotent support contains the unipotent support of $\Pi_{\mathrm{pref}}$ (see Corollary 12).

We thank Alberto Mínguez for his interest and his suggestion of considering the case of dual pairs of type II.

## 2. Unipotent representations

Let G be the group of $\mathbb{F}_q$-rational points of a connected algebraic reductive group $\mathbf{G}$ over $\bar{\mathbb{F}}_q$, defined over $\mathbb{F}_q$, and let $F\colon \mathbf{G} \to \mathbf{G}$ be the corresponding Frobenius map so that $\mathrm{G} = \mathbf{G}^F$ (fixed points by $F$).

To each G-conjugacy class of pairs $(\mathbf{T}, \theta)$ where $\mathbf{T}$ is an $F$-stable maximal torus in $\mathbf{G}$ and $\theta$ is an irreducible character of $T = \mathbf{T}^F$, Deligne and Lusztig attached a virtual character $R_{\mathbf{T}}^{\mathbf{G}}(\theta)$ of G, [**DL76**].

Recall that the *uniform class functions* on G are, by definition, the complex linear combinations of Deligne-Lusztig characters $R_{\mathbf{T}}^{\mathbf{G}}(\theta)$. Recall also that an irreducible representation of G is called *unipotent* if its character has non-zero scalar product with $R_{\mathbf{T}}^{\mathbf{G}}(1)$ for some $\mathbf{T}$. If $\mathrm{G} = \mathrm{GL}_n(\mathbb{F}_q)$, then the uniform class functions span the space of all class functions on G. For G arbitrary, it is not the case in general: for instance, the character of the cuspidal unipotent representation $\theta_{10}$ of the symplectic group $\mathrm{Sp}_4(\mathbb{F}_q)$ defined by Srinivasan is not uniform.

Because we will need to include the case of orthogonal groups, it is necessary to extend the definition of $R_{\mathbf{T}}^{\mathbf{G}}(\theta)$ to the case when $\mathbf{G}$ is a disconnected reductive algebraic group. In this case, we put $R_{\mathbf{T}}^{\mathbf{G}}(\theta) := \mathrm{Ind}_{\mathrm{G}^\circ}^{\mathrm{G}}(R_{\mathbf{T}}^{\mathbf{G}^\circ}(\theta))$, where $\mathbf{G}^\circ$ denotes the identity connected component of $\mathbf{G}$ and $\mathrm{G}^\circ := (\mathbf{G}^\circ)^F$. We will call *uniform class functions* all the linear combinations of $R_{\mathbf{T}}^{\mathbf{G}}(\theta)$. An irreducible representation of G is called *unipotent* if its character has non-zero scalar product with $R_{\mathbf{T}}^{\mathbf{G}}(1)$ for some $\mathbf{T}$.

Since the cyclic group ${\mathbb{F}_q}^*$ is of even order, $|{\mathbb{F}_q}^*/{\mathbb{F}_q}^{*2}| = 2$. Therefore there are exactly two non-equivalent non-degenerate symmetric bilinear forms on the vector space ${\mathbb{F}_q}^{2n}$, see [**Jac74**, Theorem 6.9], one is split and the other one is not split. Let $\mathrm{O}_{2n}^+(q) = \mathrm{O}_{2n}^+(\mathbb{F}_q)$ (resp. $\mathrm{O}_{2n}^-(q) = \mathrm{O}_{2n}^-(\mathbb{F}_q)$) denotes the corresponding split (resp. non-split) orthogonal group. See [**DM91**, sec. 15.3] for more details. Also, we shall write $\mathrm{Sp}_{2n}(q) := \mathrm{Sp}_{2n}(\mathbb{F}_q)$.

We recall some results from [**Lus80**]. The group $\mathrm{Sp}_{2n}(q)$ has a unipotent cuspidal irreducible representation if and only if $n$ is a triangular number, that is, $n = k^2 + k$ for some $k \in \mathbb{N}$. The group $\mathrm{Sp}_{2(k^2+k)}(q)$ has a unique unipotent cuspidal representation. Similarly, the group $\mathrm{SO}_{2n}^\varepsilon(q)$, with $\varepsilon \in \{-, +\}$, has a unipotent cuspidal irreducible representation if and only if $n$ is a square, that is, $n = k^2$ for some $k \in \mathbb{N}$. The group $\mathrm{SO}_{2k^2}^\varepsilon(q)$ has a unique unipotent cuspidal representation, say $\Pi_k$. It follows that $\mathrm{O}_{2n}^\varepsilon(q)$ admits unipotent cuspidal representations if and only if $n = k^2$ for some $k \in \mathbb{N}$, and that $\mathrm{O}_{2k^2}^\varepsilon(q)$ has exactly two unipotent cuspidal representations, $\Pi_k^{\mathrm{I}}$ and $\Pi_k^{\mathrm{II}}$. (Indeed, we have $\mathrm{Ind}_{\mathrm{SO}_{2k^2}^\varepsilon(q)}^{\mathrm{O}_{2k^2}^\varepsilon(q)} \Pi_k = \Pi_k^{\mathrm{I}} + \Pi_k^{\mathrm{II}}$. Both $\Pi_k^{\mathrm{I}}$ and $\Pi_k^{\mathrm{II}}$ have the same restriction to $\mathrm{SO}_{2k^2}^\varepsilon(q)$ and thus differ by tensoring with the determinant character of $\mathrm{O}_{2k^2}^\varepsilon(q)$.) See [**Lus77**, Theorem 8.2] or [**AM93**, Theorem 5.1] for the details.

Let $\mathbf{M} := \mathrm{Sp}_{2(k^2+k)} \times \mathbf{T}$ (resp. $\mathbf{M} := \mathrm{O}_{2k^2}^\varepsilon \times \mathbf{T}$), where $\mathbf{T}$ is a split torus of $\mathbf{G} = \mathrm{Sp}_{2n}$ (resp. $\mathbf{G} = \mathrm{O}_{2n}^\varepsilon$), and let $\Pi^{\mathbf{M}}$ be a unipotent cuspidal irreducible representation of $\mathbf{M}$. The representation $\Pi^{\mathbf{M}}$ is the tensor product of the unipotent cuspidal representation of $\mathrm{Sp}_{2(k^2+k)}$ (resp. $\Pi_k^{\mathrm{I}}$ or $\Pi_k^{\mathrm{II}}$) with the trivial representation of $\mathbf{T}$. On the other hand, $\mathbf{M}$ is an $\mathbb{F}_q$-rational Levi subgroup of an $\mathbb{F}_q$-rational parabolic subgroup $\mathbf{P}$ of $\mathbf{G}$ and the commuting algebra $\mathrm{End}_{\mathrm{G}}(\mathrm{Ind}_{\mathbf{P}}^{\mathbf{G}}(\Pi^{\mathbf{M}}))$ (where the cuspidal representation $\Pi^{\mathbf{M}}$ of $\mathbf{M}$ is trivially extended to the unipotent radical of $\mathbf{P}$, that is, $\mathrm{Ind}_{\mathbf{P}}^{\mathbf{G}}(\Pi^{\mathbf{M}})$, also denoted by $R_{\mathbf{M}}^{\mathbf{G}}(\Pi^{\mathbf{M}})$, is the usual Harish-Chandra induction) is an Iwahori-Hecke algebra of type $\mathrm{B}_{\tilde{n}}$, with $\tilde{n} := n-(k^2+k)$ (resp. $\tilde{n} := n-k^2$), see for instance [**Lus80**] or [**AMR96**, § 3.A]. Hence the irreducible representations of G which occur in $\mathrm{Ind}_{\mathbf{P}}^{\mathbf{G}}(\Pi^{\mathbf{M}})$ are in bijection with $\mathrm{Irr}(\mathsf{W}_{\tilde{n}})$, where $\mathsf{W}_{\tilde{n}} = \mathsf{W}(\mathrm{B}_{\tilde{n}}) = (\mathbb{Z}/2\mathbb{Z})^{\tilde{n}} \rtimes \mathfrak{S}_{\tilde{n}}$ (*cf.* [**Car93**, Chapter 10] or [**Lus84**, Corollary 8.7]).

We will denote by $\Pi^{\mathrm{G}}_{\Pi^{\mathrm{M}},\rho}$ the irreducible representation of G which corresponds to $\rho \in \mathrm{Irr}(\mathsf{W}_{\tilde{n}})$ by this bijection.

We put

$$\mathrm{Sp} := \{\mathrm{Sp}_{2n}(q) \,:\, n \in \mathbb{N}\} \quad \text{and} \quad \mathrm{O}^{\varepsilon} := \{\mathrm{O}^{\varepsilon}_{2n}(q) \,:\, n \in \mathbb{N}\}.$$

We call Sp (resp. $\mathrm{O}^{\varepsilon}$) a *Witt tower of symplectic* (resp. *orthogonal*) *type*. Let $\mathcal{T}$, $\mathcal{T}'$ be two Witt towers, one is of symplectic type and the other one is of orthogonal type.

For a finite group $H$ let $\mathfrak{R}(H)$ denote the free abelian group generated by the irreducible characters of $H$. Thus the subset of the irreducible characters $\mathrm{Irr}(H) \subseteq \mathfrak{R}(H)$ is a base of $\mathfrak{R}(H)$ over $\mathbb{Z}$. Let $\mathrm{G}'_{m'}$ be an element of $\mathcal{T}'$ and let $\mathrm{G}_m$ be an element of $\mathcal{T}$. Denote by $\omega_{m',m}$ the projection onto the space of the uniform class functions on $\mathrm{G}'_{m'} \times \mathrm{G}_m$ of the pullback of the character of the oscillator representation (determined by one fixed character of the field $\mathbb{F}_q$) via the map $\mathrm{G}'_{m'} \times \mathrm{G}_m \ni (g',g) \to g'g \in \mathrm{Sp}_{4m'm}(q)$. By Howe correspondence for the dual pair $(\mathrm{G}'_{m'}, \mathrm{G}_m)$ we shall understand the map

$$\theta^{\mathrm{G}_m} : \mathfrak{R}(\mathrm{G}'_{m'}) \to \mathfrak{R}(\mathrm{G}_m) \tag{1}$$

defined by

$$\omega_{m',m} = \sum_{\Pi' \in \mathrm{Irr}(\mathrm{G}'_{m'})} \Pi' \otimes \theta^{\mathrm{G}_m}(\Pi'). \tag{2}$$

(See [**AMR96**, (1.4)], where $\theta^{\mathrm{G}_m}(\Pi')$ was denoted by $\Theta_{\mathrm{G}_m}(\Pi')$.)

Let $\Pi'$ be a cuspidal irreducible representation of an element $\mathrm{G}'_{m'}$ of $\mathcal{T}'$. Then there exists $\mathrm{G}_m \in \mathcal{T}$ such that $\theta^{\mathrm{G}_m}(\Pi')$ is a cuspidal irreducible representation of $\mathrm{G}_m$, see [**AMR96**, Theorem 3.7]. Moreover (see *loc. cit.*), the image by Howe correspondence for the dual pair $(\mathrm{G}'_{m'+l'}, \mathrm{G}_{m+l})$, with $l', l \in \mathbb{N}$, of each component of the Harish-Chandra parabolic induced representation $R^{\mathbf{G}'_{m'+l'}}_{\mathbf{G}'_{m'}\times\mathbf{T}'}(\Pi')$ (where $\mathbf{T}'$ is a split torus in $\mathbf{G}'_{m'+l'}$) belongs to the Harish-Chandra parabolic induced representation $R^{\mathbf{G}_{m+l}}_{\mathbf{G}_m\times\mathbf{T}}(\theta^{\mathrm{G}_m}(\Pi'))$ (where $\mathbf{T}$ is a split torus in $\mathbf{G}_{m+l}$).

Using the description of the uniform part of the restriction of $\mathrm{G}_m \times \mathrm{G}'_{m'}$ of the Weil representation obtained by Srinivasan in [**Sri79**], Adams and Moy proved that Howe correspondence sends unipotent representations to unipotent representations, [**AM93**, Theorem 3.5], and that the unique cuspidal unipotent representation of the group $\mathrm{Sp}_{2(k^2+k)}(q)$ corresponds to the representation $\Pi^{\mathrm{II}}_k$ of $\mathrm{O}^{\varepsilon}_{2k^2}(q)$ if $\varepsilon$ is the sign of $(-1)^k$ and to the representation $\Pi^{\mathrm{I}}_{k+1}$ of $\mathrm{O}^{\varepsilon}_{2(k+1)^2}(q)$, where $\varepsilon$ is the sign of $(-1)^{k+1}$, otherwise, [**AM93**, Theorem 5.2]. (In fact this defines the representations $\Pi^{\mathrm{I}}_k$ and $\Pi^{\mathrm{II}}_k$.)

## 3. Dual pairs of type II

In this section we will consider the case of the dual pair $(\mathrm{G}',\mathrm{G}) = (\mathrm{GL}_{n'}(q), \mathrm{GL}_n(q))$. We assume that $n' \leq n$. The characters of the unipotent irreducible representations of G are in bijection with the irreducible characters of the symmetric group $\mathfrak{S}_n$, hence in bijection with the partitions of $n$. Let $\rho_\mu$ denote the irreducible representation of $\mathfrak{S}_n$ which corresponds to the partition $\mu$ of $n$. Define

$$R_{\rho_\mu} := \frac{1}{n!}\sum_{\sigma\in\mathfrak{S}_n} \rho_\mu(\sigma)\, R_{\mathbf{T}_\sigma}^{\mathrm{GL}_n}(1), \tag{3}$$

where $\mathbf{T}_\sigma$ is a maximal torus of type $\sigma$. Then $R_{\rho_\mu}$ is a unipotent irreducible character of G, and each such character is of this form for some partition $\mu$ of $n$ (see for instance [**DM91**, sec. 15.4]).

Recall that a *partition* of a positive integer $n$ is a finite sequence $\lambda = [\lambda_1 \geq \lambda_2 \geq \cdots \geq \lambda_k \geq 0]$ of integers $\lambda_i$ such that $\sum_{i=1}^k \lambda_i = n$. Let $\mathrm{ht}(\lambda)$ denote the *hight* of the partition $\lambda$ (that is, the largest $i$ with $\lambda_i \neq 0$). Flipping a Young diagram of a partition $\lambda$ of $n$ over its main diagonal (from upper left to lower right), we obtain the Young diagram of another partition ${}^t\lambda$ of $n$, which is called the *conjugate partition* of $\lambda$. Thus, for $\lambda = [\lambda_1 \geq \lambda_2 \geq \cdots \geq \lambda_k]$, we have ${}^t\lambda = [{}^t\lambda_1 \geq {}^t\lambda_2 \geq \cdots \geq {}^t\lambda_l]$, where $l = \lambda_1$ and ${}^t\lambda_j = |\{i \,:\, 1 \leq i \leq k, \lambda_i \geq j\}|$ for $1 \leq j \leq l$.

If $\lambda = [\lambda_1 \geq \lambda_2 \geq \cdots \geq \lambda_k]$ and $\mu = [\mu_1 \geq \mu_2 \geq \cdots \geq \mu_h]$ are any partitions, we write $\mu \subset \lambda$ if the followings holds: $\mathrm{ht}(\mu) \leq \mathrm{ht}(\lambda)$ and $\mu_i \leq \lambda_i$ for all $1 \leq i \leq \mathrm{ht}(\mu)$. If we identify $\lambda$ and $\mu$ with their Young diagrams, this means that the diagram of $\mu$ is contained in those of $\lambda$. Removing the boxes of $\lambda$ which belong to $\mu$, we obtain a *skew diagram* which we denote by $\lambda - \mu$.

We will also need to consider the *intersection partition* of $\lambda$ and $\mu$:

$$\lambda \cap \mu := \left[\min(\lambda_1, \mu_1), \ldots, \min(\lambda_{\min(k,h)}, \mu_{\min(k,h)})\right].$$

We have $\mu \subset \lambda$ if and only if $\lambda \cap \mu = \mu$.

Let $\nu = [\nu_1 \geq \nu_2 \geq \cdots \geq \nu_m] \subset \lambda\cap\mu$. Then we denote by $p_{\lambda=\mu}(\nu)$ the partition $(\nu_i)_{\{i:\lambda_i=\mu_i\}}$ and we put

$$\lambda \cap^{=} \mu := p_{\lambda=\mu}(\lambda\cap\mu).$$

We will say that $\lambda$ and $\mu$ are *close* if for each $i$ we have $|\lambda_i - \mu_i| \leq 1$.

For later use, we will now introduce some more notation. If $\lambda = [\lambda_1 \geq \lambda_2 \geq \cdots \geq \lambda_k]$ is a partition of $n$ and $\mu = [\mu_1 \geq \mu_2 \geq \cdots \geq \mu_h]$ is a partition of $m$, by adding zero parts if necessary we can assume that $h = k$, and we define the partition $\lambda \oplus \mu$ of $n + m$ as

$$(\lambda \oplus \mu)_i := \lambda_i + \mu_i, \quad \text{for } 1 \leq i \leq k.$$

For a partition $\lambda$ and for any integer $i$, let $n_i(\lambda)$ be the numbers of $j \geq 1$ such that $\lambda_j = i$.

Definition 1. *Let $\lambda \cup \mu$ be the unique partition of $n + m$ such that*

$$n_i(\lambda \cup \mu) = n_i(\lambda) + n_i(\mu), \quad \textit{for each } i \geq 1.$$

We observe that ${}^t(\lambda \oplus \mu) = {}^t\lambda \cup {}^t\mu$ and ${}^t(\lambda \cup \mu) = {}^t\lambda \oplus {}^t\mu$. Also, for any $L \in \mathrm{N}$, we have $(L) \cup \mu = [\mu_1 \geq \cdots \geq L \geq \cdots \geq \mu_l]$ (or $[L \geq \mu_1 \geq \cdots \geq \mu_l]$ or $[\mu_1 \geq \cdots \geq \mu_l \geq L]$).

Consider $\mathfrak{R}(\mathfrak{S}):=\bigoplus_{n\geq 0}\mathfrak{R}(\mathfrak{S}_n)$ (it is a free $\mathbb{Z}$-module with basis $\bigcup_{n\geq 0}\mathrm{Irr}(\mathfrak{S}_n)$), and define a map $\theta^{\mathfrak{S}}:\mathfrak{R}(\mathfrak{S})\to\mathfrak{R}(\mathfrak{S})$ by

$$\rho_{\mu'}\mapsto\sum_{{}^t\mu\text{ close to }{}^t\mu'} f({}^t\mu'\cap^{=}{}^t\mu)\,\rho_\mu,$$

where, if $\nu=[r^{a_1},\ldots,1^{a_r}]$, we have put $f(\nu)=\prod_i a_i$, and where the empty partition is sent by $f$ to 1 (in accordance with [**AMR96**, proof of Lemma 5.4]).

THEOREM 2. [**AMR96**, Théorème 5.5] *Howe correspondence between unipotent characters of* $\mathrm{GL}_{n'}(q)$ *and* $\mathrm{GL}_n(q)$ *is given by the map*

$$R^{\mathrm{GL}_{n'}}_{\rho_{\mu'}}\mapsto R^{\mathrm{GL}_n}_{\theta^{\mathfrak{S}}(\rho_{\mu'})}.$$

The following result is a direct consequence of Theorem 2.

THEOREM 3. *Let* $(n',n)$ *be a pair of positive integers with* $n'\leq n$. *Let* $\mu'$ *be a partition of* $n'$. *The unipotent representations of* $\mathrm{GL}_{n'}(q)$ *and* $\mathrm{GL}_n(q)$ *with characters* $R_{\rho_{\mu'}}$ *and* $R_{\rho_{\mu'\cup(n-n')}}$, *respectively, correspond by Howe correspondence.*

*Moreover, any representation of* $\mathrm{GL}_n(q)$ *which belongs to the image of* $R_{\rho_{\mu'}}$ *by Howe correspondence is of the form* $R_{\rho_\mu}$ *where* $\mu\geq(\mu'\cup(n-n'))$, *where* $\geq$ *denotes the usual order on partitions.*

PROOF. We note that the unipotent representation $R_{\rho_{\mu'\cup(n-n')}}$ occurs in the image of $R_{\rho_{\mu'}}$ by Howe correspondence. Indeed we have

$${}^t(\mu'\cup(n-n'))={}^t(\mu')\oplus 1^{n-n'},$$

the partitions ${}^t(\mu')$ and ${}^t(\mu'\cup(n-n'))$ are close, and, since

$${}^t(\mu')\cap^{=}{}^t(\mu'\cup(n-n'))=\emptyset,$$

we have $f({}^t(\mu')\cap^{=}{}^t(\mu'\cup(n-n')))=1$.

Now, ${}^t(\mu'\cup(n-n'))$ is the largest partition in the set of partitions of $n$ which are close to ${}^t(\mu')$. Hence, if $R_{\rho_\mu}$ belongs to the image of $R_{\rho_{\mu'}}$ by Howe correspondence we have ${}^t\mu\leq{}^t(\mu'\cup(n-n'))$, *i.e.*, $\mu\geq\mu'\cup(n-n')$. □

## 4. Ortho-symplectic dual pairs

**4.1. A correspondence between Weyl groups.** Let $(n_1,n_2)$ be a pair of positive integers. Let $k$ be an integer such that $0\leq k^2+k\leq n_1$ and $k^2\leq n_2$, and let $\Pi_k^{\mathrm{sp}}$ denote the unipotent cuspidal representation of $\mathrm{Sp}_{2(k^2+k)}$. We denote by $\varepsilon_k$ the sign of $(-1)^k$.

It follows that:

- Howe correspondence for the dual pair $(\mathrm{Sp}_{2n_1}(q),\mathrm{O}^{\varepsilon_k}_{2n_2}(q))$ induces a correspondence between irreducible components of $R^{\mathbf{Sp}_{2n_1}}_{\mathbf{Sp}_{2(k^2+k)}\times\mathbf{T}}(\Pi_k^{\mathrm{sp}}\otimes 1)$ and irreducible components of $R^{\mathbf{O}^{\varepsilon_k}_{2n_2}}_{\mathbf{O}^{\varepsilon_k}_{2k^2}\times\mathbf{T}}(\Pi_k^{\mathrm{II}}\otimes 1)$,
- Howe correspondence for the dual pair $(\mathrm{Sp}_{2n_1}(q),\mathrm{O}^{\varepsilon_{k+1}}_{2n_2}(q))$ induces a correspondence between irreducible components of $R^{\mathbf{Sp}_{2n_1}}_{\mathbf{Sp}_{2(k^2+k)}\times\mathbf{T}}(\Pi_k^{\mathrm{sp}}\otimes 1)$ and irreducible components of $R^{\mathbf{O}^{\varepsilon_{k+1}}_{2n_2}}_{\mathbf{O}^{\varepsilon_{k+1}}_{2(k+1)^2}\times\mathbf{T}}(\Pi_{k+1}^{\mathrm{I}}\otimes 1)$.

All these irreducible components are unipotent, see [**Lus84**, (8.5.1)].

Let $k_2 \in \{k, k+1\}$. We set

$$\Pi^{\rm or}_{k_2} := \begin{cases} \Pi^{\rm II}_k & \text{if } k_2 = k, \\ \Pi^{\rm I}_{k+1} & \text{if } k_2 = k+1, \end{cases} \tag{4}$$

$$\tilde n_1(k) := n_1 - (k^2+k), \quad \tilde n_2(k_2) := n_2 - (k_2)^2. \tag{5}$$

Then Howe correspondence for the dual pair $(\mathrm{G},\mathrm{G}') = (\mathrm{Sp}_{2n_1}(q), \mathrm{O}^{\varepsilon_{k_2}}_{2n_2}(q))$ induces a correspondence, $\Theta^{\mathrm{G},\mathrm{G}'}_k$, between $\mathrm{Irr}(\mathsf{W}_{\tilde n_1(k)})$ and $\mathrm{Irr}(\mathsf{W}_{\tilde n_2(k_2)})$, defined as follows.

DEFINITION 4. *We will say that the representations $\rho \in \mathrm{Irr}(\mathsf{W}_{\tilde n_1(k)})$ and $\rho' \in \mathrm{Irr}(\mathsf{W}_{\tilde n_2(k_2)})$ correspond by $\Theta^{\mathrm{G},\mathrm{G}'}_k$ if the character of $\Pi^{\mathrm{G}}_{\Pi^{\rm sp}_k \otimes 1,\rho} \otimes \Pi^{\mathrm{G}'}_{\Pi^{\rm or}_{k_2}\otimes 1,\rho'}$ has a non-zero scalar product with $\omega_{n_1,n_2}$.*

In particular, taking $k = k_2 = 0$, we obtain a correspondence between $\mathrm{Irr}(\mathsf{W}_{n_1})$ and $\mathrm{Irr}(\mathsf{W}_{n_2})$.

Let $\mathrm{sgn}_{\mathrm{CD},\tilde n} \colon \mathsf{W}_{\tilde n} \to \{\pm 1\}$ denote the unique character whose restriction to the normal subgroup $(\mathbb{Z}/2\mathbb{Z})^{\tilde n}$ of $\mathsf{W}_{\tilde n}$ is the product of the sign characters and that is trivial on the subgroup $\mathfrak{S}_{\tilde n}$. The kernel of $\mathrm{sgn}_{\mathrm{CD},\tilde n}$ is isomorphic to the Weyl group $\mathsf{W}(\mathrm{D}_n)$. The restriction of the character $\mathrm{sgn}_{\mathrm{CD},\tilde n}$ to the subgroup $\mathsf{W}_{\tilde n-1}$ of $\mathsf{W}_{\tilde n}$ equals the character $\mathrm{sgn}_{\mathrm{CD},\tilde n-1}$. Because of this, we will denote $\mathrm{sgn}_{\mathrm{CD},\tilde n}$ simply by $\mathrm{sgn}_{\mathrm{CD}}$.

A conjectural description of the correspondence $\Theta^{\mathrm{G},\mathrm{G}'}_k$ was stated in [**AMR96**]. It can be formulated as follows:

CONJECTURE 1. *The representations $\rho \in \mathrm{Irr}(\mathsf{W}_{\tilde n_1(k)})$ and $\rho' \in \mathrm{Irr}(\mathsf{W}_{\tilde n_2(k_2)})$ correspond by $\Theta^{\mathrm{G},\mathrm{G}'}_k$ if and only if $\rho\otimes\rho'$ has a non-zero scalar product with*

$$\sum_{0\le r\le N'} \sum_{\rho\in \mathrm{Irr}(\mathsf{W}_r)} \mathrm{Ind}^{W_{\tilde n_1(k)}}_{W_r\times W_{\tilde n_1(k)-r}}(\sigma\otimes \mathrm{sgn}_{\mathrm{CD}}) \otimes \mathrm{Ind}^{W_{\tilde n_2(k_2)}}_{W_r\times W_{\tilde n_2(k_2)-r}}(\sigma\otimes \mathrm{sgn}_{\mathrm{CD}})$$

*(resp.* $$\sum_{0\le r\le N'} \sum_{\sigma\in \mathrm{Irr}(\mathsf{W}_r)} \mathrm{Ind}^{W_{\tilde n_1(k)}}_{W_r\times W_{\tilde n_1(k)-r}}(\sigma\otimes 1) \otimes \mathrm{Ind}^{W_{\tilde n_2(k_2)}}_{W_r\times W_{\tilde n_2(k_2)-r}}(\sigma\otimes \mathrm{sgn}_{\mathrm{CD}})),$$

*where $k_2 = k$ (resp. $k_2 = k+1$).*

We put

$$N' := \min(\tilde n_1(k), \tilde n_2(k_2)) \quad N := \max(\tilde n_1(k), \tilde n_2(k_2)), \quad \text{and} \quad L := N - N'. \tag{6}$$

In Conjecture 1, G stands for a symplectic group and G′ for an orthogonal group. However we would like to consider Howe's correspondence (1) in any of the two directions. Therefore we will consider the following cases and keep in mind that Conjecture 1 applies to any of them:

**Case 1:**

(a) $\mathrm{G}' = \mathrm{O}^{\varepsilon_k}_{2(k^2+N')}(q)$ and $\mathrm{G} = \mathrm{Sp}_{2(k^2+k+N))}(q)$.
(Here we have $N' = \tilde n_2(k)$ and $N = \tilde n_1(k)$.)

(b) $\mathrm{G}' = \mathrm{Sp}_{2(k^2+k+N')}(q)$ and $\mathrm{G} = \mathrm{O}^{\varepsilon_k}_{2(k^2+N)}(q)$.
(Here we have $N' = \tilde n_1(k)$ and $N = \tilde n_2(k)$.)

**Case 2:** $\mathrm{G}' = \mathrm{Sp}_{2(k^2+k+N')}(q)$ and $\mathrm{G} = \mathrm{O}^{\varepsilon_{k+1}}_{2((k+1)^2+N)}(q)$.
(Here we have $N' = \tilde{n}_1(k)$ and $N = \tilde{n}_2(k+1)$.)
**Case 3:** $\mathrm{G}' = \mathrm{O}^{\varepsilon_{k+1}}_{2((k+1)^2+N')}(q)$ and $\mathrm{G} = \mathrm{Sp}_{2(k^2+k+N)}(q)$.
(Here we have $N' = \tilde{n}_2(k+1)$, $N = \tilde{n}_1(k)$.)

Let $(\xi, \eta)$ be a pair of partitions of $N$, *i.e.*, $\xi$ and $\eta$ are two partitions with $|\xi| + |\eta| = N$. The irreducible representations of $\mathsf{W}_N$ are parameterized by the pairs of partitions of $N$ (see [**Lus77**]). The trivial representation of $\mathsf{W}_N$ corresponds to $((N), \emptyset)$ while the sign representation corresponds to $(\emptyset, (1^N))$ and the representation afforded by the character $\mathrm{sgn}_{\mathrm{CD}} = \mathrm{sgn}_{\mathrm{CD},N}$ corresponds to $(\emptyset, (N))$.

DEFINITION 5. *We define* $\underline{\theta}^{N',N} \colon \mathrm{Irr}(\mathsf{W}_{N'}) \to \mathrm{Irr}(\mathsf{W}_N)$ *by*

$$\underline{\theta}^{N',N}(\rho_{\xi',\eta'}) := \begin{cases} \rho_{\xi',(L)\cup\eta'} & \text{in Cases 1 and 2,} \\ \rho_{(L)\cup\xi',\eta'} & \text{in Case 3.} \end{cases}$$

THEOREM 6. *If Conjecture* 1 *holds, then the representations* $\rho_{\xi',\eta'} \in \mathrm{Irr}(\mathsf{W}_{N'})$ *and* $\underline{\theta}^{N',N}(\rho_{\xi',\eta'})$ *correspond by* $\Theta_k^{\mathrm{G},\mathrm{G}'}$.

In order to prove Theorem 6, we will need to introduce some more combinatorics.

Removing the boxes of $\lambda$ which belong to $\mu$, we obtain a *skew diagram* which we denote by $\lambda - \mu$. Then a *generalized tableau* of shape $\lambda - \mu$ is a filling of the boxes of $\lambda - \mu$ with positive integers such that the entries are weakly increasing from the left to the right along each row and strictly increasing down each column. Tableaux of shape $\lambda$ are examples of generalized tableaux. Let $T$ be a generalized tableau. Let $n_i = n_i(T)$ denote the number of occurrences of the integer $i$ in $T$. The *weight* of $T$ is defined as the sequence $(n_1, n_2, \ldots)$. The *word* $w(T)$ of $T$ is the sequence obtained by reading the entries of $T$ from right to left in successive rows, starting with the top row. On the other hand, any sequence $\mathbf{a} = (a_1, a_2, \ldots, a_l)$ with $a_i \in \{1, 2, \ldots, N\}$ is called a *lattice permutation* if, for $1 \le j \le l$ and $1 \le i \le N-1$, the number of occurrences of $i$ in $(a_1, a_2, \ldots, a_j)$ is not less than the number of occurrences of $i+1$.

Let $\lambda$, $\mu$, $\nu$ be partitions such that $|\lambda| = |\mu| + |\nu|$. The *Littlewood-Richardson coefficient* $c^\lambda_{\mu,\nu}$ is defined as the number of generalized tableaux $T$ of shape $\lambda - \mu$ and weight $\nu$ such that $w(T)$ is a lattice permutation.

The Littlewood-Richardson rule (*cf.* for instance [**GP00**, 6.1.1, 6.1.6]) says that

$$\mathrm{Ind}_{\mathfrak{S}_\ell \times \mathfrak{S}_{N-\ell}}^{\mathfrak{S}_N}(\rho_\mu \otimes \rho_\nu) = \sum_\lambda c^\lambda_{\mu,\nu}\, \rho_\lambda,$$

where the sum runs over all partitions $\lambda$ of $N$.

A similar rule occurs in the group $\mathsf{W}_n = \mathsf{W}(\mathrm{B}_n)$ (*cf.* [**GP00**, 6.1.3]):

$$\mathrm{Ind}_{\mathsf{W}_\ell \times \mathsf{W}_{N-\ell}}^{\mathsf{W}_N}(\rho_{\xi_1,\eta_1} \otimes \rho_{\xi_2,\eta_2}) = \sum_{(\xi,\eta)} c^\xi_{\xi_1,\xi_2}\, c^\eta_{\eta_1,\eta_2}\, \rho_{\xi,\eta}, \tag{7}$$

where the sum runs over all pairs of partitions $(\xi, \eta)$ with $|\xi| = |\xi_1| + |\xi_2|$ and $|\eta| = |\eta_1| + |\eta_2|$.

PROPOSITION 7. *We have*

$$\mathrm{Ind}_{\mathsf{W}_\ell \times \mathsf{W}_{N-\ell}}^{\mathsf{W}_N}(\rho_{\xi_1,\eta_1} \otimes 1) = \sum_\xi \rho_{\xi,\eta_1}, \tag{8}$$

*where the sum is over all partitions $\xi$ of $N-|\eta_1| = N-\ell+|\xi_1|$ whose Young diagram is obtained from that of $\xi_1$ by adding $N-\ell$ boxes, with no two boxes in the same column. In particular $\rho_{(N-l)\cup\xi_1,\eta_1}$ occurs in (8).*

*In a similar way, we have:*

$$\mathrm{Ind}_{\mathsf{W}_\ell\times\mathsf{W}_{N-\ell}}^{\mathsf{W}_N}(\rho_{\xi_1,\eta_1}\otimes \mathrm{sgn}_{\mathrm{CD}}) = \sum_{\eta}\rho_{\xi_1,\eta}, \tag{9}$$

*where the sum is over all partition $\eta$ of $N-|\xi_1| = N-\ell+|\eta_1|$ whose Young diagram is obtained from that of $\eta_1$ by adding $N-\ell$ boxes, with no two boxes in the same column. . In particular $\rho_{\xi_1,(N-l)\cup\eta_1}$ occurs in (9).*

Proof. As already mentioned, the trivial character of $\mathsf{W}_{N-\ell}$ corresponds to the pair of partitions $((N-\ell),\emptyset)$. Hence we have to consider certain generalized tableaux $T$ of shape $\xi-\xi_1$ and weight $(N-\ell)$. The integer 1 occurs $n_1(T)=N-\ell$ times in $T$. It follows that all the entries of $T$ are equal to 1 and so the condition of $w(T)$ is empty. On the other hand, the fact that the entries of $T$ have to be strictly increasing down each column implies that there is at most one box in each column of $T$. The first equality follows. The second equality is proved in an analogous way, using the fact that $\mathrm{sgn}_{\mathrm{CD},n-\ell} = \rho_{\emptyset,(N-\ell)}$. □

The following special cases of Proposition 7 will be used in the proof of Proposition 8.

Example 1. *Assume $\ell=1$ and $N\geq 2$. We obtain*

$$\mathrm{Ind}_{\mathsf{W}_1\times\mathsf{W}_{N-1}}^{\mathsf{W}_N}(\rho_{(1),\emptyset}\otimes 1) = \rho_{(N),\emptyset}\oplus\rho_{(N-1,1),\emptyset},$$
$$\mathrm{Ind}_{\mathsf{W}_1\times\mathsf{W}_{N-1}}^{\mathsf{W}_N}(\rho_{\emptyset,(1)}\otimes 1) = \rho_{(N-1),(1)},$$
$$\mathrm{Ind}_{\mathsf{W}_1\times\mathsf{W}_{N-1}}^{\mathsf{W}_N}(\rho_{(1),\emptyset}\otimes \mathrm{sgn}_{\mathrm{CD}}) = \rho_{(1),(N-1)},$$
$$\mathrm{Ind}_{\mathsf{W}_1\times\mathsf{W}_{N-1}}^{\mathsf{W}_N}(\rho_{\emptyset,(1)}\otimes \mathrm{sgn}_{\mathrm{CD}}) = \rho_{\emptyset,(N)}\oplus\rho_{\emptyset,(N-1,1)}.$$

Example 2. *Assume $\ell=2$ and $N\geq 3$. We obtain*

$$\mathrm{Ind}_{\mathsf{W}_2\times\mathsf{W}_{N-2}}^{\mathsf{W}_N}(\rho_{(2),\emptyset}\otimes 1) = \rho_{(N),\emptyset}\oplus\rho_{(N-1,1),\emptyset}\oplus\rho_{(N-2,2),\emptyset},$$
$$\mathrm{Ind}_{\mathsf{W}_2\times\mathsf{W}_{N-2}}^{\mathsf{W}_N}(\rho_{(1^2),\emptyset}\otimes 1) = \rho_{(N-1,1),\emptyset}\oplus\rho_{(N-2,1^2),\emptyset},$$
$$\mathrm{Ind}_{\mathsf{W}_2\times\mathsf{W}_{N-2}}^{\mathsf{W}_N}(\rho_{(1),(1)}\otimes 1) = \rho_{(N-1),(1)}\oplus\rho_{(N-2,1),(1)},$$
$$\mathrm{Ind}_{\mathsf{W}_2\times\mathsf{W}_{N-2}}^{\mathsf{W}_N}(\rho_{\emptyset,(2)}\otimes 1) = \rho_{(N-2),(2)},$$
$$\mathrm{Ind}_{\mathsf{W}_2\times\mathsf{W}_{N-2}}^{\mathsf{W}_N}(\rho_{\emptyset,(1^2)}\otimes 1) = \rho_{(N-2),(1^2)},$$
$$\mathrm{Ind}_{\mathsf{W}_2\times\mathsf{W}_{N-2}}^{\mathsf{W}_N}(\rho_{(2),\emptyset}\otimes \mathrm{sgn}_{\mathrm{CD}}) = \rho_{(2),(N-2)},$$
$$\mathrm{Ind}_{\mathsf{W}_2\times\mathsf{W}_{N-2}}^{\mathsf{W}_N}(\rho_{(1^2),\emptyset}\otimes \mathrm{sgn}_{\mathrm{CD}}) = \rho_{(1^2),(N-2)},$$
$$\mathrm{Ind}_{\mathsf{W}_2\times\mathsf{W}_{N-2}}^{\mathsf{W}_N}(\rho_{(1),(1)}\otimes \mathrm{sgn}_{\mathrm{CD}}) = \rho_{(1),(N-1)}\oplus\rho_{(1),(N-2,1)},$$
$$\mathrm{Ind}_{\mathsf{W}_2\times\mathsf{W}_{N-2}}^{\mathsf{W}_N}(\rho_{\emptyset,(2)}\otimes \mathrm{sgn}_{\mathrm{CD}}) = \rho_{\emptyset,(N)}\oplus\rho_{\emptyset,(N-1,1)}\oplus\rho_{\emptyset,(N-2,2)},$$
$$\mathrm{Ind}_{\mathsf{W}_2\times\mathsf{W}_{N-2}}^{\mathsf{W}_N}(\rho_{\emptyset,(1^2)}\otimes \mathrm{sgn}_{\mathrm{CD}}) = \rho_{\emptyset,(N-1,1)}\oplus\rho_{\emptyset,(N-2,1^2)}.$$

Proof of Theorem 6. It follows easily from the description given in Conjecture 1, combined with Proposition 7. □

In the case when $N' = 2$ we will describe Conjecture 1 in a more explicit manner. For $j \in \{1,2,3\}$, and $N \geq 2$, let $\theta_j^{2,N} \colon \operatorname{Irr}(W_2) \to \mathbb{Z}\operatorname{Irr}(\mathsf{W}_N)$ be the maps defined by (where in each case, the underlined representation $\underline{\rho_{\xi,\eta}}$ is equal to $\underline{\theta}^{2,N}(\rho_{\xi',\eta'})$):

$$\theta_1^{2,2} \colon \begin{array}{rcc} \rho_{(2),\emptyset} & \mapsto & \underline{\rho_{(2),\emptyset}} \\ \rho_{(1),(1)} & \mapsto & \underline{\rho_{(1),(1)}} \oplus \rho_{(1),(1)} \\ \rho_{(1^2),\emptyset} & \mapsto & \underline{\rho_{(1^2),\emptyset}} \\ \rho_{\emptyset,(2)} & \mapsto & \underline{\rho_{\emptyset,(2)}} \oplus 2\rho_{\emptyset,(2)} \oplus \rho_{\emptyset,(1^2)} \\ \rho_{\emptyset,(1^2)} & \mapsto & \rho_{\emptyset,(2)} \oplus \underline{\rho_{\emptyset,(1^2)}} \oplus \rho_{\emptyset,(1^2)} \end{array},$$

$$\theta_2^{2,2} \colon \begin{array}{rcc} \rho_{(2),\emptyset} & \mapsto & \underline{\rho_{(2),\emptyset}} \oplus \rho_{(1),(1)} \oplus \rho_{\emptyset,(2)} \\ \rho_{(1),(1)} & \mapsto & \underline{\rho_{(1),(1)}} \oplus \rho_{\emptyset,(2)} \oplus \rho_{\emptyset,(1^2)} \\ \rho_{(1^2),\emptyset} & \mapsto & \underline{\rho_{(1^2),\emptyset}} \oplus \rho_{(1),(1)} \\ \rho_{\emptyset,(2)} & \mapsto & \underline{\rho_{\emptyset,(2)}} \\ \rho_{\emptyset,(1^2)} & \mapsto & \underline{\rho_{\emptyset,(1^2)}} \end{array},$$

$$\theta_3^{2,2} \colon \begin{array}{rcc} \rho_{(2),\emptyset} & \mapsto & \underline{\rho_{(2),\emptyset}} \\ \rho_{(1),(1)} & \mapsto & \rho_{(2),\emptyset} \oplus \underline{\rho_{(1),(1)}} \oplus \rho_{(1^2),\emptyset} \\ \rho_{(1^2),\emptyset} & \mapsto & \underline{\rho_{(1^2),\emptyset}} \\ \rho_{\emptyset,(2)} & \mapsto & \rho_{(2),\emptyset} \oplus \rho_{(1),(1)} \oplus \underline{\rho_{\emptyset,(2)}} \\ \rho_{\emptyset,(1^2)} & \mapsto & \rho_{(1),(1)} \oplus \underline{\rho_{\emptyset,(1^2)}} \end{array},$$

$$\theta_1^{2,N} \colon \begin{array}{rcc} \rho_{(2),\emptyset} & \mapsto & \underline{\rho_{(2),(N-2)}} \\ \rho_{(1),(1)} & \mapsto & 2\rho_{(1),(N-1)} \oplus \underline{\rho_{(1),(N-2,1)}} \\ \rho_{(1^2),\emptyset} & \mapsto & \underline{\rho_{(1^2),(N-2)}} \\ \rho_{\emptyset,(2)} & \mapsto & 3\rho_{\emptyset,(N)} \oplus 2\rho_{\emptyset,(N-1,1)} \oplus \underline{\rho_{\emptyset,(N-2,2)}} \\ \rho_{\emptyset,(1^2)} & \mapsto & \rho_{\emptyset,(N)} \oplus 2\rho_{\emptyset,(N-1,1)} \oplus \underline{\rho_{\emptyset,(N-2,1^2)}} \end{array}, \quad \text{if } N \geq 3,$$

$$\theta_2^{2,N} \colon \begin{array}{rcc} \rho_{(2),\emptyset} & \mapsto & \rho_{\emptyset,(N)} \oplus \underline{\rho_{(2),(N-2)}} \oplus \rho_{(1),(N-1)} \\ \rho_{(1),(1)} & \mapsto & \rho_{(1),(N-1)} \oplus \underline{\rho_{(1),(N-2,1)}} \oplus \rho_{\emptyset,(N)} \oplus \rho_{\emptyset,(N-1,1)} \\ \rho_{(1^2),\emptyset} & \mapsto & \underline{\rho_{(1^2),(N-2)}} \oplus \rho_{(1),(N-1)} \\ \rho_{\emptyset,(2)} & \mapsto & \rho_{\emptyset,(N)} \oplus \rho_{\emptyset,(N-1,1)} \oplus \underline{\rho_{\emptyset,(N-2,2)}} \\ \rho_{\emptyset,(1^2)} & \mapsto & \rho_{\emptyset,(N-1,1)} \oplus \underline{\rho_{\emptyset,(N-2,1^2)}} \end{array}, \quad \text{if } N \geq 3,$$

$$\theta_3^{2,N} \colon \begin{array}{rcc} \rho_{(2),\emptyset} & \mapsto & \rho_{(N),\emptyset} \oplus \rho_{(N-1,1),\emptyset} \oplus \underline{\rho_{(N-2,2),\emptyset}} \\ \rho_{(1),(1)} & \mapsto & \rho_{(N),\emptyset} \oplus \rho_{(N-1,1),\emptyset} \oplus \rho_{(N-1),(1)} \oplus \underline{\rho_{(N-2,1),(1)}} \\ \rho_{(1^2),\emptyset} & \mapsto & \rho_{(N-1,1),\emptyset} \oplus \underline{\rho_{(N-2,1^2),\emptyset}} \\ \rho_{\emptyset,(2)} & \mapsto & \rho_{(N),\emptyset} \oplus \rho_{(N-1),(1)} \oplus \underline{\rho_{(N-2),(2)}} \\ \rho_{\emptyset,(1^2)} & \mapsto & \rho_{(N-1),(1)} \oplus \underline{\rho_{(N-2),(1^2)}} \end{array}, \quad \text{if } N \geq 3.$$

PROPOSITION 8. *We assume that $N' = 2$, $N \geq 2$ and that Conjecture 1 holds. Then Howe correspondence for the dual pair $(\mathrm{G}', \mathrm{G})$ is given by the map*

$$\begin{cases} \theta_1^{2,N} & \text{if } \mathrm{G}' = \mathrm{O}^{\varepsilon_k}_{2(k^2+2)}(q) \text{ and } \mathrm{G} = \mathrm{Sp}_{2(k^2+k+N)}(q), \\ \theta_1^{2,N} & \text{if } \mathrm{G}' = \mathrm{Sp}_{2(k^2+k+2)}(q) \text{ and } \mathrm{G} = \mathrm{O}^{\varepsilon_k}_{2(k^2+N)}(q), \\ \theta_2^{2,N} & \text{if } \mathrm{G}' = \mathrm{Sp}_{2(k^2+k+2)}(q) \text{ and } \mathrm{G} = \mathrm{O}^{\varepsilon_{k+1}}_{2((k+1)^2+N)}(q), \\ \theta_3^{2,N} & \text{if } \mathrm{G}' = \mathrm{O}^{\varepsilon_{k+1}}_{2((k+1)^2+2)}(q) \text{ and } \mathrm{G} = \mathrm{Sp}_{2(k^2+k+N)}(q). \end{cases}$$

PROOF. We will consider the three cases listed after Conjecture 1 separately.

**Case 1:**

The combination of Conjecture 1 and Example 1 gives

$$\rho_{(2),\emptyset} \mapsto \mathrm{Ind}_{W_2\times W_{N-2}}^{W_N}(\rho_{(2),\emptyset}\otimes \mathrm{sgn}_{\mathrm{CD}})$$

$$\rho_{(1),(1)} \mapsto \mathrm{Ind}_{W_1\times W_{N-1}}^{W_N}(1\otimes \mathrm{sgn}_{\mathrm{CD}}) \oplus \mathrm{Ind}_{W_2\times W_{N-2}}^{W_N}(\rho_{(1),(1)}\otimes \mathrm{sgn}_{\mathrm{CD}})$$

$$\rho_{(1^2),\emptyset} \mapsto \mathrm{Ind}_{W_2\times W_{N-2}}^{W_N}(\rho_{(1^2),\emptyset}\otimes \mathrm{sgn}_{\mathrm{CD}})$$

$$\rho_{\emptyset,(2)} \mapsto \rho_{\emptyset,(N)}\oplus\mathrm{Ind}_{W_1\times W_{N-1}}^{W_N}(\mathrm{sgn}_{\mathrm{CD}}\otimes \mathrm{sgn}_{\mathrm{CD}})\oplus\mathrm{Ind}_{W_2\times W_{N-2}}^{W_N}(\rho_{\emptyset,(2)}\otimes \mathrm{sgn}_{\mathrm{CD}})$$

$$\rho_{\emptyset,(1^2)} \mapsto \mathrm{Ind}_{W_1\times W_{N-1}}^{W_N}(\mathrm{sgn}_{\mathrm{CD}}\otimes \mathrm{sgn}_{\mathrm{CD}}) \oplus \mathrm{Ind}_{W_2\times W_{N-2}}^{W_N}(\rho_{\emptyset,(1^2)}\otimes \mathrm{sgn}_{\mathrm{CD}}).$$

Using the computations done in Examples 1, 2, we obtain the map $\theta_1^{2,N}$.

**Case 2:**

The combination of Conjecture 1 and Example 1 gives

$$\rho_{(2),\emptyset} \mapsto \mathrm{sgn}_{\mathrm{CD}} \oplus \mathrm{Ind}_{W_1\times W_{N-1}}^{W_N}(1\otimes \mathrm{sgn}_{\mathrm{CD}}) \oplus \mathrm{Ind}_{W_2\times W_{N-2}}^{W_N}(1\otimes \mathrm{sgn}_{\mathrm{CD}})$$

$$\rho_{(1),(1)} \mapsto \mathrm{Ind}_{W_1\times W_{N-1}}^{W_N}(\mathrm{sgn}_{\mathrm{CD}}\otimes \mathrm{sgn}_{\mathrm{CD}}) \oplus \mathrm{Ind}_{W_2\times W_{N-2}}^{W_N}(\rho_{(1),(1)}\otimes \mathrm{sgn}_{\mathrm{CD}})$$

$$\rho_{(1^2),\emptyset} \mapsto \mathrm{Ind}_{W_1\times W_{N-1}}^{W_N}(1\otimes \mathrm{sgn}_{\mathrm{CD}}) \oplus \mathrm{Ind}_{W_2\times W_{N-2}}^{W_N}(\rho_{(1^2),\emptyset}\otimes \mathrm{sgn}_{\mathrm{CD}})$$

$$\rho_{\emptyset,(2)} \mapsto \mathrm{Ind}_{W_2\times W_{N-2}}^{W_N}(\rho_{\emptyset,(2)}\otimes \mathrm{sgn}_{\mathrm{CD}})$$

$$\rho_{\emptyset,(1^2)} \mapsto \mathrm{Ind}_{W_2\times W_{N-2}}^{W_N}(\rho_{\emptyset,(1^2)}\otimes \mathrm{sgn}_{\mathrm{CD}}).$$

Using the computations done in Examples 1, 2, we obtain $\theta_2^{2,N}$.

**Case 3:**

The combination of Conjecture 1 and Example 1 gives

$$\rho_{(2),\emptyset} \mapsto \mathrm{Ind}_{W_2\times W_{N-2}}^{W_N}(\rho_{(2),\emptyset}\otimes 1)$$

$$\rho_{(1),(1)} \mapsto \mathrm{Ind}_{W_1\times W_{N-1}}^{W_N}(1\otimes 1) \oplus \mathrm{Ind}_{W_2\times W_{N-2}}^{W_N}(\rho_{(1),(1)}\otimes 1)$$

$$\rho_{(1^2),\emptyset} \mapsto \mathrm{Ind}_{W_2\times W_{N-2}}^{W_N}(\rho_{(1^2),\emptyset}\otimes 1)$$

$$\rho_{\emptyset,(2)} \mapsto \rho_{(N),\emptyset} \oplus \mathrm{Ind}_{W_1\times W_{N-1}}^{W_N}(\mathrm{sgn}_{\mathrm{CD}}\otimes 1) \oplus \mathrm{Ind}_{W_2\times W_{N-2}}^{W_N}(\rho_{\emptyset,(2)}\otimes 1)$$

$$\rho_{\emptyset,(1^2)} \mapsto \mathrm{Ind}_{W_1\times W_{N-1}}^{W_N}(\mathrm{sgn}_{\mathrm{CD}}\otimes 1) \oplus \mathrm{Ind}_{W_2\times W_{N-2}}^{W_N}(\rho_{\emptyset,(1^2)}\otimes 1).$$

Using the computations done in Examples 1, 2, we get $\theta_3^{2,N}$. □

**4.2. Symbols and $u$-symbols.** We will recall part of the formalism of symbols due to Lusztig. (See [**Lus84**] and references there.)

A *symbol* is an ordered pair $\Lambda = \begin{pmatrix} A \\ B \end{pmatrix}$ of finite subsets (including the empty set $\emptyset$) of $\{0, 1, 2, \ldots\}$. The *rank* of $\Lambda$ is defined to be

$$\mathrm{rank}(\Lambda) := \sum_{a \in A} a + \sum_{b \in B} b - \left\lfloor \left( \frac{|A| + |B| - 1}{2} \right)^2 \right\rfloor,$$

where for any real number $r$ we denote by $\lfloor r \rfloor$ the largest integer not greater than $r$. The *defect* of $\Lambda$, to be denoted by $\mathrm{def}(\Lambda)$, is defined to be the absolute value of $|A| - |B|$. There is an equivalence relation on such pairs generated by the shift

$$\begin{pmatrix} A \\ B \end{pmatrix} \sim \begin{pmatrix} \{0\} \cup (A+1) \\ \{0\} \cup (B+1) \end{pmatrix}.$$

We shall identify a symbol with its equivalence class. The functions $\mathrm{rank}(\Lambda)$ and $\mathrm{def}(\Lambda)$ are invariant under the shift operation, hence are well-defined on the set of symbol classes. A symbol $\Lambda = \begin{pmatrix} A \\ B \end{pmatrix}$ is said to be *degenerate* if $A = B$, and *non-degenerate* otherwise. The entries appearing in exactly one row of $\Lambda$ are called *singles*.

There is also a notion of $u$-symbols due to Lusztig related to unipotent classes.

Let $(\xi, \eta)$ be a pair of partitions of $N$. We ensure that $\xi$ has exactly one more part than $\eta$ by adding zeros as parts where necessary. Let $m$ denote the number of parts of $\eta$. We then attach to $(\xi, \eta)$, where $\xi = (\xi_1 \geq \xi_2 \geq \cdots \geq \xi_m \geq \xi_{m+1})$ and $\eta = (\eta_1 \geq \eta_2 \geq \cdots \geq \eta_m)$, a *symbol* $\Lambda = \Lambda_{\xi,\eta}$ of defect 1 and two *$u$-symbols* $\Lambda^{\mathrm{u,sp}}_{\xi,\eta}$ and $\Lambda^{\mathrm{u,or}}_{\xi,\eta}$ to be defined by

$$\Lambda_{\xi,\eta} := \begin{pmatrix} \xi_{m+1} & & \xi_m + 1 & & \xi_{m-1} + 2 & & \cdots & \cdots & \cdots & & \xi_1 + m \\ & \eta_m & & \eta_{m-1} + 1 & & \eta_{m-2} + 2 & & \cdots & & \eta_1 + m - 1 & \end{pmatrix},$$

$$\Lambda^{\mathrm{u,sp}}_{\xi,\eta} := \begin{pmatrix} \xi_{m+1} & & \xi_m + 2 & & \xi_{m-1} + 4 & \cdots & & \cdots & & \xi_1 + 2m \\ & \eta_m + 1 & & \eta_{m-1} + 3 & & & \cdots & & \eta_1 + 2(m-1) + 1 & \end{pmatrix},$$

$$\Lambda^{\mathrm{u,or}}_{\xi,\eta} := \begin{pmatrix} \xi_{m+1} & \xi_m + 2 & \xi_{m-1} + 4 & \cdots & \cdots & \xi_1 + 2m \\ \eta_{m+1} & \eta_m + 2 & \eta_{m-1} + 4 & & \cdots & \eta_1 + 2m \end{pmatrix},$$

where in the orthogonal case we arranged for the two partitions to have the same length $m+1$. The symbol $\Lambda_{\xi,\eta}$ is called *special* if

$$\xi_{m+1} \leq \eta_m \leq \xi_m + 1 \leq \eta_{m-1} + 1 \leq \xi_{m-1} + 2 \leq \cdots \leq \eta_1 + m - 1 \leq \xi_1 + m.$$

Similarly, $\Lambda^{\mathrm{u,sp}}_{\xi,\eta}$ is called *distinguished* if

$$\xi_{m+1} \leq \eta_m + 1 \leq \xi_m + 2 \leq \eta_{m-1} + 3 \leq \cdots \leq \eta_1 + 2(m-1) + 1 \leq \xi_1 + 2m,$$

and $\Lambda^{\mathrm{u,or}}_{\xi,\eta}$ is called distinguished if

$$\xi_{m+1} \leq \eta_{m+1} \leq \xi_m + 2 \leq \eta_m + 2 \leq \cdots \leq \xi_1 + 2m \leq \eta_1 + 2m.$$

We observe that the fact that $\Lambda_{\xi,\eta}$ is special implies the distinguishness of $\Lambda^{\mathrm{u,sp}}_{\xi,\eta}$.

The set of all the symbols (resp. $u$-symbols) which contain the same entries with the same multiplicities as a given symbol (resp. $u$-symbol) is called the *similarity class* of the latter. If $\Lambda$, $\Lambda'$ belong to the same similarity class, we will write $\Lambda \sim_{\mathrm{sim}} \Lambda'$. Each similarity class of symbols (resp. $u$-symbols) contains exactly one special (resp. distinguished) element.

We will now recall the algorithm described in [**Car93**, §13.3]. To each partition $\lambda = (\lambda_1 \geq \lambda_2 \geq \cdots \geq \lambda_k)$ we attach the *sequence of $\beta$-numbers*

(10)
$$\lambda^* = (\lambda_1^* < \lambda_2^* < \cdots < \lambda_k^*), \text{ defined by } \lambda_j^* := \lambda_{k-j+1} + j - 1, \text{ for } 1 \leq j \leq k.$$

For instance, we have

$$(N)^* = (N), \quad (N-1,1)^* = (1,N), \quad (N-2,2)^* = (2,N-1), \quad (N-2,1,1)^* = (1,2,N).$$

Recall that a partition $\lambda$ is called *symplectic* (resp. *orthogonal*) if each odd (resp. even) row occurs with even multiplicity. For $N$ a given integer, let $\mathcal{P}^{\mathrm{sp}}(N)$ (resp. $\mathcal{P}^{\mathrm{or}}(N)$) denote the set of symplectic (resp. orthogonal) partitions of $N$.

Consider a symplectic or orthogonal partition $\lambda$ of $N$ and the corresponding group $\mathrm{G}_N$. We ensure that the number of parts of $\lambda$ has same parity as the defining module of $\mathrm{G}_N$, by calling the last part 0 if necessary. Thus $\lambda_1 \geq \lambda_2 \geq \cdots \geq \lambda_{2k}$ (resp. $\lambda_1 \geq \lambda_2 \geq \cdots \geq \lambda_{2k+1}$) if $\mathrm{G}_N = \mathrm{Sp}_{2N}(\bar{\mathbb{F}}_q)$ or $\mathrm{O}_{2N}(\bar{\mathbb{F}}_q)$ (resp. $\mathrm{G}_N = \mathrm{O}_{2N+1}(\bar{\mathbb{F}}_q)$). We then divide $\lambda^*$ into its odd and even parts. Let the odd parts and the even parts of $\lambda^*$ be

$$2\xi_1^* + 1 < 2\xi_2^* + 1 < \cdots < 2\xi_k^* + 1 \;\; (\text{resp. } 2\xi_{k+1}^* + 1) \quad \text{and} \quad 2\eta_1^* < 2\eta_2^* < \cdots < 2\eta_k^*,$$

respectively. Then we have

$$0 \leq \xi_1^* < \xi_2^* < \cdots < \xi_k^* \;\; (\text{resp. } \xi_{k+1}^*) \quad \text{and} \quad 0 \leq \eta_1^* < \eta_2^* < \cdots < \eta_k^*.$$

Next we define $\xi_i := \xi_{k-i+1}^* - (k-i)$ and $\eta_i := \eta_{k-i+1}^* - (k-i)$ for each $i$. We then have $\xi_i \geq \xi_{i+1} \geq 0$, $\eta_i \geq \eta_{i+1} \geq 0$, and $|\xi| + |\eta| = n$.

Thus we obtain a map

(11)
$$\varphi \colon \lambda \mapsto (\xi, \eta)$$

from $\mathcal{P}^{\mathrm{sp}}(2N)$ or $\mathcal{P}^{\mathrm{or}}(2N)$ (resp. $\mathcal{P}^{\mathrm{or}}(2N+1)$) to the set of pairs of partitions of $N$, which is injective.

A pair of partitions $(\xi_0, \eta_0)$ of $N$ is in the image of the map (11) of a symplectic partition, say $\lambda^{\mathrm{sp}}_{\xi_0,\eta_0}$ (resp. an orthogonal partition, say $\lambda^{\mathrm{or}}_{\xi_0,\eta_0}$) if and only if the $u$-symbol $\Lambda^{\mathrm{u,sp}}_{\xi_0,\eta_0}$ (resp. $\Lambda^{\mathrm{u,or}}_{\xi_0,\eta_0}$) is distinguished, see [**Car93**, page 420].

Definition 9. *If $(\xi, \eta)$ is not in the image of the map $\varphi$ defined by (11), we put $\lambda^{\mathrm{sp}}_{\xi,\eta} := \lambda^{\mathrm{sp}}_{\xi_0,\eta_0}$ (resp. $\lambda^{\mathrm{or}}_{\xi,\eta} := \lambda^{\mathrm{or}}_{\xi_0,\eta_0}$), where $\Lambda^{\mathrm{u,sp}}_{\xi_0,\eta_0}$ (resp. $\Lambda^{\mathrm{u,or}}_{\xi_0,\eta_0}$) is the distinguished $u$-symbol in the similarity class of $\Lambda^{\mathrm{u,sp}}_{\xi,\eta}$ (resp. $\Lambda^{\mathrm{u,or}}_{\xi,\eta}$).*

We will use the computations done in the following exeamples in the proof of Theorem 10.

Example 3. *Let $N \geq 2$ and $1 \leq h \leq 2$. We have*

$$\Lambda^{\mathrm{u,sp}}_{(N-h,h),\emptyset} = \begin{pmatrix} h & & N-h+2 \\ & 1 & \end{pmatrix} \quad \text{and} \quad \Lambda^{\mathrm{u,or}}_{(N-h,h),\emptyset} = \begin{pmatrix} h & N-h+2 \\ 0 & 2 \end{pmatrix}.$$

- *$\Lambda^{\mathrm{u,sp}}_{(N-1,1),\emptyset}$ is distinguished. Because $(N-1,1)^* = (1,N)$ and $\emptyset^* = (0,1)$, we obtain $\lambda^{\mathrm{sp},*}_{(N-1,1),\emptyset} = (0,2,3,2N+1)$, that gives $\lambda^{\mathrm{sp}}_{(N-1,1),\emptyset} = (2N-2,1^2)$.*
- *$\Lambda^{\mathrm{u,sp}}_{(N-2,2),\emptyset}$ is not distinguished. The distinguished $u$-symbol in its similarity class is*

$$\begin{pmatrix} 1 & & N \\ & 2 & \end{pmatrix} = \Lambda^{\mathrm{u,sp}}_{(N-2,1),(1,0)}.$$

*Because* $(N-2,1)^* = (1,N-1)$ *and* $(1,0)^* = (0,2)$*, we obtain* $\lambda^{\mathrm{sp},*}_{(N-2,1),(1,0)} = (0,3,4,2N-1)$*, that gives* $\lambda^{\mathrm{sp}}_{(N-2,2),\emptyset} = \lambda^{\mathrm{sp}}_{(N-2,1),(1)} = (2N-4,2^2)$.

- *For* $h=1,2$*, the distinguished u-symbol in the similarity class of* $\Lambda^{\mathrm{u,or}}_{(N-h,h),\emptyset}$ *is*

$$\begin{pmatrix} 0 & 2 \\ h & N-h+2 \end{pmatrix} = \Lambda^{\mathrm{u,or}}_{\emptyset,(N-h,h)}.$$

*Because* $(0^2)^* = (0,1)$ *and* $(N-h,h)^* = (h,N-h+1)$*, we obtain*

$$\lambda^{*,\mathrm{or}}_{(0^2),(N-h,h)} = \begin{cases} (1,2,3,2N) & \text{if } h=1, \\ (1,3,4,2N-2) & \text{if } h=2, \end{cases}$$

*that is,*

$$\lambda^{\mathrm{or}}_{(N-h,h),\emptyset} = \lambda^{\mathrm{or}}_{(0^2),(N-h,h)} = \begin{cases} (2N-3,1^3) & \text{if } h=1, \\ (2N-5,2^2,1) & \text{if } h=2. \end{cases}$$

EXAMPLE 4.

$$\Lambda^{\mathrm{u,sp}}_{(N-2,1^2),\emptyset} = \begin{pmatrix} 1 & & 3 & & N+2 \\ & 1 & & 3 & \end{pmatrix} \quad \text{and} \quad \Lambda^{\mathrm{u,or}}_{(N-2,1^2),\emptyset} = \begin{pmatrix} 1 & 3 & N+2 \\ 0 & 2 & 4 \end{pmatrix}.$$

- $\Lambda^{\mathrm{u,sp}}_{(N-2,1^2),\emptyset}$ *is distinguished. Because* $(N-2,1^2)^* = (1,2,N)$ *and* $(0^3)^* = (0,1,2)$*, we obtain* $\lambda^{\mathrm{sp},*}_{(N-2,1^2),\emptyset} = (0,2,3,4,5,2N+1)$*, that gives* $\lambda^{\mathrm{sp}}_{(N-2,1^2),\emptyset} = (2N-4,1^4)$.
- $\Lambda^{\mathrm{u,or}}_{(N-2,1^2),\emptyset}$ *is not distinguished. If* $N \geq 3$*, the distinguished u-symbol in its similarity class is*

$$\begin{pmatrix} 0 & 2 & 4 \\ 1 & 3 & N+2 \end{pmatrix} = \Lambda^{\mathrm{u,or}}_{\emptyset,(N-2,1^2)}.$$

*We have* $\lambda^{\mathrm{or},*}_{\emptyset,(N-2,1^2)} = (1,2,3,4,5,2N)$*, that gives* $\lambda^{\mathrm{or}}_{(N-2,1^2),\emptyset} = (2N-5,1^5)$.

EXAMPLE 5.

$$\Lambda^{\mathrm{u,sp}}_{(N-2),(1^2)} = \begin{pmatrix} 0 & & 2 & & N \\ & 2 & & 4 & \end{pmatrix} \quad \text{and} \quad \Lambda^{\mathrm{u,or}}_{(N-2),(1^2)} = \begin{pmatrix} 0 & N \\ 1 & 3 \end{pmatrix}.$$

- $\Lambda^{\mathrm{u,sp}}_{(N-2),(1^2)}$ *is distinguished if* $N \geq 4$*,* $(N-2,0)^* = (0,N-1)$*,* $(1^2)^* = (1,2)$*. Hence* $\lambda^{\mathrm{sp},*}_{(N-2),(1^2)} = (1,2,4,2N-1)$ *and* $\lambda^{\mathrm{sp}}_{(N-2),(1^2)} = (2N-4,2,1^2)$.
- $\Lambda^{\mathrm{u,or}}_{(N-2),(1^2)}$ *is not distinguished if* $N \geq 4$*. The distinguished u-symbol in its similarity class is*

$$\begin{pmatrix} 0 & 3 \\ 1 & N \end{pmatrix} = \Lambda^{\mathrm{u,or}}_{(1),(N-2,1)}.$$

$(1,0)^* = (0,2)$, $(N-2,1)^* = (1,N-1)$*. Hence* $\lambda^{\mathrm{or},*}_{(1),(N-2,1)} = (1,2,5,2N-2)$*, and* $\lambda^{\mathrm{or}}_{(N-2),(1^2)} = (2N-5,3,1^2)$.

EXAMPLE 6.

$$\Lambda^{\mathrm{u,sp}}_{(N-2),(2)} = \begin{pmatrix} 0 & & N \\ & 3 & \end{pmatrix} \quad \text{and} \quad \Lambda^{\mathrm{u,or}}_{(N-2),(2)} = \begin{pmatrix} N-2 \\ 2 \end{pmatrix}.$$

- $\lambda^{\mathrm{sp},*}_{(N-2),(2)} = (4,2N-3)$*, and* $\lambda^{\mathrm{sp}}_{(N-2),(2)} = (2N-4,4)$*, when* $N \geq 3$.

- $\lambda^{\mathrm{or},*}_{(N-2),(2)} = (5, 2N-4)$, *and* $\lambda^{\mathrm{or}}_{(N-2),(2)} = (2N-5, 5)$.

EXAMPLE 7.

$$\Lambda^{\mathrm{u,sp}}_{(N-1),(1)} = \begin{pmatrix} 0 & & N+1 \\ & 2 & \end{pmatrix} \quad \text{and} \quad \Lambda^{\mathrm{u,or}}_{(N-1),(1)} = \begin{pmatrix} N-1 \\ 1 \end{pmatrix}.$$

- $\Lambda^{\mathrm{u,sp}}_{(N-1),(1)}$ *is distinguished,* $\lambda^{\mathrm{sp},*}_{(N-1),(1)} = (2, 2N-1)$, *and* $\lambda^{\mathrm{sp}}_{(N-1),(1)} = (2N-2, 2)$.
- $\Lambda^{\mathrm{u,or}}_{(N-1),(1)}$ *is not distinguished. The distinguished u-symbol in its similarity class is* $\Lambda^{\mathrm{u,or}}_{(1),(N-1)}$. *We have* $\lambda^{\mathrm{or},*}_{(N-1),(1)} = (3, 2N-2)$, *and* $\lambda^{\mathrm{or}}_{(N-1),(1)} = (2N-3, 3)$.

EXAMPLE 8.

$$\Lambda^{\mathrm{u,sp}}_{(2),(N-2)} = \begin{pmatrix} 0 & & 4 \\ & N-1 & \end{pmatrix} \quad \text{and} \quad \Lambda^{\mathrm{u,or}}_{(2),(N-2)} = \begin{pmatrix} 2 \\ N-2 \end{pmatrix}.$$

- $\Lambda^{\mathrm{u,sp}}_{(2),(N-2)}$ *is not distinguished if* $N \geq 6$. *Then the distinguished u-symbol in its similarity class is*

$$\begin{pmatrix} 0 & & N-1 \\ & 4 & \end{pmatrix} = \Lambda^{\mathrm{u,sp}}_{(N-3),(3)}.$$

*We have* $(N-3, 0)^* = (0, N-2)$ *and* $(3, 0)^* = (0, 4)$. *Hence* $\lambda^{\mathrm{sp},*}_{(2),(N-2)} = (0, 1, 8, 2N-3)$, *and* $\lambda^{\mathrm{sp}}_{(2),(N-2)} = (2N-6, 6)$.
- $\lambda^{\mathrm{or},*}_{(2),(N-2)} = (5, 2N-4)$, *and* $\lambda^{\mathrm{or}}_{(2),(N-2)} = (2N-5, 5)$, $N \geq 5$.

EXAMPLE 9.

$$\Lambda^{\mathrm{u,sp}}_{(N-2,1),(1)} = \begin{pmatrix} 1 & & N \\ & 2 & \end{pmatrix} \quad \text{and} \quad \Lambda^{\mathrm{u,or}}_{(N-2,1),(1)} = \begin{pmatrix} 1 & N \\ 0 & 3 \end{pmatrix}.$$

- $\Lambda^{\mathrm{u,sp}}_{(N-2,1),(1)}$ *is distinguished if* $N \geq 2$, $(N-2, 1)^* = (1, N-1)$ *and* $(1, 0)^* = (0, 2)$. *Hence* $\lambda^{\mathrm{sp},*}_{(N-2,1),(1)} = (0, 3, 4, 2N-1)$, *and* $\lambda^{\mathrm{sp}}_{(N-2,1),(1)} = (2N-4, 2^2)$.
- $\Lambda^{\mathrm{u,or}}_{(N-2,1),(1)}$ *is not distinguished if* $N \geq 4$. *The distinguished u-symbol in its similarity class is*

$$\begin{pmatrix} 0 & 3 \\ 1 & N \end{pmatrix} = \Lambda^{\mathrm{u,or}}_{(1),(N-2,1)}.$$

*Hence* $\lambda^{\mathrm{or},*}_{(1),(N-2,1)} = (1, 2, 5, 2N-2)$, *and* $\lambda^{\mathrm{or}}_{(1),(N-2,1)} = (2N-5, 3, 1^2)$.

EXAMPLE 10.

$$\Lambda^{\mathrm{u,sp}}_{(1),(N-2,1)} = \begin{pmatrix} 0 & & 2 & & 5 \\ & 2 & & N+1 & \end{pmatrix}$$

*is not distinguished if* $N \geq 5$. *The distinguished u-symbol in its similarity class is*

$$\Lambda^{\mathrm{u,sp}}_{(N-3),(2,1)} = \begin{pmatrix} 0 & & 2 & & N+1 \\ & 2 & & 5 & \end{pmatrix}.$$

*We have* $\lambda^{\mathrm{sp},*}_{(N-3),(2,1)} = (1, 2, 6, 2N-3)$, *and hence* $\lambda^{\mathrm{sp}}_{(1),(N-2,1)} = (2N-6, 4, 1^2)$.

EXAMPLE 11.

$$\Lambda^{\mathrm{u,sp}}_{\emptyset,(N-2,2)} = \begin{pmatrix} 0 & & 2 & & 4 \\ & 3 & & N+1 & \end{pmatrix}$$

*is not distinguished if* $N \geq 4$. *The distinguished* $u$*-symbol in its similarity class is*

$$\Lambda^{\mathrm{u,sp}}_{(N-3,1),(1^2)} = \begin{pmatrix} 0 & & 3 & & N+1 \\ & 2 & & 4 & \end{pmatrix}.$$

*We have* $\lambda^{\mathrm{sp},*}_{(N-3,1),(1^2)} = (2,3,4,2N-3)$, *and* $\lambda^{\mathrm{sp}}_{\emptyset,(N-2,2)} = (2N-6,2^3)$.

EXAMPLE 12.

$$\Lambda^{\mathrm{u,sp}}_{\emptyset,(N-2,1^2)} = \begin{pmatrix} 0 & & 2 & & 4 & & 6 \\ & 2 & & 4 & & N+3 & \end{pmatrix}$$

*is not distinguished if* $N \geq 3$. *The distinguished* $u$*-symbol in its similarity class is*

$$\Lambda^{\mathrm{u,sp}}_{(N-3),(1^3)} = \begin{pmatrix} 0 & & 2 & & 4 & & N+3 \\ & 2 & & 4 & & 6 & \end{pmatrix}.$$

*Hence* $\lambda^{\mathrm{sp},*}_{\emptyset,(N-2,1^2)} = (1,2,3,4,6,2N-1)$, *and* $\lambda^{\mathrm{sp}}_{\emptyset,(N-2,1^2)} = (2N-6,2,1^4)$.

EXAMPLE 13.

$$\Lambda^{\mathrm{u,sp}}_{(1^2),(N-2)} = \begin{pmatrix} 1 & & 3 \\ & N-1 & \end{pmatrix}$$

*is not distinguished if* $N \geq 5$. *The distinguished* $u$*-symbol in its similarity class is* $\Lambda^{\mathrm{u,sp}}_{(N-3,1),(2)}$. *We have* $\lambda^{\mathrm{sp},*}_{(N-3,1),(2)} = (0,3,6,2N-3)$. *Hence* $\lambda^{\mathrm{sp}}_{(1^2),(N-2)} = (2N-6,4,2)$.

**4.3. Howe correspondence and unipotent orbits.** Let $\bar{\mathbb{F}}_q$ be an algebraic closure of $\mathbb{F}_q$, and let $\mathcal{N}(\mathbf{G})$ denote the set of unipotent classes of $\mathbf{G}$. This set is partially ordered by the relation $\mathcal{O}_1 \leq \mathcal{O}_2$ meaning that $\mathcal{O}_1$ is contained in $\overline{\mathcal{O}}_2$, the closure of $\mathcal{O}_2$. The unipotent orbits in the corresponding algebraic groups over $\bar{\mathbb{F}}_q$ are parameterized by partitions $\lambda$ of the dimension of the defining module. The partition $\lambda$ is symplectic (resp. orthogonal) if $G = \mathrm{Sp}_{2N}(\bar{\mathbb{F}}_q)$ (resp. $\mathrm{O}_{2N}(\bar{\mathbb{F}}_q)$ or $\mathrm{O}_{2N+1}(\bar{\mathbb{F}}_q)$).

To the representation $\rho_{\xi,\eta}$ of $W_N$ we shall associate the $u$-symbol $\Lambda^{\mathrm{u,sp}}_{\xi,\eta}$ (resp. $\Lambda^{\mathrm{u,or}}_{\xi,\eta}$) of the group $\mathbf{G} = \mathrm{Sp}_{2N}(\bar{\mathbb{F}}_q)$ (resp. $\mathbf{G} = \mathrm{O}^{\varepsilon_{k_2}}_{2N}(\bar{\mathbb{F}}_q)$).

For the groups $\mathrm{Sp}_{2N}(\bar{\mathbb{F}}_q)$ and $\mathrm{O}_{2N+1}(\bar{\mathbb{F}}_q)$, we associate the representation $\rho_{\xi,\eta}$ to the orbit $\mathcal{O}(\lambda)$, where $(\xi,\eta) := \varphi(\lambda)$, with $\varphi$ defined by (11).

Consider the group $\mathrm{O}_{2N}(\bar{\mathbb{F}}_q)$. In this case the unipotent orbits are parameterized by partitions $\lambda$ of $2N$ where the even rows occur with even multiplicities. We attach to such a partition $\lambda$ the ordered pair of partitions $(\xi,\eta)$ defined by $(\xi,\eta) := \varphi(\lambda)$. Then we associate to $\mathcal{O}(\lambda)$ the representation $\rho_{\eta,\xi}$.

Let $\mathbf{S}(\mathbf{G})$ denote the set of $u$-symbols attached to $\mathbf{G}$. Let

$$\vartheta^{2,2}_{1,a}\colon \mathbf{S}(\mathrm{O}_4(\bar{\mathbb{F}}_q)) \to \mathcal{N}(\mathrm{Sp}_4(\bar{\mathbb{F}}_q)) \quad \text{and} \quad \vartheta^{2,2}_{1,b}\colon \mathbf{S}(\mathrm{Sp}_4(\bar{\mathbb{F}}_q)) \to \mathcal{N}(\mathrm{O}_4(\bar{\mathbb{F}}_q))$$

TABLE 1. Unipotent classes and the corresponding similarity classes of the $u$-symbols for $\mathrm{Sp}_4(\bar{\mathbb{F}}_q)$

$$
\begin{array}{ccc}
\mathcal{O}(4) & \leftrightarrow & \Lambda^{\mathrm{u,sp}}_{(2),\emptyset} \\
| & & \\
\mathcal{O}(2^2) & \leftrightarrow & \{\Lambda^{\mathrm{u,sp}}_{(1),(1)}, \Lambda^{\mathrm{u,sp}}_{\emptyset,(2)}\} \\
| & & \\
\mathcal{O}(2,1^2) & \leftrightarrow & \Lambda^{\mathrm{u,sp}}_{(1^2),\emptyset} \\
| & & \\
\mathcal{O}(1^4) & \leftrightarrow & \Lambda^{\mathrm{u,sp}}_{\emptyset,(1^2)}
\end{array}
$$

TABLE 2. Unipotent classes and and the corresponding similarity classes of the $u$-symbols for $\mathrm{O}_4(\bar{\mathbb{F}}_q)$

$$
\begin{array}{ccc}
\mathcal{O}(3,1) & \leftrightarrow & \{\Lambda^{\mathrm{u,or}}_{(2),\emptyset}, \Lambda^{\mathrm{u,or}}_{\emptyset,(2)}\} \\
| & & \\
\mathcal{O}(2^2) & \leftrightarrow & \Lambda^{\mathrm{u,or}}_{(1),(1)} \\
| & & \\
\mathcal{O}(1^4) & \leftrightarrow & \{\Lambda^{\mathrm{u,or}}_{(1^2),\emptyset}, \Lambda^{\mathrm{u,or}}_{\emptyset,(1^2)}\}
\end{array}
$$

be defined by (where in each case, if the input symbol is indexed by $\xi',\eta'$, then the underlined orbit in the output corresponds to the representation $\underline{\theta}^{2,N}(\rho_{\xi',\eta'})$)

$$
\vartheta^{2,2}_{1,a}\colon
\begin{array}{ccc}
\Lambda^{\mathrm{u,or}}_{(2),\emptyset} & \mapsto & \underline{\mathcal{O}(4)} \\
\Lambda^{\mathrm{u,or}}_{(1),(1)} & \mapsto & \underline{\mathcal{O}(2^2)} \\
\Lambda^{\mathrm{u,or}}_{(1^2),\emptyset} & \mapsto & \underline{\mathcal{O}(2,1^2)} \\
\Lambda^{\mathrm{u,or}}_{\emptyset,(2)} & \mapsto & \{\underline{\mathcal{O}(2^2)}, \mathcal{O}(1^4)\} \\
\Lambda^{\mathrm{u,or}}_{\emptyset,(1^2)} & \mapsto & \{\mathcal{O}(2^2), \underline{\mathcal{O}(1^4)}\}
\end{array},
\qquad
\vartheta^{2,2}_{1,b}\colon
\begin{array}{ccc}
\Lambda^{\mathrm{u,sp}}_{(2),\emptyset} & \mapsto & \{\underline{\mathcal{O}(3,1)}, \mathcal{O}(2^2)\} \\
\Lambda^{\mathrm{u,sp}}_{(1),(1)} & \mapsto & \{\mathcal{O}(3,1), \underline{\mathcal{O}(2^2)}, \mathcal{O}(1^4)\} \\
\Lambda^{\mathrm{u,sp}}_{(1^2),\emptyset} & \mapsto & \{\mathcal{O}(2^2), \underline{\mathcal{O}(1^4)}\} \\
\Lambda^{\mathrm{u,sp}}_{\emptyset,(2)} & \mapsto & \underline{\mathcal{O}(3,1)} \\
\Lambda^{\mathrm{u,sp}}_{\emptyset,(1^2)} & \mapsto & \underline{\mathcal{O}(1^4)}
\end{array}.
$$

Let

$$
\vartheta^{2,2}_2\colon \mathbf{S}(\mathrm{Sp}_4(\bar{\mathbb{F}}_q)) \to \mathcal{N}(\mathrm{O}_4(\bar{\mathbb{F}}_q)) \quad \text{and} \quad \vartheta^{2,2}_3\colon \mathbf{S}(\mathrm{O}_4(\bar{\mathbb{F}}_q)) \to \mathcal{N}(\mathrm{Sp}_4(\bar{\mathbb{F}}_q))
$$

be defined by

$$
\vartheta^{2,2}_2\colon
\begin{array}{ccc}
\Lambda^{\mathrm{u,sp}}_{(2),\emptyset} & \mapsto & \{\underline{\mathcal{O}(3,1)}, \mathcal{O}(2^2), \mathcal{O}(3,1)\} \\
\Lambda^{\mathrm{u,sp}}_{(1),(1)} & \mapsto & \{\underline{\mathcal{O}(2^2)}, \mathcal{O}(3,1), \mathcal{O}(1^4)\} \\
\Lambda^{\mathrm{u,sp}}_{(1^2),\emptyset} & \mapsto & \{\underline{\mathcal{O}(1^4)}, \mathcal{O}(2^2)\} \\
\Lambda^{\mathrm{u,sp}}_{\emptyset,(2)} & \mapsto & \underline{\mathcal{O}(3,1)} \\
\Lambda^{\mathrm{u,sp}}_{\emptyset,(1^2)} & \mapsto & \underline{\mathcal{O}(1^4)}
\end{array},
\qquad
\vartheta^{2,2}_3\colon
\begin{array}{ccc}
\Lambda^{\mathrm{u,or}}_{(2),\emptyset} & \mapsto & \{\mathcal{O}(4), \mathcal{O}(2^2)\} \\
\Lambda^{\mathrm{u,or}}_{(1),(1)} & \mapsto & \{\mathcal{O}(4), \underline{\mathcal{O}(2^2)}, \mathcal{O}(2,1^2)\} \\
\Lambda^{\mathrm{u,or}}_{(1^2),\emptyset} & \mapsto & \{\mathcal{O}(2^2), \underline{\mathcal{O}(1^4)}\} \\
\Lambda^{\mathrm{u,or}}_{\emptyset,(2)} & \mapsto & \underline{\mathcal{O}(4)} \\
\Lambda^{\mathrm{u,or}}_{\emptyset,(1^2)} & \mapsto & \underline{\mathcal{O}(2,1^2)}
\end{array}.
$$

If $N \geq 3$, let $\vartheta^{2,N}_{1,a} : \mathbf{S}(\mathrm{O}_4(\bar{\mathbb{F}}_q)) \to \mathcal{N}(\mathrm{Sp}_{2N}(\bar{\mathbb{F}}_q))$ be defined by

$$\vartheta^{2,N}_{1,a} : \begin{array}{rcc} \Lambda^{\mathrm{u,or}}_{(2),\emptyset} & \mapsto & \underline{\mathcal{O}(2N-6,6)} \\ \Lambda^{\mathrm{u,or}}_{(1),(1)} & \mapsto & \{\mathcal{O}(2N-4,4), \underline{\mathcal{O}(2N-6,4,1^2)}\} \\ \Lambda^{\mathrm{u,or}}_{\emptyset,(2)} & \mapsto & \{\mathcal{O}(2N-2,2), \mathcal{O}(2N-4,2,1^2), \underline{\mathcal{O}(2N-6,2^3)}\} \\ \Lambda^{\mathrm{u,or}}_{(1^2),\emptyset} & \mapsto & \{\mathcal{O}(2N-2,2), \mathcal{O}(2N-4,2,1^2), \underline{\mathcal{O}(2N-6,2,1^4)}\} \\ \Lambda^{\mathrm{u,or}}_{\emptyset,(1^2)} & \mapsto & \underline{\mathcal{O}(2N-6,4,2)} \end{array}.$$

If $N \geq 3$, let $\vartheta^{2,N}_{1,b} : \mathbf{S}(\mathrm{Sp}_4(\bar{\mathbb{F}}_q)) \to \mathcal{N}(\mathrm{O}_{2N}(\bar{\mathbb{F}}_q))$ be defined by

$$\vartheta^{2,N}_{1,b} : \begin{array}{rcc} \Lambda^{\mathrm{u,sp}}_{(2),\emptyset} & \mapsto & \underline{\mathcal{O}(2N-5,5)} \\ \Lambda^{\mathrm{u,sp}}_{(1),(1)} & \mapsto & \{\mathcal{O}(2N-3,3), \underline{\mathcal{O}(2N-5,3,1^2)}\} \\ \Lambda^{\mathrm{u,sp}}_{(1^2),\emptyset} & \mapsto & \underline{\mathcal{O}(2N-5,3,1^2)} \\ \Lambda^{\mathrm{u,sp}}_{\emptyset,(2)} & \mapsto & \{\mathcal{O}(2N-1,1), \mathcal{O}(2N-3,1^3), \underline{\mathcal{O}(2N-5,2^2,1)}\} \\ \Lambda^{\mathrm{u,sp}}_{\emptyset,(1^2)} & \mapsto & \{\mathcal{O}(2N-1,1), \mathcal{O}(2N-3,1^3), \underline{\mathcal{O}(N-5,1^2)}\} \end{array}.$$

If $N \geq 3$, let $\vartheta^{2,N}_{2} : \mathbf{S}(\mathrm{Sp}_4(\bar{\mathbb{F}}_q)) \to \mathcal{N}(\mathrm{O}_{2N}(\bar{\mathbb{F}}_q))$ be defined by

$$\vartheta^{2,N}_{2} : \begin{array}{rcc} \Lambda^{\mathrm{u,sp}}_{(2),\emptyset} & \mapsto & \{\mathcal{O}(2N-1,1), \underline{\mathcal{O}(2N-5,5)}, \mathcal{O}(2N-3,3)\} \\ \Lambda^{\mathrm{u,sp}}_{(1),(1)} & \mapsto & \{\mathcal{O}(2N-3,3), \underline{\mathcal{O}(2N-5,3,1^2)}, \mathcal{O}(2N-1,1), \mathcal{O}(2N-3,1^3)\} \\ \Lambda^{\mathrm{u,sp}}_{(1^2),\emptyset} & \mapsto & \{\underline{\mathcal{O}(2N-5,3,1^2)}, \mathcal{O}(2N-3,3)\} \\ \Lambda^{\mathrm{u,sp}}_{\emptyset,(2)} & \mapsto & \{\mathcal{O}(2N-1,1), \mathcal{O}(2N-3,1^3), \underline{\mathcal{O}(2N-5,2^2,1)}\} \\ \Lambda^{\mathrm{u,sp}}_{\emptyset,(1^2)} & \mapsto & \mathcal{O}(2N-3,1^3), \underline{\mathcal{O}(2N-5,1^5)}\} \end{array}.$$

If $N \geq 3$, let $\vartheta^{2,N}_{3} : \mathbf{S}(\mathrm{O}_4(\bar{\mathbb{F}}_q)) \to \mathcal{N}(\mathrm{Sp}_{2N}(\bar{\mathbb{F}}_q))$ be defined by

$$\vartheta^{2,N}_{3} : \begin{array}{rcc} \Lambda^{\mathrm{u,or}}_{(2),\emptyset} & \mapsto & \{\mathcal{O}(2N), \mathcal{O}(2N-2,2), \underline{\mathcal{O}(2N-4,4)}\} \\ \Lambda^{\mathrm{u,or}}_{(1),(1)} & \mapsto & \{\mathcal{O}(2N), \mathcal{O}(2N-2,1^2), \mathcal{O}(2N-2,2), \underline{\mathcal{O}(2N-4,2^2)}\} \\ \Lambda^{\mathrm{u,or}}_{(1^2),\emptyset} & \mapsto & \{\mathcal{O}(2N-2,2), \underline{\mathcal{O}(2N-4,2,1^2)}\} \\ \Lambda^{\mathrm{u,or}}_{\emptyset,(2)} & \mapsto & \{\mathcal{O}(2N), \mathcal{O}(2N-2,1^2), \underline{\mathcal{O}(2N-4,2^2)}\} \\ \Lambda^{\mathrm{u,or}}_{\emptyset,(1^2)} & \mapsto & \{\mathcal{O}(2N), \underline{\mathcal{O}(2N-4,1^4)}\} \end{array}.$$

Notice that in each case, for $N \geq 3$, the underlined orbit is in the closure of any orbit in the set.

Howe correspondence over finite fields is not bijective on the level of the irreducible characters. Nevertheless, Theorem 10 below shows that it should be possible to *extract from* $\Theta^{\mathrm{G},\mathrm{G}'}_k$ *a bijective correspondence* at least when $N' = 2$, given by the map $\underline{\theta}^{2,N}$ introduced in Definition 5.

THEOREM 10. *We assume that $N' = 2$, $N \geq 2$ and that Conjecture 1 holds. Then Howe correspondence for the dual pair $(\mathrm{G}', \mathrm{G})$ induces the map*

$$\begin{cases} \vartheta^{2,N}_{1,a} & \text{if } (\mathrm{G}',\mathrm{G}) = (\mathrm{O}^{\varepsilon_k}_{2(k^2+2)}(q), \mathrm{Sp}_{2(k^2+k+N)}(q)), \\ \vartheta^{2,N}_{1,b} & \text{if } (\mathrm{G}',\mathrm{G}) = (\mathrm{Sp}_{2(k^2+k+2)}(q), \mathrm{O}^{\varepsilon_k}_{2(k^2+N)}(q)), \\ \vartheta^{2,N}_{2} & \text{if } (\mathrm{G}',\mathrm{G}) = (\mathrm{Sp}_{2(k^2+k+2)}(q), \mathrm{O}^{\varepsilon_{k+1}}_{2((k+1)^2+N)}(q)), \\ \vartheta^{2,N}_{3} & \text{if } (\mathrm{G}',\mathrm{G}) = (\mathrm{O}^{\varepsilon_{k+1}}_{2((k+1)^2+2)}(q), \mathrm{Sp}_{2(k^2+k+N)}(q)). \end{cases}$$

*Moreover, if $N \geq 3$, then the following holds:*

*Let $\rho_{\xi',\eta'} \in \mathrm{Irr}(\mathsf{W}_2)$, and let $\rho_{\xi_0,\eta_0} = \underline{\theta}^{2,N}(\rho_{\xi',\eta'})$. Then every irreducible representation $\rho_{\xi,\eta}$ of $\mathsf{W}_N$ which corresponds to $\rho_{\xi',\eta'}$ by $\Theta_k^{\mathrm{G},\mathrm{G}'}$ satisfies*

$$\mathscr{O}_{\xi_0,\eta_0} \leq \mathscr{O}_{\xi,\eta}. \tag{12}$$

REMARK 1. Proposition 8 and Theorem 10 are unconditional (since Conjecture 1 is known to be true, see [**AMR96**, § 6]) for the following triples $(\mathrm{G}', \mathrm{G}, k)$:

- in Case 1. (a):
  - $(\mathrm{O}_6^-(q), \mathrm{Sp}_{2(N+2)}(q), 1)$ where $2 \leq N \leq 9$;
  - $(\mathrm{O}_{12}^+(q), \mathrm{Sp}_{2(N+6)}(q), 2)$ where $2 \leq N \leq 5$;
- in Case 1. (b):
  - $(\mathrm{Sp}_8(q), \mathrm{O}_{2(N+1)}^-(q), 1)$ where $2 \leq N \leq 10$;
  - $(\mathrm{Sp}_{16}(q), \mathrm{O}_{2(N+4)}^+(q), 2)$ where $2 \leq N \leq 7$;
- in Case 2:
  - $(\mathrm{Sp}_8(q), \mathrm{O}_{2(N+4)}^+(q), 1)$ where $2 \leq N \leq 7$;
  - $(\mathrm{Sp}_{16}(q), \mathrm{O}_{22}^-(q), 2)$;
- in Case 3:
  - $(\mathrm{O}_{12}^+(q), \mathrm{Sp}_{2(N+2)}(q), 1)$ where $2 \leq N \leq 9$;
  - $(\mathrm{O}_{22}^-(q), \mathrm{Sp}_{2(N+6)}(q), 2)$ where $2 \leq N \leq 5$.

PROOF. We will use the examples studied in Section 4.2 and we will also need the following additional computations:

- $\Lambda^{\mathrm{u,sp}}_{(N),\emptyset} = \begin{pmatrix} N \\ - \end{pmatrix}$. We have $\lambda^{\mathrm{u,sp}}_{(N),\emptyset} = (2N)$.
- $\Lambda^{\mathrm{u,sp}}_{\emptyset,(N)} = \begin{pmatrix} 0 & & 2 \\ & N+1 & \end{pmatrix} \sim_{\mathrm{sim}} \begin{pmatrix} 0 & & N+1 \\ & 2 & \end{pmatrix} = \Lambda^{\mathrm{u,sp}}_{(N-1),(1)}$. Example 7 gives $\lambda^{\mathrm{u,sp}}_{\emptyset,(N)} = (2N-2, 2)$.
- $\Lambda^{\mathrm{u,sp}}_{(2),(N-2)} = \begin{pmatrix} 0 & & 4 \\ & N-1 & \end{pmatrix} \sim_{\mathrm{sim}} \begin{pmatrix} 0 & & N-1 \\ & 4 & \end{pmatrix} = \Lambda^{\mathrm{u,sp}}_{(N-2),(2)}$. Example 8 gives $\lambda^{\mathrm{sp}}_{(2),(N-2)} = (2N-6, 6)$.
- $\Lambda^{\mathrm{u,sp}}_{\emptyset,(N-1,1)} = \begin{pmatrix} 0 & & 2 & & 4 \\ & 2 & & N+2 & \end{pmatrix} \sim_{\mathrm{sim}} \begin{pmatrix} 0 & & 2 & & N+2 \\ & 2 & & 4 & \end{pmatrix}$ is distinguished if $N \geq 2$. Example 5 gives $\lambda^{\mathrm{sp}}_{\emptyset,(N-1,1)} = (2N-4, 2, 1^2)$.
- $\Lambda^{\mathrm{u,sp}}_{(1),(N-1)} = \begin{pmatrix} 0 & & 3 \\ & N & \end{pmatrix} \sim_{\mathrm{sim}} \begin{pmatrix} 0 & & N \\ & 3 & \end{pmatrix} = \Lambda^{\mathrm{u,sp}}_{(N-2),(2)}$. Example 6 gives $\lambda^{\mathrm{u,sp}}_{(1),(N-1)} = (2N-4, 4)$.
- $\Lambda^{\mathrm{u,or}}_{(N),\emptyset} = \begin{pmatrix} N \\ 0 \end{pmatrix} \sim_{\mathrm{sim}} \begin{pmatrix} 0 \\ N \end{pmatrix}$. Hence $\lambda^{\mathrm{or}}_{N,\emptyset} = (2N-1, 1)$.
- $\Lambda^{\mathrm{u,or}}_{\emptyset,(N)} = \begin{pmatrix} 0 \\ N \end{pmatrix}$. Hence $\lambda^{\mathrm{or}}_{\emptyset,N} = (2N-1, 1)$.
- $\Lambda^{\mathrm{u,or}}_{(2),(N-2)} = \begin{pmatrix} 2 \\ N-2 \end{pmatrix} \sim_{\mathrm{sim}} \Lambda^{\mathrm{u,or}}_{(N-2),(2)}$. Example 6 gives $\lambda^{\mathrm{or}}_{(2),(N-2)} = (2N-5, 5)$.
- $\Lambda^{\mathrm{u,or}}_{(1),(N-1)} = \begin{pmatrix} 1 \\ N-1 \end{pmatrix} \sim_{\mathrm{sim}} \Lambda^{\mathrm{u,or}}_{(N-1),(1)}$. Example 7 gives $\lambda^{\mathrm{or}}_{(1),(N-1)} = (2N-3, 3)$.
- $\Lambda^{\mathrm{u,or}}_{(1),(N-2,1)} = \begin{pmatrix} 0 & 3 \\ 1 & N \end{pmatrix}$. Example 9 gives $\lambda^{\mathrm{or}}_{(1),(N-2,1)} = (2N-5, 3, 1^2)$.

- $\Lambda^{\mathrm{u,or}}_{\emptyset,(N-1,1)} = \begin{pmatrix} 0 & 2 \\ 1 & N+1 \end{pmatrix}$. Example 3 gives $\lambda^{\mathrm{or}}_{\emptyset,(N-1,1)} = (2N-3, 1^3)$.
- $\Lambda^{\mathrm{u,or}}_{\emptyset,(N-2,2)} = \begin{pmatrix} 0 & 2 \\ 2 & N \end{pmatrix}$. Example 3 gives $\lambda^{\mathrm{or}}_{\emptyset,(N-2,2)} = (2N-5, 2^2, 1)$.
- $\Lambda^{\mathrm{u,or}}_{(1^2),(N-2)} = \begin{pmatrix} 1 & 3 \\ 0 & N \end{pmatrix} = \Lambda^{\mathrm{u,or}}_{(n-2),(1^2)}$. Example 5 gives $\lambda^{\mathrm{or}}_{(1^2),(N-2)} = (2N-5, 3, 1^2)$.
- $\Lambda^{\mathrm{u,or}}_{\emptyset,(N-2,1^2)} = \begin{pmatrix} 0 & 2 & 4 \\ 1 & 3 & N+2 \end{pmatrix}$. Example 4 gives $\lambda^{\mathrm{or}}_{\emptyset,(N-2,1^2)} = (2N-5, 1^5)$.

We will consider the four cases above separately.

Case 1 (a): The map $\theta_1^{2,N}$ induces the following correspondence between $u$-symbols:

$$\begin{array}{ccc} \Lambda^{\mathrm{u,or}}_{(2),\emptyset} & \mapsto & \underline{\Lambda^{\mathrm{u,sp}}_{(2),(N-2)}} \\ \Lambda^{\mathrm{u,or}}_{(1),(1)} & \mapsto & \{\Lambda^{\mathrm{u,sp}}_{(1),(N-1)}, \underline{\Lambda^{\mathrm{u,sp}}_{(1),(N-2,1)}}\} \\ \Lambda^{\mathrm{u,or}}_{(1^2),\emptyset} & \mapsto & \underline{\Lambda^{\mathrm{u,sp}}_{(1^2),(N-2)}} \\ \Lambda^{\mathrm{u,or}}_{\emptyset,(2)} & \mapsto & \{\Lambda^{\mathrm{u,sp}}_{\emptyset,(N)}, \Lambda^{\mathrm{u,sp}}_{\emptyset,(N-1,1)}, \underline{\Lambda^{\mathrm{u,sp}}_{\emptyset,(N-2,2)}}\} \\ \Lambda^{\mathrm{u,or}}_{\emptyset,(1^2)} & \mapsto & \{\Lambda^{\mathrm{u,sp}}_{\emptyset,(N)}, \Lambda^{\mathrm{u,sp}}_{\emptyset,(N-1,1)}, \underline{\Lambda^{\mathrm{u,sp}}_{\emptyset,(N-2,1^2)}}\} \end{array}, \quad \text{if } N \geq 3. \tag{13}$$

Here and in the rest of this proof the underlined symbols correspond to the representations $\rho_{\xi_0,\eta_0} = \underline{\theta}^{2,N}(\rho_{\xi',\eta'})$.

Combining the above computations with Example 10, Example 11, Example 12, and Example 13, we obtain the following closure order on the unipotent classes of $\mathrm{Sp}_{2N}(\bar{\mathbb{F}}_q)$ occurring in the above correspondence:

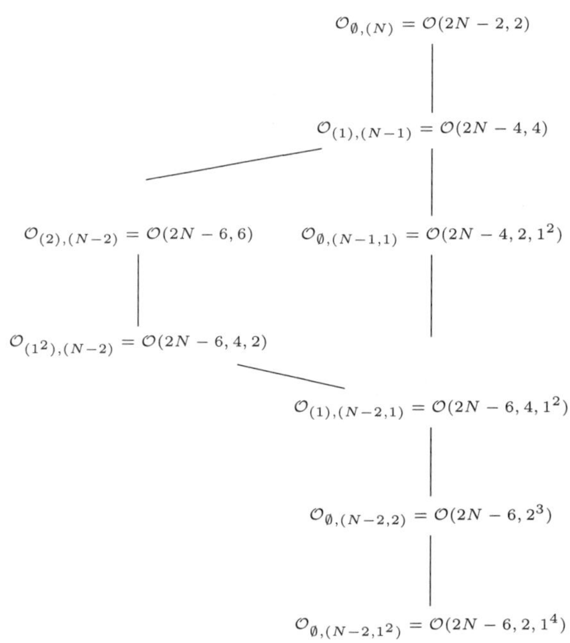

Hence (13) induces $\vartheta^{2,N}_{1,a}$ and the second assertion of the Theorem follows in the case 1 (a).

Case 1 (b): The map $\theta^{2,N}_1$ induces the following correspondence between $u$-symbols:

$$
(14)\qquad \begin{array}{ccc}
\Lambda^{\mathrm{u,sp}}_{(2),\emptyset} & \mapsto & \underline{\Lambda^{\mathrm{u,or}}_{(N-2),(2)}} \\
\Lambda^{\mathrm{u,sp}}_{(1),(1)} & \mapsto & \{\Lambda^{\mathrm{u,or}}_{(N-1),(1)}, \underline{\Lambda^{\mathrm{u,or}}_{(N-2,1),(1)}}\} \\
\Lambda^{\mathrm{u,sp}}_{(1^2),\emptyset} & \mapsto & \underline{\Lambda^{\mathrm{u,or}}_{(N-2),(1^2)}} \\
\Lambda^{\mathrm{u,sp}}_{\emptyset,(2)} & \mapsto & \{\Lambda^{\mathrm{u,or}}_{(N),\emptyset}, \Lambda^{\mathrm{u,or}}_{(N-1,1),\emptyset}, \underline{\Lambda^{\mathrm{u,or}}_{(N-2,2),\emptyset}}\} \\
\Lambda^{\mathrm{u,sp}}_{\emptyset,(1^2)} & \mapsto & \{\Lambda^{\mathrm{u,or}}_{(N),\emptyset}, \Lambda^{\mathrm{u,or}}_{(N-1,1),\emptyset}, \underline{\Lambda^{\mathrm{u,or}}_{(N-2,1^2),\emptyset}}\}
\end{array}, \quad \text{if } N \geq 3.
$$

We obtain the following closure order on the unipotent classes of $\mathrm{O}_{2N}(\bar{\mathbb{F}}_q)$ occurring in the above correspondence:

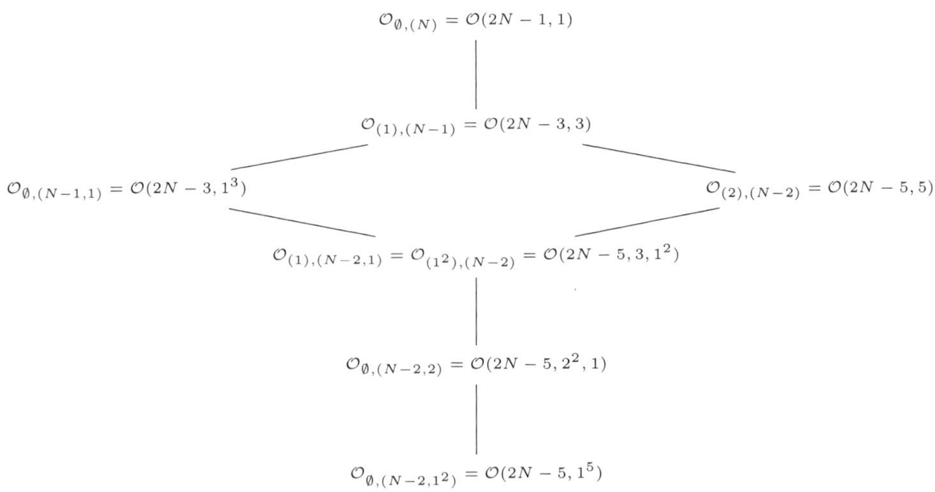


Hence (14) induces $\vartheta^{2,N}_{1,b}$ and the second assertion of the Theorem follows in the case 1 (b).

Case 2: The map $\theta^{2,N}_2$ induces the following correspondence between $u$-symbols:

$$
(15)\qquad \begin{array}{ccc}
\Lambda^{\mathrm{u,sp}}_{(2),\emptyset} & \mapsto & \{\Lambda^{\mathrm{u,or}}_{(N),\emptyset}, \underline{\Lambda^{\mathrm{u,or}}_{(N-2),(2)}}, \Lambda^{\mathrm{u,or}}_{(N-1),(1)}\} \\
\Lambda^{\mathrm{u,sp}}_{(1),(1)} & \mapsto & \{\Lambda^{\mathrm{u,or}}_{(N-1),(1)}, \underline{\Lambda^{\mathrm{u,or}}_{(N-2,1),(1)}}, \Lambda^{\mathrm{u,or}}_{(N),\emptyset}, \Lambda^{\mathrm{u,or}}_{(N-1,1),\emptyset}\} \\
\Lambda^{\mathrm{u,sp}}_{(1^2),\emptyset} & \mapsto & \{\underline{\Lambda^{\mathrm{u,or}}_{(N-2),(1^2)}}, \Lambda^{\mathrm{u,or}}_{(N-1),(1)}\} \\
\Lambda^{\mathrm{u,sp}}_{\emptyset,(2)} & \mapsto & \{\Lambda^{\mathrm{u,or}}_{(N),\emptyset}, \Lambda^{\mathrm{u,or}}_{(N-1,1),\emptyset}, \underline{\Lambda^{\mathrm{u,or}}_{(N-2,2),\emptyset}}\} \\
\Lambda^{\mathrm{u,sp}}_{\emptyset,(1^2)} & \mapsto & \Lambda^{\mathrm{u,or}}_{(N-1,1),\emptyset}, \underline{\Lambda^{\mathrm{u,or}}_{(N-2,1^2),\emptyset}}\}
\end{array}, \quad \text{if } N \geq 3.
$$

We obtain the following closure order on the unipotent classes of $\mathrm{O}_{2N}(\bar{\mathbb{F}}_q)$ occurring in the above correspondence:

$$\mathcal{O}_{\emptyset,(N)} = \mathcal{O}(2N-1,1)$$

$$\mathcal{O}_{(1),(N-1)} = \mathcal{O}(2N-3,3)$$

$$\mathcal{O}_{\emptyset,(N-1,1)} = \mathcal{O}(2N-3,1^3) \qquad\qquad \mathcal{O}_{(2),(N-2)} = \mathcal{O}(2N-5,5)$$

$$\mathcal{O}_{(1^2),(N-2)} = \mathcal{O}_{(1),(N-2,1)} = \mathcal{O}(2N-5,3,1^2)$$

$$\mathcal{O}_{\emptyset,(N-2,2)} = \mathcal{O}(2N-5,2^2,1)$$

$$\mathcal{O}_{\emptyset,(N-2,1^2)} = \mathcal{O}(2N-5,1^5)$$

Hence (15) induces $\vartheta_2^{2,N}$ and the second assertion of the Theorem follows in the case 2.

Case 3: The map $\theta_3^{2,N}$ induces the following correspondence between $u$-symbols:

$$(16)\qquad \begin{array}{rcl} \Lambda^{\mathrm{u,or}}_{(2),\emptyset} & \mapsto & \{\Lambda^{\mathrm{u,sp}}_{(N),\emptyset}, \Lambda^{\mathrm{u,sp}}_{(N-1,1),\emptyset}, \underline{\Lambda^{\mathrm{u,sp}}_{(N-2,2),\emptyset}}\} \\ \Lambda^{\mathrm{u,or}}_{(1),(1)} & \mapsto & \{\Lambda^{\mathrm{u,sp}}_{(N),\emptyset}, \Lambda^{\mathrm{u,sp}}_{(N-1,1),\emptyset}, \Lambda^{\mathrm{u,sp}}_{(N-1),(1)}, \underline{\Lambda^{\mathrm{u,sp}}_{(N-2,1),(1)}}\} \\ \Lambda^{\mathrm{u,or}}_{(1^2),\emptyset} & \mapsto & \{\Lambda^{\mathrm{u,sp}}_{(N-1,1),\emptyset}, \underline{\Lambda^{\mathrm{u,sp}}_{(N-2,1^2),\emptyset}}\} \\ \Lambda^{\mathrm{u,or}}_{\emptyset,(2)} & \mapsto & \{\Lambda^{\mathrm{u,sp}}_{(N),\emptyset}, \Lambda^{\mathrm{u,sp}}_{(N-1),(1)}, \underline{\Lambda^{\mathrm{u,sp}}_{(N-2),(2)}}\} \\ \Lambda^{\mathrm{u,or}}_{\emptyset,(1^2)} & \mapsto & \{\Lambda^{\mathrm{u,sp}}_{(N-1),(1)}, \underline{\Lambda^{\mathrm{u,sp}}_{(N-2),(1^2)}}\} \end{array} \quad , \quad \text{if } N \geq 3.$$

We obtain the following closure order on the unipotent classes of $\mathrm{Sp}_{2N}(\bar{\mathbb{F}}_q)$ occurring in the correspondence above.

$$\mathcal{O}_{(N),\emptyset} = \mathcal{O}(2N)$$

$$\mathcal{O}_{(N-1),(1)} = \mathcal{O}(2N-2,2)$$

$$\mathcal{O}_{(N-1,1),\emptyset} = \mathcal{O}(2N-2,1^2) \qquad \mathcal{O}_{(N-2),(2)} = \mathcal{O}(2N-4,4)$$

$$\mathcal{O}_{(N-2,2),\emptyset} = \mathcal{O}_{(N-2,1),(1)} = \mathcal{O}(2N-4,2^2)$$

$$\mathcal{O}_{(N-2),(1^2)} = \mathcal{O}(2N-4,2,1^2)$$

$$\mathcal{O}_{(N-2,1^2),\emptyset} = \mathcal{O}(2N-4,1^4)$$

Hence (16) induces $\vartheta_3^{2,N}$ and the second assertion of the Theorem follows in the case 3. □

Property (12) shows that the map $\underline{\theta}^{2,N}$ plays a special role in Howe correspondence. We will now restrict our attention to it and see how it relates to [**AKP13**, (19)].

Let $\widetilde{\mathrm{sgn}} := \mathrm{sgn} \otimes \mathrm{sgn}_{\mathrm{CD}}$ denote the product of the sign character $\mathrm{sgn} = \rho_{\emptyset,(1^N)}$ of the group $W_N$ by the character $\mathrm{sgn}_{\mathrm{CD}} = \rho_{\emptyset,(N)}$, *i.e.*:

$$\widetilde{\mathrm{sgn}} \otimes \rho_{\xi,\eta} = \rho_{{}^t\xi,{}^t\eta}.$$

Then, when $k_2 = k$, let $\underline{\theta}^{2,N}_{\mathrm{twist}}$ be the map defined by

$$\underline{\theta}^{2,N}_{\mathrm{twist}} := \begin{cases} \mathrm{sgn} \circ \underline{\theta}^{2,N} \circ \mathrm{sgn} & \text{if } \mathrm{G}' \text{ symplectic,} \\ \widetilde{\mathrm{sgn}} \circ \underline{\theta}^{2,N} \circ \widetilde{\mathrm{sgn}} & \text{if } \mathrm{G}' \text{ orthogonal.} \end{cases} \tag{17}$$

We obtain

$$\underline{\theta}^{2,N}_{\mathrm{twist}}(\rho_{\xi',\eta'}) = \begin{cases} \rho_{(1^2)\oplus\xi',\eta'} & \text{if } \mathrm{G}' \text{ symplectic,} \\ \rho_{\xi',(1^2)\oplus\eta'} & \text{if } \mathrm{G}' \text{ orthogonal.} \end{cases} \tag{18}$$

In the case where $k = 0$ and $\varepsilon = +$, we have $\tilde{n}_1(k) = n_1$, $\tilde{n}_2(k) = n_2$, $N = \max(n_1, n_2)$ and $N' = \min(n_1, n_2)$. Hence Howe correspondence between the irreducible components of the unipotent principal series of the groups G and G′ (where $(\mathrm{G}, \mathrm{G}') = (\mathrm{Sp}_{2n}(q), \mathrm{O}^+_{2n'}(q))$ or $(\mathrm{G}, \mathrm{G}') = (\mathrm{O}^+_{2n'}(q), \mathrm{Sp}_{2n}(q))$ is given by the correspondence $\Theta_0^{\mathrm{G},\mathrm{G}'}$ between irreducible characters of the groups $W_N$ and $W_{N'}$, and (18) coincides with [**AKP13**, (19)].

## 5. Howe correspondence and wave front set

Recall (see [**Lus92**], [**GM99**], [**AA07**]) that for every irreducible character $\chi_\Pi$ of the $\mathbb{F}_q$-points $G = \mathbf{G}^F$ of a split connected reductive group $\mathbf{G}$ defined over $\mathbb{F}_q$ (assuming that the characteristic $p$ of $\mathbb{F}_q$ is "good for $G$": for instance, if $G$ a symplectic group or a split special orthogonal group, then $p$ must be odd) there is a unique rational unipotent class $\mathcal{O}_\Pi$ in $\mathbf{G}$ which has the property that there exists $u \in \mathcal{O}_\Pi(q)$ such that $\chi_\Pi(u) \neq 0$ and $\mathcal{O}_\Pi$ has maximal dimension among classes with that property. The class $\mathcal{O}_\Pi$ is called the *unipotent support* of $\chi_\Pi$. It coincides with the class defined in [**Lus84**, §13.3].

More precisely, suppose $\Pi$ is unipotent. Then there exists an irreducible representation $\rho$ of the Weyl group $\mathsf{W}$ of $\mathbf{G}$ such that the scalar product between $\chi_\Pi$ and the *almost character* $R_\rho$ (which is defined as a certain linear combination of Deligne-Lusztig generalized characters in [**Lus84**, page 347 and (4.24.1)], and coincides with the virtual character in Eqn. (3) when $\mathrm{G} = \mathrm{GL}_n(q)$) is non-zero. Moreover, if $\rho'$ is another irreducible representation of $\mathsf{W}$ such that $\chi_\Pi$ has non-scalar product with $R_{\rho'}$, then $\rho$ and $\rho'$ belong to the same family of characters of $\mathsf{W}$ (see [**Lus84**, Theorems 5.25 and 6.17]). Thus, we can associate with $\chi_\Pi$ a unique family of characters of $\mathsf{W}$, or equivalently, a unique two-sided cell in $\mathsf{W}$. Let $\rho_{\mathrm{spe}}$ be the unique *special character* in this family (for $\mathbf{G}$ of classical type a family of characters of $\mathsf{W}$ corresponds to a similarity class of $u$-symbols, and the symbol corresponding to $\rho_{\mathrm{spe}}$ is the unique distinguished $u$-symbol in that family, [**Lus84**, (4.5.6)]). Then the class $\mathcal{O}_\Pi$ coincides with the unipotent class corresponding to $\rho_{\mathrm{spe}}$ by the Springer correspondence for the group $\mathsf{W}$. In particular the unipotent class $\mathcal{O}_\Pi$ is always special.

Moreover, every rational unipotent class $\mathcal{O}$ on which $\chi_\Pi$ does not vanish (i.e., such that there exists $u \in \mathcal{O}(q)$ with $\chi_\Pi(u) \neq 0$) satisfies

$$\mathcal{O} \leq \mathcal{O}_\Pi, \tag{19}$$

see [**AA07**, Theorem 6.1].

Suppose $\Pi$ is an irreducible unipotent representation of a split group G, such as $\mathrm{O}_{2n}^+(q)$ or $\mathrm{Sp}_{2n}(q)$, which belongs to the principal series. The algebra of the endomorphisms of the principal series which commute with the action of G is the Iwahori-Hecke algebra, whose irreducible representations coincide with the irreducible representations of the Weyl group $\mathsf{W}$. Hence, as we remarked previously, there is a one to one correspondence between the irreducible representations of $\mathsf{W}$ and the irreducible representations of G which occur in the principal series. Given an irreducible representation $\rho$ of $\mathsf{W}$ we denoted by $\Pi_\rho$ the corresponding representation of G. Furthermore, the almost character $R_\rho$ has a non-zero scalar product with the character of $\Pi_\rho$. (This follows from [**Lus84**, Theorem 4.23]. For an explicit argument see pages 297 and 298 in [**Lus84**].)

If the group $\mathbf{G}$ is disconnected and $\Pi$ be an irreducible representation of $\mathrm{G} = \mathbf{G}^F$, we define the unipotent support of $\Pi$, denoted $\mathcal{O}_\Pi$, to be the union of the rational unipotent classes $\mathcal{O} \subseteq \mathbf{G}$ of maximal dimension, such that $\mathcal{O} \cap \mathrm{G}$ has a non-empty intersection with the support of the character $\chi_\Pi$.

Let $\Pi = \Pi_{\rho_{\xi,\eta}}$ be an irreducible unipotent representation of $\mathrm{O}_{2n}^+(q)$ which belongs to the principal series, where $(\xi, \eta)$ is a pair of partitions of $n$. Two cases can occur: the restriction $\Pi_{\mathrm{SO}}$ of $\Pi$ to $\mathrm{SO}_{2n}^+(q)$ is either irreducible, or is the direct sum of two nonequivalent irreducible representations $\Pi_{\mathrm{SO}}^{\mathrm{I}}$ and $\Pi_{\mathrm{SO}}^{\mathrm{II}}$. The latter case arises if and only if the restriction of the representation $\rho_{\xi,\eta}$ to the Weyl group of the special orthogonal group splits into the sum of two inequivalent representations, i.e. the partition $(\xi, \eta)$ is such that $\xi = \eta$, see [**Car93**, Prop. 11.4.4].

$(*)$ We recall (see for instance [**Spa82**]) that any rational unipotent class $\mathcal{O}$ in $\mathrm{O}_{2n}^+$ is either a rational unipotent class in $\mathrm{SO}_{2n}^+$ or is the disjoint union of a rational unipotent class $\mathcal{O}(u)$ in $\mathrm{SO}_{2n}^+$ and a rational unipotent class $\mathcal{O}(sus^{-1})$ in $\mathrm{SO}_{2n}^+$, with $u \in \mathrm{SO}_{2n}^+(q)$ unipotent and some $s \in \mathrm{O}_{2n}(q) \setminus \mathrm{SO}_{2n}(q)$. Both these classes have the same dimension.

LEMMA 11. *If $\Pi_{\mathrm{SO}}$ is irreducible then*

$$\mathcal{O}_{\Pi} = \mathcal{O}_{\Pi_{\mathrm{SO}}}. \tag{20}$$

*If $\Pi_{\mathrm{SO}} = \Pi_{\mathrm{SO}}^{\mathrm{I}} \oplus \Pi_{\mathrm{SO}}^{\mathrm{II}}$, then*

$$\mathcal{O}_{\Pi} = \mathcal{O}_{\Pi_{\mathrm{SO}}^{\mathrm{I}}} \cup \mathcal{O}_{\Pi_{\mathrm{SO}}^{\mathrm{II}}}. \tag{21}$$

*In both cases, $\mathcal{O}_{\Pi}$ is a single unipotent class in $\mathrm{O}_{2n}^+$.*

PROOF. Let us assume first that $\Pi_{\mathrm{SO}}$ is irreducible. Then the restrictions of the two characters $\chi_{\Pi}$ and $\chi_{\Pi_{\mathrm{SO}}}$ to $\mathrm{SO}_{2n}(q)$ are equal. In particular, $\chi_{\Pi_{\mathrm{SO}}}(g) = \chi_{\Pi}(g) = \chi_{\Pi}(sgs^{-1}) = \chi_{\Pi_{\mathrm{SO}}}(sgs^{-1})$ for any $g \in \mathrm{SO}_{2n}(q)$ and $s$ as in $(*)$. Let $\mathcal{O}(u)$ be a rational unipotent class in $\mathrm{O}_{2n}^+$, as in $(*)$. We see that, with the notation of $(*)$, the restriction of $\chi_{\Pi}$ to $\mathcal{O}(u) \cap \mathrm{SO}_{2n}^+(q)$ is non-zero if and only if the restriction of $\chi_{\Pi}$ to $\mathcal{O}(sus^{-1}) \cap \mathrm{SO}_{2n}^+(q)$ is non-zero. But the classes $\mathcal{O}(sus^{-1})$ and $\mathcal{O}(u)$ have the same dimension. Therefore, if that dimension is maximal among the unipotent classes which have a non-empty intersection with the support of $\chi_{\Pi_{\mathrm{SO}}}$, we get a contradiction. Thus the unipotent support of $\Pi_{\mathrm{SO}}$ is the unipotent class in $\mathrm{O}_{2n}^+$ which is also a single unipotent class in $\mathrm{SO}_{2n}^+$. Hence, (20) follows.

Assume now that $\Pi_{\mathrm{SO}} = \Pi_{\mathrm{SO}}^{\mathrm{I}} \oplus \Pi_{\mathrm{SO}}^{\mathrm{II}}$. In this case the representations $\Pi_{\mathrm{SO}}^{\mathrm{I}}$ and $\Pi_{\mathrm{SO}}^{\mathrm{II}}$ are permuted via the action of the group element $s$, as in $(*)$, and so are their unipotent supports. More precisely, $\chi_{\Pi_{\mathrm{SO}}^{\mathrm{II}}}(u) = \chi_{\Pi_{\mathrm{SO}}^{\mathrm{I}}}(sus^{-1})$, $\mathcal{O}_{\Pi_{\mathrm{SO}}^{\mathrm{I}}} = \mathcal{O}(u)$, $\mathcal{O}_{\Pi_{\mathrm{SO}}^{\mathrm{II}}} = \mathcal{O}(sus^{-1})$ and the right hand side of (21) is a single unipotent rational class in $\mathrm{O}_{2n}$. (Since, as we noticed before, $\xi = \eta$, these classes have the same set of elementary divisors and hence the same dimension, see [**Car93**, page 399]. They are described explicitly in [**Car93**, § 13.3, Type $\mathrm{D}_l$]).

Let $u \in \mathrm{SO}_{2n}^+(q) \cap \mathcal{O}_{\Pi}$ be such that $\chi_{\Pi}(u) \neq 0$. Since $\chi_{\Pi}(u) = \chi_{\Pi_{\mathrm{SO}}^{\mathrm{I}}}(u) + \chi_{\Pi_{\mathrm{SO}}^{\mathrm{II}}}(u)$, we see that that $\chi_{\Pi_{\mathrm{SO}}^{\mathrm{I}}}(u) \neq 0$ or $\chi_{\Pi_{\mathrm{SO}}^{\mathrm{II}}}(u) \neq 0$. Hence, $\mathcal{O}_{\Pi} = \mathcal{O}_{\Pi_{\mathrm{SO}}^{\mathrm{I}}} \cup \mathcal{O}_{\Pi_{\mathrm{SO}}^{\mathrm{II}}}$. □

Since (19) holds for the representations of $\mathrm{SO}_{2n}(q)$, we see from Lemma 11 that it also holds for the representations of $\mathrm{O}_{2n}^+(q)$.

COROLLARY 12. *Let $\Pi'_{\rho_{\xi',\eta'}}$ be an irreducible unipotent representation of $\mathrm{G}' = \mathrm{Sp}_4(q)$ (resp. $\mathrm{G}' = \mathrm{O}_4^+(q)$) which belongs to the principal series of $G'$. Let $n \geq 3$, and let $(\xi_0, \eta_0) := (\xi', (n-2) \cup \eta')$.*

*Assume that Conjecture 1 holds. Then every representation of $\mathrm{G} = \mathrm{O}_{2n}^+(q)$ (resp. $\mathrm{G}' = \mathrm{Sp}_{2n}(q)$) which occurs in the image of $\Pi'$ by Howe correspondence for the dual pair $(\mathrm{G}', \mathrm{G})$ is such that the closure of its unipotent support contains the closure of the unipotent support of $\Pi_{\xi_0 \cdot \eta_0}$.*

PROOF. This follows directly from (12) and from the fact that, for the map $\vartheta_1^{2,n}$ with $n \geq 3$, the underlined orbit in the output set is contained in the closure of each orbit in that set. □

Recall Alvis-Curtis Duality $D_{\mathrm{G}} : \mathfrak{R}(\mathrm{G}) \to \mathfrak{R}(\mathrm{G})$ ([**Alv79**], [**Cur80**], [**Aub92**]), which is defined for representations of $\mathrm{G} = \mathbf{G}^F$, when $\mathbf{G}$ is connected.

Let $\Pi$ be an irreducible unipotent representation of $\mathrm{O}_{2n} := \mathrm{O}_{2n}^+(q)$, as in Lemma 11. If $\Pi_{\mathrm{SO}}$ is irreducible, define $D_{\mathrm{O}_{2n}}(\Pi)$ to be the unique irreducible representation $\tilde{\Pi}$ of $\mathrm{O}_{2n}$ such that $\tilde{\Pi}_{\mathrm{SO}} = D_{\mathrm{SO}_{2n}}(\tilde{\Pi}_{\mathrm{SO}})$. If $\Pi_{\mathrm{SO}} = \Pi_{\mathrm{SO}}^I \oplus \Pi_{\mathrm{SO}}^{II}$, let $D_{\mathrm{O}_{2n}}(\Pi)$ to be the only irreducible representation $\tilde{\Pi}$ of $\mathrm{O}_{2n}^+(q)$ such that $\tilde{\Pi}_{\mathrm{SO}} = D_{\mathrm{SO}_{2n}}(\Pi_{\mathrm{SO}}^I) \oplus D_{\mathrm{SO}_{2n}}(\Pi_{\mathrm{SO}}^{II})$. Then in both cases, $D_{\mathrm{O}_{2n}}(\Pi_{\rho_{\xi,\eta}}) = \Pi_{\rho_{{}^t\eta,{}^t\xi}}$. In other words, $D_{\mathrm{O}_{2n}}(\Pi_\rho) = \Pi_{\rho\otimes\mathrm{sgn}}$, see [**Lus84**, (6.8.6)]. Hence, $\mathcal{O}_\Pi = \mathcal{O}_{\xi,\eta}$ if and only if $\mathcal{O}_{D_{\mathrm{O}_{2n}}(\Pi)} = \mathcal{O}_{{}^t\xi,{}^t\eta}$. Also, tensoring with the sign representation of the Weyl group translates via Springer correspondence to an order reversing involution on the special unipotent orbits, see [**Car93**, pages 389, 390]. By combining this with Corollary 12, (18), (17) and [**AKP13**, Proposition 5], we deduce the following theorem.

THEOREM 13. *Consider the dual pair* $(\mathrm{G}' = \mathrm{Sp}_4(q), \mathrm{G} = \mathrm{O}_{2n}^+(q))$ *with* $n \geq 4$. *(This is a dual pair in the stable range with* $\mathrm{G}'$ *the smaller member.) Assume that Conjecture* 1 *holds.*

*Let* $\pi'$ *be an irreducible representation of* $\mathrm{G}'$ *such that* $D_{\mathrm{G}'}(\pi')$ *is unipotent and belongs to the principal series of* $\mathrm{G}'$. *Then there is a unique irreducible representation* $\pi_{\mathrm{pref}}$ *of* $\mathrm{G}$ *such that* $D_{\mathrm{G}}(\pi_{\mathrm{pref}})$ *corresponds to* $D_{\mathrm{G}'}(\pi')$ *via Howe Correspondence for the pair* $(\mathrm{G}', \mathrm{G})$ *and the unipotent support* $\mathcal{O}_{\pi_{\mathrm{pref}}}$ *of* $\pi_{\mathrm{pref}}$ *contains in its closure the unipotent support of any irreducible representation* $\pi$ *of* $\mathrm{G}$ *such that* $D_{\mathrm{G}}(\pi)$ *corresponds to* $D_{\mathrm{G}'}(\pi')$.

*Let* $\lambda'$, $\lambda'$ *be the partitions describing the rational unipotent class* $\mathcal{O}_{\pi'}$ *and* $\mathcal{O}_{\pi_{\mathrm{pref}}}$, *respectively. Then* $\lambda$ *is obtained from* $\lambda'$ *by adding a column of length* $2N-4$ *to* $\lambda'$, *as in* [**AKP13**, Theorem 1].

Lusztig has proved in [**Lus92**, Theorem 11.2], under the assumption that $p$ is large enough, that the closure of the unipotent support of $\Pi$ coincides with the wave front set (as defined by Kawanaka in [**Kaw87**]) of its dual. Recall that $D_{\mathrm{GL}_n(q)}$ maps the unipotent character $R_{\rho_\mu}$ to the unipotent character $R_{\rho_{{}^t\mu}}$. Hence Theorem 3 and Theorem 13 imply the following result.

COROLLARY 14. *Let* $(\mathrm{G}', \mathrm{G})$ *be one of the dual pairs* $(\mathrm{GL}_{n'}(q), \mathrm{GL}_n(q))$ *or* $(\mathrm{Sp}_4(q), \mathrm{O}_{2n}^+(q))$ *with* $n \geq 4$. *In the latter case, we assume that Conjecture* 1 *holds.*

*Let* $\Pi'$ *be a unipotent irreducible representation of* $\mathrm{G}'$ *that belongs to the principal series of* $\mathrm{G}'$. *Then there is a unique irreducible representation* $\Pi_{\mathrm{pref}}$ *of* $\mathrm{G}$ *such that* $\Pi_{\mathrm{pref}}$ *corresponds to* $\Pi'$ *via Howe Correspondence for the pair* $(\mathrm{G}', \mathrm{G})$ *and the wave front set of* $\Pi_{\mathrm{pref}}$ *contains the wave front set of any irreducible representation* $\Pi$ *of* $\mathrm{G}$ *such that* $\Pi$ *corresponds to* $\Pi'$.

PROOF. Let $(\mathrm{G}, \mathrm{G}') = (\mathrm{Sp}_4(q), \mathrm{O}_{2n}^+(q))$. We put $\pi' := D_{\mathrm{G}'}(\Pi')$. Since $D_{\mathrm{G}'}$ is an involution, we have $\Pi' = D_{\mathrm{G}'}(\pi')$. Then we apply Theorem 13 to the representation $\pi'$, and we set

$$\Pi_{\mathrm{pref}} := D_G(\pi_{\mathrm{pref}}).$$

From Theorem 13, it follows that $\Pi_{\mathrm{pref}}$ corresponds to $\Pi'$ by Howe correspondence, and that the unipotent support $\mathcal{O}_{\pi_{\mathrm{pref}}}$ of $\pi_{\mathrm{pref}}$ contains in its closure the unipotent support $\mathcal{O}_\pi$ of any irreducible representation $\pi$ of G such that $D_{\mathrm{G}}(\pi)$ corresponds to $D_{\mathrm{G}'}(\pi') = \Pi'$. Setting $\Pi := D_{\mathrm{G}}(\pi)$, and using the fact that the closure of $\mathcal{O}_\pi$ coincides with the wave front set of $D_G(\pi) = \Pi$, we get that the wave front set of $\Pi_{\mathrm{pref}}$ contains the wave front set of any irreducible representation $\Pi$ of G such that $\Pi$ corresponds to $\Pi'$.

A similar argument using Theorem 3 instead of 13 gives the result when $(\mathrm{G}', \mathrm{G})$ is of type II. □

## References


[AA07] P. N. Achar and A.-M. Aubert, *Supports unipotents de faisceaux caractères* (French, with English and French summaries), J. Inst. Math. Jussieu **6** (2007), no. 2, 173–207, DOI 10.1017/S1474748006000065. MR2311663 (2008c:20021)

[AKP13] A.-M. Aubert, W. Kraśkiewicz, and T. Przebinda, *Howe correspondence and Springer correspondence for real reductive dual pairs*, Manuscripta Math. **143** (2014), no. 1-2, 81–130, DOI 10.1007/s00229-013-0617-y. MR3147445

[Alv79] D. Alvis, *The duality operation in the character ring of a finite Chevalley group*, Bull. Amer. Math. Soc. (N.S.) **1** (1979), no. 6, 907–911, DOI 10.1090/S0273-0979-1979-14690-1. MR546315 (81e:20012)

[AM93] J. Adams and A. Moy, *Unipotent representations and reductive dual pairs over finite fields*, Trans. Amer. Math. Soc. **340** (1993), no. 1, 309–321, DOI 10.2307/2154558. MR1173855 (94d:20047)

[AMR96] A.-M. Aubert, J. Michel, and R. Rouquier, *Correspondance de Howe pour les groupes réductifs sur les corps finis* (French), Duke Math. J. **83** (1996), no. 2, 353–397, DOI 10.1215/S0012-7094-96-08312-X. MR1390651 (98a:20039)

[Aub92] A.-M. Aubert, *Foncteurs de Mackey et dualité de Curtis généralisés* (French, with English and French summaries), C. R. Acad. Sci. Paris Sér. I Math. **315** (1992), no. 6, 663–668. MR1183799 (93j:20096)

[Car93] R. W. Carter, *Finite groups of Lie type*, Wiley Classics Library, John Wiley & Sons, Ltd., Chichester, 1993. Conjugacy classes and complex characters; Reprint of the 1985 original; A Wiley-Interscience Publication. MR1266626 (94k:20020)

[Cur80] C. W. Curtis, *Truncation and duality in the character ring of a finite group of Lie type*, J. Algebra **62** (1980), no. 2, 320–332, DOI 10.1016/0021-8693(80)90185-4. MR563231 (81e:20011)

[DL76] P. Deligne and G. Lusztig, *Representations of reductive groups over finite fields*, Ann. of Math. (2) **103** (1976), no. 1, 103–161. MR0393266 (52 #14076)

[DM91] F. Digne and J. Michel, *Representations of finite groups of Lie type*, London Mathematical Society Student Texts, vol. 21, Cambridge University Press, Cambridge, 1991. MR1118841 (92g:20063)

[GM99] M. Geck and G. Malle, *On the existence of a unipotent support for the irreducible characters of a finite group of Lie type*, Trans. Amer. Math. Soc. **352** (2000), no. 1, 429–456, DOI 10.1090/S0002-9947-99-02210-2. MR1475683 (2000c:20064)

[GP00] M. Geck and G. Pfeiffer, *Characters of finite Coxeter groups and Iwahori-Hecke algebras*, London Mathematical Society Monographs. New Series, vol. 21, The Clarendon Press, Oxford University Press, New York, 2000. MR1778802 (2002k:20017)

[Jac74] N. Jacobson, *Basic algebra. I*, W. H. Freeman and Co., San Francisco, Calif., 1974. MR0356989 (50 #9457)

[Kaw87] N. Kawanaka, *Shintani lifting and Gel′fand-Graev representations*, The Arcata Conference on Representations of Finite Groups (Arcata, Calif., 1986), Proc. Sympos. Pure Math., vol. 47, Amer. Math. Soc., Providence, RI, 1987, pp. 147–163. MR933357 (89h:22037)

[KS05] A. C. Kable and N. Sanat, *The exterior and symmetric square of the reflection representation of $A_n(q)$ and $D_n(q)$*, J. Algebra **288** (2005), no. 2, 409–444, DOI 10.1016/j.jalgebra.2005.02.004. MR2146138 (2006e:20025)

[Lus77] G. Lusztig, *Irreducible representations of finite classical groups*, Invent. Math. **43** (1977), no. 2, 125–175. MR0463275 (57 #3228)

[Lus80] G. Lusztig, *Some problems in the representation theory of finite Chevalley groups*, The Santa Cruz Conference on Finite Groups (Univ. California, Santa Cruz, Calif., 1979), Proc. Sympos. Pure Math., vol. 37, Amer. Math. Soc., Providence, R.I., 1980, pp. 313–317. MR604598 (82i:20014)

[Lus84] G. Lusztig, *Characters of reductive groups over a finite field*, Annals of Mathematics Studies, vol. 107, Princeton University Press, Princeton, NJ, 1984. MR742472 (86j:20038)

[Lus92] G. Lusztig, *A unipotent support for irreducible representations*, Adv. Math. **94** (1992), no. 2, 139–179, DOI 10.1016/0001-8708(92)90035-J. MR1174392 (94a:20073)

[Spa82] N. Spaltenstein, *Classes unipotentes et sous-groupes de Borel* (French), Lecture Notes in Mathematics, vol. 946, Springer-Verlag, Berlin-New York, 1982. MR672610 (84a:14024)

[Sri79] B. Srinivasan, *Weil representations of finite classical groups*, Invent. Math. **51** (1979), no. 2, 143–153, DOI 10.1007/BF01390225. MR528020 (81d:20036)

Institut de Mathématiques de Jussieu Paris Rive-Gauche (U.M.R. 7586 du C.N.R.S.) and Université Pierre et Marie Curie, 4 place Jussieu, 75005 Paris, France
*E-mail address*: anne-marie.aubert@imj-prg.fr

Faculty of Mathematics, Nicolas Copernicus University, Chopina 12, 87-100 Toruń, Poland
*E-mail address*: wkras@mat.uni.torun.pl

Department of Mathematics, University of Oklahoma, Norman, Oklahoma 73019
*E-mail address*: tprzebin@gmail.com

Proceedings of Symposia in Pure Mathematics
Volume **92**, 2016
http://dx.doi.org/10.1090/pspum/092/01576

# Twisted modules for tensor product vertex operator superalgebras and permutation automorphisms of odd order

Katrina Barron

ABSTRACT. We construct and classify $(1\ 2\ \cdots\ k)$-twisted $V^{\otimes k}$-modules for $k$ odd and for $V$ a vertex operator superalgebra. In particular, we show that the category of weak $(1\ 2\ \cdots\ k)$-twisted $V^{\otimes k}$-modules for $k$ odd is isomorphic to the category of weak $V$-modules. This extends previous results of the author, along with Dong and Mason, classifying all permutation-twisted modules for tensor product vertex operator algebras, to the setting of vertex operator superalgebras for odd order cyclic permutations. We show why this construction does not extend to the case of cyclic permutations of even order in the setting of vertex operator superalgebras, and how the construction and classification in the even order case is fundamentally different than that for the odd order permutation case.

## 1. Introduction

Let $V$ be a vertex operator (super)algebra, and for a fixed positive integer $k$, consider the tensor product vertex operator (super)algebra $V^{\otimes k}$ (see [**FLM3**], [**FHL**]). Any element $g$ of the symmetric group $S_k$ acts in a natural way on $V^{\otimes k}$ as a vertex operator (super)algebra automorphism, and thus it is appropriate to consider $g$-twisted $V^{\otimes k}$-modules. This is the setting for permutation orbifold conformal field theory, and for permutation orbifold superconformal field theory if the vertex operator superalgebra is not just super, but is also supersymmetric, i.e. is a representation of a Neveu-Schwarz super-extension of the Virasoro algebra.

In [**BDM**], the author along with Dong and Mason constructed and classified the $g$-twisted $V^{\otimes k}$-modules for $V$ a vertex operator algebra and $g$ any permutation. In the present paper, we extend these results to $g$-twisted $V^{\otimes k}$-modules for $V$ a vertex operator superalgebra and $g$ a cyclic permutation of odd order. In addition, we show that the results of [**BDM**] for permutation-twisted tensor product vertex operator algebras and the results of the current paper do not extend in a straightforward way to the vertex operator superalgebra setting for permutations in the full symmetric group, but that for even order cyclic permutations the construction is necessarily fundamentally different.

2010 *Mathematics Subject Classification.* Primary 17B68, 17B69, 17B81, 81R10, 81T40, 81T60.

*Key words and phrases.* Vertex operator superalgebras, twisted sectors, permutation orbifold, superconformal field theory.

In the present paper, as our main result, we give an explicit construction and classification of $(1\ 2\ \cdots\ k)$-twisted $V^{\otimes k}$-modules for $k$ odd and $V$ any vertex operator superalgebra. In particular, we show that for $k$ odd, the category of weak $(1\ 2\ \cdots\ k)$-twisted $V^{\otimes k}$-modules is isomorphic to the category of weak $V$-modules. These results allow us to give formulas for the graded dimensions of $(1\ 2\ \cdots\ k)$-twisted $V^{\otimes k}$-modules in terms of the graded dimensions of $V$-modules, and vice versa, as in Corollary 6.5.

In this paper, we also establish certain key identities for certain operators in the case when $k$ is even. Then in [**BV2**], the author along with Vander Werf uses several of these results from this paper to construct $(1\ 2\ \cdots\ k)$-twisted $V^{\otimes k}$-modules for $k$ even. However, in this case, the category of weak $(1\ 2\ \cdots\ k)$-twisted $V^{\otimes k}$-modules is isomorphic to the category of weak *parity-twisted* $V$-modules. Here the parity map $\sigma$ on any $\mathbb{Z}_2$-graded vector space is the identity on the even subspace and $-1$ on the odd subspace. Thus this class of examples we construct and classify in this paper (i.e., for the case when $k$ is odd and $V$ is a vertex operator superalgebra) in comparison to the class of examples constructed in [**BV2**], are of fundamental importance in understanding the role of the parity map and parity-twisted modules in the theory of vertex operator superalgebras.

Twisted vertex operators were discovered and used in [**LW**]. Twisted modules for vertex operator algebras arose in the work of I. Frenkel, J. Lepowsky and A. Meurman [**FLM1**], [**FLM2**], [**FLM3**] for the case of a lattice vertex operator algebra and certain lifts of the lattice isometry $-1$, in the course of the construction of the moonshine module vertex operator algebra (see also [**Bo**]). This structure came to be understood as an "orbifold model" in the sense of conformal field theory and string theory. Twisted modules are the mathematical counterpart of "twisted sectors", which are the basic building blocks of orbifold models in conformal field theory and string theory (see [**DHVW1**], [**DHVW2**], [**DFMS**], [**DVVV**], [**DGM**], as well as [**KS**], [**FKS**], [**Ban1**], [**Ban2**], [**BHS**], [**dBHO**], [**HO**], [**GHHO**], [**Ban3**] and [**HH**]). Orbifold theory plays an important role in conformal field theory and in supersymmetric generalizations, and is also a way of constructing a new vertex operator (super)algebra from a given one.

Formal calculus arising from twisted vertex operators associated to an even lattice was systematically developed in [**Le1**], [**FLM2**], [**FLM3**] and [**Le2**], and the twisted Jacobi identity was formulated and shown to hold for these operators (see also [**DL2**]). These results led to the introduction of the notion of $g$-twisted $V$-module [**FFR**], [**D**], for $V$ a vertex operator algebra and $g$ an automorphism of $V$. This notion records the properties of twisted operators obtained in [**Le1**], [**FLM1**], [**FLM2**], [**FLM3**] and [**Le2**], and provides an axiomatic definition of the notion of twisted sectors for conformal field theory. In general, given a vertex operator algebra $V$ and an automorphism $g$ of $V$, it is an open problem as to how to construct a $g$-twisted $V$-module.

The focus of this paper is the study of permutation-twisted sectors for vertex operator superalgebras. A theory of twisted operators for integral lattice vertex operator superalgebras and finite automorphisms that are certain lifts of a lattice isometry were studied in [**DL2**] and [**X**], and the general theory of twisted modules for vertex operator superalgebras was developed by Li in [**Li**]. Certain specific examples of permutation-twisted sectors in superconformal field theory have been studied from a physical point of view in, for instance, [**FKS**], [**BHS**], [**MS1**], [**MS2**].

The main result of this paper is the explicit construction and classification of twisted sectors for permutation orbifold theory in the general setting of $V$ a vertex operator superalgebra and $g$ a cyclic permutation of odd order acting on $V^{\otimes k}$. In particular, for $g$ a $k$-cycle for $k$ odd, and $V$ any vertex operator superalgebra, we show that the categories of weak, weak admissible and ordinary $g$-twisted $V^{\otimes k}$-modules are isomorphic to the categories of weak, weak admissible and ordinary $V$-modules, respectively. (The definitions of weak, weak admissible and ordinary twisted modules are given in Section 2.3.) To construct the isomorphism between the category of weak $g$-twisted $V^{\otimes k}$-modules and the category of weak $V$-modules for $g$ a $k$-cycle, we explicitly define a weak $g$-twisted $V^{\otimes k}$-module structure on any weak $V$-module.

We show that this method of constructing the permutation-twisted modules for tensor product vertex operator superalgebras fails for cycles of even length (and thus for permutations of even order in general) by pointing out that the operators one would use to construct the twisted modules in the case of a cycle of even order live in the wrong space; see Remark 4.1. We show that these operators rather belong to a class of vertex operators which produce "generalized twisted modules" which satisfy a more general Jacobi identity such as those given by the relativized twisted operators studied in [**DL2**]; see Remark 5.1.

In [**BV2**], the author along with Vander Werf, use results from this paper to prove that for $g$ a $k$-cycle for $k$ even, and $V$ any vertex operator superalgebra, the categories of weak, weak admissible and ordinary $g$-twisted $V^{\otimes k}$-modules are isomorphic to the categories of weak, weak admissible and ordinary parity-twisted $V$-modules, respectively. Our results from this paper and from [**BV2**] can be used to construct permutation-twisted modules for any permutation acting on the multifold tensor product of a vertex operator superalgebra, as well as to calculate graded dimensions.

In addition, in [**Bar10**] we use the results of [**BV2**] based on the results of this paper, to construct and classify the mirror-twisted modules for an N=2 supersymmetric vertex operator of the form $V \otimes V$ for the signed transposition mirror map as studied in [**Bar9**]. In fact this was one of the main motivations of the current work and the work in [**BV2**]. Supersymmetric vertex operator superalgebras are vertex operator superalgebras that in addition to being a representation of the Virasoro algebra, also are representations of the Neveu-Schwarz algebra a Lie superalgebra extension of the Virasoro algebra. See, e.g., [**Bar1**]–[**Bar9**]. In this case, when $V$ is a supersymmetric vertex operator superalgebra, the parity-twisted $V$-modules are naturally a representation of the Ramond algebra. This is another extension of the Virasoro algebra to a Lie superalgebra. In physics terms, the modules for the supersymmetric vertex operator superalgebras are called the "Neveu-Schwarz sectors" and the parity-twisted modules are called the "Ramond sectors". Thus the current work, i.e. Theorem 6.4, along with the results of [**BV2**], imply that all permutation-twisted modules for tensor product supersymmetric vertex operator superalgebras are built up as tensor products of Neveu- Schwarz sectors (coming from the odd cycles) and Ramond sectors (coming from the even cycles). Furthermore, these results imply, as we show in [**Bar10**], that the category of weak mirror-twisted $(V \otimes V)$-modules is isomorphic to the category or $N = 1$ Ramond twisted sectors (i.e. parity-twisted modules) for the vertex operator superalgebra $V$.

This paper is organized as follows. In Section 2, we recall the definitions of vertex operator superalgebra, and weak, weak admissible, and ordinary twisted module, as well as some of their properties, including results from [**Bar3**]–[**Bar5**]. Although most of the notions we present are routine, we point out a clarification made first in [**BV1**] about the definition of $g$-twisted $V$-module for $V$ a vertex operator superalgebra. In particular, we point out in Remark 2.8 below that the notion of "parity-unstable $g$-twisted $V$-module", as used in for instance [**DZ1**], [**DZ2**], [**DH**], arises from a notion of $g$-twisted $V$-module that is not the natural categorical definition. In Remark 2.8, we recall our result from [**BV1**], showing that these so called "parity-unstable $g$-twisted $V$-modules" always come in pairs that together form a "parity-stable $g$-twisted $V$-module". Thus it is more appropriate to take the definition of $g$-twisted $V$-module to be a "parity-stable $g$-twisted $V$-module" in the language of these other works, and then "parity-unstable $g$-twisted $V$-modules" are simply parity-unstable *invariant subspaces* of a (properly defined) $g$-twisted $V$-module. This is the point of view we take in this paper. This fact we proved in [**BV1**] concerning the nature of parity-unstable invariant subspaces of parity-stable $g$-twisted $V$-modules can be used to clarify and simplify many aspects of past works, such as [**DZ1**], [**DZ2**], [**DH**].

In Section 3, we define the operator $\Delta_k(x)$ on a vertex operator superalgebra $V$ and prove several important properties of $\Delta_k(x)$ which are needed in subsequent sections. This is the main operator from which our twisted vertex operators will be built. The main ideas for the proofs of these identities come from the development of this operator in the nonsuper setting in [**BDM**] and the supergeometry developed in [**Bar3**]–[**Bar5**] restricted to the (non-supersummetric) vertex operator superalgebra setting, in order to extend identities involving this operator from the vertex operator subalgebra $V^{(0)}$ of the vertex operator superalgebra $V$ to all of $V = V^{(0)} \oplus V^{(1)}$.

In Section 4, we develop the setting for $(1\ 2\ \cdots\ k)$-twisted $V^{\otimes k}$-modules for $k$ a positive integer and study the vertex operators for a $V$-module modified by the orbifolding $x \to x^{1/k}$ and composing with the operator $\Delta_k(x)$. In particular we derive the supercommutator formula for these operators showing that these operators satisfy the twisted Jacobi identity for odd vectors if and only if $k$ is an odd integer.

In Section 5, we use these operators to define a weak $g = (1\ 2\ \cdots\ k)$-twisted $V^{\otimes k}$-module structure on any weak $V$-module in the case when $k$ is odd. As a result we construct a functor $T_g^k$ from the category of weak $V$-modules to the category of weak $g$-twisted $V^{\otimes k}$-modules such that $T_g^k$ maps weak admissible (resp., ordinary) $V$-modules into weak admissible (resp., ordinary) $g$-twisted $V^{\otimes k}$-modules. In addition, we show that $T_g^k$ preserves irreducible objects.

In Section 6, we define a weak $V$-module structure on any weak $g = (1\ 2\ \cdots\ k)$-twisted $V^{\otimes k}$-module, for $V$ a vertex operator superalgebra and $k$ odd. In so doing, we construct a functor $U_g^k$ from the category of weak $g$-twisted $V^{\otimes k}$-modules to the category of weak $V$-modules such that $T_g^k \circ U_g^k = id$ and $U_g^k \circ T_g^k = id$. We then use this construction and classification of $(1\ 2\ \cdots\ k)$-twisted $V^{\otimes k}$-modules in terms of $V$-modules to show in Corollary 6.5 how the graded dimensions of $(1\ 2\ \cdots\ k)$-twisted $V^{\otimes k}$-modules are given by the graded dimensions of $V$-modules under the change of variable $q \mapsto q^{1/k}$.

## 2. Vertex operator superalgebras, twisted modules and some of their properties

In this section we recall some of the formal calculus we will need, following e.g., [**LL**], [**H**], and [**Bar3**], and we recall the notions of vertex superalgebra and vertex operator superalgebra. We also recall some properties of such structures, and prove a general geometrically inspired identity using results and techniques from [**Bar3**]. Then we present the notion of $g$-twisted module for a vertex operator superalgebra and an automorphism $g$ following [**BV1**].

**2.1. Formal calculus.** Let $x, x_0, x_1, x_2$, etc., denote commuting independent formal variables. Let $\delta(x) = \sum_{n\in\mathbb{Z}} x^n$. We will use the binomial expansion convention, namely, that any expression such as $(x_1 - x_2)^n$ for $n \in \mathbb{C}$ is to be expanded as a formal power series in nonnegative integral powers of the second variable, in this case $x_2$.

For $r \in \mathbb{C}$ we have

$$x_2^{-1}\left(\frac{x_1 - x_0}{x_2}\right)^r \delta\left(\frac{x_1 - x_0}{x_2}\right) = x_1^{-1}\left(\frac{x_2 + x_0}{x_1}\right)^{-r} \delta\left(\frac{x_2 + x_0}{x_1}\right), \tag{2.1}$$

and it is easy to see that for $k$ a positive integer,

$$\sum_{p=0}^{k-1}\left(\frac{x_1 - x_0}{x_2}\right)^{p/k} x_2^{-1}\delta\left(\frac{x_1 - x_0}{x_2}\right) = x_2^{-1}\delta\left(\frac{(x_1 - x_0)^{1/k}}{x_2^{1/k}}\right). \tag{2.2}$$

Therefore, we have the $\delta$-function identity

$$x_2^{-1}\delta\left(\frac{(x_1 - x_0)^{1/k}}{x_2^{1/k}}\right) = x_1^{-1}\delta\left(\frac{(x_2 + x_0)^{1/k}}{x_1^{1/k}}\right). \tag{2.3}$$

We also have the three-term $\delta$-function identity

$$x_0^{-1}\delta\left(\frac{x_1 - x_2}{x_0}\right) - x_0^{-1}\delta\left(\frac{x_2 - x_1}{-x_0}\right) = x_2^{-1}\delta\left(\frac{x_1 - x_0}{x_2}\right). \tag{2.4}$$

Let $R$ be a ring, and let $O$ be an invertible linear operator on $R[x, x^{-1}]$. We define another linear operator $O^{x\frac{\partial}{\partial x}}$ by

$$O^{x\frac{\partial}{\partial x}} \cdot x^n = O^n x^n$$

for any $n \in \mathbb{Z}$. For example, since the formal variable $z^{1/k}$ can be thought of as an invertible linear multiplication operator from $\mathbb{C}[x, x^{-1}]$ to $\mathbb{C}[z^{1/k}, z^{-1/k}][x, x^{-1}]$, we have the corresponding operator $z^{(1/k)x\frac{\partial}{\partial x}}$ from $\mathbb{C}[x, x^{-1}]$ to $\mathbb{C}[z^{1/k}, z^{-1/k}][x, x^{-1}]$. Note that $z^{(1/k)x\frac{\partial}{\partial x}}$ can be extended to a linear operator on $\mathbb{C}[[x, x^{-1}]]$ in the obvious way.

Let $\mathbb{Z}_+$ denote the positive integers. From Proposition 2.1.1 in [**H**], we have the following lemma.

LEMMA 2.1 ([**H**]). *For any formal power series in $f(x) \in x\mathbb{C}[[x]]$, given by*

$$f(x) = \sum_{j\in\mathbb{N}} a_j x^{j+1} \quad \textit{for } a_j \in \mathbb{C}$$

*there exists a unique sequence* $\{A_j\}_{j\in\mathbb{Z}_+}$ *in* $\mathbb{C}$ *such that*

$$f(x) = \exp\left(\sum_{j\in\mathbb{Z}_+} A_j x^{j+1}\frac{d}{dx}\right) a_0^{x\frac{d}{dx}} x. \tag{2.5}$$

Let $\varphi$ be a formal anti-commuting variable, that is, commuting with $x$ but satisfying $\varphi^2 = 0$. From Proposition 3.5 in [**Bar3**], we have the following lemma.

LEMMA 2.2 ([**Bar3**]). *For any formal power series in* $f(x) \in x\mathbb{C}[[x]]$ *given as in Lemma* 2.1, *and a choice of* $a_0^{1/2}$, *we have that*

$$\exp\left(-\sum_{j\in\mathbb{Z}_+} A_j L_j(x,\varphi)\right)(a_0^{1/2})^{-2L_0(x,\varphi)}(x,\varphi) = \left(f(x), \varphi\sqrt{f'(x)}\right) \tag{2.6}$$

*where*

$$L_j(x,\varphi) = -\left(x^{j+1}\frac{\partial}{\partial x} + \left(\frac{j+1}{2}\right)\varphi x^j \frac{\partial}{\partial \varphi}\right),$$

*and* $\sqrt{f'(x)}$ *is the unique series in* $\mathbb{C}[[x]]$ *satisfying* $\sqrt{f'(x)}^2 = f'(x)$ *and such that* $\sqrt{f'(x)}|_{x=0} = a_0^{1/2}$.

Note that the formal power series in $x\mathbb{C}[[x]]\oplus\varphi\mathbb{C}[[x]]$ given by (2.6) is superconformal in the sense of [**Bar3**], but is not the most general form of a superconformal formal power series vanishing at zero. In particular, the infinitesimal superconformal transformations involved, in this case $L_j(x,\varphi)$, only give a representation of the Virasoro algebra and not a super extension of the Virasoro algebra such as the Neveu-Schwarz algebra. This reflects the fact that throughout this work, we will be assuming that we are working with a vertex operator superalgebra, i.e. one that has supercommuting properties with respect to a $\mathbb{Z}_2$-grading as well as a $\frac{1}{2}\mathbb{Z}$-grading, but is not necessarily supersymmetric, meaning it will not necessarily be a representation of any super extension of the Virasoro algebra.

**2.2. Vertex superalgebras, vertex operator superalgebras, and some of their properties.** A *vertex superalgebra* is a vector space which is $\mathbb{Z}_2$-graded (by *sign* or *parity*)

$$V = V^{(0)} \oplus V^{(1)} \tag{2.7}$$

equipped with a linear map

$$\begin{aligned} V &\longrightarrow (\mathrm{End}\, V)[[x,x^{-1}]] \\ v &\mapsto Y(v,x) = \sum_{n\in\mathbb{Z}} v_n x^{-n-1} \end{aligned} \tag{2.8}$$

such that $v_n \in (\mathrm{End}\, V)^{(j)}$ for $v \in V^{(j)}$, $j \in \mathbb{Z}_2$, and equipped with a distinguished vector $\mathbf{1} \in V^{(0)}$, (the *vacuum vector*), satisfying the following conditions for $u, v \in V$:

$$u_n v = 0 \quad \text{for } n \text{ sufficiently large}; \tag{2.9}$$

$$Y(\mathbf{1},x) = Id_V; \tag{2.10}$$

$$Y(v,x)\mathbf{1} \in V[[x]] \quad \text{and} \quad \lim_{x\to 0} Y(v,x)\mathbf{1} = v; \tag{2.11}$$

and for $u, v \in V$ of homogeneous sign, the *Jacobi identity* holds

$$x_0^{-1}\delta\left(\frac{x_1 - x_2}{x_0}\right) Y(u, x_1)Y(v, x_2) - (-1)^{|u||v|}x_0^{-1}\delta\left(\frac{x_2 - x_1}{-x_0}\right) Y(v, x_2)Y(u, x_1)$$

$$= x_2^{-1}\delta\left(\frac{x_1 - x_0}{x_2}\right) Y(Y(u, x_0)v, x_2) \tag{2.12}$$

where $|v| = j$ if $v \in V^{(j)}$ for $j \in \mathbb{Z}_2$.

This completes the definition. We denote the vertex superalgebra just defined by $(V, Y, \mathbf{1})$, or briefly, by $V$.

Note that as a consequence of the definition, we have that there exists a distinguished endomorphism $T \in (\mathrm{End}\, V)^{(0)}$ defined by

$$T(v) = v_{-2}\mathbf{1} \quad \text{for } v \in V$$

such that

$$[T, Y(v, x)] = Y(T(v), x) \;=\; \frac{d}{dx}Y(v, x),$$

(cf. [**LL**], [**Bar6**], [**Bar7**]).

A *vertex operator superalgebra* is a vertex superalgebra with a distinguished vector $\omega \in V_2$ (the *conformal element*) satisfying the following conditions:

$$[L(m), L(n)] = (m - n)L(m + n) + \frac{1}{12}(m^3 - m)\delta_{m+n,0}c \tag{2.13}$$

for $m, n \in \mathbb{Z}$, where

$$L(n) = \omega_{n+1} \quad \text{for } n \in \mathbb{Z}, \qquad \text{i.e., } Y(\omega, x) = \sum_{n\in\mathbb{Z}} L(n)x^{-n-2} \tag{2.14}$$

and $c \in \mathbb{C}$ (the *central charge* of $V$);

$$T = L(-1) \quad \text{i.e., } \frac{d}{dx}Y(v, x) = Y(L(-1)v, x) \text{ for } v \in V; \tag{2.15}$$

$V$ is $\frac{1}{2}\mathbb{Z}$-graded (by *weight*)

$$V = \coprod_{n\in\frac{1}{2}\mathbb{Z}} V_n \tag{2.16}$$

such that

$$L(0)v \;=\; nv \;=\; (\mathrm{wt}\, v)v \quad \text{for } n \in \tfrac{1}{2}\mathbb{Z} \text{ and } v \in V_n; \tag{2.17}$$

$$\dim V_n \;<\; \infty; \tag{2.18}$$

$$V_n \;=\; 0 \quad \text{for } n \text{ sufficiently negative}; \tag{2.19}$$

and $V^{(j)} = \coprod_{n\in\mathbb{Z}+\frac{j}{2}} V_n$ for $j \in \mathbb{Z}_2$.

This completes the definition. We denote the vertex operator superalgebra just defined by $(V, Y, \mathbf{1}, \omega)$, or briefly, by $V$.

REMARK 2.3. For the purposes of this paper we do not assume any supersymmetric properties of a vertex operator superalgebra. That is we do not assume that $V$ is necessarily a representation for any super extension of the Virasoro algebra. However one of the motivations for constructing and classifying permutation-twisted modules for tensor product vertex operator superalgebras is the application to constructing mirror-twisted sectors for N=2 supersymmetric vertex operator superalgebras as discussed in [**Bar8**], [**Bar9**], and [**BV2**]. This construction of mirror-twisted sectors is presented in [**Bar10**] as an application of the results of this paper and [**BV2**].

REMARK 2.4. Note that if $(V, Y, \mathbf{1})$ and $(V', Y', \mathbf{1}')$ are two vertex superalgebras, then $(V \otimes V',\ Y \otimes Y',\ \mathbf{1} \otimes \mathbf{1}')$ is a vertex superalgebra where

$$(Y \otimes Y')(u \otimes u', x)(v \otimes v') = (-1)^{|u'||v|} Y(u,x)v \otimes Y'(u',x)v'. \tag{2.20}$$

If in addition, $V$ and $V'$ are vertex operator superalgebras with conformal vectors $\omega$ and $\omega'$ respectively, then $V \otimes V'$ is a vertex operator superalgebra with conformal vector $\omega \otimes \mathbf{1}' + \mathbf{1} \otimes \omega'$.

REMARK 2.5. As a consequence of the definition of vertex operator superalgebra, independent of the requirement that as a vertex superalgebra we should have $v_n \in (\mathrm{End}\, V)^{(|v|)}$, we have that $\mathrm{wt}(v_n u) = \mathrm{wt} u + \mathrm{wt} v - n - 1$, for $u, v \in V$ and $n \in \mathbb{Z}$. This implies that $v_n \in (\mathrm{End}\, V)^{(j)}$ if and only if $v \in V^{(j)}$ for $j \in \mathbb{Z}_2$, without us having to assume this as an axiom.

Next we present some change of variables formulas and identities, generalizing [**H**] in the nonsuper setting, and reducing the results and techniques of [**Bar3**], [**Bar4**], [**Bar5**] to the case of vertex operator superalgebras that are not necessarily supersymmetric.

Let $f_{t^{1/2}}(x) \in t^{-1}x\mathbb{C}[[x]]$ be a formal power series given by

$$f_{t^{1/2}}(x) = \exp\Bigg( - \sum_{j \in \mathbb{Z}_+} A_j L_j(x, \varphi) \Bigg) (t^{-1/2} a_0^{1/2})^{-2L_0(x,\varphi)} \cdot x \tag{2.21}$$

for $A_j, a_0^{1/2} \in \mathbb{C}$ for $j \in \mathbb{Z}_+$, and for $t^{-1/2}$ a formal commuting variable. In particular, we have

$$\begin{aligned} \exp\Bigg( - \sum_{j \in \mathbb{Z}_+} A_j L_j(x, \varphi) \Bigg) (t^{-1/2} a_0^{1/2})^{-2L_0(x,\varphi)} \cdot \varphi \Bigg|_{x=0} &= (t^{-1/2} a_0^{1/2})^{-2L_0(x,\varphi)} \cdot \varphi \\ &= t^{-1/2} a_0^{1/2} \varphi. \end{aligned}$$

Define $\Theta_j = \Theta_j(t^{-1/2} a_0^{1/2}, \{A_n\}_{n \in \mathbb{Z}_+}, x) \in \mathbb{C}[x][[t^{1/2}]]$, for $j \in \mathbb{N}$, by

$$\begin{aligned} &f_{t^{1/2}}(t a_0^{-1} w + f^{-1}_{t^{1/2}}(x)) - x \\ &\qquad = \exp\Bigg( - \sum_{j \in Z_+} \Theta_j L_j(w, \rho) \Bigg) \exp(-\Theta_0 2 L_0(w, \rho)) \cdot w \end{aligned} \tag{2.22}$$

for $\rho$ an anticommuting formal variable. That is, in particular, we define $\Theta_0$ such that

$$e^{-\Theta_0 2 L_0(w,\rho)} \cdot \rho \Big|_{w=0} = e^{-\Theta_0} \rho.$$

By Corollary 3.44 of [**Bar3**], the formal series $\Theta_j$ are indeed well defined and in $\mathbb{C}[x][[t^{1/2}]]$.

PROPOSITION 2.6. *Let $V = (V, Y, \mathbf{1}, \omega)$ be a vertex operator superalgebra. Then for $t^{1/2}$ a formal variable, $A_j \in \mathbb{C}$ for $j \in \mathbb{Z}_+$, and $a_0^{1/2} \in \mathbb{C}$, we have that*

$$
\begin{aligned}
(2.23)\quad & e^{-\sum_{j\in\mathbb{Z}_+} A_j L(j)} \cdot (t^{-\frac{1}{2}} a_0^{\frac{1}{2}})^{-2L(0)} Y(u,x) \cdot (t^{-\frac{1}{2}} a_0^{\frac{1}{2}})^{2L(0)} \cdot e^{\sum_{j\in\mathbb{Z}_+} A_j L(j)} \\
&= Y\left( (t^{-\frac{1}{2}} a_0^{\frac{1}{2}})^{-2L(0)} e^{-\sum_{j\in\mathbb{Z}_+} \Theta_j L(j)} \cdot e^{-2\Theta_0 L(0)} u, f^{-1}_{t^{1/2}}(x) \right)
\end{aligned}
$$

*where*

$$
(2.24)\qquad f^{-1}_{t^{1/2}}(x) = (t^{-\frac{1}{2}} a_0^{\frac{1}{2}})^{-2x\frac{\partial}{\partial x}} \cdot \exp\Big( - \sum_{j\in\mathbb{Z}_+} A_j x^{j+1} \frac{\partial}{\partial x} \Big) \cdot x
$$

*and the $\Theta_j = \Theta_j(t^{-1/2} a_0^{1/2}, \{A_n\}_{n\in\mathbb{Z}_+}, x)$, for $j \in \mathbb{N}$, are defined by (2.22).*

PROOF. By the proof of Equation (5.4.10)[1] in [**H**] extended to the case of a vertex operator superalgebra or analogously by the proof of Equation (7.17) in [**Bar4**], restricted to a not necessarily supersymmetric vertex operator superalgebra (i.e., by assuming all $G(j-1/2)$ terms, for $j \in \mathbb{Z}$ are zero) the result follows. □

**2.3. The notion of twisted module.** Let $(V, Y, \mathbf{1})$ and $(V', Y', \mathbf{1}')$ be vertex superalgebras. A *homomorphism of vertex superalgebras* is a linear map $g : V \longrightarrow V'$ of $\mathbb{Z}_2$-graded vector spaces such that $g(\mathbf{1}) = \mathbf{1}'$ and

$$
(2.25)\qquad gY(v,x) = Y'(gv,x)g
$$

for $v \in V$. Note that this implies that $g \circ T = T' \circ g$. If in addition, $V$ and $V'$ are vertex operator superalgebras with conformal elements $\omega$ and $\omega'$, respectively, then a *homomorphism of vertex operator superalgebras* is a homomorphism of vertex superalgebras $g$ such that $g(\omega) = \omega'$. In particular $gV_n \subset V'_n$ for $n \in \frac{1}{2}\mathbb{Z}$.

An *automorphism* of a vertex (operator) superalgebra $V$ is a bijective vertex (operator) superalgebra homomorphism from $V$ to $V$.

If $g$ is an automorphism of a vertex (operator) superalgebra $V$ such that $g$ has finite order, then $V$ is a direct sum of the eigenspaces $V^j$ of $g$,

$$
(2.26)\qquad V = \coprod_{j\in\mathbb{Z}/k\mathbb{Z}} V^j,
$$

where $k \in \mathbb{Z}_+$ and $g^k = 1$, and

$$
(2.27)\qquad V^j = \{v \in V \mid gv = \eta^j v\},
$$

for $\eta$ a fixed primitive $k$-th root of unity. We denote the projection of $v \in V$ onto the $j$-th eigenspace, $V^j$, by $v_{(j)}$.

Let $(V, Y, \mathbf{1})$ be a vertex superalgebra and $g$ an automorphism of $V$ of period $k$. A *$g$-twisted $V$-module* is a $\mathbb{Z}_2$-graded vector space $M = M^{(0)} \oplus M^{(1)}$ equipped with a linear map

$$
\begin{aligned}
(2.28)\qquad V &\rightarrow (\mathrm{End}\, M)[[x^{1/k}, x^{-1/k}]] \\
v &\mapsto Y_g(v,x) = \sum_{n\in\frac{1}{k}\mathbb{Z}} v_n^g x^{-n-1},
\end{aligned}
$$

[1]There is a typo in this equation in [**H**]. $A^{(0)}$ in the first line of Equation (5.4.10) should be $A^{(1)}$ which is the infinite series $\{A_j^{(1)}\}_{j\in\mathbb{Z}_+}$, where $A_j^{(1)} \in \mathbb{C}$.

with $v_n^g \in (\mathrm{End}\, M)^{(|v|)}$, such that for $u, v \in V$ and $w \in M$ the following hold:

$$v_n^g w = 0 \text{ if } n \text{ is sufficiently large;} \tag{2.29}$$

$$Y_g(\mathbf{1}, x) = Id_M; \tag{2.30}$$

the twisted Jacobi identity holds: for $u, v \in V$ of homogeneous sign

$$x_0^{-1}\delta\left(\frac{x_1 - x_2}{x_0}\right) Y_g(u, x_1)Y_g(v, x_2) - (-1)^{|u||v|}x_0^{-1}\delta\left(\frac{x_2 - x_1}{-x_0}\right) Y_g(v, x_2)Y_g(u, x_1)$$

$$= \frac{x_2^{-1}}{k} \sum_{j \in \mathbb{Z}/k\mathbb{Z}} \delta\left(\eta^j \frac{(x_1 - x_0)^{1/k}}{x_2^{1/k}}\right) Y_g(Y(g^j u, x_0)v, x_2). \tag{2.31}$$

This completes the definition of $g$-twisted $V$-module for a vertex superalgebra. We denote the $g$-twisted $V$-module just defined by $(M, Y_g)$ or just by $M$ for short.

We note that the generalized twisted Jacobi identity (2.31) is equivalent to

$$x_0^{-1}\delta\left(\frac{x_1 - x_2}{x_0}\right) Y_g(u, x_1)Y_g(v, x_2) - (-1)^{|u||v|}x_0^{-1}\delta\left(\frac{x_2 - x_1}{-x_0}\right) Y_g(v, x_2)Y_g(u, x_1)$$

$$= x_2^{-1}\left(\frac{x_1 - x_0}{x_2}\right)^{-r/k} \delta\left(\frac{x_1 - x_0}{x_2}\right) Y_g(Y(u, x_0)v, x_2) \tag{2.32}$$

for $u \in V^r$, $r = 0, \dots, k-1$. In addition, this implies that for $v \in V^r$,

$$Y_g(v, x) = \sum_{n \in r/k + \mathbb{Z}} v_n^g x^{-n-1}, \tag{2.33}$$

and for $v \in V$,

$$Y_g(gv, x) = \lim_{x^{1/k} \to \eta^{-1}x^{1/k}} Y_g(v, x). \tag{2.34}$$

If $g = 1$, then a $g$-twisted $V$-module is a $V$-module for the vertex superalgebra $V$. If $(V, Y, \mathbf{1}, \omega)$ is a vertex operator superalgebra and $g$ is a vertex operator superalgebra automorphism of $V$, then since $\omega \in V^0$, we have that $Y_g(\omega, x)$ has component operators which satisfy the Virasoro algebra relations and $Y_g(L(-1)u, x) = \frac{d}{dx}Y_g(u, x)$. In this case, a $g$-twisted $V$-module as defined above, viewed as a vertex superalgebra module, is called a *weak $g$-twisted $V$-module* for the vertex operator superalgebra $V$.

A *weak admissible* $g$-twisted $V$-module is a weak $g$-twisted $V$-module $M$ which carries a $\frac{1}{2k}\mathbb{N}$-grading

$$M = \coprod_{n \in \frac{1}{2k}\mathbb{N}} M(n) \tag{2.35}$$

such that $v_m^g M(n) \subseteq M(n + \mathrm{wt}\ v - m - 1)$ for homogeneous $v \in V$, $n \in \frac{1}{2k}\mathbb{N}$, and $m \in \frac{1}{k}\mathbb{Z}$. We may assume that $M(0) \neq 0$ if $M \neq 0$. If $g = 1$, then a weak admissible $g$-twisted $V$-module is called a weak admissible $V$-module.

REMARK 2.7. Note that if $k$ is even where $k$ is the order of $g$, then the grading of a weak admissible $g$-twisted $V$-module can be assumed to be a $\frac{1}{k}\mathbb{N}$ grading.

An (ordinary) $g$-twisted $V$-module is a weak $g$-twisted $V$-module $M$ graded by $\mathbb{C}$ induced by the spectrum of $L(0)$. That is, we have

$$M = \coprod_{\lambda \in \mathbb{C}} M_\lambda \tag{2.36}$$

where $M_\lambda = \{w \in M \mid L(0)^g w = \lambda w\}$, for $L(0)^g = \omega_1^g$. Moreover we require that $\dim M_\lambda$ is finite and $M_{n/2k+\lambda} = 0$ for fixed $\lambda$ and for all sufficiently small integers $n$. If $g = 1$, then a $g$-twisted $V$-module is a $V$-module.

A *homomorphism of weak $g$-twisted $V$-modules*, $(M, Y_g)$ and $(M', Y_g')$, is a linear map $f : M \longrightarrow M'$ satisfying

$$f(Y_g(v, x)w) = Y_g'(v, x)f(w), \tag{2.37}$$

for $v \in V$, and $w \in M$. If in addition, $M$ and $M'$ are weak admissible $g$-twisted $V$-modules, then a *homomorphism of weak admissible $g$-twisted $V$-modules*, is a homomorphism of weak $g$-twisted $V$-modules such that $f(M(n)) \subseteq M'(n)$. And if $M$ and $M'$ are ordinary $g$-twisted $V$-modules, then a *homomorphism of $g$-twisted $V$-modules*, is a homomorphism of weak $g$-twisted $V$-modules such that $f(M_\lambda) \subseteq M'_\lambda$.

We note here that an example of an automorphism of a vertex operator superalgebra is the *parity map*

$$\begin{aligned} \sigma : V &\longrightarrow V \\ v &\mapsto (-1)^{|v|} v. \end{aligned} \tag{2.38}$$

REMARK 2.8. In many works on vertex superalgebras, e.g. [**Li**], [**DZ1**], [**DZ2**], [**DH**], [**Bar8**], [**Bar9**], the condition that $v_n^g \in (\mathrm{End}\, M)^{(|v|)}$ for $v \in V$, is not given as one of the axioms of a $g$-twisted $V$-module $M$ for a vertex superalgebra $V$. That is, it is not assumed that the $\mathbb{Z}_2$-grading of $V$ is compatible with the $\mathbb{Z}_2$-grading of $M$ via the action of $V$ as super-endomorphisms acting on $M$. Then in, for instance, [**DZ1**], [**DZ2**], [**DH**], the notion of "parity-stable $g$-twisted $V$-module" is introduced for those modules that are representative of the $\mathbb{Z}_2$-grading of $V$, and modules that do not have this property are called "parity-unstable". Thus a "parity-unstable $g$-twisted $V$-module" is a vector space $M$ that satisfies all the axioms of our notion of $g$-twisted $V$-module except for $v_n^g \in (\mathrm{End}\, M)^{(|v|)}$. That is there exists no $\mathbb{Z}_2$-grading on $M$ such that the operators $v_n^g$ act as even or odd endomorphisms on $M$ according to the sign (or parity) of $v$. However in [**BV1**], we prove that any so called "parity-unstable $g$-twisted $V$-module" can always be realized as a subspace of a $g$-twisted $V$-module in the sense of the notion of $g$-twisted $V$-module we give above. In particular, in [**BV1**] we proved the following (reworded to fit our current setting):

THEOREM 2.9 ([**BV1**]). *Let $V$ be a vertex superalgebra and $g$ an automorphism. Suppose $(M, Y_M)$ is a "parity-unstable $g$-twisted $V$-module" (in the sense of* [**DZ1**]*). Then $(M, Y_M \circ \sigma_V)$ is a "parity-unstable $g$-twisted $V$-module" which, in general, is not isomorphic to $(M, Y_M)$. Moreover $(M, Y_M) \oplus (M, Y_M \circ \sigma_V)$ is a "parity-stable $g$-twisted $V$-module" (in the sense of* [**DZ1**]*), i.e., a $g$-twisted $V$ module in terms of the definition given above in this paper.*

Requiring weak twisted modules to be "parity stable" as part of the definition gives the more canonical notion of twisted module from a categorical point of view, for instance, for the purpose of defining a $(V_1 \otimes V_2)$-module structure on $M_1 \otimes M_2$ for $M_j$ a $V_j$-module, for $j = 1, 2$. (See e.g., (2.20)). In particular, the notion of a $g$-twisted $V$-module corresponding to a *representation* of $V$ as a vertex superalgebra only holds for "parity-stable $g$-twisted $V$-modules" in the sense of [**DZ1**], in that the vertex operators acting on a $g$-twisted $V$-module have coefficients in $\mathrm{End}\, M$ such that, the operators $v_m^g$ have a $\mathbb{Z}_2$-graded structure compatible with that of $V$. For instance the operators $v_0^g$, for $v \in V$, give a representation of the Lie superalgebra generated by $v_0$ in $\mathrm{End}\, V$ if and only if $M$ is "parity stable". This corresponds to $V$ acting (via the modes of the vertex operators) as endomorphisms on $M$ in the category of vector spaces (i.e., via even or odd endomorphisms) rather than in the category of $\mathbb{Z}_2$-graded vectors spaces (i.e., as grade-preserving and thus strictly even endomorphisms). However, it is interesting to note that, as is shown in [**BV1**], for a lift of a lattice isometry, the "twisted modules" for a lattice vertex operator superalgebra constructed following [**DL2**], [**X**], naturally sometimes give rise to pairs of parity-unstable invariant subspaces in the language of the current paper, i.e., to pairs of "parity-unstable $g$-twisted modules" in the sense of [**DZ1**], that then must be taken as a direct sum to realize the actual $g$-twisted module that is constructed.

**2.4. Parity-twisted $V$-modules.** A crucial example in the study of $g$-twisted $V$-modules for $V$ a vertex superalgebra is that of parity-twisted $V$-modules. (Not to be confused with the notion discussed above of "parity-stable" or "parity-unstable" modules.) Above in (2.38), we define the parity automorphism, denoted $\sigma$, for any vertex superalgebra. Thus we have the notion of a *parity-twisted $V$-module*, also denoted by *$\sigma$-twisted $V$-module.*

REMARK 2.10. Note that it follows from the definitions, that any weak admissible $\sigma$-twisted module for a vertex operator superalgebra is $\mathbb{N}$-graded.

If $V$, in addition to being a vertex operator superalgebra is N=1 or N=2 supersymmetric, i.e., is also a representation of the N=1 or N=2 Neveu-Schwarz algebra super extension of the Virasoro algebra, then a $\sigma$-twisted $V$-module is naturally a representation of the N=1 or N=2 Ramond algebra, respectively, cf. [**Bar8**], [**Bar9**], [**Bar10**] and references therein.

## 3. The operator $\Delta_k(x)$

In this section, following and generalizing [**BDM**], we define an operator $\Delta_k(x) = \Delta_k^V(x)$ on a vertex operator superalgebra $V$ for a fixed positive integer $k$. In Sections 4 and 5, we will use $\Delta_k(x)$ to construct a $g$-twisted $V^{\otimes k}$-module from a $V$-module where $g$ is a certain $k$-cycle with $k$ odd.

Let $x$, $y$, $z$, and $z_0$ be formal variables commuting with each other. Consider the polynomial

$$\frac{1}{k}(1+x)^k - \frac{1}{k} \in x\mathbb{C}[x].$$

By Lemma 2.1, for $k \in \mathbb{Z}_+$, we can define $a_j \in \mathbb{C}$ for $j \in \mathbb{Z}_+$, by

$$\exp\left(-\sum_{j\in\mathbb{Z}_+} a_j x^{j+1}\frac{\partial}{\partial x}\right)\cdot x = \frac{1}{k}(1+x)^k - \frac{1}{k}. \tag{3.1}$$

For example, $a_1 = (1-k)/2$ and $a_2 = (k^2-1)/12$.

Let

$$\begin{aligned} f(x) &= z^{1/k}\exp\left(-\sum_{j\in\mathbb{Z}_+} a_j x^{j+1}\frac{\partial}{\partial x}\right)\cdot x \\ &= \exp\left(-\sum_{j\in\mathbb{Z}_+} a_j x^{j+1}\frac{\partial}{\partial x}\right)\cdot z^{(1/k)x\frac{\partial}{\partial x}}\cdot x \\ &= \frac{z^{1/k}}{k}(1+x)^k - \frac{z^{1/k}}{k} \qquad \in z^{1/k}x\mathbb{C}[x]. \end{aligned}$$

Then the compositional inverse of $f(x)$ in $x\mathbb{C}[z^{-1/k}, z^{1/k}][[x]]$ is given by

$$\begin{aligned} f^{-1}(x) &= z^{-(1/k)x\frac{\partial}{\partial x}}\exp\left(\sum_{j\in\mathbb{Z}_+} a_j x^{j+1}\frac{\partial}{\partial x}\right)\cdot x \\ &= z^{-1/k}\exp\left(\sum_{j\in\mathbb{Z}_+} a_j z^{-j/k} x^{j+1}\frac{\partial}{\partial x}\right)\cdot x \\ &= (1+kz^{-1/k}x)^{1/k} - 1 \end{aligned}$$

where the last line is considered as a formal power series in $z^{-1/k}x\mathbb{C}[z^{-1/k}][[x]]$, i.e., we are expanding about $x=0$ taking $1^{1/k}=1$.

Now let $\Theta_j = \Theta_j(z^{1/2k}, \{-a_n\}_{n\in\mathbb{Z}_+}, x) \in \mathbb{C}[x][[z^{-1/2k}]]$, for $j\in\mathbb{N}$, where the $\Theta_j(t^{-1/2}a_0^{1/2}, \{A_n\}_{n\in\mathbb{Z}_+}, x)$ are defined by (2.22), and the series $\{-a_j\}_{j\in\mathbb{Z}_+}$ are defined by (3.1). That is

$$f(z^{-1/k}w + f^{-1}(x)) - x = e^{\left(-\sum_{j\in Z_+}\Theta_j L_j(w,\rho)\right)} e^{-\Theta_0 2L_0(w,\rho)}\cdot w, \tag{3.2}$$

and $e^{-\Theta_0 2L_0(w,\rho)}\cdot\rho = e^{\Theta_0}\rho$.

Proposition 3.1. *The series $\Theta_j(z^{1/2k}, \{-a_n\}_{n\in\mathbb{Z}_+}, \frac{1}{k}z^{1/k-1}z_0)$, for $j\in\mathbb{N}$, is a well-defined series with terms in $\mathbb{C}[z_0][[z^{-1/k}]]$. Furthermore*

$$\Theta_j\left(z^{\frac{1}{2k}}, \{-a_n\}_{n\in\mathbb{Z}_+}, k^{-1}z^{\frac{1}{k}-1}z_0\right) = -a_j(z+z_0)^{-\frac{j}{k}} \tag{3.3}$$

*for $j\in\mathbb{Z}_+$, and*

$$\exp\left(z^{\frac{1}{2k}}, \Theta_0(\{-a_n\}_{n\in\mathbb{Z}_+}, k^{-1}z^{\frac{1}{k}-1}z_0)\right) = z^{-\frac{1}{2k}(k-1)}(z+z_0)^{\frac{1}{2k}(k-1)}, \tag{3.4}$$

*where $(z+z_0)^{-r/2k}$ is understood to be expanded in nonnegative integral powers of $z_0$.*

Proof. From (3.2), the formal series $\Theta_j(z^{1/2k}, \{-a_n\}_{n\in\mathbb{Z}_+}, x)$ for $j\in\mathbb{Z}_+$ are the same as the series $\Theta_j(\{-a_n\}_{n\in\mathbb{Z}_+}, z^{1/k}, x)$ of [**BDM**]. We also have that the square of the exponential of $\Theta_0(z^{1/2k}, \{-a_n\}_{n\in\mathbb{Z}_+}, x)$ is equal to the exponential of the series $\Theta_0(\{-a_n\}_{n\in\mathbb{Z}_+}, z^{1/k}, x)$ of [**BDM**]. Thus by Proposition 2.1 in [**BDM**],

Equation (3.3) follows. In particular, in $w\mathbb{C}[z_0][[z^{-1/k}]][[w]]$, we have

$$\begin{aligned}
&\exp\left(\sum_{j\in\mathbb{Z}_+} a_j(z+z_0)^{-\frac{j}{k}} L_j(w,\rho)\right) z^{\frac{1}{2k}(k-1)2L_0(w,\rho)}(z+z_0)^{-\frac{1}{2k}(k-1)2L_0(w,\rho)}\cdot w\\
&\quad = z^{\frac{1}{k}-1}(z+z_0)^{-\frac{1}{k}+1}\exp\left(-\sum_{j\in\mathbb{Z}_+} a_j(z+z_0)^{-\frac{j}{k}} w^{j+1}\frac{\partial}{\partial w}\right)\cdot w\\
&\quad = f(f^{-1}(x)+z^{-1/k}w)-x\Big|_{x=\frac{1}{k}z^{1/k-1}z_0}\\
&\quad = \exp\left(-\sum_{j\in\mathbb{Z}_+}\Theta_j(z^{1/2k},\{-a_n\}_{n\in\mathbb{Z}_+},x)L_j(w,\rho)\right)\\
&\qquad\quad \exp\left(-\Theta_0(z^{1/2k},\{-a_n\}_{n\in\mathbb{Z}_+},x)2L_0(w,\rho)\right)\cdot w\Big|_{x=\frac{1}{k}z^{1/k-1}z_0}.
\end{aligned}$$

In addition, we have that in $\rho\mathbb{C}[z_0][[z^{-1/2k}]][[w]]$,

$$\begin{aligned}
&\exp\left(\sum_{j\in\mathbb{Z}_+} a_j(z+z_0)^{-\frac{j}{k}} L_j(w,\rho)\right) z^{\frac{1}{2k}(k-1)2L_0(w,\rho)}(z+z_0)^{-\frac{1}{2k}(k-1)2L_0(w,\rho)}\cdot \rho\Big|_{w=0}\\
&\qquad = z^{-\frac{1}{2k}(k-1)}(z+z_0)^{\frac{1}{2k}(k-1)}\rho\\
&\qquad = \exp\left(\Theta_0(z^{1/2k},\{-a_n\}_{n\in\mathbb{Z}_+},x)\right)\cdot\rho\Big|_{x=\frac{1}{k}z^{1/k-1}z_0},
\end{aligned}$$

giving Equation (3.4). □

Let $V=(V,Y,\mathbf{1},\omega)$ be a vertex operator superalgebra. Note that for $k$ a positive integer, and $k^{1/2}$ a fixed square root of $k$, we have that $(k^{1/2})^{-2L(0)}$ is a well-defined element of End $V$ and for $z^{1/2k}$ a formal commuting variable, $(z^{1/2k})^{(1-k)2L(0)}$ is a well-defined element of $(\mathrm{End}\,V)[[z^{1/2k},z^{-1/2k}]]$.

Define the operator $\Delta_k^V(z)\in(\mathrm{End}\,V)[[z^{1/2k},z^{-1/2k}]]$, by

$$\Delta_k^V(z)=\exp\left(\sum_{j\in\mathbb{Z}_+} a_j z^{-\frac{j}{k}}L(j)\right)(k^{\frac{1}{2}})^{-2L(0)}\left(z^{\frac{1}{2k}(k-1)}\right)^{-2L(0)}. \tag{3.5}$$

PROPOSITION 3.2. *In* $((\mathrm{End}\,V)[[z^{1/2k},z^{-1/2k}]])[[z_0,z_0^{-1}]]$, *we have*

$$\Delta_k^V(z)Y(u,z_0)\Delta_k^V(z)^{-1}=Y(\Delta_k^V(z+z_0)u,(z+z_0)^{1/k}-z^{1/k}),$$

*for all* $u\in V$.

PROOF. By Proposition 2.6, where in our case, $A_j^{(1)}=-a_j$, and by Proposition 3.1 above, we have that the steps of the proof of Proposition 2.2 in [**BDM**] all hold in this setting. That is, we have

$$
\begin{aligned}
&\Delta_k^V(z)Y(u,z_0)\Delta_k^V(z)^{-1}\\
&= \exp\Bigg(\sum_{j\in\mathbb{Z}_+} a_j z^{-\frac{j}{k}} L(j)\Bigg) Y((k^{\frac{1}{2}})^{-2L(0)}\left(z^{\frac{1}{2k}(k-1)}\right)^{-2L(0)} u, k^{-1}z^{1/k-1}z_0)\cdot\\
&\qquad \cdot \exp\Bigg(-\sum_{j\in\mathbb{Z}_+} a_j z^{-\frac{j}{k}} L(j)\Bigg)\\
&= (z^{\frac{1}{2k}})^{2L(0)} \exp\Bigg(\sum_{j\in\mathbb{Z}_+} a_j L(j)\Bigg)(z^{\frac{1}{2k}})^{-2L(0)} Y((k^{\frac{1}{2}})^{-2L(0)} z^{-\frac{1}{2k}(k-1)2L(0)},\\
&\qquad k^{-1}z^{1/k-1}z_0)(z^{\frac{1}{2k}})^{2L(0)} \exp\Bigg(-\sum_{j\in\mathbb{Z}_+} a_j L(j)\Bigg)(z^{\frac{1}{2k}})^{-2L(0)}\\
&= (z^{\frac{1}{2k}})^{2L(0)} Y\Bigg((z^{\frac{1}{2k}})^{-2L(0)} \exp\Bigg(-\sum_{j\in\mathbb{Z}_+} \Theta_j(z^{\frac{1}{2k}}, \{-a_n\}_{n\in\mathbb{Z}_+}, k^{-1}z^{\frac{1}{k}-1}z_0)L(j)\Bigg)\cdot\\
&\qquad \exp\left(-\Theta_0(z^{\frac{1}{2k}}, \{-a_n\}_{n\in\mathbb{Z}_+}, k^{-1}z^{\frac{1}{k}-1}z_0)2L(0)\right)(k^{\frac{1}{2}})^{-2L(0)} z^{-\frac{1}{2k}(k-1)2L(0)} u,\\
&\qquad f^{-1}(k^{-1}z^{1/k-1}z_0)\Bigg)(z^{\frac{1}{2k}})^{-2L(0)}\\
&= (z^{\frac{1}{2k}})^{2L(0)} Y\Bigg((z^{\frac{1}{2k}})^{-2L(0)} \exp\Bigg(\sum_{j\in\mathbb{Z}_+} a_j (z+z_0)^{-\frac{j}{k}} L(j)\Bigg) z^{\frac{1}{2k}(k-1)2L(0)}\\
&\qquad (z+z_0)^{-\frac{1}{2k}(k-1)2L(0)}(k^{\frac{1}{2}})^{-2L(0)} z^{-\frac{1}{2k}(k-1)2L(0)} u, \left(1+z^{-1}z_0\right)^{\frac{1}{k}} - 1\Bigg)\\
&\qquad (z^{\frac{1}{2k}})^{-2L(0)}\\
&= (z^{\frac{1}{2k}})^{2L(0)} Y\left((z^{\frac{1}{2k}})^{-2L(0)}\Delta_k^V(z+z_0)u, \left(1+z^{-1}z_0\right)^{\frac{1}{k}} - 1\right)(z^{\frac{1}{2k}})^{-2L(0)}\\
&= Y\left(\Delta_k^V(z+z_0)u, (z+z_0)^{\frac{1}{k}} - z^{\frac{1}{k}}\right)
\end{aligned}
$$

as desired. □

REMARK 3.3. In [**BDM**] ,we proved Proposition 3.2 for the special case when $V = V^{(0)}$. In the current setting, this coincides with the case when $\Delta_k^V(z) \in (\mathrm{End}\, V)[[z^{1/2k}, z^{-1/2k}]]$ is restricted to $\Delta_k^V(z)|_{V^{(0)}} \in (\mathrm{End}\, V^{(0)})[[z^{1/k}, z^{-1/k}]]$, we restrict to $u \in V^{(0)}$, and the operator $Y(u,z_0) \in (\mathrm{End}\, V)[[z_0, z_0^{-1}]]$ is restricted to $Y(u,z_0)|_{V^{(0)}} \in (\mathrm{End}\, V^{(0)})[[z_0, z_0^{-1}]]$. Thus in this paper, Proposition 3.2 extends this result to the full operator $\Delta_k^V(z)$ and the operator $Y(u,z_0)$ for $u \in V^{(0)} \oplus V^{(1)}$ acting on $V = V^{(0)} \oplus V^{(1)}$.

Define $\Delta_k^{(x,\varphi)}(z) \in (\mathrm{End}\, \mathbb{C}[x,x^{-1}][\varphi])[[z^{1/2k}, z^{-1/2k}]]$ by

$$
\Delta_k^{(x,\varphi)}(z) = \exp\Bigg(\sum_{j\in\mathbb{Z}_+} a_j z^{-\frac{j}{k}} L_j(x,\varphi)\Bigg)(k^{\frac{1}{2}})^{-2L_0(x,\varphi)}(z^{\frac{1}{2k}(k-1)})^{-2L_0(x,\varphi)},
$$

that is

$$\begin{aligned}\Delta_k^{(x,\varphi)}(z)\cdot(x,\varphi) &= \left(z^{\frac{1}{k}}\right)^{L_0(x,\varphi)}\cdot\left(kz^{1-\frac{1}{k}}f(x),\ \varphi k^{\frac{1}{2}}z^{\frac{1}{2}-\frac{1}{2k}}\sqrt{f'(x)}\right)\\ &= \left((z^{\frac{1}{k}}+x)^k - z,\ \varphi k^{\frac{1}{2}}(z^{\frac{1}{k}}+x)^{\frac{1}{2}(k-1)}\right).\end{aligned}$$

PROPOSITION 3.4. *In* $(\mathrm{End}\,\mathbb{C}[x,x^{-1}][\varphi])[[z^{1/2k},z^{-1/2k}]]$, *we have*

$$-\Delta_k^{(x,\varphi)}(z)\frac{\partial}{\partial x} + k^{-1}z^{\frac{1}{k}-1}\frac{\partial}{\partial x}\Delta_k^{(x,\varphi)}(z) = \frac{\partial}{\partial z}\Delta_k^{(x,\varphi)}(z), \tag{3.6}$$

$$-\Delta_k^{(x,\varphi)}(z)^{-1}\frac{\partial}{\partial x} + kz^{-\frac{1}{k}+1}\frac{\partial}{\partial x}\Delta_k^{(x,\varphi)}(z)^{-1} = kz^{-\frac{1}{k}+1}\frac{\partial}{\partial z}\Delta_k^{(x,\varphi)}(z)^{-1}. \tag{3.7}$$

PROOF. Since $L_j(x,\varphi)\cdot x^n = -x^{j+1}\frac{\partial}{\partial x}\cdot x^n$ for $n\in\mathbb{Z}$, we have from Proposition 2.3 in [**BDM**] that Equations (3.6) and (3.7) hold in $(\mathrm{End}\,\mathbb{C}[x,x^{-1}])[[z^{1/k},z^{-1/k}]] \subset (\mathrm{End}\,\mathbb{C}[x,x^{-1}][\varphi])[[z^{1/2k},z^{-1/2k}]]$.

Next we observe that in $\mathbb{C}[x,x^{-1}][\varphi][[z^{1/2k},z^{-1/2k}]]$, we have

$$\begin{aligned}&-\Delta_k^{(x,\varphi)}(z)\frac{\partial}{\partial x}\cdot\varphi + k^{-1}z^{\frac{1}{k}-1}\frac{\partial}{\partial x}\Delta_k^{(x,\varphi)}(z)\cdot\varphi\\ &\quad= k^{-1}z^{\frac{1}{k}-1}\frac{\partial}{\partial x}\varphi k^{\frac{1}{2}}(z^{\frac{1}{k}}+x)^{\frac{1}{2}(k-1)}\\ &\quad= z^{\frac{1}{k}-1}\varphi k^{-\frac{1}{2}}\frac{1}{2}(k-1)(z^{\frac{1}{k}}+x)^{\frac{1}{2}(k-3)}\\ &\quad= \frac{\partial}{\partial z}\varphi k^{\frac{1}{2}}(z^{\frac{1}{k}}+x)^{\frac{1}{2}(k-1)}\\ &\quad= \frac{\partial}{\partial z}\Delta_k^x(z)\cdot\varphi.\end{aligned}$$

By Proposition 3.11 in [**Bar3**] and Proposition 2.3 in [**BDM**], we have

$$\begin{aligned}\Delta_k^{(x,\varphi)}(z)\cdot(\varphi x^n) &= (\Delta_k^{(x,\varphi)}(z)\cdot\varphi)(\Delta_k^{(x,\varphi)}(z)\cdot x^n)\\ &= (\Delta_k^{(x,\varphi)}(z)\cdot\varphi)(\Delta_k^{(x,\varphi)}(z)\cdot x)^n\end{aligned}$$

for all $n\in\mathbb{Z}$. Therefore

$$\begin{aligned}&-\Delta_k^{(x,\varphi)}(z)\frac{\partial}{\partial x}\cdot\varphi x^n + k^{-1}z^{\frac{1}{k}-1}\frac{\partial}{\partial x}\Delta_k^{(x,\varphi)}(z)\cdot\varphi x^n\\ &= -n\Delta_k^{(x,\varphi)}(z)\cdot\left(\varphi x^{n-1}\frac{\partial}{\partial x}\cdot x\right) + k^{-1}z^{\frac{1}{k}-1}\frac{\partial}{\partial x}(\Delta_k^{(x,\varphi)}(z)\cdot\varphi)(\Delta_k^{(x,\varphi)}(z)\cdot x)^n\\ &= -n\left(\Delta_k^{(x,\varphi)}(z)\cdot\varphi\right)\left(\Delta_k^{(x,\varphi)}(z)\cdot x\right)^{n-1}\left(\Delta_k^{(x,\varphi)}(z)\frac{\partial}{\partial x}\cdot x\right)\\ &\quad +k^{-1}z^{\frac{1}{k}-1}\left(\frac{\partial}{\partial x}(\Delta_k^{(x,\varphi)}(z)\cdot\varphi)\right)(\Delta_k^{(x,\varphi)}(z)\cdot x)^n\\ &\quad +k^{-1}z^{\frac{1}{k}-1}(\Delta_k^{(x,\varphi)}(z)\cdot\varphi)n(\Delta_k^{(x,\varphi)}(z)\cdot x)^{n-1}\frac{\partial}{\partial x}(\Delta_k^{(x,\varphi)}(z)\cdot x)\end{aligned}$$

$$
\begin{aligned}
&= \left(-\Delta_k^{(x,\varphi)}(z)\frac{\partial}{\partial x}\cdot\varphi + k^{-1}z^{\frac{1}{k}-1}\frac{\partial}{\partial x}\Delta_k^{(x,\varphi)}(z)\cdot\varphi\right)(\Delta_k^{(x,\varphi)}(z)\cdot x)^n \\
&\quad +(\Delta_k^{(x,\varphi)}(z)\cdot\varphi)n(\Delta_k^{(x,\varphi)}(z)\cdot x)^{n-1}\left(-\Delta_k^{(x,\varphi)}(z)\frac{\partial}{\partial x}\cdot x\right. \\
&\quad \left. +k^{-1}z^{\frac{1}{k}-1}\frac{\partial}{\partial x}\Delta_k^{(x,\varphi)}(z)\cdot x\right) \\
&= \left(\frac{\partial}{\partial z}(\Delta_k^{(x,\varphi)}(z)\cdot\varphi)\right)(\Delta_k^{(x,\varphi)}(z)\cdot x)^n \\
&\quad +(\Delta_k^{(x,\varphi)}(z)\cdot\varphi)n(\Delta_k^{(x,\varphi)}(z)\cdot x)^{n-1}\left(\frac{\partial}{\partial z}(\Delta_k^{(x,\varphi)}(z)\cdot x)\right) \\
&= \frac{\partial}{\partial z}\left((\Delta_k^{(x,\varphi)}(z)\cdot\varphi)(\Delta_k^{(x,\varphi)}(z)\cdot x)^n\right) \\
&= \frac{\partial}{\partial z}\left(\Delta_k^{(x,\varphi)}(z)\cdot\varphi x^n\right)
\end{aligned}
$$

for all $n\in\mathbb{Z}$. Equation (3.6) follows by linearity in $\mathbb{C}[x,x^{-1}][\varphi][[z^{1/2k},z^{-1/2k}]]$.

Similarly, noting that

$$
\Delta_k^{(x,\varphi)}(z)^{-1}(x,\varphi) = \left((x+z)^{1/k}-z^{1/k},\ \varphi k^{-\frac{1}{2}}(x+z)^{\frac{1}{2k}(1-k)}\right)
$$

we have

$$
\begin{aligned}
&-\Delta_k^{(x,\varphi)}(z)^{-1}\frac{\partial}{\partial x}\cdot\varphi + kz^{-\frac{1}{k}+1}\frac{\partial}{\partial x}\Delta_k^{(x,\varphi)}(z)^{-1}\cdot\varphi \\
&= kz^{-\frac{1}{k}+1}\frac{\partial}{\partial x}\varphi k^{-\frac{1}{2}}(x+z)^{\frac{1}{2k}(1-k)} \\
&= z^{-\frac{1}{k}+1}\varphi k^{-\frac{1}{2}}\frac{1}{2}(1-k)(x+z)^{\frac{1}{2k}(1-3k)} \\
&= kz^{-\frac{1}{k}+1}\frac{\partial}{\partial z}\varphi k^{-\frac{1}{2}}(x+z)^{\frac{1}{2k}(1-k)} \\
&= kz^{-\frac{1}{k}+1}\frac{\partial}{\partial z}\Delta_k^{(x,\varphi)}(z)^{-1}\cdot\varphi.
\end{aligned}
$$

The proof of identity (3.7) acting on $\varphi x^n$ for $n\in\mathbb{Z}$ is analogous to the proof of identity (3.6) on $\varphi x^n$ for $n\in\mathbb{Z}$. Identity (3.7) then follows by linearity. □

Let $\mathfrak{L}$ be the Virasoro algebra with basis $L_j$, $j\in\mathbb{Z}$, and central charge $d\in\mathbb{C}$. The identities (3.6) and (3.7) can be thought of as identities for the representation of the Virasoro algebra on $\mathbb{C}[x,x^{-1}][\varphi]$ given by $L_j\mapsto L_j(x,\varphi)$, for $j\in\mathbb{Z}$, with central charge equal to zero. We want to prove the corresponding identity for certain other representations of the Virasoro algebra, in particular for vertex operator superalgebras. We do this by following the method of proof used in Chapter 4 of [**H**], and extending [**BDM**] with the main point being extending to a $\frac{1}{2}\mathbb{Z}$-graded representation of the Virasoro algebra with the grading given by eigenspaces under the action of $L_0$. Let $\kappa^{1/2}$ be another formal commuting variable. We first prove the identity in $\mathcal{U}_\Pi(\mathfrak{L})[[z^{1/2k},z^{-1/2k}]][[\kappa^{1/2},\kappa^{-1/2}]]$ where $\mathcal{U}_\Pi(\mathfrak{L})$ is a certain extension of the universal enveloping algebra for the Virasoro algebra, and then letting $\kappa^{1/2}=k^{1/2}$, the identity will follow in $(\mathrm{End}\,V)[[z^{1/2k},z^{-1/2k}]]$ where $V$ is a certain type of module for the Virasoro algebra.

We want to construct an extension of $\mathcal{U}(\mathfrak{L})$, the universal enveloping algebra for the Virasoro algebra, in which the operators $(\kappa^{1/2})^{-2L_0}$ and $(z^{1/2k})^{(1-k)2L_0}$ can be defined. Let $V_\Pi$ be a vector space over $\mathbb{C}$ with basis $\{P_j \mid j \in \frac{1}{2}\mathbb{Z}\}$. Let $\mathcal{T}(\mathfrak{L} \oplus V_\Pi)$ be the tensor algebra generated by the direct sum of $\mathfrak{L}$ and $V_\Pi$, and let $\mathcal{I}$ be the ideal of $\mathcal{T}(\mathfrak{L} \oplus V_\Pi)$ generated by

$$\Big\{L_i \otimes L_j - L_j \otimes L_i - [L_i, L_j],\ L_i \otimes d - d \otimes L_i,\ P_r \otimes P_s - \delta_{rs}P_r, \quad P_r \otimes d - d \otimes P_r,\ P_r \otimes L_j - L_j \otimes P_{r+j} \mid i,j \in \mathbb{Z},\ r,s \in \frac{1}{2}\mathbb{Z}\Big\}. \tag{3.8}$$

Define $\mathcal{U}_\Pi(\mathfrak{L}) = \mathcal{T}(\mathfrak{L} \oplus V_\Pi)/\mathcal{I}$. For any formal commuting variable $t^{1/2}$ and for $n \in \mathbb{Z}$, we define

$$(t^{1/2})^{n2L_0} = \sum_{j \in \frac{1}{2}\mathbb{Z}} P_j t^{nj} \in \mathcal{U}_\Pi(\mathfrak{L})[[t^{1/2}, t^{-1/2}]].$$

Note then that $(\kappa^{1/2})^{-2L_0}$ and $(z^{1/2k})^{(1-k)2L_0}$ are well-defined elements of

$$U_\Pi(\mathfrak{L})[[z^{1/2k}, z^{-1/2k}]][[\kappa^{1/2}, \kappa^{-1/2}]].$$

In $\mathcal{U}_\Pi(\mathfrak{L})[[z^{1/2k}, z^{-1/2k}]][[\kappa^{1/2}, \kappa^{-1/2}]]$, define

$$\Delta_k^{\mathfrak{L}}(z) = \exp\Bigg(\sum_{j \in \mathbb{Z}_+} a_j z^{-j/k} L_j\Bigg)(\kappa^{1/2})^{-2L_0}(z^{1/2k})^{(1-k)2L_0}.$$

PROPOSITION 3.5. *In $\mathcal{U}_\Pi(\mathfrak{L})[[z^{1/2k}, z^{-1/2k}]][[\kappa^{1/2}, \kappa^{-1/2}]]$, we have*

$$\Delta_k^{\mathfrak{L}}(z)L_{-1} - \kappa^{-1}z^{\frac{1}{k}-1}L_{-1}\Delta_k^{\mathfrak{L}}(z) = \frac{\partial}{\partial z}\Delta_k^{\mathfrak{L}}(z), \tag{3.9}$$

$$\Delta_k^{\mathfrak{L}}(z)^{-1}L_{-1} - \kappa z^{-\frac{1}{k}+1}L_{-1}\Delta_k^{\mathfrak{L}}(z)^{-1} = kz^{-\frac{1}{k}+1}\frac{\partial}{\partial z}\Delta_k^{\mathfrak{L}}(z)^{-1}. \tag{3.10}$$

PROOF. In $\mathcal{U}_\Pi(\mathfrak{L})[[z^{1/2k}, z^{-1/2k}]][[\kappa^{1/2}, \kappa^{-1/2}]]$, we have

$$\begin{aligned}
&\Delta_k^{\mathfrak{L}}(z)L_{-1} - \kappa^{-1}z^{\frac{1}{k}-1}L_{-1}\Delta_k^{\mathfrak{L}}(z)\\
&= \kappa^{-1}z^{\frac{1}{k}-1}\left[e^{\sum_{j\in\mathbb{Z}_+} a_j z^{-j/k}L_j}, L_{-1}\right](\kappa^{1/2})^{-2L_0}(z^{1/2k})^{(1-k)2L_0}\\
&= \kappa^{-1}z^{\frac{1}{k}-1}\sum_{n\in\mathbb{Z}_+}\frac{1}{n!}\Bigg(\sum_{j_1,\dots,j_n\in\mathbb{Z}_+} a_{j_1}\cdots a_{j_n} z^{-(j_1+\cdots+j_n)/k}\\
&\qquad \Bigg(\sum_{i=1,\dots,n} L_{j_1}L_{j_2}\cdots L_{j_{i-1}}[L_{j_i}, L_{-1}]L_{j_{i+1}}\cdots L_{j_n}\Bigg)\Bigg)(\kappa^{1/2})^{-2L_0}(z^{1/2k})^{(1-k)2L_0}
\end{aligned}$$

which is a well-defined element of $\mathcal{U}_\Pi(\mathfrak{L})[[z^{1/2k}, z^{-1/2k}]][[\kappa^{1/2}, \kappa^{-1/2}]]$ involving only elements $L_j$ with $j \in \mathbb{N}$. The right-hand side of (3.9) also involves only $L_j$ for $j \in \mathbb{N}$. Thus comparing with the identity (3.6) for the representation $L_j \mapsto L_j(x, \varphi)$, the identity (3.9) must hold. The proof of (3.10) is analogous. □

Let $V$ be a module for the Virasoro algebra satisfying $V = \coprod_{n \in \frac{1}{2}\mathbb{Z}} V_n$. For $j \in \mathbb{Z}$, let $L(j) \in \text{End}\, V$ and $c \in \mathbb{C}$ be the representation images of $L_j$ and $d$, respectively, for the Virasoro algebra. Assume that for $v \in V_n$, we have $L(0)v = nv$. For any formal variable $t^{1/2}$, define $(t^{1/2})^{j2L(0)} \in (\text{End}\, V)[[t^{1/2}, t^{-1/2}]]$ by

$$(t^{1/2})^{j2L(0)}v = t^{jn}v$$

for $v \in V_n$. Or equivalently, let $P(n) : V \to V_n$ be the projection from $V$ to the homogeneous subspace of weight $n$ for $n \in \frac{1}{2}\mathbb{Z}$. Then

$$(t^{1/2})^{j2L(0)}v = \sum_{n\in\mathbb{Z}} t^{jn}P(n)v$$

for $v \in V$. The elements $P(n) \in \mathrm{End}\, V$ can be thought of as the representation images of $P_n$ in the algebra $\mathcal{U}_\Pi(\mathfrak{L})$.

As in (3.5), in $(\mathrm{End}\, V)[[z^{1/2k}, z^{-1/2k}]]$, let

$$\Delta_k^V(z) = \exp\left(\sum_{j\in\mathbb{Z}_+} a_j z^{-j/k}L(j)\right)(k^{\frac{1}{2}})^{-2L(0)}(z^{\frac{1}{2k}})^{(1-k)2L(0)}.$$

From Proposition 4.1.1 in [**H**] extended to this setting of a $\frac{1}{2}\mathbb{Z}$-graded $\mathcal{L}$-module, or equivalently, from Proposition 3.32 of [**Bar3**] restricted to the Virasoro subalgebra of the N=1 Neveu-Schwarz algebra, and from Proposition 3.5, we obtain the following corollary.

COROLLARY 3.6. *In* $(\mathrm{End}\, V)[[z^{1/2k}, z^{-1/2k}]]$, *for a vertex operator superalgebra* $V$, *we have*

$$\Delta_k^V(z)L(-1) - \frac{1}{k}z^{1/k-1}L(-1)\Delta_k^V(z) = \frac{\partial}{\partial z}\Delta_k^V(z), \tag{3.11}$$

$$\Delta_k^V(z)^{-1}L(-1) - kz^{-1/k+1}L(-1)\Delta_k^V(z)^{-1} = kz^{-1/k+1}\frac{\partial}{\partial z}\Delta_k^V(z)^{-1}. \tag{3.12}$$

REMARK 3.7. In [**BDM**] we proved the identities (3.11) and (3.12) restricted to $\Delta_k^V(z)|_{V^{(0)}} \in (\mathrm{End}\, V^{(0)})[[z^{1/k}, z^{-1/k}]]$, and $L(-1)|_{V^{(0)}}$. Thus Corollary 3.6 extends these identities to operators in $(\mathrm{End}\, V)[[z^{1/2k}, z^{-1/2k}]]$ acting on $V = V^{(0)} \oplus V^{(1)}$.

## 4. The setting of $(1\ 2\ \cdots k)$-twisted $V^{\otimes k}$-modules and the operators $Y_M(\Delta_k(x)u, x^{1/k})$ for a $V$-module $(M, Y_M)$

Now we turn our attention to tensor product vertex operator superalgebras. Let $V = (V, Y, \mathbf{1}, \omega)$ be a vertex operator superalgebra, and let $k$ be a fixed positive integer. Then by Remark 2.4, $V^{\otimes k}$ is also a vertex operator superalgebra, and the symmetric group $S_k$ acts naturally as vertex superalgebra automorphisms on $V^{\otimes k}$, namely as signed permutations of the tensor factors. That is $(j\ j+1)\cdot(v_1 \otimes v_2 \otimes \cdots \otimes v_k) = (-1)^{|v_j||v_{j+1}|}(v_1 \otimes v_2 \otimes \cdots v_{j-1} \otimes v_{j+1} \otimes v_j \otimes v_{j+2} \otimes \cdots \otimes v_k)$, and we take this to be a right action so that, for instance

$$\begin{aligned}(1\,2\cdots k) : V \otimes V \otimes \cdots \otimes V &\longrightarrow V \otimes V \otimes \cdots \otimes V \\ v_1 \otimes v_2 \otimes \cdots \otimes v_k &\mapsto (-1)^{|v_1|(|v_2|+\cdots+|v_k|)}v_2 \otimes v_3 \otimes \cdots \otimes v_k \otimes v_1.\end{aligned} \tag{4.1}$$

(Note that in [**BDM**], this action was given as a left action. For convenience we make the change here to a right action as in [**BHL**]).

Let $g = (1\ 2\ \cdots\ k)$. In the next section we will, construct a functor $T_g^k$ from the category of weak $V$-modules to the category of weak $g$-twisted modules for $V^{\otimes k}$ for the case when $k$ is odd. This construction will be based on the operators $Y_M(\Delta_k(x)u, x^{1/k})$ for a $V$-module $(M, Y_M)$, where $\Delta_k(x)$ is the operator on $V$ defined and studied in the previous section. In this section, we establish several properties of these operators. We will also show why our construction of $g$-twisted $V^{\otimes k}$-modules given in this paper will not follow through for $k$ even. Rather, as shown in [**BV2**], one needs to incorporate parity-twisted $V$-modules.

For $v \in V$, and $k$ any positive integer, denote by $v^j \in V^{\otimes k}$, for $j = 1, \ldots, k$, the vector whose $j$-th tensor factor is $v$ and whose other tensor factors are $\mathbf{1}$. Then $gv^j = v^{j-1}$ for $j = 1, \ldots, k$ where 0 is understood to be $k$.

Suppose that $W$ is a weak $g$-twisted $V^{\otimes k}$-module, and let $\eta$ be a fixed primitive $k$-the root of unity. We first make some general observations for this setting following and extending [**BDM**]. First, it follows from the definition of twisted module (cf. (2.34)) that the $g$-twisted vertex operators on $W$ satisfy

$$Y_g(v^{j+1}, x) = Y_g(g^{-j}v^1, x) = \lim_{x^{1/k} \to \eta^j x^{1/k}} Y_g(v^1, x). \tag{4.2}$$

Since $V^{\otimes k}$ is generated by $v^j$ for $v \in V$ and $j = 1, \ldots, k$, the twisted vertex operators $Y_g(v^1, x)$ for $v \in V$ determine all the twisted vertex operators $Y_g(u, x)$ on $W$ for any $u \in V^{\otimes k}$. This observation is very important in our construction of twisted modules.

Secondly, if $u, v \in V$ are of homogeneous sign, then by (2.31) the twisted Jacobi identity for $Y_g(u^1, x_1)$ and $Y_g(v^1, x_2)$ is

$$\begin{aligned} x_0^{-1}\delta\left(\frac{x_1 - x_2}{x_0}\right) & Y_g(u^1, x_1)Y_g(v^1, x_2) \\ & - (-1)^{|u||v|}x_0^{-1}\delta\left(\frac{x_2 - x_1}{-x_0}\right) Y_g(v^1, x_2)Y_g(u^1, x_1) \\ & = \frac{1}{k}x_2^{-1}\sum_{j=0}^{k-1}\delta\left(\eta^j\frac{(x_1 - x_0)^{1/k}}{x_2^{1/k}}\right)Y_g(Y(g^j u^1, x_0)v^1, x_2). \end{aligned} \tag{4.3}$$

Since $g^{-j}u^1 = u^{j+1}$, we see that $Y(g^{-j}u^1, x_0)v^1$ only involves nonnegative integer powers of $x_0$ unless $j = 0 \pmod k$. Thus the we have the supercommutator

$$\begin{aligned} & [Y_g(u^1, x_1), Y_g(v^1, x_2)] \\ & \qquad = \mathrm{Res}_{x_0}\frac{1}{k}x_2^{-1}\delta\left(\frac{(x_1 - x_0)^{1/k}}{x_2^{1/k}}\right)Y_g(Y(u^1, x_0)v^1, x_2). \end{aligned} \tag{4.4}$$

This shows that the component operators of $Y_g(u^1, x)$ for $u \in V$ on $W$ form a Lie superalgebra.

For $u \in V$ and $\Delta_k(z) = \Delta_k^V(x)$ given by (3.5), and $(M, Y_M)$ a weak $V$-module, define

$$\bar{Y}(u, x) = Y_M(\Delta_k(x)u, x^{1/k}). \tag{4.5}$$

For example, as in [**BDM**], taking $u = \omega$, and recalling that $a_2 = (k^2 - 1)/12$, we have

$$\begin{aligned} \bar{Y}(\omega, x) & = Y_M\left(\frac{x^{2(1/k-1)}}{k^2}\left(\omega + a_2\frac{c}{2}x^{-2/k}\right), x^{1/k}\right) \\ & = \frac{x^{2(1/k-1)}}{k^2}Y(\omega, x^{1/k}) + \frac{(k^2 - 1)c}{24k^2}x^{-2}, \end{aligned} \tag{4.6}$$

where $c$ is the central charge of $V$.

REMARK 4.1. Note that

$$\Delta_k(x)u \in \begin{cases} V^{(0)}[[x^{1/k}, x^{-1/k}]] & \text{if } u \text{ is even, and } k \in \mathbb{Z}_+ \\ V^{(1)}[[x^{1/k}, x^{-1/k}]] & \text{if } u \text{ is odd, and } k \text{ is odd} \\ x^{1/2k}V^{(1)}[[x^{1/k}, x^{-1/k}]] & \text{if } u \text{ is odd, and } k \text{ is even.} \end{cases} \tag{4.7}$$

Thus since $Y_M(v, x) \in (\mathrm{End}\, M)[[x, x^{-1}]]$, we have

$$Y_M(\Delta_k(x)u, x^{1/k}) \in \begin{cases} (\mathrm{End}\, M)[[x^{1/k}, x^{-1/k}]] & \text{if } k \text{ is odd} \\ x^{|v|/2k}(\mathrm{End}\, M)[[x^{1/k}, x^{-1/k}]] & \text{if } k \text{ is even.} \end{cases} \tag{4.8}$$

When we put a weak $g$-twisted $V^{\otimes k}$-module structure on $M$, this operator $\bar{Y}(u, x) = Y_M(\Delta_k(x)u, x^{1/k})$ will be the twisted vertex operator acting on $M$ associated to $u^1$, when $k$ is odd. However, since this operator contains powers of $x^{1/2k}$ for $k$ even, it cannot be the twisted vertex operator associated to $u^1$ in this case. It is this crucial fact that distinguishes the $k$ even versus $k$ odd cases, making the construction of twisted modules fundamentally different. As noted in Remark 5.1 below, the operator $\bar{Y}$ is a type of "generalized" twisted vertex operator, generalizing the notion of "relativized" twisted vertex operator for lattice vertex operator superalgebras as constructed in [**DL2**]. Alternately, $\bar{Y}$ can be realized as parity-twisted $V$-modules as is done in [**BV2**].

Next we study the properties of the operators $\bar{Y}(u, x)$, following and generalizing the results of [**BDM**].

LEMMA 4.2. *For $u \in V$*

$$\bar{Y}(L(-1)u, x) = \frac{d}{dx}\bar{Y}(u, x).$$

PROOF. By Corollary 3.6, the analogous proof given in [**BDM**] (Lemma 3.2 of [**BDM**]) in the nonsuper setting follows in the setting of vertex operator superalgebras. That is, we have

$$\begin{aligned}
\bar{Y}(L(-1)u, x) &= Y_M(\Delta_k(x)L(-1)u, x^{1/k}) \\
&= Y_M(\frac{d}{dx}\Delta_k(x)u, x^{1/k}) + k^{-1}x^{1/k-1}Y_M(L(-1)\Delta_k(x)u, x^{1/k}) \\
&= Y_M(\frac{d}{dx}\Delta_k(z)u, x^{1/k}) + k^{-1}x^{1/k-1}\frac{d}{dy}Y_M(\Delta_k(x)u, y)\Big|_{y=x^{1/k}} \\
&= Y_M(\frac{d}{dx}\Delta_k(x)u, x^{1/k}) + \frac{d}{dy}Y_M(\Delta_k(x)u, y^{1/k})\Big|_{y=x} \\
&= \frac{d}{dx}Y_M(\Delta_k(x)u, x^{1/k}) \\
&= \frac{d}{dx}\bar{Y}(u, x)
\end{aligned}$$

as desired. □

LEMMA 4.3. *For $u, v \in V$ of homogeneous sign, we have the supercommutator*

$$(4.9)\quad [\bar{Y}(u,x_1), \bar{Y}(v,x_2)] = \mathrm{Res}_{x_0} \frac{x_2^{-1}}{k} \delta\left(\frac{(x_1-x_0)^{1/k}}{x_2^{1/k}}\right) \bar{Y}(Y(u,x_0)v,x_2) \left(\frac{x_1-x_0}{x_2}\right)^{\frac{|u|}{2k}(1-k)},$$

*or equivalently*

$$(4.10)\quad [\bar{Y}(u,x_1), \bar{Y}(v,x_2)] = \begin{cases} \mathrm{Res}_{x_0} \frac{x_2^{-1}}{k} \delta\left(\frac{(x_1-x_0)^{1/k}}{x_2^{1/k}}\right) \bar{Y}(Y(u,x_0)v,x_2) & \text{if } k \text{ is odd} \\ \mathrm{Res}_{x_0} \frac{x_2^{-1}}{k} \delta\left(\frac{(x_1-x_0)^{1/k}}{x_2^{1/k}}\right) \bar{Y}(Y(u,x_0)v,x_2) \left(\frac{x_1-x_0}{x_2}\right)^{\frac{|u|}{2k}} & \text{if } k \text{ is even.} \end{cases}$$

PROOF. The supercommutator formula for the weak $V$-module $M$ is given by

$$(4.11)\qquad [Y_M(u,x_1), Y_M(v,x_2)] = \mathrm{Res}_x x_2^{-1} \delta\left(\frac{x_1-x}{x_2}\right) Y_M(Y(u,x)v,x_2)$$

which is a consequence of the Jacobi identity on $M$. Replacing $Y_M(u,x_1)$ and $Y_M(v,x_2)$ by $Y_M(\Delta_k(x_1)u, x_1^{1/k})$ and $Y_M(\Delta_k(x_2)v, x_2^{1/k})$, respectively, in the supercommutator formula, we have the supercommutator

$$(4.12)\quad [\bar{Y}(u,x_1), \bar{Y}(v,x_2)] = \mathrm{Res}_x x_2^{-1/k} \delta\left(\frac{x_1^{1/k}-x}{x_2^{1/k}}\right) Y_M(Y(\Delta_k(x_1)u,x)\Delta_k(x_2)v, x_2^{1/k}).$$

We want to make the change of variable $x = x_1^{1/k} - (x_1-x_0)^{1/k}$ where by $x_1^{1/k} - (x_1-x_0)^{1/k}$ we mean the power series expansion in positive powers of $x_0$. For $n \in \mathbb{Z}$, it was shown in [**BDM**] that

$$(4.13)\qquad (x_1^{1/k}-x)^n \Big|_{x=x_1^{1/k}-(x_1-x_0)^{1/k}} = (x_1-x_0)^{n/k}.$$

Thus substituting $x = x_1^{1/k} - (x_1-x_0)^{1/k}$ into

$$x_2^{-1/k} \delta\left(\frac{x_1^{1/k}-x}{x_2^{1/k}}\right) Y_M(Y(\Delta_k(x_1)u,x)\Delta_k(x_2)v, x_2^{1/k}),$$

we have a well-defined power series given by

$$x_2^{-1/k} \delta\left(\frac{(x_1-x_0)^{1/k}}{x_2^{1/k}}\right) Y_M(Y(\Delta_k(x_1)u, x_1^{1/k} - (x_1-x_0)^{1/k})\Delta_k(x_2)v, x_2^{1/k}).$$

Let $f(z_1,z_2,x)$ be a complex analytic function in the complex variables $z_1, z_2$, and $x$, and let $h(z_1,z_2,z_0)$ be a complex analytic function in $z_1, z_2$, and $z_0$. Then if $f(z_1,z_2,h(z_1,z_2,z_0))$ is well defined, and thinking of $z_1$ and $z_2$ as fixed ( i.e., considering $f(z_1,z_2,h(z_1,z_2,z_0))$ as a Laurent series in $z_0$) by the residue theorem of complex analysis, we have

$$(4.14)\qquad \mathrm{Res}_x f(z_1,z_2,x) = \mathrm{Res}_{z_0} \left(\frac{\partial}{\partial z_0} h(z_1,z_2,z_0)\right) f(z_1,z_2,h(z_1,z_2,z_0)).$$

This of course remains true for $f$ and $h$ formal power series in their respective variables. Thus making the change of variable $x = h(x_1, x_2, x_0) = x_1^{1/k} - (x_1 - x_0)^{1/k}$, using (4.12), (4.14), the $\delta$-function identity (2.3), (4.8), and Proposition 3.2, we obtain

$$\begin{aligned}
&[\bar{Y}(u, x_1), \bar{Y}(v, x_2)] \\
&= \mathrm{Res}_{x_0} \frac{1}{k} x_2^{-1/k} (x_1 - x_0)^{1/k - 1} \delta\left( \frac{(x_1 - x_0)^{1/k}}{x_2^{1/k}} \right) \\
&\qquad Y_M(Y(\Delta_k(x_1)u, x_1^{1/k} - (x_1 - x_0)^{1/k}) \Delta_k(x_2)v, x_2^{1/k}) \\
&= \mathrm{Res}_{x_0} \frac{1}{k} x_2^{-1} \delta\left( \frac{(x_1 - x_0)^{1/k}}{x_2^{1/k}} \right) \\
&\qquad Y_M(Y(\Delta_k(x_1)u, x_1^{1/k} - (x_1 - x_0)^{1/k}) \Delta_k(x_2)v, x_2^{1/k}) \\
&= \mathrm{Res}_{x_0} \frac{1}{k} x_1^{-1} \delta\left( \frac{(x_2 + x_0)^{1/k}}{x_1^{1/k}} \right) \\
&\qquad Y_M(Y(\Delta_k(x_1)u, x_1^{1/k} - (x_1 - x_0)^{1/k}) \Delta_k(x_2)v, x_2^{1/k}).
\end{aligned}$$

Now we observe that

$$(4.15)\quad Y_M(Y(\Delta_k(x_1)u, x_1^{1/k} - (x_1 - x_0)^{1/k}) \Delta_k(x_2)v, x_2^{1/k})$$
$$\in \begin{cases} x_1^{|u|/2k} x_2^{|v|/2k} (\mathrm{End}\, M)[[x_0]][[x_1^{1/k}, x_1^{-1/k}]][[x_2^{1/k}, x_2^{-1/k}]] & \text{if } k \text{ is even} \\ (\mathrm{End}\, M)[[x_0]][[x_1^{1/k}, x_1^{-1/k}]][[x_2^{1/k}, x_2^{-1/k}]] & \text{if } k \text{ is odd.} \end{cases}$$

Thus using the $\delta$-function substitution property (see e.g. [**LL**]) and Proposition 3.2, we obtain

$$\begin{aligned}
&[\bar{Y}(u, x_1), \bar{Y}(v, x_2)] = \\
&= \mathrm{Res}_{x_0} \frac{1}{k} x_1^{-1} \delta\left( \frac{(x_2 + x_0)^{1/k}}{x_1^{1/k}} \right) \left( \frac{x_2 + x_0}{x_1} \right)^{\frac{|u|}{2k}(k-1)} \\
&\qquad Y_M(Y(\Delta_k(x_2 + x_0)u, (x_2 + x_0)^{1/k} - x_2^{1/k}) \Delta_k(x_2)v, x_2^{1/k}) \\
&= \mathrm{Res}_{x_0} \frac{1}{k} x_2^{-1} \delta\left( \frac{(x_1 - x_0)^{1/k}}{x_2^{1/k}} \right) \left( \frac{x_1 - x_0}{x_2} \right)^{-\frac{|u|}{2k}(k-1)} \\
&\qquad Y_M(Y(\Delta_k(x_2 + x_0)u, (x_2 + x_0)^{1/k} - x_2^{1/k}) \Delta_k(x_2)v, x_2^{1/k}) \\
&= \mathrm{Res}_{x_0} \frac{1}{k} x_2^{-1} \delta\left( \frac{(x_1 - x_0)^{1/k}}{x_2^{1/k}} \right) Y_M(\Delta_k(x_2) Y(u, x_0)v, x_2^{1/k}) \left( \frac{x_1 - x_0}{x_2} \right)^{\frac{|u|}{2k}(1-k)} \\
&= \mathrm{Res}_{x_0} \frac{1}{k} x_2^{-1} \delta\left( \frac{(x_1 - x_0)^{1/k}}{x_2^{1/k}} \right) \bar{Y}(Y(u, x_0)v, x_2) \left( \frac{x_1 - x_0}{x_2} \right)^{\frac{|u|}{2k}(1-k)},
\end{aligned}$$

giving (4.9). Equation (4.10) follows from the properties of the $\delta$-function. □

## 5. The construction of a weak $(1\ 2\ \cdots\ k)$-twisted $V^{\otimes k}$-module structure on a weak $V$-module $(M, Y_M)$ for $k$ odd

Let $M = (M, Y_M)$ be a weak $V$-module. Now we begin our construction of a weak $g$-twisted $V^{\otimes k}$-module structure on $M$ when $k$ is an odd positive integer and $g = (1\ 2\ \cdots\ k)$. Now that we have established the properties of $\Delta_k(x)$ as in Section 3 in the setting of vertex operator superalgebras and proved the supercommutator (4.9), our construction of a weak $g$-twisted $V^{\otimes k}$-module structure on $M$ in the case when $k$ is odd follows the same spirit as the construction in the nonsuper case given in [**BDM**], but now using the full power of local systems for twisted operators in the setting of vertex operator superalgebras as established by Li in [**Li**].

For $k$ odd, we construct these weak $g$-twisted $V^{\otimes k}$-modules by first defining $g$-twisted vertex operators on a weak $V$-module $M$ for a set of generators which are mutually local (see [**Li**]). These $g$-twisted vertex operators generate a local system which is a vertex superalgebra. We then construct a homomorphism of vertex superalgebras from $V^{\otimes k}$ to this local system which thus gives a weak $g$-twisted $V^{\otimes k}$-module structure on $M$.

For $u \in V$ and $j = 0, \dots, k-1$, set

$$Y_g(u^1, x) = \bar{Y}(u, x) \quad \text{and} \quad Y_g(u^{j+1}, x) = \lim_{x^{1/k} \to \eta^j x^{1/k}} Y_g(u^1, x). \tag{5.1}$$

REMARK 5.1. From the supercommutator (4.9) for $\bar{Y}$, we see that defining $g$-twisted operators as above for the case when $k$ is even, can not result in a twisted module structure on $M$ due to appearance of the extra term involving $(x_2^{-1}(x_1 - x_0))^{|u|/2k}$. In particular, the most we could hope for would be a type of "generalized" $g$-twisted $V^{\otimes k}$-module structure in the spirit of [**DL1**] and the "relativized" twisted vertex operators for lattice vertex operator superalgebras as constructed in [**DL2**]. In fact, as is proved in [**BV2**], these should be interpreted as parity-twisted vertex operators, thus showing the importance of this example of permutation-twisted orbifold superconformal field theory for fully understanding the role of the parity map in the general problem of how to construct and classify twisted sectors.

Note that $Y_g(u^j, x) = \sum_{p=0}^{k-1} Y_g^p(u^j, x)$ where $Y_g^p(u^j, x) = \sum_{n \in \frac{p}{k} + \mathbb{Z}} u_n^j x^{-n-1}$.

LEMMA 5.2. *Let $u, v \in V$ of homogeneous sign. Then we have the supercommutator*

$$\begin{aligned}(5.2)\quad & [Y_g(u^j, x_1), Y_g(v^m, x_2)] \\ &= \mathrm{Res}_{x_0} \frac{1}{k} x_2^{-1} \delta\left( \frac{\eta^{j-m}(x_1 - x_0)^{1/k}}{x_2^{1/k}} \right) Y_g((Y(u, x_0)v)^m, x_2)\end{aligned}$$

*where $(Y(u, x_0)v)^m = \sum_{n \in \mathbb{Z}} (u_n v)^m x_0^{-n-1}$, and*

$$\begin{aligned}(5.3)\quad & [Y_g^p(u^j, x_1), Y_g(v^m, x_2)] \\ &= \mathrm{Res}_{x_0} \frac{1}{k} x_2^{-1} \eta^{(m-j)p} \left( \frac{x_1 - x_0}{x_2} \right)^{-p/k} \delta\left( \frac{x_1 - x_0}{x_2} \right) Y_g((Y(u, x_0)v)^m, x_2).\end{aligned}$$

PROOF. By Lemma 4.3, Equation (5.2) holds if $j = m = 1$ and $k$ odd. Then using (5.1), we obtain Equation (5.2) for any $j, m = 1, \dots, k$. Equation (5.3) is a direct consequence of Equation (5.2). □

By Lemma 5.2 for $u, v \in V$ of homogeneous sign, there exists a positive integer $N$ such that

$$(x_1 - x_2)^N [Y_g(u^j, x_1), Y_g(v^m, x_2)] = 0. \tag{5.4}$$

Taking the limit $x^{1/k} \longrightarrow \eta^{j-1} x^{1/k}$ in Lemma 4.2, for $j = 1, \ldots, k$, we have

$$Y_g(L(-1)u^j, x) = \frac{d}{dx} Y_g(u^j, x). \tag{5.5}$$

Thus the operators $Y_g(u^j, x)$ for $u \in V$, and for $j = 1, \ldots, k$, are mutually local and generate a local system $A$ of weak twisted vertex operators on $(M, L(-1))$ in the sense of [**Li**].

Let $\rho$ be a map from $A$ to $A$ such that $\rho Y_g(u^j, x) = Y_g(u^{j-1}, x)$ for $u \in V$ and $j = 1, \ldots, k$. By Theorem 3.14 of [**Li**][2], the local system $A$ generates a vertex superalgebra we denote by $(A, Y_A)$, and $\rho$ extends to an automorphism of $A$ of order $k$ such that $M$ is a natural weak generalized $\rho$-twisted $A$-module in the sense that $Y_A(\alpha(x), x_1) = \alpha(x_1)$ for $\alpha(x) \in A$ are $\rho$-twisted vertex operators on $M$.

REMARK 5.3. $\rho$ is given by

$$\rho a(x) = \lim_{x^{1/k} \to \eta^{-1} x^{1/k}} a(x) \tag{5.6}$$

for $a(x) \in A$; see [**Li**].

Let $A^j = \{c(x) \in A \mid \rho c(x) = \eta^j c(x)\}$, and let $a(x) \in A^j$ be of homogeneous sign in $A$. Then for any integer $n$ and $b(x) \in A$ of homogeneous sign, the operator $a(x)_n b(x)$ is an element of $A$ given by

$$a(x)_n b(x) = \mathrm{Res}_{x_1} \mathrm{Res}_{x_0} \left( \frac{x_1 - x_0}{x} \right)^{j/k} x_0^n \cdot X \tag{5.7}$$

where

$$X = x_0^{-1} \delta \left( \frac{x_1 - x}{x_0} \right) a(x_1) b(x) - (-1)^{|a||b|} x_0^{-1} \delta \left( \frac{x - x_1}{-x_0} \right) b(x) a(x_1).$$

Or, equivalently, $a(x)_n b(x)$ is defined by:

$$\sum_{n \in \mathbb{Z}} (a(x)_n b(x))\, x_0^{-n-1} = \mathrm{Res}_{x_1} \left( \frac{x_1 - x_0}{x} \right)^{j/k} \cdot X. \tag{5.8}$$

Thus following [**Li**], for $a(z) \in A^j$, we define $Y_A(a(z), x)$ by setting $Y_A(a(z), x_0) b(z)$ equal to (5.8).

LEMMA 5.4. *For $u, v \in V$ of homogeneous sign, we have the supercommutator*

$$[Y_A(Y_g(u^j, x), x_1), Y_A(Y_g(v^m, x), x_2)] = 0$$

*for $j, m = 1, \ldots, k$, with $j \neq m$.*

PROOF. The proof is analogous to the proof of Lemma 3.6 in [**BDM**] where we use the vertex superalgebra structure of $A$ rather than just the vertex algebra structure and we use the supercommutators of Lemma 5.2. $\square$

[2]There is a typo in the statement of Theorem 3.14 in [**Li**]. The $V$ in the theorem should be $A$. That is, the main result of the theorem is that the local system $A$ of the theorem has the structure of a vertex superalgebra.

Let $Y_g(u^i, z)_n$ for $n \in \mathbb{Z}$ denote the coefficient of $x^{-n-1}$ in the vertex operator $Y_A(Y_g(u^i, z), x)$ for $u \in V$, and $i = 1, \dots, k$. That is

$$Y_A(Y_g(u^i, z), x) = \sum_{n \in \mathbb{Z}} Y_g(u^i, z)_n \, x^{-n-1} \in (\mathrm{End}\, A)[[x, x^{-1}]].$$

LEMMA 5.5. *For $u_1, \dots, u_k \in V$, we have*

$$\begin{aligned} &Y_A(Y_g(u_1^1, z)_{-1} \cdots Y_g(u_{k-1}^{k-1}, z)_{-1} Y_g(u_k^k, z), x) \\ &\qquad = Y_A(Y_g(u_1^1, z), x) \cdots Y_A(Y_g(u_{k-1}^{k-1}, z), x) Y_A(Y_g(u_k^k, z), x). \end{aligned}$$

PROOF. From the Jacobi identity on $A$, and Lemma 5.4, we have, for $1 \le i < j \le k$,

$$\begin{aligned} &Y_A(Y_g(u^i, z)_{-1} Y_g(v^j, z), x) \\ &= \mathrm{Res}_{x_1} \mathrm{Res}_{x_0} x_0^{-1} \left( x_0^{-1} \delta\left(\frac{x_1 - x}{x_0}\right) Y_A(Y_g(u^i, z), x_1) Y_A(Y_g(v^j, z), x) \right. \\ &\qquad \left. -(-1)^{|u||v|} x_0^{-1} \delta\left(\frac{x - x_1}{-x_0}\right) Y_A(Y_g(v^j, z), x) Y_A(Y_g(u^i, z), x_1) \right) \\ &= \mathrm{Res}_{x_1} \Big( (x_1 - x)^{-1} Y_A(Y_g(u^i, z), x_1) Y_A(Y_g(v^j, z), x) \\ &\qquad -(-1)^{|u||v|} (x - x_1)^{-1} Y_A(Y_g(v^j, z), x) Y_A(Y_g(u^i, z), x_1) \Big) \\ &= \sum_{n<0} Y_g(u^i, z)_n x^{-n-1} Y_A(Y_g(v^j, z), x) \\ &\qquad -(-1)^{|u||v|} Y_A(Y_g(v^j, z), x) \sum_{n \ge 0} Y_g(u^i, z)_n x^{-n-1} \\ &= \sum_{n \in \mathbb{Z}} Y_g(u^i, z)_n x^{-n-1} Y_A(Y_g(v^j, z), x) \\ &= Y_A(Y_g(u^i, z), x) Y_A(Y_g(v^j, z), x). \end{aligned}$$

The result follows by induction. □

Define the map $f : V^{\otimes k} \longrightarrow A$ by

$$\begin{aligned} f : V^{\otimes k} &\longrightarrow A \\ u_1 \otimes \cdots \otimes u_k = (u_1^1)_{-1} \cdots (u_{k-1}^{k-1})_{-1} u_k^k &\mapsto Y_g(u_1^1, z)_{-1} \cdots Y_g(u_{k-1}^{k-1}, z)_{-1} Y_g(u_k^k, z) \end{aligned}$$

for $u_1, \dots u_k \in V$. Then $f(u^j) = Y_g(u^j, z)$.

LEMMA 5.6. *$f$ is a homomorphism of vertex superalgebras.*

PROOF. We need to show that

$$fY(u_1 \otimes \cdots \otimes u_k, x) = Y_A(Y_g(u_1^1, z)_{-1} \cdots Y_g(u_{k-1}^{k-1}, z)_{-1} Y_g(u_k^k, z), x) f$$

for $u_i \in V$. Take $v_i \in V$ for $i = 1, \dots, k$. Then

$$\begin{aligned} &fY(u_1 \otimes \cdots \otimes u_k, x)(v_1 \otimes \cdots \otimes v_k) \\ &= (-1)^s f(Y(u_1, x) v_1 \otimes \cdots Y(u_k, x) v_k) \\ &= (-1)^s Y_g(Y(u_1^1, x) v_1^1, z)_{-1} \cdots Y_g(Y(u_{k-1}^{k-1}, x) v_{k-1}^{k-1}, z)_{-1} Y_g(Y(u_k^k, x) v_k^k, z) \end{aligned}$$

for $s = \sum_{j=1}^{k-1} |v_j| \sum_{i=j+1}^{k} |u_i|$.

By Lemma 5.5, we have

$$\begin{aligned}&Y_A(Y_g(u^1_1, z)_{-1} \cdots Y_g(u^{k-1}_{k-1}, z)_{-1} Y_g(u^k_k, z), x) f(v_1 \otimes \cdots \otimes v_k)\\&\quad = Y_A(Y_g(u^1_1, z), x) \cdots Y_A(Y_g(u^{k-1}_{k-1}, z), x) Y_A(Y_g(u^k_k, z), x) Y_g(v^1_1, z)_{-1}\\&\qquad\qquad \cdots Y_g(v^{k-1}_{k-1}, z)_{-1} Y_g(v^k_k, z).\end{aligned}$$

By Lemma 5.4, it is enough to show that

$$Y_g(Y(u^j, x)v^j, z) = Y_A(Y_g(u^j, z), x) Y_g(v^j, z)$$

for $u, v \in V$ and $j = 1, \ldots, k$. In fact, in view of the relation between $Y(u^1, z)$ and $Y(u^j, z)$ for $u \in V$, we only need to prove the case $j = 1$.

By Proposition 3.2,

$$\begin{aligned}Y_g(Y(u^1, x_0)v^1, x_2) &= Y_M(\Delta_k(x_2) Y(u, x_0)v, x_2^{1/k})\\&= Y_M(Y(\Delta_k(x_2 + x_0)u, (x_2 + x_0)^{1/k} - x_2^{1/k}) \Delta_k(x_2)v, x_2^{1/k}).\end{aligned}$$

On the other hand,

$$Y_A(Y_g(u^1, x_2), x_0) Y_g(v^1, x_2) = \sum_{p=0}^{k-1} \mathrm{Res}_{x_1} \left(\frac{x_1 - x_0}{x_2}\right)^{p/k} X$$

where

$$X = x_0^{-1}\delta\left(\frac{x_1 - x_2}{x_0}\right) Y_g(u^1, x_1) Y_g(v^1, x_2) - (-1)^{|u||v|} x_0^{-1}\delta\left(\frac{x_2 - x_1}{-x_0}\right) Y_g(v^1, x_2) Y_g(u^1, x_1). \tag{5.9}$$

By Equation (5.4), there exists a positive integer $N$ such that

$$(x_1 - x_2)^N Y_g(u^1, x_1) Y_g(v^1, x_2) = (-1)^{|u||v|} (x_1 - x_2)^N Y_g(v^1, x_2) Y_g(u^1, x_1).$$

Thus, using the three-term $\delta$-function identity (2.4), we have

$$\begin{aligned}X &= x_0^{-1}\delta\left(\frac{x_1 - x_2}{x_0}\right) Y_g(u^1, x_1) Y_g(v^1, x_2)\\&\quad -(-1)^{|u||v|} x_0^{-1}\delta\left(\frac{x_2 - x_1}{-x_0}\right) x_0^{-N} (x_1 - x_2)^N Y_g(v^1, x_2) Y_g(u^1, x_1)\\&= x_0^{-1}\delta\left(\frac{x_1 - x_2}{x_0}\right) x_0^{-N} \left((x_1 - x_2)^N Y_g(u^1, x_1) Y_g(v^1, x_2)\right)\\&\quad -(-1)^{|u||v|} x_0^{-1}\delta\left(\frac{x_2 - x_1}{-x_0}\right) x_0^{-N} \left((x_1 - x_2)^N Y_g(u^1, x_1) Y_g(v^1, x_2)\right)\\&= x_2^{-1} x_0^{-N} \delta\left(\frac{x_1 - x_0}{x_2}\right) \left((x_1 - x_2)^N Y_g(u^1, x_1) Y_g(v^1, x_2)\right).\end{aligned}$$

Therefore using the $\delta$-function relation (2.2), we have

$$\begin{aligned}&Y_A(Y_g(u^1, x_2), x_0) Y_g(v^1, x_2)\\&\quad = \mathrm{Res}_{x_1} x_0^{-N} x_2^{-1} \delta\left(\frac{(x_1 - x_0)^{1/k}}{x_2^{1/k}}\right) \left((x_1 - x_2)^N Y_g(u^1, x_1) Y_g(v^1, x_2)\right).\end{aligned}$$

And the rest of the proof is analogous to the corresponding part of the proof of Lemma 3.8 in [**BDM**]. □

Let $(M, Y)$ be a weak $V$-module, $k$ a positive odd integer, and $g = (1\ 2\ \cdots\ k)$. Define $T_g^k(M, Y) = (T_g^k(M), Y_g) = (M, Y_g)$. That is $T_g^k(M, Y)$ is $M$ as the underlying vector space and the vertex operator $Y_g$ is given by (5.1).

Now we state our first main theorem of the paper.

THEOREM 5.7. *$(T_g^k(M), Y_g)$ is a weak $g$-twisted $V^{\otimes k}$-module such that $T_g^k(M) = M$, and $Y_g$, defined by (5.1), is the linear map from $V^{\otimes k}$ to $(\mathrm{End}\, T_g^k(M))[[x^{1/k}, x^{-1/k}]]$ defining the twisted module structure. Moreover,*

*(1) $(M, Y)$ is an irreducible weak $V$-module if and only if $(T_g^k(M), Y_g)$ is an irreducible weak $g$-twisted $V^{\otimes k}$-module.*

*(2) $M$ is a weak admissible $V$-module if and only if $T_g^k(M)$ is a weak admissible $g$-twisted $V^{\otimes k}$-module.*

*(3) $M$ is an ordinary $V$-module if and only if $T_g^k(M)$ is an ordinary $g$-twisted $V^{\otimes k}$-module.*

PROOF. It is immediate from Lemma 5.6 that $T_g^k(M) = M$ is a weak $g$-twisted $V^{\otimes k}$-module with $Y_g(u^1, x) = \bar{Y}(u, x)$. Note that with

$$\Delta_k(x)^{-1} = (x^{1/2k})^{-(1-k)2L(0)}(k^{1/2})^{2L(0)} \exp\left( - \sum_{j \in \mathbb{Z}_+} a_j x^{-j/k} L(j) \right), \tag{5.10}$$

we have

$$Y_g((\Delta_k(x^k)^{-1}u)^1, x) = \bar{Y}(\Delta_k(x)^{-1}u, x) = Y_M(u, x^{1/k}), \tag{5.11}$$

and all twisted vertex operators $Y_g(v, x)$ for $v \in V^{\otimes k}$ can be generated from $Y_g(u^1, x)$ for $u \in V$. It is clear now that $M$ is an irreducible weak $V$-module if and only if $T_g^k(M)$ is an irreducible weak $g$-twisted $V^{\otimes k}$-module, proving statement (1).

For statement (2), we first assume that $M$ is a weak admissible $V$-module, i.e. we have $M = \coprod_{n \in \frac{1}{2}\mathbb{N}} M(n)$ such that for $m \in \mathbb{Z}$, the component operator $u_m$ of $Y_M(u, z)$, satisfies $u_m M(n) \subset M(\mathrm{wt}\ u - m - 1 + n)$ if $u \in V$ is of homogeneous weight.

Define a $\frac{1}{2k}\mathbb{N}$-gradation on $T_g^k(M)$ such that $T_g^k(M)(n/k) = M(n)$ for $n \in \frac{1}{2}\mathbb{Z}$. Recall that $Y_g(v, x) = \sum_{m \in \frac{1}{k}\mathbb{Z}} v_m^g x^{-m-1}$ for $v \in V^{\otimes k}$. We need to show that $v_m^g T_g^k(M)(n) \subset T_g^k(M)(\mathrm{wt}\ v - m - 1 + n)$ for $m \in \frac{1}{k}\mathbb{Z}$, and $n \in \frac{1}{2k}\mathbb{N}$. Since all twisted vertex operators $Y_g(v, x)$ for $v \in V^{\otimes k}$ can be generated from $Y_g(u^1, x)$ for $u \in V$, it is enough to show $(u^1)_m^g T_g^k(M)(n) \subset T_g^k(M)(\mathrm{wt}\ u - m - 1 + n)$.

Let $u \in V_p$ for $p \in \frac{1}{2}\mathbb{Z}$. Then we can write

$$\Delta_k(x)u = \sum_{j=0}^{\infty} u(j) x^{p/k - p - j/k}$$

where $u(j) \in V_{p-j}$. Thus

$$Y_g(u^1, x) = Y_M(\Delta_k(x)u, x^{1/k}) = \sum_{j=0}^{\infty} Y_M(u(j), x^{1/k}) x^{p/k - p - j/k},$$

and thus for $m \in \frac{1}{k}\mathbb{Z}$, we have

$$(u^1)_m^g = \sum_{j=0}^{\infty} u(j)_{(1-k)p - j - 1 + km + k}.$$

Since the weight of $u(j)_{(1-k)p-j-1+km+k}$ is $k(p-m-1)$, we see that for $n \in \frac{1}{2k}\mathbb{N}$, we have $(u^1)^g_m T^k_g(M)(n) = (u^1)^g_m M(kn) \subset M(k(p-m-1+n)) = T^k_g(M)(p-m-1+n)$, showing that $T^k_g(M)$ is a weak admissible $g$-twisted $V^{\otimes k}$-module.

Conversely, we assume that $T^k_g(M)$ is a weak admissible $g$-twisted $V^{\otimes k}$-module, i.e., we have $T^k_g(M) = \coprod_{n\in\frac{1}{2k}\mathbb{N}} T^k_g(M)(n)$ such that for $m \in \frac{1}{k}\mathbb{Z}$, the component operator $u^g_m$ of $Y_g(u,x)$ satisfies $u^g_m T^k_g(M)(n) \subset T^k_g(M)(\mathrm{wt}\ u - m - 1 + n)$ if $u \in V^{\otimes k}$ is of homogeneous weight. Define a $\frac{1}{2}\mathbb{N}$-gradation on $M$ such that $M(n) = T^k_g(M)(n/k)$ for $n \in \frac{1}{2}\mathbb{Z}$.

Note that by again letting $u \in V_p$ for $p \in \frac{1}{2}\mathbb{Z}$, then we can write

$$\Delta_k(x)^{-1}u = \sum_{j=0}^{\infty} u[j]x^{p-p/k-j}$$

where $u[j] \in V_{p-j}$. Thus Equation (5.11) implies

$$Y_M(u,x) = Y_g((\Delta_k(x^k)^{-1}u)^1, x^k) = \sum_{j=0}^{\infty} Y_g(u[j]^1, x^k)x^{pk-p-jk},$$

and thus for $m \in \mathbb{Z}$, we have

$$u_m = \sum_{j=0}^{\infty} (u[j]^1)^g_{\frac{1}{k}((k-1)p-jk-k+m+1)}.$$

The weight of $(u[j]^1)^g_{\frac{1}{k}((k-1)p-jk-k+m+1)}$ is $\frac{1}{k}(p-m-1)$. Therefore for the weak $V$-module $M$, we have for $n \in \mathbb{N}$, that $u_m M(n) = u_m T^k_g(M)(n/k) \subset T^k_g(M)(\frac{1}{k}(p - m - 1 + n)) = M(p-m-1+n)$, finishing the proof of (2).

In order to prove (3) we write $Y_g(\bar{\omega}, x) = \sum_{n\in\mathbb{Z}} L^g(n)x^{-n-2}$ where $\bar{\omega} = \sum_{j=1}^k \omega^j$. We have

$$Y_g(\bar{\omega}, x) = \sum_{j=0}^{k-1} \lim_{x^{1/k}\mapsto\eta^{-j}x^{1/k}} Y_g(\omega^1, x).$$

It follows from (4.6) that

$$L^g(0) = \frac{1}{k}L(0) + \frac{(k^2-1)c}{24k}, \tag{5.12}$$

immediately implying (3). □

Let $V$ be an arbitrary vertex operator superalgebra and $g$ an automorphism of $V$ of finite order. We denote the categories of weak, weak admissible and ordinary generalized $g$-twisted $V$-modules by $\mathcal{C}^g_w(V)$, $\mathcal{C}^g_a(V)$ and $\mathcal{C}^g(V)$, respectively. If $g = 1$, we habitually remove the index $g$.

Now again consider the vertex operator superalgebra $V^{\otimes k}$ and the $k$-cycle $g = (1\ 2\ \cdots\ k)$ for $k$ odd. Define

$$\begin{array}{rcl} T^k_g : \mathcal{C}_w(V) & \longrightarrow & \mathcal{C}^g_w(V^{\otimes k}) \\ (M,Y) & \mapsto & (T^k_g(M), Y_g) = (M, Y_g) \\ f & \mapsto & T^k_g(f) = f \end{array}$$

for $(M, Y_M)$ an object and $f$ a morphism in $\mathcal{C}_w(V)$.

The following corollary to Theorem 5.7 follows immediately.

COROLLARY 5.8. *If $k$ is odd, then $T_g^k$ is a functor from the category $\mathcal{C}_w(V)$ of weak $V$-modules to the category $\mathcal{C}_w^g(V^{\otimes k})$ of weak $g = (1\ 2\ \cdots\ k)$-twisted $V^{\otimes k}$-modules such that: (1) $T_g^k$ preserves irreducible objects; (2) The restrictions of $T_g^k$ to $\mathcal{C}_a(V)$ and $\mathcal{C}(V)$ are functors from $\mathcal{C}_a(V)$ and $\mathcal{C}(V)$ to $\mathcal{C}_a^g(V^{\otimes k})$ and $\mathcal{C}^g(V^{\otimes k})$, respectively.*

In the next section we will construct a functor $U_g^k$, in the case when $k$ is odd, from the category $\mathcal{C}_w^g(V^{\otimes k})$ to the category $\mathcal{C}_w(V)$ such that $U_g^k \circ T_g^k = id_{\mathcal{C}_w(V)}$ and $T_g^k \circ U_g^k = id_{\mathcal{C}_w^g(V^{\otimes k})}$.

## 6. Constructing a weak $V$-module structure on a weak $g = (1\,2\ \cdots\ k)$-twisted $V^{\otimes k}$-module for $k$ odd

For $k \in \mathbb{Z}_+$ and $g = (1\ 2\ \cdots\ k)$, let $M = (M, Y_g)$ be a weak $g$-twisted $V^{\otimes k}$-module. Motivated by the construction of weak $g$-twisted $V^{\otimes k}$-modules from weak $V$-modules in Section 5, we consider

$$Y_g((\Delta_k(x^k)^{-1}u)^1, x^k) \tag{6.1}$$

for $u \in V$ where $\Delta_k(x)^{-1} = \Delta_k^V(x)^{-1}$ is given by (5.10). Note that (6.1) is multi-valued since $Y_g((\Delta_k(x)^{-1}u)^1, x) \in (\mathrm{End}\, M)[[x^{1/2k}, x^{-1/2k}]]$. Thus we define

$$Y_M(u, x) = Y_g((\Delta_k(x^k)^{-1}u)^1, x^k) \tag{6.2}$$

to be the unique formal Laurent series in $(\mathrm{End}\, M)[[x^{1/2}, x^{-1/2}]]$ given by taking $(x^k)^{1/k} = x$. Note that if $k$ is odd, then $Y_M(u, x) \in (\mathrm{End}\, M)[[x, x^{-1}]]$.

Our goal in this section is to construct a functor $U_g^k : \mathcal{C}_w^g(V^{\otimes k}) \to \mathcal{C}_w(V)$ with $U_g^k(M_g, Y_g) = (U_g^k(M_g), Y_M) = (M_g, Y_M)$ for the case when $k$ is odd. If we instead define $Y_M$ by taking $(x^k)^{1/k} = \eta^j x$ for $\eta$ a fixed primitive $k$-th root of unity for $j = 1, \ldots, k-1$, then $(M_g, Y_M)$ will not be a weak $V$-module. Further note that this implies that if we allow $x = z$ to be a complex number and if we define $z^{1/k}$ using the principal branch of the logarithm, then much of our work in this section is valid if and only if $-\pi/k < \arg z < \pi/k$.

LEMMA 6.1. *For $u \in V$, we have*

$$\begin{aligned} Y_M(L(-1)u, x) &= \left(\frac{d}{dx}((x^k)^{1/k})\right) \frac{d}{dx} Y_M(u, x) \\ &= \frac{d}{dx} Y_M(u, x) \end{aligned}$$

*on $M_g$. Thus the $L(-1)$-derivative property holds for $Y_M$ on $M_g$.*

PROOF. The proof is similar to that of Lemma 4.2, and follows from Corollary 3.6 via the analogous proof of Lemma 4.1 in [**BDM**]. □

LEMMA 6.2. *Let $u, v \in V$. Then on $M_g$, we have the supercommutator*

$$\begin{aligned} &[Y_M(u, x_1), Y_M(v, x_2)] \\ &\quad = \mathrm{Res}_{x_0} x_2^{-1} \delta\left(\frac{x_1 - x_0}{x_2}\right) \left(\frac{x_1 - x_0}{x_2}\right)^{\frac{1}{2}(1-k)|u|} Y_M(Y(u, x_0)v, x_2), \end{aligned} \tag{6.3}$$

*i.e.*,

$$(6.4)\quad [Y_M(u,x_1),Y_M(v,x_2)] = \begin{cases} \mathrm{Res}_{x_0} x_2^{-1}\delta\left(\frac{x_1-x_0}{x_2}\right)\left(\frac{x_1-x_0}{x_2}\right)^{\frac{|u|}{2}} Y_M(Y(u,x_0)v,x_2) & \text{if } k \text{ is even} \\ \mathrm{Res}_{x_0} x_2^{-1}\delta\left(\frac{x_1-x_0}{x_2}\right) Y_M(Y(u,x_0)v,x_2) & \text{if } k \text{ is odd.} \end{cases}$$

*Therefore, the operators $Y_M$ can satisfy the Jacobi identity only if $k$ is odd.*

Proof. The proof is similar to the proof of Lemma 4.3 and is analogous to the proof of Lemma 4.2 in [**BDM**]. From the twisted Jacobi identity on $(M_g,Y_g)$, we have

$$(6.5)\quad [Y_g(u^1,x_1),Y_g(v^1,x_2)] = \mathrm{Res}_{x_0}\frac{1}{k}x_2^{-1}\delta\left(\frac{(x_1-x_0)^{1/k}}{x_2^{1/k}}\right)Y_g(Y(u^1,x_0)v^1,x_2).$$

Therefore,

$$\begin{aligned} &[Y_M(u,x_1),Y_M(v,x_2)] \\ &= [Y_g((\Delta_k(x_1^k)^{-1}u)^1,x_1^k),Y_g((\Delta_k(x_2^k)^{-1}v)^1,x_2^k)] \\ &= \mathrm{Res}_x\frac{1}{k}x_2^{-k}\delta\left(\frac{(x_1^k-x)^{1/k}}{x_2}\right)Y_g(Y((\Delta_k(x_1^k)^{-1}u)^1,x)(\Delta_k(x_2^k)^{-1}v)^1,x_2^k). \end{aligned}$$

We want to make the change of variable $x = x_1^k-(x_1-x_0)^k$ where we choose $x_0$ such that $((x_1-x_0)^k)^{1/k} = x_1-x_0$. Then noting that $(x_1^k-x)^{n/k}|_{x=x_1^k-(x_1-x_0)^k} = (x_1-x_0)^n$ for all $n\in\mathbb{Z}$, and using (4.14), we have

$$\begin{aligned} &[Y_M(u,x_1),Y_M(v,x_2)] \\ &= \mathrm{Res}_{x_0}x_2^{-k}(x_1-x_0)^{k-1}\delta\left(\frac{x_1-x_0}{x_2}\right)Y_g(Y((\Delta_k(x_1^k)^{-1}u)^1,x_1^k-(x_1-x_0)^k) \\ &\qquad (\Delta_k(x_2^k)^{-1}v)^1,x_2^k) \\ &= \mathrm{Res}_{x_0}x_2^{-1}\delta\left(\frac{x_1-x_0}{x_2}\right)Y_g(Y((\Delta_k(x_1^k)^{-1}u)^1,x_1^k-(x_1-x_0)^k) \\ &\qquad (\Delta_k(x_2^k)^{-1}v)^1,x_2^k) \\ &= \mathrm{Res}_{x_0}x_1^{-1}\delta\left(\frac{x_2+x_0}{x_1}\right)Y_g(Y((\Delta_k(x_1^k)^{-1}u)^1,x_1^k-(x_1-x_0)^k) \\ &\qquad (\Delta_k(x_2^k)^{-1}v)^1,x_2^k) \\ &= \mathrm{Res}_{x_0}x_1^{-1}\delta\left(\frac{x_2+x_0}{x_1}\right)\left(\frac{x_2+x_0}{x_1}\right)^{\frac{1}{2}(1-k)|u|}Y_g((Y(\Delta_k((x_2+x_0)^k)^{-1}u, \\ &\qquad (x_2+x_0)^k-x_2^k)\Delta_k(x_2^k)^{-1}v)^1,x_2^k) \\ &= \mathrm{Res}_{x_0}x_2^{-1}\delta\left(\frac{x_1-x_0}{x_2}\right)\left(\frac{x_1-x_0}{x_2}\right)^{\frac{1}{2}(k-1)|u|}Y_g((Y(\Delta_k((x_2+x_0)^k)^{-1}u, \\ &\qquad (x_2+x_0)^k-x_2^k)\Delta_k(x_2^k)^{-1}v)^1,x_2^k). \end{aligned}$$

Thus the proof is reduced to proving

$$Y(\Delta_k((x_2+x_0)^k)^{-1}u,(x_2+x_0)^k-x_2^k)\Delta_k(x_2^k)^{-1} = \Delta_k(x_2^k)^{-1}Y\left(u,x_0\right),$$

i.e., proving

$$\Delta_k(x_2^k)Y(\Delta_k((x_2+x_0)^k)^{-1}u, (x_2+x_0)^k - x_2^k)\Delta_k(x_2^k)^{-1} = Y(u, x_0)\,. \tag{6.6}$$

In Proposition 3.2, substituting $u$, $z$ and $z_0$ by $\Delta_k((x_2+x_0)^k)^{-1}u$, $x_2^k$ and $(x_2+x_0)^k - x_2^k$, respectively, gives Equation (6.6). □

Let $(M_g, Y_g)$ be a weak $g$-twisted $V$-module, for $k$ a positive odd integer, and $g = (1\ 2\ \cdots\ k)$. Define $U_g^k(M_g, Y_g) = (U_g^k(M_g), Y_M) = (M_g, Y_M)$. That is $U_g^k(M_g, Y_g)$ is $M_g$ as the underlying vector space and the untwisted vertex operator $Y_M$ on the space $M = M_g$ is given by (6.2).

THEOREM 6.3. *Given a weak $g = (1\ 2\ \cdots\ k)$-twisted $V$-modules $(M_g, Y_g)$, with the notations as above, for $k$ odd, $U_g^k(M_g, Y_g) = (U_g^k(M_g), Y_M) = (M_g, Y_M)$ is a weak $V$-module.*

PROOF. Since the $L(-1)$-derivative property has been shown to hold for $Y_M$ in Lemma 6.1, we only need to prove the Jacobi identity. The Jacobi identity is equivalent to the supercommutator formula given by Lemma 6.2, where we are making the crucial restriction to the case when $k$ is odd, and the associator formula. The associator formula states that for $u, v \in V$ and $w \in U_g^k(M_g)$ there exists a positive integer $n$ such that

$$(x_0+x_2)^n Y_M(u, x_0+x_2)Y_M(v, x_2)w = (x_2+x_0)^n Y_M(Y(u, x_0)v, x_2)w.$$

Write $u^1 = \sum_{j=0}^{k-1} u^1_{(j)}$ where $gu^1_{(j)} = \eta^j u^1_{(j)}$. Then from the twisted Jacobi identity, we have the following associator: There exists a positive integer $m$ such that for $n \geq m$, and $j = 0, \ldots, k-1$,

$$(x_0+x_2)^{j/k+n}Y_g(u^1_{(j)}, x_0+x_2)Y_g(v^1, x_2)w = (x_2+x_0)^{j/k+n}Y_g(Y(u^1_{(j)}, x_0)v^1, x_2)w.$$

Note that $\Delta_k(x)^{-1}u \in (x^{\frac{1}{2k}})^{(k-1)2\mathrm{wt}\, u}V[x^{-1}]$. Thus for $k$ odd, we have that $\Delta_k(x^k)^{-1}u \in V[x, x^{-1}]$. From this fact, the rest of the proof is analogous to the proof of Theorem 4.3 in [**BDM**] where we use Proposition 3.2 above instead of Proposition 2.2 in [**BDM**]. □

THEOREM 6.4. *For $k$ an odd positive integer, and $g = (1\ 2\ \cdots\ k)$, the map $U_g^k$ is a functor from the category $\mathcal{C}_w^g(V^{\otimes k})$ of weak $g$-twisted $V^{\otimes k}$-modules to the category $\mathcal{C}_w(V)$ of weak $V$-modules such that $T_g^k \circ U_g^k = id_{\mathcal{C}_w^g(V^{\otimes k})}$ and $U_g^k \circ T_g^k = id_{\mathcal{C}_w(V)}$. In particular, the categories $\mathcal{C}_w^g(V^{\otimes k})$ and $\mathcal{C}_w(V)$ are isomorphic. Moreover,*

*(1) The restrictions of $T_g^k$ and $U_g^k$ to the category of admissible $V$-modules $\mathcal{C}_a(V)$ and to the category of admissible $g$-twisted $V^{\otimes k}$-modules $\mathcal{C}_a^g(V^{\otimes k})$, respectively, give category isomorphisms. In particular, $V^{\otimes k}$ is $g$-rational if and only if $V$ is rational.*

*(2) The restrictions of $T_g^k$ and $U_g^k$ to the category of ordinary $V$-modules $\mathcal{C}(V)$ and to the category of ordinary $g$-twisted $V^{\otimes k}$-modules $\mathcal{C}^g(V^{\otimes k})$, respectively, give category isomorphisms.*

PROOF. It is trivial to verify $T_g^k \circ U_g^k = id_{\mathcal{C}_w^g(V^{\otimes k})}$ and $U_g^k \circ T_g^k = id_{\mathcal{C}_w(V)}$ from the definitions of the functors $T_g^k$ and $U_g^k$. Parts (1) and (2) follow from Theorem 5.7. □

Using the functor $T_g^k$ giving the isomorphism between the categories $\mathcal{C}(V)$ and $\mathcal{C}(V^{\otimes k})$ as well as the actual construction of $g$-twisted $V^{\otimes k}$-modules from

$V$-modules, we have a correspondence between graded traces of modules in $\mathcal{C}(V)$ and modules in $\mathcal{C}(V^{\otimes k})$. In particular, from (5.12), we have the following corollary.

COROLLARY 6.5. *Let $g = (1\ 2\ \cdots\ k)$ for $k$ odd. Then $(M, Y_M)$ is an ordinary $V$-module with graded dimension*

$$\dim_q M = tr_M q^{-c/24 + L(0)} = q^{-c/24} \sum_{\lambda \in \mathbb{C}} \dim(M_\lambda) q^\lambda$$

*if and only if $(T_g^k(M), Y_g)$ is an ordinary $(1\ 2\ \cdots\ k)$-twisted $V^{\otimes k}$-module with graded dimension*

$$\begin{aligned} \dim_q T_g^k(M) &= tr_{T_g^k(M)} q^{-kc/24 + L^g(0)} = tr_{T_g^k(M)} q^{-kc/24 + \frac{1}{k}L(0) + (k^2-1)c/24k} \\ &= \dim_{q^{1/k}} M. \end{aligned}$$

## References


[Ban1] P. Bántay, *Algebraic aspects of orbifold models*, Internat. J. Modern Phys. A **9** (1994), no. 9, 1443–1456, DOI 10.1142/S0217751X94000649. MR1270363 (95g:81169)

[Ban2] P. Bantay, *Characters and modular properties of permutation orbifolds*, Phys. Lett. B **419** (1998), no. 1-4, 175–178, DOI 10.1016/S0370-2693(97)01464-0. MR1620424 (98m:81165)

[Ban3] P. Bantay, *Permutation orbifolds and their applications*, Vertex operator algebras in mathematics and physics (Toronto, ON, 2000), Fields Inst. Commun., vol. 39, Amer. Math. Soc., Providence, RI, 2003, pp. 13–23. MR2029789 (2005b:81169)

[Bar1] K. Barron, *A supergeometric interpretation of vertex operator superalgebras*, Internat. Math. Res. Notices **9** (1996), 409–430, DOI 10.1155/S107379289600027X. MR1399408 (98a:17043)

[Bar2] K. Barron, *$N = 1$ Neveu-Schwarz vertex operator superalgebras over Grassmann algebras and with odd formal variables*, Representations and quantizations (Shanghai, 1998), China High. Educ. Press, Beijing, 2000, pp. 9–35. MR1802167 (2001k:17041)

[Bar3] K. Barron, *The moduli space of $N = 1$ superspheres with tubes and the sewing operation*, Mem. Amer. Math. Soc. **162** (2003), no. 772, vi+135, DOI 10.1090/memo/0772. MR1954484 (2004b:32010)

[Bar4] K. Barron, *The notion of $N = 1$ supergeometric vertex operator superalgebra and the isomorphism theorem*, Commun. Contemp. Math. **5** (2003), no. 4, 481–567, DOI 10.1142/S0219199703001051. MR2003210 (2004j:17034)

[Bar5] K. Barron, *Superconformal change of variables for $N = 1$ Neveu-Schwarz vertex operator superalgebras*, J. Algebra **277** (2004), no. 2, 717–764, DOI 10.1016/j.jalgebra.2004.02.031. MR2067629 (2005c:17043)

[Bar6] K. Barron, *Alternate notions of $N = 1$ superconformality and deformations of $N = 1$ vertex superalgebras*, Vertex operator algebras and related areas, Contemp. Math., vol. 497, Amer. Math. Soc., Providence, RI, 2009, pp. 33–51, DOI 10.1090/conm/497/09767. MR2568397 (2011d:17046)

[Bar7] K. Barron, *Axiomatic aspects of $N = 2$ vertex superalgebras with odd formal variables*, Comm. Algebra **38** (2010), no. 4, 1199–1268, DOI 10.1080/00927870902828900. MR2656574 (2011i:17040)

[Bar8] K. Barron, *On twisted modules for* N = 2 *supersymmetric vertex operator superalgebras*, Lie theory and its applications in physics, Springer Proc. Math. Stat., vol. 36, Springer, Tokyo, 2013, pp. 411–420, DOI 10.1007/978-4-431-54270-4_29. MR3070668

[Bar9] K. Barron, Twisted modules for N=2 supersymmetric vertex operator superalgebras arising from finite automorphisms of the N=2 Neveu-Schwarz algebra, arXiv:1110.0229v3.

[Bar10] K. Barron, On the correspondence between mirror-twisted sectors for N=2 supersymmetric vertex operator superalgebras of the form $(V \otimes V)$ and N=1 Ramond sectors of $V$, in: "Proceedings of the Xth International Workshop on Lie Theory and Its Applications in Physics", June 2013, Varna, Bulgaria. ed. V. Dobrev. Springer Proceedings in Mathematics & Statistics, Vol. 111, 2014.

[BDM] K. Barron, C. Dong, and G. Mason, *Twisted sectors for tensor product vertex operator algebras associated to permutation groups*, Comm. Math. Phys. **227** (2002), no. 2, 349–384, DOI 10.1007/s002200200633. MR1903649 (2003f:17034)

[BHL] K. Barron, Y.-Z. Huang, and J. Lepowsky, *An equivalence of two constructions of permutation-twisted modules for lattice vertex operator algebras*, J. Pure Appl. Algebra **210** (2007), no. 3, 797–826, DOI 10.1016/j.jpaa.2006.12.005. MR2324608 (2008f:17043)

[BV1] K. Barron and N. Vander Werf, On permutation-twisted free fermion vertex operator superalgebras and two conjectures, in: "Proceedings of the XXIst International Conference on Integrable Systems and Quantum Symmetries", June 2013, Prague, Czech Republic; ed. C. Burdik, O. Navratil and S. Posta; Jour. of Physics: Conference Series, **474** (2013), 012009; 35 pages.

[BV2] K. Barron and N. Vander Werf, *Permutation-twisted modules for even order cycles acting on tensor product vertex operator superalgebras*, Internat. J. Math. **25** (2014), no. 2, 1450018, 35, DOI 10.1142/S0129167X14500189. MR3189774

[Bo] R. E. Borcherds, *Vertex algebras, Kac-Moody algebras, and the Monster*, Proc. Nat. Acad. Sci. U.S.A. **83** (1986), no. 10, 3068–3071, DOI 10.1073/pnas.83.10.3068. MR843307 (87m:17033)

[BHS] L. Borisov, M. B. Halpern, and C. Schweigert, *Systematic approach to cyclic orbifolds*, Internat. J. Modern Phys. A **13** (1998), no. 1, 125–168, DOI 10.1142/S0217751X98000044. MR1606821 (99i:81068)

[dBHO] J. de Boer, M. Halpern and N. Obers, The operator algebra and twisted KZ equations of WZW orbifolds, *J. High Energy Phys.* **10** (2001), 71 pages.

[DVVV] R. Dijkgraaf, C. Vafa, E. Verlinde, and H. Verlinde, *The operator algebra of orbifold models*, Comm. Math. Phys. **123** (1989), no. 3, 485–526. MR1003430 (91c:81132)

[DFMS] L. Dixon, D. Friedan, E. Martinec, and S. Shenker, *The conformal field theory of orbifolds*, Nuclear Phys. B **282** (1987), no. 1, 13–73, DOI 10.1016/0550-3213(87)90676-6. MR869937 (88f:81120)

[DHVW1] L. Dixon, J. A. Harvey, C. Vafa, and E. Witten, *Strings on orbifolds*, Nuclear Phys. B **261** (1985), no. 4, 678–686, DOI 10.1016/0550-3213(85)90593-0. MR818423 (87k:81104a)

[DHVW2] L. Dixon, J. Harvey, C. Vafa, and E. Witten, *Strings on orbifolds. II*, Nuclear Phys. B **274** (1986), no. 2, 285–314, DOI 10.1016/0550-3213(86)90287-7. MR851703 (87k:81104b)

[DGM] L. Dolan, P. Goddard, and P. Montague, *Conformal field theory of twisted vertex operators*, Nuclear Phys. B **338** (1990), no. 3, 529–601, DOI 10.1016/0550-3213(90)90644-S. MR1063589 (92a:17032)

[D] C. Dong, *Twisted modules for vertex algebras associated with even lattices*, J. Algebra **165** (1994), no. 1, 91–112, DOI 10.1006/jabr.1994.1099. MR1272580 (95i:17032)

[DH] C. Dong and J. Han, *Some finite properties for vertex operator superalgebras*, Pacific J. Math. **258** (2012), no. 2, 269–290, DOI 10.2140/pjm.2012.258.269. MR2981954

[DL1] C. Dong and J. Lepowsky, *Generalized vertex algebras and relative vertex operators*, Progress in Mathematics, vol. 112, Birkhäuser Boston, Inc., Boston, MA, 1993. MR1233387 (95b:17032)

[DL2] C. Dong and J. Lepowsky, *The algebraic structure of relative twisted vertex operators*, J. Pure Appl. Algebra **110** (1996), no. 3, 259–295, DOI 10.1016/0022-4049(95)00095-X. MR1393116 (98e:17036)

[DZ1] C. Dong and Z. Zhao, *Modularity in orbifold theory for vertex operator superalgebras*, Comm. Math. Phys. **260** (2005), no. 1, 227–256, DOI 10.1007/s00220-005-1418-2. MR2175996 (2007a:17041)

[DZ2] C. Dong and Z. Zhao, *Twisted representations of vertex operator superalgebras*, Commun. Contemp. Math. **8** (2006), no. 1, 101–121, DOI 10.1142/S0219199706002040. MR2208812 (2007b:17038)

[FFR] A. J. Feingold, I. B. Frenkel, and J. F. X. Ries, *Spinor construction of vertex operator algebras, triality, and* $E_8^{(1)}$, Contemporary Mathematics, vol. 121, American Mathematical Society, Providence, RI, 1991. MR1123265 (92k:17041)

[FHL] I. B. Frenkel, Y.-Z. Huang, and J. Lepowsky, *On axiomatic approaches to vertex operator algebras and modules*, Mem. Amer. Math. Soc. **104** (1993), no. 494, viii+64, DOI 10.1090/memo/0494. MR1142494 (94a:17007)

[FLM1] I. B. Frenkel, J. Lepowsky, and A. Meurman, *A natural representation of the Fischer-Griess Monster with the modular function J as character*, Proc. Nat. Acad. Sci. U.S.A. **81** (1984), no. 10, Phys. Sci., 3256–3260, DOI 10.1073/pnas.81.10.3256. MR747596 (85e:20018)

[FLM2] I. B. Frenkel, J. Lepowsky, and A. Meurman, *Vertex operator calculus*, Mathematical aspects of string theory (San Diego, Calif., 1986), Adv. Ser. Math. Phys., vol. 1, World Sci. Publishing, Singapore, 1987, pp. 150–188. MR915822

[FLM3] I. Frenkel, J. Lepowsky, and A. Meurman, *Vertex operator algebras and the Monster*, Pure and Applied Mathematics, vol. 134, Academic Press, Inc., Boston, MA, 1988. MR996026 (90h:17026)

[FKS] J. Fuchs, A. Klemm, and M. G. Schmidt, *Orbifolds by cyclic permutations in Gepner type superstrings and in the corresponding Calabi-Yau manifolds*, Ann. Physics **214** (1992), no. 2, 221–257, DOI 10.1016/S0003-4916(05)80001-6. MR1154598 (93b:81247)

[GHHO] O. Ganor, M. B. Halpern, C. Helfgott, and N. A. Obers, *The outer-automorphic WZW orbifolds on* $\mathfrak{so}(2n)$*, including five triality orbifolds on* $\mathfrak{so}(8)$, J. High Energy Phys. **12** (2002), 019, 45, DOI 10.1088/1126-6708/2002/12/019. MR1960316 (2004f:81199)

[HH] M. B. Halpern and C. Helfgott, *The general twisted open WZW string*, Internat. J. Modern Phys. A **20** (2005), no. 5, 923–992, DOI 10.1142/S0217751X05020628. MR2123426 (2006b:81273)

[HO] M. B. Halpern and N. A. Obers, *Two large examples in orbifold theory: abelian orbifolds and the charge conjugation orbifold on* $\mathfrak{su}(n)$, Internat. J. Modern Phys. A **17** (2002), no. 27, 3897–3961, DOI 10.1142/S0217751X02010832. MR1932323 (2004a:81112)

[H] Y.-Z. Huang, *Two-dimensional conformal geometry and vertex operator algebras*, Progress in Mathematics, vol. 148, Birkhäuser Boston, Inc., Boston, MA, 1997. MR1448404 (98i:17037)

[KS] A. Klemm and M. G. Schmidt, *Orbifolds by cyclic permutations of tensor product conformal field theories*, Phys. Lett. B **245** (1990), no. 1, 53–58, DOI 10.1016/0370-2693(90)90164-2. MR1068695 (91m:81254)

[Le1] J. Lepowsky, *Calculus of twisted vertex operators*, Proc. Nat. Acad. Sci. U.S.A. **82** (1985), no. 24, 8295–8299, DOI 10.1073/pnas.82.24.8295. MR820716 (88f:17030)

[Le2] J. Lepowsky, *Perspectives on vertex operators and the Monster*, The mathematical heritage of Hermann Weyl (Durham, NC, 1987), Proc. Sympos. Pure Math., vol. 48, Amer. Math. Soc., Providence, RI, 1988, pp. 181–197, DOI 10.1090/pspum/048/974335. MR974335 (90f:17031)

[LL] J. Lepowsky and H. Li, *Introduction to vertex operator algebras and their representations*, Progress in Mathematics, vol. 227, Birkhäuser Boston, Inc., Boston, MA, 2004. MR2023933 (2004k:17050)

[LW] J. Lepowsky and R. L. Wilson, *Construction of the affine Lie algebra* $A_1(1)$, Comm. Math. Phys. **62** (1978), no. 1, 43–53. MR0573075 (58 #28089)

[Li] H.-S. Li, *Local systems of twisted vertex operators, vertex operator superalgebras and twisted modules*, Moonshine, the Monster, and related topics (South Hadley, MA, 1994), Contemp. Math., vol. 193, Amer. Math. Soc., Providence, RI, 1996, pp. 203–236, DOI 10.1090/conm/193/02373. MR1372724 (96m:17050)

[MS1] M. Maio and A. N. Schellekens, *Permutation orbifolds of* $N = 2$ *supersymmetric minimal models*, Nuclear Phys. B **845** (2011), no. 2, 212–245, DOI 10.1016/j.nuclphysb.2010.12.005. MR2753566 (2012a:81249)

[MS2] M. Maio and A. N. Schellekens, *Permutation orbifolds of heterotic Gepner models*, Nuclear Phys. B **848** (2011), no. 3, 594–628, DOI 10.1016/j.nuclphysb.2011.03.012. MR2793326 (2012i:81243)

[X] X. Xu, *Introduction to vertex operator superalgebras and their modules*, Mathematics and its Applications, vol. 456, Kluwer Academic Publishers, Dordrecht, 1998. MR1656671 (2000h:17019)



Department of Mathematics, University of Notre Dame
*E-mail address*: kbarron@nd.edu

Proceedings of Symposia in Pure Mathematics
Volume **92**, 2016
http://dx.doi.org/10.1090/pspum/092/01579

# Third cohomology for Frobenius kernels and related structures

Christopher P. Bendel, Daniel K. Nakano, and Cornelius Pillen

Abstract. Let $G$ be a simple simply connected group scheme defined over $\mathbb{F}_p$ and $k$ be an algebraically closed field of characteristic $p > 0$. Moreover, let $B$ be a Borel subgroup of $G$ and $U$ be the unipotent radical of $B$. In this paper the authors compute the third cohomology group for $B$ and its Frobenius kernels, $B_r$, with coefficients in a one-dimensional representation. These computations hold with relatively mild restrictions on the characteristic of the field. As a consequence of our calculations, the third ordinary Lie algebra cohomology group for $\mathfrak{u} = \text{Lie } U$ with coefficients in $k$ is determined, as well as the third $G_r$-cohomology with coefficients in the induced modules $H^0(\lambda)$.

## 1. Introduction

**1.1.** Over the last 40 years, one of the central questions in the representation theory of algebraic groups that remains open is to understand the structure and vanishing behavior of the line bundle cohomology $H^n(\lambda) := \mathcal{H}^n(G/B, \mathcal{L}(\lambda))$ for $n \geq 0$ where $G/B$ is the flag variety and $\mathcal{L}(\lambda)$ is a line bundle over $G/B$. For $n = 0$, the characters are given by Weyl's character formula, but the composition factors are not known when $k = \overline{\mathbb{F}}_p$. Andersen computed the socle of $H^1(\lambda)$, but more general results other than calculations for low rank examples remain elusive. A fundamental computation which is related to understanding line bundle cohomology is the calculation of rational $B$-cohomology, $\mathrm{H}^\bullet(B, \lambda)$, where $B$ is a Borel subgroup for the reductive group $G$ and $\lambda \in X(T)$ is a weight regarded as a one-dimensional representation of $B$.

In [**BNP2**] the authors calculated $\mathrm{H}^2(B, \lambda)$ for $p \geq 3$. Our methods involved weaving other relevant $\mathrm{H}^2$-calculations for Frobenius kernels and Lie algebras into the picture. These fundamental cohomology groups and the status of their computation are presented in the following table.[1] The references are [**Kos**], [**FP2**], [**PT**], [**AJ**], [**KLT**], [**Jan2**], [**BNP1,BNP2**], [**AR**], [**W**], [**UGA**].

---

2010 *Mathematics Subject Classification.* Primary 17B50, 17B56, 20G05, 20G10.

Research of the first author was supported in part by a Simons Foundation Collaboration Grant #317062.

Research of the second author was supported in part by NSF grant DMS-1402271.

Research of the third author was supported in part by a Simons Foundation Collaboration Grant #245236.

[1] See the following section for the notation.

| | Cohomology Group | Known Results |
|---|---|---|
| (1) | $\mathrm{H}^n(\mathfrak{u}, k)$ | $[p \geq h-1,\, n \geq 0]$; $[p \geq 2,\, n = 0, 1]$; $[p \geq 3,\, n = 2]$ |
| (2) | $\mathrm{H}^n(U_1, k)$ | $[p \geq h,\, n \geq 0]$; $[p \geq 2,\, n = 0, 1, 2]$ |
| (3) | $\mathrm{H}^n(B_1, \lambda)$ | $[p \geq h,\, n \geq 0]$; $[p \geq 2,\, n = 0, 1, 2]$ |
| (4) | $\mathrm{H}^n(B_r, \lambda)$ | $[p \geq 2,\, n = 0, 1, 2]$ |
| (5) | $\mathrm{H}^n(B, \lambda)$ | $[p \geq 2,\, n = 0, 1, 2]$; $[p > h,\, n = 3]$ |
| (6) | $\mathrm{H}^n(G_r, H^0(\lambda))$ | $[p > h,\, r = 1,\, n \geq 0]$; $[p \geq 2,\, r \geq 1,\, n = 0, 1, 2]$ |

Recently, for $SL_2$, Ngo [**N**] has computed (4) and (6) for all primes and all $n$.

The calculation of the aforementioned cohomology groups is formidable and of general interest. For example, the complete calculation of (1) and (2) would yield a general version (for *all* characteristics) of the celebrated theorem of Kostant [**Kos**]. For (6) (when $\lambda = 0$), the computation of $\mathrm{H}^\bullet(G_r, k)$ for $r \geq 2$ presents a major challenge and is geometrically related to the variety of commuting $r$-tuples of $p$-nilpotent matrices [**SFB1**, **SFB2**]. The goal in this paper is to expand the computations of (1)-(6) to the case when $n = 3$.

The paper is organized as follows. In Section 2 we provide a preliminary analysis on possible weights that can occur in the ordinary Lie algebra cohomology group $\mathrm{H}^3(\mathfrak{u}, k)$. At the beginning of Section 3, further results are provided on constraints involving root sums. These results are crucial throughout the paper in considering differentials in various spectral sequences. Later in this section, we provide a realization of $\mathrm{H}^3(U_1, k)$ via the ordinary cohomology groups $\mathrm{H}^j(\mathfrak{u}, k)$ for $j = 1, 3$.

At the beginning of Section 4, we introduce assumptions on the characteristic of the field that will be used during the remainder of the paper. Several key ideas from Andersen [**And**] involving the $B$-cohomology are then employed to give an explicit description of $\mathrm{H}^3(\mathfrak{u}, k)$. The section concludes by establishing another crucial calculation of ordinary Lie algebra cohomology, namely $\mathrm{H}^1(\mathfrak{u}, \mathfrak{u}^*)$.

In Section 5, with these prior computations and some arguments using spectral sequences, we determine $\mathrm{H}^3(B_r, \lambda)$ for $\lambda \in X(T)$. As a consequence of these results, the $B$-cohomology groups, $\mathrm{H}^3(B, \lambda)$, for $\lambda \in X(T)$ are determined which extends the prior work of Andersen and Rian [**AR**]. Finally, in Section 6, we give a description of $\mathrm{H}^3(G_r, H^0(\lambda))^{(-r)}$ and demonstrate that these cohomology groups (as $G$-modules) admit a good filtration. An application is provided at the end of the section, which uses the $B$-cohomology calculations to give linear bounds on the third cohomology of finite Chevalley groups. The Appendix at the end of the paper contains the computations of certain weights that appear in the calculations of $B_1$-cohomology groups.

**1.2. Notation:** Throughout this paper, we will follow the basic conventions provided in [**Jan1**].

(1) $k$: an algebraically closed field of characteristic $p > 0$.
(2) $G$: a simple, simply connected algebraic group which is defined and split over the finite prime field $\mathbb{F}_p$ of characteristic $p$. The assumption that $G$ is simple (equivalently, its root system $\Phi$ is irreducible) is largely one of convenience. All the results of this paper extend easily to the semisimple, simply connected case.

(3) $F : G \to G$: the Frobenius morphism.
(4) $G_r = \ker F^r$: the $r$th Frobenius kernel of $G$.
(5) $G^{(r)}$: the $r$th Frobenius twist of $G$; $G^{(r)} \cong G/G_r$.
(6) $G(\mathbb{F}_q)$: the associated finite Chevalley group where $\mathbb{F}_q$ is the field with $q = p^r$ elements.
(7) $T$: a maximal split torus in $G$.
(8) $\Phi$: the corresponding (irreducible) root system associated to $(G, T)$. When referring to short and long roots, when a root system has roots of only one length, all roots shall be considered as both short and long.
(9) $\Pi = \{\alpha_1, \ldots, \alpha_n\}$: the set of simple roots. We will adhere to the ordering of the simple roots as given in [**Jan2**] (following Bourbaki). In particular, for type $B_n$, $\alpha_n$ denotes the unique short simple root and for type $C_n$, $\alpha_n$ denotes the unique long simple root.
(10) $\Phi^{\pm}$: the positive (respectively, negative) roots.
(11) $\alpha_0$: the maximal short root.
(12) $B$: a Borel subgroup containing $T$ corresponding to the negative roots.
(13) $U$: the unipotent radical of $B$.
(14) $\mathbb{E}$: the Euclidean space spanned by $\Phi$ with inner product $\langle\, ,\, \rangle$ normalized so that $\langle \alpha, \alpha \rangle = 2$ for $\alpha \in \Phi$ any short root.
(15) $\alpha^\vee = 2\alpha/\langle \alpha, \alpha \rangle$: the coroot of $\alpha \in \Phi$.
(16) $\rho$: the Weyl weight defined by $\rho = \frac{1}{2}\sum_{\alpha \in \Phi^+} \alpha$.
(17) $h$: the Coxeter number of $\Phi$, given by $h = \langle \rho, \alpha_0^\vee \rangle + 1$.
(18) $W = \langle s_{\alpha_1}, \ldots, s_{\alpha_n} \rangle \subset \mathbb{O}(\mathbb{E})$: the Weyl group of $\Phi$, generated by the orthogonal reflections $s_{\alpha_i}$, $1 \leq i \leq n$. For $\alpha \in \Phi$, $s_\alpha : \mathbb{E} \to \mathbb{E}$ is the orthogonal reflection in the hyperplane $H_\alpha \subset \mathbb{E}$ of vectors orthogonal to $\alpha$.
(19) $\ell : W \to \mathbb{Z}$: the usual length function on $W$; for $w \in w$, $\ell(w)$ is the minimum number of simple reflections required to express $w$ as a product of simple reflections.
(20) $X(T) = \mathbb{Z}\omega_1 \oplus \cdots \oplus \mathbb{Z}\omega_n$: the weight lattice, where the fundamental dominant weights $\omega_i \in \mathbb{E}$ are defined by $\langle \omega_i, \alpha_j^\vee \rangle = \delta_{ij}$, $1 \leq i, j \leq n$.
(21) $X(T)_+ = \mathbb{N}\omega_1 + \cdots + \mathbb{N}\omega_n$: the cone of dominant weights.
(22) $X_r(T) = \{\lambda \in X(T)_+ : 0 \leq \langle \lambda, \alpha^\vee \rangle < p^r, \forall \alpha \in \Pi\}$: the set of $p^r$-restricted dominant weights.
(23) $M^{(s)}$: the module obtained by composing the underlying representation for a rational $G$-module $M$ with $F^s$.
(24) $H^0(\lambda) := \operatorname{ind}_B^G \lambda$, $\lambda \in X(T)_+$: the induced module whose character is provided by Weyl's character formula.

## 2. Observations on $\mathfrak{u}$-cohomology

**2.1.** We begin by recalling the definition of the ordinary Lie algebra cohomology of a Lie algebra $L$ over $k$. The ordinary Lie algebra cohomology $\mathrm{H}^i(L, k)$ may be computed as the cohomology of the complex

$$k \xrightarrow{d_0} L^* \xrightarrow{d_1} \Lambda^2(L)^* \xrightarrow{d_2} \Lambda^3(L)^* \to \cdots .$$

The differentials are given as follows: $d_0 = 0$ and $d_1 : L^* \to \Lambda^2(L)^*$ with

$$(d_1\phi)(x \wedge y) = -\phi([x, y])$$

where $\phi \in L^*$ and $x, y \in L$. For the higher differentials, we identify $\Lambda^n(L)^* \cong \Lambda^n(L^*)$. Then the differentials are determined by the following product rule (see [**Jan1**, I.9.17]):

$$d_{i+j}(\phi \wedge \psi) = d_i(\phi) \wedge \psi + (-1)^i \phi \wedge d_j(\psi).$$

Using this formula, one can use induction to obtain the following formula.

LEMMA 2.1.1. *Let $n \geq 1$ be an integer. Consider $d_n : \Lambda^n(L^*) \to \Lambda^{n+1}(L^*)$. Let*

$$x = \bigwedge_{i=1}^{n} \phi_i \in \Lambda^n(L^*).$$

*Then*

$$d_n(x) = \sum_{j=1}^{n} (-1)^{j+1} \bigwedge_{i=1}^{n} \phi_i^j$$

*where* $\phi_i^j = \begin{cases} \phi_i & \text{if } i \neq j \\ d_1(\phi_i) & \text{if } i = j. \end{cases}$

*In other words,*

$$\begin{aligned} d_n(x) = d_1(\phi_1) \wedge \phi_2 \wedge \phi_3 \wedge \cdots \wedge \phi_n - \phi_1 \wedge d_1(\phi_2) \wedge \phi_3 \wedge \cdots \wedge \phi_n \\ + \cdots + (-1)^{n+1} \phi_1 \wedge \phi_2 \wedge \cdots \wedge \phi_{n-1} \wedge d_1(\phi_n). \end{aligned}$$

For the remainder of the paper we will be primarily interested in the ordinary Lie algebra cohomology of the Lie algebra $\mathfrak{u} = \text{Lie}(U)$ where $U$ is the unipotent radical of a Borel subgroup of $G$.

**2.2. First cohomology of $\mathfrak{u}$:** Now let us consider the case when $L = \mathfrak{u} = \text{Lie}\, U$. A basis for $\mathfrak{u}$ is given by a basis of negative root vectors $\{x_\alpha : \alpha \in \Phi^-\}$. Let $\{\phi_\alpha : \alpha \in \Phi^+\}$ be the dual basis in $\mathfrak{u}^*$ with $\phi_\alpha(x_\beta) = \delta_{-\alpha,\beta}$ for all $\alpha \in \Phi^+$ and $\beta \in \Phi^-$. It is well known that the first cohomology (for any Lie algebra) is

$$\text{H}^1(\mathfrak{u}, k) = \ker d_1 = (\mathfrak{u}/[\mathfrak{u}, \mathfrak{u}])^*.$$

For large primes, the simple roots give a basis for $\text{H}^1(\mathfrak{u}, k)$. Specifically, we recall the following result of Jantzen [**Jan2**].

PROPOSITION 2.2.1. *Assume $p \geq 3$.*

(a) *Assume further that $\Phi$ is not of type $G_2$ if $p = 3$. Then a $T$-basis for $\text{H}^1(\mathfrak{u}, k)$ is $\{\phi_\alpha : \alpha \in \Pi\}$.*
(b) *If $p=3$ and $\Phi$ is of type $G_2$, a $T$-basis for $\text{H}^1(\mathfrak{u}, k)$ is $\{\phi_{\alpha_1}, \phi_{\alpha_2}, \phi_{3\alpha_1+\alpha_2}\}$.*

**2.3. Second cohomology of $\mathfrak{u}$:** For $p \geq 3$, the second cohomology groups $\text{H}^2(\mathfrak{u}, k)$ were computed by the authors in [**BNP2**, Thm. 4.4]. We remind the reader of the results here.

THEOREM 2.3.1. *Let $p \geq 3$ and $\pi = \{-w \cdot 0 : w \in W, \ell(w) = 2\}$. As a $T$-module*

$$\text{H}^2(\mathfrak{u}, k) \cong \bigoplus_{\lambda \in \pi \cup \pi'} \lambda$$

*where $\pi'$ is given below. Moreover, if $\lambda = -w \cdot 0$ with $w = s_\alpha s_\beta$, then the corresponding cohomology class is represented by $\phi_\alpha \wedge \phi_{-\langle \beta, \alpha^\vee \rangle \alpha + \beta}$.*

(a) *$p > 3$ or $\Phi$ is of type $A_n$, $B_2 = C_2$, $D_n$, or $E_n$: $\pi' = \emptyset$.*

(b) $p = 3$ *and* $\Phi$ *is of type* $B_n$, $n \geq 3$: $\pi' = \{\alpha_{n-2} + 2\alpha_{n-1} + 3\alpha_n\}$ *corresponding to the cohomology class*

$$\phi_{\alpha_n} \wedge \phi_{\alpha_{n-2}+2\alpha_{n-1}+2\alpha_n} - \phi_{\alpha_{n-1}+\alpha_n} \wedge \phi_{\alpha_{n-2}+\alpha_{n-1}+2\alpha_n} + \phi_{\alpha_{n-2}+\alpha_{n-1}+\alpha_n} \wedge \phi_{\alpha_{n-1}+2\alpha_n}.$$

(c) $p = 3$ *and* $\Phi$ *is of type* $C_n$, $n \geq 3$: $\pi' = \{\alpha_{n-2}+3\alpha_{n-1}+\alpha_n\}$ *corresponding to the cohomology class*

$$\phi_{\alpha_{n-1}} \wedge \phi_{\alpha_{n-2}+2\alpha_{n-1}+\alpha_n} - \phi_{\alpha_{n-2}+\alpha_{n-1}} \wedge \phi_{2\alpha_{n-1}+\alpha_n}.$$

(d) $p = 3$ *and* $\Phi$ *is of type* $F_4$: $\pi' = \{\alpha_1 + 2\alpha_2 + 3\alpha_3, \alpha_2 + 3\alpha_3 + \alpha_4\}$ *corresponding to the cohomology classes*

$$\phi_{\alpha_3} \wedge \phi_{\alpha_1+2\alpha_2+2\alpha_3} - \phi_{\alpha_2+\alpha_3} \wedge \phi_{\alpha_1+\alpha_2+2\alpha_3} + \phi_{\alpha_1+\alpha_2+\alpha_3} \wedge \phi_{\alpha_2+2\alpha_3},$$

$$\phi_{\alpha_3} \wedge \phi_{\alpha_2+2\alpha_3+\alpha_4} - \phi_{\alpha_3+\alpha_4} \wedge \phi_{\alpha_2+2\alpha_3}.$$

(e) $p = 3$ *and* $\Phi$ *is of type* $G_2$: $\pi' = \{3\alpha_1+\alpha_2, 3\alpha_1+3\alpha_2, 6\alpha_1+3\alpha_2, 4\alpha_1+2\alpha_2\}$ *corresponding to the cohomology classes*

$$\phi_{\alpha_1} \wedge \phi_{2\alpha_1+\alpha_2}, \phi_{\alpha_2} \wedge \phi_{3\alpha_1+2\alpha_2}, \phi_{3\alpha_1+\alpha_2} \wedge \phi_{3\alpha_1+2\alpha_2},$$

$$\phi_{\alpha_1} \wedge \phi_{3\alpha_1+2\alpha_2} + \phi_{\alpha_1+\alpha_2} \wedge \phi_{3\alpha_1+\alpha_2}.$$

**2.4. General results:** We present a general observation about weight spaces of $\Lambda^n(\mathfrak{u}^*)$.

PROPOSITION 2.4.1 ([**FP2**, Prop. 2.2], [**BNP2**, Prop. 2.3]). *Let* $w \in W$. *Then*

$$\dim_k \Lambda^n(\mathfrak{u}^*)_{-w\cdot 0} = \begin{cases} 1 & \text{if } n = \ell(w) \\ 0 & \text{otherwise.} \end{cases}$$

*Let* $x_w \in \Lambda^{\ell(w)}(\mathfrak{u}^*)$ *be an element of weight* $-w\cdot 0$. *Then* $x_w$ *represents a cohomology class in* $\mathrm{H}^{\ell(w)}(\mathfrak{u}, k)$.

Over characteristic zero, $\mathrm{H}^\bullet(\mathfrak{u}, k)$ was computed by Kostant [**Kos**]. The cohomology classes in the preceding proposition in fact yield a $T$-basis. In prime characteristic, it is known for $p \geq h - 1$ by work in [**FP2**], [**PT**], and [**UGA**] that the formal characters of these cohomology groups are the same as in characteristic zero. Our goal will be to show that this holds for $\mathrm{H}^3(\mathfrak{u}, k)$ when the prime $p$ is not too small. See Theorem 4.3.1.

**2.5.** We first investigate the nature of weights $-w \cdot 0$. Let $w \in W$ have $\ell(w) = m$. Then $w$ can be expressed in reduced form as $w = s_1 s_2 \cdots s_m$ where $s_i = s_\beta$ for some $\beta \in \Pi$. Inductively, one obtains the following.

PROPOSITION 2.5.1. *Given* $w = s_1 s_2 \cdots s_m$, *let* $\beta_i$ *denote the simple root corresponding to the simple reflection* $s_i$. *Then*

$$\begin{aligned} -w \cdot 0 &= -s_1 s_2 \cdots s_m(\beta_m) - s_1 s_2 \cdots s_{m-1}(\beta_{m-1}) - \cdots - s_1 s_2(\beta_2) - s_1(\beta_1) \\ &= -s_1 s_2 \cdots s_{m-1}(-\beta_m) - s_1 s_2 \cdots s_{m-2}(-\beta_{m-1}) - \cdots - s_1(-\beta_2) - (-\beta_1) \\ &= s_1 s_2 \cdots s_{m-1}(\beta_m) + s_1 s_2 \cdots s_{m-2}(\beta_{m-1}) + \cdots + s_1(\beta_2) + \beta_1. \end{aligned}$$

*Furthermore, each of the summands lies in* $\Phi^+$.

In particular, suppose that $w \in W$ has length 3. Write $w = s_{\beta_1}s_{\beta_2}s_{\beta_3}$. Then

$$\begin{aligned}
-w \cdot 0 &= -s_{\beta_1}s_{\beta_2}s_{\beta_3} \cdot 0 \\
&= s_{\beta_1}s_{\beta_2}(\beta_3) + s_{\beta_1}(\beta_2) + \beta_1 \\
&= [\beta_3 - \langle \beta_3, \beta_2^\vee\rangle\beta_2 - (\langle \beta_3, \beta_1^\vee\rangle - \langle \beta_3, \beta_2^\vee\rangle\langle \beta_2, \beta_1^\vee\rangle)\beta_1] \\
&\quad + [\beta_2 - \langle \beta_2, \beta_1^\vee\rangle\beta_1] + [\beta_1]
\end{aligned}$$

where the roots in brackets are each positive roots.

One can now conclude the following.

COROLLARY 2.5.2. *Let $\beta_1, \beta_2, \beta_3 \in \Pi$ such that $w = s_{\beta_1}s_{\beta_2}s_{\beta_3}$ has length 3. Then $\phi_{\beta_1} \wedge \phi_{s_{\beta_1}(\beta_2)} \wedge \phi_{s_{\beta_1}s_{\beta_2}(\beta_3)}$ has weight $-w \cdot 0$ and represents a cohomology class in $\mathrm{H}^3(\mathfrak{u}, k)$.*

Note for $w = s_{\beta_1}s_{\beta_2}s_{\beta_3} \in W$ with $\ell(w) = 3$, $-w \cdot 0$ involves at most three simple roots. Indeed, if $\beta_1$, $\beta_2$, and $\beta_3$ are all distinct, then $-w \cdot 0 = i\beta_1 + j\beta_2 + \beta_3$ for some $i, j > 0$ (with a more precise formula determined as above). On the other hand, suppose the simple roots are not distinct. In order to be of length three, the only possibility is that $\beta_1 = \beta_3$ and $\beta_2$ must be adjacent to $\beta_1$. Then we have that $-w \cdot 0 = -s_{\beta_1}s_{\beta_2}s_{\beta_1} \cdot 0 = i\beta_1 + j\beta_2$ for some $i \geq j > 0$.

**2.6.** We now identify some limitations on which other wedge products $\phi_\alpha \wedge \phi_\beta \wedge \phi_\gamma$ or linear combinations thereof can represent cohomology classes. Since the differentials are additive and preserve the action of $T$, of interest are linear combinations of wedge products that have the same weight. To avoid "trivial" linear combinations, we say that an expression $\sum c_{\alpha,\beta,\gamma}\phi_\alpha \wedge \phi_\beta \wedge \phi_\gamma \in \Lambda^3(\mathfrak{u}^*)$ is in *reduced* form if a triple $(\alpha, \beta, \gamma)$ appears at most once and each $c_{\alpha,\beta,\gamma} \neq 0$.

While we are interested particularly in degree 3, we make the following general observation which extends [**BNP2**, Prop. 2.4].

PROPOSITION 2.6.1. *Let $x = \sum_j c_j \bigwedge_{i=1}^n \phi_{\sigma_{i,j}}$ be an element in $\Lambda^n(\mathfrak{u}^*)$ in reduced form of weight $\gamma$ for some $\gamma \in X(T)$. If $d_n(x) = 0$, then $d_1(\phi_{\sigma_{i,j}}) = 0$ for at least one $\sigma_{i,j}$ appearing in the sum.*

PROOF. Fix an ordering on $\Phi^+$ such that $\Phi^+$ may be identified as $\Phi^+ = \{\gamma_i : 1 \leq i \leq |\Phi^+|\}$ with $\mathrm{ht}(\gamma_i) \leq \mathrm{ht}(\gamma_{i+1})$ for all $i$. By reordering if necessary, one may assume that in each wedge $\bigwedge_{i=1}^n \phi_{\sigma_{i,j}}$, we have $\sigma_{i,j} \prec_{\mathrm{ht}} \sigma_{i+1,j}$. In particular, $\mathrm{ht}(\sigma_{i,j}) \leq \mathrm{ht}(\sigma_{i+1,j})$.

Using the ordering given by $\prec_{\mathrm{ht}}$, order the wedge products found in $x$ lexicographically based on $\bigwedge_{i=2}^n \phi_{\sigma_{i,j}}$. That is, we ignore the first element in the wedge when forming this ordering. Note that, for each $j$, since $\gamma = \sum_{i=1}^n \sigma_{i,j}$, $\sigma_{1,j}$ is determined if one knows $\gamma$ and $\{\sigma_{i,j} : 2 \leq i \leq n\}$. In this lexicographical ordering, choose the wedge product that is maximal. For simplicity, denote this simply by $x_{max} = \bigwedge_{i=1}^n \phi_{\sigma_i}$. By our construction, for all $j$,

$$\mathrm{ht}\left(\bigwedge_{i=2}^n \phi_{\sigma_i}\right) := \sum_{i=2}^n \mathrm{ht}(\sigma_i) \geq \sum_{i=2}^n \mathrm{ht}(\sigma_{i,j}) = \mathrm{ht}\left(\bigwedge_{i=2}^n \phi_{\sigma_{i,j}}\right).$$

Note that equality is possible. Although in such a case, the actual roots must differ in some manner since $x$ is assumed to be in reduced form.

We claim that in this element $x_{max}$, $d_1(\phi_{\sigma_1}) = 0$. By assumption, $d_n(x) = 0$ and the coefficient of $x_{max}$ is non-trivial. From Lemma 2.1.1,

$$d_n(x_{max}) = d_1(\phi_{\sigma_1}) \wedge \bigwedge_{i=2}^{n} \phi_{\sigma_i} + \sum_{k=2}^{n}(-1)^{k+1} \bigwedge_{i=1}^{n} \phi_{\sigma_i}^k.$$

For each other summand of $x$, we get a similar expression for the result of applying $d_n$.

Consider the term $d_1(\phi_{\sigma_1}) \wedge \bigwedge_{i=2}^n \phi_{\sigma_i}$. Either this is zero or it must cancel with another term of the form

$$\phi_{\sigma_{1,j}} \wedge \cdots \wedge \phi_{\sigma_{l-1,j}} \wedge d_1(\phi_{\sigma_{l,j}}) \wedge \phi_{\sigma_{l+1,j}} \wedge \cdots \wedge \phi_{\sigma_{n,j}} \tag{2.6.1}$$

for some integers $j, l$. We show that the latter cannot happen.

Suppose on the contrary that it did. Then, for each $2 \leq i \leq n$, $\phi_{\sigma_i}$ would have to appear within a wedge as in (2.6.1). This could happen by involving zero, one, or two terms from the double wedge $d_1(\phi_{\sigma_{l,j}})$.

If it involved no components of $d_1(\phi_{\sigma_{l,j}})$, this would mean that

$$\bigwedge_{i=2}^{n} \phi_{\sigma_i} = \pm \bigwedge_{i \neq l} \phi_{\sigma_{i,j}}.$$

As noted above, by weight considerations, this would imply that $\sigma_1 = \sigma_{l,j}$ and moreover that

$$\bigwedge_{i=1}^{n} \phi_{\sigma_i} = \pm \bigwedge_{i=1}^{n} \phi_{\sigma_{i,j}}$$

which contradicts the fact that $x$ is in reduced form.

For the remaining cases, note that for any $\eta \in \Phi^+$, if $d_1(\phi_\eta) = \sum c_{\alpha,\beta}\phi_\alpha \wedge \phi_\beta$, then $\mathrm{ht}(\alpha) < \mathrm{ht}(\eta)$ and $\mathrm{ht}(\beta) < \mathrm{ht}(\eta)$.

The second case would be that $\bigwedge_{i=2}^n \phi_{\sigma_i}$ is (up to a sign) the wedge of all but one of the $\phi_{\sigma_{i,j}}$ (with $i \neq l$) along with a $\phi_\alpha$ appearing in $d_1(\phi_{\sigma_{l,j}})$. Since $\mathrm{ht}(\alpha) < \mathrm{ht}(\sigma_{l,j})$ and $\mathrm{ht}(\sigma_{1,j}) \leq \mathrm{ht}(\sigma_{i,j})$ for $2 \leq i \leq n$,

$$\mathrm{ht}\left(\bigwedge_{i=2}^{n} \phi_{\sigma_i}\right) < \mathrm{ht}\left(\bigwedge_{i=2}^{n} \phi_{\sigma_{i,j}}\right) \leq \mathrm{ht}\left(\bigwedge_{i=2}^{n} \phi_{\sigma_i}\right).$$

We can conclude that this is not possible.

Similarly, in the third case, we would have that $\bigwedge_{i=2}^n \phi_{\sigma_i}$ is (up to a sign) the wedge of all but two of the $\phi_{\sigma_{i,j}}$ (with $i \neq l$) along with a wedge $\phi_\alpha \wedge \phi_\beta$ appearing in $d_1(\phi_{\sigma_{l,j}})$. Note that $\alpha + \beta = \sigma_{l,j}$. Hence,

$$\mathrm{ht}\left(\bigwedge_{i=2}^{n} \phi_{\sigma_i}\right) \leq \mathrm{ht}\left(\bigwedge_{i=3}^{n} \phi_{\sigma_{i,j}}\right) < \mathrm{ht}\left(\bigwedge_{i=2}^{n} \phi_{\sigma_{i,j}}\right) \leq \mathrm{ht}\left(\bigwedge_{i=2}^{n} \phi_{\sigma_i}\right).$$

Consequently, this is also not possible. Thus we must have $d_1(\phi_{\sigma_1}) \wedge \bigwedge_{i=2}^n \phi_{\sigma_i} = 0$.

If $d_1(\phi_{\sigma_1}) \neq 0$, then $d_1(\phi_{\sigma_1}) = \sum_{\alpha+\beta=\sigma_1} c_{\alpha,\beta}\phi_\alpha \wedge \phi_\beta$. As above, we have $\mathrm{ht}(\alpha) < \mathrm{ht}(\sigma_1) \leq \mathrm{ht}(\sigma_i)$ for $2 \leq i \leq n$. The analogous condition holds for $\beta$ as well. Hence, no $\alpha$ or $\beta$ can equal a $\sigma_i$, and we must have $d_1(\phi_{\sigma_1}) = 0$. □

**2.7.** Combining this with Proposition 2.2.1, we get the following.

COROLLARY 2.7.1. *Assume that $p \geq 3$. For $p = 3$, assume further that $\Phi$ is not of type $G_2$.*

(a) *Let $x \in \mathrm{H}^n(\mathfrak{u}, k)$ be a representative cohomology class in reduced form having weight $\gamma$ for some $\gamma \in X(T)$. Then one of the components of $x$ is of the form $\bigwedge_{i=1}^n \phi_{\sigma_i}$ for distinct positive roots $\sigma_i \in \Phi^+$ with $\gamma = \sum_{i=1}^n \sigma_i$ and at least one $\sigma_i$ being simple.*

(b) *Suppose $\phi_\alpha \wedge \phi_\beta \wedge \phi_\delta$ represents a cohomology class in $\mathrm{H}^3(\mathfrak{u}, k)$. By reordering if necessary, we have the following conditions on $\alpha$, $\beta$, and $\delta$:*
   (i) *$\alpha$ is a simple root.*
   (ii) *Either $\beta$ is a simple root or $\beta = \alpha + \sigma$ for some $\sigma \in \Phi^+$, and this is the unique decomposition of $\beta$ as a sum of positive roots.*
   (iii) *Either $\delta$ is a simple root (in which case $\beta$ is also) or $\delta = \alpha + \sigma_1$ or $\delta = \beta + \sigma_2$ or both for some $\sigma_i \in \Phi^+$, and those are the only possible decompositions of $\delta$ as a sum of positive roots.*

PROOF. Part (a) follows immediately from Propositions 2.2.1 and 2.6.1. For part (b), by reordering as needed, we may assume that $\mathrm{ht}(\alpha) \leq \mathrm{ht}(\beta) \leq \mathrm{ht}(\delta)$. From part (a), $\alpha$ must be simple.

By assumption $d_3(\phi_\alpha \wedge \phi_\beta \wedge \phi_\delta) = 0$. We have

$$\begin{aligned} d_3(\phi_\alpha \wedge \phi_\beta \wedge \phi_\delta) &= d_1(\phi_\alpha) \wedge \phi_\beta \wedge \phi_\delta - \phi_\alpha \wedge d_1(\phi_\beta) \wedge \phi_\delta + \phi_\alpha \wedge \phi_\beta \wedge d_1(\phi_\delta) \\ &= -\phi_\alpha \wedge d_1(\phi_\beta) \wedge \phi_\delta + \phi_\alpha \wedge \phi_\beta \wedge d_1(\phi_\delta). \end{aligned}$$

In order for this to be zero, either both terms are independently zero or the terms cancel each other out. However, the first wedge product contains $\phi_\delta$ and the second wedge product can never contain $\phi_\delta$ since the roots $\alpha, \beta, \delta$ are necessarily distinct. Thus the latter scenario is impossible. In other words, we must have both

$$\phi_\alpha \wedge d_1(\phi_\beta) \wedge \phi_\delta = 0 \qquad \text{and} \qquad \phi_\alpha \wedge \phi_\beta \wedge d_1(\phi_\delta) = 0.$$

Consider the first wedge. If $d_1(\phi_\beta) = 0$, then by Proposition 2.2.1, $\beta$ is simple. Otherwise,

$$d_1(\phi_\beta) = \sum_{\sigma_1+\sigma_2=\beta} c_{\sigma_1,\sigma_2}\phi_{\sigma_1} \wedge \phi_{\sigma_2}$$

where the sum is over all distinct decompositions of $\beta$ as a sum of positive roots. Note further that under the hypotheses, all coefficients $c_{\sigma_1,\sigma_2}$ are non-zero mod $p$. Therefore, the only way to have $\phi_\alpha \wedge d_1(\phi_\beta) \wedge \phi_\delta = 0$ is if $d_1(\phi_\beta)$ involves $\phi_\alpha$ or $\phi_\delta$. However, by our height assumption, the latter case is not possible. That is, there is a unique decomposition of $\beta$ into a sum of positive roots and it has the form $\beta = \alpha + \sigma$ for some $\sigma \in \Phi^+$. An analogous argument gives the constraints on $\delta$. □

**2.8.** We make some further observations about weights that arise in $\mathrm{H}^3(\mathfrak{u}, k)$. Recall that our goal is to show that, for sufficiently large primes, all weights of $\mathrm{H}^3(\mathfrak{u}, k)$ have the form $-w \cdot 0$ for $w \in W$ with $\ell(w) = 3$. We have also noted above that weights $-w \cdot 0$ with $\ell(w) = 3$ have the form $i\beta_1 + j\beta_2 + \beta_3$ or $i\beta_1 + j\beta_2$ for some $\beta_i \in \Pi$ and positive integers $i, j$. In this section, we show conversely that if such a weight arises as the weight of a cohomology class, then it must indeed equal $-w \cdot 0$.

To show this, we first need some elementary computations for rank 3 root systems about the differential $d_1 : \mathfrak{u}^* \to \Lambda^2(\mathfrak{u}^*)$. Note that these computations also hold as appropriate for rank 2 subsystems. We leave these computations as a straightforward exercise. Note that these results are unique only up to a consistent sign change.

PROPOSITION 2.8.1. *Assume that $\Phi$ is of type $A_3$, $B_3$, or $C_3$. Let the simple roots be $\alpha_1, \alpha_2, \alpha_3$ ordered in the natural way. So $\alpha_3$ is the short simple root in type $B_3$ and the long simple root in type $C_3$. Then the following holds for $d_1 : \mathfrak{u}^* \to \Lambda^2(\mathfrak{u}^*)$.*

(a) *All types:*
- $d_1(\phi_{\alpha_1+\alpha_2}) = \phi_{\alpha_1} \wedge \phi_{\alpha_2}$
- $d_1(\phi_{\alpha_2+\alpha_3}) = \phi_{\alpha_2} \wedge \phi_{\alpha_3}$
- $d_1(\phi_{\alpha_1+\alpha_2+\alpha_3}) = \phi_{\alpha_1} \wedge \phi_{\alpha_2+\alpha_3} + \phi_{\alpha_1+\alpha_2} \wedge \phi_{\alpha_3}$

(b) *Type $B_3$:*
- $d_1(\phi_{\alpha_1+2\alpha_2+2\alpha_3}) = \phi_{\alpha_2} \wedge \phi_{\alpha_1+\alpha_2+2\alpha_3} + \phi_{\alpha_2+2\alpha_3} \wedge \phi_{\alpha_1+\alpha_2} + 2\cdot\phi_{\alpha_2+\alpha_3} \wedge \phi_{\alpha_1+\alpha_2+\alpha_3}$
- $d_1(\phi_{\alpha_1+\alpha_2+2\alpha_3}) = \phi_{\alpha_1} \wedge \phi_{\alpha_2+2\alpha_3} + 2\cdot\phi_{\alpha_3} \wedge \phi_{\alpha_1+\alpha_2+\alpha_3}$
- $d_1(\phi_{\alpha_2+2\alpha_3}) = 2\cdot\phi_{\alpha_3} \wedge \phi_{\alpha_2+\alpha_3}$

(c) *Type $C_3$:*
- $d_1(\phi_{2\alpha_1+2\alpha_2+\alpha_3}) = 2\cdot\phi_{\alpha_1} \wedge \phi_{\alpha_1+2\alpha_2+\alpha_3} + 2\cdot\phi_{\alpha_1+\alpha_2} \wedge \phi_{\alpha_1+\alpha_2+\alpha_3}$
- $d_1(\phi_{\alpha_1+2\alpha_2+\alpha_3}) = \phi_{\alpha_1} \wedge \phi_{2\alpha_2+\alpha_3} + \phi_{\alpha_2} \wedge \phi_{\alpha_1+\alpha_2+\alpha_3} + \phi_{\alpha_1+\alpha_2} \wedge \phi_{\alpha_2+\alpha_3}$
- $d_1(\phi_{2\alpha_2+\alpha_3}) = 2\cdot\phi_{\alpha_2} \wedge \phi_{\alpha_2+\alpha_3}$

Note that $\Phi$ necessarily has rank at least 3 in the following. In particular, type $G_2$ is not under consideration.

PROPOSITION 2.8.2. *Assume that $p \geq 3$. If $p = 3$, assume further that $\Phi$ is not of type $B_n$, $C_n$, or $F_4$. Suppose $\gamma \in X(T)$ is a weight of $\mathrm{H}^3(\mathfrak{u}, k)$ with $\gamma = i\beta_1 + j\beta_2 + \beta_3$ for distinct $\beta_1, \beta_2, \beta_3 \in \Pi$ and $i \geq j > 0$. Then $\gamma = -(s_{\beta_1}s_{\beta_2}s_{\beta_3})\cdot 0 = s_{\beta_1}s_{\beta_2}(\beta_3) + s_{\beta_1}(\beta_2) + \beta_1$.*

PROOF. The proof employs a case-by-case analysis. By assumption, $\gamma = \sigma_1 + \sigma_2 + \sigma_3$ where $\{\sigma_i\}$ consists of distinct positive roots. Furthermore, for each $i$, $\sigma_i = a\beta_1 + b\beta_2 + c\beta_3$ for some $a, b \geq 0$, and $c \in \{0, 1\}$.

**Types $A_n$, $D_n$, and $E_n$:** For these types, the coefficients $a, b$ must also lie in the set $\{0, 1\}$. The possible sums of three distinct positive roots giving the desired form are as follows:

I. $i = 3$, $j = 2$: $\gamma = (\beta_1 + \beta_2 + \beta_3) + (\beta_1 + \beta_2) + \beta_1$

II. $i = 3$, $j = 1$: $\gamma = (\beta_1 + \beta_2) + (\beta_1 + \beta_3) + \beta_1$

III. $i = 2$, $j = 2$:
  (a) $\gamma = (\beta_1 + \beta_2 + \beta_3) + \beta_2 + \beta_1$
  (b) $\gamma = (\beta_1 + \beta_2) + (\beta_1 + \beta_3) + \beta_2$
  (c) $\gamma = (\beta_1 + \beta_2) + (\beta_2 + \beta_3) + \beta_1$

IV. $i = 2$, $j = 1$:
  (a) $\gamma = (\beta_1 + \beta_2) + \beta_3 + \beta_1$
  (b) $\gamma = (\beta_1 + \beta_3) + \beta_2 + \beta_1$

V. $i = 1$, $j = 1$: $\gamma = \beta_3 + \beta_2 + \beta_1$

**Case I.** Since there is only one way that the weight $\gamma$ can arise, the corresponding cohomology class must be represented by $\phi_{\beta_1+\beta_2+\beta_3} \wedge \phi_{\beta_1+\beta_2} \wedge \phi_{\beta_1}$. In particular, this element must be a cocycle. From Lemma 2.1.1, Proposition 2.2.1, and Proposition 2.8.1,

$$\begin{aligned}0 &= d_3(\phi_{\beta_1+\beta_2+\beta_3} \wedge \phi_{\beta_1+\beta_2} \wedge \phi_{\beta_1})\\ &= d_1(\phi_{\beta_1+\beta_2+\beta_3}) \wedge \phi_{\beta_1+\beta_2} \wedge \phi_{\beta_1} - \phi_{\beta_1+\beta_2+\beta_3} \wedge d_1(\phi_{\beta_1+\beta_2}) \wedge \phi_{\beta_1}\\ &\qquad + \phi_{\beta_1+\beta_2+\beta_3} \wedge \phi_{\beta_1+\beta_2} \wedge d_1(\phi_{\beta_1})\\ &= d_1(\phi_{\beta_1+\beta_2+\beta_3}) \wedge \phi_{\beta_1+\beta_2} \wedge \phi_{\beta_1} \pm \phi_{\beta_1+\beta_2+\beta_3} \wedge \phi_{\beta_1} \wedge \phi_{\beta_2} \wedge \phi_{\beta_1}\\ &= d_1(\phi_{\beta_1+\beta_2+\beta_3}) \wedge \phi_{\beta_1+\beta_2} \wedge \phi_{\beta_1}.\end{aligned}$$

Since $\beta_1 + \beta_2$ is a positive root, the roots $\beta_1$ and $\beta_2$ are adjacent. Further, since $\beta_1 + \beta_2 + \beta_3$ is a positive root, $\beta_3$ is adjacent to either $\beta_1$ or $\beta_2$. That is, the roots correspond to nodes of the Dynkin diagram in one of two ways:

$$\beta_3 \leftrightarrow \beta_1 \leftrightarrow \beta_2 \qquad \text{or} \qquad \beta_1 \leftrightarrow \beta_2 \leftrightarrow \beta_3.$$

In the first case, by Lemma 2.1.1,

$$\begin{aligned}d_1(\phi_{\beta_1+\beta_2+\beta_3}) \wedge \phi_{\beta_1+\beta_2} \wedge \phi_{\beta_1} &= \phi_{\beta_3} \wedge \phi_{\beta_1+\beta_2} \wedge \phi_{\beta_1+\beta_2} \wedge \phi_{\beta_1}\\ &\quad + \phi_{\beta_3+\beta_1} \wedge \phi_{\beta_2} \wedge \phi_{\beta_1+\beta_2} \wedge \phi_{\beta_1}\\ &= \phi_{\beta_3+\beta_1} \wedge \phi_{\beta_2} \wedge \phi_{\beta_1+\beta_2} \wedge \phi_{\beta_1} \neq 0,\end{aligned}$$

which is a contradiction. Hence, the latter situation must hold. Then one can readily check that $3\beta_1 + 2\beta_2 + \beta_3 = -s_{\beta_1}s_{\beta_2}s_{\beta_3} \cdot 0$.

**Case II.** As in Case I, since the weight is unique, it must correspond to the element $\phi_{\beta_1+\beta_2} \wedge \phi_{\beta_1+\beta_3} \wedge \phi_{\beta_1}$. Since $\beta_1 + \beta_2$ and $\beta_1 + \beta_3$ are positive roots, $\beta_1$ is adjacent to both $\beta_2$ and $\beta_3$. That is, $\beta_2 \leftrightarrow \beta_1 \leftrightarrow \beta_3$. With that condition, one can check that $3\beta_1 + \beta_2 + \beta_3 = -s_{\beta_1}s_{\beta_2}s_{\beta_3} \cdot 0$.

**Case III.** Here, there are potentially three ways in which the weight $\gamma$ could arise. Notice that for (a), $\beta_1$, $\beta_2$, and $\beta_3$ lie in a row in some order. For (b), we must have $\beta_2 \leftrightarrow \beta_1 \leftrightarrow \beta_3$, and for (c) we have $\beta_1 \leftrightarrow \beta_2 \leftrightarrow \beta_3$. So cases (b) and (c) cannot occur simultaneously. There are three scenarios to consider up to a flip of the Dynkin diagram.

Suppose first that $\beta_1 \leftrightarrow \beta_3 \leftrightarrow \beta_2$. Then neither case (b) or (c) holds and $\gamma$ arises uniquely corresponding to $\phi_{\beta_1+\beta_2+\beta_3} \wedge \phi_{\beta_2} \wedge \phi_{\beta_1}$. Moreover, one can check that $2\beta_1 + 2\beta_2 + \beta_3 = -s_{\beta_1}s_{\beta_2}s_{\beta_3} \cdot 0$.

Suppose next that $\beta_2 \leftrightarrow \beta_1 \leftrightarrow \beta_3$. Then $\gamma$ can arise in two ways - (a) or (b). Notice however that $\gamma \neq -s_{\beta_1}s_{\beta_2}s_{\beta_3} \cdot 0$ (indeed, $-s_{\beta_1}s_{\beta_2}s_{\beta_3} \cdot 0$ is the weight arising in Case II). So we need to show that in fact there is no cohomology class having weight $\gamma$. If there was, $\gamma$ would have to correspond to a linear combination $a\phi_{\beta_1+\beta_2+\beta_3} \wedge \phi_{\beta_2} \wedge \phi_{\beta_1} + b\phi_{\beta_1+\beta_2} \wedge \phi_{\beta_1+\beta_3} \wedge \phi_{\beta_2}$. Then (rewriting to match the ordering of the roots)

$$\begin{aligned}&d_3(a\phi_{\beta_2+\beta_1+\beta_3} \wedge \phi_{\beta_2} \wedge \phi_{\beta_1} + b\phi_{\beta_2+\beta_1} \wedge \phi_{\beta_1+\beta_3} \wedge \phi_{\beta_2})\\ &\qquad = a\phi_{\beta_2} \wedge \phi_{\beta_1+\beta_3} \wedge \phi_{\beta_2} \wedge \phi_{\beta_1} + a\phi_{\beta_2+\beta_1} \wedge \phi_{\beta_3} \wedge \phi_{\beta_2} \wedge \phi_{\beta_1}\\ &\qquad\quad + b\phi_{\beta_2} \wedge \phi_{\beta_1} \wedge \phi_{\beta_1+\beta_3} \wedge \phi_{\beta_2} - b\phi_{\beta_2+\beta_1} \wedge \phi_{\beta_1} \wedge \phi_{\beta_3} \wedge \phi_{\beta_2}\\ &\qquad = a\phi_{\beta_2+\beta_1} \wedge \phi_{\beta_3} \wedge \phi_{\beta_2} \wedge \phi_{\beta_1} - b\phi_{\beta_2+\beta_1} \wedge \phi_{\beta_1} \wedge \phi_{\beta_3} \wedge \phi_{\beta_2}\\ &\qquad = (a-b)\phi_{\beta_2+\beta_1} \wedge \phi_{\beta_3} \wedge \phi_{\beta_2} \wedge \phi_{\beta_1}.\end{aligned}$$

So this is a cocyle if and only if $a = b$. That is the potential cohomology class would be represented by $\phi_{\beta_2+\beta_1+\beta_3} \wedge \phi_{\beta_2} \wedge \phi_{\beta_1} + \phi_{\beta_2+\beta_1} \wedge \phi_{\beta_1+\beta_3} \wedge \phi_{\beta_2}$. Notice however that

$$\begin{aligned}
d_2(&\phi_{\beta_2+\beta_1+\beta_3} \wedge \phi_{\beta_2+\beta_1}) \\
&= \phi_{\beta_2} \wedge \phi_{\beta_1+\beta_3} \wedge \phi_{\beta_2+\beta_1} + \phi_{\beta_2+\beta_1} \wedge \phi_{\beta_3} \wedge \phi_{\beta_2+\beta_1} - \phi_{\beta_2+\beta_1+\beta_3} \wedge \phi_{\beta_2} \wedge \phi_{\beta_1} \\
&= \phi_{\beta_2} \wedge \phi_{\beta_1+\beta_3} \wedge \phi_{\beta_2+\beta_1} - \phi_{\beta_1+\beta_2+\beta_3} \wedge \phi_{\beta_2} \wedge \phi_{\beta_1} \\
&= -\phi_{\beta_1+\beta_2+\beta_3} \wedge \phi_{\beta_2} \wedge \phi_{\beta_1} - \phi_{\beta_2+\beta_1} \wedge \phi_{\beta_1+\beta_3} \wedge \phi_{\beta_2}.
\end{aligned}$$

And so the above class is a coboundary. Hence, there is no cohomology class of weight $\gamma$ as claimed.

Suppose finally that $\beta_1 \leftrightarrow \beta_2 \leftrightarrow \beta_3$. Then $\gamma$ can arise in two ways - (a) or (c). As in the preceding case, $\gamma \neq -s_{\beta_1}s_{\beta_2}s_{\beta_3} \cdot 0$ (indeed, $-s_{\beta_1}s_{\beta_2}s_{\beta_3} \cdot 0$ is the weight arising in Case I). So we need to show that in fact there is no cohomology class having weight $\gamma$. If there was, $\gamma$ would have to correspond to a linear combination $a\phi_{\beta_1+\beta_2+\beta_3} \wedge \phi_{\beta_2} \wedge \phi_{\beta_1} + b\phi_{\beta_1+\beta_2} \wedge \phi_{\beta_2+\beta_3} \wedge \phi_{\beta_1}$. Then (rewriting to match the ordering of the roots)

$$\begin{aligned}
d_3(a\phi_{\beta_1+\beta_2+\beta_3} \wedge \phi_{\beta_1} \wedge \phi_{\beta_2} + b\phi_{\beta_1+\beta_2} \wedge \phi_{\beta_2+\beta_3} \wedge \phi_{\beta_1}) \\
= a\phi_{\beta_1} \wedge \phi_{\beta_2+\beta_3} \wedge \phi_{\beta_1} \wedge \phi_{\beta_2} + a\phi_{\beta_1+\beta_2} \wedge \phi_{\beta_3} \wedge \phi_{\beta_1} \wedge \phi_{\beta_2} \\
+ b\phi_{\beta_1} \wedge \phi_{\beta_2} \wedge \phi_{\beta_2+\beta_3} \wedge \phi_{\beta_1} - b\phi_{\beta_1+\beta_2} \wedge \phi_{\beta_2} \wedge \phi_{\beta_3} \wedge \phi_{\beta_1} \\
= a\phi_{\beta_1+\beta_2} \wedge \phi_{\beta_3} \wedge \phi_{\beta_1} \wedge \phi_{\beta_2} - b\phi_{\beta_1+\beta_2} \wedge \phi_{\beta_2} \wedge \phi_{\beta_3} \wedge \phi_{\beta_1} \\
= (a-b)\phi_{\beta_1+\beta_2} \wedge \phi_{\beta_1} \wedge \phi_{\beta_2} \wedge \phi_{\beta_3}.
\end{aligned}$$

So this is a cocyle if and only if $a = b$. That is the potential cohomology class would be represented by $\phi_{\beta_1+\beta_2+\beta_3} \wedge \phi_{\beta_1} \wedge \phi_{\beta_2} + \phi_{\beta_1+\beta_2} \wedge \phi_{\beta_2+\beta_3} \wedge \phi_{\beta_1}$. Notice however that

$$\begin{aligned}
d_2(&\phi_{\beta_1+\beta_2+\beta_3} \wedge \phi_{\beta_1+\beta_2}) \\
&= \phi_{\beta_1} \wedge \phi_{\beta_2+\beta_3} \wedge \phi_{\beta_1+\beta_2} + \phi_{\beta_1+\beta_2} \wedge \phi_{\beta_3} \wedge \phi_{\beta_1+\beta_2} - \phi_{\beta_1+\beta_2+\beta_3} \wedge \phi_{\beta_1} \wedge \phi_{\beta_2} \\
&= \phi_{\beta_1} \wedge \phi_{\beta_2+\beta_3} \wedge \phi_{\beta_1+\beta_2} - \phi_{\beta_1+\beta_2+\beta_3} \wedge \phi_{\beta_1} \wedge \phi_{\beta_2} \\
&= -\phi_{\beta_1+\beta_2+\beta_3} \wedge \phi_{\beta_1} \wedge \phi_{\beta_2} - \phi_{\beta_1+\beta_2} \wedge \phi_{\beta_2+\beta_3} \wedge \phi_{\beta_1}.
\end{aligned}$$

And so the above class is a coboundary. Hence, there is no cohomology class of weight $\gamma$ as claimed.

**Case IV.** Notice that in case (a) $\beta_1$ is adjacent to $\beta_2$, and in case (b) $\beta_1$ is adjacent to $\beta_3$. Notice also that both $\phi_{\beta_1+\beta_2} \wedge \phi_{\beta_1} \wedge \phi_{\beta_3}$ and $\phi_{\beta_1+\beta_3} \wedge \phi_{\beta_1} \wedge \phi_{\beta_2}$ are cocycles.

Suppose first that $\beta_1$ is adjacent to both $\beta_2$ and $\beta_3$. Then $\beta_2 \leftrightarrow \beta_1 \leftrightarrow \beta_3$ and $\gamma \neq -s_{\beta_1}s_{\beta_2}s_{\beta_3} \cdot 0$. That would be Case II. So we want to show that there is no cohomology class having weight $\gamma$. Indeed, observe that

$$\begin{aligned}
d_2(&\phi_{\beta_2+\beta_1+\beta_3} \wedge \phi_{\beta_1} + \phi_{\beta_2+\beta_1} \wedge \phi_{\beta_1+\beta_3}) \\
&= \phi_{\beta_2} \wedge \phi_{\beta_1+\beta_3} \wedge \phi_{\beta_1} + \phi_{\beta_2+\beta_1} \wedge \phi_{\beta_3} \wedge \phi_{\beta_1} + \phi_{\beta_2} \wedge \phi_{\beta_1} \wedge \phi_{\beta_1+\beta_3} \\
&\quad - \phi_{\beta_2+\beta_1} \wedge \phi_{\beta_1} \wedge \phi_{\beta_3} \\
&= -2\phi_{\beta_2+\beta_1} \wedge \phi_{\beta_1} \wedge \phi_{\beta_3}.
\end{aligned}$$

Hence, $\phi_{\beta_2+\beta_1} \wedge \phi_{\beta_1} \wedge \phi_{\beta_3}$ is a coboundary. Similarly,

$$d_2(\phi_{\beta_2+\beta_1+\beta_3} \wedge \phi_{\beta_1} - \phi_{\beta_2+\beta_1} \wedge \phi_{\beta_1+\beta_3}) = -2\phi_{\beta_2+\beta_1} \wedge \phi_{\beta_1} \wedge \phi_{\beta_3}.$$

And so $\phi_{\beta_2+\beta_1} \wedge \phi_{\beta_1} \wedge \phi_{\beta_3}$ is also a coboundary. Hence, $\gamma$ cannot be the weight of a cohomology class.

Next assume that $\beta_1$ is adjacent to $\beta_2$ but not adjacent to $\beta_3$. Then we need only consider case (a) in which $\gamma$ must correspond to $\phi_{\beta_1+\beta_2} \wedge \phi_{\beta_1} \wedge \phi_{\beta_3}$. There are still two cases to consider depending upon whether $\beta_3$ is adjacent to $\beta_2$. If it is, then $\beta_1 \leftrightarrow \beta_2 \leftrightarrow \beta_3$ and $\gamma \neq -s_{\beta_1}s_{\beta_2}s_{\beta_3} \cdot 0$. This should be Case I. However,

$$\begin{aligned} d_2(\phi_{\beta_1+\beta_2+\beta_3} \wedge \phi_{\beta_1}) &= \phi_{\beta_1} \wedge \phi_{\beta_2+\beta_3} \wedge \phi_{\beta_1} + \phi_{\beta_1+\beta_2} \wedge \phi_{\beta_3} \wedge \phi_{\beta_1} \\ &= -\phi_{\beta_1+\beta_2} \wedge \phi_{\beta_1} \wedge \phi_{\beta_3}. \end{aligned}$$

And so $\phi_{\beta_1+\beta_2} \wedge \phi_{\beta_1} \wedge \phi_{\beta_3}$ is indeed a coboundary. On the other hand, if $\beta_3$ is not adjacent to $\beta_2$ (nor $\beta_1$), then one can check that $2\beta_1 + \beta_2 + \beta_3 = -s_{\beta_1}s_{\beta_2}s_{\beta_3} \cdot 0$.

Finally, assume that $\beta_1$ is adjacent to $\beta_3$ but not to $\beta_2$. Then we only need to consider case (b) in which $\gamma$ must correspond to $\phi_{\beta_1+\beta_3} \wedge \phi_{\beta_1} \wedge \phi_{\beta_2}$. Similar to the preceding case, if $\beta_2$ is adjacent to $\beta_3$ (i.e., $\beta_1 \leftrightarrow \beta_3 \leftrightarrow \beta_2$), one sees that $\phi_{\beta_1+\beta_3} \wedge \phi_{\beta_1} \wedge \phi_{\beta_2}$ is a coboundary. On the other hand, if $\beta_2$ is not adjacent to $\beta_3$ (nor $\beta_1$), then $2\beta_1 + \beta_2 + \beta_3 = -s_{\beta_1}s_{\beta_2}s_{\beta_3} \cdot 0$.

**Case V.** Here $\gamma$ must correspond to $\phi_{\beta_1} \wedge \phi_{\beta_2} \wedge \phi_{\beta_3}$ which is evidently a cocycle. Observe that if any of the roots $\beta_1$, $\beta_2$, and $\beta_3$ are adjacent, then this class is a coboundary. For example, suppose $\beta_1$ and $\beta_2$ are adjacent. Then

$$d_2(\phi_{\beta_1+\beta_2} \wedge \phi_{\beta_3}) = \phi_{\beta_1} \wedge \phi_{\beta_2} \wedge \phi_{\beta_3}.$$

The other cases are similar. Hence, for this to represent a cohomology class, the three roots must be completely disjoint. Under that condition, one indeed has $\beta_1 + \beta_2 + \beta_3 = -s_{\beta_1}s_{\beta_2}s_{\beta_3} \cdot 0$ as claimed.

**Type $B_n$:** If the none of the roots $\beta_1$, $\beta_2$, or $\beta_3$ is equal to the short root $\alpha_n$, then the roots lie in the natural root subsystem of type $A_{n-1}$ and the result follows from above. Next, if $\beta_i \in \{\alpha_{n-2}, \alpha_{n-1}, \alpha_n\}$ for each $i$, then the problem reduces to a Lie subalgebra of type $B_3$. For type $B_3$, the Lie algebra cohomology can be computed directly by hand (with a fair amount of work) or using computer programs and the software MAGMA [**BC**, **BCP**]. Such programs have been written by students at the University of Wisconsin-Stout and by the VIGRE Algebra Group at the University of Georgia. Alternatively, since $h = 6$, one can apply the general theory of [**FP2**], [**PT**], [**UGA**]. One finds that for $p \geq 5$, each cohomology class in $\mathrm{H}^3(\mathfrak{u}, k)$ has weight $-w \cdot 0$ for some $w \in W$ with $\ell(w) = 3$. Hence, the result follows. Note that for $p = 3$, there do exist three "extra" cohomology classes, two of which indeed have weight as given in the lemma.

It remains to consider the case when one of the $\beta_i$s is $\alpha_n$ and another is $\alpha_j$ for some $j < n-2$. Recall that the weight must arise as a sum $\sigma_1 + \sigma_2 + \sigma_3$ of distinct positive roots. We have the following possibilities:

(i) $\alpha_j + (\alpha_{n-1} + 2\alpha_n) + (\alpha_{n-1} + \alpha_n)$ for some $1 \leq j \leq n-3$
(ii) $\alpha_j + (\alpha_{n-1} + 2\alpha_n) + \alpha_n$ for some $1 \leq j \leq n-3$
(iii) $\alpha_j + (\alpha_{n-1} + 2\alpha_n) + \alpha_{n-1}$ for some $1 \leq j \leq n-3$
(iv) $\alpha_j + (\alpha_{n-1} + \alpha_n) + \alpha_n$ for some $1 \leq j \leq n-3$
(v) $\alpha_j + (\alpha_{n-1} + \alpha_n) + \alpha_{n-1}$ for some $1 \leq j \leq n-3$
(vi) $\alpha_j + \alpha_{n-1} + \alpha_n$ for some $1 \leq j \leq n-3$
(vii) $\alpha_j + \alpha_{n-2} + \alpha_n$ for some $1 \leq j \leq n-4$
(viii) $\alpha_{n-3} + \alpha_{n-2} + \alpha_n$
(ix) $\sigma_1 + \sigma_2 + \alpha_n$ where $\sigma_1$ and $\sigma_2$ together involve precisely two simple roots $\alpha_j$ with $1 \leq j \leq n-3$.

For the first eight cases, the given weight arises uniquely as a sum of three distinct positive roots. Hence, the corresponding cohomology class would have to be $\phi_{\sigma_1} \wedge \phi_{\sigma_2} \wedge \phi_{\sigma_3}$. In case (i), the corresponding element is not a cocyle. Indeed, $\phi_{\alpha_j} \wedge \phi_{\alpha_{n-1}+2\alpha_n} \wedge \phi_{\alpha_{n-1}+\alpha_n} \mapsto \phi_{\alpha_j} \wedge \phi_{\alpha_{n-1}+2\alpha_n} \wedge \phi_{\alpha_{n-1}} \wedge \phi_{\alpha_n}$. In case (ii), the weight equals $-s_{\alpha_n} s_{\alpha_{n-1}} s_{\alpha_j} \cdot 0$ as needed. In case (iii), the corresponding element is not a cocycle. Indeed, $\phi_{\alpha_j} \wedge \phi_{\alpha_{n-1}+2\alpha_n} \wedge \phi_{\alpha_{n-1}} \mapsto -2\phi_{\alpha_j} \wedge \phi_{\alpha_n} \wedge \phi_{\alpha_{n-1}+\alpha_n} \wedge \phi_{\alpha_{n-1}}$. In case (iv), the corresponding element is a cocycle but is also a coboundary. Indeed, $\phi_{\alpha_j} \wedge \phi_{\alpha_{n-1}+2\alpha_n} \mapsto 2\phi_{\alpha_j} \phi_{\alpha_{n-1}+\alpha_n} \wedge \phi_{\alpha_n}$. In case (v), the weight equals $-s_{\alpha_{n-1}} s_{\alpha_n} s_{\alpha_j} \cdot 0$ as needed. In case (vi), the corresponding element is a cocycle but is also a coboundary. Indeed, $\phi_{\alpha_j} \wedge \phi_{\alpha_{n-1}+\alpha_n} \mapsto -\phi_{\alpha_j} \wedge \phi_{\alpha_{n-1}} \wedge \phi_{\alpha_n}$. In case (vii), the weight equals $-s_{\alpha_n} s_{\alpha_{n-2}} s_{\alpha_j} \cdot 0$ as needed. In case (viii), the corresponding element is a cocyle but is also a colbouncary. Indeed, $\phi_{\alpha_{n-3}+\alpha_{n-2}} \wedge \phi_{\alpha_n} \mapsto \phi_{\alpha_{n-3}} \wedge \phi_{\alpha_{n-2}} \wedge \phi_{\alpha_n}$.

For case (ix), there are several subcases to consider:

(a) $\sigma_1 = \alpha_j$ and $\sigma_2 = \alpha_{j+1}$ with $1 \leq j \leq n-4$
(b) $\sigma_1 = \alpha_i$ and $\sigma_2 = \alpha_j$ with $1 \leq i < j-1 \leq n-4$
(c) $\sigma_1 = \alpha_j$ and $\sigma_2 = \alpha_j + \alpha_{j+1}$ with $1 \leq j \leq n-4$
(d) $\sigma_1 = \alpha_j$ and $\sigma_2 = \alpha_{j-1} + \alpha_j$ with $2 \leq j \leq n-3$.

Again, these weights arise uniquely and all correpsond to cocycles. However, in case (a), the corresponding element is also a coboundary. Indeed, $\phi_{\alpha_j+\alpha_{j+1}} \wedge \phi_{\alpha_n} \mapsto \phi_{\alpha_j} \wedge \phi_{\alpha_{j+1}} \wedge \phi_{\alpha_n}$. In the latter three cases, the weight equals $-w \cdot 0$ as needed. In order, the words are $s_{\alpha_i} s_{\alpha_j} s_{\alpha_n}$, $s_{\alpha_j} s_{\alpha_{j+1}} s_{\alpha_n}$, and $s_{\alpha_j} s_{\alpha_{j-1}} s_{\alpha_n}$. Thus the claim holds for type $B_n$.

**Type $C_n$:** For type $C_n$ one can argue similarly to type $B_n$.

**Type $F_4$:** Note that in this case $\mathrm{H}^3(\mathfrak{u}, k)$ could be computed with the aid of a computer. We present the general argument anyhow. If the simple roots $\{\beta_1, \beta_2, \beta_3\}$ form a root subsystem of type $B_3$ or $C_3$, we are done by above. That leaves the case of type $A_2 \times A_1$. The possible root sums are

(i) $\alpha_1 + \alpha_3 + \alpha_4$
(ii) $\alpha_1 + (\alpha_3 + \alpha_4) + \alpha_4$
(iii) $\alpha_1 + \alpha_3 + (\alpha_3 + \alpha_4)$
(iv) $\alpha_1 + \alpha_2 + \alpha_4$
(v) $\alpha_1 + (\alpha_1 + \alpha_2) + \alpha_4$
(vi) $(\alpha_1 + \alpha_2) + \alpha_2 + \alpha_4$.

Note that all of these weights arise uniquely. Cases (ii), (iii), (v), and (vi) all have the desired form $-w \cdot 0$. Specifically, in order, they are $-s_1 s_4 s_3 \cdot 0$, $-s_1 s_3 s_4 \cdot 0$, $-s_1 s_2 s_4 \cdot 0$, and $-s_2 s_1 s_4 \cdot 0$. In the remaining two cases, the corresponding elements are cocycles but also coboundaries. Indeed, we have $\phi_{\alpha_1} \wedge \phi_{\alpha_3+\alpha_4} \mapsto -\phi_{\alpha_1} \wedge \phi_{\alpha_3} \wedge \phi_{\alpha_4}$ and $\phi_{\alpha_1+\alpha_2} \wedge \phi_{\alpha_4} \mapsto \phi_{\alpha_1} \wedge \phi_{\alpha_2} \wedge \phi_{\alpha_4}$. □

PROPOSITION 2.8.3. *Assume that $p \geq 3$. Assume that $\Phi$ is not of type $G_2$. Suppose $\gamma \in X(T)$ is a weight of $\mathrm{H}^3(\mathfrak{u}, k)$ with $\gamma = i\beta_1 + j\beta_2$ for $\beta_1, \beta_2 \in \Pi$ and $i \geq j > 0$. Then $\gamma = -(s_{\beta_1} s_{\beta_2} s_{\beta_1}) \cdot 0 = s_{\beta_1} s_{\beta_2}(\beta_1) + s_{\beta_1}(\beta_2) + \beta_1$.*

PROOF. As in the proof of the preceding lemma, the weight $\gamma$ must equal the sum of three distinct positive roots. For types $A_n$, $D_n$, and $E_n$, the only way such a weight can arise is if $\gamma = \alpha_i + \alpha_{i+1} + (\alpha_i + \alpha_{i+1}) = -s_{\alpha_i} s_{\alpha_{i+1}} s_{\alpha_i} \cdot 0 = -s_{\alpha_{i+1}} s_{\alpha_i} s_{\alpha_{i+1}} \cdot 0$ as claimed. In types $B_n$, $C_n$, and $F_4$, we have additional cases to consider when working within a type $B_2$ (or equivalently $C_2$) root system. For type $B_2$, there are four cases:

(i) $\alpha_1 + \alpha_2 + (\alpha_1 + \alpha_2)$
(ii) $\alpha_1 + \alpha_2 + (\alpha_1 + 2\alpha_2)$
(iii) $\alpha_1 + (\alpha_+ + \alpha_2) + (\alpha_1 + 2\alpha_2)$
(iv) $\alpha_2 + (\alpha_1 + \alpha_2) + (\alpha_1 + 2\alpha_2)$.

Each of these weights arise uniquely. In the first two cases, the corresponding elements are coboundaries. Indeed, we have $\phi_{\alpha_1} \wedge \phi_{\alpha_1+2\alpha_2} \mapsto -2\phi_{\alpha_1} \wedge \phi_{\alpha_2} \wedge \phi_{\alpha_1+\alpha_2}$ and $\phi_{\alpha_1+\alpha_2} \wedge \phi_{\alpha_1+2\alpha_2} \mapsto \phi_{\alpha_1} \wedge \phi_{\alpha_2} \wedge \phi_{\alpha_1+2\alpha_2}$. The latter two weights are of the desired form: $-s_1s_2s_1 \cdot 0$ and $-s_2s_1s_2 \cdot 0$ respectively. □

REMARK 2.8.4. In the case that $\Phi$ is of type $G_2$, all cohomology groups $\mathrm{H}^i(\mathfrak{u}, k)$ may be computed by hand. For $p \geq 5$, $\mathrm{H}^3(\mathfrak{u}, k)$ is two dimensional with weights corresponding to $-w \cdot 0$ for the two elements of length three. However, for $p = 3$, there are six "extra" cohomology classes whose weights are not of the form $-w \cdot 0$ for $\ell(w) = 3$.

## 3. Relationship between ordinary and restricted cohomology

**3.1.** In this section, we investigate a spectral sequence relating $\mathrm{H}^3(\mathfrak{u}, k)$ to $\mathrm{H}^3(U_1, k)$. In Section 4, this will be further related to $B_1$-cohomology, from which we will be able to make a precise determination of $\mathrm{H}^3(\mathfrak{u}, k)$ for primes that are not too small. See Theorem 4.3.1.

By Corollary 2.7.1, a weight $\gamma$ of $\mathrm{H}^3(\mathfrak{u}, k)$ can be expressed as the sum of three distinct positive roots, at least one of which is simple. In studying these relationships between $\mathfrak{u}$-, $U_1$-, and $B_1$-cohomology, we make use of the fact that such root sums cannot take certain forms. This will be the focus of the next two subsections.

**3.2. Root sums I:** The following result shows that weights of $\mathrm{H}^3(\mathfrak{u}, k)$ cannot lie in $pX(T)$ for $p \geq 5$.

PROPOSITION 3.2.1. *Assume $p \geq 5$. Let $\alpha, \sigma_1, \sigma_2 \in \Phi^+$ be distinct positive roots with $\alpha \in \Pi$. Then $\alpha + \sigma_1 + \sigma_2 \notin pX(T)$.*

PROOF. Set $\gamma := \alpha + \sigma_1 + \sigma_2$ and suppose that $\gamma = p\nu$ for some $\nu \in X(T)$. Clearly $\nu \neq 0$. Express $\gamma$ as a sum of fundamental dominant weights: $\gamma = \sum c_i\omega_i$ for integers $c_i$. By direct calculation, one finds the following constraints on the $c_i$ for any $i$.

| $\Phi$ | $A_n, D_n, E_n$ | $B_n$, $F_4$ | $C_n$ | $G_2$ |
|---|---|---|---|---|
| Bounds on $c_i$ | $-3 \leq c_i \leq 4$ | $-6 \leq c_i \leq 6$ | $-4 \leq c_i \leq 5$ | $-4 \leq c_i \leq 6$ |

For example, in types $A_n$, $D_n$, or $E_n$, when a positive root is expressed as a sum of fundamental dominant weights, the coefficients are $-1, 0, 1$, or $2$, with $2$ occurring only for a simple root. Since we have a sum of three distinct roots, for any given $\omega_i$, the coefficient is at most $2 + 1 + 1 = 4$.

Since $\nu \neq 0$, in a similar expression for $p\nu$, at least one $\omega_i$ has a coefficient that is a multiple of $p$. From the above table, this is clearly not possible for $p \geq 7$. Furthermore, for types $A_n, D_n$, and $E_n$ this is not possible for $p \geq 5$. We are left to consider types $B_n$, $C_n$, $F_4$, and $G_2$ when $p = 5$. In these cases, since $p\nu$ lies in the root lattice, $\nu$ necessarily also lies in the root lattice. In fact, $\nu$ must lie in the positive root lattice.

Similar to above, express $\gamma$ as a sum of simple roots: $\gamma = \sum m_i\alpha_i$ with $m_i \geq 0$. To have $\gamma = 5\nu$, we must have each $m_i$ being divisible by 5. For type $B_n$, the largest a coefficient can be is 5, and this occurs only if $\alpha = \alpha_n$ and $\sigma_1$, $\sigma_2$ both contain $2\alpha_n$. But then the coefficient of $\alpha_{n-1}$ is non-zero (it is at least two) and at most 4. So this is impossible. In type $C_n$, one similarly could have $\alpha = \alpha_{n-1}$ and $\sigma_1, \sigma_2$ both containing $2\alpha_{n-1}$. But then the coefficient of $\alpha_n$ is non-zero (it is at least two) and not divisible by 5. One can argue similarly in types $F_4$ and $G_2$ or compute all possible values of $\gamma$ to verify the claim. □

REMARK 3.2.2. For $p = 2$ or $p = 3$, it is possible for such weights $\gamma$ to lie in $pX(T)$. For example, in type $B_2$, $\alpha_1 + (\alpha_1 + \alpha_2) + (\alpha_1 + 2\alpha_2) = 3(\alpha_1 + \alpha_2)$.

**3.3. Root sums II:** Next we consider whether weights of $\mathrm{H}^3(\mathfrak{u}, k)$ can have the form $\sigma + p\nu$ for $\sigma$ in the positive root lattice and $\nu \neq 0$. For sufficiently large primes, this is not possible. For later computations, slightly more general statements are proven than just for the case that $\alpha, \sigma_1, \sigma_2$ are distinct positive roots.

LEMMA 3.3.1. *Let $\alpha \in \Pi$ and $\sigma_1, \sigma_2 \in \Phi^+ \cup \{0\}$. Let $\sigma \in \mathbb{Z}\Phi^+$ with $\sigma = \sum n_i\alpha_i$ and $0 \leq n_i < p$. Make the following assumption on $p$ dependent on the root system:*

| $\Phi$ | $A_n$ | $B_n$ | $C_n$ | $D_n$ | $E_6$ | $E_7$ | $E_8$ | $F_4$ | $G_2$ |
|---|---|---|---|---|---|---|---|---|---|
| $p \geq$ | 5 | 7 | 7 | 7 | 11 | 11 | 17 | 11 | 11 |

*Suppose $\alpha + \sigma_1 + \sigma_2 = \sigma + p\nu$ for some $\nu \in X(T)$. Suppose further that in type $A_n$ either $\nu$ lies in the root lattice or $p$ does not divide $n+1$. Then $\nu = 0$.*

PROOF. First observe that $p\nu$ necessarily lies in the root lattice. By the assumptions on $p$, this implies that $\nu$ lies in the root lattice.

Set $\gamma := \alpha + \sigma_1 + \sigma_2$ and express $\gamma = \sum m_i\alpha_i$ as a sum of simple roots (with $m_i \geq 0$). The equation $\gamma = \sigma + p\nu$ becomes $\sum m_i\alpha_i = \sum n_i\alpha_i + p\nu$. Since $0 \leq n_i < p$ for each $i$, if $p > m_i$ for all $i$, then we necessarily have $\nu = 0$. We recall in the following table the maximum value of a coefficient when a root is expressed as a sum of simple roots.

| $\Phi$ | $A_n$ | $B_n$ | $C_n$ | $D_n$ | $E_6$ | $E_7$ | $E_8$ | $F_4$ | $G_2$ |
|---|---|---|---|---|---|---|---|---|---|
| max coefficient | 1 | 2 | 2 | 2 | 3 | 4 | 6 | 4 | 3 |

Keeping in mind that $\alpha$ is simple, the assumptions on $p$ indeed give $p > m_i$ for all $i$. □

For our purposes, the weight $\sigma$ appearing in Lemma 3.3.1 will have more specific form. The next two results present cases where one can lower the condition on the prime.

PROPOSITION 3.3.2. *Let $\alpha \in \Pi$, $\sigma_1, \sigma_2 \in \Phi^+ \cup \{0\}$, and assume that all non-zero roots are distinct. Suppose that*

$$\alpha + \sigma_1 + \sigma_2 = \beta + \sigma_3 + p\nu$$

*for distinct roots $\beta, \sigma_3 \in \Phi^+$ with $\beta \in \Pi$ and $\nu \in \mathbb{Z}\Phi$. Make the following assumption on $p$ dependent on the root system:*

| $\Phi$ | $A_n$ | $B_n$ | $C_n$ | $D_n$ | $E_n$ | $F_4$ | $G_2$ |
|---|---|---|---|---|---|---|---|
| $p \geq$ | 5 | 7 | 5 | 5 | 5 | 7 | 7 |

*Then* $\nu = 0$.

PROOF. By Lemma 3.3.1, we only need to consider the following cases: types $C_n$, $D_n$ and $E_n$ for $p = 5$, types $E_n$, $F_4$, and $G_2$ for $p = 7$, and type $E_8$ for $p = 11, 13$. We proceed in a case-by-case basis, and, as in the proof of Lemma 3.3.1, express the left and right hand sides of the given equation as sums of simple roots.

Consider type $C_n$ with $p = 5$. To have a non-trivial solution, one would need to have $\alpha = \alpha_{n-1}$, both $\sigma_1$ and $\sigma_2$ containing $2\alpha_{n-1}$, $\nu = \alpha_{n-1}$, and neither $\beta$ nor $\sigma_3$ containing $\alpha_{n-1}$. Since $\sigma_3$ does not contain an $\alpha_{n-1}$, it cannot contain *both* an $\alpha_{n-2}$ and an $\alpha_n$, nor does $\beta$ (since $\beta$ is simple). So the sum of the coefficients of $\alpha_{n-2}$ and $\alpha_n$ in $\beta + \sigma_3$ is at most two. On the other hand, both $\sigma_1$ and $\sigma_2$ must contain an $\alpha_n$ and, since they are distinct, at least one contains an $\alpha_{n-2}$. Therefore, the sum of the coefficients of $\alpha_{n-2}$ and $\alpha_n$ in $\sigma_1 + \sigma_2$ is at least three. So no solution is possible.

For type $D_n$ with $p = 5$, similar to the type $C_n$ case, to have a non-trivial solution, we must have $\alpha = \alpha_{n-2}$, $\sigma_1, \sigma_2$ each containing $2\alpha_{n-2}$, and neither $\beta$ nor $\sigma_3$ containing an $\alpha_{n-2}$. Then $\sigma_1 + \sigma_2$ contains $2\alpha_{n-1} + 2\alpha_n$, while $\beta$ and $\sigma_3$ can each contain at most one of $\alpha_{n-1}$ or $\alpha_n$. Therefore, no solution is possible.

For the exceptional cases, the claim has been verified by using MAGMA [**BC**, **BCP**] to compute all possible sums over $\alpha, \sigma_1, \sigma_2, \beta, \sigma_3$. One can also argue in a manner similar to above (with many more cases). □

REMARK 3.3.3. When $p = 2, 3$, there exists cases with $\nu \neq 0$ in all types. For $p = 5$, arguing as above, one finds that the *only* non-trivial solutions are as follows:

(i) Type $B_n$: $n \geq 3$, and $1 \leq i \leq n-2$,
- $\alpha_n + (\alpha_{n-1} + 2\alpha_n) + (\alpha_i + \cdots + \alpha_{n-1} + 2\alpha_n) = \alpha_{n-1} + (\alpha_i + \cdots + \alpha_{n-1}) + 5\alpha_n$
- For $i < n-2$, the weight $\gamma := \alpha_i + \cdots + \alpha_{n-2} + 2\alpha_{n-1} + 5\alpha_n$ does not have the form $-w \cdot 0$ for $\ell(w) = 3$. Moreover, the only way that $\gamma$ can arise as the sum of three distinct positive roots is as $\gamma = \alpha_n + (\alpha_{n-1} + 2\alpha_n) + (\alpha_i + \cdots + \alpha_{n-1} + 2\alpha_n)$. As such, the corresponding cohomology class would have to be represented by $\phi_{\alpha_n} \wedge \phi_{\alpha_{n-1}+2\alpha_n} \wedge \phi_{\alpha_i+\cdots+\alpha_{n-1}+2\alpha_n} \in \Lambda^3(\mathfrak{u}^*)$. However, one can directly check that the differential is non-zero on this element, and so it does not represent a cohomology class in $\mathrm{H}^3(\mathfrak{u}, k)$.

(ii) Type $F_4$:
- $\alpha_3 + (\alpha_2 + 2\alpha_3) + (\alpha_1 + \alpha_2 + 2\alpha_3) = \alpha_2 + (\alpha_1 + \alpha_2) + 5\alpha_3$
- $\alpha_3 + (\alpha_2 + 2\alpha_3) + (\alpha_1 + 2\alpha_2 + 4\alpha_3 + 2\alpha_2) = \alpha_2 + (\alpha_1 + 2\alpha_2 + 2\alpha_3 + 2\alpha_4) + 5\alpha_3$

(iii) Type $G_2$:
- $\alpha_1 + (2\alpha_1 + \alpha_2) + (3\alpha_1 + \alpha_2) = \alpha_2 + (\alpha_1 + \alpha_2) + 5\alpha_1.$

PROPOSITION 3.3.4. *Let* $\alpha, \sigma_1, \sigma_2 \in \Phi^+$ *be distinct positive roots with* $\alpha \in \Pi$. *Suppose that*

$$\alpha + \sigma_1 + \sigma_2 = i_1\beta_1 + i_2\beta_2 + i_3\beta_3 + p\nu$$

*for distinct $\beta_j \in \Pi$ with $0 \leq i_1, i_2 < p$, $0 \leq i_3 \leq 1$ and $\nu \in X(T)$. Make the following assumption on $p$ dependent on the root system:*

| $\Phi$ | $A_n, n \neq 4, 6$ | $A_4$ | $A_6$ | $B_2 = C_2$ | $B_n, n \geq 3$ | $C_n, n \geq 3$ | $D_n$ | $E_n$ | $F_4$ | $G_2$ |
|---|---|---|---|---|---|---|---|---|---|---|
| $p \geq$ | 5 | 7 | $5(\neq 7)$ | 5 | 7 | 7 | 5 | 5 | 11 | 11 |

*Then $\nu = 0$.*

PROOF. In type $A_n$ when $p$ does not divide $n+1$, type $B_n$ ($n \geq 3$), type $C_n$ ($n \geq 3$), type $F_4$, and type $G_2$, we are done by Lemma 3.3.1. Set $\gamma := \alpha + \sigma_1 + \sigma_2$.

For type $B_2$ (or equivalently $C_2$) when $p = 5$, when expressed as a sum of simple roots, the maximal coefficient of $\gamma$ is 4. Hence, no solution is possible. In type $D_n$, we only need to consider the case $p = 5$. We may assume that $n \geq 5$ since in $D_4$ only one root has a coefficient of 2 when expressed as a sum of simple roots. As in the proof of Proposition 3.3.2, to have a non-trivial solution, one would need $\alpha = \alpha_{n-2}$, $\sigma_1, \sigma_2$ each containing $2\alpha_{n-2}$, and $\nu = \alpha_{n-2}$. Then $\gamma - 5\alpha_{n-2}$ would contain at least four simple roots, and we are allowed at most 3. For type $E_n$, the claim has again been verified by direct computation using MAGMA [**BC**, **BCP**].

Lastly, consider the case that $\Phi$ is of type $A_n$ with $p$ dividing $n+1$. We argue as in [**BNP2**, Section 3.4]. In this situation, $X(T)/\mathbb{Z}\Phi = \{t\omega_1 + \mathbb{Z}\Phi : t = 0, 1, \dots, n\}$ and

$$t\omega_1 = \frac{t}{n+1}(n\alpha_1 + (n-1)\alpha_2 + \cdots + \alpha_n).$$

If $\nu$ lies in the root lattice, then we are done, so we may assume that $\nu = t\omega_1 + \nu'$ for some $t \neq 0$ and $\nu' \in \mathbb{Z}\Phi$. Our equation becomes

$$\gamma = i_1\beta_1 + i_2\beta_2 + i_3\beta_3 + pt\omega_1 + p\nu'.$$

We may further assume that $pt/(n+1)$ is not congruent to zero mod $p$, or we would be done as previously. Express both sides as sums of simple roots. On the left, the coefficients appearing in $\gamma$ can be $0, 1, 2$, or 3. In other words, there are at most three distinct non-zero coefficients. On the right, notice that, mod $p$, the numbers $1, 2, \dots, p-1$ all appear as coefficients in $pt\omega_1$. At most three of those can be "cancelled" by the $\beta_j$ terms. Hence, there are at least $p - 4$ distinct non-zero coefficients mod $p$. So we have a contradiction if $p - 4 > 3$ or $p > 7$. For $p = 7$, if $n+1 \geq 14$, notice that the numbers $1, 2, \dots, p-1$ appear at least *twice* as coefficients (from distinct simple roots) in $pt\omega_1$, which again leads to a contradiction. One has a similar situation for $p = 5$ and $n + 1 \geq 10$. That leaves the two "bad" cases of type $A_6$ when $p = 7$ and $A_4$ when $p = 5$.

Note that there are non-trivial solutions in these "bad" type $A$ cases. For example, in type $A_4$ (with $p = 5$) some solutions (not all) appear in Remark 3.3.7 below (which gives solutions to a special case of the equation under consideration here). □

REMARK 3.3.5. For the following root systems and primes, the following are the *only* non-trivial solutions:

(i) Type $C_n$, $n \geq 3$, $p = 5$
- $\alpha_{n-1} + (c_1\alpha_{n-2} + 2\alpha_{n-1} + \alpha_n) + (c_2\alpha_{n-3} + c_3\alpha_{n-2} + 2\alpha_{n-1} + \alpha_n) = c_2\alpha_{n-3} + (c_1 + c_3)\alpha_{n-2} + 2\alpha_n + 5\alpha_{n-1}$ for appropriate $c_1 \in \{0, 1, 2\}$, $c_2 \in \{0, 1\}$, $c_3 \in \{1, 2\}$ (9 cases for $n \geq 4$, 3 cases for $n = 3$)
- $\alpha_{n-2} + (2\alpha_{n-2} + 2\alpha_{n-1} + \alpha_n) + (\alpha_{n-3} + 2\alpha_{n-2} + 2\alpha_{n-1} + \alpha_n) = \alpha_{n-3} + 4\alpha_{n-1} + 2\alpha_n + 5\alpha_{n-2}$ ($n \geq 4$)

- Note that all weights occur within the type $C_4$ root subsystem.
- Type $F_4$, $p = 7$
  - $\alpha_3 + (\alpha_2 + 2\alpha_3 + c\alpha_4) + (\alpha_1 + 2\alpha_2 + 4\alpha_3 + 2\alpha_4) = \alpha_1 + 3\alpha_2 + (c+2)\alpha_4 + 7\alpha_3$ for $c \in \{0, 1, 2\}$
  - $\alpha_3 + (\alpha_2 + 2\alpha_3 + c\alpha_4) + (\alpha_1 + 3\alpha_2 + 4\alpha_3 + 2\alpha_4) = \alpha_1 + 4\alpha_2 + (c+2)\alpha_4 + 7\alpha_3$ for $c \in \{0, 1, 2\}$
  - $\alpha_2 + (\alpha_1 + 3\alpha_2 + 4\alpha_3 + 2\alpha_4) + (2\alpha_1 + 3\alpha_2 + 4\alpha_3 + 2\alpha_4) = 3\alpha_1 + \alpha_3 + 4\alpha_4 + 7(\alpha_2 + \alpha_3)$
- Type $G_2$, $p = 7$
  - $\alpha_1 + (3\alpha_1 + \alpha_2) + (3\alpha_1 + 2\alpha_2) = 3\alpha_2 + 7\alpha_1$
- Denote the above expressions by $\gamma = \sigma + p\nu$. Note that in each case neither $\gamma$ nor $\sigma$ has the form $-w \cdot 0$ for $\ell(w) = 3$.

COROLLARY 3.3.6. *Assume $p \geq 5$. If $\Phi$ is of type $A_4$, assume $p \geq 7$. Let $\alpha, \sigma_1, \sigma_2 \in \Phi^+$ be distinct positive roots with $\alpha \in \Pi$. Then*

$$\alpha + \sigma_1 + \sigma_2 \neq \beta + p\nu$$

*for any $\beta \in \Pi$ and $\nu \in X(T)$.*

PROOF. First observe that we could not have $\nu = 0$. In types $A_n$ (except $A_6$ with $p = 7$), $D_n$, and $E_n$, the claim follows immediately from Proposition 3.3.4. By direct calculation using MAGMA [**BC**, **BCP**], the claim can be verified for type $A_6$ with $p = 7$ and types $F_4$ and $G_2$ when $p = 5, 7$ (with larger primes following from the proposition). It remains to consider types $B_n$ and $C_n$ ($n \geq 3$) for $p = 5$. In those cases, we argue as in the proof of Proposition 3.3.2 supposing that there was a non-zero $\nu$ for which the equation held. As seen there, in type $B_n$ (respectively, $C_n$), for that to be possible, the left side would contain between $2\alpha_{n-1}$ and $4\alpha_{n-1}$ (respectively, $2\alpha_n$ and $4\alpha_n$) which would not match with the single simple root $\beta$ on the right. □

REMARK 3.3.7. For type $A_4$ with $p = 5$, there are precisely two solutions to the equation $\alpha + \sigma_1 + \sigma_2 = \beta + p\nu$.

- $-s_3 s_2 s_1 \cdot 0 = \alpha_3 + (\alpha_2 + \alpha_3) + (\alpha_1 + \alpha_2 + \alpha_3) = \alpha_4 + 5(\omega_3 - \omega_4)$
- $-s_2 s_3 s_4 \cdot 0 = \alpha_2 + (\alpha_2 + \alpha_3) + (\alpha_2 + \alpha_3 + \alpha_4) = \alpha_1 + 5(-\omega_1 + \omega_2)$.

**3.4. Relating $\mathfrak{u}$ and $U_1$:** We can relate $\mathrm{H}^3(\mathfrak{u}, k)$ to $\mathrm{H}^3(U_1, k)$ via the first quadrant spectral sequence introduced by Friedlander and Parshall [**FP1**] for $p \geq 3$ (and later generalized by Andersen and Jantzen [**AJ**] and Friedlander and Parshall [**FP2**]; cf. also [**Jan1**, I.9.20]):

$$E_2^{2i,j} = S^i(\mathfrak{u}^*)^{(1)} \otimes \mathrm{H}^j(\mathfrak{u}, k) \Rightarrow \mathrm{H}^{2i+j}(U_1, k). \tag{3.4.1}$$

The spectral sequence admits an action of the maximal torus $T$. Under this action, all the differentials are $T$-homomorphisms. Since the odd columns are all zero, the only terms that can contribute to $\mathrm{H}^3(U_1, k)$ are $E_2^{2,1} = (\mathfrak{u}^*)^{(1)} \otimes \mathrm{H}^1(\mathfrak{u}, k)$ and $E_2^{0,3} = \mathrm{H}^3(\mathfrak{u}, k)$.

The first question to consider is whether or not these terms (or what portion of them) survive in the spectral sequence. For the first term, we must consider the differential

$$d_2 : (\mathfrak{u}^*)^{(1)} \otimes \mathrm{H}^1(\mathfrak{u}, k) = E_2^{2,1} \to E_2^{4,0} = S^2(\mathfrak{u}^*)^{(1)}.$$

We need to identify the kernel of this map. From Proposition 2.2.1, as long as $p \geq 3$ or $\Phi$ is not of type $G_2$ when $p = 3$, then elements in $(\mathfrak{u}^*)^{(1)} \otimes \mathrm{H}^1(\mathfrak{u}, k)$ have weight

$p\sigma+\alpha$ for some $\sigma\in\Phi^+$ and some simple root $\alpha\in\Pi$. On the other hand, any weight of the righthand side is of the form $p\lambda$ for $\lambda=\sigma_1+\sigma_2$ with $\sigma_i\in\Phi^+$. This implies that $\alpha=p\tau$ for some $\tau\in X(T)$ which is impossible for $p\geq 3$. Therefore, all elements must map to zero. When $p=3$ and $\Phi$ is of type $G_2$, the "extra" weight of $\mathrm{H}^1(\mathfrak{u},k)$ is $3\alpha_1+\alpha_2$ which would similarly lead to the condition that $\alpha_2=3\tau$ which is impossible. So for $p\geq 3$, $\ker(d_2)=(\mathfrak{u}^*)^{(1)}\otimes\mathrm{H}^1(\mathfrak{u},k)$.

Next, we consider the image of

$$d_2:\mathrm{H}^2(\mathfrak{u},k)=E_2^{0,2}\to E_2^{2,1}=(\mathfrak{u}^*)^{(1)}\otimes\mathrm{H}^1(\mathfrak{u},k).$$

This differential was studied in [**BNP2**]. For $p>3$, it was shown [**BNP2**, Prop. 4.1] that the map is zero. For $p=3$, it was found that the map could be non-zero for elements having certain weight in types $B_n$ ($\lambda=\alpha_{n-1}+3\alpha_n$), $C_n$ ($\lambda=3\alpha_{n-1}+\alpha_n$), $F_4$ ($\lambda=\alpha_2+3\alpha_3$), and $G_2$ ($\lambda=3\alpha_1+\alpha_2$). Moreover, these weights arise uniquely on each side, and using [**BNP2**, Prop. 5.2] along with the fact that $\mathrm{H}^2(B_1,\lambda)=\mathrm{H}^2(U_1,\lambda)^{T_1}=(\mathrm{H}^2(U_1,k)\otimes\lambda)^{T_1}$, one finds that the map is indeed non-zero on representative elements corresponding to these weights.

As noted in [**BNP2**], the spectral sequence can be refined. Since weight spaces are preserved by the differentials, and all modules for $T$ are completely reducible, for each $\lambda\in X(T)$, one obtains a spectral sequence:

$$E_2^{2i,j}=[S^i(\mathfrak{u}^*)^{(1)}\otimes\mathrm{H}^j(\mathfrak{u},k)]_\lambda\Rightarrow\mathrm{H}^{2i+j}(U_1,k)_\lambda. \tag{3.4.2}$$

From our analysis above, we obtain the following result.

PROPOSITION 3.4.1. *let $p\geq 3$ and $\lambda\in X(T)$. As a $T$-module,*

$$\mathrm{H}^3(U_1,k)_\lambda\supseteq((\mathfrak{u}^*)^{(1)}\otimes\mathrm{H}^1(\mathfrak{u},k))_\lambda \tag{3.4.3}$$

*except for the following weights*

(i) $p=3$, $\Phi$ *is of type* $B_n$: $\lambda=\alpha_{n-1}+3\alpha_n$,
(ii) $p=3$, $\Phi$ *is of type* $C_n$: $\lambda=3\alpha_{n-1}+\alpha_n$,
(iii) $p=3$, $\Phi$ *is of type* $F_4$: $\lambda=\alpha_2+3\alpha_3$,
(iv) $p=3$, $\Phi$ *is of type* $G_2$: $\lambda=3\alpha_1+\alpha_2$.

*In the exceptional cases,* $\mathrm{H}^3(U_1,k)_\lambda\subseteq\mathrm{H}^3(\mathfrak{u},k)_\lambda$ *as $T$-modules.*

**3.5.** We next turn to the investigation of the term $E_2^{0,3}=\mathrm{H}^3(\mathfrak{u},k)$. In order to determine its contribution to $\mathrm{H}^3(U_1,k)$, there are two differentials to consider:

$$d_2:\mathrm{H}^3(\mathfrak{u},k)=E_2^{0,3}\to E_2^{2,2}=(\mathfrak{u}^*)^{(1)}\otimes\mathrm{H}^2(\mathfrak{u},k)$$

and then

$$d_4:\mathrm{H}^3(\mathfrak{u},k)=E_2^{0,3}\to E_2^{4,0}=S^2(\mathfrak{u}^*)^{(1)}.$$

From Corollary 2.7.1, the weights of $\mathrm{H}^3(\mathfrak{u},k)$ have the form $\alpha+\sigma_1+\sigma_2$ for distinct positive roots $\alpha,\sigma_1,\sigma_2$ with $\alpha$ being simple.

Consider the differential $d_2$. On the right hand side, by Theorem 2.3.1, the weights have the form $p\nu+\beta+\sigma_3$ for positive roots $\nu,\beta,\sigma_3$ with $\beta$ simple and $\beta\neq\sigma_3$. To have a non-trivial differential, we must have $\alpha+\sigma_1+\sigma_2=p\nu+\beta+\sigma_3$ (with $\nu\neq 0$). Applying Proposition 3.3.2, this is impossible for the primes given therein.

For those classes that survive to the $E_4$-page, consider now $d_4$. Here, the image has to have weight of the form $p(\sigma_3+\sigma_4)$ for positive roots $\sigma_3,\sigma_4$. In other words, $\alpha+\sigma_1+\sigma_2=p(\sigma_3+\sigma_4)$. By Proposition 3.2.1, this is not possible for $p\geq 5$. We conclude the following.

THEOREM 3.5.1. *Let*

(a) $p \geq 5$ *when* $\Phi$ *is of types* $A_n$, $C_n$, $D_n$, $E_n$,
(b) $p \geq 7$ *when* $\Phi$ *is of type* $B_n$ $(n \geq 3)$, $F_4$, *or* $G_2$.

*Then, as* $T$*-modules,*

$$\mathrm{H}^3(U_1, k) \cong \mathrm{H}^3(\mathfrak{u}, k) \oplus ((\mathfrak{u}^*)^{(1)} \otimes \mathrm{H}^1(\mathfrak{u}, k)).$$

REMARK 3.5.2. In the cases for $p = 5$ when $\Phi$ is of type $B_n$ $(n \geq 3)$, $F_4$, or $G_2$, one still has $\mathrm{H}^3(U_1, k) \supset (\mathfrak{u}^*)^{(1)} \otimes \mathrm{H}^1(\mathfrak{u}, k)$, however, $\mathrm{H}^3(U_1, k)_\lambda$ may not contain $\mathrm{H}^3(\mathfrak{u}, k)_\lambda$ for the following weights (cf. Remark 3.3.3):

(i) Type $B_n$ $(n \geq 3)$: $\lambda = \alpha_{n-2} + 2\alpha_{n-1} + 5\alpha_n = -s_n s_{n-1} s_{n-2} \cdot 0$,
(ii) Type $F_4$: $\lambda = \alpha_1 + 2\alpha_2 + 5\alpha_3 = -s_3 s_2 s_1 \cdot 0$ or $\lambda = \alpha_1 + 3\alpha_2 + 7\alpha_3 + 2\alpha_4$,
(iii) Type $G_2$: $\lambda = 6\alpha_1 + 2\alpha_2$.

## 4. Utilizing the $B$-cohomology

**4.1.** From this point on, unless otherwise specified, we will make the following assumptions on the characteristic of the field $k$.

ASSUMPTION 4.1.1. *Let* $p$ *be the characteristic of the field* $k$. *We impose the condition that*

(a) $p \geq 5$ *when* $\Phi$ *is of types* $A_n$, $C_n$, $D_n$, *or* $E_n$,
(b) $p \neq 5$ *when* $\Phi$ *is of type* $A_4$, $p \neq 7$ *when* $\Phi$ *is of type* $A_6$,
(c) $p \geq 7$ *when* $\Phi$ *is of type* $B_n$ $(n \geq 3)$, $F_4$, *or* $G_2$.

**4.2. Relating $\mathfrak{u}$ and $B_1$:** The following result relates the $B_1$-cohomology with the $\mathfrak{u}$-cohomology.

PROPOSITION 4.2.1. *Let* $p$ *satisfy Assumption* 4.1.1 *and* $\lambda \in X(T)$.

(a) *As a* $T$*-module,*

$$\mathrm{H}^3(B_1, \lambda) \cong \bigoplus_{\nu \in X(T)} p\nu^{\dim \mathrm{H}^3(\mathfrak{u},k)_{-\lambda+p\nu} + \dim[((\mathfrak{u}^*)^{(1)} \otimes \mathrm{H}^1(\mathfrak{u},k)]_{-\lambda+p\nu}}$$

(b) $\mathrm{H}^3(B_1, k) = 0$.

PROOF. (a) The same argument as in [**BNP2**, Prop. 4.2] gives that

$$\mathrm{H}^3(B_1, \lambda) \cong \bigoplus_{\nu \in X(T)} p\nu^{\dim \mathrm{H}^3(U_1,k)_{-\lambda+p\nu}}.$$

The claim now follows from Theorem 3.5.1.

(b) We have from Theorem 3.5.1 that $\mathrm{H}^3(B_1, k) \cong \mathrm{H}^3(U_1, k)^{T_1} \cong \mathrm{H}^3(\mathfrak{u}, k)^{T_1}$. By Corollary 2.7.1 and Proposition 3.2.1, $\mathrm{H}^3(\mathfrak{u}, k)^{T_1} = 0$. □

REMARK 4.2.2. The proposition holds also for types $A_4$ and $A_6$ for all $p \geq 5$.

**4.3. $\mathfrak{u}$-cohomology:** In [**BNP2**], the authors made use of known information on $\mathrm{H}^j(B, \lambda)$ $j = 0, 1, 2$ in order to compute $\mathrm{H}^2(\mathfrak{u}, k)$. This strategy will be used in the following theorem. Information about $\mathrm{H}^j(B, k)$ for $j = 0, 1, 2, 3$ will be used to determine $\mathrm{H}^3(\mathfrak{u}, k)$. The following theorem extends the work of [**PT**, **FP2**, **UGA**] which required $p \geq h - 1$.

THEOREM 4.3.1. *Let* $p$ *be an odd prime. In addition, assume that* $p \geq 5$ *when* $\Phi$ *is of type* $A_n$ $(n \geq 4)$, $B_3$, $C_n$ $(n \geq 3)$, $D_n$, $E_n$, *and* $G_2$, *and that* $p \geq 7$ *when* $\Phi$ *is of type* $B_n$ $(n \geq 4)$ *and* $F_4$.

(a) *If* $\gamma \in X(T)$ *is a weight of* $\mathrm{H}^3(\mathfrak{u}, k)$, *then* $\gamma = -w \cdot 0$ *for some* $w \in W$ *with* $\ell(w) = 3$.
(b) *As a* $T$*-module,*

$$\mathrm{H}^3(\mathfrak{u}, k) \cong \bigoplus_{w\in W,\ \ell(w)=3} -w \cdot 0$$

Proof. By direct computation of $\mathrm{H}^3(\mathfrak{u}, k)$, with the aid of MAGMA [**BC**, **BCP**], (or by [**PT**] when $p \geq h-1$) the theorem can be verified for types $A_1$, $A_2$, $A_3$, and $B_2$ (or $C_2$) when $p = 3$, type $A_4$ when $p = 5$, type $A_6$ when $p = 7$, type $F_4$ when $p \geq 7$, and type $G_2$ when $p \geq 5$. We exclude those cases for the remainder of the proof. Assume also for the moment that $\Phi$ is not of type $C_n$ $(n \geq 3)$ when $p = 5$.

Let $\gamma$ be a weight of $\mathrm{H}^3(\mathfrak{u}, k)$. Then $\gamma = \alpha + \sigma_1 + \sigma_2$ where $\alpha \in \Pi$ and $\sigma_1, \sigma_2 \in \Phi^+$ by Corollary 2.7.1. Now by Proposition 4.2.1, we have that $\mathrm{H}^3(B_1, -\gamma) \neq 0$. Consider the Lyndon-Hochschild-Serre (LHS) spectral sequence applied to $B_1 \trianglelefteq B$ with $-\gamma + p\nu \in X(T)$:

$$E_2^{i,j} = \mathrm{Ext}^i_{B/B_1}(-p\nu, \mathrm{H}^j(B_1, -\gamma)) \Rightarrow \mathrm{H}^{i+j}(B, -\gamma + p\nu).$$

First observe that $-\gamma \notin pX(T)$ by Proposition 4.2.1, thus $E_2^{i,0} = 0$ for $i \geq 0$. Next suppose that $E_2^{i,1} \neq 0$ for some $i \geq 0$. Then $\mathrm{H}^1(B_1, -\gamma) \neq 0$. By [**Jan2**], $\gamma = p\sigma + \delta$ for some $\delta \in \Pi$ and $\sigma \in X(T)$. By Corollary 3.3.6, this is not possible. Thus $E_2^{i,1} = 0$ for $i \geq 0$. Finally, by [**BNP2**, Thm 5.3], if $\mathrm{H}^2(B_1, -\gamma) \neq 0$, then $\gamma = -w \cdot 0 + p\sigma$ where $\ell(w) = 2$ and $\sigma \in X(T)$. From Proposition 2.5.1, $-w \cdot 0 = c\delta_1 + \delta_2$ for $\delta_i \in \Pi$ and $c \geq 1$. By Proposition 3.3.4, we must have $\sigma = 0$, and so $\gamma = c\delta_1 + \delta_2$, which is not possible. Therefore, $E_2^{i,2} = 0$ for $i \geq 2$. We can now conclude that

$$\mathrm{H}^3(B, -\gamma + p\nu) \cong E_2^{0.3} \cong \mathrm{Hom}_{B/B_1}(-p\nu, \mathrm{H}^3(B_1, -\gamma))) \neq 0 \tag{4.3.1}$$

for some $\nu \in X(T)$.

From the proof in [**And**, 2.9] there exists a simple root $\beta_1$ such that

$$0 \neq \mathrm{H}^3(B, -\gamma+p\nu) \cong \mathrm{H}^2(P_{\beta_1}, \mathrm{H}^1(P_{\beta_1}/B, -\gamma+p\nu)) \cong \mathrm{H}^2(B, \mathrm{H}^1(P_{\beta_1}/B, -\gamma+p\nu)),$$

where $P_{\beta_1} \supset B$ is the standard parabolic subgroup associated to $\beta_1$. Therefore, there exists a weight $\mu$ of $\mathrm{H}^1(P_{\beta_1}/B, -\gamma + p\nu))$ with $\mathrm{H}^2(B, \mu) \neq 0$. According to [**BNP2**, Thm. 5.8(a)], $\mu = -i_2\beta_2 - i_3\beta_3$ for some $\beta_2, \beta_3 \in \Pi$ where $i_2, i_3 \geq 0$ and without loss of generality $i_3 = p^l$ for $l \geq 0$. On the other hand, $\mu = -\gamma + p\nu + i_1\beta_1$ for $i_1 \geq 0$ by [**And**, (2.8)]. Therefore,

$$\gamma = i_1\beta_1 + i_2\beta_2 + i_3\beta_3 + p\nu.$$

If any $i_j \geq p$, the $p$-portion can be combined with the $p\nu$ term to re-express this as

$$\gamma = i_1'\beta_1 + i_2'\beta_2 + i_3'\beta_3 + p\nu'$$

where $0 \leq i_j' \leq p-1$ for $j \in \{1, 2\}$ and $0 \leq i_3' \leq 1$. By Proposition 3.3.4, $\nu' = 0$. From Proposition 2.8.2 and Proposition 2.8.3, it follows that $\gamma = -w \cdot 0$ with $\ell(w) = 3$.

Lastly, we need to consider the case of type $C_n$ with $p = 5$. In this setting, Proposition 3.3.4, which has been applied twice, might fail to hold. Then $\gamma$ would need to be one of the weights listed in Remark 3.3.5. As noted there, none of those weights have the form $-w \cdot 0$ with $\ell(w) = 3$, and they all lie in a type $C_4$

root subsystem. By direct calculation, the theorem holds for type $C_4$ (or $C_3$). In particular, none of the weights from Remark 3.3.5 can appear in $\mathrm{H}^3(\mathfrak{u},k)$. Since they do not appear in type $C_4$, they also cannot appear as weights of $\mathrm{H}^3(\mathfrak{u},k)$ for any $C_n$ for $n \geq 4$. Hence, the claim holds. □

REMARK 4.3.2. For odd primes, the conditions on the prime given in Theorem 4.3.1 are necessary and sufficient to obtain the analogue of Kostant's Theorem in positive characteristic. Specifically, one can show by direct calculation that there exist "extra" cohomology classes in $\mathrm{H}^3(\mathfrak{u},k)$ when $p=3$ in types $A_4$, $D_4$, and $G_2$, . Such classes will further give rise to "extra" cohomology classes in types $A_n$ and $D_n$ when $n>4$ and types $E_n$. Similarly, when $p=3,5$, there exist "extra" cohomology classes in type $B_n$ $(n \geq 4)$ and $F_4$.

**4.4. Other $u$-cohomology calculations:** In this section, we prove some results about $\mathrm{H}^i(\mathfrak{u},\mathfrak{u}^*)$, $i=0,1$, that will be used to compute $B_r$-cohomology.

PROPOSITION 4.4.1. *Let $p$ satisfy Assumption 4.1.1.*

(a) $\dim \mathrm{H}^0(\mathfrak{u},\mathfrak{u}^*)_\lambda = \begin{cases} 1 & \text{if } \lambda=\alpha,\ \alpha\in\Pi, \\ 0 & \text{else.}\end{cases}$
(b) *Let $\lambda \in X(T)$. If $\mathrm{H}^1(\mathfrak{u},\mathfrak{u}^*)_\lambda \neq 0$, then $\lambda=\alpha+\beta$ where $\alpha\in\Pi$ and $\beta\in\Phi^+$.*
(c) *As $T$-modules, $\mathrm{H}^1(U_1,\mathfrak{u}^*)\cong \mathrm{H}^1(\mathfrak{u},\mathfrak{u}^*)$.*
(d) *Let $\lambda\in X(T)$. If $\lambda\neq\alpha+\beta$ where $\alpha\in\Pi$ and $\beta\in\Phi^+$, then*
$$\mathrm{H}^1(B/U_1, \mathrm{Hom}_{U_1}(k,\mathfrak{u}^*\otimes -\lambda)) \cong \mathrm{H}^1(B,\mathfrak{u}^*\otimes -\lambda).$$
(e) *If $\lambda=\alpha+\beta$ where $\alpha\in\Pi$ and $\beta\in\Phi^+$, then*
$$\mathrm{H}^1(\mathfrak{u},\mathfrak{u}^*)_\lambda \cong \mathrm{H}^1(B,\mathfrak{u}^*\otimes -\lambda).$$

PROOF. Part (a) follows from the fact that with our assumptions on $p$ the $\mathfrak{u}$-socle of $\mathfrak{u}^*$ has basis $\{\phi_\alpha : \alpha\in\Pi\}$. For part (b), we recall the fact that
$$\mathrm{H}^1(\mathfrak{u},\mathfrak{u}^*)\cong \mathrm{Der}(\mathfrak{u},\mathfrak{u}^*)/\mathrm{Inn}(\mathfrak{u},\mathfrak{u}^*)$$
where $\mathrm{Der}(\mathfrak{u},\mathfrak{u}^*)$ (resp. $\mathrm{Inn}(\mathfrak{u},\mathfrak{u}^*)$) are both $T$-modules. Since a derivation is completely determined by its values on the generators of $\mathfrak{u}$, if $\mathrm{H}^1(\mathfrak{u},\mathfrak{u}^*)_\lambda\neq 0$ then $\lambda=\alpha+\beta$ where $\alpha\in\Pi$ and $\beta\in\Phi^+$.

For part (c), there exists a spectral sequence when $p\geq 3$ (cf. [**FP2**], [**AJ**], [**Jan1**, I.9.20]):
$$E_2^{2i,j} = S^i(\mathfrak{u}^*)^{(1)}\otimes \mathrm{H}^j(\mathfrak{u},\mathfrak{u}^*) \Rightarrow \mathrm{H}^{2i+j}(U_1,\mathfrak{u}^*).$$
We show that the differential $d_2 : \mathrm{H}^1(\mathfrak{u},\mathfrak{u}^*)\to(\mathfrak{u}^*)^{(1)}$ is zero. As shown above, any weight of $\mathrm{H}^1(\mathfrak{u},\mathfrak{u}^*)$ can be expressed as $\alpha+\beta$ where $\alpha\in\Pi$ and $\beta\in\Phi^+$. For the differential $d_2$ to be non-zero, there would need to be a $\gamma\in\Phi^+$ such that $\alpha+\beta=p\gamma$. It was shown in [**BNP2**, Prop. 3.1(A)] (under conditions weaker than Assumption 4.1.1) that this is not possible if $\alpha\neq\beta$. However, it is straightforward to see that this is also impossible when $\alpha=\beta$, particularly as $\alpha$ is simple. Therefore, $\mathrm{H}^1(U_1,\mathfrak{u}^*)\cong\mathrm{H}^1(\mathfrak{u},\mathfrak{u}^*)$.

For parts (d) and (e), apply the LHS spectral sequence for $U_1 \trianglelefteq B$ and $\lambda\in X(T)$:
$$E_2^{i,j} = \mathrm{H}^i(B/U_1, \mathrm{H}^j(U_1,\mathfrak{u}^*\otimes -\lambda)) \Rightarrow \mathrm{H}^{i+j}(B,\mathfrak{u}^*\otimes -\lambda). \tag{4.4.1}$$

First observe that, if $\lambda \neq \alpha+\beta$ where $\alpha \in \Pi$ and $\beta \in \Phi^+$, then (using part (b))

$$E_2^{0,1} = \operatorname{Hom}_{B/U_1}(\lambda, \mathrm{H}^1(U_1, \mathfrak{u}^*)) \cong \operatorname{Hom}_{B/U_1}(\lambda, \mathrm{H}^1(\mathfrak{u}, \mathfrak{u}^*)) = 0.$$

Therefore, $E_2^{1,0} \cong \mathrm{H}^1(B, \mathfrak{u} \otimes -\lambda)$ which proves (d).

Using the fact that $U/U_1 \trianglelefteq B/U_1$ with $(B/U_1)/(U/U_1) \cong T$, one has

$$\begin{aligned} E_2^{i,0} &= \mathrm{H}^i(B/U_1, \mathrm{H}^0(U_1, \mathfrak{u}^* \otimes -\lambda)) \\ &\cong \mathrm{H}^i(U/U_1, \mathrm{H}^0(U_1, \mathfrak{u}^* \otimes -\lambda))^T \\ &\cong [\mathrm{H}^i(U/U_1, k) \otimes \mathrm{H}^0(U_1, \mathfrak{u}^*) \otimes -\lambda]^T \\ &\cong [\mathrm{H}^i(U, k)^{(1)} \otimes \mathrm{H}^0(\mathfrak{u}, \mathfrak{u}^*) \otimes -\lambda]^T. \end{aligned}$$

By part (a), if $E_2^{1,0} \neq 0$, then there exists $\nu \neq 0$ with $p\nu + \gamma - \lambda = 0$ for some $\gamma \in \Pi$. Suppose that $\lambda = \alpha + \beta$ for $\alpha \in \Pi$ and $\beta \in \Phi^+$. Then $\lambda = \alpha + \beta = \gamma + p\nu$. If $\alpha \neq \beta$, this cannot happen under Assumption 4.1.1 by [**BNP2**, Prop. 3.1(B)]. If $\alpha = \beta \in \Pi$, one can argue in a similar manner that this is impossible. Therefore,

$$\mathrm{H}^1(B, \mathfrak{u} \otimes -\lambda) \cong E_2^{0,1} \cong \operatorname{Hom}_{B/U_1}(k, \mathrm{H}^1(\mathfrak{u}, \mathfrak{u}^*) \otimes -\lambda)$$

for such $\lambda$. To finish off the proof, we need to show that $\mathrm{H}^1(\mathfrak{u}, \mathfrak{u}^*)$ is semisimple as $B/U_1$-module. Observe that, for any $\sigma \in X(T)$,

$$\operatorname{Ext}^1_{B/U_1}(k, \sigma) \cong \mathrm{H}^1(U/U_1, \sigma)^T \cong [\mathrm{H}^1(U/U_1, k) \otimes \sigma]^T \cong [\mathrm{H}^1(U, k)^{(1)} \otimes \sigma]^T.$$

Consequently $\operatorname{Ext}^1_{B/U_1}(k, \sigma) \neq 0$ implies that $\sigma = -p^l\gamma$ where $l > 0$ and $\gamma \in \mathbb{Z}\Phi$. Suppose we have two one-dimensional representations in $\mathrm{H}^1(\mathfrak{u}, \mathfrak{u}^*)$ which extend one another. Viewed as weights, let these be represented by $\delta_1 + \beta_1$ and $\delta_2 + \beta_2$ for $\delta_1, \delta_2 \in \Pi$ and $\beta_1, \beta_2 \in \Phi^+$. Then $\delta_1 + \beta_1 = \delta_2 + \beta_2 + p\nu$ with $\nu \in \mathbb{Z}\Phi$ and $\nu \neq 0$. We can rule this out by using Proposition 3.3.2 (with $\alpha = \delta_1$, $\sigma_1 = \beta_1$, and $\sigma_2 = 0$). □

THEOREM 4.4.2. *Let $p$ satisfy Assumption 4.1.1. Then*

$$\dim \mathrm{H}^1(\mathfrak{u}, \mathfrak{u}^*)_\lambda = \begin{cases} 2 & \text{if } \lambda = \alpha + \beta,\ \alpha, \beta \in \Pi, \alpha \neq \beta, \alpha + \beta \notin \Phi^+, \\ 1 & \text{if } \lambda = \alpha + \beta,\ \alpha, \beta \in \Pi, \alpha + \beta \in \Phi^+, \\ 1 & \text{if } \lambda = 2\alpha,\ \alpha \in \Pi, \\ 1 & \text{if } \lambda = -s_\alpha s_\beta \cdot 0,\ \alpha, \beta \in \Pi, \alpha + \beta \in \Phi^+, \\ 0 & \text{else.} \end{cases}$$

PROOF. We use induction on the rank of the root system. First one can verify the statement directly for root systems of rank less than or equal to two (i.e., $\Phi = A_1, A_1 \times A_1, A_2, B_2, G_2$) by using the restriction on the allowable weights of $\mathrm{H}^1(\mathfrak{u}, \mathfrak{u}^*)$ given in part (b) of Proposition 4.4.1. Alternatively, we can also appeal to [**AR**, Prop. 4.3] since our assumptions on $p$ imply $p > h$ for these small rank cases.

The next step is to show that one can reduce from $\mathfrak{u}$ to a subalgebra corresponding to a subset of simple roots $J$ that is properly contained in $\Pi$. Let $J \subseteq \Pi$. Consider the parabolic subalgebra $\mathfrak{p}_J = \mathfrak{l}_J \ltimes \mathfrak{u}_J$ of Lie$(G)$ where $\mathfrak{l}_J$ is the Levi subalgebra associated to $J$. Then one can express $\mathfrak{u} \cong \mathfrak{a}_J \ltimes \mathfrak{u}_J$. We have the LHS spectral sequence:

$$E_2^{i,j} = \mathrm{H}^i(\mathfrak{a}_J, \mathrm{H}^j(\mathfrak{u}_J, \mathfrak{u}^*)) \Rightarrow \mathrm{H}^{i+j}(\mathfrak{u}, \mathfrak{u}^*).$$

The five term exact sequence yields

$$0 \to \mathrm{H}^1(\mathfrak{a}_J, \mathrm{H}^0(\mathfrak{u}_J, \mathfrak{u}^*)) \hookrightarrow \mathrm{H}^1(\mathfrak{u}, \mathfrak{u}^*) \to [\mathrm{H}^1(\mathfrak{u}_J, \mathfrak{u}^*)]^{\mathfrak{a}_J} \to \tag{4.4.2}$$

Note this is an exact sequence of $T$-modules. If $\lambda \in \mathbb{Z}\Phi_J$ then

$$\mathrm{H}^1(\mathfrak{a}_J, \mathrm{H}^0(\mathfrak{u}_J, \mathfrak{u}^*))_\lambda \cong \mathrm{H}^1(\mathfrak{u}, \mathfrak{u}^*)_\lambda \tag{4.4.3}$$

because all the weights of $\mathrm{H}^1(\mathfrak{u}_J, \mathfrak{u}^*)$ are positive linear combinations of simple roots with at least one simple root in $\Pi - J$. Now the short exact sequence

$$0 \to \mathfrak{a}_J^* \to \mathfrak{u}^* \to \mathfrak{u}_J^* \to 0$$

yields a short exact sequence

$$0 \to \mathrm{H}^0(\mathfrak{u}_J, \mathfrak{a}_J^*) \to \mathrm{H}^0(\mathfrak{u}_J, \mathfrak{u}^*) \to Z \to 0$$

or equivalently,

$$0 \to \mathfrak{a}_J^* \to \mathrm{H}^0(\mathfrak{u}_J, \mathfrak{u}^*) \to Z \to 0,$$

where $Z$ is a submodule of $\mathrm{H}^0(\mathfrak{u}_J, \mathfrak{u}_J^*)$. Observe that any weight in $Z$ is not in $\mathbb{Z}\Phi_J$. This means that any weight of $\mathrm{H}^j(\mathfrak{a}_J, Z)$ is not in $\mathbb{Z}\Phi_J$ for $j \geq 0$. Consequently, for $\lambda \in \mathbb{Z}\Phi_J$,

$$\mathrm{H}^1(\mathfrak{a}_J, \mathfrak{a}_J^*)_\lambda \cong \mathrm{H}^1(\mathfrak{a}_J, \mathrm{H}^0(\mathfrak{u}_J, \mathfrak{u}^*))_\lambda. \tag{4.4.4}$$

Combining this with (4.4.3) yields

$$\mathrm{H}^1(\mathfrak{a}_J, \mathfrak{a}_J^*)_\lambda \cong \mathrm{H}^1(\mathfrak{u}, \mathfrak{u}^*)_\lambda \tag{4.4.5}$$

for $\lambda \in \mathbb{Z}\Phi_J$.

As seen in Proposition 4.4.1, every weight of $\mathrm{H}^1(\mathfrak{u}, \mathfrak{u}^*)$ is of the form $\alpha + \beta$ where $\alpha \in \Pi$ and $\beta \in \Phi^+$. Set $\lambda = \alpha + \beta$. If $\lambda \in \mathbb{Z}\Phi_J$ where $J$ is properly contained in $\Pi$ then one can use the induction hypothesis and (4.4.5) to get the claim.

It remains to show that $\mathrm{H}^1(\mathfrak{u}, \mathfrak{u}^*)_\lambda = 0$ when $\lambda = \alpha + \beta$ where $\lambda$ is a positive linear combination with every simple root occurring and $|\Pi| \geq 3$. If this occurs, by Proposition 4.4.1, $\mathrm{H}^1(B, \mathfrak{u}^* \otimes -\lambda) \neq 0$. In [**AR**, Prop. 4.3], a complete computation of $\mathrm{H}^1(B, \mathfrak{u}^* \otimes -\lambda)$ is given under the condition $p > h$. That proof consists of two parts: a determination of those $\lambda$ for which the cohomology is non-zero and then a computation of those cohomology groups. For our purposes, we only need the determination portion of that proof. In that portion, it is shown that if $\mathrm{H}^1(B, \mathfrak{u}^* \otimes -\lambda) \neq 0$, then $\lambda = a\beta_1 + b\beta_2$ for simple roots $\beta_1, \beta_2$ and non-negative integers $a, b$, which contradicts our assumption that $\lambda$ involves at least three simple roots, and hence completes the proof. The necessary portion of the argument in [**AR**] only requires $p \geq 5$ (in order to guarantee that $\langle \eta + \delta, \gamma^\vee \rangle < p$ for simple roots $\eta, \gamma$ and a positive root $\delta$). In particular, it holds under our Assumption 4.1.1. □

REMARK 4.4.3. An alternative, non-inductive, proof of the aforementioned result can also be obtained by arguing along the lines used in [**BNP2**] to compute $\mathrm{H}^2(\mathfrak{u}, k)$, and again using [**AR**, Prop. 4.3].

## 5. $B_r$-cohomology

**5.1.** In this section, we compute $\mathrm{H}^3(B_r,\lambda)$ for all $\lambda \in X(T)$. Recall that the first cohomology groups $\mathrm{H}^1(B_1,\lambda)$ were computed for all primes and all weights $\lambda \in X(T)$ by Jantzen [**Jan2**]. For higher $r$, $\mathrm{H}^1(B_r,\lambda)$ is computed by the authors in [**BNP1**], and $\mathrm{H}^2(B_r,\lambda)$ is computed by the authors [**BNP2**] and Wright [**W**].

We investigate the third cohomology $\mathrm{H}^3(B_r,\lambda)$ by starting out with $B_1$. Note that for $\lambda \in X(T)$, we may write $\lambda = \lambda_0 + p\lambda_1$ for unique weights $\lambda_0, \lambda_1$ with $\lambda_0 \in X_1(T)$. Then

$$\mathrm{H}^3(B_1,\lambda) = \mathrm{H}^3(B_1,\lambda_0 + p\lambda_1) \cong \mathrm{H}^3(B_1,\lambda_0) \otimes p\lambda_1. \tag{5.1.1}$$

Hence, it suffices to compute $\mathrm{H}^3(B_1,\lambda)$ for $\lambda \in X_1(T)$. To this end, we will be interested in $\lambda$ of the form $\lambda = w\cdot 0 + p\nu \in X_1(T)$ with $w \in W$ and $\ell(w) = 3$. Given such a $w$, there exists a unique weight $\nu \in X(T)$ such that $\lambda = w\cdot 0 + p\nu \in X_1(T)$. Such weights $\nu$ are summarized in Lemma 7.1.1 (in the Appendix).

**5.2.** We can now present the computation of $\mathrm{H}^3(B_1,\lambda)$ for $\lambda \in X_1(T)$. The weight $\gamma_w$ given in the statement of the theorem is identified in Lemma 7.1.1. The precise identification of $\gamma_w$ is not necessary for the proof.

THEOREM 5.2.1. *Let $p$ satisfy Assumption* 4.1.1 *and $\lambda \in X_1(T)$. Then as $B/B_1$-modules*

$$\mathrm{H}^3(B_1,\lambda) \cong \begin{cases} \gamma_w^{(1)} & \text{if } \lambda = w\cdot 0 + p\gamma_w \text{ with } \ell(w) = 3,\\ (\mathfrak{u}^*)^{(1)} \otimes \omega_\alpha^{(1)} & \text{if } \lambda = s_\alpha\cdot 0 + p\omega_\alpha \text{ where } \alpha \in \Pi,\\ 0 & \text{otherwise.}\end{cases}$$

PROOF. From Proposition 3.5.1, we have the following isomorphisms as $T/T_1$-modules:

$$\begin{aligned}\mathrm{H}^3(B_1,\lambda) &\cong [\mathrm{H}^3(U_1,k)\otimes\lambda]^{T_1}\\ &\cong [\mathrm{H}^3(\mathfrak{u},k)\otimes\lambda]^{T_1} \oplus \left((\mathfrak{u}^*)^{(1)} \otimes [\mathrm{H}^1(\mathfrak{u},k)\otimes\lambda]^{T_1}\right).\end{aligned}$$

Note that the first isomorphism holds as $B/B_1$-modules and the second isomorphism is obtained from a spectral sequence on which $B/B_1$ acts, preserving the differentials. We will argue below that these two summands cannot occur simultaneously, and hence the identification of $\mathrm{H}^3(B_1,\lambda)$ holds as a $B/B_1$-module.

Therefore, if $p\nu$ is a weight of $\mathrm{H}^3(B_1,\lambda)$, then, by Theorem 4.3.1 and Proposition 2.2.1, either (i) $p\nu = -w\cdot 0 + \lambda$ for some $w \in W$ with $\ell(w) = 3$ or (ii) $p\nu = p\sigma + \beta + \lambda$ with $\sigma \in \Phi^+$, $\beta \in \Pi$. In the first case, since $\lambda \in X_1(T)$, $\nu = \gamma_w$. In the second case, $\lambda = -\beta + p(\nu - \sigma) = s_\beta\cdot 0 + p(\nu-\sigma)$ and we must have $\nu - \sigma = \omega_\beta$.

Given $\lambda \in X_1(T)$, we consider whether it can simultaneously take form (i) and (ii). This would imply that $w\cdot 0 + p\gamma_w = -\beta + p\omega_\beta$ where $\ell(w) = 3$ and $\beta \in \Pi$. Equivalently, we would have $-w\cdot 0 = \beta + p(\gamma_w - \omega_\beta)$. By Corollary 3.3.6 (with $-w\cdot 0$ playing the role of $\alpha + \sigma_1 + \sigma_2$), this is impossible.

Lastly, it remains to check whether the weights of type (i) or (ii) can occur in multiple ways, something that would lead to doubling of the cohomological dimension. For $p \geq 5$, it is straightforward to check (or see [**BNP1**]) that if $-\beta_1 + p\omega_{\beta_1} = -\beta_2 + p\omega_{\beta_2}$, then $\beta_1 = \beta_2$. Similarly, suppose that $w_1\cdot 0 + p\nu_1 = w_2\cdot 0 + p\nu_2 \in X_1(T)$ for some $w_1$, $w_2$ with $\ell(w_1) = \ell(w_2) = 3$. Such an equation can be rewritten as $-w_1\cdot 0 = -w_2\cdot 0 + p(\nu_1 - \nu_2)$. We may apply Proposition 3.3.4 (with $-w_1\cdot 0$

playing the role of $\alpha+\sigma_1+\sigma_2$ and $-w_2\cdot 0$ playing the role of $i_1\beta_1+i_2\beta_2+\beta_3$) to conclude that $\nu_1-\nu_2=0$ or $\nu_1=\nu_2$, from which it follows that $w_1=w_2$.

This application of Proposition 3.3.4 potentially fails in types $C_n$ for $p=5$, $F_4$ for $p=7$, and $G_2$ for $p=7$. However, from Remark 3.3.5, none of the potential "bad" weights have the form $-w\cdot 0$ for $\ell(w)=3$, and so the claim follows. □

REMARK 5.2.2. For type $A_4$, when $p=5$, and type $A_6$, when $p=7$, the structure of $\mathrm{H}^3(B_1,\lambda)$ may be more complex. In these cases there exist pairs $w_1,w_2\in W$, both of length at most 3, with $w_1\cdot 0=w_2\cdot 0+p\nu$ for some $\nu\in X(T)$. Following the discussion in [**AJ**, 6.1 - 6.3] one observes that this happens if and only if the pair $w_1,w_2$ appears in the same left coset of the cyclic subgroup of $W$ generated by $s_1s_2\dots s_n$. For example, in type $A_4$, let $w_1=s_4$ and $w_2=s_3s_2s_1$ then $\lambda=s_4\cdot 0+5\omega_4=s_3s_2s_1\cdot 0+5\omega_3$. The weight $\lambda$ takes both forms as given in the theorem and the two cases combine. For type $A_6$ there is just the pair, $w_1=s_3s_2s_1, w_2=s_4s_5s_6$. Note that these are the "bad" cases referred to in Proposition 3.3.4, see also Remark 3.3.7. From length considerations one can see that no such pairs occur in ranks greater than 6 and cohomological degree 3.

**5.3.** The preceding calculations can be used to compute $\mathrm{H}^3(B_r,\lambda)$ for any $r$ and $\lambda\in X(T)$. We first need to review the known computations of $\mathrm{H}^j(B_1,\lambda_0)$ for $j=0,1,2$ as a $B/B_1$-module where $\lambda_0\in X_1(T)$ when $p$ satisfies Assumption 4.1.1:

$$\mathrm{H}^0(B_1,\lambda_0)\cong\begin{cases}k & \text{if } \lambda_0=0,\\ 0 & \text{else,}\end{cases}\tag{5.3.1}$$

$$\mathrm{H}^1(B_1,\lambda_0)\cong\begin{cases}\omega_\alpha^{(1)} & \text{if } \lambda_0=-\alpha+p\omega_\alpha, \alpha\in\Pi,\\ 0 & \text{else,}\end{cases}\tag{5.3.2}$$

$$\mathrm{H}^2(B_1,\lambda_0)\cong\begin{cases}(\mathfrak{u}^*)^{(1)} & \text{if } \lambda_0=0,\\ \gamma_w^{(1)} & \text{if } \lambda_0=w\cdot 0+p\gamma_w, \ell(w)=2,\\ 0 & \text{else.}\end{cases}\tag{5.3.3}$$

We also note that more generally, for $\lambda\in X(T)$,

$$\mathrm{H}^0(B_r,\lambda)\cong\begin{cases}\nu^{(r)} & \text{if } \lambda=p^r\nu,\\ 0 & \text{else.}\end{cases}\tag{5.3.4}$$

Lastly, we need the following observation.

LEMMA 5.3.1. *Let $p$ satisfy Assumption 4.1.1 and $\alpha,\beta\in\Pi$.*
(a) *If $\lambda=-\alpha$, then the zero weight space of $\mathrm{H}^3(B_r,\lambda)$ is zero.*
(b) *If $\lambda\in\{-\alpha-p^i\beta\mid 0\le i\le r-1\}$, then the zero weight space of $\mathrm{H}^4(B_r,\lambda)$ is zero.*
(c) *If $\lambda=s_\alpha s_\beta\cdot 0$ and $\alpha+\beta\in\Phi^+$, then the zero weight space of $\mathrm{H}^4(B_r,\lambda)$ is zero.*

PROOF. Recall that $\mathrm{H}^i(B_r,\lambda)\cong\mathrm{H}^i(U_r,\lambda)^{T_r}$ and consider the spectral sequence of [**Jan1**, I.9.14] abutting to $\mathrm{H}^\bullet(U_r,\lambda)$ (whose differentials preserve the action of $T$). The only terms that can contribute to $\mathrm{H}^3(U_r,\lambda)$ have the form $\lambda\otimes(\mathfrak{u}^*)^{(i)}\otimes(\mathfrak{u}^*)^{(j)}$ or $\lambda\otimes\Lambda^3(\mathfrak{u}^*)^{(j)}$ for $1\le i\le r$ and $0\le j\le r-1$. For $\lambda=-\alpha$, neither term contains the zero weight. Similarly the only terms that contribute to $\mathrm{H}^4(U_r,\lambda)$ are $\lambda\otimes S^2(\mathfrak{u}^*)^{(i)}$,

$\lambda\otimes(\mathfrak{u}^*)^{(i_1)}\otimes(\mathfrak{u}^*)^{(i_2)}$, $\lambda\otimes(\mathfrak{u}^*)^{(i)}\otimes\Lambda^2(\mathfrak{u}^*)^{(j)}$, $\lambda\otimes(\mathfrak{u}^*)^{(i)}\otimes(\mathfrak{u}^*)^{(j_1)}\otimes(\mathfrak{u}^*)^{(j_2)}$, or $\lambda\otimes\Lambda^4(\mathfrak{u}^*)^{(j)}$ for $1\leq i,i_1,i_2\leq r$, $0\leq j\leq r-1$, and $0\leq j_1<j_2\leq r-1$. □

The following proposition provides a recursive algorithm to compute $\mathrm{H}^3(B_r,\lambda)$.

PROPOSITION 5.3.2. *Let $p$ satisfy Assumption 4.1.1, $r\geq 2$, and $\lambda\in X(T)$. Then*

(a) *If $\lambda_0\neq 0$, then*

$$\mathrm{H}^3(B_r,\lambda)\cong\begin{cases}\nu^{(r)} & \text{if } \lambda=p^r\nu-p^i\alpha+w\cdot 0,\ \ell(w)=2,\\ & \quad 1\leq i\leq r-1,\alpha\in\Pi,\nu\in X(T),\\ \nu^{(r)} & \text{if } \lambda=p^r\nu+p^iw\cdot 0-\alpha,\ \ell(w)=2,\\ & \quad 1\leq i\leq r-1,\alpha\in\Pi,\nu\in X(T),\\ \nu^{(r)} & \text{if } \lambda=p^r\nu+w\cdot 0,\ \ell(w)=3,\nu\in X(T),\\ \nu^{(r)} & \text{if } \lambda=p^r\nu-p^i\beta-\alpha,\\ & \quad 1\leq i\leq r-1,\alpha,\beta\in\Pi,\nu\in X(T),\\ \nu^{(r)} & \text{if } \lambda=p^r\nu-p^l\gamma-p^i\beta-\alpha,\\ & \quad 1\leq i<l\leq r-1,\alpha,\beta,\gamma\in\Pi,\nu\in X(T),\\ (\mathfrak{u}^*\otimes\nu)^{(r)} & \text{if } \lambda=p^r\nu-\alpha,\alpha\in\Pi,\nu\in X(T),\\ 0 & \text{otherwise.}\end{cases}$$

(b) *If $\lambda_0=0$, then*

$$\mathrm{H}^3(B_r,\lambda)\cong\begin{cases}\nu^{(r)}\oplus\nu^{(r)} & \text{if } \lambda=p^r\nu-p^i\beta-p\alpha,\\ & \quad 2\leq i\leq r-1,\alpha,\beta\in\Pi,\nu\in X(T),\\ \nu^{(r)}\oplus\nu^{(r)} & \text{if } \lambda=p^r\nu-p\beta-p\alpha,\\ & \quad \alpha,\beta\in\Pi,\alpha\neq\beta,\alpha+\beta\notin\Phi^+,\nu\in X(T),\\ \nu^{(r)} & \text{if } \lambda=p^r\nu-p\beta-p\alpha,\\ & \quad \alpha,\beta\in\Pi,\alpha+\beta\in\Phi^+,\nu\in X(T),\\ \nu^{(r)} & \text{if } \lambda=p^r\nu-2p\alpha,\ \alpha\in\Pi,\nu\in X(T),\\ \nu^{(r)} & \text{if } \lambda=p^r\nu+ps_\alpha s_\beta\cdot 0,\\ & \quad \alpha,\beta\in\Pi,\alpha+\beta\in\Phi^+,\nu\in X(T),\\ \mathrm{H}^3(B_{r-1},\lambda_1)^{(1)} & \text{otherwise.}\end{cases}$$

PROOF. Consider the LHS spectral sequence

$$E_2^{i,j}=\mathrm{H}^i(B_r/B_1,\mathrm{H}^j(B_1,\lambda_0)\otimes p\lambda_1)\Rightarrow\mathrm{H}^{i+j}(B_r,\lambda)$$

where $\lambda=\lambda_0+p\lambda_1$, $\lambda_0\in X_1(T)$ and $\lambda_1\in X(T)$. The possible terms that can contribute to $\mathrm{H}^3(B_r,\lambda)$ are $E_2^{i,j}$ where $i+j=3$, and this can be non-zero only if $\mathrm{H}^j(B_1,\lambda_0)\neq 0$ for some $j=0,1,2,3$. These cases will be analyzed using Theorem 5.2.1 and Equations (5.3.1)-(5.3.3). Note that, by Proposition 3.2.1, Proposition 3.3.2, Corollary 3.3.6, and [**BNP2**, Prop. 3.1(A)(B)], the four possibilities given below for $\lambda_0$ cannot happen simultaneously.

*Case 1*: $\lambda_0 = w \cdot 0 + p\gamma_w$, $\ell(w) = 2$

In this case, we have $\mathrm{H}^j(B_1, \lambda_0) = 0$ for $j = 0, 1, 3$ and $\mathrm{H}^2(B_1, \lambda_0) \cong \gamma_w^{(1)}$. Therefore,

$$\mathrm{H}^3(B_r, \lambda) \cong E_2^{1,2} \cong \mathrm{H}^1(B_r/B_1, p(\gamma_w + \lambda_1)) \cong \mathrm{H}^1(B_{r-1}, \gamma_w + \lambda_1)^{(1)}.$$

According to [**BNP1**, Thm. 2.8(A)], $\mathrm{H}^1(B_{r-1}, \gamma_w + \lambda_1)^{(1)} \cong \nu^{(r)}$ if $\gamma_w + \lambda_1 = p^{r-1}\nu - p^i\alpha$ for some $\alpha \in \Pi$, $0 \leq i \leq r-2$, otherwise it is zero. This translates to $\lambda = w \cdot 0 + p^r\nu - p^i\alpha$ where $\alpha \in \Pi$ and $1 \leq i \leq r-1$.

*Case 2*: $\lambda_0 = w \cdot 0 + p\gamma_w$, $\ell(w) = 3$

This condition implies that $\mathrm{H}^j(B_1, \lambda_0) = 0$ for $j = 0, 1, 2$. Thus,

$$\mathrm{H}^3(B_r, \lambda) \cong E_2^{0,3} = \mathrm{H}^0(B_r/B_1, \mathrm{H}^3(B_1, w \cdot 0 + p\gamma_w) \otimes p\lambda_1) \cong \mathrm{H}^0(B_{r-1}, \gamma_w + \lambda_1)^{(1)}$$

where the last isomorphism follows from Theorem 5.2.1. Moreover,

$$\mathrm{H}^0(B_{r-1}, \gamma_w + \lambda_1) \cong \begin{cases} \nu^{(r-1)} & \text{if } \gamma_w + \lambda_1 = p^{r-1}\nu, \nu \in X(T), \\ 0 & \text{if } \gamma_w + \lambda_1 \notin p^{r-1}X(T). \end{cases}$$

The non-vanishing case translates to $\lambda = w \cdot 0 + p^r\nu$.

*Case 3*: $\lambda_0 = -\alpha + p\omega_\alpha$

We have $\mathrm{H}^j(B_1, \lambda_0) = 0$ for $j = 0, 2$ and $E_2^{2,1} \hookrightarrow \mathrm{H}^3(B_r, \lambda)$. In this case we have a possible non-zero differential at the $E_3$-level: $d_3 : E_3^{0,3} \to E_3^{3,1}$. Note that $E_2^{0,3} = E_3^{0,3}$ and $E_2^{3,1} = E_3^{3,1}$. One can immediately conclude that (as $T/T_r$-modules) $\mathrm{H}^3(B_r, \lambda) \cong E_2^{2,1} \oplus \ker d_3$. One of the goals will be to show that $d_3 = 0$ so that

$$\mathrm{H}^3(B_r, \lambda) \cong E_2^{2,1} \oplus E_2^{0,3}. \tag{5.3.5}$$

In the following, we apply Theorem 5.2.1 and Proposition 4.4.1(a) to obtain

$$\begin{aligned} E_2^{0,3} &\cong \mathrm{H}^0(B_r/B_1, \mathrm{H}^3(B_1, \lambda_0) \otimes p\lambda_1) \\ &\cong \mathrm{H}^0(B_r/B_1, (\mathfrak{u}^*)^{(1)} \otimes p(\omega_\alpha + \lambda_1)) \\ &\cong \mathrm{H}^0(B_{r-1}, \mathfrak{u}^* \otimes (\omega_\alpha + \lambda_1))^{(1)} \\ &\cong [(\mathrm{H}^0(U_{r-1}, \mathfrak{u}^*) \otimes (\omega_\alpha + \lambda_1))^{T_{r-1}}]^{(1)} \\ &\cong [(\mathrm{H}^0(\mathfrak{u}, \mathfrak{u}^*) \otimes (\omega_\alpha + \lambda_1))^{T_{r-1}}]^{(1)} \\ &\cong \begin{cases} \nu^{(r)} & \text{if } \lambda_1 = p^{r-1}\nu - \beta - \omega_\alpha, \text{ for } \beta \in \Pi, \nu \in X(T), \\ 0 & \text{else} \end{cases} \\ &\cong \begin{cases} \nu^{(r)} & \text{if } \lambda = p^r\nu - p\beta - \alpha, \text{ for } \alpha, \beta \in \Pi, \nu \in X(T), \\ 0 & \text{else.} \end{cases} \end{aligned}$$

On the other hand,

$$E_2^{3,1} \cong \mathrm{H}^3(B_r/B_1, \mathrm{H}^1(B_1,\lambda_0)\otimes p\lambda_1) \cong \mathrm{H}^3(B_{r-1},\omega_\alpha+\lambda_1)^{(1)}.$$

Now suppose that $E_2^{0,3}\neq 0$. Then $\omega_\alpha+\lambda_1 = p^{r-1}\nu-\beta$ for some $\beta\in\Pi,\nu\in X(T)$. It follows that

$$E_2^{3,1} \cong \mathrm{H}^3(B_{r-1},\omega_\alpha+\lambda_1)^{(1)} \cong \mathrm{H}^3(B_{r-1},-\beta)^{(1)}\otimes\nu^{(r)}.$$

By Lemma 5.3.1, $\mathrm{H}^3(B_{r-1},-\beta)^{(1)}$ does not have a zero weight space. Since the differential $d_3$ must preserve $T/T_r$ weight spaces, it follows that $d_3=0$. Consequently, (5.3.5) holds.

If $E_2^{0,3}\neq 0$, then $\beta+\omega_\alpha+\lambda_1 = p^{r-1}\nu$ for some $\beta\in\Pi,\nu\in X(T)$. When this occurs

$$E_2^{2,1} = \mathrm{H}^2(B_r/B_1, p(\omega_\alpha+\lambda_1)) \cong \mathrm{H}^2(B_{r-1},-\beta)^{(1)}\otimes\nu^{(r)} = 0$$

by [**BNP2**, Lemma 5.6]. Hence, under the given assumption on $\lambda_0$, if $\mathrm{H}^3(B_r,\lambda)\neq 0$, then $\mathrm{H}^3(B_r,\lambda)\cong E_2^{0,3}$ or $\mathrm{H}^3(B_r,\lambda)\cong E_2^{2,1}\cong \mathrm{H}^2(B_{r-1},\omega_\alpha+\lambda_1)^{(1)}$.

In the latter case one obtains from [**BNP2**, Thm. 5.7] that

$$\mathrm{H}^3(B_r,\lambda)\cong\begin{cases}(\mathfrak{u}^*\otimes\nu)^{(r)} & \text{if } \lambda=p^r\nu-\alpha, \alpha\in\Pi,\nu\in X(T),\\ \nu^{(r)} & \text{if } \lambda=p^r\nu+p^iw\cdot 0-\alpha, \ell(w)=2,\\ & \quad 1\le i\le r-1, \alpha\in\Pi,\nu\in X(T),\\ \nu^{(r)} & \text{if } \lambda=p^r\nu-p^i\beta-\alpha,\\ & \quad 2\le i\le r-1,\alpha,\beta\in\Pi,\nu\in X(T),\\ \nu^{(r)} & \text{if } \lambda=p^r\nu-p^l\gamma-p^i\beta-\alpha,\\ & \quad 1\le i<l\le r-1,\alpha,\beta,\gamma\in\Pi,\nu\in X(T),\\ 0 & \text{otherwise.}\end{cases}$$

Combining these gives the remaining conditions on $\lambda$ for non-vanishing in part (a) of the theorem and completes the proof of the theorem for the case $\lambda_0\neq 0$.

*Case 4*: $\lambda_0=0$

In this case, we have $\mathrm{H}^j(B_1,\lambda_0)=0$ for $j=1,3$. We have

$$E_3^{1,2}=E_2^{1,2}\cong \mathrm{H}^1(B_{r-1},\mathfrak{u}^*\otimes\lambda_1)^{(1)}$$

and

$$E_3^{3,0}=E_2^{3,0}\cong \mathrm{H}^3(B_{r-1},\lambda_1)^{(1)}.$$

We will show that (as $T/T_r$-modules)

$$\mathrm{H}^3(B_r,\lambda)\cong E_3^{3,0}\oplus E_3^{1,2}.$$

To do so, we will show that the differentials $d_3: E_3^{0,2}\to E_3^{3,0}$ and $d_3: E_3^{1,2}\to E_3^{4,0}$ are both zero.

In the first case, we have

$$E_3^{0,2} = \mathrm{H}^0(B_r/B_1,\mathrm{H}^2(B_1,k)\otimes p\lambda_1)\cong \mathrm{H}^0(B_{r-1},\mathfrak{u}^*\otimes\lambda_1)^{(1)}.$$

As seen in Case 3, this is non-zero only if $\lambda_1 + \beta = p^{r-1}\nu$ for some $\beta \in \Pi$ and $\nu \in X(T)$. In that case, $E_3^{0,2} \cong \nu^{(r)}$ and we have

$$E_3^{3,0} \cong \mathrm{H}^3(B_{r-1}, -\beta + p^{r-1}\nu)^{(1)} \cong \mathrm{H}^3(B_{r-1}, -\beta)^{(1)} \otimes \nu^{(r)}.$$

Lemma 5.3.1 now implies that $\mathrm{H}^3(B_{r-1}, -\beta)$ has no zero weight space. Hence, the differential is the zero map.

To consider the second differential, we first need to compute $E_3^{1,2}$. For $r = 1$, Theorem 4.4.2 can be used to see that

$$\begin{aligned}
\mathrm{H}^1(B_1, \mathfrak{u}^* \otimes \lambda_1) &\cong [\mathrm{H}^1(U_1, \mathfrak{u}^*) \otimes \lambda_1]^{T_1} \\
&\cong [\mathrm{H}^1(\mathfrak{u}, \mathfrak{u}^*) \otimes \lambda_1]^{T_1} \\
&\cong \begin{cases} \nu^{(1)} \oplus \nu^{(1)} & \text{if } \lambda_1 = p\nu - \alpha - \beta, \text{ for } \alpha, \beta \in \Pi, \\ & \quad \alpha \neq \beta,\ \alpha + \beta \notin \Phi^+,\ \nu \in X(T), \\ \nu^{(1)} & \text{if } \lambda_1 = p\nu - \alpha - \beta, \text{ for } \alpha, \beta \in \Pi, \\ & \quad \alpha + \beta \in \Phi^+,\ \nu \in X(T), \\ \nu^{(1)} & \text{if } \lambda_1 = p\nu - 2\alpha, \text{ for } \alpha \in \Pi,\ \nu \in X(T), \\ \nu^{(1)} & \text{if } \lambda_1 = p\nu + s_\alpha s_\beta \cdot 0, \text{ for } \alpha, \beta \in \Pi, \\ & \quad \alpha + \beta \in \Phi^+,\ \nu \in X(T), \\ 0 & \text{else.} \end{cases}
\end{aligned} \tag{5.3.6}$$

Similarly, one obtains

$$\mathrm{H}^0(B_1, \mathfrak{u}^* \otimes \lambda_1) \cong \begin{cases} \nu^{(1)} & \text{if } \lambda_1 = p\nu - \alpha, \text{ for } \alpha \in \Pi,\ \nu \in X(T), \\ 0 & \text{else.} \end{cases} \tag{5.3.7}$$

One can apply the LHS spectral sequence (for $B_1 \trianglelefteq B_{r-1}$) abutting to $\mathrm{H}^\bullet(B_{r-1}, \mathfrak{u}^* \otimes \lambda_1)$ to obtain an exact sequence

$$0 \to E_1^{1,0} \to E^1 \to E_2^{0,1} \to E_2^{2,0}$$

which is equivalent to

$$\begin{aligned}
0 &\to \mathrm{H}^1(B_{r-2}, \mathrm{H}^0(B_1, \mathfrak{u}^* \otimes \lambda_1)^{(-1)})^{(1)} \to \mathrm{H}^1(B_{r-1}, \mathfrak{u}^* \otimes \lambda_1) \\
&\to \mathrm{H}^0(B_{r-2}, \mathrm{H}^1(B_1, \mathfrak{u}^* \otimes \lambda_1)^{(-1)})^{(1)} \to \mathrm{H}^2(B_{r-2}, \mathrm{H}^0(B_1, \mathfrak{u}^* \otimes \lambda_1)^{(-1)})^{(1)}.
\end{aligned}$$

Note that Assumption 4.1.1 does not allow for both $\mathrm{H}^0(B_1, \mathfrak{u}^* \otimes \lambda_1)$ and $\mathrm{H}^1(B_1, \mathfrak{u}^* \otimes \lambda_1)$ to be non-zero.

First we consider the case $\mathrm{H}^0(B_1, \mathfrak{u}^* \otimes \lambda_1) \neq 0$. By (5.3.7), $\lambda_1 = p\kappa - \alpha$ for some $\alpha \in \Pi$ and $\kappa \in X(T)$. Since $\mathrm{H}^1(B_1, \mathfrak{u}^* \otimes \lambda_1) = 0$ one concludes

$$\mathrm{H}^1(B_{r-1}, \mathfrak{u}^* \otimes \lambda_1) \cong \mathrm{H}^1(B_{r-2}, \mathrm{H}^0(B_1, \mathfrak{u}^* \otimes \lambda_1)^{(-1)})^{(1)} \cong \mathrm{H}^1(B_{r-2}, \kappa)^{(1)}.$$

Generalizing (5.3.2), from [**BNP1**, Thm. 2.8(A)], we have that $\mathrm{H}^1(B_{r-2},\kappa)^{(1)} \neq 0$ if and only if $\kappa = p^{r-2}\nu - p^i\beta$ for some $\beta \in \Pi$ with $0 \le i \le r-3$. This translates to $\lambda_1 = p^{r-1}\nu - p^{i+1}\beta - \alpha$, in which case we have $\lambda = p^r\nu - p^i\beta - p\alpha$ with $2 \le i \le r-1$ and $E_3^{1,2} \cong \nu^{(r)}$.

Next we consider the case $\mathrm{H}^0(B_1, \mathfrak{u}^* \otimes \lambda_1) = 0$ and $\mathrm{H}^1(B_1, \mathfrak{u} \otimes \lambda_1) \neq 0$. By (5.3.6), $\lambda_1 = p\kappa - \alpha - \beta$, $\lambda_1 = p\kappa - 2\alpha$, or $\lambda_1 = p\kappa + s_\alpha s_\beta \cdot 0$ for some $\alpha, \beta \in \Pi$ and $\kappa \in X(T)$. Moreover, from the above exact sequence,

$$\mathrm{H}^1(B_{r-1}, \mathfrak{u}^* \otimes \lambda_1)^{(1)} \cong \mathrm{H}^0(B_{r-2}, \mathrm{H}^1(B_1, \mathfrak{u}^* \otimes \lambda_1)^{(-1)})^{(1)}.$$

For the latter to be non-zero, by (5.3.6) and (5.3.4), we need $\kappa = p^{r-2}\nu$ for some $\nu \in X(T)$. Combining these with our earlier observation we conclude that

$$E_3^{1,2} \cong \mathrm{H}^1(B_{r-1}, \mathfrak{u}^* \otimes \lambda_1)^{(1)}$$

$$(5.3.8) \qquad \cong \begin{cases} \nu^{(r)} & \text{if } \lambda_1 = p^{r-1}\nu - p^i\beta - \alpha \text{ for } \alpha, \beta \in \Pi, \alpha + \beta \in \Phi^+, \\ & \quad 1 \le i \le r-2, \nu \in X(T), \\ \nu^{(r)} \oplus \nu^{(r)} & \text{if } \lambda_1 = p^{r-1}\nu - \alpha - \beta \text{ for } \alpha, \beta \in \Pi, \alpha \neq \beta, \\ & \quad \alpha + \beta \notin \Phi^+, \nu \in X(T), \\ \nu^{(r)} & \text{if } \lambda_1 = p^{r-1}\nu - \alpha - \beta \text{ for } \alpha, \beta \in \Pi, \alpha + \beta \in \Phi^+, \\ & \quad \nu \in X(T), \\ \nu^{(r)} & \text{if } \lambda_1 = p^{r-1}\nu - 2\alpha \text{ for } \alpha \in \Pi,\ \nu \in X(T), \\ \nu^{(r)} & \text{if } \lambda_1 = p^{r-1}\nu + s_\alpha s_\beta \cdot 0 \text{ for } \alpha, \beta \in \Pi, \alpha + \beta \in \Phi^+, \\ & \quad \nu \in X(T), \\ 0 & \text{otherwise.} \end{cases}$$

Now it follows from Lemma 5.3.1 that at least one of the terms $E_3^{1,2}$ and $E_3^{4,0}$ vanishes. Hence $d_3 : E_3^{1,2} \to E_3^{4,0}$ is the zero map and $\mathrm{H}^3(B_r, \lambda) = E_3^{1,2} \oplus E_3^{3,0}$.

Finally, we consider whether $E_3^{1,2}$ and $E_3^{3,0}$ can simultaneously be non-zero. Suppose that $E_3^{1,2} \neq 0$. Then $\lambda = p\lambda_1$ for one of the weights given in (5.3.8). Note that in all (non-zero) cases $\lambda_1 \notin pX(T)$. On the other hand, for $\lambda = p\lambda_1$, $E_3^{3,0} \cong \mathrm{H}^3(B_{r-1}, \lambda_1)^{(1)}$. Inductively, applying the theorem to $r-1$, $E_3^{3,0}$ can be non-zero only for those weights listed in part (a) of the theorem. Comparing lists, we see that $E_3^{1,2}$ and $E_3^{3,0}$ are both non-zero if and only if $\lambda_1 = p^r\nu - p^i\beta - \alpha$ with $1 \le i \le r-2$ (necessarily requiring $r \ge 3$). In this case $E_3^{3,0} = E_3^{1,2} = \nu^{(r)}$. The statement of the theorem in the $\lambda_0 = 0$ case follows from our above analysis. □

We now state the results of Proposition 5.3.2 in closed form.

THEOREM 5.3.3. *Let $p$ satisfy Assumption 4.1.1, $r \geq 2$, and $\lambda \in X(T)$.*

$$
\mathrm{H}^3(B_r,\lambda) \cong \begin{cases}
(\mathfrak{u}^* \otimes \nu)^{(r)} & \text{if } \lambda = p^r\nu - p^l\alpha, \\
& \quad 0 \leq l \leq r-1,\ \alpha \in \Pi,\ \nu \in X(T), \\
\nu^{(r)} & \text{if } \lambda = p^r\nu + p^l w\cdot 0,\ \ell(w) = 3, \\
& \quad 0 \leq l \leq r-1,\ \nu \in X(T), \\
\nu^{(r)} & \text{if } \lambda = p^r\nu - p^m\alpha + p^l w\cdot 0,\ \ell(w) = 2, \\
& \quad 0 \leq l < m \leq r-1,\ \alpha \in \Pi,\ \nu \in X(T), \\
\nu^{(r)} & \text{if } \lambda = p^r\nu + p^m w\cdot 0 - p^l\alpha,\ \ell(w) = 2, \\
& \quad 0 \leq l < m \leq r-1,\ \alpha \in \Pi,\ \nu \in X(T), \\
\nu^{(r)} & \text{if } \lambda = p^r\nu - p^l\beta - \alpha, \\
& \quad 1 \leq l \leq r-1,\ \alpha,\beta \in \Pi,\ \nu \in X(T), \\
\nu^{(r)} \oplus \nu^{(r)} & \text{if } \lambda = p^r\nu - p^m\beta - p^l\alpha, \\
& \quad 1 \leq l < m \leq r-1,\ \alpha,\beta \in \Pi,\ \nu \in X(T), \\
\nu^{(r)} & \text{if } \lambda = p^r\nu - p^n\gamma - p^m\beta - p^l\alpha, \\
& \quad 0 \leq l < m < n \leq r-1,\ \alpha,\beta,\gamma \in \Pi,\ \nu \in X(T), \\
\nu^{(r)} \oplus \nu^{(r)} & \text{if } \lambda = p^r\nu - p^l(\alpha+\beta), \\
& \quad 1 \leq l \leq r-1,\ \alpha,\beta \in \Pi,\ \alpha \neq \beta,\ \alpha+\beta \notin \Phi^+,\ \nu \in X(T), \\
\nu^{(r)} & \text{if } \lambda = p^r\nu - 2p^l\alpha, \\
& \quad 1 \leq l \leq r-1,\ \alpha \in \Pi, \nu \in X(T), \\
\nu^{(r)} & \text{if } \lambda = p^r\nu - p^l(\alpha+\beta), \\
& \quad 1 \leq l \leq r-1,\ \alpha,\beta \in \Pi,\ \alpha+\beta \in \Phi^+,\ \nu \in X(T), \\
\nu^{(r)} & \text{if } \lambda = p^r\nu + p^l s_\alpha s_\beta\cdot 0, \\
& \quad 1 \leq l \leq r-1,\ \alpha,\beta \in \Pi,\ \alpha+\beta \in \Phi^+,\ \nu \in X(T), \\
0 & \text{otherwise.}
\end{cases}
$$

**5.4. $B$-cohomology:** From [**CPS**, Cor. 7.2], we have $\mathrm{H}^3(B,\lambda) \cong \varprojlim \mathrm{H}^3(B_r,\lambda)$. If there is $\lambda \in X(T)$ with $\mathrm{H}^3(B,\lambda) \neq 0$, then there exists $s > 0$ such that the restriction map $\mathrm{H}^3(B,\lambda) \to \mathrm{H}^3(B_r,\lambda)$ is non-zero for all $r \geq s$. In particular, we must have $\mathrm{H}^3(B_r,\lambda) \neq 0$ for all $r \geq s$. From Theorem 5.3.3, one can readily determine those $\lambda$ for which $\mathrm{H}^3(B,\lambda) \neq 0$ along with the dimensions of these groups. These are given in Theorem 5.4.1. Since $B$ acts trivially on $\mathrm{H}^\bullet(B,\lambda)$, this can then be used to compute $\mathrm{H}^3(B,\lambda)$ for all $\lambda \in X(T)$ satisfying Assumption 4.1.1. This recovers and extends the work of Andersen and Rian [**AR**, Thm. 5.2] who computed these groups for $p > h$.

THEOREM 5.4.1. *Let $p$ satisfy Assumption 4.1.1 and $\lambda \in X(T)$. Then*

$$\dim \mathrm{H}^3(B,\lambda) \cong \begin{cases} 1 & \text{if } \lambda = p^l w\cdot 0,\ \ell(w)=3,\ l\geq 0, \\ 1 & \text{if } \lambda = -p^m\alpha + p^l w\cdot 0,\ \ell(w)=2,\ m>l\geq 0,\ \alpha\in\Pi, \\ 1 & \text{if } \lambda = p^m w\cdot 0 - p^l\alpha,\ \ell(w)=2,\ m>l\geq 0,\ \alpha\in\Pi, \\ 1 & \text{if } \lambda = -p^l\beta-\alpha,\ l\geq 1,\ \alpha,\beta\in\Pi, \\ 2 & \text{if } \lambda = -p^m\beta - p^l\alpha,\ m>l\geq 1,\ \alpha,\beta\in\Pi, \\ 1 & \text{if } \lambda = -p^n\gamma - p^m\beta - p^l\alpha,\ n>m>l\geq 0,\ \alpha,\beta,\gamma\in\Pi, \\ 2 & \text{if } \lambda = -p^l(\alpha+\beta),\ l\geq 1,\ \alpha,\beta\in\Pi,\ \alpha\neq\beta,\ \alpha+\beta\notin\Phi^+, \\ 1 & \text{if } \lambda = -p^l(\alpha+\beta),\ l\geq 1,\ \alpha,\beta\in\Pi,\ \alpha+\beta\in\Phi^+, \\ 1 & \text{if } \lambda = -2p^l\alpha,\ l\geq 1,\ \alpha\in\Pi, \\ 1 & \text{if } \lambda = p^l s_\alpha s_\beta\cdot 0,\ l\geq 1,\ \alpha,\beta\in\Pi,\ \alpha+\beta\in\Phi^+, \\ 0 & \text{otherwise.} \end{cases}$$

## 6. $G_r$-cohomology

**6.1.** The computation of $B_r$-cohomology can now be used to determine the $G_r$-cohomology of induced modules $H^0(\lambda)$ for some $\lambda \in X(T)_+$. For $i = 1, 2$, one has the isomorphism [**Jan1**, II.12.2], [**BNP2**, Thm. 6.1]

$$\mathrm{H}^i(G_r, H^0(\lambda))^{(-r)} \simeq \mathrm{ind}_B^G(\mathrm{H}^i(B_r,\lambda)^{(-r)})$$

for any $\lambda \in X(T)_+$. This isomorphism holds independently of the prime and was used in [**BNP1**,**BNP2**] to give explicit descriptions of $\mathrm{H}^i(G_r, H^0(\lambda))$ for all primes. The following theorem uses the calculations done by the authors in [**BNP2**] and Wright [**W**] to show that this isomorphism can be extended further to degree three for good primes. Recall that a prime $p$ is good if $p$ does not divide any coefficient of a root when expressed as a sum of simple roots.

THEOREM 6.1.1. *Let $\lambda \in X(T)_+$.*

(a) *Let $p$ be a good prime. Then*

$$\mathrm{H}^3(G_r, H^0(\lambda))^{(-r)} \simeq \mathrm{ind}_B^G\left(\mathrm{H}^3(B_r,\lambda)^{(-r)}\right).$$

(b) *Let $p$ satisfy Assumption 4.1.1. Then $\mathrm{H}^3(G_r, H^0(\lambda))^{(-r)}$ has a good filtration.*

PROOF. Consider the spectral sequence (cf. [**Jan1**, II.12.2])

$$E_2^{i,j} = R^i\,\mathrm{ind}_B^G\left(\mathrm{H}^j(B_r,\lambda)^{(-r)}\right) \Rightarrow \mathrm{H}^{i+j}(G_r, \mathrm{ind}_B^G\lambda)^{(-r)} = \mathrm{H}^{i+j}(G_r, H^0(\lambda))^{(-r)}.$$

We would like to show that

$$E_2^{i,0} = R^i\,\mathrm{ind}_B^G\left(\mathrm{Hom}_{B_r}(k,\lambda)^{(-r)}\right),$$
$$E_2^{i,1} = R^i\,\mathrm{ind}_B^G\left(\mathrm{H}^1(B_r,\lambda)^{(-r)}\right), \text{ and}$$
$$E_2^{i,2} = R^i\,\mathrm{ind}_B^G\left(\mathrm{H}^2(B_r,\lambda)^{(-r)}\right)$$

vanish for all $i > 0$. This would imply that $E^3 \cong E_2^{0,3}$.

In the proof of [**BNP2**, Thm. 6.1], it was shown that $E_2^{i,0} = 0$ and $E_2^{i,1} = 0$ for all $\lambda \in X(T)_+$ and all primes $p$. For $E_2^{i,2}$, we need to consider $\mathrm{H}^2(B_r, \lambda)^{(-r)}$ which was computed in [**BNP2**, Thm. 5.3, 5.7] for $p \geq 3$ and [**W**] for $p = 2$.

A careful analysis of these results shows that, as a $B$-module, and for a dominant weight $\lambda$, $\mathrm{H}^2(B_r, \lambda)^{(-r)}$ always has a $B$-filtration with factors of the form $S$ where either

(i) $S$ is one-dimensional of weight $\mu$ with $\langle \mu, \alpha^\vee \rangle \geq -1$ for all $\alpha \in \Pi$, or
(ii) $S = \mathfrak{u}^* \otimes \mu$ where $\mu$ satisfies (i).

In all the cases $R^i\mathrm{ind}_B^G S = 0$ for $i > 0$ using [**Jan1**, II.5.4] and [**KLT**, Thm. 2] (where the good prime requirement is needed). Therefore, $E_2^{i,2} = 0$ for $i > 0$. This proves part (a).

For a dominant weight $\lambda$, Theorem 5.3.3 shows that $\mathrm{H}^3(B_r, \lambda)^{(-r)}$ also has a $B$-filtration whose factors satisfy the same conditions (i) or (ii). Part (b) now follows from [**KLT**, Thm. 7]. □

**6.2. An application to $G(\mathbb{F}_q)$-cohomology:** Let $G(\mathbb{F}_q)$ be the finite Chevalley group obtained from $G$ by taking the $\mathbb{F}_q$-rational points, and $kG(\mathbb{F}_q)$ be its group algebra. With Theorem 5.3.3 we can extend the results given in [**BBDNPPW**, Thm. 4.3.2]. The latter theorem required $p > h$, a condition which was needed to guarantee that $\dim \mathrm{H}^3(B, \lambda) \leq 2$ for any weight $\lambda$. From Theorem 5.3.3 this dimension condition holds under Assumption 4.1.1.

THEOREM 6.2.1. *Suppose $p$ satisfies Assumption 4.1.1. Then there exists a constant $D(\Phi)$, depending on $\Phi$, such that if $r \geq D(\Phi)$ and if $q = p^r$, then, for each finite-dimensional $kG(\mathbb{F}_q)$-module $V$, one has*

$$\dim \mathrm{H}^3(G(\mathbb{F}_q), V) \leq 2 \cdot \dim V.$$

## 7. Appendix

**7.1.** For each $w \in W$, there exists a unique weight $\gamma_w$ such that $w \cdot 0 + p\gamma_w$ lies in the restricted region $X_1(T)$. The lemma below provides an identification of all such weights $\gamma_w$ when $\ell(w) = 3$ for $p \geq 3$.

We first introduce some notation. For two simple roots $\alpha$, $\beta$, we write $\alpha \sim \beta$ for adjacent roots and $\alpha \nsim \beta$ for non-ajacent roots. Given simple roots $\alpha$ and $\beta$, if there exists a third simple root $\gamma$ with $\alpha \sim \gamma$ and $\gamma \sim \beta$ (i.e., we have a subgraph of the Dynkin diagram of the form $\alpha \leftrightarrow \gamma \leftrightarrow \beta$ up to a flip), we write $\alpha \approx \beta$. Furthermore, we write $\omega_{\alpha,\beta}$ for $\omega_\gamma$.

LEMMA 7.1.1. *Let $p \geq 3$. For $w = s_{\alpha_i} s_{\alpha_j} s_{\alpha_k} \in W$ with $\ell(w) = 3$, we define $\gamma_w$ as follows. Then $w \cdot 0 + p\gamma_w \in X_1(T)$.*

(I) *Suppose $k = i$ and $\alpha_i \sim \alpha_j$. Then*

$$\gamma_w = \omega_i + \omega_j$$

*except in the following cases, where*

$$\gamma_w = \begin{cases} \omega_{n-2}+\omega_{n-1}-\omega_n & \text{if } p=3,\ \Phi \text{ is of type } B_n, \text{ and } w = s_{\alpha_{n-2}}s_{\alpha_{n-1}}s_{\alpha_{n-2}}, \\ \omega_{n-1} & \text{if } p\geq 5,\ \Phi \text{ is of type } B_n, \text{ and } w = s_{\alpha_{n-1}}s_{\alpha_n}s_{\alpha_{n-1}}, \\ \omega_{n-1}-\omega_{n-2} & \text{if } p=3,\ \Phi \text{ is of type } B_n, \text{ and } w = s_{\alpha_{n-1}}s_{\alpha_n}s_{\alpha_{n-1}}, \\ \omega_n & \text{if } p\geq 5,\ \Phi \text{ is of type } B_n, \text{ and } w = s_{\alpha_n}s_{\alpha_{n-1}}s_{\alpha_n}, \\ 2\omega_n & \text{if } p=3,\ \Phi \text{ is of type } B_n, \text{ and } w = s_{\alpha_n}s_{\alpha_{n-1}}s_{\alpha_n}, \\ \omega_{n-1} & \text{if } p\geq 5,\ \Phi \text{ is of type } C_n, \text{ and } w = s_{\alpha_{n-1}}s_{\alpha_n}s_{\alpha_{n-1}}, \\ 2\omega_{n-1}-\omega_{n-2} & \text{if } p=3,\ \Phi \text{ is of type } C_n, \text{ and } w = s_{\alpha_{n-1}}s_{\alpha_n}s_{\alpha_{n-1}}, \\ \omega_n & \text{if } p\geq 5,\ \Phi \text{ is of type } C_n, \text{ and } w = s_{\alpha_n}s_{\alpha_{n-1}}s_{\alpha_n}, \\ \omega_n-\omega_{n-2} & \text{if } p=3,\ \Phi \text{ is of type } C_n, \text{ and } w = s_{\alpha_n}s_{\alpha_{n-1}}s_{\alpha_n}, \\ \omega_1+\omega_2-\omega_3 & \text{if } p=3,\ \Phi \text{ is of type } F_4, \text{ and } w = s_{\alpha_1}s_{\alpha_2}s_{\alpha_1}, \\ \omega_2 & \text{if } p\geq 5,\ \Phi \text{ is of type } F_4, \text{ and } w = s_{\alpha_2}s_{\alpha_3}s_{\alpha_2}, \\ \omega_2-\omega_1-\omega_4 & \text{if } p=3,\ \Phi \text{ is of type } F_4, \text{ and } w = s_{\alpha_2}s_{\alpha_3}s_{\alpha_2}, \\ \omega_3+\omega_2-\omega_4 & \text{if } p=3,\ \Phi \text{ is of type } F_4, \text{ and } w = s_{\alpha_3}s_{\alpha_2}s_{\alpha_3}, \\ \omega_1 & \text{if } p\geq 7,\ \Phi \text{ is of type } G_2, \text{ and } w = s_{\alpha_1}s_{\alpha_2}s_{\alpha_1}, \\ 2\omega_1 & \text{if } p=3,5,\ \Phi \text{ is of type } G_2, \text{ and } w = s_{\alpha_1}s_{\alpha_2}s_{\alpha_1}, \\ \omega_2 & \text{if } p\geq 5,\ \Phi \text{ is of type } G_2, \text{ and } w = s_{\alpha_2}s_{\alpha_1}s_{\alpha_2}, \\ 2\omega_2-\omega_1 & \text{if } p=3,\ \Phi \text{ is of type } G_2, \text{ and } w = s_{\alpha_2}s_{\alpha_1}s_{\alpha_2}. \end{cases}$$

*Suppose from now on that $\alpha_i$, $\alpha_j$, and $\alpha_k$ are distinct.*

(II) *Suppose none of the simple roots $\alpha_i$, $\alpha_j$, and $\alpha_k$ are adjacent to each other. Then*

$$\gamma_w = \omega_i + \omega_j + \omega_k$$

*except in the following $p = 3$ cases, where*

$$\gamma_w = \begin{cases} \omega_n+\omega_{n-2}+\omega_k+\omega_{n-1} & \text{if } \Phi \text{ is of type } C_n, \text{ and } w = s_{\alpha_n}s_{\alpha_{n-2}}s_{\alpha_k} \\ & \quad \text{with } k \leq n-4, \\ \omega_n+\omega_{n-1}+\omega_{n-3}-\omega_{n-2} & \text{if } \Phi \text{ of type } D_n, \text{ and } w = s_{\alpha_n}s_{\alpha_{n-1}}s_{\alpha_{n-3}}, \\ \omega_2+\omega_3+\omega_5-\omega_4 & \text{if } \Phi \text{ is of type } E_n \text{ and } w = s_{\alpha_2}s_{\alpha_3}s_{\alpha_5}. \end{cases}$$

(III) *Suppose that precisely one pair of simple roots from $\alpha_i$, $\alpha_j$, and $\alpha_k$ are adjacent. By rewriting $w$ if necessary, we may assume that $\alpha_i \sim \alpha_j$, $\alpha_i \nsim \alpha_k$, and $\alpha_j \nsim \alpha_k$. Then*

$$\gamma_w = \omega_i + \omega_k$$

*unless $p = 3$ and $\alpha_i \approx \alpha_k$, in which case*

$$\gamma_w = \omega_i + \omega_k - \omega_{i,k}$$

*except in the following cases, where*

$$\gamma_w = \begin{cases} 2\omega_n + \omega_k & \text{if } p = 3,\ \Phi \text{ is of type } B_n, \text{ and } w = s_{\alpha_n} s_{\alpha_{n-1}} s_{\alpha_k} \\ & \quad \text{with } k \leq n-3, \\ 2\omega_{n-1} + \omega_k - \omega_{n-2} & \text{if } p = 3,\ \Phi \text{ is of type } C_n, \text{ and } w = s_{\alpha_{n-1}} s_{\alpha_n} s_{\alpha_k} \\ & \quad \text{with } k \leq n-3, \\ \omega_1 + \omega_4 - \omega_3 & \text{if } p = 3,\ \Phi \text{ is of type } F_4, \text{ and } w = s_{\alpha_1} s_{\alpha_2} s_{\alpha_4}, \\ \omega_2 + \omega_4 - \omega_3 & \text{if } p = 5,\ \Phi \text{ is of type } F_4, \text{ and } w = s_{\alpha_2} s_{\alpha_1} s_{\alpha_4}. \end{cases}$$

(IV) *Suppose* $\alpha_i \sim \alpha_j$ *and* $\alpha_j \sim \alpha_k$ *(i.e., there is a subgraph of the form* $\alpha_i \leftrightarrow \alpha_j \leftrightarrow \alpha_k$*). Then*

$$\gamma_w = \begin{cases} \omega_i & \text{if } p \geq 5, \\ 2\omega_i - \sum_{\alpha_\ell \sim \alpha_i, \ell \neq j} \omega_\ell & \text{if } p = 3, \end{cases}$$

*except in the following cases, where*

$$\gamma_w = \begin{cases} \omega_{n-1} - \omega_n & \text{if } p \geq 5, \Phi \text{ is of type } B_n, \text{ and } w = s_{\alpha_{n-1}} s_{\alpha_{n-2}} s_{\alpha_{n-3}}, \\ 2\omega_{n-1} - 2\omega_n & \text{if } p = 3, \Phi \text{ is of type } B_n, \text{ and } w = s_{\alpha_{n-1}} s_{\alpha_{n-2}} s_{\alpha_{n-3}}, \\ 2\omega_n & \text{if } p \geq 5, \Phi \text{ is of type } B_n, \text{ and } w = s_{\alpha_n} s_{\alpha_{n-1}} s_{\alpha_{n-2}}, \\ 2\omega_n - \omega_{n-1} & \text{if } p = 3, \Phi \text{ is of type } C_n, \text{ and } w = s_{\alpha_n} s_{\alpha_{n-1}} s_{\alpha_{n-2}}, \\ 2\omega_2 - \omega_1 - \omega_3 & \text{if } p = 3, \Phi \text{ is of type } F_4, \text{ and } w = s_{\alpha_2} s_{\alpha_3} s_{\alpha_4}. \end{cases}$$

(V) *Suppose* $\alpha_i \sim \alpha_j$ *and* $\alpha_i \sim \alpha_k$ *(i.e., there is a subgraph of the form* $\alpha_j \leftrightarrow \alpha_i \leftrightarrow \alpha_k$*). Then*

$$\gamma_w = \begin{cases} \omega_i & \text{if } p \geq 5, \\ 2\omega_i - \sum_{\alpha_\ell \sim \alpha_i, \ell \neq j,k} \omega_\ell & \text{if } p = 3, \end{cases}$$

*except in the following cases, where*

$$\gamma_w = \begin{cases} \omega_{n-1} - \omega_n & \text{if } p = 5, \Phi \text{ is of type } B_n, \text{ and } w = s_{\alpha_{n-1}} s_{\alpha_{n-2}} s_{\alpha_n}, \\ 2\omega_{n-1} - 2\omega_n & \text{if } p = 3, \Phi \text{ is of type } B_n, \text{ and } w = s_{\alpha_{n-1}} s_{\alpha_{n-2}} s_{\alpha_n}, \\ 2\omega_{n-1} & \text{if } p = 5, \Phi \text{ is of type } C_n, \text{ and } w = s_{\alpha_{n-1}} s_{\alpha_{n-2}} s_{\alpha_n}, \\ 2\omega_{n-1} - \omega_n & \text{if } p = 3, \Phi \text{ is of type } C_n, \text{ and } w = s_{\alpha_{n-1}} s_{\alpha_{n-2}} s_{\alpha_n}, \\ \omega_2 - \omega_3 & \text{if } p = 5, \Phi \text{ is of type } F_4, \text{ and } w = s_{\alpha_2} s_{\alpha_1} s_{\alpha_3}, \\ 2\omega_2 - 2\omega_3 & \text{if } p = 3, \Phi \text{ is of type } F_4, \text{ and } w = s_{\alpha_2} s_{\alpha_1} s_{\alpha_3}, \\ 2\omega_3 & \text{if } p = 5, \Phi \text{ is of type } F_4, \text{ and } w = s_{\alpha_3} s_{\alpha_4} s_{\alpha_2}, \\ 2\omega_3 - \omega_2 & \text{if } p = 3, \Phi \text{ is of type } F_4, \text{ and } w = s_{\alpha_3} s_{\alpha_4} s_{\alpha_2}. \end{cases}$$

(VI) *Suppose* $\alpha_i \sim \alpha_k$ *and* $\alpha_j \sim \alpha_k$ *(i.e., there is a subgraph of the form* $\alpha_i \leftrightarrow \alpha_k \leftrightarrow \alpha_j$*). Then*

$$\gamma_w = \omega_i + \omega_j$$

*except in the following cases when* $p = 3$, *where*

$$\gamma_w = \begin{cases} \omega_{n-1} + \omega_{n-3} - \omega_n & \text{if } \Phi \text{ is of type } B_n \text{ and } w = s_{\alpha_{n-1}} s_{\alpha_{n-3}} s_{\alpha_{n-2}}, \\ 2\omega_n + \omega_{n-2} - \omega_{n-1} & \text{if } \Phi \text{ is of type } B_n \text{ and } w = s_{\alpha_n} s_{\alpha_{n-2}} s_{\alpha_{n-1}}, \\ \omega_n + \omega_{n-2} - \omega_{n-1} & \text{if } \Phi \text{ is of type } C_n \text{ and } w = s_{\alpha_n} s_{\alpha_{n-2}} s_{\alpha_{n-1}}, \\ \omega_1 + \omega_3 - \omega_2 - \omega_4 & \text{if } \Phi \text{ is of type } F_4 \text{ and } w = s_{\alpha_1} s_{\alpha_3} s_{\alpha_2}, \\ \omega_2 + \omega_4 - \omega_3 & \text{if } \Phi \text{ is of type } F_4 \text{ and } w = s_{\alpha_2} s_{\alpha_4} s_{\alpha_3}. \end{cases}$$

## References


[And] H. H. Andersen, *Extensions of modules for algebraic groups*, Amer. J. Math. **106** (1984), no. 2, 489–504, DOI 10.2307/2374311. MR737781 (86g:20056)

[AJ] H. H. Andersen and J. C. Jantzen, *Cohomology of induced representations for algebraic groups*, Math. Ann. **269** (1984), no. 4, 487–525, DOI 10.1007/BF01450762. MR766011 (86g:20057)

[AR] H. H. Andersen and T. Rian, *B-cohomology*, J. Pure Appl. Algebra **209** (2007), no. 2, 537–549, DOI 10.1016/j.jpaa.2006.07.009. MR2293326 (2007k:20100)

[BBDNPPW] C. P. Bendel, B. D. Boe, C. M. Drupieski, D. K. Nakano, B. J. Parshall, C. Pillen, and C. B. Wright, *Bounding the dimensions of rational cohomology groups*, Developments and retrospectives in Lie theory, Dev. Math., vol. 38, Springer, Cham, 2014, pp. 51–69, DOI 10.1007/978-3-319-09804-3_2. MR3308777

[BC] W. Bosma, J. Cannon, *Handbook on Magma Functions*, Sydney University, 1996.

[BCP] W. Bosma, J. Cannon, and C. Playoust, *The Magma algebra system. I. The user language*, J. Symbolic Comput. **24** (1997), no. 3-4, 235–265, DOI 10.1006/jsco.1996.0125. Computational algebra and number theory (London, 1993). MR1484478

[BNP1] C. P. Bendel, D. K. Nakano, and C. Pillen, *Extensions for Frobenius kernels*, J. Algebra **272** (2004), no. 2, 476–511, DOI 10.1016/j.jalgebra.2003.04.003. MR2028069 (2004m:20089)

[BNP2] C. P. Bendel, D. K. Nakano, and C. Pillen, *Second cohomology groups for Frobenius kernels and related structures*, Adv. Math. **209** (2007), no. 1, 162–197, DOI 10.1016/j.aim.2006.05.001. MR2294220 (2008c:20085)

[CPS] E. Cline, B. Parshall, and L. Scott, *Cohomology, hyperalgebras, and representations*, J. Algebra **63** (1980), no. 1, 98–123, DOI 10.1016/0021-8693(80)90027-7. MR568566 (81k:20060)

[FP1] E. M. Friedlander and B. J. Parshall, *On the cohomology of algebraic and related finite groups*, Invent. Math. **74** (1983), no. 1, 85–117, DOI 10.1007/BF01388532. MR722727 (85d:20035)

[FP2] E. M. Friedlander and B. J. Parshall, *Cohomology of infinitesimal and discrete groups*, Math. Ann. **273** (1986), no. 3, 353–374, DOI 10.1007/BF01450727. MR824427 (87e:22026)

[Jan1] J. C. Jantzen, *Representations of algebraic groups*, 2nd ed., Mathematical Surveys and Monographs, vol. 107, American Mathematical Society, Providence, RI, 2003. MR2015057 (2004h:20061)

[Jan2] J. C. Jantzen, *First cohomology groups for classical Lie algebras*, Representation theory of finite groups and finite-dimensional algebras (Bielefeld, 1991), Progr. Math., vol. 95, Birkhäuser, Basel, 1991, pp. 289–315. MR1112165 (92e:17024)

[Kos] B. Kostant, *Lie algebra cohomology and the generalized Borel-Weil theorem*, Ann. of Math. (2) **74** (1961), 329–387. MR0142696 (26 #265)

[KLT] S. Kumar, N. Lauritzen, and J. F. Thomsen, *Frobenius splitting of cotangent bundles of flag varieties*, Invent. Math. **136** (1999), no. 3, 603–621, DOI 10.1007/s002220050320. MR1695207 (2000g:20088)

[N] N. V. Ngo, *Cohomology for Frobenius kernels of* $SL_2$, J. Algebra **396** (2013), 39–60, DOI 10.1016/j.jalgebra.2013.07.033. MR3108071

[OHal] J. O'Halloran, *A vanishing theorem for the cohomology of Borel subgroups*, Comm. Algebra **11** (1983), no. 14, 1603–1606, DOI 10.1080/00927878308822921. MR702726 (84j:20047)

[PT] P. Polo and J. Tilouine, *Bernstein-Gelfand-Gelfand complexes and cohomology of nilpotent groups over $\mathbb{Z}_{(p)}$ for representations with p-small weights* (English, with English and French summaries), Astérisque **280** (2002), 97–135. Cohomology of Siegel varieties. MR1944175 (2003j:17027)

[SFB1] A. Suslin, E. M. Friedlander, and C. P. Bendel, *Infinitesimal 1-parameter subgroups and cohomology*, J. Amer. Math. Soc. **10** (1997), no. 3, 693–728, DOI 10.1090/S0894-0347-97-00240-3. MR1443546 (98h:14055b)

[SFB2] A. Suslin, E. M. Friedlander, and C. P. Bendel, *Support varieties for infinitesimal group schemes*, J. Amer. Math. Soc. **10** (1997), no. 3, 729–759, DOI 10.1090/S0894-0347-97-00239-7. MR1443547 (98h:14055c)

[UGA] University of Georgia VIGRE Algebra Group, On Kostant's theorem for Lie algebra cohomology, *Cont. Math.*, 478, (2009), 39–60.

[W] C. B. Wright, *Second cohomology groups for algebraic groups and their Frobenius kernels*, J. Algebra **330** (2011), 60–75, DOI 10.1016/j.jalgebra.2011.01.013. MR2774617 (2012g:20090)

Department of Mathematics, Statistics and Computer Science, University of Wisconsin-Stout, Menomonie, Wisconsin 54751

*E-mail address*: bendelc@uwstout.edu

Department of Mathematics, University of Georgia, Athens, Georgia 30602

*E-mail address*: nakano@math.uga.edu

Department of Mathematics and Statistics, University of South Alabama, Mobile, Alabama 36688

*E-mail address*: pillen@southalabama.edu

Proceedings of Symposia in Pure Mathematics
Volume **92**, 2016
http://dx.doi.org/10.1090/pspum/092/01585

# Invariant theory for quantum Weyl algebras under finite group action

S. Ceken, J. H. Palmieri, Y.-H. Wang, and J. J. Zhang

ABSTRACT. We study the invariant theory of a class of quantum Weyl algebras under group actions and prove that the fixed subrings are always Gorenstein. We also verify the Tits alternative for the automorphism groups of these quantum Weyl algebras.

## Introduction

Fix a field $k$. For $n \geq 2$, let $W_n$ be the $(-1)$-*quantum Weyl algebra*: this is the $k$-algebra generated by $x_1, \ldots, x_n$ subject to the relations $x_i x_j + x_j x_i = 1$ for all $i \neq j$.

THEOREM 1. *Assume that* $\operatorname{char} k = 0$. *Let $n$ be an even integer $\geq 4$ and let $G$ be a group acting on $W_n$. Then the fixed subring $W_n^G$ under the $G$-action is filtered Artin-Schelter Gorenstein.*

The above theorem was announced in [**CPWZ1**, Theorem 2] without proof. The first purpose of this paper is to provide a proof of this, using some earlier results from noncommutative invariant theory [**JiZ**, **KKZ1**]. Secondly, we discuss the automorphism group of $W_n$ when $n \geq 3$ is odd [Theorem 2]. We also want to correct some small errors in [**CPWZ1**]: see Remarks 1.5(2) and 2.7. Finally, we give a criterion for an isomorphism question for a class of $(-1)$-quantum Weyl algebras [Theorem 3].

One aspect of invariant theory is to study homological properties of fixed subrings (also called invariant subrings) under group actions. When $A$ is regular (or has finite global dimension), the fixed subring $A^G$ under the action of a finite group $G$ is Cohen-Macaulay when $\operatorname{char} k$ does not divide the order of $G$. One interesting question is *when a fixed subring $A^G$ is Gorenstein.* The famous Watanabe theorem [**Wa**] answers such a question for commutative polynomial rings. Theorem 1 above provides a solution for $W_n$ when $n$ is even. However, this question is open for $W_n^G$ when $n$ is odd (even when $n = 3$ and $G$ is finite). Another interesting question is *when a fixed subring $A^G$ is regular.* The classical result of Shephard-Todd-Chevalley [**ST**, **Ch**] answers this question for commutative polynomial rings. The question is open for $W_n$ for all $n \geq 3$, though we have an easy partial result:

2010 *Mathematics Subject Classification.* Primary 16W20, 16E65.
*Key words and phrases.* Automorphism group, quantum Weyl algebra, Artin-Schelter Gorenstein property, free subgroup.

see Proposition 1.8. A recent survey on invariant theory of Artin-Schelter regular algebras is given by Kirkman in [**Ki**].

Another aspect of invariant theory is to study the structure of the automorphism group $\mathrm{Aut}(A)$ of an algebra $A$. There is a long history and an extensive study of the automorphism groups of algebras. Determining the full automorphism group of an algebra is generally a notoriously difficult problem. Recently, significant progress has been made in finding the full automorphism groups of some noncommutative algebras. For example, during the last two years, Yakimov has proved the Andruskiewitsch-Dumas conjecture and the Launois-Lenagan conjecture by using a rigidity theorem for quantum tori [**Y1**, **Y2**]. The automorphism groups of generalized or quantum Weyl algebras have been studied by several researchers [**AD**, **BJ**, **SAV**]. The authors used the discriminant method to determine the automorphism groups of some noncommutative algebras [**CPWZ1**, **CPWZ2**]. When $n$ is even, $\mathrm{Aut}(W_n)$ was worked out by the authors in [**CPWZ1**, Theorem 1]. Unfortunately, we have not been able to determine the full automorphism group of $W_n$ when $n$ is odd [**CPWZ1**, Example 5.10 and Question 5.13].

The second theorem concerns $\mathrm{Aut}(W_n)$ when $n$ is odd. We will consider a slightly more general setting. Let $\mathcal{A} := \{a_{ij} \in k \mid 1 \le i < j \le n\}$ be a set of scalars. Define a modified $(-1)$-quantum Weyl algebra $V_n(\mathcal{A})$ [**CPWZ1**, Section 4] to be the $k$-algebra generated by $\{x_1, \ldots, x_n\}$ subject to the relations

$$x_i x_j + x_j x_i = a_{ij}, \quad \forall\, 1 \le i < j \le n.$$

As a special case, note that if $a_{ij} = 1$ for all $i < j$, then $V_n(\mathcal{A}) = W_n$. Another special case is when $a_{ij} = 0$ for all $i < j$, in which case $V_n(\mathcal{A})$ is just the skew polynomial ring $k_{-1}[x_1, \ldots, x_n]$.

THEOREM 2. *Let $n \ge 3$ be an odd integer. Suppose* $\mathrm{char}\, k$ *does not divide $(n-1)!$. Then* $\mathrm{Aut}(V_n(\mathcal{A}))$ *contains a free group on two generators. As a consequence, it contains a free group on countably many generators.*

Note that [**CPWZ1**, Theorem 1] implies that, when $n$ is even, $\mathrm{Aut}(W_n)$ is finite (and so virtually solvable). A more general statement for $\mathrm{Aut}(V_n(\mathcal{A}))$ is Lemma 1.1. Combining Theorem 2 with Lemma 1.1(2,3), we obtain that in characteristic 0, $\mathrm{Aut}(V_n(\mathcal{A}))$ either is virtually solvable or contains a free subgroup of rank two. This can be viewed as a version of the Tits alternative [**Ti**] for the automorphism groups of $V_n(\mathcal{A})$. The original Tits alternative states that every finitely generated linear group is either virtually solvable or contains a free subgroup of rank two. We also have a version of the Tits alternative for the class of the automorphism groups of skew polynomial rings in a separate paper [**CPWZ3**].

In Section 3, we use the discriminant to prove the following criterion for when two $V_n(\mathcal{A})$s are isomorphic (in the case when $n$ is even). For simplicity, let $a_{ij} = a_{ji}$ if $i > j$.

THEOREM 3. *Suppose* $\mathrm{char}\, k \neq 2$ *and let $n$ be an even integer. Let $\mathcal{A}' := \{a'_{ij} \mid 1 \le i < j \le n\}$ be another set of scalars in $k$. Then $V_n(\mathcal{A}) \cong V_n(\mathcal{A}')$ if and only if there are a permutation $\sigma \in S_n$ and nonzero scalars $\lambda_i$ for $i = 1, \ldots, n$ such that $a'_{ij} = \lambda_i \lambda_j a_{\sigma(i)\sigma(j)}$ for all $i$ and $j$.*

As a consequence of Theorem 3, when $n \ge 4$ is even, there are infinitely many non-isomorphic $V_n(\mathcal{A})$s.

In Section 4, we give some examples of $\mathrm{Aut}(V_n(\mathcal{A}))$ for $n = 4, 6$ and list some questions about $V_n(\mathcal{A})$.

## 1. Proof of Theorem 1

Throughout let $k$ be a base commutative domain. Modules, vector spaces, algebras, and morphisms are over $k$. In this section we further assume that $k$ is a field.

As above, for each $n \geq 2$, let $V_n(\mathcal{A})$ denote the algebra generated by $\{x_i\}_{i=1}^n$ subject to the relations $x_i x_j + x_j x_i = a_{ij}$ for all $i \neq j$. This is a filtered Artin-Schelter regular algebra in the following sense. First we recall the definition of Artin-Schelter regularity [**AS**, p.171]. A connected graded algebra $A$ is called *Artin-Schelter Gorenstein* (or *AS Gorenstein* for short) of dimension $d$ if the following conditions hold:

(1) $A$ has injective dimension $d < \infty$ on the left and on the right,
(2) $\mathrm{Ext}^i_A({}_A k, {}_A A) = \mathrm{Ext}^i_A(k_A, A_A) = 0$ for all $i \neq d$, and
(3) $\mathrm{Ext}^d_A({}_A k, {}_A A) \cong \mathrm{Ext}^d_A(k_A, A_A) \cong k(l)$ for some $l$.

If, in addition,

(4) $A$ has finite global dimension, and
(5) $A$ has finite Gelfand-Kirillov dimension,

then $A$ is called *Artin-Schelter regular* (or *AS regular* for short) of dimension $d$. Now let $A$ be an ungraded algebra and let $F := \{F_n \subseteq A \mid n \geq 0\}$ be an increasing filtration on $A$ satisfying

(a) $F_0 = k$,
(b) $F_n F_m \subseteq F_{n+m}$ for all $n, m \geq 0$,
(c) $A = \bigcup_n F_n$.

The associated graded algebra with respect to $F$, denoted by $\mathrm{gr}_F A$, is defined to be $\mathrm{gr}_F A = \bigoplus_{i=0}^{\infty} F_i/F_{i-1}$. This algebra is connected graded by condition (a). An algebra $A$ is called *filtered Artin-Schelter regular* (resp. *filtered Artin-Schelter Gorenstein*) if there is a filtration $F$ such that $\mathrm{gr}_F A$ is Artin-Schelter regular (resp. Artin-Schelter Gorenstein). In our case, we filter $V_n(\mathcal{A})$ by setting $F_0 = k$, $F_1 = k + \sum_{s=1}^n k x_s$, and $F_i = (F_1)^i$ for all $i \geq 2$. Then it is easy to see that $\mathrm{gr}_F V_n(\mathcal{A}) \cong k_{-1}[x_1, \ldots, x_n] (= V_n(\{0\}))$, the skew polynomial ring generated by $\{x_i\}_{i=1}^n$ subject to the relations $x_j x_i = -x_i x_j$ for all $i < j$. Since $k_{-1}[x_1, \ldots, x_n]$ is Artin-Schelter regular, $V_n(\mathcal{A})$ is filtered Artin-Schelter regular. The symmetric group $S_n$ acts on the set $\{x_i\}_{i=1}^n$ naturally. It is easy to see that this action extends to an $S_n$-action on the algebras $V_n(\mathcal{A})$ and $k_{-1}[x_1, \ldots, x_n]$ uniquely; note that the invariant theory of $S_n$-actions on $k_{-1}[x_1, \ldots, x_n]$ was studied in [**KKZ1**].

The group of all affine automorphisms of $V_n(\mathcal{A})$, denoted by $\mathrm{Aut}_{\mathrm{af}}(V_n(\mathcal{A}))$, can be worked out by using easy combinatorics [**CPWZ1**, Lemma 4.3]. Let $G(\mathcal{A})$ be the set of automorphisms of $V_n(\mathcal{A})$ of the form

$$g: \quad x_i \to r_i x_{\sigma(i)}, \ \forall\, i \tag{E1.0.1}$$

where $\sigma \in S_n$ and $r_i \in k^\times = k \setminus \{0\}$. It is easy to see that $g$ is an automorphism of $V_n(\mathcal{A})$ if and only if

$$a_{ij} = r_i r_j a_{\sigma(i)\sigma(j)} \tag{E1.0.2}$$

for all $i \neq j$. Such an automorphism $g$ of $V_n(\mathcal{A})$ is denoted by $g(\sigma, \{r_1, \ldots, r_n\})$. The following lemma is easy. We will use another obvious automorphism of $V_n(\mathcal{A})$, which sends $x_i$ to $-x_i$ for all $i = 1, \ldots, n$. We denote this automorphism by $-1$, while the identity is just denoted by 1.

LEMMA 1.1. *Suppose that $n \geq 2$.*

(1) [**CPWZ1**, Lemma 4.3] *$G(\mathcal{A})$ is a group, and $\operatorname{Aut}_{\text{af}}(V_n(\mathcal{A})) = G(\mathcal{A})$.*

(2) *There is a short exact sequence*

$$1 \to Z \to G(\mathcal{A}) \to S_n$$

*where $Z$ is a subgroup of $(k^\times)^n$. As a consequence, $G(\mathcal{A})$ is virtually abelian (so virtually solvable).*

(3) [**CPWZ1**, Theorem 4.9(3)] *If $n$ is even, then $\operatorname{Aut}(V_n(\mathcal{A})) = \operatorname{Aut}_{\text{af}}(V_n(\mathcal{A})) = G(\mathcal{A})$.*

*In parts (4–5) we further assume that $a_{ij} \neq 0$ for all $i < j$.*

(4) *If $n \geq 3$, there is a short exact sequence*

$$1 \to \{\pm 1\} \to G(\mathcal{A}) \to S_n.$$

(5) *If $n \geq 4$ is even, then $\operatorname{Aut}(V_n(\mathcal{A})) = G(\mathcal{A})$ is finite.*

PROOF. (1) By a direct computation, $G(\mathcal{A}) \subseteq \operatorname{Aut}_{\text{af}}(V_n(\mathcal{A}))$.

Now fix $g \in \operatorname{Aut}_{\text{af}}(V_n(\mathcal{A}))$. By [**CPWZ1**, Lemma 4.3], there is a permutation $\sigma \in S_n$ and scalars $r_i \in k^\times$ such that $g(x_i) = r_i x_{\sigma(i)}$ for all $i$. Applying $g$ to the relations of $V_n(\mathcal{A})$, one obtains that $a_{ij} = r_i r_j a_{\sigma(i)\sigma(j)}$ for all $i \neq j$. So $g \in G(\mathcal{A})$. Thus $\operatorname{Aut}_{\text{af}}(V_n(\mathcal{A})) \subseteq G(\mathcal{A})$.

(2) The map from $G(\mathcal{A})$ to $S_n$ takes $g(\sigma, \{r_1, \ldots, r_n\})$ to $\sigma$. So the kernel $Z$ of this map consists of all $g(Id, \{r_1, \ldots, r_n\})$ where $a_{ij} = r_i r_j a_{ij}$ for all $i \neq j$. Since each $r_i$ is in $k^\times$, $Z$ is a subgroup of $(k^\times)^n$. Since $Z$ is abelian, $G(\mathcal{A})$ is virtually abelian by definition.

(3) By [**CPWZ1**, Theorem 4.9(3)], $\operatorname{Aut}(V_n(\mathcal{A})) = \operatorname{Aut}_{\text{af}}(V_n(\mathcal{A}))$. The assertion follows from part (1).

(4) By the proof of part (2), $Z$ consists of all $g(Id, \{r_1, \ldots, r_n\})$ where $\{r_i\}_{i=1}^n$ satisfy $a_{ij} = r_i r_j a_{ij}$ for all $i \neq j$. Since $a_{ij} \neq 0$, we have $r_i r_j = 1$ for all $i < j$. Since $n \geq 3$, we obtain that $r_i = r_j$ for all $i < j$ and so $r_i = 1$ or $r_i = -1$.

(5) This follows from parts (3,4). □

Let $G_1(\mathcal{A})$ denote the subgroup of $G(\mathcal{A})$ consisting of all $g(\sigma, \{r_1, \ldots, r_n\})$ such that $\prod_{i=1}^n r_i = 1$. If $\mathcal{A} = \{1\}_{i<j}$ (i.e., the $W_n$ case) and $n \geq 1$ is even, then $G(\mathcal{A}) = G_1(\mathcal{A})$ by [**CPWZ1**, Theorem 1]. We give an example for which $G(\mathcal{A}) \neq G_1(\mathcal{A})$ when $n = 4$.

EXAMPLE 1.2. Let $n = 4$ and $(a_{ij})_{4\times 4} = \begin{pmatrix} \emptyset & 1 & -1 & 1 \\ 1 & \emptyset & 1 & -a^2 \\ -1 & 1 & \emptyset & -a \\ 1 & -a^2 & -a & \emptyset \end{pmatrix}$ where $a^3 = -1$ and $a \neq -1$. In the definition of the algebra $V_n(\mathcal{A})$, we do not need the entries $a_{ii}$ in the above matrix, which are denoted by $\emptyset$ for simplicity. It is easy to check that $g((123), \{1, 1, -1, a\})$ is in $G(\mathcal{A})$, but not in $G_1(\mathcal{A})$.

Next we recall the definitions of Auslander-Gorenstein and GKdim-Macaulay, see also [**JiZ**, Definition 0.6]. Let $A$ be a noetherian ring.

(1) The *grade* of an $A$-module $M$ is defined to be

$$j(M) = \min\{i \mid \operatorname{Ext}^i_A(M, A) \neq 0\}$$

or $\infty$ if no such $i$ exists.

(2) We say $A$ is *Auslander-Gorenstein* if the following two conditions hold:
   (a) $A$ has finite left and right injective dimension,
   (b) for every noetherian $A$-module $M$ and for all $i \geq 0$ and all submodules $N \subseteq \operatorname{Ext}^i_A(M, A)$, one has $j(N) \geq i$.

   We say $A$ is *Auslander regular* if $A$ is Auslander-Gorenstein and $A$ has finite global dimension.

(3) We say $A$ is GKdim-*Macaulay* (where GKdim denotes the Gelfand-Kirillov dimension) if

$$\operatorname{GKdim} M + j(M) = \operatorname{GKdim} A < \infty$$

   for every noetherian $A$-module $M$.

Another concept that we will use is the homological determinant defined in [**JoZ**, **JiZ**, **KKZ3**]. We only recall the definition for connected graded AS Gorenstein algebras using [**KKZ3**, Proposition 5.3(d)]. Since $A$ is AS Gorenstein, $\operatorname{Ext}^d_A(k, A)$ is 1-dimensional where $d$ is the injective dimension of $A$. Let $\mathfrak{e}$ be any nonzero element in $\operatorname{Ext}^d_A(k, A)$. For any graded algebra automorphism $g$ of $A$, it acts on $\operatorname{Ext}^d_A(k, A)$, see [**KKZ3**, Proposition 5.3(d)]. Thus $g(\mathfrak{e}) = \eta(g)\mathfrak{e}$ where $\eta(g)$ is a nonzero scalar in $k$. The *homological determinant* of $g$ is defined to be $\eta(g)^{-1}$. We refer to [**JiZ**] for the definition of the homological determinant in the filtered case.

THEOREM 1.3. *Let $G$ be a finite subgroup of $G_1(\mathcal{A})$. Suppose that* $\operatorname{char} k$ *does not divide $|G|$. Then the fixed subring $V_n(\mathcal{A})^G$ is filtered Artin-Schelter Gorenstein, Auslander-Gorenstein and* GKdim-*Macaulay.*

PROOF. First of all, we note that both $k_{-1}[x_1, \dots, x_n]$ and $V_n(\mathcal{A})$ are filtered Artin-Schelter Gorenstein, Auslander-Gorenstein and GKdim-Macaulay. We claim that, for each $g \in G \subset G_1(\mathcal{A})$, its homological determinant, denoted by $\operatorname{hdet}_{V_n(\mathcal{A})} g$, is 1. If the claim holds, then the assertion follows from [**JiZ**, Theorem 3.5].

Now we prove the claim. By definition [**JiZ**, p. 677], $\operatorname{hdet}_{V_n(\mathcal{A})} g = \operatorname{hdet}_{\operatorname{gr} V_n(\mathcal{A})} g$. To show the claim, we consider the algebra $\operatorname{gr} V_n(\mathcal{A}) \cong k_{-1}[x_1, \dots, x_n] := B$ and identify $G$ with a subgroup of $\operatorname{Aut}(B)$ in the natural way. For any element $g = g(\sigma, \{r_1, \dots, r_n\}) \in G$, there is a decomposition

$$g(\sigma, \{r_1, \dots, r_n\}) = g(\sigma, \{1, \dots, 1\}) \circ g(Id, \{r_1, \dots, r_n\})$$

as in $\operatorname{Aut}(B)$. Then the homological determinant of $g(\sigma, \{1, \dots, 1\})$ is 1 by the proof of [**KKZ1**, Theorem 1.5]. The homological determinant of $g(Id, \{r_1, \dots, r_n\})$ is $\prod_{i=1}^n r_i$, which is 1 by the definition of $G_1(\mathcal{A})$. Combining these, we see that $\operatorname{hdet}_B g = 1$ for all $g \in G$. Thus we have proved the claim, and therefore the theorem. □

Now we are ready to prove Theorem 1.

THEOREM 1.4. *Let $n$ be an even integer $\geq 4$ and $G$ a group acting on $W_n$ faithfully. Suppose that* $\operatorname{char} k$ *does not divide $2|G|$. Then the fixed subring $W_n^G$ is filtered Artin-Schelter Gorenstein, Auslander-Gorenstein, and* GKdim-*Macaulay.*

Proof. Since $n$ is even and $\operatorname{char} k \neq 2$, by [**CPWZ1**, Theorem 1], $\operatorname{Aut}(W_n) = S_n \times \{\pm 1\}$. So $G$ is a subgroup of $S_n \times \{\pm 1\}$, which is $G_1(\{1\}_{i<j})$ when $n$ is even. The assertion follows from Theorem 1.3. □

Remark 1.5. (1) Theorem 1 is clearly a consequence of Theorem 1.4: if the group $G$ acts on $W_n$, then $W_n^G = W_n^{\operatorname{im} G}$, where $\operatorname{im} G$ is the image of $G$ in $\operatorname{Aut} W_n$. Note that $\operatorname{im} G$ acts faithfully.

(2) The hypothesis that "$\operatorname{char} k$ does not divide $2|G|$" is missing from [**CPWZ1**, Theorem 2]. We do not know whether $W_n^G$ is filtered Artin-Schelter Gorenstein when $\operatorname{char} k$ divides $2|G|$.

(3) By [**CWWZ**, Example 3.4(2) for $D_4$], the fixed subring $W_2^{S_2 \times \{\pm 1\}}$ is a commutative Gorenstein algebra with isolated singularities. So we expect that, in general, $W_n^G$ is not regular. (Note that [**CWWZ**, Example 3.4(2)] was stated in terms of the ring $R = k\langle u, v\rangle/(u^2 + v^2 - 1)$, but this is isomorphic to $W_2$.)

For completeness we have the following assertion for $n = 2$; see also [**CWWZ**, Theorem 0.1].

Proposition 1.6. *Let $G$ be a finite group acting on $W_2$ faithfully. Suppose that $\operatorname{char} k$ does not divide $2|G|$. Then the fixed subring $W_2^G$ is filtered Artin-Schelter Gorenstein, Auslander-Gorenstein, and* GKdim-*Macaulay.*

Proof. By [**AD**], $\operatorname{Aut}(W_2) = S_2 \ltimes k^\times$. So $G$ is a finite subgroup of $S_2 \ltimes k^\times$. Note that the $G$-action preserves the given filtration of $W_2$. By [**CWWZ**, Lemma 2.7], the homological determinant of the $G$-action is trivial (namely, $\operatorname{hdet}_{W_2} g = 1$ for all $g \in G$). The assertion follows from [**JiZ**, Theorem 3.5]. □

By [**KKZ1**, Theorem 1.5(2)], if $G$ is a subgroup of $S_n$ other than $\{1\}$, then $k_{-1}[x_1, \ldots, x_n]^G$ is not regular: its global dimension is infinite. When $G$ is a special subgroup of $S_n$, then $k_{-1}[x_1, \ldots, x_n]^G$ is a classical complete intersection [**KKZ1**, Theorem 3.12]. Some version of this should be true for $W_n$. Here we only give one simple example.

Example 1.7. Some of the arguments in this example were motivated by [**KKZ1**, Example 3.1].

Let $S_2$ act on $W_2$ in the obvious way. Then $A := W_2^{S_2}$ is a complete intersection, in the sense that it is isomorphic to $B/(f)$ for some regular algebra $B$ and some normal nonzerodivisor $f \in B$. To see this, we note that $A$ is generated by $u = x_1 + x_2$ and $v = x_1^3 + x_2^3$ satisfying two relations $[u^2, v] = [u, v^2] = 0$. By a straightforward computation, we have

$$r_3 := 2u^6 - 3u^3v - 3vu^3 + 4v^2 - 5u^4 + 3u^2 + 4(uv + vu) = 0.$$

One can check that $A$ is isomorphic to $B/(r_3)$, where $B$ is the down-up algebra $k\langle u, v\rangle/([u^2, v] = [u, v^2] = 0)$.

Proposition 1.8. *Let $G$ be the subgroup $\{\pm 1\}$ of* $\operatorname{Aut}(V_n(\mathcal{A}))$. *If $a_{ij} \neq 0$ for some $i < j$, then $V_n(\mathcal{A})^G$ is regular.*

Proof. Define $\deg x_i = 1 \in \mathbb{Z}/(2)$. This makes $V_n(\mathcal{A})$ into a $\mathbb{Z}/(2)$-graded algebra. Since the automorphism $-1$ will act on a length $s$ monomial $m = x_{i_1} \ldots x_{i_s}$ by multiplication by $(-1)^s$, such a monomial is in $V_n(\mathcal{A})^G$ if and only if $s$ is even, which holds if and only if $m$ is in the degree 0 part $(V_n(\mathcal{A}))_0$ of $V_n(\mathcal{A})$. Thus

$V_n(\mathcal{A})^G = (V_n(\mathcal{A}))_0$. Since $x_ix_j + x_jx_i = a_{ij} \neq 0$ for some $i < j$, $V_n(\mathcal{A})$ is strongly $\mathbb{Z}/(2)$-graded. It is well known that $V_n(\mathcal{A})$ is regular, so $(V_n(\mathcal{A}))_0$ is regular of the same global dimension by a theorem of Dade's [**Da**, Theorem 2.8] that the categories of $(V_n(\mathcal{A}))_0$-modules and $\mathbb{Z}/(2)$-graded $V_n(\mathcal{A})$-modules are equivalent. □

## 2. Proof of Theorem 2

In this section we assume that $k$ is a commutative domain.

Let $A$ be an algebra generated over $k$ by $x_1, x_2, \ldots, x_t$ for some $t \geq 3$, and let $R$ be the subalgebra generated by $x_3, \ldots, x_t$. Suppose we have an algebra automorphism of $A$ of the following form:

$$g(x_i) = \begin{cases} x_1 + a_0 + a_1x_2 & i = 1, \\ x_i & i \neq 1, \end{cases} \tag{E2.0.1}$$

where $a_0, a_1$ are in the subalgebra $R$, and in particular are fixed by $g$. The following lemma is easy.

LEMMA 2.1. *Let $n$ be an integer. Then*

$$g^n(x_i) = \begin{cases} x_1 + na_0 + na_1x_2 & i = 1, \\ x_i & i \neq 1. \end{cases}$$

Assume that, when $n$ and $m$ are any elements in $k$, the following are algebra isomorphisms of $A$:

$$g^n(x_i) = \begin{cases} x_1 + na_0 + na_1x_2 & i = 1, \\ x_i & i \neq 1, \end{cases} \tag{E2.1.1}$$

and

$$h^m(x_i) = \begin{cases} x_2 + mb_0 + mb_1x_1 & i = 2, \\ x_i & i \neq 2. \end{cases} \tag{E2.1.2}$$

This is automatic if $\operatorname{char} k = 0$. We abuse notation and denote these by $g^n$ and $h^m$, even if $n$ and $m$ are not integers. It is easy to check that $g^ng^{n'} = g^{n+n'}$.

By symmetry, if we also assume that we have an algebra automorphism

$$h(x_i) = \begin{cases} x_2 + b_0 + b_1x_1 & i = 2, \\ x_i & i \neq 2, \end{cases} \tag{E2.1.3}$$

where $b_0, b_1 \in R$, then for any $m \in k$, we have

$$h^m(x_i) = \begin{cases} x_2 + mb_0 + mb_1x_1 & i = 2, \\ x_i & i \neq 2. \end{cases}$$

LEMMA 2.2. *For any $m, n \in k$,*

$$h^mg^n(x_i) = \begin{cases} x_1 + (na_0 + na_1mb_0) + na_1x_2 + na_1mb_1x_1 & i = 1, \\ x_2 + mb_0 + mb_1x_1 & i = 2, \\ x_i & i \neq 1, 2, \end{cases}$$

*and*

$$g^n h^m(x_i) = \begin{cases} x_1 + na_0 + na_1x_2 & i = 1, \\ x_2 + (mb_0 + mb_1na_0) + mb_1x_1 + mb_1na_1x_2 & i = 2, \\ x_i & i \neq 1, 2. \end{cases}$$

Let $z$ be a positive integer and $n_1, \ldots, n_z, m_1, \ldots, m_z$ nonzero elements in $k$. Define automorphisms

$$\tau_z = g^{n_z} h^{m_{z-1}} g^{n_{z-1}} \cdots h^{m_1} g^{n_1},$$
$$\sigma_z = h^{m_z} \tau_z = h^{m_z} g^{n_z} h^{m_{z-1}} g^{n_{z-1}} \cdots h^{m_1} g^{n_1}.$$

LEMMA 2.3. *Retain the above notation. We have*

$$\tau_z(x_1) = d_0 + d_1x_1 + d_2x_2 + \left(\prod_{s=1}^{z-1} m_s\right)\left(\prod_{s=1}^{z} n_s\right) a_1(b_1a_1)^{z-1}x_2$$

*where $d_0, d_1, d_2$ are elements in $R$. Further,*

$$d_1 = \sum_{s=0}^{z-1} \alpha_s (a_1b_1)^s \quad and \quad d_2 = \sum_{s=0}^{z-2} \beta_s a_1 (b_1a_1)^s$$

*where $\alpha_s, \beta_s$ are elements in the subring of $k$ generated by $n_1, \ldots, n_z, m_1, \ldots, m_z$.*

PROOF. When $z = 1$, this is Lemma 2.1. Now assume that $z > 1$. By definition, $\tau_z = g^{n_z} h^{m_{z-1}} \tau_{z-1}$. By a direct computation and the induction hypothesis,

$$\begin{aligned}
\tau_z(x_1) &= g^{n_z} h^{m_{z-1}} \tau_{z-1}(x_1) \\
&= g^{n_z} h^{m_{z-1}} \left(d'_0 + d'_1x_1 + d'_2x_2 + \left(\prod_{s=1}^{z-2} m_s\right)\left(\prod_{s=1}^{z-1} n_s\right) a_1(b_1a_1)^{z-2}x_2\right) \\
&= d'_0 + d'_1(x_1 + n_za_0 + n_za_1x_2) \\
&\quad + d'_2(x_2 + (m_{z-1}b_0 + m_{z-1}b_1n_za_0) + m_{z-1}b_1x_1 + m_{z-1}b_1n_za_1x_2) \\
&\quad + \left(\prod_{s=1}^{z-2} m_s\right)\left(\prod_{s=1}^{z-1} n_s\right) a_1(b_1a_1)^{z-2}. \\
&\qquad (x_2 + (m_{z-1}b_0 + m_{z-1}b_1n_za_0) + m_{z-1}b_1x_1 + m_{z-1}b_1n_za_1x_2) \\
&= d_0 + d_1x_1 + d_2x_2 + \left(\prod_{s=1}^{z-1} m_s\right)\left(\prod_{s=1}^{z} n_s\right) a_1(b_1a_1)^{z-1}x_2
\end{aligned}$$

where

$$\begin{aligned}
d_0 &= d'_0 + d'_1n_za_0 + d'_2(m_{z-1}b_0 + m_{z-1}b_1n_za_0) \\
&\quad + \left(\prod_{s=1}^{z-2} m_s\right)\left(\prod_{s=1}^{z-1} n_s\right) a_1(b_1a_1)^{z-2}(m_{z-1}b_0 + m_{z-1}b_1n_za_0), \\
d_1 &= d'_1 + d'_2m_{z-1}b_1 + \left(\prod_{s=1}^{z-2} m_s\right)\left(\prod_{s=1}^{z-1} n_s\right) a_1(b_1a_1)^{z-2}m_{z-1}b_1, \\
d_2 &= d'_1n_za_1 + d'_2 + d'_2m_{z-1}b_1n_za_1 + \left(\prod_{s=1}^{z-2} m_s\right)\left(\prod_{s=1}^{z-1} n_s\right) a_1(b_1a_1)^{z-2}.
\end{aligned}$$

The formulas for $d_1$ and $d_2$ follow easily by induction. □

In the next proposition, the free product $G * H$ of two groups $G$ and $H$ will be used. It is well known that every element in $G * H$ can be expressed uniquely in the form $g_1h_1g_2h_2\cdots g_sh_s$ for some $s$ and for some $g_i \in G$, $h_i \in H$ with $h_i \neq 1$ for $i \neq s$ and $g_i \neq 1$ for $i \neq 1$. The following lemma is well known.

LEMMA 2.4. *Let $G$ be a group of order at least 3. Then $G * G$ contains a free subgroup of rank two.*

PROOF. If $x \in G$ has infinite order, then $\langle x\rangle * \langle x\rangle$ is a free group of rank two contained in $G * G$.

If $x$ has order $n \geq 3$, then $\langle x\rangle * \langle x\rangle$ contains a free subgroup of rank $(n-1)^2$ [**AP**, Theorem 1].

If every element $x$ in $G$ has order 2, then $G$ is abelian and $G$ contains $(\mathbb{Z}/(2))^{\oplus 2}$. It is well known that $(\mathbb{Z}/(2))^{\oplus 2} * (\mathbb{Z}/(2))^{\oplus 2}$ contains a free subgroup of rank $(4-1)^2$ [**AP**, Theorem 1]. □

Let $k_0$ be a subring of $k$. We say an element $f \in A$ is *transcendental over* $k_0$ if there are no $r_i \in k_0$ with $r_n \neq 0$ such that $\sum_{i=0}^n r_i f^i = 0$.

PROPOSITION 2.5. *Retain the hypotheses and notation as above; in particular, assume that $g^n$ and $h^m$ from* (E2.1.1) *and* (E2.1.2) *are automorphisms of $A$ for all $n, m \in k$. Assume that $R + Rx_1 + Rx_2$ is a free left $R$-module with basis $\{1, x_1, x_2\}$ and that the element $a_1b_1 \in A$ is transcendental over a subring $k_0 \subseteq k$.*

(1) *Suppose $m_s$ and $n_s$ are nonzero elements in $k_0$ for all $s$. Then neither $\sigma_z$ nor $\tau_z$ is the identity automorphism.*
(2) $\mathrm{Aut}(A)$ *contains the free product $(k_0, +) * (k_0, +)$. As a consequence, if $k_0 \neq \mathbb{Z}/(2)$, then* $\mathrm{Aut}(A)$ *contains a free subgroup of rank two.*

PROOF. (1) Since $a_1b_1$ is transcendental over $k_0$,

$$d_2 + (\prod_{s=1}^{z-1} m_s)(\prod_{s=1}^{z} n_s)a_1(b_1a_1)^{z-1}$$

is not zero. Since $R + Rx_1 + Rx_2$ is free, $\tau_z(x_1) \neq x_1$ by Lemma 2.3. So $\tau_z$ is not the identity. Since $h^{-m_z}(x_1) = x_1$, we deduce that $\tau_z \neq h^{-m_z}$, so $\sigma_z$ is not the identity.

(2) There is a natural group map $(k_0, +) * (k_0, +) \to \mathrm{Aut}(A)$ sending $m_z * n_z * m_{z-1} * n_{z-1} * \cdots * m_1 * n_1$ to $h^{m_z}g^{n_z}h^{m_{z-1}}g^{n_{z-1}}\cdots h^{m_1}g^{n_1}$. By part (1) the kernel of this map is $\{1\}$. So the main assertion follows. The consequence follows from Lemma 2.4. □

Here is a version of [**CPWZ1**, Lemma 5.11] for the algebra $V_n(\mathcal{A})$. First recall from [**CPWZ1**, Example 5.10(2)],

$$\Omega(x_1, \ldots, x_t) := \sum_{\sigma \in S_t} (-1)^{|\sigma|} x_{\sigma(1)} \cdots x_{\sigma(t)}.$$

LEMMA 2.6. *For any $t \leq n$, consider the element $\Omega_t = \Omega(x_1, \ldots, x_t)$ in $V_n(\mathcal{A})$. Then $x_i\Omega_t = (-1)^{n-1}\Omega_t x_i$ for all $i = 1, \ldots, t$.*

PROOF. The proof is similar to that of [**CPWZ1**, Lemma 5.11] and is therefore omitted. □

Now we are ready to prove Theorem 2.

PROOF OF THEOREM 2. Let $n \geq 3$ be odd. By Lemma 2.6, we have two automorphisms of $V_n(\mathcal{A})$, as in [**CPWZ1**, Example 5.10(2)]:

$$g : x_i \to \begin{cases} x_1 + \Omega(x_2, x_3 \dots, x_n) & i = 1, \\ x_i & 2 \leq i \leq n, \end{cases}$$

and

$$h : x_i \to \begin{cases} x_2 + \Omega(x_1, x_3, \dots, x_n) & i = 2, \\ x_i & i = 1 \text{ or } 3 \leq i \leq n. \end{cases}$$

Using the notation in (E2.0.1) and (E2.1.3), both $a_1$ and $b_1$ have leading term equal to $(-1)^{n-2}(n-1)! \prod_{i=3}^{n} x_i$. It is easy to see that $R + Rx_1 + Rx_2$ is a free left $R$-module with basis $\{1, x_1, x_2\}$, and it is clear that $a_1 b_1$ is transcendental over $k$. Since $k_0 = k \neq \mathbb{Z}/(2)$, the assertion follows from Proposition 2.5(2). $\square$

REMARK 2.7. It was claimed in [**CPWZ1**, Example 5.10] that $\operatorname{Aut}(W_n)$ is not affine when $n$ is odd. This is true when $\operatorname{char} k$ does not divide $(n-1)!$; it is not known what happens when $\operatorname{char} k$ divides $(n-1)!$. The hypothesis that "$\operatorname{char} k$ does not divide $(n-1)!$" was missing from [**CPWZ1**, Example 5.10].

We conclude with a corollary of Proposition 2.5 which deals with skew polynomial rings. See [**CPWZ3**] for more results along these lines. Let $\{p_{ij} \in k^\times : 1 \leq i < j \leq n\}$ be a set of parameters. The *skew polynomial ring*, denoted by $k_{p_{ij}}[x_1, \dots, x_n]$, is defined to be the algebra generated by $x_1, \dots, x_n$ subject to the relations $x_j x_i = p_{ij} x_i x_j$ for all $i < j$. We set $p_{ji} = p_{ij}^{-1}$ and $p_{ii} = p_{jj} = 1$ for all $i < j$. For any $1 \leq s \leq n$, let

$$T_s := \left\{ (d_1, \dots, \widehat{d_s}, \dots, d_n) \in \mathbb{N}^{n-1} : \prod_{\substack{j=1 \\ j \neq s}}^{n} p_{ij}^{d_j} = p_{is} \quad \forall i \neq s \right\}.$$

In [CPWZ2, Theorem 3.8], it was proved that if each $p_{ij}$ is a root of unity and $T_s = \emptyset$ for all $s$, then every automorphism of $k_{p_{ij}}[x_1, \dots, x_n]$ is affine. Note also that if each $p_{ij}$ is a root of unity and some $T_s$ is non-empty, then $T_s$ is in fact infinite. If we drop the assumption on the $p_{ij}$'s and we allow at most one $T_s$ to be infinite, then we can still understand the automorphism group of $k_{p_{ij}}[x_1, \dots, x_n]$ – see [CPWZ2,Theorem 4]. The following result concerns $\operatorname{Aut}(k_{p_{ij}}[x_1, \dots, x_n])$ in another case.

COROLLARY 2.8. *Let $n \geq 3$ be an integer and let $A := k_{p_{ij}}[x_1, \dots, x_n]$ be a skew polynomial ring. Suppose that $T_1$ contains an element of the form $(1, d_3, \dots, d_n)$ and $T_2$ contains an element of the form $(1, d'_3, \dots, d'_n)$. Suppose that at least one of $d_3, \dots, d_n, d'_3, \dots, d'_n$ is nonzero. Then $\operatorname{Aut}(A)$ contains $(k, +) * (k, +)$. If, further, $k \neq \mathbb{Z}/(2)$, then $\operatorname{Aut}(A)$ contains a free group on two generators.*

PROOF. For $d = (1, d_3, \dots, d_n) \in T_1$ and $d' = (1, d'_3, \dots, d'_n) \in T_2$ as in the statement, define $f_d := x_2 x_3^{d_3} \dots x_n^{d_n}$, $f_{d'} := x_1 x_3^{d'_3} \dots x_n^{d'_n}$. The maps, defined in [CPWZ2, (2.10.1)],

$$g(f_d, 1) : x_1 \mapsto \begin{cases} x_i & \text{if } i \neq 1, \\ x_1 + f_d & \text{if } i = 1, \end{cases} \qquad g(f_{d'}, 2) : x_i \mapsto \begin{cases} x_i & \text{if } i \neq 2, \\ x_2 + f_{d'} & \text{if } i = 2, \end{cases}$$

extend to algebra automorphisms of $A$. Using the notation (E2.0.1) and (E2.1.3), $a_1 = \prod_{j=3}^n p_{j,2}^{d_j} x_3^{d_3} \ldots x_n^{d_n}$ and $b_1 = \prod_{j=3}^n p_{j,1}^{d'_j} x_3^{d'_3} \ldots x_n^{d'_n}$. It is clear that $R + Rx_1 + Rx_2$ is a free left $R$-module with basis $\{1, x_1, x_2\}$ and that $a_1 b_1$ is transcendental over $k$. The assertion follows from Proposition 2.5(2). □

Note that this applies to the ordinary polynomial ring: if $p_{ij} = 1$ for all $i, j$, then each set $T_s$ is all of $\mathbb{N}^{n-1}$.

## 3. Proof of Theorem 3

In this short section we assume that $k$ is a commutative domain. The main idea is to use the discriminant to solve isomorphism questions. Given two algebras with similar properties, it is generally very difficult to determine whether they are isomorphic. When the discriminant controls the the automorphism groups of these algebras, it also controls the isomorphisms between them.

Given an algebra $A$, we write $d(A/Z(A))$ for the discriminant of $A$ over its center, as defined in [**CPWZ1**] or [**CPWZ2**]. The following lemma is clear.

LEMMA 3.1. *Suppose $A$ and $B$ are PI algebras, and let $g : A \to B$ be an algebra isomorphism. Then $g(d(A/Z(A))) = c\, d(B/Z(B))$ for some unit $c \in Z(B)$.*

We keep the notation for $V_n(\mathcal{A})$ and $V_n(\mathcal{A}')$ as above, except that it is convenient to distinguish between their generators, so we use $\{x_i\}$ for the generators of $V_n(\mathcal{A})$ and $\{x'_i\}$ for the generators of $V_n(\mathcal{A}')$.

PROOF OF THEOREM 3. First of all, if there exist $\sigma \in S_n$ and invertible scalars $\lambda_i$ such that $a'_{ij} = \lambda_i \lambda_j a_{\sigma(i)\sigma(j)}$, then the map $g$ determined by $x'_i \mapsto \lambda_i x_{\sigma(i)}$ for $i = 1, \ldots, n$ extends to an algebra isomorphism $V_n(\mathcal{A}') \to V_n(\mathcal{A})$.

Conversely, let $g : V_n(\mathcal{A}') \to V_n(\mathcal{A})$ be an algebra isomorphism. By [**CPWZ1**, Theorem 4.9],

$$d := d(V_n(\mathcal{A})/Z(V_n(\mathcal{A}))) = \left(\prod_{i=1}^n x_i^2\right)^{2^{n-1}} + (\text{cwlt}),$$

where (cwlt) ("component-wise less than") is a linear combination of terms with degree strictly less than the degree of $(\prod_{i=1}^n x_i^2)^{2^{n-1}}$. By symmetry,

$$d' := d(V_n(\mathcal{A}')/Z(V_n(\mathcal{A}'))) = \left(\prod_{i=1}^n (x'_i)^2\right)^{2^{n-1}} + (\text{cwlt}).$$

By Lemma 3.1, $g(d') = cd$ for some unit $c \in Z(V_n(\mathcal{A})) = k[x_1^2, \ldots, x_n^2]$. So $c$ is a unit in $k^\times$. The equation $g(d') = cd$ implies that

$$\text{(E3.1.1)} \qquad \left(\prod_{i=1}^n (g(x'_i))^2\right)^{2^{n-1}} + g((\text{cwlt})) = c\left(\prod_{i=1}^n x_i^2\right)^{2^{n-1}} + (\text{cwlt}).$$

Note that $g(x'_i)$ has degree at least 1 since it is not in $k$. So each term in $g((\text{cwlt}))$ has degree strictly less than the degree of $(\prod_{i=1}^n (g(x'_i))^2)^{2^{n-1}}$. This implies that the degree of the left-hand term in (E3.1.1) is equal to the degree of $(\prod_{i=1}^n (g(x'_i))^2)^{2^{n-1}}$. The degree of the right-hand term of (E3.1.1) is $n2^n$. Therefore $\deg g(x'_i) = 1$ for all $i$. Thus if we use the standard filtration on both $V_n(\mathcal{A}')$ and $V_n(\mathcal{A})$, then $g$ preserves the filtration. Consider the graded algebra isomorphism $\operatorname{gr} g : \operatorname{gr} V_n(\mathcal{A}') \to$

$\operatorname{gr} V_n(\mathcal{A})$, recalling that $\operatorname{gr} V_n(\mathcal{A}) \cong k_{-1}[x_1, \ldots, x_n]$. By [**CPWZ1**, Lemma 4.3], there are units $r_i \in k$ and a permutation $\sigma \in S_n$ such that $\operatorname{gr} g(x_i') = r_i x_{\sigma(i)}$. Going back to $g$, there must be $a_i \in k$ such that $g(x_i') = r_i x_{\sigma(i)} + a_i$ for all $i$. Using the second half of the proof of [**CPWZ1**, Lemma 4.3], one sees that $a_i = 0$ for all $i$. The assertion follows. □

## 4. Examples and Questions

In this final section, we give some examples and list some questions concerning the algebra $V_n(\mathcal{A})$. For the sake of convenience, we assume that $k$ is an algebraically closed field of characteristic zero.

LEMMA 4.1. *If the order of* $\operatorname{Aut}(V_n(\mathcal{A}))$ *is* $2n!$, *then* $n$ *is an even integer* $\geq 4$ *and* $V_n(\mathcal{A})$ *is isomorphic to* $W_n$. *As a consequence,* $\operatorname{Aut}(V_n(\mathcal{A})) \cong S_n \times \{\pm 1\}$.

PROOF. By Theorem 2, $n$ is even. If $a_{ij} = 0$ for all $i < j$, then $\operatorname{Aut}(V_n(\mathcal{A})) \cong S_n \ltimes (k^\times)^n$ which does not have order $2n!$. So $a_{ij} \neq 0$ for some $i < j$.

If $n = 2$, then $a_{12}$ must be nonzero, in which case $V_2(\mathcal{A})$ is isomorphic to $W_2$. By [**AD**], $\operatorname{Aut}(W_2) = S_2 \ltimes k^\times$, so $n \neq 2$. Hence $n$ is an even integer larger than or equal to 4.

Given that $a_{ij} \neq 0$ for some $i < j$, we claim that $a_{st} \neq 0$ for all $s < t$. By Lemma 1.1, $|\operatorname{Aut}(V_n(\mathcal{A}))| = 2m$, where $m$ is the order of the image of $\operatorname{Aut}(V_n(\mathcal{A}))$ in $S_n$. Since $|\operatorname{Aut}(V_n(\mathcal{A}))|$ is $2n!$, the map $\operatorname{Aut}(V_n(\mathcal{A})) \to S_n$ must be surjective, so there is a short exact sequence

$$1 \to \{\pm 1\} \to \operatorname{Aut}(V_n(\mathcal{A})) \to S_n \to 1.$$

This means that for each $\sigma \in S_n$, there is a corresponding element $g(\sigma, \{r_i\}) \in \operatorname{Aut}(V_n(\mathcal{A}))$. In particular, given $s$ and $t$ with $1 \leq s < t \leq n$, there is a permutation $\sigma \in S_n$ such that $(\sigma(i), \sigma(j)) = (s, t)$. By (E1.0.2), $a_{ij} = r_i r_j a_{\sigma(i)\sigma(j)} = r_i r_j a_{st}$. Since $a_{ij} \neq 0$, we obtain that $a_{st} \neq 0$.

Since elements in $k$ have square roots in $k$, after replacing $x_i$ by $b_i x_i$ for $i = 1, 2, 3$ for appropriate $b_1, b_2, b_3$, we may assume that $a_{12} = a_{13} = a_{23} = 1$. For each $i > 3$, by replacing $x_i$ by $a_{1i}^{-1} x_i$, we may assume that $a_{1i} = 1$ for all $i > 3$. Summarizing, we have $a_{1i} = 1$ for all $i > 1$ and $a_{23} = 1$.

Next we claim that $a_{ij} = 1$ for all $i < j$. First we show that $a_{2i} = 1$ for all $i \neq 2$. Taking an element $g((12), \{r_i\})$ in $\operatorname{Aut}(V_n(\mathcal{A}))$, (E1.0.2) implies that

$$\begin{aligned} a_{12} &= r_1 r_2 a_{21}, \\ a_{13} &= r_1 r_3 a_{23}, \\ a_{23} &= r_2 r_3 a_{13}. \end{aligned}$$

Since $a_{12} = a_{13} = a_{23} = 1$, $r_1 = r_2 = r_3 = \pm 1$. By (E1.0.2),

$$a_{3i} = r_3 r_i a_{3i}, \qquad \forall\, i > 3.$$

Hence $r_i = r_3 = \pm 1$ for all $i > 3$. So $r_i = r_1 = \pm 1$ for all $i$. By (E1.0.2) again, $a_{2i} = r_2 r_i a_{1i} = 1$ for all $i$. By symmetry, $a_{3i} = 1$ for all $i \neq 3$. Secondly, we show that $a_{ij} = 1$ for all $3 < i < j$. Nothing needs to be proved if $n = 4$. For $n \geq 6$, it suffices to show $a_{45} = 1$ by symmetry. Taking $g((14), \{r_i\})$ in $\operatorname{Aut}(V_n(\mathcal{A}))$, (E1.0.2) implies that

$$\begin{aligned}1 &= a_{12} = r_1 r_2 a_{42} = r_1 r_2,\\ 1 &= a_{13} = r_1 r_3 a_{43} = r_1 r_3,\\ 1 &= a_{23} = r_2 r_3 a_{23} = r_2 r_3,\\ 1 &= a_{34} = r_3 r_4 a_{31} = r_3 r_4.\end{aligned}$$

Then $r_1 = r_2 = r_3 = r_4 = \pm 1$. By (E1.0.2),

$$a_{3i} = r_3 r_i a_{3i}, \qquad \forall\, i > 4.$$

Hence $r_i = r_3 = \pm 1$ for all $i > 4$. So $r_i = r_1 = \pm 1$ for all $i$. By (E1.0.2) again, $a_{45} = r_4 r_5 a_{15} = 1$, as desired.

The consequence follows from [**CPWZ1**, Theorem 1]. □

There are cases when $\mathrm{Aut}(V_n(\mathcal{A}))$ is smaller than $S_n \times \{\pm 1\}$, as the next example shows.

EXAMPLE 4.2. Let $n = 4$.

(1) Let $q$ be transcendental over $\mathbb{Q} \subseteq k$. Let $a_{12} = q$, $a_{13} = q^2$, $a_{14} = q^4$, $a_{23} = q^8$, $a_{24} = q^{16}$, and $a_{34} = q^{32}$. For any $\{i_1, \ldots, i_4\} = \{j_1, \ldots, j_4\} = \{1, \ldots, 4\}$, unless $i_s = j_s$ for all $s$, the element $a_{i_1 i_2} a_{i_3 i_4} a_{j_1 j_2}^{-1} a_{j_3 j_4}^{-1}$ is a non-trivial power of $q$, which is not a root of unity. Using (E1.0.1), (E1.0.2) and the fact that the homological determinant of $g$ is $r_1 r_2 r_3 r_4$, one can show that

$$G(\mathcal{A})/\{\pm 1\} \cong \{Id, (12)(34), (13)(24), (14)(23)\} \cong (\mathbb{Z}/(2))^{\oplus 2}.$$

In general, one can show that when $n = 4$ and $a_{ij} \neq 0$ for all $i < j$, then $G(\mathcal{A})/\{\pm 1\}$ always contains the subgroup $\{Id, (12)(34), (13)(24), (14)(23)\}$.

(2) Let $q$ be transcendental over $\mathbb{Q} \subseteq k$. Let $a_{12} = q$, $a_{13} = q^2$, $a_{14} = q^4$, $a_{23} = q^8$, and $a_{24} = a_{34} = 0$. We claim that $\mathrm{Aut}(V_n(\mathcal{A})) \cong S_2 \times \{\pm 1\}$. If $g = g(\sigma, \{r_i\})$ is in $\mathrm{Aut}(V_n(\mathcal{A}))$, then $\sigma$ fixes 1 and 4, as 1 is the only index so that $a_{1i} \neq 0$ for all $i$ and 4 is the only index so that $a_{i4} = 0$ for two different $i$. If $g$ is neither 1 nor $-1$, $g$ must be of the form $g((23), \{r_i\})$. In fact, one can check that $G(\mathcal{A})/\{\pm 1\} \cong \{(23)\} \cong S_2$.

(3) Let $(a_{ij})_{4\times 4} = \begin{pmatrix} \emptyset & 1 & -1 & 1\\ 1 & \emptyset & 1 & -a^2\\ -1 & 1 & \emptyset & -a\\ 1 & -a^2 & -a & \emptyset \end{pmatrix}$ where $a^3 = -1$ and $a \neq -1$. This is the same as Example 1.2. One can show that $G(\mathcal{A})/\{\pm 1\} \cong A_4$.

EXAMPLE 4.3. Let $n = 6$.

(1) Suppose that $a_{15} = a_{45} = a_{i6} = 0$ for all $i = 2, 3, 4, 5$, that all other $a_{ij}$ are nonzero, and that $a_{12} a_{34} \neq a_{13} a_{24}$. Then $\mathrm{Aut}(V_n(\mathcal{A})) \cong \{\pm 1\}$. To see this, we first note that 6 is the only index so that there are 4 different $i$ such that $a_{i6} = 0$; thus if $g(\sigma, \{r_i\})$ is in $\mathrm{Aut}(V_n(\mathcal{A}))$, $\sigma$ fixes 6. Similarly, $\sigma$ fixes 1, 4 and 5. The only possible nontrivial $\sigma$ is (23). In the case of $g((23), \{r_i\})$, we have

$$\begin{aligned} a_{12} &= r_1 r_2 a_{13},\\ a_{13} &= r_1 r_3 a_{12},\\ a_{23} &= r_2 r_3 a_{23},\\ a_{14} &= r_1 r_4 a_{14},\\ a_{24} &= r_2 r_4 a_{34}.\end{aligned}$$

By the first three equations, we have $r_1^2 = 1$ and $r_2 r_3 = 1$. So $r_1 = \pm 1$. By the fourth equation, $r_4 = r_1$. Then the first and fifth equations contradict the hypothesis that $a_{12}a_{34} \neq a_{13}a_{24}$. Therefore the only $g \in \mathrm{Aut}(V_n(\mathcal{A}))$ is of the form $\pm 1$, so $\mathrm{Aut}(V_n(\mathcal{A})) \cong \{\pm 1\}$.

(2) Suppose that $a_{12} = 0$ and that $a_{ij} = 1$ for all $i < j$ except for $a_{12}$. Then $\mathrm{Aut}(V_n(\mathcal{A}))/\{\pm 1\} \cong S_2 \times S_4$ and therefore $\mathrm{Aut}(V_n(\mathcal{A})) \cong (S_2 \times S_4) \times \{\pm 1\}$.

QUESTION 4.4. Which finite groups can be realized as $\mathrm{Aut}(V_n(\mathcal{A}))$? For which of those groups $G$ can one classify the algebras $V_n(\mathcal{A})$ such that $\mathrm{Aut}(V_n(\mathcal{A})) \cong G$?

Similar to Theorem 1, we have the following. Recall that $G_1(\mathcal{A})$ is defined just after Lemma 1.1.

LEMMA 4.5. *Suppose there is a short exact sequence*

$$1 \to \{\pm 1\} \to \mathrm{Aut}(V_n(\mathcal{A})) \to H \to 1$$

*where $H$ is a subgroup of $S_n$ such that $H = [H, H]$ (for example, $H = A_n$ when $n \geq 6$). Then*

(1) *$G(\mathcal{A}) = G_1(\mathcal{A})$.*
(2) *For each subgroup $G \subseteq \mathrm{Aut}(V_n(\mathcal{A}))$, the fixed subring $V_n(\mathcal{A})^G$ is filtered AS Gorenstein.*

PROOF. Part (2) is a consequence of part (1) and Theorem 1.3, so we only prove part (1). Since $n$ must be even, the homological determinant $\mathrm{hdet} : \mathrm{Aut}(V_n(\mathcal{A})) \to k^\times$ maps $\pm 1$ to 1. Hence it induces a group homomorphism $\mathrm{hdet}' : H \to k^\times$. Since $H = [H, H]$ and $k^\times$ is abelian, $\mathrm{hdet}'$ is the trivial map. So the image of hdet is $\{1\}$. This is equivalent to saying that $G(\mathcal{A}) = G_1(\mathcal{A})$. □

Related to the above and Theorem 1, we have the following question.

QUESTION 4.6. Classify all $V_n(\mathcal{A})$ (when $n$ is even) such that $V_n(\mathcal{A})^G$ is Gorenstein for all subgroups $G \subseteq \mathrm{Aut}(V_n(\mathcal{A}))$.

Theorem 3 suggests the following question.

QUESTION 4.7. When $n$ is odd, determine when two $V_n(\mathcal{A})$s are isomorphic.

A related question is the following.

QUESTION 4.8. If $V_n(\mathcal{A})$ is Morita equivalent to $V_n(\mathcal{A}')$, is $V_n(\mathcal{A})$ isomorphic to $V_n(\mathcal{A}')$?

Some Hopf algebra actions on $W_2$ are given in [**CWWZ**, Example 3.4]. It would be very interesting to work out all possible Hopf algebra actions.

QUESTION 4.9. Suppose $n$ is even. Classify all finite dimensional Hopf algebras that can act on $W_n$ inner-faithfully.

**Acknowledgments.** S. Ceken was supported by the Scientific and Technological Research Council of Turkey (TUBITAK), Science Fellowships and Grant Programmes Department (Programme no. 2214). Y. H. Wang was supported by the Natural Science Foundation of China (grant nos. 10901098 and 11271239), the Foundation of Shanghai Science and Technology Committee (no. 14511107202), the Scientific Research Starting Foundation for the Returned Overseas Chinese Scholars, Ministry of Education of China and the Innovation program of Shanghai Municipal Education Commission. J. J. Zhang was supported by the US National Science Foundation (nos. DMS-0855743 and DMS-1402863).

## References


[AD] J. Alev and F. Dumas, *Rigidité des plongements des quotients primitifs minimaux de* $U_q(\mathrm{sl}(2))$ *dans l'algèbre quantique de Weyl-Hayashi* (French), Nagoya Math. J. **143** (1996), 119–146. MR1413010 (97h:17014)

[AP] M. Anshel and R. Prener, *On free products of finite abelian groups*, Proc. Amer. Math. Soc. **34** (1972), 343–345. MR0302768 (46 #1911)

[AS] M. Artin and W. F. Schelter, *Graded algebras of global dimension* 3, Adv. in Math. **66** (1987), no. 2, 171–216, DOI 10.1016/0001-8708(87)90034-X. MR917738 (88k:16003)

[BJ] V. V. Bavula and D. A. Jordan, *Isomorphism problems and groups of automorphisms for generalized Weyl algebras*, Trans. Amer. Math. Soc. **353** (2001), no. 2, 769–794, DOI 10.1090/S0002-9947-00-02678-7. MR1804517 (2002g:16040)

[CPWZ1] S. Ceken, J. H. Palmieri, Y.-H. Wang, and J. J. Zhang, *The discriminant controls automorphism groups of noncommutative algebras*, Adv. Math. **269** (2015), 551–584, DOI 10.1016/j.aim.2014.10.018. MR3281142

[CPWZ2] S. Ceken, J. H. Palmieri, Y.-H. Wang, and J. J. Zhang, *The discriminant criterion and automorphism groups of quantized algebras*, Adv. Math. **286** (2016), 754–801, DOI 10.1016/j.aim.2015.09.024. MR3415697

[CPWZ3] S. Ceken, J. Palmieri, Y.-H. Wang and J.J. Zhang, Tits alternative for automorphisms and derivations, in preparation (2015).

[CWWZ] K. Chan, C. Walton, Y. H. Wang, and J. J. Zhang, *Hopf actions on filtered regular algebras*, J. Algebra **397** (2014), 68–90, DOI 10.1016/j.jalgebra.2013.09.002. MR3119216

[Ch] C. Chevalley, *Invariants of finite groups generated by reflections*, Amer. J. Math. **77** (1955), 778–782. MR0072877 (17,345d)

[Da] E. C. Dade, *Group-graded rings and modules*, Math. Z. **174** (1980), no. 3, 241–262, DOI 10.1007/BF01161413. MR593823 (82c:16028)

[JiZ] N. Jing and J. J. Zhang, *Gorensteinness of invariant subrings of quantum algebras*, J. Algebra **221** (1999), no. 2, 669–691, DOI 10.1006/jabr.1999.8023. MR1728404 (2001i:16076)

[JoZ] P. Jørgensen and J. J. Zhang, *Gourmet's guide to Gorensteinness*, Adv. Math. **151** (2000), no. 2, 313–345, DOI 10.1006/aima.1999.1897. MR1758250 (2001d:16023)

[Ki] E. Kirkman, Invariant theory of Artin-Schelter regular algebras: a survey, preprint, (2015), arXiv:1506.06121.

[KKZ1] E. Kirkman, J. Kuzmanovich, and J. J. Zhang, *Invariants of* $(-1)$*-skew polynomial rings under permutation representations*, Recent advances in representation theory, quantum groups, algebraic geometry, and related topics, Contemp. Math., vol. 623, Amer. Math. Soc., Providence, RI, 2014, pp. 155–192, DOI 10.1090/conm/623/12463. MR3288627

[KKZ2] E. Kirkman, J. Kuzmanovich, and J. J. Zhang, *Invariant theory of finite group actions on down-up algebras*, Transform. Groups **20** (2015), no. 1, 113–165, DOI 10.1007/s00031-014-9279-4. MR3317798

[KKZ3] E. Kirkman, J. Kuzmanovich, and J. J. Zhang, *Gorenstein subrings of invariants under Hopf algebra actions*, J. Algebra **322** (2009), no. 10, 3640–3669, DOI 10.1016/j.jalgebra.2009.08.018. MR2568355 (2011b:16115)

[ST] G. C. Shephard and J. A. Todd, *Finite unitary reflection groups*, Canadian J. Math. **6** (1954), 274–304. MR0059914 (15,600b)

[SAV] M. Suárez-Alvarez and Q. Vivas, *Automorphisms and isomorphisms of quantum generalized Weyl algebras*, J. Algebra **424** (2015), 540–552, DOI 10.1016/j.jalgebra.2014.08.045. MR3293233

[Ti] J. Tits, *Free subgroups in linear groups*, J. Algebra **20** (1972), 250–270. MR0286898 (44 #4105)

[Wa] K. Watanabe, *Certain invariant subrings are Gorenstein. I, II*, Osaka J. Math. **11** (1974), 1–8; ibid. 11 (1974), 379–388. MR0354646 (50 #7124)

[Y1] M. Yakimov, *Rigidity of quantum tori and the Andruskiewitsch-Dumas conjecture*, Selecta Math. (N.S.) **20** (2014), no. 2, 421–464, DOI 10.1007/s00029-013-0145-3. MR3177924

[Y2] M. Yakimov, *The Launois-Lenagan conjecture*, J. Algebra **392** (2013), 1–9, DOI 10.1016/j.jalgebra.2013.07.001. MR3085018

Department of Mathematics–Computer Programming, Faculty of Art & Sciences, İstanbul Aydin University, İstanbul, Turkey
*E-mail address*: cekensecil@gmail.com

Department of Mathematics, Box 354350, University of Washington, Seattle, Washington 98195
*E-mail address*: palmieri@math.washington.edu

School of Mathematics, Shanghai University of Finance and Economics – and – Shanghai Key Laboratory of Financial Information Technology, Shanghai 200433, China
*E-mail address*: yhw@mail.shufe.edu.cn

Department of Mathematics, Box 354350, University of Washington, Seattle, Washington 98195
*E-mail address*: zhang@math.washington.edu

Proceedings of Symposia in Pure Mathematics
Volume **92**, 2016
http://dx.doi.org/10.1090/pspum/092/01593

# Bounded highest weight modules over $\mathfrak{osp}(1, 2n)$

Thomas Ferguson, Maria Gorelik, and Dimitar Grantcharov

ABSTRACT. We classify all simple bounded highest weight modules of the ortosymplectic superalgebras $\mathfrak{osp}(1, 2n)$. The classification is obtained in two independent ways: using equivalence of categories of $\mathfrak{osp}(1, 2n)$-modules and $\mathfrak{osp}(1, 2n)_{\bar{0}}$-modules, and by finding primitive vectors in tensor products of bounded and finite-dimensional $\mathfrak{osp}(1, 2n)$-modules. We also obtain character formulae for the simple bounded highest weight modules of $\mathfrak{osp}(1, 2n)$.

## 1. Introduction

Weight modules are modules that are semisimple as modules over a fixed Cartan subalgebra. Examples of weight modules include quotients of parabolically induced modules (in particular, highest weight modules) and some generalized Harish-Chandra modules. The classification of all simple weight $\mathfrak{g}$-modules with finite weight multiplicities over finite-dimensional simple Lie superalgebras $\mathfrak{g}$ is not completed yet. This classification for Lie algebras $\mathfrak{g}$ was completed in the breakthrough paper [**M**] by classifying all simple cuspidal $\mathfrak{g}$-modules, i.e. modules on which all root elements of $\mathfrak{g}$ act bijectively. In the Lie superalgebra case, the classification was obtained in [**DMP**] for all $\mathfrak{g}$ except for the Lie superalgebra series $\mathfrak{osp}(m; 2n)$, $m = 1, 3, 4, 5, 6$; $\mathfrak{psq}(n)$, $D(2, 1, \alpha)$, and the Cartan series of type $S$ and $H$. It is interesting to note that the Lie superalgebras $\mathfrak{osp}(m, 2n)$, $m \geq 7$, are not in the list because their even parts do not have cuspidal modules. For the classical Lie superalgebras $\mathfrak{g}$, the classification of simple weight modules with finite weight multiplicities was reduced to the classification of the so-called bounded highest weight modules, [**Gr**]. The latter classification was obtained for $\mathfrak{g} = \mathfrak{psq}(n)$ and $\mathfrak{g} = D(2, 1, \alpha)$ in [**GG**] and [**H**], respectively, leaving the orthosymplectic series the only remaining classical Lie superalgebras to consider.

Fix a triangular decomposition of $\mathfrak{g}$ and denote by $L(\lambda)$ the simple highest weight $\mathfrak{g}$-module of highest weight $\lambda$. If the set of weight multiplicities of $L(\lambda)$ is uniformly bounded we call the module $L(\lambda)$ *bounded* and the weight $\lambda$ $\mathfrak{g}$-*bounded*. In this paper we make the first step towards the classification of the bounded highest weight modules $L(\lambda)$ (and, hence, of the simple weight modules with finite

---

2010 *Mathematics Subject Classification.* Primary 17B10.
*Key words and phrases.* Weight modules, orthosymplectic superalgebras, character formulae.
The second author was supported in part by BSF Grant No. 711623. The third author was supported in part by NSA grant H98230-13-1-0245.

weight multiplicities) of the orthosymplectic superalgebras - we solve the problem for $\mathfrak{osp}(1,2n)$. Our result can be written in short as follows.

**Theorem.** *An infinite-dimensional module $L(\lambda)$ is $\mathfrak{osp}(1,2n)$-bounded if and only if $\lambda$ and $\lambda-\delta_n$ are $\mathfrak{sp}(2n)$-bounded.*

We believe that our classification will play crucial role in the classification of the simple bounded highest weight modules for the remaining four orthosymplectic superlagebras series. In addition to the classification of the bounded $\mathfrak{osp}(1,2n)$-modules $L(\lambda)$, we obtain character formulae of $L(\lambda)$ in terms of characters of simple finite-dimensional $\mathfrak{so}(2n)$-modules, see (2).

We present two alternative proofs of the classifcation of the bounded modules $L(\lambda)$. The first is based on the equivalence of categories of graded $\mathfrak{osp}(1,2n)$-modules and of $\mathfrak{sp}(2n)$-modules established in [**G2**]. The second relies on finding primitive vectors of tensor products of the Weyl module $L(-\frac{1}{2}\sum_{i=1}^{n}\delta_i)$ and finite-dimensional modules.

The paper is organized as follows. The main result together with character formulae for bounded $L(\lambda)$ are presented in Section 3. The alternative proof of the classification of bounded highest weight modules is included in Section 4. Some important facts on the equivalences of categories and character formulae are collected in the Appendix.

## 2. Notation and conventions

Except for the appendix, throughout the paper, $\mathfrak{g}=\mathfrak{osp}(1,2n)$. We fix a triangular decomposition of $\mathfrak{g}$, hence of $\mathfrak{g}_{\bar{0}}$, and by $\mathfrak{h}$ we denote the Cartan subalgebra of $\mathfrak{g}$ and $\mathfrak{g}_{\bar{0}}$.

The root system of $(\mathfrak{g},\mathfrak{h})$ is $\Delta=\Delta_{\bar{0}}\cup\Delta_{\bar{1}}$, where $\Delta_{\bar{1}}=\{\pm\delta_i \mid i=1,\dots,n\}$, $\Delta_{\bar{0}}=\{\pm\delta_i\pm\delta_j,\pm 2\delta_i \mid 1\leq i<j\leq n\}$. We will also use the notation $\Delta_C=\Delta_{\bar{0}}$ (root system of $C_n=\mathfrak{sp}(2n)$) and $\Delta_D=\{\pm\delta_i\pm\delta_j \mid 1\leq i<j\leq n\}$ (root system of $D_n=\mathfrak{so}(2n)$). By $\Delta^+$ we will denote the set of all positive roots of $\mathfrak{g}$ and $W$ will stand for the Weyl group. Fix $(-,-)$ to be the symmetric bilinear form on $\mathfrak{h}^*$ such that $(\delta_i,\delta_j)=\delta_{i,j}$.

In most of the paper we will fix $\Pi=\{\delta_1-\delta_2,\dots,\delta_{n-1}-\delta_n,\delta_n\}$ to be the base of $\Delta$. We will also use $\widetilde{\Pi}=\{-\delta_1,\delta_1-\delta_2,\dots,\delta_{n-1}-\delta_n\}$. Denote by $\theta$ the automorphism of $\Delta$ defined by $\delta_i\mapsto-\delta_{n+1-i}$. This automorphism extends to an automorphism of $\mathfrak{g}$ that we will denote by the same letter. Note that $\widetilde{\Pi}=\theta(\Pi)$.

We will call a vector $v$ in a $\mathfrak{g}$-module primitive (respectively, $\widetilde{\Pi}$-primitive), if the elements of the $\alpha$-root spaces of $\mathfrak{g}$ for all $\alpha\in\Pi$ (respectively, $\alpha\in\widetilde{\Pi}$) annihilate $v$.

We denote by $W(C)$ and $W(D)$ the Weyl groups of $C_n$ and $D_n$, respectively, and by $\rho_C,\rho_D$ the corresponding half sums of positive roots. Also, by $\rho=\rho_0-\rho_1$ we denote the difference of the half sums of the positive even roots and the positive odd roots of $\mathfrak{g}$. We have:

$$\rho=\rho_C-\frac{1}{2}\sum_{i=1}^{n}\delta_i=\rho_D+\frac{1}{2}\sum_{i=1}^{n}\delta_i.$$

We denote by $L(\lambda)$ (resp., $M(\lambda)$) the irreducible (respectively, the Verma) $\mathfrak{g}$-module of highest weight $\lambda$. Also, by $L_{\widetilde{\Pi}}(\lambda)$ we denote the simple $\widetilde{\Pi}$-highest weight module with highest weight $\lambda$. We denote by $L_C(\lambda)$ (resp., $L_D(\lambda)$) an irreducible

$C_n$ (resp., $D_n$) module of the highest weight $\lambda$. We consider two shifted actions of $W(C)$ to $\mathfrak{h}^*$:

$$w \circ \lambda := w(\lambda+\rho)-\rho, \quad w \circ_C \lambda := w(\lambda+\rho_C)-\rho_C,$$

and the shifted action of $W(D)$: $w \circ_D \lambda := w(\lambda+\rho_D)-\rho_D$. For each $\lambda \in \mathfrak{h}^*$, $W(\lambda)$ stands for the corresponding integral Weyl group, i.e. the subgroup of $W$ generated by the reflection $r_\alpha$ for the even roots $\alpha$ satisfying $2(\lambda,\alpha) \in \mathbb{Z}(\alpha,\alpha)$.

We say that a $\mathfrak{g}$-module $M$ is a weight module if $M = \bigoplus_{\lambda\in\mathfrak{h}^*} M_\lambda$, where $M_\lambda = \{m \in M \mid hm = \lambda(h)m \text{ for every } h \in \mathfrak{h}\}$ is the $\lambda$-weight space. We call a weight module $M$ bounded if there is $C$ such that $\dim M_\lambda < C$ for all $\lambda \in \mathfrak{h}^*$. A weight $\lambda$ is called $\mathfrak{g}$-bounded, or simply bounded, if $L(\lambda)$ is a bounded module. Analogously, we introduce the notions of $\mathfrak{sp}(2n)$-bounded modules and bounded weights.

For each root $\alpha$ we introduce $\alpha^\vee := 2\alpha/(\alpha,\alpha)$. Recall that for a simple Lie algebra an irreducible module of a highest weight module $\lambda$ is finite-dimensional if and only if $(\lambda+\rho,\alpha^\vee) \in \mathbb{Z}_{>0}$ for each positive root $\alpha$. A $\mathfrak{g}$-module $L(\lambda)$ is finite-dimensional if and only if $L_C(\lambda)$ is finite-dimensional, see Theorem 8 in [**K1**]. For each $\lambda \in \mathfrak{h}^*$ we set $\Delta(\lambda) = \{\alpha \in \Delta_C \mid (\lambda,\alpha^\vee) \in \mathbb{Z}\}$ and $\Delta(\lambda)^+ = \Delta(\lambda) \cap \Delta^+$.

## 3. Bounded highest weight modules of $\mathfrak{osp}(1,2n)$

In this section we classify the bounded highest weight $\mathfrak{osp}(1,2n)$-modules and obtain their character formulae.

**3.1.** By [**M**], Lemma 9.1, $L_C(\lambda)$ is bounded if and only if

(i) $(\lambda,\delta_i-\delta_{i+1}) \in \mathbb{Z}_{\geq 0}$ for $i = 1,\dots,n-1$, and

(ii) either $(\lambda,\delta_n) \in \mathbb{Z}_{\geq 0}$ (then $L_C(\lambda)$ is finite-dimensional), or $(\lambda,\delta_n) \in \mathbb{Z}+1/2$ and $(\lambda,\delta_{n-1}+\delta_n) \in \mathbb{Z}_{\geq -2}$.

The above conditions can be rewritten as follows: $L_C(\lambda)$ is bounded if and only if $(\lambda+\rho_C,\alpha^\vee) \in \mathbb{Z}_{>0}$ for each $\alpha \in \Delta(\lambda)^+$ and $\Delta(\lambda) = \Delta_C$ (then $\dim L_C(\lambda) < \infty$) or $\Delta(\lambda) = \Delta_D$.

**3.2. Lemma.** *If $(\lambda+\rho_C,\alpha^\vee) \in \mathbb{Z}_{>0}$ and $\Delta(\lambda) = \Delta_D$ (equivalently, if $L_C(\lambda)$ is infinite-dimensional $\mathfrak{sp}(2n)$-bounded module), then*

$$e^{\rho_C} \operatorname{ch} L_C(\lambda) = \prod_{i=1}^{n}(1-e^{-2\delta_i})^{-1} e^{\rho_D} \operatorname{ch} L_D(\lambda+\rho_C-\rho_D), \quad \dim L_D(\lambda+\rho_C-\rho_D) < \infty.$$

PROOF. Since $(\lambda+\rho_C,\alpha^\vee) \in \mathbb{Z}_{>0}$ for each $\alpha \in \Delta(\lambda)^+$, we have

$$\operatorname{ch} L_C(\lambda) = \sum_{w\in W(\Delta(\lambda))} (-1)^{l(w)} \operatorname{ch} M_C(y \circ_C \lambda). \tag{1}$$

If $\Delta(\lambda) = \Delta_D$, then $W(\Delta(\lambda)) = W(D)$. The condition $(\lambda+\rho,\alpha^\vee) \in \mathbb{Z}_{>0}$ for each $\alpha \in \Delta(\lambda)^+$ is equivalent to $\dim L_D(\lambda+\rho_C-\rho_D) < \infty$. Combining (1), the Weyl character formula for $L_D(\lambda+\rho_C-\rho_D)$ and the formula $\operatorname{ch} M_C(\nu) = \prod_{i=1}^n (1-e^{-2\delta_i})^{-1} \operatorname{ch} M_D(\nu)$ we obtain the desired identity. □

**3.3.** Lemma 3.2 implies that for each $\mu \in \mathfrak{h}^*$

$$\dim L_C(\lambda)_\mu \leq \dim L_D(\lambda+\rho_C-\rho_D).$$

**3.4. Lemma.** *If $L(\lambda)$ is an infinite-dimensional bounded module, then $L_C(\lambda)$ and $L_C(\lambda-\delta_n)$ are bounded $\mathfrak{sp}(2n)$-modules.*

PROOF. Since $L_C(\lambda)$ is an $\mathfrak{sp}(2n)$-subquotient of $L(\lambda)$, $L_C(\lambda)$ is $\mathfrak{sp}(2n)$-bounded. It remains to prove that $L_C(\lambda-\delta_n)$ is $\mathfrak{sp}(2n)$-bounded. Since $L(\lambda)$ is an infinite-dimensional bounded module, we have that $\Delta(\lambda)=\Delta_D$, in particular $(\lambda,\delta_n)\neq 0$. But then one easily checks that $X_{-\delta_n}v$ is a nonzero $\mathfrak{g}_{\bar{0}}$-primitive vector in $L(\lambda)$, where $v$ is a highest weight vector of $L(\lambda)$ and $X_{-\delta_n}$ is in $\mathfrak{g}_{-\delta_n}$. Hence $L(\lambda)$ has a $\mathfrak{g}_{\bar{0}}$-subquotient isomorphic to $L_C(\lambda-\delta_n)$. □

**3.5. Proposition.**

(i) *The module $L(\lambda)$ is bounded if and only if $\dim L(\lambda)<\infty$ or $\Delta(\lambda)=\Delta_D$ and $(\lambda+\rho,\alpha^\vee)\in\mathbb{Z}_{>0}$ for each $\alpha\in\Delta(\lambda)^+$. In other words, $L(\lambda)$ is bounded if and only if $L_C(\lambda)$ is bounded and $(\lambda,\delta_{n-1}+\delta_n)\neq -2$.*

(ii) *If $L(\lambda)$ is bounded and infinite-dimensional, then $\dim L_D(\lambda+\rho-\rho_D)<\infty$ and*

$$e^{\rho}\operatorname{ch} L(\lambda)=\prod_{i=1}^{n}(1-e^{-\delta_i})^{-1}e^{\rho_D}\operatorname{ch} L_D(\lambda+\rho-\rho_D). \tag{2}$$

*In particular, for each $\mu\in\mathfrak{h}^*$ one has*

$$\dim L(\lambda)_\mu\leq \dim L_D(\lambda+\rho-\rho_D).$$

PROOF. Assume that $L(\lambda)$ is bounded.

Since $L_C(\lambda)$ is a submodule of $L(\lambda)$, we conclude that $\Delta(\lambda)$ is $\Delta_C$ or $\Delta_D$. If $\Delta(\lambda)=\Delta_C$, then $\dim L_C(\lambda)<\infty$, which is equivalent to $\dim L(\lambda)<\infty$.

Consider the case $\Delta(\lambda)=\Delta_D$, in particular, $L(\lambda)$ is infinite-dimensional. Since $L_C(\lambda)$ is bounded, for $i=1,\dots,n-1$ we have

$$(\lambda+\rho,\delta_i-\delta_{i+1})=(\lambda+\rho_C,\delta_i-\delta_{i+1})\in\mathbb{Z}_{>0}.$$

Since $L(\lambda)$ is infinite dimensional, by Lemma 3.4, $L_C(\lambda-\delta_n)$ is bounded. This gives

$$(\lambda+\rho,\delta_{n-1}+\delta_n)=(\lambda-\delta_n+\rho_C,\delta_{n-1}+\delta_n)\in\mathbb{Z}_{>0}.$$

Since $\delta_i-\delta_{i+1}, i=1,\dots,n-1$ and $\delta_{n-1}+\delta_n$ are simple roots of $\Delta(\lambda)=\Delta_D$, we have $(\lambda+\rho,\alpha^\vee)\in\mathbb{Z}_{>0}$ for each $\alpha\in\Delta(\lambda)^+$.

Now let $\lambda$ be such that $\Delta(\lambda)=\Delta_D$ and $(\lambda+\rho,\alpha^\vee)\in\mathbb{Z}_{>0}$ for each $\alpha\in\Delta(\lambda)^+$. This means that $W(\lambda)=W(D)$, $\lambda$ is maximal in its $W(\lambda)$-orbit $W(\lambda)\circ\lambda$, and the stabilizer of $\lambda$ is trivial. In particular, $\dim L_D(\lambda+\rho-\rho_D)<\infty$ and the Weyl character formula gives

$$e^{\rho_D}\operatorname{ch} L_D(\lambda+\rho-\rho_D)=\sum_{w\in W(D)}\operatorname{sgn} w\operatorname{ch} M_D(w(\lambda+\rho)).$$

On the other hand, by Corollary 5.5 we obtain

$$e^{\rho}\operatorname{ch} L(\lambda)=\sum_{w\in W(D)}\operatorname{sgn} w\operatorname{ch} M(w(\lambda+\rho)).$$

Combining the last two identities with $\operatorname{ch} M(\nu)=\prod_{i=1}^{n}(1-e^{-\delta_i})^{-1}\operatorname{ch} M_D(\nu)$ leads to (2). This implies $\dim L(\lambda)_\mu\leq\dim L_D(\lambda+\rho-\rho_D)$ as required. □

## 4. Bounded highest weight modules of $\mathfrak{osp}(1,2n)$: an alternative approach

In this section we establish the classification of bounded weights of $\mathfrak{g}$ in an alternative way. Namely, we will present every bounded $L(\lambda)$ as a subquotient of a tensor product of a bounded module and a finite-dimensional module.

By $\mathcal{D}(n)$ we will denote the Weyl algebra $\mathbb{C}[x_1,\dots,x_n;\partial_1,\dots,\partial_n]$ generated by $x_i,\partial_j$ subject to the relations $x_ix_j - x_jx_i = \partial_i\partial_j - \partial_j\partial_i = 0$; $x_i\partial_j - \partial_jx_i = \delta_{ij}$. We consider $\mathcal{D}(n)$ as an associative superalgebra letting $x_i$ and $\partial_i$ to be odd. There are several ways to define a homomorphism $U(\mathfrak{g}) \to \mathcal{D}(n)$. In this paper we will use a presentation for which the $\delta_i$- and $(-\delta_i)$-root vectors of $\mathfrak{g}$ act as $\frac{1}{\sqrt{2}}x_i$ and $\frac{1}{\sqrt{2}}\partial_i$, respectively.

Fix for convenience elements $X_\alpha$ in the $\alpha$-root space of $\mathfrak{g}$ so that $[X_{\delta_i}, X_{\pm\delta_j}] = X_{\delta_i\pm\delta_j}$, $[X_{-\delta_i}, X_{-\delta_j}] = -X_{-\delta_i-\delta_j}$, $i \neq j$, $[X_{\pm\delta_i}, X_{\pm\delta_i}] = \pm 2X_{\pm 2\delta_i}$. The complete list of relations $[X_\alpha, X_\beta] = c_{\alpha,\beta}X_{\alpha+\beta}$ can be found in §4, [**F**]. We also fix elements $h_{\delta_i-\delta_j} = [X_{\delta_i}, X_{-\delta_j}]$ and $h_{2\delta_i} = [X_{\delta_i}, X_{-\delta_i}] = [X_{2\delta_i}, X_{-2\delta_i}]$ in $\mathfrak{h}$.

The following proposition can be verified with a direct computation, [**F**].

**4.1. Proposition.** *The following correspondences define a homomorphism $\phi : U(\mathfrak{g}) \longmapsto \mathcal{D}(n)$ of associative superalgebras:*

$$\begin{aligned}
X_{\delta_i-\delta_j} &\longmapsto x_i\partial_j, i \neq j;\\
X_{2\delta_i} &\longmapsto \frac{1}{2}x_i^2;\\
X_{-2\delta_i} &\longmapsto -\frac{1}{2}\partial_i^2;\\
X_{\delta_i+\delta_j} &\longmapsto x_ix_j, i \neq j;\\
X_{-\delta_i-\delta_j} &\longmapsto -\partial_i\partial_j, i \neq j;\\
h_{\delta_i-\delta_j} &\longmapsto x_i\partial_i - x_j\partial_j, i \neq j;\\
h_{2\delta_i} &\longmapsto x_i\partial_i + \frac{1}{2};\\
X_{\delta_i} &\longmapsto \frac{1}{\sqrt{2}}x_i;\\
X_{-\delta_i} &\longmapsto \frac{1}{\sqrt{2}}\partial_i.
\end{aligned}$$

From the above proposition we easily find that the $\mathcal{D}(n)$-module $\mathbb{C}[x_1,\dots,x_n]$, when considered as a $\mathfrak{g}$-module though the homomorphism $\phi$, is isomorphic to $L_{\widetilde{\Pi}}(\frac{1}{2}(\delta_1 + \cdots + \delta_n))$.

**4.2. Lemma.** *Let $N$ be positive integer and $v$ be a highest weight vector of $L_{\widetilde{\Pi}}(-N(\delta_1 + \cdots + \delta_n))$. Then the vector*

$$u = x_1^{2N} \otimes v + \sum_{k=1}^{2N} c_{2N-k}x_1^{2N-k} \otimes X_{\delta_1}^k(v)$$

*is a primitive vector of* $L_{\widetilde{\Pi}}(\frac{1}{2}(\delta_1 + \cdots + \delta_n)) \otimes L_{\widetilde{\Pi}}(-N(\delta_1 + \cdots + \delta_n))$*, where the scalars* $c_i$ *are defined as follows:*

$$\begin{aligned} c_{2N} &= 1 \\ c_{2N-2j} &= \frac{(2N-1)(2N-3)\dots(2N-(2j-1))}{j!}, j > 0 \\ c_{2N-(2j+1)} &= -\sqrt{2}c_{2N-2j}, j \geq 0. \end{aligned}$$

PROOF. The identity $X_{\delta_i - \delta_{i+1}} u = 0$ is straightforward. It remains to show that $X_{-\delta_1} u = 0$. We easily check that

$$X_{-\delta_1}(x_1^{2N-k} \otimes X_{\delta_1}^k(v))$$
$$= \begin{cases} \frac{2N-k}{\sqrt{2}} x_1^{2N-k-1} \otimes X_{\delta_1}^k(v) + \frac{k}{2} x_1^{2N-k} \otimes X_{\delta_1}^{k-1}(v) \text{ if } k \text{ is even,} \\ \frac{2N-k}{\sqrt{2}} x_1^{2N-k-1} \otimes X_{\delta_1}^k(v) - \left(\frac{k-1}{2} - N\right) x_1^{2N-k} \otimes X_{\delta_1}^{k-1}(v) \text{ if } k \text{ is odd.} \end{cases}$$

Using the above identities we easily complete the proof. □

**4.3. Corollary.** *If* $\lambda$ *is such that* $(\lambda, \delta_n) \in \frac{1}{2} + \mathbb{Z}$ *and* $(\lambda, \delta_{n-1} + \delta_n) \in \mathbb{Z}_{\geq -1}$*, then* $\lambda$ *is bounded.*

PROOF. We first prove that for every nonnegative integer $N$, the weight

$$\lambda_N = \left(N - \frac{1}{2}\right)\delta_1 + \cdots + \left(N - \frac{1}{2}\right)\delta_{n-1} + \left(-N - \frac{1}{2}\right)\delta_n$$

is bounded. For this we note that the $\mathfrak{g}$-module $L(\lambda_N)^\theta$ obtained by twisting $L(\lambda_N)$ by $\theta$ is isomorphic to $L_{\widetilde{\Pi}}\left(\left(N + \frac{1}{2}\right)\delta_1 + \left(\frac{1}{2} - N\right)\delta_2 + \cdots + \left(\frac{1}{2} - N\right)\delta_n\right)$. Then, the latter by Lemma 4.2 is a subquotient of the tensor product of the bounded $\mathfrak{g}$-module $L_{\widetilde{\Pi}}(\frac{1}{2}(\delta_1 + \cdots + \delta_n))$ and the finite-dimensional $\mathfrak{g}$-module $L_{\widetilde{\Pi}}(-N(\delta_1 + \cdots + \delta_n))$. Therefore $L(\lambda_N)^\theta$ is bounded and hence $\lambda_N$ is bounded.

Now let $\lambda$ be a weight for which $(\lambda, \delta_n) \in \frac{1}{2} + \mathbb{Z}$ and $(\lambda, \delta_{n-1} + \delta_n) \in \mathbb{Z}_{\geq -1}$ and let $\mu = \lambda + \frac{1}{2}\sum_{i=1}^n \delta_i$. Also, set for simplicity $\mu_i = (\mu, \delta_i)$. Then we have that $\mu_i \in \mathbb{Z}$, $\mu_1 \geq \mu_2 \geq \cdots \geq \mu_n$ and $\mu_{n-1} + \mu_n \geq 0$. In particular $\mu_1 \geq 0$ and hence $\lambda' = \left(\mu_1 - \frac{1}{2}\right)\sum_{i=1}^{n-1}\delta_i - \left(\mu_1 + \frac{1}{2}\right)\delta_n$ is a bounded weight. But then $\lambda = \lambda' + \lambda''$ for the bounded weight $\lambda'$ and for the dominant integral weight $\lambda'' = \sum_{i=2}^n (\mu_i - \mu_1)\delta_i$. Thus $\lambda$ is bounded. □

The above corollary together with Lemma 3.4 leads to an alternative proof of the classification of bounded weights in Proposition 3.5(i).

## 5. Appendix: Characters of some highest weight modules

**5.1. Conventions.** In this appendix $\mathfrak{g} = \mathfrak{g}_0 \oplus \mathfrak{g}_1$ will be a basic classical Lie superalgebra with a fixed triangular decomposition $\mathfrak{g} = \mathfrak{n}_- \oplus \mathfrak{h} \oplus \mathfrak{n}$. By $\mathfrak{h}$ we denote the Cartan subalgebra of $\mathfrak{g}$, $W$ will be the Weyl group, and $\Delta = \Delta_0 \cup \Delta_1$, $\Delta^+$ will be the set of all roots and all positive roots, respectively. We fix an even non-degenerate bilinear invariant form $(-,-)$ on $\mathfrak{h}^*$. We write $\nu \geq \mu$ for weights $\nu, \mu$ in $\mathfrak{h}^*$ if $\nu - \mu \in \mathbb{Z}_{\geq 0}\Delta^+$. Like in the case $\mathfrak{g} = \mathfrak{osp}(1, 2n)$, for the shifted action of $W$ on $\mathfrak{h}^*$ we write

$$w \circ \lambda := w(\lambda + \rho) - \rho,$$

where $\rho = \rho_0 - \rho_1$ is the difference of the half sums of the positive even roots and the positive odd roots.

By $\operatorname{sgn} : W \to \{\pm 1\}$ we denote the sign homomorphism. For each $\lambda \in \mathfrak{h}^*$, like one the case $\mathfrak{g} = \mathfrak{osp}(1,2n)$, $W(\lambda)$ denotes the corresponding integral Weyl group.

As before, by $M(\nu)$ (resp., $L(\nu)$) we denote the Verma (resp., irreducible) module of highest weight $\nu$. We denote by $\dot{M}(\nu)$ (resp., $\dot{L}(\nu)$) the corresponding Verma (resp., irreducible) $\mathfrak{g}_0$-modules. We introduce another shifted action of the Weyl group $W$ on $\mathfrak{h}^*$ by

$$w \circ_{\mathfrak{g}_0} \lambda = w(\lambda + \rho_0) - \rho_0.$$

**5.2. Character formulae for typical highest weight modules.** Let $\lambda$ be a maximal element in its $W(\lambda)$ orbit $W(\lambda) \circ \lambda$ (i.e., $\lambda \geq w \circ \lambda$ for each $w \in W$). Recall that if $\mathfrak{g}$ is a semisimple Lie algebra, then for each $w \in W(\lambda)$ the character of an irreducible highest weight module $L(w \circ \lambda)$ is given by the Kazhdan-Lusztig character formula:

$$\operatorname{ch} L(w \circ \lambda) = \sum_{y \in W(\lambda)} a_y^w \operatorname{ch} M(y \circ \lambda), \tag{3}$$

and the coefficients $a_y^w$ are given by in terms of the inverse Kazhdan-Lusztig polynomial for the Weyl group $W(\lambda)$ and the stabilizer

$$\operatorname{Stab}_W(\lambda + \rho) = \{w \in W |\ w(\lambda + \rho) = \lambda + \rho\} = \{w \in W(\lambda) |\ w \circ \lambda = \lambda\}.$$

Note that any weight $\nu \in \mathfrak{h}^*$ is of the form $w \circ \lambda$, where $\lambda$ is the maximal element in $W(\nu) \circ \nu$ (in this case $\lambda$ is maximal in $W \circ \lambda$) and $w \in W(\nu) = W(\lambda)$. Hence, (3) gives the character of any irreducible highest weight module.

The character formula (3) also holds for a basic classical Lie superalgebra $\mathfrak{g}$ in the case when $\lambda$ is strongly typical (i.e., $(\lambda + \rho, \beta) \neq 0$ for each $\beta \in \Delta_1$) or some weakly atypical weights, see §5.5 below; this gives a character formula for all strongly typical highest weight modules. The coefficients $a_y^w$ are determined by the same formulae as for the Lie algebras case (they depend on $W(\lambda)$ and $\operatorname{Stab}_W(\lambda+\rho)$). This result easily follows from the equivalence of categories established in [**PS1**], [**PS2**] and [**G2**], see details below in §5.4 and §5.5.

If $\lambda$ is strongly typical, maximal in $W(\lambda) \circ \lambda$, and has the trivial stabilizer, then $a_y^e = \operatorname{sgn} y$. Hence (3) takes the form

$$\operatorname{ch} L(\lambda) = \sum_{y \in W(\lambda)} \operatorname{sgn}(y) \operatorname{ch} M(y \circ \lambda). \tag{4}$$

This holds, in particular, if $L(\lambda)$ is typical and finite-dimensional. In the latter case $W(\lambda) = W$ and (3) becomes the Weyl-Kac character formula established in [**K2**].

**5.3. Typicality and strong typicality.** Recall that $\lambda \in \mathfrak{h}^*$ is *typical* if $(\lambda + \rho, \beta) \neq 0$ for each isotropic root $\beta$. We call $\lambda \in \mathfrak{h}^*$ *strongly typical* if $(\lambda + \rho, \beta) \neq 0$ for each odd root $\beta$.

A central character $\chi : \mathcal{Z}(\mathcal{U}(\mathfrak{g})) \to \mathbb{C}$ is called typical (resp., strongly typical) if it is a central character of $L(\lambda)$ with typical (resp., strongly typical) $\lambda$. Note that this definition does not depend on the triangular decomposition in the following sense. If $\mathfrak{g} = \mathfrak{n}_- \oplus \mathfrak{h} \oplus \mathfrak{n} = \mathfrak{n}'_- \oplus \mathfrak{h} \oplus \mathfrak{n}'$ are two triangular decompositions such that $\mathfrak{n} \cap \mathfrak{g}_0 = \mathfrak{n}' \cap \mathfrak{g}_0$, and if $\lambda$ is typical, then the Verma module $M(\lambda, \mathfrak{n})$ relative to the first triangular decomposition is isomorphic to a Verma module $M(\lambda', \mathfrak{n}')$ relative to the second triangular decomposition; in this case $\lambda + \rho = \lambda' + \rho'$. In particular, if $\lambda$ is such that $(\lambda+\rho,\beta)\neq 0$ for all isotropic roots $\beta$, then $L(\lambda,\mathfrak{n})=L(\lambda',\mathfrak{n}')$, where $\lambda+\rho=\lambda'+\rho'$, so $(\lambda'+\rho',\beta)\neq 0$ for all isotropic roots $\beta$.

We call a $\mathfrak{g}$-module strongly typical if it has a strongly typical central character. The set of highest weights of irreducible highest weight modules with a fixed central character forms a single $W$-orbit if and only if this central character is typical: if $\chi:\mathcal{Z}(\mathcal{U}(\mathfrak{g}))\to\mathbb{C}$ is a central character, then $(\operatorname{Ker}\chi)L(\lambda)=(\operatorname{Ker}\chi)L(\lambda')=0$ implies $\lambda'+\rho\in W(\lambda+\rho)$ if and only if $\chi$ is typical.

**5.4. Strongly typical case.** Take any strongly typical central character $\chi$. By [**G2**] Theorem 3.3.1, there exists a $\mathfrak{g}_0$-central character $\dot\chi:\mathcal{Z}(\mathcal{U}(\mathfrak{g}_0))\to\mathbb{C}$ such that the map $\Psi: N\mapsto N_{\dot\chi}:=\{v\in N|\ (\operatorname{Ker}\dot\chi)v=0\}$ provides an equivalence between the category of $\mathfrak{g}$-modules with the central character $\chi$ and the category of $\mathfrak{g}_0$-modules with the central character $\dot\chi$. The map $\Psi$ maps a Verma $\mathfrak{g}$ module to a Verma $\mathfrak{g}_0$-module. Recall that $M(\nu),M(\lambda)$ (resp., $\dot M(\nu),\dot M(\lambda)$) have the same typical central character (resp., the same central character) if and only if $\nu\in W\circ\lambda$ (resp., $\nu\in W\circ_{\mathfrak{g}_0}\lambda$).

5.4.1. **Lemma.** *If $\Psi(M(\nu))=\dot M(\nu')$, then for each $w\in W$ one has*

$$\Psi(M(w\circ\nu))=\dot M(w\circ_{\mathfrak{g}_0}\nu').$$

PROOF. Recall that $M(\lambda)$ has a filtration of $\mathfrak{g}_0$-modules with the factors $\{\dot M(\lambda-\gamma)\}_{\gamma\in\Gamma}$, where

$$\Gamma:=\left\{\sum_{\beta\in X}\beta\mid X\subset\Delta_1^+\right\}.$$

By [**G1**], Lemma 8.3.4(i), for each $M(\lambda)$ with central character $\chi$, there exists a unique $\gamma\in\Gamma$ such that $\dot M(\lambda-\gamma)$ has the central character $\dot\chi$, and such that $\Psi(M(\lambda))=M(\lambda-\gamma)$. Thus it is enough to verify that for each $w\in W$ one has $w\circ\nu-w\circ_{\mathfrak{g}_0}\nu'\in\Gamma$ if $\nu-\nu'\in\Gamma$. Observe that

$$w\circ\nu-w\circ_{\mathfrak{g}_0}\nu'=w(\nu-\nu'-\rho_1)+\rho_1.$$

One readily sees that

$$\Gamma-\rho_1=\{\gamma-\rho_1|\ \gamma\in\Gamma\}=\left\{\sum_{\beta\in Y}\beta/2\mid Y\subset\Delta_1,\Delta_1=Y\sqcup(-Y)\right\}.$$

Since $\Delta_1$ is $W$-invariant, $\Gamma-\rho_1$ is also $W$-invariant and thus $\nu-\nu'\in\Gamma$ implies $w\circ\nu-w\circ_{\mathfrak{g}_0}\nu'\in\Gamma$ as required. $\square$

5.4.2. **Proposition.** *Let $\mu$ be strongly typical and a minimal element in its orbit $W(\mu)\circ\mu$. Let $w_0$ be the longest element in $W(\mu)$ and $\nu:=w_0\circ\mu$. Then the character formula (3) holds for $\lambda=\nu$, i.e.*

$$\operatorname{ch}L(w\circ\nu)=\sum_{y\in W(\nu)}a_y^w\operatorname{ch}M(y\circ\nu).$$

PROOF. Since $M(\mu)$ is an irreducible Verma module, the $\mathfrak{g}_0$-Verma module $\Psi(M(\mu))=\dot M(\mu')$ is also irreducible, that is $\mu'$ is minimal in its orbit $W(\mu')\circ_{\mathfrak{g}_0}\mu'$. Note that for each $\alpha\in\Delta_0$ one has $2(\alpha,\beta)/(\alpha,\alpha)\in\mathbb{Z}$ for all $\beta\in\Delta$ and thus for all $\beta\in\Gamma$. Hence $W(\mu')=W(\mu)$. Then $\nu=w_0\circ\mu$ (resp., $\nu':=w_0\circ_{\mathfrak{g}_0}\mu'$) is a (unique) maximal element in the orbit $W(\mu)\circ\mu$ (resp., $W(\mu)\circ_{\mathfrak{g}_0}\mu$). We have $W(\nu)=W(\mu)=W(\mu')=W(\nu')$. By (3), for each $w\in W(\nu)$ we have

$$\operatorname{ch}\dot{L}(w\circ_{\mathfrak{g}_0}\nu')=\sum_{y\in W(\nu)} a_y^w \operatorname{ch}\dot{M}(y\circ_{\mathfrak{g}_0}\nu').$$

Since $\Psi$ is an equivalence of categories, $\Psi(M(y\circ\nu))=\dot{M}(y\circ_{\mathfrak{g}_0}\nu')$ for each $y\in W$ (by Lemma 5.4.1) and so $\Psi(L(y\circ\nu))=\dot{L}(y\circ_{\mathfrak{g}_0}\nu')$. A standard reasoning (see [**BGG**]) leads to

$$\operatorname{ch}L(w\circ\nu)=\sum_{y\in W(\nu)} a_y^w \operatorname{ch}M(y\circ\nu)$$

as needed. □

**5.5. The special case $\mathfrak{g}=\mathfrak{osp}(1,2n)$.** Let now $\mathfrak{g}:=\mathfrak{osp}(1,2n)$. Let $\nu\in\mathfrak{h}^*$ be such that $(\nu+\rho,\beta)=0$ for a unique odd positive root $\beta$ and $\nu$ be the maximal element in its $W(\nu)$-orbit $W(\nu)\circ\nu$ (note that if $\lambda\in\mathfrak{h}^*$ is such that $(\lambda+\rho,\beta)=0$ for a unique odd positive root $\beta$, then $\lambda=w\circ\nu$, where $\nu$ is as above and $w\in W(\nu)$). We show that (3) holds for $\lambda=\nu$.

Let $\chi:\mathcal{Z}(\mathcal{U}(\mathfrak{g}))\to\mathbb{C}$ be the central character of $M(\nu)$. By [**G2**], Theorem 4.3 there exists a $\mathfrak{g}_0$-central character $\dot{\chi}:\mathcal{Z}(\mathcal{U}(\mathfrak{g}_0))\to\mathbb{C}$ such that the map $\Psi_0:N\mapsto N_{\bar{0},\dot{\chi}}:=\{v\in N_{\bar{0}}|\ (\operatorname{Ker}\dot{\chi})v=0\}$ provides an equivalence between the category of graded $\mathfrak{g}$-modules ($N=N_{\bar{0}}\oplus N_{\bar{1}}$) with the central character $\chi$ and the category of $\mathfrak{g}_0$-modules with the central character $\dot{\chi}$. The map $\Psi_0$ maps a graded Verma $\mathfrak{g}$-module to a Verma $\mathfrak{g}_0$-module.

Let $\mu$ be the minimal element in $W(\nu)\circ\nu$ (i.e., $\mu=w_0\circ\nu$, where $w_0$ is the longest element in $W(\nu)$). View $M(\mu)$ as a graded Verma module (with one of two possible gradings); note that $M(\mu)$ is irreducible, so $\Psi(M(\mu))=\dot{M}(\mu')$ is irreducible. Thus $\mu'$ is minimal in its $W(\mu')$-orbit $W(\mu')\circ_{\mathfrak{g}_0}\mu'$-orbit and $W(\nu)=W(\mu)=W(\mu')$. Arguing as in Lemma 5.4.1, we obtain that $\Psi^{-1}(\dot{M}(w\circ_{\mathfrak{g}_0}\mu'))$ is the Verma module $M(w\circ\mu)$ with one of two possible gradings (such that $M(w\circ\mu)_{w\circ_{\mathfrak{g}_0}\mu'}\subset M(w\circ\mu)_{\bar{0}}$). Now the argument in the proof of Proposition 5.4.2 shows that (3) holds for $\lambda=\nu$.

**Corollary.** *Let $\mathfrak{g}=\mathfrak{osp}(1,2n)$, $\lambda$ be maximal in its $W(\lambda)$-orbit $W(\lambda)\circ\lambda$, and the stabilizer $\operatorname{Stab}_W(\lambda+\rho)$ be trivial. Then the character formula (4) holds for $\lambda$.*

## References


[BGG] I. N. Bernšteĭn, I. M. Gel′fand, and S. I. Gel′fand, *Structure of representations that are generated by vectors of highest weight* (Russian), Funckcional. Anal. i Priložen. **5** (1971), no. 1, 1–9. MR0291204 (45 #298)

[DMP] I. Dimitrov, O. Mathieu, and I. Penkov, *On the structure of weight modules*, Trans. Amer. Math. Soc. **352** (2000), no. 6, 2857–2869, DOI 10.1090/S0002-9947-00-02390-4. MR1624174 (2000j:17008)

[F] T. L. Ferguson, *Weight Modules of Orthosymplectic Lie Superalgebras*, ProQuest LLC, Ann Arbor, MI, 2015. Thesis (Ph.D.)–The University of Texas at Arlington. MR3407351

[G1] M. Gorelik, *Annihilation theorem and separation theorem for basic classical Lie superalgebras*, J. Amer. Math. Soc. **15** (2002), no. 1, 113–165 (electronic), DOI 10.1090/S0894-0347-01-00382-4. MR1862799 (2002j:17003)

[G2] M. Gorelik, *Strongly typical representations of the basic classical Lie superalgebras*, J. Amer. Math. Soc. **15** (2002), no. 1, 167–184 (electronic), DOI 10.1090/S0894-0347-01-00381-2. MR1862800 (2002j:17004)

[GG] M. Gorelik, D. Grantcharov, *Bounded highest weight modules of* $\mathfrak{q}(n)$, Int. Math. Res. Not. (2013), Id: rnt147.

[Gr] D. Grantcharov, *Explicit realizations of simple weight modules of classical Lie superalgebras*, Groups, rings and group rings, Contemp. Math., vol. 499, Amer. Math. Soc., Providence, RI, 2009, pp. 141–148, DOI 10.1090/conm/499/09797. MR2581932 (2011b:17016)

[H] C. Hoyt, *Weight modules of* $D(2,1,\alpha)$, Advances in Lie superalgebras, Springer INdAM Ser., vol. 7, Springer, Cham, 2014, pp. 91–100, DOI 10.1007/978-3-319-02952-8_6. MR3205083

[K1] V. G. Kac, *Lie superalgebras*, Advances in Math. **26** (1977), no. 1, 8–96. MR0486011 (58 #5803)

[K2] V. G. Kac, *Characters of typical representations of classical Lie superalgebras*, Comm. Algebra **5** (1977), no. 8, 889–897. MR0444725 (56 #3075)

[M] O. Mathieu, *Classification of irreducible weight modules* (English, with English and French summaries), Ann. Inst. Fourier (Grenoble) **50** (2000), no. 2, 537–592. MR1775361 (2001h:17017)

[PS1] I. Penkov and V. Serganova, *Representations of classical Lie superalgebras of type* I, Indag. Math. (N.S.) **3** (1992), no. 4, 419–466, DOI 10.1016/0019-3577(92)90020-L. MR1201236 (93k:17006)

[PS2] I. Penkov and V. Serganova, *Generic irreducible representations of finite-dimensional Lie superalgebras*, Internat. J. Math. **5** (1994), no. 3, 389–419, DOI 10.1142/S0129167X9400022X. MR1274125 (95c:17015)

Department of Mathematics, University of Texas at Arlington, Arlington, Texas 76019

*E-mail address*: thomas.ferguson@mavs.uta.edu

Department of Mathematics, Weizmann Institute of Science, Rehovot 76100, Israel
*E-mail address*: maria.gorelik@weizmann.ac.il

Department of Mathematics, University of Texas at Arlington, Arlington, Texas 76019

*E-mail address*: grandim@uta.edu

Proceedings of Symposia in Pure Mathematics
Volume **92**, 2016
http://dx.doi.org/10.1090/pspum/092/01577

# A combinatorial description of the affine Gindikin-Karpelevich formula of type $A_n^{(1)}$

Seok-Jin Kang, Kyu-Hwan Lee, Hansol Ryu, and Ben Salisbury

ABSTRACT. The classical Gindikin-Karpelevich formula appears in Langlands' calculation of the constant terms of Eisenstein series on reductive groups and in Macdonald's work on $p$-adic groups and affine Hecke algebras. The formula has been generalized in the work of Garland to the affine Kac-Moody case, and the affine case has been geometrically constructed in a recent paper of Braverman, Finkelberg, and Kazhdan. On the other hand, there have been efforts to write the formula as a sum over Kashiwara's crystal basis or Lusztig's canonical basis, initiated by Brubaker, Bump, and Friedberg. In this paper, we write the affine Gindikin-Karpelevich formula as a sum over the crystal of generalized Young walls when the underlying Kac-Moody algebra is of affine type $A_n^{(1)}$. The coefficients of the terms in the sum are determined explicitly by the combinatorial data from Young walls.

## 0. Introduction

The classical Gindikin-Karpelevich formula originated from a certain integration on real reductive groups [**GK62**]. When Langlands calculated the constant terms of Eisenstein series on reductive groups [**Lan71**], he considered a $p$-adic analogue of the integration and called the resulting formula the *Gindikin-Karpelevich formula*. In the case of $\mathrm{GL}_{n+1}$, the formula can be described as follows: let $F$ be a $p$-adic field with residue field of $q$ elements and let $N_-$ be the maximal unipotent subgroup of $\mathrm{GL}_{n+1}(F)$ with maximal torus $T$. Let $f^\circ$ denote the standard spherical vector corresponding to an unramified character $\chi$ of $T$, let $T(\mathbf{C})$ be the maximal torus in the $L$-group $\mathrm{GL}_{n+1}(\mathbf{C})$ of $\mathrm{GL}_{n+1}(F)$, and let $\boldsymbol{z} \in T(\mathbf{C})$ be the element corresponding to $\chi$ via the Satake isomorphism. Then the Gindikin-Karpelevich

---

2010 *Mathematics Subject Classification.* Primary 17B37; Secondary 05E10.

*Key words and phrases.* Crystal, Gindikin-Karpelevich formula, generalized Young wall.

The work of the first author was supported by NRF Grant # 2011-0017937 and NRF Grant # 2011-0027952.

The work of the second author was partially supported by a grant from the Simons Foundation (#318706).

The work of the third author was supported by BK21 Mathematical Sciences Division and NRF Grant # 2011-0027952, and NRF 2014R1A2A1A11050917.

The work of the fourth author was partially supported by NSF grant DMS 0847586.

formula is given by

$$\int_{N_-(F)} f^\circ(\boldsymbol{n})\, \mathrm{d}\boldsymbol{n} = \prod_{\alpha\in\Delta^+} \frac{1-q^{-1}\boldsymbol{z}^\alpha}{1-\boldsymbol{z}^\alpha}, \tag{0.1}$$

where $\Delta^+$ is the set of positive roots of $\mathrm{GL}_{n+1}(\mathbf{C})$. The formula appears in Macdonald's study on $p$-adic groups and affine Hecke algebras as well [**Mac71**], and the product side of (0.1) is also known as Macdonald's $c$-function.

In the paper [**Gar04**], Garland generalized Langlands' calculation to affine Kac-Moody groups and obtained an affine Gindikin-Karpelevich formula as a product over $\Delta^+ \cap w^{-1}(\Delta^-)$ for each $w \in W$, where $\Delta^+$ (resp. $\Delta^-$) is the set of positive (resp. negative) roots of the corresponding affine Kac-Moody algebra and $W$ is the Weyl group. In a recent paper of Braverman, Finkelberg, and Kazhdan [**BFK12**], the authors interpreted the classical Gindikin-Karpelevich formula in a geometric way, and generalized the formula to affine Kac-Moody groups and obtained another version of affine Gindikin-Karpelevich formula, which has an additional "correction factor" in the product side.

On the other hand, in the works of Brubaker, Bump and Friedberg [**BBF11**], Bump and Nakasuji [**BN10**], and McNamara [**McN11**], the product side of the classical Gindikin-Karpelevich formula in type $A_n$ was written as a sum over the crystal $\mathcal{B}(\infty)$. (For the definition of a crystal, see [**HK02**, **Kas02**].) More precisely, they proved

$$\prod_{\alpha\in\Delta^+} \frac{1-q^{-1}\boldsymbol{z}^\alpha}{1-\boldsymbol{z}^\alpha} = \sum_{b\in\mathcal{B}(\infty)} G_{\mathbf{i}}^{(e)}(b) q^{\langle \mathrm{wt}(b),\rho\rangle} \boldsymbol{z}^{-\mathrm{wt}(b)},$$

where $\rho$ is the half-sum of the positive roots, $\mathrm{wt}(b)$ is the weight of $b$, and the coefficients $G_{\mathbf{i}}^{(e)}(b)$ are defined using so-called BZL paths or Kashiwara's parametrization. As shown in [**KL11**] by H. Kim and K.-H. Lee, one can also choose a reduced word for the longest element of the Weyl group and use Lusztig's parametrization of canonical bases ([**Lus90**, **Lus91**]), and the product can be written as

$$\prod_{\alpha\in\Delta^+} \frac{1-q^{-1}\boldsymbol{z}^\alpha}{1-\boldsymbol{z}^\alpha} = \sum_{b\in\mathcal{B}(\infty)} (1-q^{-1})^{\mathscr{N}(\phi_{\mathbf{i}}(b))} \boldsymbol{z}^{-\mathrm{wt}(b)}, \tag{0.2}$$

where $\mathscr{N}(\phi_{\mathbf{i}}(b))$ is the number of nonzero entries in Lusztig's parametrization $\phi_{\mathbf{i}}(b)$. The equation (0.2) was proved for all finite roots systems $\Delta$, and was generalized in a subsequent paper [**KL12**] to the affine Kac-Moody case using the results of Beck, Chari, and Pressley [**BCP99**] and Beck and Nakajima [**BN04**] on PBW-type bases. The parametrizations of basis elements in simply-laced affine cases can be found in [**BCP99**, Theorem 3]. We will call them *canonical parametrizations.*

The use of crystals connects the Gindikin-Karpelevich formula to combinatorial representation theory, since much work has been done on realizations of crystals through various combinatorial objects (e.g., [**Kam10**, **Kan03**, **KN94**, **KS97**, **Lit95**]). Indeed, for type $A_n$, K.-H. Lee and Salisbury [**LS12**] expressed the right side of (0.2) as a sum over marginally large Young tableaux using J. Hong and H. Lee's [**HL08**] description of $\mathcal{B}(\infty)$ and the coefficients were determined by a simple statistic $\mathrm{seg}(b)$ of the tableau $b$. Furthermore, the meaning of $\mathrm{seg}(b)$ was studied in the frameworks of Kamnitzer's MV polytope model [**Kam10**] and Kashiwara-Saito's geometric realization [**KS97**] of the crystal $\mathcal{B}(\infty)$. The segment statistic was then generalized to types $B_n$, $C_n$, $D_n$, and $G_2$ in [**LS14**].

The goal of this paper is to extend this approach to affine type $A_n^{(1)}$ through generalized Young walls. The notion of a Young wall was first introduced by Kang [**Kan03**] in his extensive study of affine crystals. In the case of $\mathcal{B}(\infty)$ in type $A_n^{(1)}$, J.-A. Kim and D.-U. Shin [**KS10**] considered a set of *generalized* Young walls to obtain a realization of $\mathcal{B}(\infty)$, while H. Lee [**Lee07**] established a different realization. These constructions in type $A_n^{(1)}$ are closely related to Zelevinsky's *multisegments* [**Zel80**] and Lusztig's *aperiodic* multisegments [**Lus91**], whose crystal structure was studied by Leclerc, Thibon and Vasserot [**LTV99**]. In this paper, we will adopt Kim and Shin's realization and prove (Theorem 3.23)

$$\prod_{\alpha\in\Delta^+}\left(\frac{1-q^{-1}\boldsymbol{z}^{\alpha}}{1-\boldsymbol{z}^{\alpha}}\right)^{\mathrm{mult}(\alpha)} = \sum_{Y\in\mathcal{Y}(\infty)} (1-q^{-1})^{\mathscr{N}(Y)}\boldsymbol{z}^{-\mathrm{wt}(Y)},$$

where $\mathcal{Y}(\infty)$ is the set of reduced proper generalized Young walls and $\mathscr{N}(Y)$ is a certain statistic on $Y \in \mathcal{Y}(\infty)$.

There are two main constructions in the proof. The first one is to establish natural bijections starting from $\mathcal{Y}(\infty)$ so that we may assign a Kostant partition to an element $Y$ of $\mathcal{Y}(\infty)$. The second is to develop an algorithm to calculate the number $\mathscr{N}(Y)$ of distinct parts in the Kostant partition corresponding to $Y$. Note that if one can read off a canonical parametrization established by Beck, Chari, Nakajima and Pressley, directly from $Y$, then the corresponding Kostant partition is readily obtained. However, to the authors' knowledge, an efficient way to read off a canonical parametrization from $Y$ in the affine setting is not known. Instead, our construction uses the more combinatorial nature of $\mathcal{Y}(\infty)$ and produces an explicit correspondence between $\mathcal{Y}(\infty)$ and the set of Kostant partitions. Our method then assigns a canonical-type parametrization to $Y$ through the corresponding Kostant partition. We do not know at the moment whether our parametrization coincides with a canonical parametrization of Beck, Chari, Nakajima and Pressley.

In type $A_n^{(1)}$, the correction factor in the formula of Braverman, Finkelberg, and Kazhdan, mentioned above is given by

$$\prod_{i=1}^{n}\prod_{j=1}^{\infty}\frac{1-q^{-i}\boldsymbol{z}^{j\delta}}{1-q^{-(i+1)}\boldsymbol{z}^{j\delta}}, \tag{0.3}$$

where $\delta$ is the minimal positive imaginary root. In the last section we will write this correction factor as a sum over a subset of reduced proper generalized Young walls (Proposition 4.4), obtain an expansion of the whole product as a sum over pairs of reduced proper generalized Young walls (Corollary 4.5), and derive a combinatorial formula for the number of points in the intersection $T^{-\gamma}\cap S^0$ of certain orbits $T^{-\gamma}$ and $S^0$ in the (double) affine Grassmannian (Corollary 4.6).

**Acknowledgements.** The authors are grateful to A. Braverman for helpful comments. They also thank the referee for useful comments.

## 1. General definitions

Let $I = \{0, 1, \dots, n\}$ be an index set and let $(A, \Pi, \Pi^\vee, P, P^\vee)$ be a Cartan datum of type $A_n^{(1)}$; i.e.,

- $A = (a_{ij})_{i,j\in I}$ is a generalized Cartan matrix of type $A_n^{(1)}$,
- $\Pi = \{\alpha_i : i \in I\}$ is the set of simple roots,

- $\Pi^\vee = \{h_i : i \in I\}$ is the set of simple coroots,
- $P^\vee = \mathbf{Z}h_1 \oplus \cdots \oplus \mathbf{Z}h_n \oplus \mathbf{Z}d$ is the dual weight lattice,
- $\mathfrak{h} = \mathbf{C} \otimes_{\mathbf{Z}} P^\vee$ is the Cartan subalgebra,
- and $P = \{\lambda \in \mathfrak{h}^* : \lambda(P^\vee) \subset \mathbf{Z}\}$ is the weight lattice.

In addition to the above data, we have a bilinear pairing $\langle\ ,\ \rangle \colon P^\vee \times P \longrightarrow \mathbf{Z}$ defined by $\langle h_i, \alpha_j\rangle = a_{ij}$ and $\langle d, \alpha_j\rangle = \delta_{0,j}$.

Let $\mathfrak{g}$ be the affine Kac-Moody algebra associated with this Cartan datum, and denote by $U_v(\mathfrak{g})$ the quantized universal enveloping algebra of $\mathfrak{g}$. We denote the generators of $U_v(\mathfrak{g})$ by $e_i$, $f_i$ $(i \in I)$, and $v^h$ $(h \in P^\vee)$. The subalgebra of $U_v(\mathfrak{g})$ generated by $f_i$ $(i \in I)$ will be denoted by $U_v^-(\mathfrak{g})$.

A $U_v(\mathfrak{g})$-crystal is a set $\mathcal{B}$ together with maps

$$\widetilde{e}_i, \widetilde{f}_i \colon \mathcal{B} \longrightarrow \mathcal{B} \sqcup \{0\}, \qquad \varepsilon_i, \varphi_i \colon \mathcal{B} \longrightarrow \mathbf{Z} \sqcup \{-\infty\}, \qquad \mathrm{wt} \colon \mathcal{B} \longrightarrow P$$

satisfying certain conditions (see [**HK02**, **Kas95**]). The negative part $U_v^-(\mathfrak{g})$ has a crystal base (see [**Kas91**]) which is a $U_v(\mathfrak{g})$-crystal. We denote this crystal by $\mathcal{B}(\infty)$, and denote its highest weight element by $u_\infty$.

Finally, we will describe the set of roots $\Delta$ for $\mathfrak{g}$. Since we are fixing $\mathfrak{g}$ to be of type $A_n^{(1)}$, we may make this explicit. Define

$$\begin{aligned} \Delta_{\mathrm{cl}} &= \{\pm(\alpha_i + \cdots + \alpha_j) : 1 \le i \le j \le n\}, \\ \Delta_{\mathrm{cl}}^+ &= \{\alpha_i + \cdots + \alpha_j : 1 \le i \le j \le n\} \end{aligned}$$

to be set of classical roots and positive classical roots; *i.e.*, roots in the root system of $\mathfrak{g}_{\mathrm{cl}} = \mathfrak{sl}_{n+1}$. The minimal imaginary root is $\delta = \alpha_0 + \alpha_1 + \cdots + \alpha_n$. Then

$$\Delta_{\mathrm{Im}} = \{m\delta : m \in \mathbf{Z} \setminus \{0\}\}, \qquad \Delta_{\mathrm{Im}}^+ = \{m\delta : m \in \mathbf{Z}_{>0}\}.$$

We have $\Delta = \Delta_{\mathrm{Re}} \sqcup \Delta_{\mathrm{Im}}$ and $\Delta^+ = \Delta_{\mathrm{Re}}^+ \sqcup \Delta_{\mathrm{Im}}^+$, where

$$\begin{aligned} \Delta_{\mathrm{Re}} &= \{\alpha + m\delta : \alpha \in \Delta_{\mathrm{cl}},\ m \in \mathbf{Z}\} \\ \Delta_{\mathrm{Re}}^+ &= \{\alpha + m\delta : \alpha \in \Delta_{\mathrm{cl}},\ m \in \mathbf{Z}_{>0}\} \cup \Delta_{\mathrm{cl}}^+. \end{aligned}$$

Recall $\mathrm{mult}(\alpha) = 1$ for any $\alpha \in \Delta_{\mathrm{Re}}$ and $\mathrm{mult}(\alpha) = n$ for any $\alpha \in \Delta_{\mathrm{Im}}$. For notational convenience, since $\mathrm{mult}(m\delta) = n$, we write

$$\Delta_{\mathrm{Im}}^+ = \{m_1\delta_1, \dots, m_n\delta_n : m_1, \dots, m_n \in \mathbf{Z}_{>0}\},$$

where each $\delta_j$ is a copy of the imaginary root $\delta$.

## 2. Generalized Young walls

In this section we describe generalized Young walls. We refer the reader to [**Kan03**, **Zel80**, **Lus91**, **LTV99**] for related constructions and background. We start by defining the board on which all generalized Young walls will be built.

Define

(2.1)

| | | | | | | | |
|---|---|---|---|---|---|---|---|
| | $\vdots$ | $\vdots$ | $\vdots$ | | $\vdots$ | $\vdots$ | $\vdots$ |
| $\cdots$ | 0 | 1 | 2 | $\cdots$ | 0 | 1 | $(n+3)$rd row |
| $\cdots$ | $n$ | 0 | 1 | $\cdots$ | $n$ | 0 | $(n+2)$nd row |
| $\cdots$ | $n-1$ | $n$ | 0 | $\cdots$ | $n-1$ | $n$ | $(n+1)$st row |
| | $\vdots$ | $\vdots$ | $\vdots$ | | $\vdots$ | $\vdots$ | $\vdots$ |
| $\cdots$ | 0 | 1 | 2 | $\cdots$ | 0 | 1 | 2nd row |
| $\cdots$ | $n$ | 0 | 1 | $\cdots$ | $n$ | 0 | 1st row |

In particular, the color of the $j$th site from the bottom of the $i$th column from the right in (2.1) is $j - i \bmod n+1$.

DEFINITION 2.1. A *generalized Young wall* is a finite collection of $i$-colored boxes $\boxed{i}$ $(i \in I)$ on the board (2.1) satisfying the following building conditions.

(1) The colored boxes should be located according to the colors of the sites on the board (2.1).
(2) The colored boxes are put in rows; that is, one stacks boxes from right to left in each row.

For a generalized Young wall $Y$, we define the *weight* $\mathrm{wt}(Y)$ of $Y$ to be

$$\mathrm{wt}(Y) = -\sum_{i\in I} m_i(Y)\alpha_i,$$

where $m_i(Y)$ is the number of $i$-colored boxes in $Y$.

DEFINITION 2.2. A generalized Young wall is called *proper* if for any $k > \ell$ and $k - \ell \equiv 0 \bmod n+1$, the number of boxes in the $k$th row from the bottom is less than or equal to that of the $\ell$th row from the bottom.

DEFINITION 2.3. Let $Y$ be a generalized Young wall and let $Y_k$ be the $k$th column of $Y$ from the right. Set $a_i(k)$, with $i \in I$ and $k \geq 1$, to be the number of $i$-colored boxes in the $k$th column $Y_k$.

(1) We say $Y_k$ contains a *removable* $\delta$ if we may remove one $i$-colored box for all $i \in I$ from $Y_k$ and still obtain a generalized Young wall. In other words, $Y_k$ contains a removable $\delta$ if $a_{i-1}(k+1) < a_i(k)$ for all $i \in I$.
(2) $Y$ is said to be *reduced* if no column $Y_k$ of $Y$ contains a removable $\delta$.

Let $\mathcal{Y}(\infty)$ denote the set of all reduced proper generalized Young walls. In [**KS10**], Kim and Shin defined a crystal structure on $\mathcal{Y}(\infty)$ and proved the following theorem. We refer the reader to [**KS10**] for the details.

THEOREM 2.4 ([**KS10**]). *We have $\mathcal{B}(\infty) \cong \mathcal{Y}(\infty)$ as crystals.*

## 3. Kostant partitions

Let

$$\alpha_i^{(\ell)} = \alpha_i + \alpha_{i-1} + \cdots + \alpha_{i-\ell+1}, \qquad i \in I,\ 1 \le \ell \le n,$$

where the indices are understood mod $n+1$.

EXAMPLE 3.1. Let $n = 2$. Then

$$\alpha_0^{(1)} = \alpha_0 \qquad \alpha_1^{(1)} = \alpha_1 \qquad \alpha_2^{(1)} = \alpha_2$$

$$\alpha_0^{(2)} = \alpha_0 + \alpha_2 \qquad \alpha_1^{(2)} = \alpha_1 + \alpha_0 \qquad \alpha_2^{(2)} = \alpha_2 + \alpha_1.$$

Let

$$S_1 = \left\{ (m_k\delta_k),\ (c_{i,\ell}\,\delta + \alpha_i^{(\ell)}) : {\scriptstyle m_k > 0,\ 1 \le k \le n, \atop c_{i,\ell} \ge 0,\ i \in I,\ 1 \le \ell \le n} \right\}.$$

We introduce the generator $\boldsymbol{\delta}^{(m)}$ for $m \in \mathbf{Z}_{>0}$ and set

$$S_2 = \{\boldsymbol{\delta}^{(m)} : m \in \mathbf{Z}_{>0}\}.$$

Let $\widetilde{\mathcal{G}}$ be the free abelian group generated by $S_1 \cup S_2$. Consider the subgroup $L$ of $\widetilde{\mathcal{G}}$ generated by the elements: for $m > 0$,

$$\text{(3.1)} \qquad \begin{cases} \boldsymbol{\delta}^{(m)} - \sum_{i \in I} (k\delta + \alpha_i^{(\ell)}), & m = (n+1)k + \ell,\ 1 \le \ell \le n; \\ \boldsymbol{\delta}^{(m)} - \boldsymbol{\delta}^{(k)} - \sum_{i=1}^{n} (k\delta_i), & m = (n+1)k. \end{cases}$$

We set $\mathcal{G} = \widetilde{\mathcal{G}}/L$ and let $\mathcal{G}^+$ be the $\mathbf{Z}_{\ge 0}$-span of $S_1 \cup S_2$ in $\mathcal{G}$. The following observation will play an important role.

REMARK 3.2. If we slightly abuse language, we may say that, in $\mathcal{G}$, the element $\boldsymbol{\delta}^{(m)}$ is equal to the sum of $n+1$ distinct positive "roots" of equal length $m$ whose total weight is $m\delta$. In particular, if $m = (n+1)k + \ell$ $(1 \le \ell \le n)$, then $\boldsymbol{\delta}^{(m)}$ is equal to the sum of $n+1$ distinct positive real roots of equal length $m$, and if $m = (n+1)k$, then $\boldsymbol{\delta}^{(m)}$ is equal to the sum of $(k\delta_1), \dots, (k\delta_n), \boldsymbol{\delta}^{(k)}$ of equal length $m$.

EXAMPLE 3.3. Let $n = 2$. Then in $\mathcal{G}$,

$$\boldsymbol{\delta}^{(1)} = (\alpha_0) + (\alpha_1) + (\alpha_2)$$

$$\boldsymbol{\delta}^{(2)} = (\alpha_0 + \alpha_2) + (\alpha_1 + \alpha_0) + (\alpha_2 + \alpha_1)$$

$$\boldsymbol{\delta}^{(3)} = \boldsymbol{\delta}^{(1)} + (\delta_1) + (\delta_2) = (\alpha_0) + (\alpha_1) + (\alpha_2) + (\delta_1) + (\delta_2)$$

$$\boldsymbol{\delta}^{(4)} = (\delta + \alpha_0) + (\delta + \alpha_1) + (\delta + \alpha_2)$$

$$\boldsymbol{\delta}^{(5)} = (\delta + \alpha_0 + \alpha_2) + (\delta + \alpha_1 + \alpha_0) + (\delta + \alpha_2 + \alpha_1)$$

$$\boldsymbol{\delta}^{(6)} = \boldsymbol{\delta}^{(2)} + (2\delta_1) + (2\delta_2) = (\alpha_0 + \alpha_2) + (\alpha_1 + \alpha_0) + (\alpha_2 + \alpha_1) + (2\delta_1) + (2\delta_2)$$

$$\vdots$$

$$\boldsymbol{\delta}^{(9)} = \boldsymbol{\delta}^{(3)} + (3\delta_1) + (3\delta_2) = (\alpha_0) + (\alpha_1) + (\alpha_2) + (\delta_1) + (\delta_2) + (3\delta_1) + (3\delta_2)$$

$$\vdots$$

DEFINITION 3.4. Let $\mathfrak{p} \in \mathcal{G}^+$, and write $\mathfrak{p}$ as a $\mathbf{Z}_{\ge 0}$-linear combination of elements in $S_1 \cup S_2$.

(1) We say an expression of $\mathfrak{p}$ contains a *removable* $\delta$ if it contains some parts that can be replaced by $\boldsymbol{\delta}^{(k)}$ for some $k > 0$.
(2) We say an expression of $\mathfrak{p}$ is *reduced* if it does not contain a removable $\delta$.

Let $\mathcal{K}(\infty)$ denote the set of reduced expressions of elements in $\mathcal{G}^+$. We define the set $\mathcal{K}$ of *Kostant partitions* to be the $\mathbf{Z}_{\geq 0}$-span of the set $S_1$ in $\mathcal{G}^+$. Notice that the set $S_1$ is linearly independent.

DEFINITION 3.5. For $\mathfrak{p} \in \mathcal{K}$, we denote by $\mathscr{N}(\mathfrak{p})$ the number of distinct parts in $\mathfrak{p}$.

EXAMPLE 3.6. If $\mathfrak{p} = 2(\alpha_0 + \alpha_1) + 5(\alpha_2 + \alpha_1) + 2(\delta_1) + (\delta_2) + (\alpha_0) + 4(\alpha_1)$, then $\mathscr{N}(\mathfrak{p}) = 6$.

Define a *reduction map* $\psi\colon \mathcal{K} \longrightarrow \mathcal{K}(\infty)$ as follows: Given $\mathfrak{p} \in \mathcal{K}$, write it as a $\mathbf{Z}_{\geq 0}$-linear combination of elements in $S_1$. Replace $k_1 \sum_{i \in I}(\alpha_i^{(1)})$ in the expression, where $k_1$ is the largest possible, with $k_1 \boldsymbol{\delta}^{(1)}$. The resulting expression is denoted by $\mathfrak{p}^{(1)}$. Next, replace $k_2 \sum_{i \in I}(\alpha_i^{(2)})$ (or $k_2(\boldsymbol{\delta}^{(1)} + (\delta_1))$ if $n = 1$), where $k_2$ is the largest possible, with $k_2 \boldsymbol{\delta}^{(2)}$. The result is denoted by $\mathfrak{p}^{(2)}$. Continue this process with $\boldsymbol{\delta}^{(k)}$ ($k \geq 3$) using the relations in (3.1). The process stops with $\mathfrak{p}^{(s)}$ for some $s$. By construction, $\mathfrak{p}^{(s)} \in \mathcal{K}(\infty)$, and we define $\psi(\mathfrak{p}) = \mathfrak{p}^{(s)}$.

Conversely, we define the *unfolding map* $\phi\colon \mathcal{K}(\infty) \longrightarrow \mathcal{K}$ by unfolding the $\boldsymbol{\delta}^{(k)}$'s consecutively. That is, given $\mathfrak{q} \in \mathcal{K}(\infty)$, find $\boldsymbol{\delta}^{(r)}$ with the largest $r$ and replace it with the corresponding sum from (3.1). The resulting expression is denoted by $\mathfrak{q}^{(r)}$. Next, replace $\boldsymbol{\delta}^{(r-1)}$ with the corresponding sum from (3.1). The result is denoted by $\mathfrak{q}^{(r-1)}$. Continue this process until we replace $\boldsymbol{\delta}^{(1)}$ with $\sum_{i \in I}(\alpha_i^{(1)})$ and obtain $\mathfrak{q}^{(1)}$. By construction, $\mathfrak{q}^{(1)} \in \mathcal{K}$, and we define $\phi(\mathfrak{q}) = \mathfrak{q}^{(1)}$.

It is clear from the definitions that $\psi$ and $\phi$ are inverses to each other. Hence, we have proven the following lemma.

LEMMA 3.7. *The reduction map $\psi\colon \mathcal{K} \longrightarrow \mathcal{K}(\infty)$ is a bijection, whose inverse is the unfolding map $\phi\colon \mathcal{K}(\infty) \longrightarrow \mathcal{K}$.*

For later use, we need to describe the unfolding map $\phi$ more explicitly.

LEMMA 3.8. *For $p \in \mathbf{Z}_{\geq 0}$ and $q \in \mathbf{Z}_{>0}$, we have*

$$\phi\big(\boldsymbol{\delta}^{((n+1)^p q)}\big) = \sum_{j=1}^{n+1}(r\delta + \alpha_{j-1}^{(s)}) + \sum_{i=0}^{p-1}\Big(\sum_{j=1}^{n}\big((n+1)^i q\, \delta_j\big)\Big), \tag{3.2}$$

*where we write $q = (n+1)r + s$, $1 \leq s \leq n$. In particular, $\boldsymbol{\delta}^{((n+1)^p q)}$ has $n+1+np$ parts.*

PROOF. We use induction on $p$. Assume that $p = 0$. Then it follows from (3.1) that

$$\phi\big(\boldsymbol{\delta}^{(q)}\big) = \sum_{j=1}^{n+1}(r\delta + \alpha_{j-1}^{(s)}).$$

Now assume that $p \geq 1$. From (3.1) and the induction hypothesis, we obtain

$$\begin{aligned}\phi\big(\boldsymbol{\delta}^{((n+1)^p q)}\big) &= \phi\big(\boldsymbol{\delta}^{((n+1)^{p-1} q)}\big) + \sum_{j=1}^{n} \big((n+1)^{p-1} q\, \delta_j\big) \\ &= \sum_{j=1}^{n+1} (r\delta + \alpha_{j-1}^{(s)}) + \sum_{i=0}^{p-2} \Big( \sum_{j=1}^{n} \big((n+1)^i q\, \delta_j\big)\Big) + \sum_{j=1}^{n} \big((n+1)^{p-1} q\, \delta_j\big) \\ &= \sum_{j=1}^{n+1} (r\delta + \alpha_{j-1}^{(s)}) + \sum_{i=0}^{p-1} \Big( \sum_{j=1}^{n} \big((n+1)^i q\, \delta_j\big)\Big).\end{aligned}$$

■

In what follows, we will establish a bijection between $\mathcal{Y}(\infty)$ and $\mathcal{K}(\infty)$. For $Y \in \mathcal{Y}(\infty)$, we define $N_k(Y)$ $(k \geq 1)$ to be the number of boxes in the $k$th row of $Y$. We first define a map $\Psi\colon \mathcal{Y}(\infty) \longrightarrow \mathcal{K}(\infty)$ by describing how the blocks in a reduced proper generalized Young wall $Y$ contribute to the parts in a reduced Kostant partition. For $Y \in \mathcal{Y}(\infty)$, $1 \leq j \leq n+1$ and $m \geq 0$, define $\Psi(Y; j, m)$ by

$$\Psi(Y; j, m) = \tag{3.3} \begin{cases} (k\delta_j) & \text{if } 1 \leq j \leq n \text{ and} \\ & N_{(n+1)m+j}(Y) = (n+1)k \text{ for some } k > 0, \\ (k\delta + \alpha_{j-1}^{(\ell)}) & \text{if } 1 \leq j \leq n \text{ and} \\ & N_{(n+1)m+j}(Y) = (n+1)k + \ell \text{ for some } 1 \leq \ell \leq n, k \geq 0, \\ \boldsymbol{\delta}^{(k)} & \text{if } j = n+1 \text{ and} \\ & N_{(n+1)(m+1)}(Y) = (n+1)k \text{ for some } k > 0, \\ (k\delta + \alpha_n^{(\ell)}) & \text{if } j = n+1 \text{ and} \\ & N_{(n+1)(m+1)}(Y) = (n+1)k + \ell \text{ for some } 1 \leq \ell \leq n, k \geq 0. \end{cases}$$

Then

$$\Psi(Y) = \sum_{m \geq 0} \sum_{j=1}^{n+1} \Psi(Y; j, m).$$

LEMMA 3.9. *For any $Y \in \mathcal{Y}(\infty)$, we have $\Psi(Y) \in \mathcal{K}(\infty)$.*

PROOF. Let $\mathfrak{p} = \Psi(Y)$. It is clear that $\mathfrak{p} \in \mathcal{G}^+$, so it remains to show the expression of $\mathfrak{p}$ is reduced. On the contrary, assume that $\mathfrak{p}$ contains a removable $\delta$. By Remark 3.2, the expression of $\mathfrak{p}$ contains a sum of $n+1$ distinct positive "roots" of equal length, and the sum corresponds through (3.3) to a collection of rows of $Y$ with equal length in non-congruent positions. Then $Y$ contains a removable $\delta$, which is a contradiction. Thus $\mathfrak{p}$ does not contain a removable $\delta$, so $\mathfrak{p}$ is reduced. ■

EXAMPLE 3.10. Let $Y = \widetilde{f}_2^3 \widetilde{f}_0^2 \widetilde{f}_1^2 \widetilde{f}_2 \widetilde{f}_1 \widetilde{f}_0 Y_\infty$. That is, let

$Y =$ 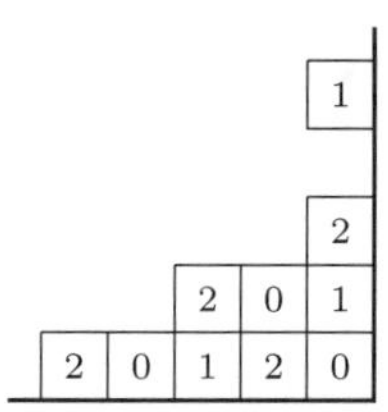

.

Then $\Psi(Y) = (\delta + \alpha_0 + \alpha_2) + (\delta_2) + (\alpha_2) + (\alpha_1)$.

Now define a function $\Phi\colon \mathcal{K}(\infty) \longrightarrow \mathcal{Y}(\infty)$ in the following way. Let $\mathfrak{p}$ be a reduced Kostant partition. To each part of the partition, we assign a row of a generalized Young wall using the following prescription. For $1 \le j \le n$ and $1 \le \ell \le n$,

$$\Phi\colon \begin{cases} (k\delta_j) & \mapsto (n+1)k \text{ boxes in row} \equiv j \pmod{n+1}, \\ (k\delta + \alpha_{j-1}^{(\ell)}) & \mapsto (n+1)k+\ell \text{ boxes in row} \equiv j \pmod{n+1}, \\ \boldsymbol{\delta}^{(k)} & \mapsto (n+1)k \text{ boxes in row} \equiv 0 \pmod{n+1}, \\ (k\delta + \alpha_n^{(\ell)}) & \mapsto (n+1)k+\ell \text{ boxes in row} \equiv 0 \pmod{n+1}. \end{cases} \tag{3.4}$$

To construct the Young wall $\Phi(\mathfrak{p})$ from this data, we arrange the rows so that the number of boxes in each row of the form $(n+1)k+j$, for a fixed $j$, is weakly decreasing as $k$ increases. Hence $\Phi(\mathfrak{p})$ is proper.

LEMMA 3.11. *For any $\mathfrak{p} \in \mathcal{K}(\infty)$, we have $\Phi(\mathfrak{p}) \in \mathcal{Y}(\infty)$.*

PROOF. We set $Y = \Phi(\mathfrak{p})$. Since $\mathfrak{p}$ is reduced, $\mathfrak{p}$ does not contain a removable $\delta$. Using a similar argument as in the proof of Lemma 3.9, we see that a removable $\delta$ of $Y$ corresponds to a removable $\delta$ of $\mathfrak{p}$. Thus $Y$ does not contain a removable $\delta$, so $Y \in \mathcal{Y}(\infty)$. ∎

EXAMPLE 3.12. Let $\mathfrak{p} = (\alpha_0) + (2\delta + \alpha_1 + \alpha_0) + \boldsymbol{\delta}^{(3)} + (\alpha_2)$. Then

$\Phi(\mathfrak{p}) =$ 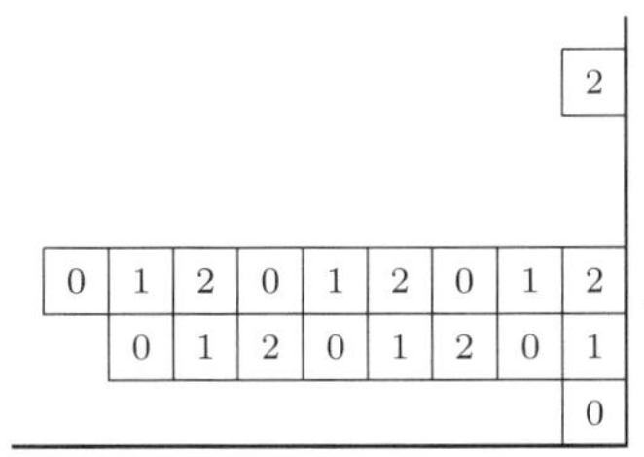

.

PROPOSITION 3.13. *The maps $\Psi$ and $\Phi$ are bijections which are inverses to each other. In particular, we have $\mathcal{Y}(\infty) \cong \mathcal{K}(\infty)$ as sets.*

The existence of a bijection is guaranteed by the theory of Kostant partitions and crystal bases. The importance of the proposition is that we have constructed an explicit, combinatorial description of a bijection.

PROOF. Assume that $Y \in \mathcal{Y}(\infty)$. It is enough to check that a row $j$ of $Y$ is mapped onto the same stack of boxes in a row $\equiv j \pmod{n+1}$ by $\Phi \circ \Psi$, since the rows are arranged uniquely so that the number of boxes in each row of the form $(n+1)k+j$ for a fixed $j$ is weakly decreasing as $k$ increases. It follows from (3.3) and (3.4) that a row $j$ of $Y$ is mapped onto the same stack of boxes in a row $\equiv j \pmod{n+1}$.

Conversely, assume that $\mathfrak{p} \in \mathcal{K}(\infty)$. It is enough to check that each part of $\mathfrak{p}$ is mapped onto itself through $\Psi \circ \Phi$. Using (3.3) and (3.4), we see that it is the case. ∎

REMARK 3.14. While one may define a crystal structure on $\mathcal{K}(\infty)$ directly in order to show that the bijection in Proposition 3.13 is a crystal isomorphism, the bijection given is very explicit and easily understood, so one my simply pull back the crystal structure on $\mathcal{Y}(\infty)$ to $\mathcal{K}(\infty)$ in order to obtain a crystal isomorphism.

For $1 \le j \le n+1$ and $Y \in \mathcal{Y}(\infty)$, define $S_j(Y)$ be the set of distinct $N_{(n+1)m+j}(Y)$'s for $m \ge 0$; *i.e.*, set

$$S_j(Y) = \bigcup_{m \ge 0} \{N_{(n+1)m+j}(Y)\}.$$

When $j = n+1$, for each $m \ge 0$, define $(p_m, q_m) \in \mathbf{Z}_{\ge 0} \times \mathbf{Z}_{\ge 0}$ by

$$N_{(n+1)(m+1)}(Y) = (n+1)^{p_m} q_m,$$

with $q_m$ not divisible by $n+1$. If $N_{(n+1)(m+1)}(Y) = 0$, then we put $(p_m, q_m) = (0,0)$. We set

$$\mathscr{Q}(Y) = \Big( \bigcup_{m \ge 0} \{(n+1)^s q_m : s = 0, 1, \dots, p_m - 1\} \Big) \cup \{0\},$$

and let

$$\mathscr{P}(Y) = \sum_{\substack{t \ge 1 \\ (n+1) \nmid t}} \max \{ p_m : q_m = t,\ m \ge 0 \}.$$

Define

$$\mathscr{N}(Y) = n\mathscr{P}(Y) + \sum_{j=1}^{n+1} \#\big(S_j(Y) \setminus \mathscr{Q}(Y)\big). \tag{3.5}$$

PROPOSITION 3.15. *Assume that $Y \in \mathcal{Y}(\infty)$, and let $\mathfrak{p} = (\phi \circ \Psi)(Y) \in \mathcal{K}$, where $\phi$ is the unfolding map defined in the proof of Lemma 3.7. Then $\mathscr{N}(Y)$ is equal to the number of distinct parts in the Kostant partition $\mathfrak{p}$; i.e., we have $\mathscr{N}(Y) = \mathscr{N}(\mathfrak{p})$.*

Before we prove this proposition, we provide a pair of examples. In the first example, we do not have $\boldsymbol{\delta}^{(k)}$ in $\Psi(Y)$, and in the second example, we have $\boldsymbol{\delta}^{(k)}$ in $\Psi(Y)$. We will see how the formula for $\mathscr{N}(Y)$ works.

EXAMPLE 3.16. Suppose that

$Y =$ 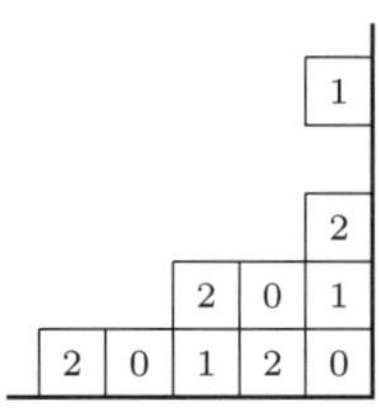

.

Then $\mathfrak{p} = (\phi \circ \Psi)(Y) = (\delta + \alpha_0 + \alpha_1) + (\delta_2) + (\alpha_2) + (\alpha_1)$, and the number of distinct parts is 4. On the other hand,

$$S_1(Y) = \{5, 0\}, \quad S_2(Y) = \{3, 1, 0\}, \quad S_3(Y) = \{1, 0\}.$$

Now setting $N_{3(m+1)}(Y) = 3^{p_m} q_m$ implies $(p_0, q_0) = (0, 1)$ and $(p_m, q_m) = (0, 0)$ for $m \ge 1$. Thus $\mathscr{Q}(Y) = \{0\}$ and $\mathscr{P}(Y) = 0$. So

$$\mathscr{N}(Y) = 1 + 2 + 1 + 2 \cdot 0 = 4.$$

EXAMPLE 3.17. Suppose that

$Y =$ 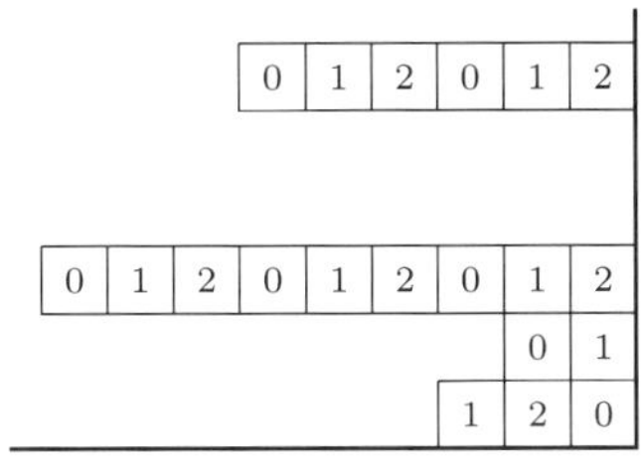 .

Then we have

$$\begin{aligned}\mathfrak{p} = (\phi \circ \Psi)(Y) &= \phi\Big((\delta_1) + (\alpha_0 + \alpha_1) + \boldsymbol{\delta}^{(3)} + \boldsymbol{\delta}^{(2)}\Big) \\ &= (\delta_1) + (\alpha_0 + \alpha_1) + (\alpha_0) + (\alpha_1) + (\alpha_2) + (\delta_1) + (\delta_2) + (\alpha_0 + \alpha_2) \\ &\qquad + (\alpha_1 + \alpha_0) + (\alpha_2 + \alpha_1) \\ &= 2(\alpha_1 + \alpha_0) + (\alpha_0 + \alpha_2) + (\alpha_2 + \alpha_1) + 2(\delta_1) + (\delta_2) + (\alpha_0) + (\alpha_1) + (\alpha_2).\end{aligned}$$

Hence the number of distinct parts is 8. On the other hand, we get

$$S_1(Y) = \{3, 0\}, \quad S_2(Y) = \{2, 0\}, \quad S_3(Y) = \{9, 6, 0\}.$$

From $N_{3(m+1)}(Y) = 3^{p_m} q_m$, we obtain $(p_0, q_0) = (2, 1)$, $(p_1, q_1) = (1, 2)$ and $(p_m, q_m) = (0, 0)$ for $m \geq 2$. Then $\mathscr{Q}(Y) = \{1, 3, 2, 0\}$ and $\mathscr{P}(Y) = 2 + 1 = 3$. So

$$\mathscr{N}(Y) = 0 + 0 + 2 + 2 \cdot 3 = 8.$$

PROOF OF PROPOSITION 3.15.

Step 1: Assume that $p_m = 0$ for all $m \geq 0$. Then $\Psi(Y)$ has no $\boldsymbol{\delta}^{(k)}$, or equivalently, $Y$ is such that $N_{(n+1)(m+1)}(Y) \neq (n+1)k$ for any $m \geq 0$ and $k \geq 1$. Then $(\phi \circ \Psi)(Y) = \Psi(Y)$ as $\Psi(Y)$ does not have a $\boldsymbol{\delta}^{(k)}$. On the other hand, since $p_m = 0$ for all $m \geq 0$, we have $\mathscr{Q}(Y) = \{0\}$ and $\mathscr{P}(Y) = 0$. Hence

$$\mathscr{N}(Y) = \sum_{j=1}^{n+1} \#\big(S_j(Y) \setminus \{0\}\big).$$

For each $1 \leq j \leq n+1$, define $R_j(Y)$ to be the collection of $k$th rows of $Y$ with $k \equiv j \pmod{n+1}$. From (3.3), we see that two nonempty rows $y_1, y_2 \in R_j(Y)$ correspond to distinct parts in $\Psi(Y)$ if and only if the lengths of $y_1$ and $y_2$ are different. Since $\#\left(S_j(Y) \setminus \{0\}\right)$ is the number of distinct nonzero lengths of rows in $R_j(Y)$, it is equal to the number of distinct parts in $\Psi(Y)$ corresponding to $R_j(Y)$. Furthermore, if $j \neq j'$, then $y \in R_j(Y)$ and $y' \in R_{j'}(Y)$ correspond to distinct parts in $\Psi(Y)$. Thus $\mathscr{N}(Y)$ is the total number of distinct parts in $\Psi(Y) = (\phi \circ \Psi)(Y)$, as required.

Step 2: Now assume that $p_m \geq 1$ for some $m$ and $p_{m'} = 0$ for all $m' \neq m$. From the definition $N_{(n+1)(m+1)}(Y) = (n+1)^{p_m} q_m$, we see that the row $(n+1)(m+1)$ has $(n+1)^{p_m} q_m$ boxes, and the corresponding part in $\Psi(Y)$ is $\boldsymbol{\delta}^{((n+1)^{p_m - 1} q_m)}$. We obtain from Lemma 3.8

$$\phi\big(\boldsymbol{\delta}^{((n+1)^{p_m-1} q_m)}\big) = \sum_{j=1}^{n+1} (r_m \delta + \alpha_{j-1}^{(s_m)}) + \sum_{i=0}^{p_m - 2} \Big( \sum_{j=1}^{n} \big((n+1)^i q_m \, \delta_j\big)\Big), \tag{3.6}$$

where we write $q_m = (n+1) r_m + s_m$, $1 \leq s_m \leq n$. Thus $\phi\big(\boldsymbol{\delta}^{((n+1)^{p_m-1} q_m)}\big)$ has $n p_m + 1$ distinct parts, some of which may be the same as other parts in $\Psi(Y)$. It follows from (3.3) that the part $(r_m\delta + \alpha_{j-1}^{(s_m)})$ corresponds to $q_m$ boxes in a row $\equiv j \pmod{n+1}$ for $1 \le j \le n+1$. Similarly, the part $((n+1)^i q_m\,\delta_j)$ corresponds to $(n+1)^{i+1}q_m$ boxes in a row $\equiv j \pmod{n+1}$ for $1 \le j \le n$ and $0 \le i \le p_m - 2$. Then the number of distinct parts in $(\phi\circ\Psi)(Y)$ is

(3.7)

$$np_m+1+\sum_{j=1}^{n}\#\big(S_j(Y)\setminus\{0,(n+1)^i q_m\}_{0\le i\le p_m-1}\big)+\#\big(S_{n+1}(Y)\setminus\{0,q_m,(n+1)^{p_m}q_m\}\big)$$
$$= np_m + \sum_{j=1}^{n}\#\big(S_j(Y)\setminus\{0,(n+1)^i q_m\}_{0\le i\le p_m-1}\big) + \#\big(S_{n+1}(Y)\setminus\{0,q_m\}\big).$$

Since $S_{n+1}(Y)$ does not contain $(n+1)^i q_m$, $1\le i\le p_m-1$, by the assumption, the expression (3.7) is equal to

$$\begin{aligned}
&np_m + \sum_{j=1}^{n}\#\big(S_j(Y)\setminus\{0,(n+1)^i q_m\}_{0\le i\le p_m-1}\big)\\
&\qquad\qquad + \#\big(S_{n+1}(Y)\setminus\{0,(n+1)^i q_m\}_{0\le i\le p_m-1}\big)\\
&\qquad = np_m + \sum_{j=1}^{n+1}\#\big(S_j(Y)\setminus\{0,(n+1)^i q_m\}_{0\le i\le p_m-1}\big)\\
&\qquad = np_m + \sum_{j=1}^{n+1}\#\big(S_j(Y)\setminus\mathscr{Q}(Y)\big)\\
&\qquad = \mathscr{N}(Y).
\end{aligned}$$

Thus the number of distinct parts in $(\phi\circ\Psi)(Y)$ is $\mathscr{N}(Y)$.

Step 3: Next we assume $p_m = \max\{p_{m'} : m' \ge 0\}$ and $q_m = q_{m'}$ for any $p_{m'} \ge 1$. We have $\boldsymbol{\delta}^{((n+1)^{p_{m'}-1}q_{m'})}$ in $\Psi(Y)$ for each $p_{m'}\ge 1$, and each $\phi(\boldsymbol{\delta}^{((n+1)^{p_{m'}-1}q_{m'})})$ yields $np_{m'}+1$ parts as in (3.6). However, we can see from (3.6) that $\phi(\boldsymbol{\delta}^{(n+1)^{p_m-1}q_m})$ with the maximal $p_m$ generates all the distinct parts including those from other $p_{m'}$, since $q_m = q_{m'}$ for all $p_{m'}\ge 1$ by the assumption. Then the number of distinct parts in $(\phi\circ\Psi)(Y)$ is given by

$$\begin{aligned}
&np_m + 1 + \sum_{j=1}^{n}\#\big(S_j(Y)\setminus\{0,(n+1)^i q_m\}_{0\le i\le p_m-1}\big)\\
&\qquad\qquad + \#\big(S_{n+1}(Y)\setminus\{0,q_m,(n+1)^{p_{m'}}q_m\}_{1\le p_{m'}\le p_m}\big)\\
&\qquad = np_m + \sum_{j=1}^{n}\#\big(S_j(Y)\setminus\{0,(n+1)^i q_m\}_{0\le i\le p_m-1}\big)\\
&\qquad\qquad + \#\big(S_{n+1}(Y)\setminus\{0,(n+1)^i q_m\}_{0\le i\le p_m-1}\big)\\
&\qquad = np_m + \sum_{j=1}^{n+1}\#\big(S_j(Y)\setminus\mathscr{Q}(Y)\big) = \mathscr{N}(Y).
\end{aligned}$$

Step 4: Finally we consider the general case. We group $p_m$'s using the rule that $p_m$ and $p_{m'}$ are in the same group if and only if $q_m = q_{m'}$. For each of such groups, we use the result in Step 3, and see that the number of distict parts in $(\phi\circ\Psi)(Y)$

is equal to

$$n\mathscr{P}(Y) + \sum_{j=1}^{n+1} \#\big(S_j(Y) \setminus \mathscr{Q}(Y)\big),$$

recalling the definitions

$$\mathscr{P}(Y) = \sum_{\substack{t\geq 1\\ (n+1)\nmid t}} \max\{p_m : q_m = t,\ m \geq 0\},$$

$$\mathscr{Q}(Y) = \Big(\bigcup_{m\geq 0}\{(n+1)^s q_m : s = 0,1,\dots,p_m - 1\}\Big) \cup \{0\}.$$

Hence the number of distinct parts in $(\phi \circ \Psi)(Y)$ is $\mathscr{N}(Y)$. ■

The rule for calculating the number $\mathscr{N}(Y)$, for $Y \in \mathcal{Y}(\infty)$, may be reinterpreted using the following algorithm. For this algorithm, we say that two rows in $Y$ are *distinct* if their rightmost boxes are different or if their rightmost boxes are equal but they have an unequal number of boxes.

ALGORITHM 3.18. Define a map $\psi_{\mathcal{Y}}$ on $\mathcal{Y}(\infty)$ as follows.

(1) If $Y$ has no row with rightmost box $n$ and length $\equiv 0 \bmod n+1$, then $\psi_{\mathcal{Y}}(Y) := Y$.
(2) If $Y$ has at least one row with rightmost box $n$ and length $(n+1)\ell$, then replace any row with maximal such $\ell$ with $n+1$ distinct rows of length $\ell$. Rearrange all rows (if necessary) so that it is proper. This gives $\psi_{\mathcal{Y}}^{(\ell)}(Y)$.
(3) Apply Step 2 with $\ell$ replaced by $\ell - 1$ and $Y$ replaced by $\psi_{\mathcal{Y}}^{(\ell)}(Y)$. This gives $\psi_{\mathcal{Y}}^{(\ell-1)}(Y)$.
(4) Iterate this process until $\ell = 1$. Then $\psi_{\mathcal{Y}}(Y) = \psi_{\mathcal{Y}}^{(1)}(Y)$.

Note that $\psi_{\mathcal{Y}}(Y)$ is proper, but need not be reduced, so $\psi_{\mathcal{Y}}(Y) \notin \mathcal{Y}(\infty)$ in general. Then $\mathscr{N}(Y)$ is the number of distinct rows in $\psi_{\mathcal{Y}}(Y)$.

EXAMPLE 3.19. Let $n = 2$ and let $Y$ be as in Example 3.17. Then

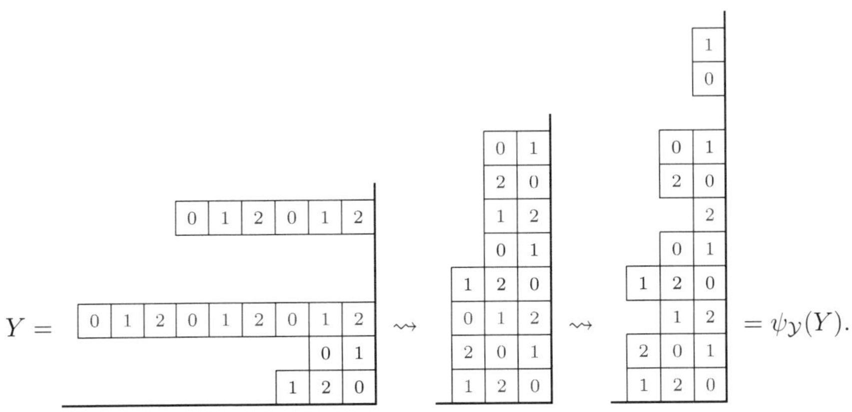

Counting the number of distinct rows gives $8 = \mathscr{N}(Y)$.

Let $W$ be the Weyl group of $\mathfrak{g}$ and $s_i$ $(i \in I)$ be the simple reflections. We fix $\boldsymbol{h} = (\dots, i_{-1}, i_0, i_1, \dots)$ as in Section 3.1 in [**BN04**]. Then for any integers

$m < k$, the product $s_{i_m} s_{i_{m+1}} \cdots s_{i_k} \in W$ is a reduced expression, so is the product $s_{i_k} s_{i_{k-1}} \cdots s_{i_m} \in W$. We set

$$\beta_k = \begin{cases} s_{i_0} s_{i_{-1}} \cdots s_{i_{k+1}}(\alpha_{i_k}) & \text{if } k \leq 0, \\ s_{i_1} s_{i_2} \cdots s_{i_{k-1}}(\alpha_{i_k}) & \text{if } k > 0. \end{cases} \tag{3.8}$$

Let $T_i = T''_{i,1}$ be the automorphism of $U_v(\mathfrak{g})$ as in Section 37.1.3 of [**Lus93**], and let

$$\boldsymbol{c}_+ = (c_0, c_{-1}, c_{-2}, \dots) \in \mathbf{Z}_{\geq 0}^{\mathbf{Z}_{\leq 0}} \quad \text{and} \quad \boldsymbol{c}_- = (c_1, c_2, \dots) \in \mathbf{Z}_{\geq 0}^{\mathbf{Z}_{>0}}$$

be functions (or sequences) that are almost everywhere zero. We denote by $\mathscr{C}_>$ (resp. by $\mathscr{C}_<$) the set of such functions $\boldsymbol{c}_+$ (resp. $\boldsymbol{c}_-$). For an element $\boldsymbol{c}_+ = (c_0, c_{-1}, \dots) \in \mathscr{C}_>$ (resp. $\boldsymbol{c}_- = (c_1, c_2, \dots) \in \mathscr{C}_>$), we define

$$E_{\boldsymbol{c}_+} = E_{i_0}^{(c_0)} T_{i_0}^{-1}(E_{i_{-1}}^{(c_{-1})}) T_{i_0}^{-1} T_{i_{-1}}^{-1}(E_{i_{-2}}^{(c_{-2})}) \cdots$$

and

$$E_{\boldsymbol{c}_-} = \cdots T_{i_1} T_{i_2}(E_{i_3}^{(c_3)}) T_{i_1}(E_{i_2}^{(c_2)}) E_{i_1}^{(c_1)}.$$

We also define $\mathscr{N}(\boldsymbol{c}_+)$ (resp. $\mathscr{N}(\boldsymbol{c}_-)$) to be the number of nonzero $c_i$'s in $\boldsymbol{c}_+$ (resp. $\boldsymbol{c}_-$).

Let $\boldsymbol{c}_0 = (\rho^{(1)}, \rho^{(2)}, \dots, \rho^{(n)})$ be a multi-partition with $n$ components; *i.e.*, each component $\rho^{(i)}$ is a partition. We denote by $\mathcal{P}(n)$ the set of all multi-partitions with $n$ components. Let $S_{\boldsymbol{c}_0}$ be defined as in [**BN04**, p. 352] for $\boldsymbol{c}_0 \in \mathcal{P}(n)$. For a partition $\boldsymbol{p} = (1^{m_1} 2^{m_2} \cdots r^{m_r} \cdots)$, we define

$$\mathscr{N}(\boldsymbol{p}) = \#\{r : m_r \neq 0\} \quad \text{and} \quad |\boldsymbol{p}| = m_1 + 2m_2 + 3m_3 + \cdots. \tag{3.9}$$

Then for a multi-partition $\boldsymbol{c}_0 = (\rho^{(1)}, \rho^{(2)}, \dots, \rho^{(n)}) \in \mathcal{P}(n)$, we set

$$\mathscr{N}(\boldsymbol{c}_0) = \mathscr{N}(\rho^{(1)}) + \mathscr{N}(\rho^{(2)}) + \cdots + \mathscr{N}(\rho^{(n)}).$$

Let $\mathscr{C} = \mathscr{C}_> \times \mathcal{P}(n) \times \mathscr{C}_<$. We denote by $\mathscr{B}$ the Kashiwara-Lusztig canonical basis for $U_v^+(\mathfrak{g})$, the positive part of the quantum affine algebra.

THEOREM 3.20 ([**BCP99**, **BN04**]). *There is a bijection $\eta\colon \mathscr{B} \longrightarrow \mathscr{C}$ such that for each $\boldsymbol{c} = (\boldsymbol{c}_+, \boldsymbol{c}_0, \boldsymbol{c}_-) \in \mathscr{C}$, there exists a unique $b = \eta^{-1}(\boldsymbol{c}) \in \mathscr{B}$ satisfying*

$$b \equiv E_{\boldsymbol{c}_+} S_{\boldsymbol{c}_0} E_{\boldsymbol{c}_-} \mod v^{-1}\mathbf{Z}[v^{-1}]. \tag{3.10}$$

Now the number $\mathscr{N}(\boldsymbol{c})$ is defined by $\mathscr{N}(\boldsymbol{c}) = \mathscr{N}(\boldsymbol{c}_+) + \mathscr{N}(\boldsymbol{c}_0) + \mathscr{N}(\boldsymbol{c}_-)$ for each $\boldsymbol{c} \in \mathscr{C}$. Using the canonical basis $\mathscr{B}$, H. Kim and K.-H. Lee expanded the product side of the Gindikin-Karpelevich formula as a sum, and obtained the following theorem.

THEOREM 3.21 ([**KL12**]). *We have*

$$\prod_{\alpha \in \Delta^+} \left( \frac{1 - q^{-1} \boldsymbol{z}^\alpha}{1 - \boldsymbol{z}^\alpha} \right)^{\mathrm{mult}(\alpha)} = \sum_{b \in \mathscr{B}} (1 - q^{-1})^{\mathscr{N}(\eta(b))} \boldsymbol{z}^{\mathrm{wt}(b)}. \tag{3.11}$$

In the rest of this section, we will prove a combinatorial description of the formula (3.11) using the set $\mathcal{Y}(\infty)$ of reduced proper generalized Young walls.

We define a map $\theta\colon \mathcal{P}(n) \longrightarrow \mathcal{K}$ as follows. For $\boldsymbol{c}_0 = (\rho^{(1)}, \rho^{(2)}, \dots, \rho^{(n)}) \in \mathcal{P}(n)$, we define

$$\theta(\boldsymbol{c}_0) = \sum_{i=1}^{n} m_{1,i}(\delta_i) + m_{2,i}(2\delta_i) + \cdots + m_{r,i}(r\delta_i) + \cdots,$$

where $\rho^{(i)} = (1^{m_{1,i}}2^{m_{2,i}}\cdots r^{m_{r,i}}\cdots)$ for $i = 1, 2, \ldots, n$. Then we define a map $\Theta\colon \mathscr{C} \longrightarrow \mathcal{K}$ by

$$\Theta(\boldsymbol{c}) = \theta(\boldsymbol{c}_0) + \sum_{i\in\mathbf{Z}} c_i(\beta_i),$$

where $\boldsymbol{c} = (\boldsymbol{c}_+, \boldsymbol{c}_0, \boldsymbol{c}_-)$, $\boldsymbol{c}_+ = (c_0, c_{-1}, c_{-2}, \ldots)$, $\boldsymbol{c}_- = (c_1, c_2, \ldots)$ and $\beta_i$ is given by (3.8) with $(\beta_i) \in \mathcal{K}$.

COROLLARY 3.22. *The map $\Theta\colon \mathscr{C} \longrightarrow \mathcal{K}$ is a bijection, and for $\boldsymbol{c} \in \mathscr{C}$, the number of distinct parts in $\mathfrak{p} = \Theta(\boldsymbol{c})$ is the same as $\mathscr{N}(\boldsymbol{c})$; i.e., $\mathscr{N}(\Theta(\boldsymbol{c})) = \mathscr{N}(\boldsymbol{c})$.*

PROOF. By Theorem 3.20, the set $\mathscr{C}$ parametrizes a PBW type basis of $U_v^+(\mathfrak{g})$. Thus the set $\mathscr{C}$ also parametrizes a PBW basis of the universal enveloping algebra $U^+(\mathfrak{g})$. Now the first assertion follows from the fact that the Kostant partitions parametrize the elements in a PBW basis of $U^+(\mathfrak{g})$ and that the function $\Theta$ is defined according to these correspondences. The second assertion follows from the definitions of $\mathscr{N}$ for $\mathscr{C}$ and $\mathcal{K}$, respectively. ∎

THEOREM 3.23. *Let $\mathfrak{g}$ be an affine Kac-Moody algebra of type $A_n^{(1)}$. Then*

$$\prod_{\alpha\in\Delta^+}\left(\frac{1-q^{-1}\boldsymbol{z}^\alpha}{1-\boldsymbol{z}^\alpha}\right)^{\mathrm{mult}(\alpha)} = \sum_{Y\in\mathcal{Y}(\infty)} (1-q^{-1})^{\mathscr{N}(Y)}\boldsymbol{z}^{-\mathrm{wt}(Y)}, \tag{3.12}$$

*where $\mathscr{N}(Y)$ is defined in* (3.5).

PROOF. By Lemma 3.7, Proposition 3.13, Theorem 3.20 and Corollary 3.22, we have bijections

$$\mathscr{B} \xrightarrow{\eta} \mathscr{C} \xrightarrow{\Theta} \mathcal{K} \xrightarrow{\psi} \mathcal{K}(\infty) \xrightarrow{\Phi} \mathcal{Y}(\infty).$$

For $b \in \mathscr{B}$, we write $Y = (\Phi\circ\psi\circ\Theta\circ\eta)(b) \in \mathcal{Y}(\infty)$. Then, by Proposition 3.15 and Corollary 3.22, we have $\mathscr{N}(\eta(b)) = \mathscr{N}(Y)$. We also see from the constructions that $\mathrm{wt}(b) = -\mathrm{wt}(Y)$. Now the equality (3.12) follows from Theorem 3.21. ∎

## 4. Connection to Braverman-Finkelberg-Kazhdan's formula

We briefly recall the framework of the paper [**BFK12**]. Let $G$ (resp. $\widehat{G}$) be the minimal (resp. formal) Kac-Moody group functor attached to a symmetrizable Kac-Moody root datum and let $\mathfrak{g}$ be the corresponding Lie algebra. There is a natural imbedding $G \hookrightarrow \widehat{G}$. The group $G$ has the closed subgroup functors $U \subset B$, $U_- \subset B_-$ such that the quotients $B/U$ and $B_-/U_-$ are naturally isomorphic to the Cartan subgroup $H$ of $G$. We denote by $\widehat{B}$ and $\widehat{U}$ the closures of $B$ and $U$ in $\widehat{G}$, respectively. We will denote the coroot lattice of $G$ by $\Lambda$ and the set of positive coroots by $R^+ \subset \Lambda$. The subsemigroup of $\Lambda$ generated by $R^+$ will be denoted by $\Lambda^+$. For an element $\gamma = \sum a_i\alpha_i^\vee \in \Lambda^+$ with simple coroots $\alpha_i^\vee$, we write $|\gamma| = \sum a_i$. We assume that $G$ is "simply connected"; *i.e.*, the lattice $\Lambda$ is equal to the cocharacter lattice of $H$.

We set $\mathcal{F} = \mathbf{F}_q((t))$ and $\mathcal{O} = \mathbf{F}_q[[t]]$, where $\mathbf{F}_q$ is the finite field with $q$ elements. We let $\mathrm{Gr} = \widehat{G}(\mathcal{F})/\widehat{G}(\mathcal{O})$. Each $\lambda \in \Lambda$ defines a homomorphism $\mathcal{F}^* \longrightarrow H(\mathcal{F})$. We will denote the image of $t$ under this homomorphism by $t^\lambda$, and its image in Gr will also be denoted by $t^\lambda$. We set

$$S^\lambda = \widehat{U}(\mathcal{F})\cdot t^\lambda \subset \mathrm{Gr} \quad\text{and}\quad T^\lambda = U_-(\mathcal{F})\cdot t^\lambda \subset \mathrm{Gr}.$$

In a recent paper [**BFK12**], Braverman, Finkelberg and Kazhdan defined the generating function

$$I_{\mathfrak{g}}(q) = \sum_{\gamma \in \Lambda^+} \#(T^{-\gamma} \cap S^0) q^{-|\gamma|} \boldsymbol{z}^{\gamma}, \tag{4.1}$$

and computed this sum as a product using a geometric method. Now we assume that the set of positive coroots $R^+$ forms a root system of type $A_n^{(1)}$, and we identify $R^+$ with the set of positive roots $\Delta^+$ in the previous sections of this paper. In this case, the resulting product in [**BFK12**] is

$$I_{\mathfrak{g}}(q) = \prod_{i=1}^{n} \prod_{j=1}^{\infty} \frac{1 - q^{-i} \boldsymbol{z}^{j\delta}}{1 - q^{-(i+1)} \boldsymbol{z}^{j\delta}} \prod_{\alpha \in \Delta^+} \left( \frac{1 - q^{-1} \boldsymbol{z}^{\alpha}}{1 - \boldsymbol{z}^{\alpha}} \right)^{\mathrm{mult}(\alpha)}.$$

We separate the factor

$$\prod_{i=1}^{n} \prod_{j=1}^{\infty} \frac{1 - q^{-i} \boldsymbol{z}^{j\delta}}{1 - q^{-(i+1)} \boldsymbol{z}^{j\delta}},$$

and call it the *correction factor*. Our goal of this section is to write this correction factor and the function $I_{\mathfrak{g}}(q)$ as sums over reduced proper generalized Young walls.

Let $\mathcal{Y}_0$ denote the subset of $\mathcal{Y}(\infty)$ consisting of the reduced proper generalized Young walls with empty rows in positions $\equiv 0 \bmod n+1$. We define a map $\xi\colon \mathcal{P}(n) \longrightarrow \mathcal{Y}_0$ by the following assignment. If $\boldsymbol{p} = (\rho^{(1)}, \dots, \rho^{(n)})$ is a multi-partition, then the parts of the partition $\rho^{(j)}$ give the lengths of the rows $\equiv j \bmod n+1$ in a reduced proper generalized Young wall $\xi(\boldsymbol{p}) = Y \in \mathcal{Y}_0$. The following is clear from the definition.

LEMMA 4.1. *The map $\xi\colon \mathcal{P}(n) \longrightarrow \mathcal{Y}_0$ defined above is a bijection.*

EXAMPLE 4.2. Let $n = 2$. If

$\boldsymbol{p} =$ 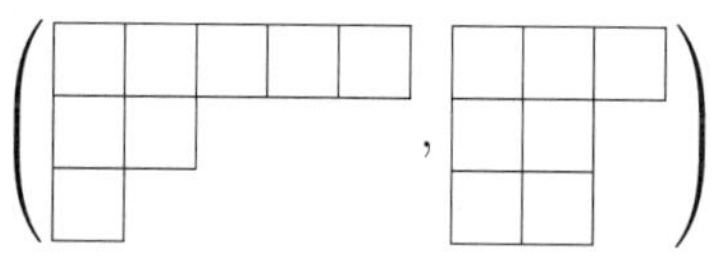 ,

then the corresponding element in $\mathcal{Y}_0$ is

$Y = \xi(\boldsymbol{p}) =$ 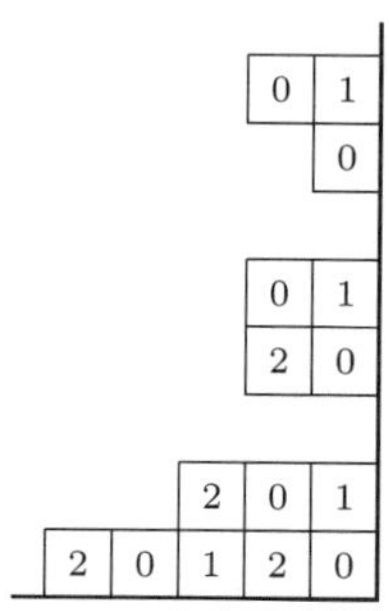 .



For $Y \in \mathcal{Y}_0$, define

$$\mathscr{M}(Y) = \sum_{i=1}^{n} (i+1) M_i(Y),$$

where $M_i(Y)$ is the number of nonempty rows $\equiv i \bmod n+1$ in $Y$. Moreover, we define $|Y|$ to be the total number of blocks in $Y$.

EXAMPLE 4.3. Let $Y$ be as in Example 4.2. Then

$$\mathscr{M}(Y) = 2\cdot 3 + 3\cdot 3 = 15 \quad \text{and} \quad |Y| = 15.$$

Let us consider $\mathscr{N}(Y)$ for $Y \in \mathcal{Y}_0$, where $\mathscr{N}(Y)$ is defined in (3.5). Since $Y$ has empty rows in positions $\equiv 0 \bmod n+1$, we have $(p_m, q_m) = (0,0)$ for all $m \geq 0$, and obtain $\mathscr{Q}(Y) = \{0\}$ and $\mathscr{P}(Y) = 0$. Hence we have

$$\mathscr{N}(Y) = \sum_{j=1}^{n} \#(S_j(Y) \setminus \{0\}) \qquad \text{for } Y \in \mathcal{Y}_0. \tag{4.2}$$

PROPOSITION 4.4. *Let $\mathfrak{g}$ be an affine Kac-Moody algebra of type $A_n^{(1)}$. Then*

$$\prod_{i=1}^{n}\prod_{j=1}^{\infty} \frac{1-q^{-i}\boldsymbol{z}^{j\delta}}{1-q^{-(i+1)}\boldsymbol{z}^{j\delta}} = \sum_{Y\in\mathcal{Y}_0} (1-q)^{\mathscr{N}(Y)} q^{-\mathscr{M}(Y)} \boldsymbol{z}^{|Y|\delta}.$$

PROOF. We have

$$\begin{aligned}\prod_{j=1}^{\infty} \frac{1-q^{-i}\boldsymbol{z}^{j\delta}}{1-q^{-(i+1)}\boldsymbol{z}^{j\delta}} &= \prod_{j=1}^{\infty}\left(1+\sum_{k=1}^{\infty}(1-q)q^{-k(i+1)}\boldsymbol{z}^{kj\delta}\right) \\ &= \sum_{\rho^{(i)}\in\mathcal{P}(1)} (1-q)^{\mathscr{N}(\rho^{(i)})} q^{-(i+1)M(\rho^{(i)})} \boldsymbol{z}^{|\rho^{(i)}|\delta},\end{aligned}$$

where $\mathscr{N}(\rho^{(i)}) = \#\{r : m_r \neq 0\}$ and $|\rho^{(i)}| = m_1 + 2m_2 + \cdots$ are defined in (3.9) and we set $M(\rho^{(i)}) = m_1 + m_2 + \cdots$ for $\rho^{(i)} = (1^{m_1}2^{m_2}\cdots) \in \mathcal{P}(1)$. For a multi-partition $\rho = (\rho^{(1)}, \ldots, \rho^{(n)}) \in \mathcal{P}(n)$, define

$$\mathscr{N}(\rho) = \sum_{i=1}^{n} \mathscr{N}(\rho^{(i)}), \quad |\rho| = \sum_{i=1}^{n} |\rho^{(i)}| \quad \text{and} \quad \mathscr{M}(\rho) = \sum_{i=1}^{n}(i+1)M(\rho^{(i)}).$$

Then we have

$$\begin{aligned}\prod_{i=1}^{n}\prod_{j=1}^{\infty} \frac{1-q^{-i}\boldsymbol{z}^{j\delta}}{1-q^{-(i+1)}\boldsymbol{z}^{j\delta}} &= \prod_{i=1}^{n}\sum_{\rho^{(i)}\in\mathcal{P}(1)} (1-q)^{\mathscr{N}(\rho^{(i)})} q^{-(i+1)M(\rho^{(i)})} \boldsymbol{z}^{|\rho^{(i)}|\delta} \\ &= \sum_{\rho\in\mathcal{P}(n)} (1-q)^{\mathscr{N}(\rho)} q^{-\mathscr{M}(\rho)} \boldsymbol{z}^{|\rho|\delta}.\end{aligned} \tag{4.3}$$

Using the map $\xi$ in Lemma 4.1, one can see that $\mathscr{N}(\rho) = \mathscr{N}(\xi(\rho))$, $\mathscr{M}(\rho) = \mathscr{M}(\xi(\rho))$ and $|\rho| = |\xi(\rho)|$ for $\rho \in \mathcal{P}(n)$, and the proposition follows from (4.3). ∎

The following formula provides a combinatorial description of the affine Gindikin-Karpelevich formula proved by Braverman, Finkelberg and Kazhdan.

COROLLARY 4.5. *When $\mathfrak{g}$ is an affine Kac-Moody algebra of type $A_n^{(1)}$, we have*

$$\begin{aligned}I_{\mathfrak{g}}(q) &= \prod_{i=1}^{n}\prod_{j=1}^{\infty} \frac{1-q^{-i}\boldsymbol{z}^{j\delta}}{1-q^{-(i+1)}\boldsymbol{z}^{j\delta}} \prod_{\alpha\in\Delta^+}\left(\frac{1-q^{-1}\boldsymbol{z}^{\alpha}}{1-\boldsymbol{z}^{\alpha}}\right)^{\mathrm{mult}(\alpha)} \\ &= \sum_{(Y_1,Y_2)\in\mathcal{Y}(\infty)\times\mathcal{Y}_0} (1-q^{-1})^{\mathscr{N}(Y_1)}(1-q)^{\mathscr{N}(Y_2)} q^{-\mathscr{M}(Y_2)} \boldsymbol{z}^{-\mathrm{wt}(Y_1)+|Y_2|\delta}.\end{aligned} \tag{4.4}$$

Furthermore, comparing (4.4) with (4.1), we obtain a combinatorial formula for the number of points in the intersection $T^{-\gamma} \cap S^0$:

Corollary 4.6. *We have*

$$\#(T^{-\gamma} \cap S^0) = \sum_{\substack{(Y_1,Y_2)\in\mathcal{Y}(\infty)\times\mathcal{Y}_0 \\ -\mathrm{wt}(Y_1)+|Y_2|\delta=\gamma}} (1-q^{-1})^{\mathscr{N}(Y_1)}(1-q)^{\mathscr{N}(Y_2)}q^{|\gamma|-\mathscr{M}(Y_2)},$$

*where $\gamma \in \Lambda^+$ is identified with the corresponding element of the root lattice of $\mathfrak{g}$.*

Example 4.7. Assume $n = 1$ and $\gamma = \delta$. Then we have

$$(Y_1, Y_2) = \left(\varnothing\,,\ \boxed{0}\right),\quad \left(\boxed{1\,|\,0}\,,\varnothing\right),\quad \text{or}\quad \left(\boxed{0\,|\,1}\,,\varnothing\right).$$

From the first pair, we get $(1-q^{-1})^0(1-q)^1q^{2-2} = 1-q$. The second yields $(1-q^{-1})^1(1-q)^0q^{2-0} = q^2-q$, and the third $(1-q^{-1})^2(1-q)^0q^{2-0} = (q-1)^2$. Thus we have

$$\#(T^{-\gamma} \cap S^0) = 1-q+q^2-q+(q-1)^2 = 2(q-1)^2.$$

## Appendix A. Implementation in Sage

Together with Lucas Roesler and Travis Scrimshaw, the fourth named author has implemented generalized Young walls and the statistics developed here in the open-source mathematical software Sage [**SCc08**, **S$^+$14**]. We conclude with some examples using our package.

First we may verify examples given above. To verify Example 3.16, we have the following, where Y.number_of_parts() refers to $\mathscr{N}(Y)$.

```
sage: Yinf = crystals.infinity.GeneralizedYoungWalls(2)
sage: Y = Yinf([[0,2,1,0,2],[1,0,2],[2],[],[1]])
sage: Y.pp()
        1|
         |
        2|
    2|0|1|
2|0|1|2|0|
sage: Y.number_of_parts()
4
```

Similarly, to see Examples 3.17 and 3.19 using Sage, use the following commands.

```
sage: Yinf = crystals.infinity.GeneralizedYoungWalls(2)
sage: row1 = [0,2,1]
sage: row2 = [1,0]
sage: row3 = [2,1,0,2,1,0,2,1,0]
sage: row6 = [2,1,0,2,1,0]
sage: Y = Yinf([row1,row2,row3,[],[],row6])
```

```
sage: Y.pp()
          0|1|2|0|1|2|
                     |
                     |
 0|1|2|0|1|2|0|1|2|
                0|1|
              1|2|0|
sage: Y.number_of_parts()
8
```

Note that the remaining crystal structure pertaining to generalized Young walls has also been implemented. We continue using the $Y$ from the previous example.

```
sage: Y.weight(root_lattice=True)
-7*alpha[0] - 7*alpha[1] - 6*alpha[2]
sage: Y.f(1).pp()
          0|1|2|0|1|2|
                    1|
                     |
 0|1|2|0|1|2|0|1|2|
                0|1|
              1|2|0|
sage: Y.e(0).pp()
            1|2|0|1|2|
                     |
                     |
 0|1|2|0|1|2|0|1|2|
                0|1|
              1|2|0|
sage: Y.content()
20
```

One may also generate the top part of the crystal graph.

```
sage: Yinf = crystals.infinity.GeneralizedYoungWalls(2)
sage: S = Yinf.subcrystal(max_depth=4)
sage: G = Yinf.digraph(subset=S)
sage: view(G, tightpage=True)
```

We conclude by mentioning that highest weight crystals realized by generalized Young walls have also been implemented in Sage, following Theorem 4.1 of [**KS10**].

```
sage: Delta = RootSystem(['A',3,1])
sage: P = Delta.weight_lattice(extended=True)
sage: La = P.fundamental_weights()
sage: YLa = crystals.GeneralizedYoungWalls(3,La[0])
sage: S = YLa.subcrystal(max_depth=6)
sage: G = YLa.digraph(subset=S)
sage: view(G, tightpage=True)
```

## References


[BBF11] B. Brubaker, D. Bump, and S. Friedberg, *Weyl group multiple Dirichlet series: type A combinatorial theory*, Annals of Mathematics Studies, vol. 175, Princeton University Press, Princeton, NJ, 2011. MR2791904

[BCP99] J. Beck, V. Chari, and A. Pressley, *An algebraic characterization of the affine canonical basis*, Duke Math. J. **99** (1999), no. 3, 455–487, DOI 10.1215/S0012-7094-99-09915-5. MR1712630 (2000g:17013)

[BFK12] A. Braverman, M. Finkelberg, and D. Kazhdan, *Affine Gindikin-Karpelevich formula via Uhlenbeck spaces*, Contributions in analytic and algebraic number theory, Springer Proc. Math., vol. 9, Springer, New York, 2012, pp. 17–29, DOI 10.1007/978-1-4614-1219-9_2. MR3060455

[BN04] J. Beck and H. Nakajima, *Crystal bases and two-sided cells of quantum affine algebras*, Duke Math. J. **123** (2004), no. 2, 335–402. MR2066942 (2005e:17020)

[BN10] D. Bump and M. Nakasuji, *Integration on p-adic groups and crystal bases*, Proc. Amer. Math. Soc. **138** (2010), no. 5, 1595–1605, DOI 10.1090/S0002-9939-09-10206-X. MR2587444 (2011f:22015)

[Gar04] H. Garland, *Certain Eisenstein series on loop groups: convergence and the constant term*, Algebraic groups and arithmetic, Tata Inst. Fund. Res., Mumbai, 2004, pp. 275–319. MR2094114 (2005i:11118)

[GK62] S. G. Gindikin and F. I. Karpelevič, *Plancherel measure for symmetric Riemannian spaces of non-positive curvature* (Russian), Dokl. Akad. Nauk SSSR **145** (1962), 252–255. MR0150239 (27 #240)

[HK02] J. Hong and S.-J. Kang, *Introduction to quantum groups and crystal bases*, Graduate Studies in Mathematics, vol. 42, American Mathematical Society, Providence, RI, 2002. MR1881971 (2002m:17012)

[HL08] J. Hong and H. Lee, *Young tableaux and crystal $\mathcal{B}(\infty)$ for finite simple Lie algebras*, J. Algebra **320** (2008), no. 10, 3680–3693, DOI 10.1016/j.jalgebra.2008.06.008. MR2457716 (2009j:17008)

[Kam10] J. Kamnitzer, *Mirković-Vilonen cycles and polytopes*, Ann. of Math. (2) **171** (2010), no. 1, 245–294, DOI 10.4007/annals.2010.171.245. MR2630039 (2011g:20070)

[Kan03] S.-J. Kang, *Crystal bases for quantum affine algebras and combinatorics of Young walls*, Proc. London Math. Soc. (3) **86** (2003), no. 1, 29–69, DOI 10.1112/S0024611502013734. MR1971463 (2004c:17028)

[Kas91] M. Kashiwara, *On crystal bases of the q-analogue of universal enveloping algebras*, Duke Math. J. **63** (1991), no. 2, 465–516, DOI 10.1215/S0012-7094-91-06321-0. MR1115118 (93b:17045)

[Kas95] M. Kashiwara, *On crystal bases*, Representations of groups (Banff, AB, 1994), CMS Conf. Proc., vol. 16, Amer. Math. Soc., Providence, RI, 1995, pp. 155–197. MR1357199 (97a:17016)

[Kas02] M. Kashiwara, *Bases cristallines des groupes quantiques* (French), Cours Spécialisés [Specialized Courses], vol. 9, Société Mathématique de France, Paris, 2002. Edited by Charles Cochet. MR1997677 (2004g:17013)

[KL11] H. H. Kim and K.-H. Lee, *Representation theory of p-adic groups and canonical bases*, Adv. Math. **227** (2011), no. 2, 945–961, DOI 10.1016/j.aim.2011.02.017. MR2793028 (2012i:22025)

[KL12] H. H. Kim and K.-H. Lee, *Quantum affine algebras, canonical bases, and q-deformation of arithmetical functions*, Pacific J. Math. **255** (2012), no. 2, 393–415, DOI 10.2140/pjm.2012.255.393. MR2928558

[KN94] M. Kashiwara and T. Nakashima, *Crystal graphs for representations of the q-analogue of classical Lie algebras*, J. Algebra **165** (1994), no. 2, 295–345, DOI 10.1006/jabr.1994.1114. MR1273277 (95c:17025)

[KS97] M. Kashiwara and Y. Saito, *Geometric construction of crystal bases*, Duke Math. J. **89** (1997), no. 1, 9–36, DOI 10.1215/S0012-7094-97-08902-X. MR1458969 (99e:17025)

[KS10] J.-A. Kim and D.-U. Shin, *Generalized Young walls and crystal bases for quantum affine algebra of type A*, Proc. Amer. Math. Soc. **138** (2010), no. 11, 3877–3889, DOI 10.1090/S0002-9939-2010-10428-8. MR2679610 (2011k:17030)

[Lan71] R. P. Langlands, *Euler products*, Yale University Press, New Haven, Conn.-London, 1971. A James K. Whittemore Lecture in Mathematics given at Yale University, 1967; Yale Mathematical Monographs, 1. MR0419366 (54 #7387)

[Lee07] H. Lee, *Realizations of crystal $\mathcal{B}(\infty)$ using Young tableaux and Young walls*, J. Algebra **308** (2007), no. 2, 780–799, DOI 10.1016/j.jalgebra.2006.05.015. MR2295089 (2008f:17027)

[Lit95] P. Littelmann, *Paths and root operators in representation theory*, Ann. of Math. (2) **142** (1995), no. 3, 499–525, DOI 10.2307/2118553. MR1356780 (96m:17011)

[LS12] K.-H. Lee and B. Salisbury, *A combinatorial description of the Gindikin-Karpelevich formula in type A*, J. Combin. Theory Ser. A **119** (2012), no. 5, 1081–1094, DOI 10.1016/j.jcta.2012.01.011. MR2891384

[LS14] K.-H. Lee and B. Salisbury, *Young tableaux, canonical bases, and the Gindikin-Karpelevich formula*, J. Korean Math. Soc. **51** (2014), no. 2, 289–309, DOI 10.4134/JKMS.2014.51.2.289. MR3178585

[LTV99] B. Leclerc, J.-Y. Thibon, and E. Vasserot, *Zelevinsky's involution at roots of unity*, J. Reine Angew. Math. **513** (1999), 33–51, DOI 10.1515/crll.1999.062. MR1713318 (2001f:20011)

[Lus90] G. Lusztig, *Canonical bases arising from quantized enveloping algebras*, J. Amer. Math. Soc. **3** (1990), no. 2, 447–498, DOI 10.2307/1990961. MR1035415 (90m:17023)

[Lus91] G. Lusztig, *Quivers, perverse sheaves, and quantized enveloping algebras*, J. Amer. Math. Soc. **4** (1991), no. 2, 365–421, DOI 10.2307/2939279. MR1088333 (91m:17018)

[Lus93] G. Lusztig, *Introduction to quantum groups*, Progress in Mathematics, vol. 110, Birkhäuser Boston, Inc., Boston, MA, 1993. MR1227098 (94m:17016)

[Mac71] I. G. Macdonald, *Spherical functions on a group of p-adic type*, Ramanujan Institute, Centre for Advanced Study in Mathematics,University of Madras, Madras, 1971. Publications of the Ramanujan Institute, No. 2. MR0435301 (55 #8261)

[McN11] P. J. McNamara, *Metaplectic Whittaker functions and crystal bases*, Duke Math. J. **156** (2011), no. 1, 1–31, DOI 10.1215/00127094-2010-064. MR2746386 (2012a:11060)

[S+14] W. A. Stein et al., *Sage Mathematics Software (Version 7.0)*, The Sage Developers, 2016, `http://www.sagemath.org`.

[SCc08] The Sage-Combinat community, *Sage-Combinat: enhancing Sage as a toolbox for computer exploration in algebraic combinatorics*, 2008, `http://combinat.sagemath.org`.

[Zel80] A. V. Zelevinsky, *Induced representations of reductive $\mathfrak{p}$-adic groups. II. On irreducible representations of* GL($n$), Ann. Sci. École Norm. Sup. (4) **13** (1980), no. 2, 165–210. MR584084 (83g:22012)



Gwanak Wiberpolis 101-1601, Gwanak-Ro 195, Gwanak-gu, Seoul 151-811, Korea

Department of Mathematics, University of Connecticut, 196 Auditorium Road, Unit 3009, Storrs, Connecticut 06269-3009

*E-mail address*: `khlee@math.uconn.edu`

Department of Mathematical Sciences, Seoul National University, Gwanak-ro 599, Gwanak-gu, Seoul 151-747, Korea

*E-mail address*: `sol8586@snu.ac.kr`

Department of Mathematics, Central Michigan University, Mount Pleasant, Michigan 48859

*E-mail address*: `ben.salisbury@cmich.edu`

Proceedings of Symposia in Pure Mathematics
Volume **92**, 2016
http://dx.doi.org/10.1090/pspum/092/01565

# Canonical bases of Cartan-Borcherds type, II: Constructible functions on singular supports

Yiqiang Li

ABSTRACT. The singular supports of the simple perverse sheaves studied are shown to be contained in certain varieties of $\Lambda$ (3.3.1) similar to Lusztig's nilpotent quiver varieties in the negative part of the universal enveloping algebra of a generalized Kac-Moody Lie algebra is realized inside the algebra of constructible functions on $\Lambda$.

## 1. Introduction

Let $\hat{U}^-$ be the negative part of the universal enveloping algebra of the Kac-Moody Lie algebra associated to a graph, $\Gamma$, without loops. Let $\hat{\mathbf{U}}^-$ be the quantized version of $\hat{U}^-$.

In [**L91**], Lusztig studied a geometric algebra $\hat{\mathcal{K}}$ spanned by the set, $\hat{\mathcal{P}}$, of isomorphism classes of certain simple perverse sheaves on the representation varieties $\hat{\mathbf{E}}$ of an oriented graph of $\Gamma$. He showed the following results among others.

(a) The algebra $\hat{\mathbf{U}}^-$ is isomorphic to $\hat{\mathcal{K}}$. Moreover, if one identifies $\hat{\mathbf{U}}^-$ with $\hat{\mathcal{K}}$, the basis $\hat{\mathcal{P}}$, is the so-called canonical basis of $\hat{\mathbf{U}}^-$, having many remarkable properties.

(b) The singular supports of elements in $\hat{\mathcal{P}}$ are contained in certain closed subvariety $\hat{\Lambda}$ of the cotangent bundle of $\hat{\mathbf{E}}$ defined by the Gelfand-Ponomarev relation (or ADHM-equation) and the nilpotent condition.

(c) The algebra $\hat{U}^-$ is isomorphic to the subalgebra generated by the simplest possible elements in the algebra of constructible functions on $\hat{\Lambda}$.

One may wonder to what extent the results (a), (b) and (c) in [**L91**] can be generalized when $\Gamma$ has loops.

The analogous algebra to $\hat{U}^-$ (resp. $\hat{\mathbf{U}}^-$) in the graph with loops case is the negative part, $U^-$, (resp. $\mathbf{U}^-$) of the universal (resp. quantized) enveloping algebra of the *generalized* Kac-Moody Lie algebra in [**B88**] (resp. [**K95**]).

Similar to the pair $(\hat{\mathcal{K}}, \hat{\mathcal{P}})$, a pair $(\mathcal{K}, \mathcal{P})$ is obtained in the graph-with-loops case in [**L93b**]. It is shown in [**KS06**] that the algebra $\mathbf{U}^-$ is naturally a subalgebra of $\mathcal{K}$, and inside $\mathbf{U}^-$ there is a basis, $\mathcal{B}$, consisting of certain semisimple perverse sheaves on the representation variety $\mathbf{E}$ of an oriented graph with loops. It is further shown in [**LL09**] that the construction of the basis $\mathcal{B}$ is independent of the choices of the graph, and under certain condition on the graph, the algebra $\mathcal{K}$ is generated

*Key words and phrases.* Singular support, canonical basis, Lusztig's nilpotent quiver variety.

by the simplest possible elements in $\mathcal{K}$ as was conjectured in [**L93b**]. In short, the extension of the result (a) in [**L91**] to the graph-with-loops case is almost complete by the work [**L93b**], [**KS06**] and [**LL09**].

The purpose of this article is to extend the results (b) and (c) in [**L91**] to the graph-with-loops case. More precisely, we show that the singular supports for the elements in $\mathcal{P}$ are contained in a closed subvariety $\Lambda$ of the cotangent bundle of $\mathbf{E}$. Similar to Lusztig's nilpotent quiver variety $\hat{\Lambda}$, the variety $\Lambda$ is defined by the Galfand-Ponomarev relation and a stable relation, in replace of the nilpotent condition. (The stable condition coincides with the nilpotent condition if there is no edge loops in the graph.) The variety $\Lambda$ has been appeared implicitly in [**KKS08**]. We then show that the algebra $U^-$ is isomorphic to a certain subalgebra, $\mathscr{U}$, in the algebra of constructible functions on $\Lambda$.

In [**L92**] and [**L00**], the so-called semicanonical basis of $\hat{U}^-$ is constructed in the setting of the result (c). One expects to have a similar basis for $U^-$ arising from $\mathscr{U}$ in view of the work [**KKS08**] and [**KKS09**].

## 2. Jordan Quiver

**2.1.** Let $V$ be a finite dimensional vector space over $\mathbb{C}$, the field of complex numbers. Let $\mathbf{G}_V = \mathrm{GL}(V)$, the general linear group, and $\mathfrak{g}_V = \mathrm{End}(V)$ its Lie algebra. Let

$$\mathbf{E}_{\Omega,V}=\mathrm{End}(V), \quad \mathbf{E}_{\bar{\Omega},V}=\mathrm{End}(V) \quad \text{and} \quad \mathbf{E}_V=\mathbf{E}_{\Omega,V}\oplus\mathbf{E}_{\bar{\Omega},V}=\mathrm{End}(V)\oplus\mathrm{End}(V).$$

Elements in $\mathbf{E}_V$ will be denoted by $x=(x_h,x_{\bar{h}})$ with $x_h\in\mathbf{E}_{\Omega,V}$ and $x_{\bar{h}}\in\mathbf{E}_{\bar{\Omega}}$. We define a non degenerate symplectic form

$$\langle,\rangle:\mathbf{E}_V\times\mathbf{E}_V\to\mathbb{C}$$

on $\mathbf{E}_V$ by

$$\langle x,y\rangle=\mathrm{tr}(x_hy_{\bar{h}}-x_{\bar{h}}y_h),$$

where "tr" is the trace map. Observe that $\mathbf{E}_{V,\Omega}$ is a lagrangian subspace of $\mathbf{E}_V$ and the restriction of the symplectic form $\langle,\rangle$ to $\mathbf{E}_{\Omega,V}\times\mathbf{E}_{\bar{\Omega},V}$ is a non degenerate pairing. Thus we can identify the cotangent bundle of $\mathbf{E}_{\Omega,V}$ with $\mathbf{E}_V$.

DEFINITION 2.1.1. Let $\Lambda_V$ be the variety of all elements $(x_h,x_{\bar{h}})\in\mathbf{E}_V$ such that $x_hx_{\bar{h}}-x_{\bar{h}}x_h=0$ and $x_h$ is a nilpotent morphism.

The group $\mathbf{G}_V$ acts on $\mathbf{E}_{\Omega,V}$, $\mathbf{E}_{\bar{\Omega},V}$ and $\mathbf{E}_V$ by conjugation. It is clear that $\Lambda_V$ is stable under the $\mathbf{G}_V$-action.

LEMMA 2.1.2. *Let $O$ be the $\mathbf{G}_V$-orbit of a nilpotent element $x_h$ in $\mathbf{E}_{\Omega,V}$. Then the conormal bundle $T^*_O\mathbf{E}_{\Omega,V}$ of $O$ in $\mathbf{E}_V$ is contained in $\Lambda_V$.*

PROOF. By [**H75**], the tangent space to $O$ at $x_h$ consists of all elements of the form $[g,x_h]:=gx_h-x_hg$, where $g\in\mathfrak{g}_V$. Thus, the conormal space to $O$ at $x_h$ consists of all elements $y_{\bar{h}}\in\mathbf{E}_{\bar{\Omega},V}$ such that $\langle y_{\bar{h}},[g,x_h]\rangle=0$ for all $g\in\mathfrak{g}_V$. By definition,

$$\langle y_{\bar{h}},[g,x_h]\rangle=\mathrm{tr}([g,x_h]y_{\bar{h}})=\mathrm{tr}(gx_hy_{\bar{h}}-x_hgy_{\bar{h}})=\mathrm{tr}(g(x_hy_{\bar{h}}-y_{\bar{h}}x_h)).$$

Thus the condition that $\langle y_{\bar{h}},[g,x_h]\rangle=0$ for all $g\in\mathfrak{g}_V$ is equivalent to the condition that $x_hy_{\bar{h}}-y_{\bar{h}}x_h=0$, i.e., $(x_h,y_{\bar{h}})\in\Lambda_V$. Lemma follows. □

From Lemma 2.1.2, we see that $\Lambda_V$ is the union of the conormal bundle of nilpotent orbits in $\mathbf{E}_{\Omega,V}$. Now that we know these conormal bundles are irreducible and of the same dimension and there are only finitely many nilpotent orbits in $\mathbf{E}_{\Omega,V}$, we see that the closure of each conormal bundle is an irreducible component of $\Lambda_V$, and moreover any irreducible component are of the form. Summing up, we have

COROLLARY 2.1.3. *The irreducible components of $\Lambda_V$ are of the same dimension, and they are the closures of the conormal bundle of the nilpotent orbits in $\mathbf{E}_{\Omega,V}$. Moreover, $\Lambda_V$ is lagrangian.*

**2.2.** We refer to [**KS90**, V] for the definition of the singular support of a given complex of constructible sheaves on a variety over $\mathbb{C}$. We will recall a result from [**KS90**] that we need in this paper.

Let $f : X \to Y$ be a morphism of smooth, connected varieties. It then induces a map

$$df^* : T^*Y \to T^*X$$

from the cotangent bundle $T^*Y$ of $Y$ to the cotangent bundle $T^*X$ of $X$. We set

$$\tilde{X} = \{(x,\xi) \in X \times_Y T^*Y | (df)^*_x(\xi) = 0\}$$

where $(df)^*_x$ is the induced cotangent map from $f$ at $x$. Let $\tilde{Y}$ be the image of $\tilde{X}$ under the second projection. Let us recall a result from [**KS90**, Prop. 5.4.4].

PROPOSITION 2.2.1 ([**KS90**, Prop. 5.4.4]). *Let $\mathbb{C}_X$ be the constant sheaf on $X$, regarded as a complex concentrated on the zeroth degree. If, moreover, $f$ is proper, then the singular support, $SS(f_!(\mathbb{C}_X))$, of the semisimple complex $f_!(\mathbb{C}_X)$ is contained in $\tilde{Y}$.*

**2.3.** Let $\mathbf{a} = (a_1, \cdots, a_n)$ be a sequence of nonnegative integers such that $a_1 + \cdots + a_n = \dim V$. Let $\mathcal{F}_{\mathbf{a}}$ denote the variety consisting of all flags of the form

$$F = (V = F^0 \supseteq F^1 \supseteq \cdots \supseteq F^n = 0)$$

such that $\dim F^i/F^{i+1} = a_{i+1}$ for $i = 0, \cdots, n-1$. Given any pair $(F, x_h) \in \mathcal{F}_{\mathbf{a}} \times \mathbf{E}_{\Omega,V}$, we say that $F$ is *$x_h$-stable* if $x_h(F^i) \subseteq F^{i+1}$ for $i = 0, \cdots, n-1$. Such a pair $(F, x_h)$ will be called a *stable pair*. We say that $F$ is *$x$-quasi-stable* if $x_h(F^i) \subseteq F^i$ for $i = 0, \cdots, n-1$.

Similar notions can be defined if the element $x_h$ is replaced by an element $x_{\bar{h}} \in \mathbf{E}_{\bar{\Omega},V}$ or $x \in \mathbf{E}_V$. Notice that if there is an $F \in \mathcal{F}_{\mathbf{a}}$ such that $F$ is $x_h$-stable, then $x_h$ is nilpotent.

Let $\tilde{\mathcal{F}}_{\mathbf{a}}$ be the closed subvariety of $\mathcal{F}_{\mathbf{a}} \times \mathbf{E}_{\Omega,V}$ consisting of all stable pairs; and

$$\pi_{\mathbf{a}} : \tilde{\mathcal{F}}_{\mathbf{a}} \to \mathbf{E}_{\Omega,V}$$

be the second projection.

**2.4.** Let $F_0$ be a flag of type $\mathbf{a}$. Let $P = \mathrm{stab}_{\mathbf{G}_V}(F_0)$ be the stabilizer of $F_0$ in $\mathbf{G}_V$, $\mathfrak{p}$ its Lie algebra in $\mathfrak{g}_V$, and $\mathfrak{n} = \mathfrak{n}_{\mathfrak{p}}$ the nilpotent radical of $\mathfrak{p}$. Note that $\mathfrak{n}$ consists of all elements $a$ in $\mathfrak{g}_V$ such that $F_0$ is $a$-stable.

Let $P$ act on $\mathbf{G}_V \times \mathfrak{n}$ by $g_1(g,n) = (gg_1^{-1}, g_1ng_1^{-1})$ for all $g_1 \in P$, $g \in G_V$ and $n \in \mathfrak{n}$. This action is free, and so the quotient $\mathbf{G}_V \times^P \mathfrak{n}$ of $\mathbf{G}_V \times \mathfrak{n}$ by $P$ exists. Moreover,

$$\mathbf{G}_V \times^P \mathfrak{n} \simeq \tilde{\mathcal{F}}_{\mathbf{a}}, \quad (g,n) \mapsto (gF_0, gng^{-1}).$$

Let us fix a stable pair $(F_0, x_0)$ in $\tilde{\mathcal{F}}_{\mathbf{a}}$. Consider the following diagram

$$\begin{array}{ccc} P & \xrightarrow{\iota} & P \\ \downarrow & & \downarrow \\ \mathbf{G}_V \times \mathfrak{n} & \xrightarrow{\phi} & \mathbf{G}_V \times \mathbf{E}_{\Omega,V} \\ \downarrow & & \downarrow \\ \mathbf{G}_V \times^P \mathfrak{n} & \longrightarrow & \mathbf{G}_V/P \times \mathbf{E}_{\Omega,V}, \end{array}$$

where the maps in the first column is defined by $p \mapsto (p^{-1}, px_0p^{-1})$ and $(g,n) \mapsto (g,n)$, the maps in the second column is defined by $p \mapsto (p, x_0)$ and $(g,n) \mapsto (gP, n)$, $\iota$ is the inverse maps, $\phi$ is defined by $\phi(g,n) = (g, gng^{-1})$, and the maps on the third row is induced by $\phi$.

If we identify $\mathbf{G}_V/P$ with $\mathcal{F}_{\mathbf{a}}$, via $gP \mapsto gF_0$, then the third row in the diagram gets identified with the natural inclusion $\tilde{\mathcal{F}}_{\mathbf{a}} \to \mathcal{F}_{\mathbf{a}} \times \mathbf{E}_{\Omega,V}$.

By taking the differentials of the above maps at the point $e$ for $P$, $(e, x_0)$ for $\mathbf{G}_V \times \mathfrak{n}$, etc., we have the following diagram

$$\begin{array}{ccccc} \mathfrak{p} & \xrightarrow{-\mathrm{id}} & \mathfrak{p} & & \\ \downarrow & & \downarrow & & \\ \mathfrak{g}_V \oplus \mathfrak{n} & \xrightarrow{d\phi} & \mathfrak{g}_V \oplus \mathbf{E}_{\Omega,V} & \xrightarrow{d\psi} & \mathbf{E}_{\Omega,V}/\mathfrak{n} \\ \downarrow & & \downarrow & & \\ (\mathfrak{g}_V \oplus \mathfrak{n})/\mathfrak{p} & \longrightarrow & (\mathfrak{g}_V/\mathfrak{p}) \oplus \mathbf{E}_{\Omega,V} & & \\ \| & & \| & & \\ T_{(F_0,x_0)}\tilde{\mathcal{F}}_{\mathbf{a}} & \longrightarrow & T_{(F_0,x_0)}(\mathcal{F}_{\mathbf{a}} \times \mathbf{E}_{\Omega,V}), & & \end{array}$$

where $d\phi(a,n) = (a, n + [a, x_0])$ and $d\psi(a,x) = x - [a, x_0]$. Observe that the second row is an exact sequence and the map $d\psi$ factors through $(\mathfrak{g}_V/\mathfrak{p}) \oplus \mathbf{E}_{\Omega,V}$. Thus we have a complex

$$(1) \qquad \begin{array}{c} (\mathfrak{g}_V \oplus \mathfrak{n})/\mathfrak{p} \to (\mathfrak{g}_V/\mathfrak{p}) \oplus \mathbf{E}_{\Omega,V} \to \mathbf{E}_{\Omega,V}/\mathfrak{n} \quad \text{or} \\ T_{F_0,x_0}\tilde{\mathcal{F}}_{\mathbf{a}} \to T_{F_0,x_0}(\mathcal{F}_{\mathbf{a}} \times \mathbf{E}_{\Omega,V}) \to \mathbf{E}_{\Omega,V}/\mathfrak{n}. \end{array}$$

This is a short exact sequence by observing that the first map is injective and the second term is surjective and comparing the dimensions of the three terms.

Through the pairing $\langle , \rangle : \mathbf{E}_{\Omega,V} \times \mathbf{E}_{\bar{\Omega},V} \to \mathbb{C}$, we may identify $\mathfrak{p}$, regarded as a subspace in $\mathbf{E}_{\bar{\Omega},V}$, with the dual space of $\mathbf{E}_{\Omega,V}/\mathfrak{n}$; $\mathfrak{n}$ with the dual of $\mathfrak{g}_V/\mathfrak{p}$. Under these identifications, the transpose of the short exact sequence (1) is

$$\mathfrak{p} \xrightarrow{c} \mathfrak{n} \oplus \mathbf{E}_{\bar{\Omega},V} \to T^*_{(F_0,x_0)}\tilde{\mathcal{F}}_{\mathbf{a}}$$

where $c(p) = ([p, x_0], p)$ , for any $p \in \mathfrak{p}$. Thus, $T^*_{(F_0,x_0)}\tilde{\mathcal{F}}_{\mathbf{a}}$ can be identified with the cokernel of $c$.

Observe that $\pi_{\mathbf{a}}$ factors through $\mathcal{F}_{\mathbf{a}} \times \mathbf{E}_{\Omega,V}$:

$$\pi_{\mathbf{a}} : \tilde{\mathcal{F}}_{\mathbf{a}} \to \mathcal{F}_{\mathbf{a}} \times \mathbf{E}_{\Omega,V} \to \mathbf{E}_{\Omega,V}$$

where the first map is the embedding and the second map is the second projection. Thus the cotangent map $(d\pi_{\mathbf{a}})^*$ at $(F_0, x_0)$ can be decomposed as

$$(d\pi_{\mathbf{a}})^*_{(F_0,x_0)} : \mathbf{E}_{\bar{\Omega},V} \to \mathfrak{n} \oplus \mathbf{E}_{\bar{\Omega},V} \to (\mathfrak{n} \oplus E_{\bar{\Omega},V})/\mathfrak{p},$$

where the first map is $x_{\bar{h}} \mapsto (0, x_{\bar{h}})$ and the second map is the cokernel map of $c$. From this, we have

LEMMA 2.4.1. *For any $x_{\bar{h}} \in \mathbf{E}_{\bar{\Omega},V}$, $(d\pi_{\mathbf{a}})^*_{(x_0,F_0)}(x_{\bar{h}}) = 0$ if and only if $x_{\bar{h}} \in \mathfrak{p}$ and $[x_{\bar{h}}, x_0] = 0$, if and only if $F_0$ is $x_{\bar{h}}$-quasi-stable and $[x_{\bar{h}}, x_0] = 0$.*

Let us regard $\pi_{\mathbf{a}}$ as the map $f$ in section 2.2. Thus $\tilde{Y}$ and $\tilde{X}$ are defined. The following proposition follows from Lemma 2.4.1.

PROPOSITION 2.4.2. *We have*

$$\tilde{X} \simeq \{(x_h, x_{\bar{h}}, F) \in \mathbf{E}_V \times \mathcal{F}_{\mathbf{a}} | [x_h, x_{\bar{h}}] = 0, \exists\, F \in \tilde{\mathcal{F}}_{\mathbf{a}} \text{ that is } x_h\text{-stable, } x_{\bar{h}}\text{-quasi-stable}\}.$$

*$\tilde{Y}$ is contained in $\Lambda_V$, consisting of all elements $(x_h, x_{\bar{h}})$ such that there is $F \in \mathcal{F}_{\mathbf{a}}$ $x_h$-stable, and $x_{\bar{h}}$-quasi-stable.*

Let IC$(O)$ be the intersection cohomology complex ([**BBD82**]) attached to a nilpotent orbit $O$ in $\mathbf{E}_{\Omega,V}$. We have

THEOREM 2.4.3. *The singular support of* IC$(O)$ *is contained in $\Lambda_V$.*

PROOF. It is well known that there is a sequence $\mathbf{a}$, say $\mathbf{a} = (1, \cdots, 1)$, such that IC$(O)$ is a direct summand of the semisimple perverse sheaf $\pi_{\mathbf{a}!}(\mathbb{C}_{\tilde{\mathcal{F}}_{\mathbf{a}}})[\dim \mathbb{C}_{\tilde{\mathcal{F}}_{\mathbf{a}}}]$. The theorem follows from Proposition 2.2.1 and Proposition 2.4.2. □

## 3. Arbitrary Quivers

**3.1.** Let $\Gamma$ be a *graph*. It consists of a pair $(I, H)$ of sets and a triple $(', '' : H \to I; ^- : H \to H)$ of maps such that "$^-$" is a fixed-point-free involution and $h' = (\bar{h})''$ for any $h \in H$.

The set $I$ is called the *vertex* set, while $H$ the *edge* set. The map "$'$" is called the *source* map, while "$''$" the *target* map. Notice that we allow edge loops, i.e., there may be some $h \in H$ such that $h' = h''$.

We fix an *orientation* $\Omega$ in $\Gamma$, i.e., $\Omega$ is a subset of $H$ such that $\Omega \sqcup \bar{\Omega} = H$ and $\Omega \cap \bar{\Omega} = \emptyset$. Let

$$\epsilon : H \to \{\pm 1\}$$

be the map defined by $\epsilon(h) = 1$ if $h \in \Omega$ and $\epsilon(h) = -1$ if $h \in \bar{\Omega}$.

Let $H^- = \{h \in H | h' = h''\}$ and $H^+ = H \backslash H^-$. We set $\Omega^+ = \Omega \cap H^+$ and $\Omega^- = \Omega \cap H^-$.

Let $I^- = \{i \in I | \exists h \in H^- \text{ s.t. } h' = i = h''\}$ and $I^+ = I \backslash I^-$.

**3.2.** To a finite dimensional $I$-graded vector space $V$ over $\mathbb{C}$, we attach the algebraic group

$$\mathbf{G}_V = \prod_{i\in I} \mathrm{GL}(V_i)$$

and the spaces

$$\mathbf{E}_{\Omega,V} = \oplus_{h\in\Omega}\mathrm{Hom}(V_{h'}, V_{h''}),\ \mathbf{E}_{\bar{\Omega},V} = \oplus_{h\in\bar{\Omega}}\mathrm{Hom}(V_{h'}, V_{h''}) \text{ and } \mathbf{E}_V = \mathbf{E}_{\Omega,V} \oplus \mathbf{E}_{\bar{\Omega},V}.$$

The group $\mathbf{G}_V$ acts on $\mathbf{E}_V$ by conjugation: $g.x = y$, $y_h = g_{h''}x_h g_{h'}^{-1}$ for any $g \in \mathbf{G}_V$, $x \in \mathbf{E}_V$ and $h \in H$. This action induces a $\mathbf{G}_V$-action on $\mathbf{E}_{\Omega,V}$ and $\mathbf{E}_{\bar{\Omega},V}$, respectively.

The Lie algebra $\mathfrak{g}_V = \oplus_{i\in I}\mathfrak{gl}_{V_i}$ of $\mathbf{G}_V$ acts on $\mathbf{E}_{\Omega,V}$ by $g.x = [g,x]$, where

$$[g,x]_h = g_{h''}x_h - x_h g_{h'}, \quad \forall h \in \Omega.$$

Given any $x \in \mathbf{E}_V$, let $x_\Omega$ and $x_{\bar{\Omega}}$ denote the components of $x$ in $\mathbf{E}_{\Omega,V}$ and in $\mathbf{E}_{\bar{\Omega},V}$, respectively.

To the space $\mathbf{E}_V$, we associate a non-degenerate symplectic form

$$\langle,\rangle : \mathbf{E}_V \times \mathbf{E}_V \to \mathbb{C}$$

defined by $\langle x,y\rangle = \sum_{h\in H} \epsilon(h)\mathrm{tr}(x_h y_{\bar{h}})$. ("tr" is the trace map.) Via this form, we can identify $\mathbf{E}_V$ with the cotangent bundle of $\mathbf{E}_{\Omega,V}$.

**3.3.** Let $(\mathbf{i},\mathbf{a})$ be a pair of sequences

$$\mathbf{i} = (i_1,\cdots,i_n), \quad \mathbf{a} = (a_1,\cdots,a_n), \quad \forall i_1\cdots,i_n \in I,\ a_1,\cdots,a_n \in \mathbb{N}$$

such that $a_1 i_1 + \cdots + a_n i_n = \sum_{i\in I} \dim V_i\, i$. Let $\mathcal{F}_{\mathbf{i},\mathbf{a}}$ be the variety of all flags of the form:

$$F = (V = F^0 \supseteq F^1 \supseteq \cdots \supseteq F^n = 0)$$

such that $\dim F^l/F^{l+1} = a_{l+1}i_{l+1}$ for $l = 0,\cdots,n-1$.

A pair $(F,x) \in \mathcal{F}_{\mathbf{i},\mathbf{a}} \times \mathbf{E}_{\Omega,V}$ is called *stable* if $x_h(F^l_{h'}) \subseteq F^{l+1}_{h''}$ for any $h \in H$ and $l = 0,\cdots,n-1$. The element $x$ in a stable pair $(F,x)$ will be called a *nilpotent* element. A pair $(F,x) \in \mathcal{F}_{\mathbf{i},\mathbf{a}} \times \mathbf{E}_{\Omega,V}$ is called *quasi-stable* if $x_h(F^l_{h'}) \subseteq F^l_{h''}$ for any $h \in H$ and $l = 0,\cdots,n-1$.

Let $\tilde{\mathcal{F}}_{\mathbf{i},\mathbf{a}}$ be the variety of all stable pairs in $\mathcal{F}_{\mathbf{i},\mathbf{a}} \times \mathbf{E}_{\Omega,V}$, and

$$\pi_{\mathbf{i},\mathbf{a}} : \tilde{\mathcal{F}}_{\mathbf{i},\mathbf{a}} \to \mathbf{E}_{\Omega,V}$$

the second projection.

DEFINITION 3.3.1. Let $\Lambda_V$ be the closed subvariety of $\mathbf{E}_V$ consisting of all elements $x$ such that

$$\mu_i(x) := \sum_{h\in H:h''=i} \epsilon(h)x_h x_{\bar{h}} = 0, \quad \forall i \in I; \tag{2}$$

and there is an $F \in \mathcal{F}_{\mathbf{i},\mathbf{a}}$ for some pair $(\mathbf{i},\mathbf{a})$ such that $F$ is $x_\Omega$-stable and $x_{\bar{\Omega}}$-quasi-stable.

REMARK 3.3.2. (1). If $H$ has no loops, $\Lambda_V$ is Lusztig's quiver variety defined in [**L91**, 12].

(2). When $H$ is the Jordan graph, $\Lambda_V$ is the variety defined in Definition 2.1.1.

**3.4.** Let us fix a stable pair $(F_0, x_0) \in \tilde{\mathcal{F}}_{\mathbf{i},\mathbf{a}}$. We shall compute the cotangent map $(d\pi_{\mathbf{i},\mathbf{a}})^*_{(F_0,x_0)}$ of $\pi_{\mathbf{i},\mathbf{a}}$ at the stable pair $(F_0, x_0)$. The procedure goes exactly the same as that of $\pi_{\mathbf{a}}$ in Section 2.4.

Let $P = \mathrm{Stab}_{\mathbf{G}_V}(F_0)$ be the stabilizer of $F_0$ in $\mathbf{G}_V$. Let $\Xi$ be the subvariety of all $x \in \mathbf{E}_{\Omega,V}$ such that $F_0$ is $x$-stable. Define a $P$-action on $\mathbf{G}_V \times \Xi$ by

$$g'.(g,\xi) = (g(g')^{-1}, g'.\xi), \quad \forall g' \in P, g \in \mathbf{G}_V, \xi \in \Xi.$$

The $P$-action on $\mathbf{G}_V \times \Xi$ is free, and so the quotient, $\mathbf{G}_V \times^P \Xi$, of $\mathbf{G}_V \times \Xi$ by $P$ exists. We have an isomorphism of varieties

$$\mathbf{G}_V \times^P \Xi \to \tilde{\mathcal{F}}_{\mathbf{i},\mathbf{a}}$$

given by $(g,\xi) \mapsto (gF_0, g.\xi)$. Consider the following diagram

$$\begin{array}{ccc} P & \xrightarrow{\iota} & P \\ \downarrow & & \downarrow \\ \mathbf{G}_V \times \Xi & \xrightarrow{\phi} & \mathbf{G}_V \times \mathbf{E}_{\Omega,V} \\ \downarrow & & \downarrow \\ \mathbf{G}_V \times^P \Xi & \longrightarrow & \mathbf{G}_V/P \times \mathbf{E}_{\Omega,V}, \end{array}$$

where the maps in the first column is defined by $p \mapsto (p^{-1}, p.x_0)$ and $(g,\xi) \mapsto (g,\xi)$, the maps in the second column is defined by $p \mapsto (p, x_0)$ and $(g,\xi) \mapsto (gP,\xi)$, $\iota$ is the inverse maps, $\phi$ is defined by $\phi(g,\xi) = (g, g.\xi)$, and the maps on the third row is induced by $\phi$.

The differentials of the maps in the above diagram at $e \in P$, $(e, x_0) \in \mathbf{G}_V \times \Xi$, etc, can be computed as follows:

$$\begin{array}{ccccc} \mathfrak{p} & \xrightarrow{-\mathrm{id}} & \mathfrak{p} & & \\ \downarrow & & \downarrow & & \\ \mathfrak{g}_V \oplus \Xi & \xrightarrow{d\phi} & \mathfrak{g}_V \oplus \mathbf{E}_{\Omega,V} & \xrightarrow{d\psi} & \mathbf{E}_{\Omega,V}/\Xi \\ \downarrow & & \downarrow & & \\ (\mathfrak{g}_V \oplus \Xi)/\mathfrak{p} & \longrightarrow & (\mathfrak{g}_V/\mathfrak{p}) \oplus \mathbf{E}_{\Omega,V} & & \\ \| & & \| & & \\ T_{(F_0,x_0)}\tilde{\mathcal{F}}_{\mathbf{i},\mathbf{a}} & \longrightarrow & T_{(F_0,x_0)}(\mathcal{F}_{\mathbf{i},\mathbf{a}} \times \mathbf{E}_{\Omega,V}), & & \end{array}$$

where $\mathfrak{p}$ is the Lie algebra of $P$, $\mathfrak{g}_V$ the Lie algebra of $\mathbf{G}_V$, $d\phi(a,n) = (a, n+[a,x_0])$ and $d\psi$ is defined by $d\psi(a,x) = x - [a, x_0]$. Observe that the second row is an exact sequence and the map $d\psi$ factors through $(\mathfrak{g}_V/\mathfrak{p}) \oplus \mathbf{E}_{\Omega,V}$. Thus we have a complex

$$(\mathfrak{g}_V \oplus \Xi)/\mathfrak{p} \to (\mathfrak{g}_V/\mathfrak{p}) \oplus \mathbf{E}_{\Omega,V} \to \mathbf{E}_{\Omega,V}/\Xi \tag{3}$$

or

$$T_{F_0,x_0}\tilde{\mathcal{F}}_{\mathbf{i},\mathbf{a}} \to T_{F_0,x_0}(\mathcal{F}_{\mathbf{i},\mathbf{a}} \times \mathbf{E}_{\Omega,V}) \to \mathbf{E}_{\Omega,V}/\Xi.$$

This is a short exact sequence by observing that the first map is injective and the second term is surjective and comparing the dimensions of the three terms.

Let $\Xi^{qs}$ be the subvariety of $\mathbf{E}_{\bar{\Omega},V}$ consisting of all elements $x$ such that $F_0$ is $x$-quasi-stable. Via the non-degenerate pairing $\langle,\rangle : \mathbf{E}_{\Omega,V} \times \mathbf{E}_{\bar{\Omega},V} \to \mathbb{C}$, we may identify $\Xi^{qs}$ with the dual of $\mathbf{E}_{\Omega,V}/\Xi$.

The pairing $\mathrm{tr} : \mathfrak{g}_V \times \mathfrak{g}_V \to \mathbb{C}$ defined by $\mathrm{tr}(g,g') = \sum_{i\in I} tr(g_i g_i')$ is non degenerate. Via this pairing, we may identify $\mathfrak{n}$, the nilradical of $\mathfrak{p}$, with the space $\mathfrak{g}_V/\mathfrak{p}$.

Via the above identifications, the transpose of the short exact sequence (3) is the short exact sequence:

$$\Xi^{qs} \xrightarrow{c} \mathfrak{n} \oplus \mathbf{E}_{\bar{\Omega},V} \to T^*_{(F_0,x_0)}\tilde{\mathcal{F}}_{\mathbf{i},\mathbf{a}}$$

where $c(\xi) = (n,\xi)$ , with $n_i = \sum_{h\in H:h''=i} \xi_h x_{0,\bar{h}} - x_{0,h}\xi_{\bar{h}} = \mu_i(\xi + x_0)$, for any $i \in I$ and $\xi \in \Xi^{qs}$. Here we embed $\mathbf{E}_{\Omega,V}$ into $\mathbf{E}_V$, so $x_{0,h} = 0$ if $h \in \bar{\Omega}$. Thus the map

$$T^*_{(F_0,x_0)}\mathbf{E}_{\Omega,V} \times \mathcal{F}_{\mathbf{i},\mathbf{a}} \to T^*_{(F_0,x_0)}\tilde{\mathcal{F}}_{\mathbf{i},\mathbf{a}}$$

can be identified with the cokernel map of $c$.

Observe that $\pi_{\mathbf{i},\mathbf{a}}$ factors through $\mathcal{F}_{\mathbf{i},\mathbf{a}} \times \mathbf{E}_{\Omega,V}$:

$$\pi_{\mathbf{i},\mathbf{a}} : \tilde{\mathcal{F}}_{\mathbf{i},\mathbf{a}} \to \mathcal{F}_{\mathbf{i},\mathbf{a}} \times \mathbf{E}_{\Omega,V} \to \mathbf{E}_{\Omega,V}$$

where the first map is the embedding and the second map is the first projection. Thus the cotangent map $(d\pi_{\mathbf{i},\mathbf{a}})^*$ at $(F_0, x_0)$ can be decomposed as

$$(d\pi_{\mathbf{a}})^*_{(F_0,x_0)} : \mathbf{E}_{\bar{\Omega},V} \to \mathfrak{n} \oplus \mathbf{E}_{\bar{\Omega},V} \to (\mathfrak{n} \oplus E_{\bar{\Omega},V})/\Xi^{qs},$$

where the first map is $x \mapsto (0,x)$ and the second map is the cokernel map of $c$. From this, we have

LEMMA 3.4.1. *For any $x \in \mathbf{E}_{\bar{\Omega},V}$, $(d\pi_{\mathbf{i},\mathbf{a}})^*_{(x_0,F_0)}(x) = 0$ if and only if $x \in \Xi^{qs}$ and $\mu_i(x+x_0) = 0$ for any $i \in I$; if and only if $F_0$ is $x$-quasi-stable and $\mu_i(x+x_0) = 0$ for any $i \in I$.*

By taking the morphism $f$ in Section 2.2 to be $\pi_{\mathbf{i},\mathbf{a}}$, the varieties $\tilde{Y}$ and $\tilde{X}$ are then defined. By Lemma 3.4.1, we have

PROPOSITION 3.4.2. *We have*

$$\tilde{X} \simeq \{(x,F) \in \mathbf{E}_V \times \mathcal{F}_{\mathbf{i},\mathbf{a}} | \mu_i(x) = 0, \forall i \in I,\ F \textit{ is } x_\Omega\textit{-stable, } x_{\bar{\Omega}}\textit{-quasi-stable}\}.$$

*$\tilde{Y}$ is contained in $\Lambda_V$, consisting of all elements $x$ such that there is a flag $F \in \mathcal{F}_{\mathbf{i},\mathbf{a}}$ $x_\Omega$-stable, and $x_{\bar{\Omega}}$-quasi-stable.*

Let $S$ be a simple perverse sheaf appearing as an irreducible summand in the semisimple complex $\pi_{\mathbf{i},\mathbf{a}!}(\mathbb{C}_{\tilde{\mathcal{F}}_{\mathbf{i},\mathbf{a}}})[z]$ for some $z \in \mathbb{Z}$. By Proposition 3.4.2 and Proposition 2.2.1, we have

THEOREM 3.4.3. *The singular support of $S$ is contained in $\Lambda_V$.*

Let $\Omega' = \Omega^+ \sqcup \bar{\Omega}^-$. Let $\tilde{\mathbb{F}}_{\mathbf{i},\mathbf{a}}$ be the subvariety of $\mathcal{F}_{\mathbf{i},\mathbf{a}} \times \mathbf{E}_{\Omega',V}$ consisting of all *quasi-stable* pairs. Let

$$\varpi_{\mathbf{i},\mathbf{a}} : \tilde{\mathbb{F}}_{\mathbf{i},\mathbf{a}} \to \mathbf{E}_{\Omega',V}$$

be the second projection.

COROLLARY 3.4.4. *The singular supports of the simple perverse sheaves appearing in the semisimple complex $\varpi_{\mathbf{i},\mathbf{a}!}(\mathbb{C}_{\tilde{\mathbb{F}}_{\mathbf{i},\mathbf{a}}})[z]$ for some $z \in \mathbb{Z}$ are contained in $\Lambda_V$.*

This corollary can be proved in a similar way as the proof of Theorem 3.4.3.

REMARK 3.4.5. Note that the simple perverse sheaves $S$ in Theorem 3.4.3 are those used in [**KS06**] and [**LL09**], while the ones in Corollary 3.4.4 are those used in [**L93b**].

QUESTION 3.4.6. Is $\Lambda_V$ lagrangian?

## 4. Constructible functions on $\Lambda_V$

We would like to investigate the algebra of constructible functions on the $\Lambda_V$'s.

**4.1.** We shall recall some basics of constructible functions. We refer to [**M74**], [**KS90**] and [**D04**] for the proofs of the results.

Let $X$ be an algebraic variety over $\mathbb{C}$. A subset $X_1$ in $X$ is called *constructible* if $X_1$ can be obtained by finitely many set-theoretic operations on the subvarieties of $X$. A function $\phi : X \to \mathbb{C}$ is called *constructible* if $X$ has a partition $X = \sqcup_{i=1}^n X_i$ of constructible subsets such that $\phi$ is constant on each $X_i$, $i = 1, \cdots, n$. For any constructible function $\phi : X \to \mathbb{C}$, the *integral* of $\phi$ with respect to the Euler characteristic $\chi$ is defined by

$$\int_{x\in X} \phi(x) = \sum_{c\in\mathbb{C}} c\chi(\phi^{-1}(c)).$$

Let $\mathscr{C}'(X)$ be the space of $\mathbb{C}$-valued constructible functions on $X$. If, furthermore, there is a linear algebraic group $G$ acting on $X$, we define $\mathscr{C}'_G(X)$ to be the subspace of $\mathscr{C}'(X)$ consisting of all $G$-invariant constructible functions.

Let $f : Y \to X$ be a morphism of varieties. We define the linear maps

$$f^* : \mathscr{C}'(X) \to \mathscr{C}'(Y) \quad \text{and} \quad f_* : \mathscr{C}'(Y) \to \mathscr{C}'(X)$$

by

$$f^*(\phi)(y) = \phi(f(y)) \quad \text{and} \quad f_*(\psi)(x) = \int_{y\in f^{-1}(x)} \psi(y),$$

for $x \in X$, $y \in Y$, $\phi \in \mathscr{C}'(X)$, and $\psi \in \mathscr{C}'(Y)$. If, further, $H$ and $G$ acts from the left on $Y$ and $X$, respectively, and there is a morphism of linear algebraic group $m : H \to G$ such that $f(h.y) = m(h)f(y)$ for any $h \in H$ and $y \in Y$, we say that $f$ *is compatible with the $H$-action and $G$-action on $Y$ and $X$, respectively.* Then the linear maps $f^*$ and $f_*$ give rise to the linear maps

$$f^* : \mathscr{C}'_G(X) \to \mathscr{C}'_H(Y) \quad \text{and} \quad f_* : \mathscr{C}'_H(Y) \to \mathscr{C}'_G(X). \tag{4}$$

**4.2.** To each $\alpha \in \mathbb{N}[I]$, we fix an $I$-graded vector space $V$ such that $\dim V = \alpha$. We consider the space

$$\mathscr{C}' = \oplus_{\alpha\in\mathbb{N}[I]} \mathscr{C}'_{\mathbf{G}_V}(\Lambda_V). \tag{5}$$

Following [**L91**, 12], we associate to $\mathscr{C}'$ an associative algebra structure as follows.

We call a tuple $(x^1, x^2, x^3, f^1, f^2)$, where $(x^1, x^2, x^3) \in \Lambda_{V^1} \times \Lambda_{V^2} \times \Lambda_{V^3}$ and $f^1 : V^1 \to V^2$ and $f^2 : V^2 \to V^3$ are $I$-graded linear maps, a *short exact sequence* if the sequence $V^1 \xrightarrow{f^1} V^2 \xrightarrow{f^2} V^3$ is a short exact sequence of $I$-graded vector spaces and $f^1_{h''} x^1_h = x^2_h f^1_{h''}$ and $x^2_h f^2_{h''} = f^2_{h'} x^3_h$ for all $h \in \Omega$. Let $Z_{V^1,V^2,V^3}$ be the variety of all short exact sequences. Let

$$p_i : Z_{V^1,V^2,V^3} \to \Lambda_{V^i}, \quad \forall i = 1, 2, 3$$

be the $i$-th projection. Let $\mathbf{H} = \mathbf{G}_{V^1} \times \mathbf{G}_{V^2} \times \mathbf{G}_{V^3}$. Then we have a natural projection $m_i : \mathbf{H} \to \mathbf{G}_{V^i}$. There is a natural $\mathbf{H}$-action on $Z_{V^1,V^2,V^3}$ such that the morphism $p_i$ is compatible with the $\mathbf{H}$-action on $Z(V^1, V^2, V^3)$ and the $\mathbf{G}$-action on $\Lambda_{V^i}$, i.e., $p_i(h.\mathbf{z}) = m_i(h)p_i(\mathbf{z})$ for any $h \in \mathbf{H}$ and $\mathbf{z} \in Z(V^1, V^2, V^3)$. Thus $p_i$ induces the following linear maps as in (4):

$$p_i^* : \mathscr{C}'_{\mathbf{G}_{V^i}}(\Lambda_{V^i}) \to \mathscr{C}'_{\mathbf{H}}(Z_{V^1,V^2,V^3}) \quad \text{and} \quad p_{i*} : \mathscr{C}'_{\mathbf{H}}(Z_{V^1,V^2,V^3}) \to \mathscr{C}'_{\mathbf{G}_{V^i}}(\Lambda_{V^i}).$$

Note that $\mathbf{G}_{V^1} \times \mathbf{G}_{V^3}$ acts freely on $Z_{V^1,V^2,V^3}$. So $p_2$ can be decomposed as follows:

$$Z_{V^1,V^2,V^3} \xrightarrow{q_1} (\mathbf{G}_{V^1} \times \mathbf{G}_{V^3})\backslash Z_{V^1,V^2,V^3} \xrightarrow{q_2} \Lambda_{V^2},$$

where $q_1$ is the quotient map, and $q_2$ is the projection induced by $p_2$. To any elements $\phi_i \in \mathscr{C}'_{\mathbf{G}_{V^i}}(\Lambda_{V^i})$ for $i = 1, 3$, we associate an element $\phi_1 \star \phi_3 \in \mathscr{C}'_{\mathbf{G}_{V^2}}(\Lambda_{V^2})$ as follows. There is a unique element $\phi_2$ in $\mathscr{C}'_{\mathbf{G}_{V^2}}((\mathbf{G}_{V^1} \times \mathbf{G}_{V^3})\backslash Z_{V^1,V^2,V^3})$ such that $p_1^*(\phi_1) \otimes p_3^*(\phi_3) = q_1^*(\phi_2)$, since $q_1$ is a $\mathbf{G}_{V^1} \times \mathbf{G}_{V^3}$-principal bundle. We define

$$\phi_1 \star \phi_3 := q_{2*}(\phi_2).$$

Thus, we have a bilinear map

$$\star : \mathscr{C}'_{\mathbf{G}_{V^1}}(\Lambda_{V^1}) \times \mathscr{C}'_{\mathbf{G}_{V^3}}(\Lambda_{V^3}) \to \mathscr{C}'_{\mathbf{G}_{V^2}}(\Lambda_{V^2}).$$

By [**L91**, 12], we have

LEMMA 4.2.1. *The space $\mathscr{C}'$ equipped with the operation $\star$ is an associative algebra.*

To each vertex $i \in I^+$ and $n \in \mathbb{N}$, the variety $\Lambda_V$ ($\dim V = ni$) consists of exactly one element 0. Let $\mathbf{F}_i^{(n)}$ denote the characteristic function of the variety $\Lambda_V$. We write $\mathbf{F}_i$ for $\mathbf{F}_i^{(1)}$.

To each vertex $i \in I^-$ and $n \in \mathbb{N}$, let $Y_0$ be the subvariety of $\Lambda_V$, with $\dim V = ni$, consisting of all elements in $\lambda_V$, whose $\Omega$-components are zero. Notice that $Y_0$ is an irreducible component in $\Lambda_V$, i.e., the conormal bundle to the orbit 0 in $\mathbf{E}_\Omega$. Let $\mathbf{F}_i^{(n)}$ denote the characteristic function of $Y_0$.

DEFINITION 4.2.2. Let $\mathscr{C}$ be the subalgebra of $\mathscr{C}'$ generated by the $\mathbf{F}_i^{(n)}$'s for $i \in I$ and $n \in \mathbb{N}$. Let $\mathscr{U}$ be the subalgebra of $\mathscr{C}'$ generated by the $\mathbf{F}_i^{(n)}$'s for $i \in I^+$ and $n \in \mathbb{N}$, and the $\mathbf{F}_i$'s for $i \in I^-$.

Let $U^-$ be the negative part of the universal enveloping algebra of the generalized Kac-Moody algebra attached to the graph $\Gamma$. Recall that $U^-$ is a $\mathbb{C}$-algebra generated by the symbols $F_i$, $\forall i \in I$ subject to the following relations

$$\sum_{t=0}^{n_{ij}} (-1)^t F_i^{(n_{ij}-t)} F_j F_i^{(t)} = 0, \quad \forall i \in I^+, j \in I \tag{6}$$

where $n_{ij} = 1 + \#\{h \in H | h' = i, h'' = j\}$, and $F_i^{(m)} = \frac{F_i^m}{m!}$, $\forall m \in \mathbb{N}$.

PROPOSITION 4.2.3. *The assignments* $F_i \mapsto \mathbf{F}_i$, $\forall i \in I$, *define an algebra isomorphism*

$$\Phi : U^- \to \mathscr{U}.$$

PROOF. We show that the assignments $F_i \mapsto \mathbf{F}_i$ define an algebra homomorphism

$$\Phi : U^- \to \mathscr{U}. \tag{7}$$

This is to show that the elements $\mathbf{F}_i$ satisfy the Serre relations (6). When $i$ and $j$ are in $I^+$, it is shown in [**L91**, Lem. 12.11]. Thus we only need to consider the case when $j$ is in $I^-$. Observe that the values of the functions $\mathbf{F}_i^{(n_{ij}-t)}\mathbf{F}_j\mathbf{F}_i^{(t)}$ at a point $x \in \Lambda_V$ ($\dim V = n_{ij}i + j$) are zero unless $x_h = 0$ for any $h \in \Omega^-$ by the definition of $\mathbf{F}_j$.

For any given $x \in \Lambda_V$ such that $x_h = 0$ for any $h \in \Omega^-$, the values of the functions $\mathbf{F}_i^{(n_{ij}-t)}\mathbf{F}_j\mathbf{F}_i^{(t)}$ at $x$ are independent of the values of $x_h$ for any $h \in H^-$ (the set of loops). Thus we may assume that $x_h = 0$ for any $h \in H^-$. In other words, the values of the functions $\mathbf{F}_i^{(n_{ij}-t)}\mathbf{F}_j\mathbf{F}_i^{(t)}$ can be calculated over the varieties $\Lambda_V^+$ defined over the graph obtained from $\Gamma$ by throwing away all loops. Again by [**L91**, Lem. 12.11], we see that the Serre relations (6) hold in this case. Therefore, we have the algebra homomorphism $\Phi$.

Finally, we show that $\Phi$ is an isomorphism. Let $\mathbf{N}_{\Omega,V}$ be the subvariety of $\mathbf{E}_{\Omega,V}$ consisting of all nilpotent elements (see 3.2). Then we have a closed embedding

$$\iota : \mathbf{N}_{\Omega,V} \to \Lambda_{\Omega,V}, \quad x_\Omega \mapsto (x_\Omega, 0).$$

Similar to $\mathscr{C}'$, an associative algebra structure is attached to the space

$$\mathscr{C}'_\Omega = \oplus_{\alpha\in\mathbb{N}[I]}\mathscr{C}'_{\mathbf{G}_V}(\mathbf{N}_{\Omega,V}).$$

Moreover, the morphisms $\iota^* : \mathscr{C}'_{\mathbf{G}_V}(\Lambda_V) \to \mathscr{C}'_{\mathbf{G}_V}(\mathbf{N}_{\Omega,V})$ induce an algebra homomorphism

$$\iota^* : \mathscr{C}' \to \mathscr{C}'_\Omega. \tag{8}$$

We still use the same notations for the images of the $\mathbf{F}_i^{(n)}$'s under $\iota^*$. Let $\mathscr{U}_\Omega$ be the subalgebra of $\mathscr{C}'_\Omega$ generated by the $\mathbf{F}_i$'s. Then by specializing $v$ to 1 in [**KS06**, Thm. 2.2], we get an isomorphism $U^- \to \mathscr{U}_\Omega$ sending $F_i$ to $\mathbf{F}_i$ for all $i \in I$. By this fact, (7) and (8), we see that $\Phi$ is an isomorphism. □

REMARK 4.2.4. Surprisingly, the algebra $\mathscr{C}$ is NOT isomorphic to the classical Hall algebra when the graph is the Jordan graph. In particular, the commutative relations $F_i^{(m)}F_i^{(n)} = F_i^{(n)}F_i^{(m)}$ fail. It seems that Borel-Moore homology is a more appropriate language for realizing the classical Hall algebra by using $\Lambda_V$ in the Jordan quiver case.

**4.3.** Recall from Proposition 3.4.2 that $\tilde{X}$, which we shall denote by $\tilde{X}_{\mathbf{i},\mathbf{a}}$ in this subsection, is the closed subvariety of $\mathbf{E}_V \times \tilde{\mathcal{F}}_{\mathbf{i},\mathbf{a}}$ consisting of all pairs $(x, F)$ such that $\mu_i(x) = 0$ for any $i \in I$ and $F$ is $x_\Omega$-stable and $x_{\bar{\Omega}}$-quasi-stable. We denote by

$$\Pi_{\mathbf{i},\mathbf{a}} : \tilde{X}_{\mathbf{i},\mathbf{a}} \to \Lambda_V \tag{9}$$

the projection to the first component. Let

$$\mathscr{L}_{\mathbf{i},\mathbf{a}} = \Pi_{\mathbf{i},\mathbf{a}!}(1_{\tilde{X}_{\mathbf{i},\mathbf{a}}}), \tag{10}$$

where $1_{\tilde{X}_{\mathbf{i},\mathbf{a}}}$ is the characteristic function of $\tilde{X}_{\mathbf{i},\mathbf{a}}$. Then we have

$$\mathscr{L}_{\mathbf{i},\mathbf{a}} \star \mathscr{L}_{\mathbf{j},\mathbf{b}} = \mathscr{L}_{\mathbf{ij},\mathbf{ab}}, \quad \forall \text{pairs } (\mathbf{i},\mathbf{a}), (\mathbf{j},\mathbf{b}).$$

Let $\mathscr{U}_{\mathbb{Z}}$ be the $\mathbb{Z}$-subalgebra of $\mathscr{U}$ spanned by the $\mathscr{L}_{\mathbf{i},\mathbf{a}}$'s for various $(\mathbf{i},\mathbf{a})$. Let $U_{\mathbb{Z}}^-$ be the integral form of $U^-$, i.e., the subalgebra of $U^-$ over $\mathbb{Z}$ generated by $F_i^{(n)}$ for $i \in I^+$ and $F_i$ for $i \in I^-$. Then the assignments, $F_i^{(n)} \mapsto \mathbf{F}_i^{(n)}$, $\forall i \in I^+$; $F_i \mapsto \mathbf{F}_i$, $\forall i \in I^-$ induce an isomorphism of algebras over $\mathbb{Z}$:

$$U_{\mathbb{Z}}^- \simeq \mathscr{U}_{\mathbb{Z}}. \tag{11}$$

**Acknowledgment.** We thank W.L. Gan for answering several questions related to the variety $\Lambda_V$ associated to the Jordan quiver.

## References

[BBD82] A. A. Beĭlinson, J. Bernstein, and P. Deligne, *Faisceaux pervers* (French), Analysis and topology on singular spaces, I (Luminy, 1981), Astérisque, vol. 100, Soc. Math. France, Paris, 1982, pp. 5–171. MR751966 (86g:32015)

[B88] R. Borcherds, *Generalized Kac-Moody algebras*, J. Algebra **115** (1988), no. 2, 501–512, DOI 10.1016/0021-8693(88)90275-X. MR943273 (89g:17004)

[D04] A. Dimca, *Sheaves in topology*, Universitext, Springer-Verlag, Berlin, 2004. MR2050072 (2005j:55002)

[H75] J. E. Humphreys, *Linear algebraic groups*, Springer-Verlag, New York-Heidelberg, 1975. Graduate Texts in Mathematics, No. 21. MR0396773 (53 #633)

[K95] S.-J. Kang, *Quantum deformations of generalized Kac-Moody algebras and their modules*, J. Algebra **175** (1995), no. 3, 1041–1066, DOI 10.1006/jabr.1995.1226. MR1341758 (96k:17023)

[KS06] S.-J. Kang and O. Schiffmann, *Canonical bases for quantum generalized Kac-Moody algebras*, Adv. Math. **200** (2006), no. 2, 455–478, DOI 10.1016/j.aim.2005.01.002. MR2200853 (2006i:17021)

[KS07] S.-J. Kang and O. Schiffmann, *Canonical bases for quantum generalized Kac-Moody algebras*, Adv. Math. **200** (2006), no. 2, 455–478, DOI 10.1016/j.aim.2005.01.002. MR2200853 (2006i:17021)

[KKS08] S.-J. Kang, M. Kashiwara, and O. Schiffmann, *Geometric construction of crystal bases for quantum generalized Kac-Moody algebras*, Adv. Math. **222** (2009), no. 3, 996–1015, DOI 10.1016/j.aim.2009.05.015. MR2553376 (2010h:17016)

[KKS09] S.-J. Kang, M. Kashiwara, and O. Schiffmann, *Geometric construction of highest weight crystals for quantum generalized Kac-Moody algebras*, Math. Ann. **354** (2012), no. 1, 193–208, DOI 10.1007/s00208-011-0725-5. MR2957624

[KS90] M. Kashiwara and P. Schapira, *Sheaves on manifolds*, Grundlehren der Mathematischen Wissenschaften [Fundamental Principles of Mathematical Sciences], vol. 292, Springer-Verlag, Berlin, 1990. With a chapter in French by Christian Houzel. MR1074006 (92a:58132)

[LL09] Y. Li and Z. Lin, *Canonical bases of Borcherds-Cartan type*, Nagoya Math. J. **194** (2009), 169–193. MR2536530 (2010j:17029)

[L90] G. Lusztig, *Canonical bases arising from quantized enveloping algebras*, J. Amer. Math. Soc. **3** (1990), no. 2, 447–498, DOI 10.2307/1990961. MR1035415 (90m:17023)

[L91] G. Lusztig, *Quivers, perverse sheaves, and quantized enveloping algebras*, J. Amer. Math. Soc. **4** (1991), no. 2, 365–421, DOI 10.2307/2939279. MR1088333 (91m:17018)

[L92] G. Lusztig, *Affine quivers and canonical bases*, Inst. Hautes Études Sci. Publ. Math. **76** (1992), 111–163. MR1215594 (94h:16021)

[L93a] G. Lusztig, *Introduction to quantum groups*, Progress in Mathematics, vol. 110, Birkhäuser Boston, Inc., Boston, MA, 1993. MR1227098 (94m:17016)

[L93b] G. Lusztig, *Tight monomials in quantized enveloping algebras*, Quantum deformations of algebras and their representations (Ramat-Gan, 1991/1992; Rehovot, 1991/1992), Israel Math. Conf. Proc., vol. 7, Bar-Ilan Univ., Ramat Gan, 1993, pp. 117–132. MR1261904 (95i:17016)

[L00] G. Lusztig, *Semicanonical bases arising from enveloping algebras*, Adv. Math. **151** (2000), no. 2, 129–139, DOI 10.1006/aima.1999.1873. MR1758244 (2001e:17033)

[M74] R. D. MacPherson, *Chern classes for singular algebraic varieties*, Ann. of Math. (2) **100** (1974), 423–432. MR0361141 (50 #13587)

Department of Mathematics, University at Buffalo, SUNY, Buffalo, NY 14260

*E-mail address*: `yqiang@buffalo.edu`

Proceedings of Symposia in Pure Mathematics
Volume **92**, 2016
http://dx.doi.org/10.1090/pspum/092/01581

# Krichever–Novikov type algebras. An introduction

Martin Schlichenmaier

ABSTRACT. Krichever–Novikov type algebras are generalizations of the Witt, Virasoro, affine Lie algebras, and their relatives to Riemann surfaces of arbitrary genus. We give the most important results about their structure, almost-grading and central extensions. This contribution is based on a sequence of introductory lectures delivered by the author at the Southeast Lie Theory Workshop 2012 in Charleston, U.S.A.

## 1. Introduction

The Witt algebra and its universal central extension the Virasoro algebra respectively are in some sense the simplest non-trivial examples of infinite dimensional Lie algebras[1]. Nevertheless, they already exhibit a very rich algebraic theory of representations. Furthermore, they appear prominently as symmetry algebras of a number of systems with infinitely many independent degrees of freedom. Their appearance in Conformal Field Theory (CFT) [**4**], [**106**] is well-known. But this is not their only application. At many other places in- and outside of mathematics they play an important role.

The algebras can be given by meromorphic objects on the Riemann sphere (genus zero) with possible poles only at $\{0, \infty\}$. For the Witt algebra these objects are vector fields. More generally, one obtains its central extension the Virasoro algebra, the current algebras and their central extensions the affine Kac-Moody algebras. For Riemann surfaces of higher genus, but still only for two points where poles are allowed, they were generalized by Krichever and Novikov [**56**], [**57**], [**58**] in 1986. In 1990 the author [**77**], [**78**], [**79**], [**80**] extended the approach further to the general multi-point case.

These extensions were not at all straight-forward. The main point was to introduce a replacement of the graded algebra structure present in the "classical" case. Krichever and Novikov found that an almost-grading, see Definition 5.1 below, will be enough to allow for the standard constructions in representation theory. In [**79**], [**80**] it was realized that a splitting of the set $A$ of points where poles are

2010 *Mathematics Subject Classification.* Primary 17B65; Secondary 14H15, 17B56, 17B66, 17B67, 17B68, 30F30, 81R10, 81T40.

Partial support by the Internal Research Project GEOMQ11, University of Luxembourg, and in the frame of the OPEN scheme of the Fonds National de la Recherche (FNR) with the project QUANTMOD O13/570706 is gratefully acknowledged.

[1]For a discussion about the correct naming, see the book [**37**].

allowed, into two disjoint non-empty subsets $A = I \cup O$ is crucial for introducing an almost-grading. The corresponding almost-grading was explicitly given. A Krichever-Novikov (KN) type algebra is an algebra of meromorphic objects with an almost-grading coming from such a splitting. In the classical situation there is only one such splitting (up to inversion) possible, Hence, there is only one almost-grading, which is indeed a grading.

From the algebraic point of view these KN type algebras are of course infinite dimensional Lie algebras, but they are still defined as (algebraic-)geometric objects. This point of view will be very helpful for further examinations.

As already mentioned above the crucial property is the almost-grading which replaces the honest grading in the Witt and Virasoro case. For a number of representation theoretic constructions the almost-grading will be enough. In contrast to the classical situation, where there is only one grading, we will have a finite set of non-equivalent gradings and new interesting phenomena show up. This is already true for the genus zero case (i.e. the Riemann sphere case) with more than two points where poles are allowed. These algebras will be only almost-graded, see e.g. [**81**], [**29**], [**30**].

Quite recently the book *Krichever–Novikov type algebras. Theory and applications* [**91**] by the author appeared. It gives a more or less complete treatment of the state of the art of the theory of KN type algebras including some applications. For more applications in direction of integrable systems and description of the Lax operator algebras see also the recent book *Current algebras on Riemann surfaces* [**103**] by Sheinman.

The goal of the lectures at the workshop and of this extended write-up was to give a gentle introduction to the KN type algebras in the multi-point setting, their definitions and main properties. Proofs are mostly omitted. They all can be found in [**91**], beside in the original works. KN type algebras carry a very rich representation theory. We have Verma modules, highest weight representations, Fermionic and Bosonic Fock representations, semi-infinite wedge forms, $b-c$ systems, Sugawara representations and vertex algebras. Due to space limitations as far as these representations are concerned we are very short here and have to refer to [**91**] too. There also a quite extensive list of references can be found, including articles published by physicists on applications in the field-theoretical context. This review extends and updates in certain respects the previous review [**89**] and has consequently some overlap with it.

It is a pleasure for me to thank the organisers Iana Anguelova, Ben Cox, Elizabeth Jurisich, and Oleg Smirnov of the Southeast Lie Theory Workshop 2012 in Charleston, for the invitation to this very inspiring activity. Particular thanks also goes to the audience for their lively feed-back. I acknowledge partial support from Internal Research Project GEOMQ11, University of Luxembourg, and in the frame of the OPEN scheme of the Fonds National de la Recherche (FNR) with the project QUANTMOD O13/570706.

## 2. A Reminder of the Virasoro Algebra and its Relatives

These algebras are in some sense the simplest non-trivial infinite dimensional Lie algebras. For the convenience of the reader we will start by recalling their conventional algebraic definitions.

**2.1. The Witt algebra.** The *Witt algebra* $\mathcal{W}$, sometimes also called Virasoro algebra without central term, is the Lie algebra generated as vector space over $\mathbb{C}$ by the basis elements $\{e_n \mid n \in \mathbb{Z}\}$ with Lie structure

$$[e_n, e_m] = (m-n)e_{n+m}, \quad n, m \in \mathbb{Z}. \tag{2.1}$$

The algebra $\mathcal{W}$ is more than just a Lie algebra. It is a graded Lie algebra. If we set for the degree $\deg(e_n) := n$ then

$$\mathcal{W} = \bigoplus_{n \in \mathbb{Z}} \mathcal{W}_n, \qquad \mathcal{W}_n = \langle e_n \rangle_{\mathbb{C}}. \tag{2.2}$$

Obviously, $\deg([e_n, e_m]) = \deg(e_n) + \deg(e_m)$.

Algebraically $\mathcal{W}$ can also be given as Lie algebra of derivations of the algebra of Laurent polynomials $\mathbb{C}[z, z^{-1}]$.

REMARK 2.1. In the purely algebraic context our field of definition $\mathbb{C}$ can be replaced by an arbitrary field $\mathbf{K}$ of characteristics 0. This concerns all cases in this section.

**2.2. The Virasoro algebra.** For the Witt algebra the universal central extension is the *Virasoro algebra* $\mathcal{V}$. As vector space it is the direct sum $\mathcal{V} = \mathbb{C} \oplus \mathcal{W}$. If we set for $x \in \mathcal{W}$, $\hat{x} := (0, x)$, and $t := (1, 0)$ then its basis elements are $\hat{e}_n$, $n \in \mathbb{Z}$ and $t$ with the Lie product [2].

$$[\hat{e}_n, \hat{e}_m] = (m-n)\hat{e}_{n+m} + \frac{1}{12}(n^3 - n)\delta_n^{-m}\, t, \qquad [\hat{e}_n, t] = [t, t] = 0, \tag{2.3}$$

for all $n, m \in \mathbb{Z}$. By setting $\deg(\hat{e}_n) := \deg(e_n) = n$ and $\deg(t) := 0$ the Lie algebra $\mathcal{V}$ becomes a graded algebra. The algebra $\mathcal{W}$ will only be a subspace, not a subalgebra of $\mathcal{V}$. But it will be a quotient. Up to equivalence of central extensions and rescaling the central element $t$, this is beside the trivial (splitting) central extension the only central extension of $\mathcal{W}$.

**2.3. The affine Lie algebra.** Given $\mathfrak{g}$ a finite-dimensional Lie algebra (i.e. a finite-dimensional simple Lie algebra) then the tensor product of $\mathfrak{g}$ with the associative algebra of Laurent polynomials $\mathbb{C}[z, z^{-1}]$ carries a Lie algebra structure via

$$[x \otimes z^n, y \otimes z^m] := [x, y] \otimes z^{n+m}. \tag{2.4}$$

This algebra is called *current algebra* or *loop algebra* and denoted by $\overline{\mathfrak{g}}$. Again we consider central extensions. For this let $\beta$ be a symmetric, bilinear form for $\mathfrak{g}$ which is invariant (e.g. $\beta([x, y], z) = \beta(x, [y, z])$ for all $x, y, z \in \mathfrak{g}$). Then a central extension is given by

$$[\widehat{x \otimes z^n}, \widehat{y \otimes z^m}] := \widehat{[x, y] \otimes z^{n+m}} - \beta(x, y) \cdot m\, \delta_n^{-m} \cdot t. \tag{2.5}$$

This algebra is denoted by $\widehat{\mathfrak{g}}$ and called *affine Lie algebra*. With respect to the classification of Kac-Moody Lie algebras, in the case of a simple $\mathfrak{g}$ they are exactly the Kac-Moody algebras of affine type, [**47**], [**48**], [**68**].

[2]Here $\delta_k^l$ is the Kronecker delta which is equal to 1 if $k = l$, otherwise zero.

**2.4. The Lie superalgebra.** To complete the description I will introduce the Lie superalgebra of Neveu-Schwarz type. The centrally extended superalgebra has as basis (we drop the $\hat{}$)

$$e_n,\ n \in \mathbb{Z}, \quad \varphi_m,\ m \in \mathbb{Z} + \frac{1}{2}, \quad t \tag{2.6}$$

with structure equations

$$\begin{aligned} [e_n, e_m] &= (m-n)e_{m+n} + \frac{1}{12}(n^3 - n)\,\delta_n^{-m}\,t, \\ [e_n, \varphi_m] &= (m - \frac{n}{2})\,\varphi_{m+n}, \\ [\varphi_n, \varphi_m] &= e_{n+m} - \frac{1}{6}\,(n^2 - \frac{1}{4})\,\delta_n^{-m}\,t. \end{aligned} \tag{2.7}$$

By "setting $t = 0$" we obtain the non-extended superalgebra. The elements $e_n$ (and $t$) are a basis of the subspace of even elements, the elements $\varphi_m$ a basis of the subspace of odd elements.

These algebras are indeed Lie superalgebras. For completeness I recall their definition here.

REMARK 2.2 (Definition of a Lie superalgebra). Let $\mathcal{S}$ be a vector space which is decomposed into even and odd elements $\mathcal{S} = \mathcal{S}_{\bar{0}} \oplus \mathcal{S}_{\bar{1}}$, i.e. $\mathcal{S}$ is a $\mathbb{Z}/2\mathbb{Z}$-graded vector space. Furthermore, let $[.,.]$ be a $\mathbb{Z}/2\mathbb{Z}$-graded bilinear map $\mathcal{S} \times \mathcal{S} \to \mathcal{S}$ such that for elements $x, y$ of pure parity

$$[x, y] = -(-1)^{\bar{x}\bar{y}}[y, x]. \tag{2.8}$$

Here $\bar{x}$ is the parity of $x$, etc. These conditions say that

$$[\mathcal{S}_{\bar{0}}, \mathcal{S}_{\bar{0}}] \subseteq \mathcal{S}_{\bar{0}}, \qquad [\mathcal{S}_{\bar{0}}, \mathcal{S}_{\bar{1}}] \subseteq \mathcal{S}_{\bar{1}}, \qquad [\mathcal{S}_{\bar{1}}, \mathcal{S}_{\bar{1}}] \subseteq \mathcal{S}_{\bar{0}}, \tag{2.9}$$

and that $[x, y]$ is symmetric for $x$ and $y$ odd, otherwise anti-symmetric. Now $\mathcal{S}$ is a *Lie superalgebra* if in addition the *super-Jacobi identity* (for $x, y, z$ of pure parity)

$$(-1)^{\bar{x}\bar{z}}[x, [y, z]] + (-1)^{\bar{y}\bar{x}}[y, [z, x]] + (-1)^{\bar{z}\bar{y}}[z, [x, y]] = 0 \tag{2.10}$$

is valid. As long as the type of the arguments is different from *(even, odd, odd)* all signs can be put to $+1$ and we obtain the form of the usual Jacobi identity. In the remaining case we get

$$[x, [y, z]] + [y, [z, x]] - [z, [x, y]] = 0. \tag{2.11}$$

By the definitions $\mathcal{S}_0$ is a Lie algebra.

## 3. The Geometric Picture

In the previous section I gave the Virasoro algebra and its relatives by purely algebraic means, i.e. by basis elements and structure equations. The full importance and strength will become more visible in a geometric context. Also from this geometric realization the need for a generalization as obtained via the Krichever–Novikov type algebras will become evident.

**3.1. The geometric realizations of the Virasoro algebra.** One of its realizations is a complexification of the Lie algebra of polynomial vector fields $Vect_{pol}(S^1)$ on the circle $S^1$. This is a subalgebra of $Vect(S^1)$, the Lie algebra of all $C^\infty$ vector fields on the circle. In this realization the basis elements are

$$e_n := -\mathrm{i} \exp \mathrm{i}\, n\, \varphi \frac{d}{d\varphi}, \qquad n \in \mathbb{Z}\,, \tag{3.1}$$

where $\varphi$ is the angle variable. The Lie product is the usual Lie bracket of vector fields.

If we extend analytically these generators to the whole punctured complex plane we obtain $e_n = z^{n+1}\frac{d}{dz}$. This gives another realization of the Witt algebra as the algebra of those meromorphic vector fields on the Riemann sphere $S^2 = \mathbb{P}^1(\mathbb{C})$ which are holomorphic outside $\{0\}$ and $\{\infty\}$.

Let $z$ be the (quasi) global coordinate $z$ (quasi, because it is not defined at $\infty$). Let $w = 1/z$ be the local coordinate at $\infty$. A global meromorphic vector field $v$ on $\mathbb{P}^1(\mathbb{C})$ will be given on the corresponding subsets where $z$ respectively $w$ are defined as

$$v = \left(v_1(z)\frac{d}{dz}\,,\; v_2(w)\frac{d}{dw}\right), \qquad v_2(w) = -v_1(z(w))w^2. \tag{3.2}$$

The function $v_1$ will determine the vector field $v$. Hence, we will usually just give $v_1$ and in fact identify the vector field $v$ with its local representing function $v_1$, which we will denote by the same letter.

For the Lie bracket we calculate

$$[v, u] = \left(v\frac{d}{dz}u - u\frac{d}{dz}v\right)\frac{d}{dz}. \tag{3.3}$$

The space of all meromorphic vector fields constitute a Lie algebra.

The subspace of those meromorphic vector fields which are holomorphic outside of $\{0, \infty\}$ is a Lie subalgebra. Its elements can be given as

$$v(z) = f(z)\frac{d}{dz} \tag{3.4}$$

where $f$ is a meromorphic function on $\mathbb{P}^1(\mathbb{C})$, which is holomorphic outside $\{0, \infty\}$. Those are exactly the Laurent polynomials $\mathbb{C}[z, z^{-1}]$. Consequently, this subalgebra has the set $\{e_n, n \in \mathbb{Z}\}$ as basis. The Lie product is the same and the subalgebra can be identified with the Witt algebra $\mathcal{W}$. The subalgebra of global holomorphic vector fields is $\langle e_{-1}, e_0, e_1\rangle_{\mathbb{C}}$. It is isomorphic to the Lie algebra $\mathfrak{sl}(2, \mathbb{C})$.

Similarly, the algebra $\mathbb{C}[z, z^{-1}]$ can be given as the algebra of meromorphic functions on $S^2 = \mathbb{P}(\mathbb{C})$ holomorphic outside of $\{0, \infty\}$.

**3.2. Arbitrary genus generalizations.** In the geometric setup for the Virasoro algebra the objects are defined on the Riemann sphere and might have poles at most at two fixed points. For a global operator approach to conformal field theory and its quantization this is not sufficient. One needs Riemann surfaces of arbitrary genus. Moreover, one needs more than two points were singularities are allowed [3]. Such a generalizations were initiated by Krichever and Novikov [**56**], [**57**], [**58**], who considered arbitrary genus and the two-point case. As far as the current algebras

[3]The singularities correspond to points where free fields are entering the region of interaction or leaving it. In particular from the very beginning there is a natural decomposition of the set of points into two disjoint subsets.

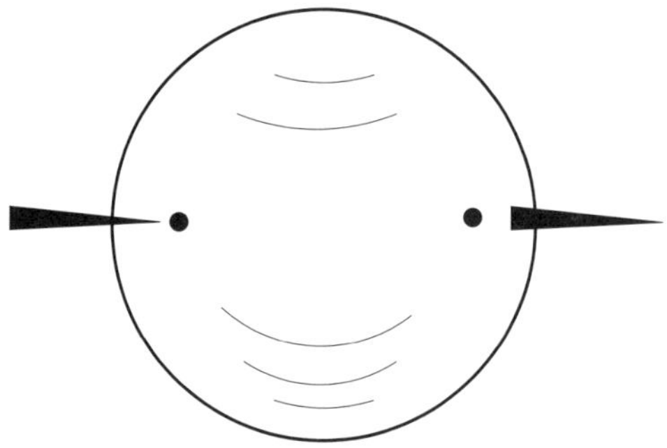

FIGURE 1. Riemann surface of genus zero with one incoming and one outgoing point.

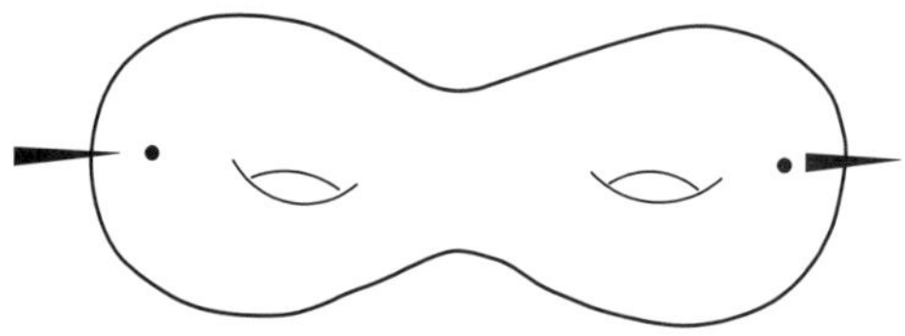

FIGURE 2. Riemann surface of genus two with one incoming and one outgoing point.

are concerned see also Sheinman [**97**], [**98**], [**99**], [**100**]. The multi-point case was systematically examined by the author [**77**], [**78**], [**79**], [**80**], [**81**] [**82**], [**83**], [**84**]. For some related approach see also Sadov [**76**].

For the whole contribution let $\Sigma$ be a compact Riemann surface without any restriction for the genus $g = g(\Sigma)$. Furthermore, let $A$ be a finite subset of $\Sigma$. Later we will need a splitting of $A$ into two non-empty disjoint subsets $I$ and $O$, i.e. $A = I \cup O$. Set $N := \#A$, $K := \#I$, $M := \#O$, with $N = K + M$. More precisely, let

$$I = (P_1, \dots, P_K), \quad \text{and} \quad O = (Q_1, \dots, Q_M) \tag{3.5}$$

be disjoint ordered tuples of distinct points ("marked points", "punctures") on the Riemann surface. In particular, we assume $P_i \neq Q_j$ for every pair $(i, j)$. The points in $I$ are called the *in-points*, the points in $O$ the *out-points*. Occasionally, we consider $I$ and $O$ simply as sets.

Sometimes we refer to the classical situation. By this we understand

$$\Sigma = \mathbb{P}^1(\mathbb{C}) = S^2, \quad I = \{z = 0\}, \quad O = \{z = \infty\}, \tag{3.6}$$

and the situation considered in Section 3.1.

The figures should indicate the geometric picture. Figure 1 shows the classical situation. Figure 2 is genus 2, but still two-point situation. Finally, in Figure 3 the case of a Riemann surface of genus 2 with two incoming points and one outgoing point is visualized.

REMARK 3.1. We stress the fact, that these generalizations are needed also in the case of genus zero if one considers more than two points. Even there in the case of three points interesting algebras show up. See also [**92**].

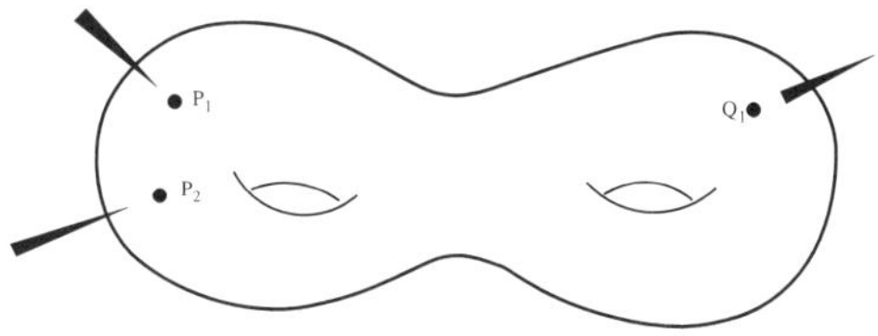


FIGURE 3. Riemann surface of genus two with two incoming points and one outgoing point.

**3.3. Meromorphic forms.** To introduce the elements of the generalized algebras we first have to discuss forms of certain (conformal) weights. Recall that $\Sigma$ is a compact Riemann surface of genus $g \ge 0$. Let $A$ be a fixed finite subset of $\Sigma$. In fact we could allow for this and the following sections (as long as we do not talk about almost-grading) that $A$ is an arbitrary subset. This includes the extremal cases $A = \emptyset$ or $A = \Sigma$.

Let $\mathcal{K} = \mathcal{K}_\Sigma$ be the canonical line bundle of $\Sigma$. Its local sections are the local holomorphic differentials. If $P \in \Sigma$ is a point and $z$ a local holomorphic coordinate at $P$ then a local holomorphic differential can be written as $f(z)dz$ with a local holomorphic function $f$ defined in a neighborhood of $P$. A global holomorphic section can be described locally in coordinates $(U_i, z_i)_{i\in J}$ by a system of local holomorphic functions $(f_i)_{i\in J}$, which are related by the transformation rule induced by the coordinate change map $z_j = z_j(z_i)$ and the condition $f_i dz_i = f_j dz_j$. This yields

$$f_j = f_i \cdot \left(\frac{dz_j}{dz_i}\right)^{-1}. \tag{3.7}$$

A meromorphic section of $\mathcal{K}$, i.e. a *meromorphic differential* is given as a collection of local meromorphic functions $(h_i)_{i\in J}$ with respect to a coordinate covering for which the transformation law (3.7) is true. We will not make any distinction between the canonical bundle and its sheaf of sections, which is a locally free sheaf of rank 1.

In the following $\lambda$ is either an integer or a half-integer. If $\lambda$ is an integer then
(1) $\mathcal{K}^\lambda = \mathcal{K}^{\otimes\lambda}$ for $\lambda > 0$,
(2) $\mathcal{K}^0 = \mathcal{O}$, the trivial line bundle, and
(3) $\mathcal{K}^\lambda = (\mathcal{K}^*)^{\otimes(-\lambda)}$ for $\lambda < 0$.
Here $\mathcal{K}^*$ denotes the dual line bundle of the canonical line bundle. This is the holomorphic tangent line bundle, whose local sections are the holomorphic tangent vector fields $f(z)(d/dz)$.

If $\lambda$ is a half-integer, then we first have to fix a "square root" of the canonical line bundle, sometimes called a *theta characteristics*. This means we fix a line bundle $L$ for which $L^{\otimes 2} = \mathcal{K}$. After such a choice of $L$ is done we set $\mathcal{K}^\lambda := \mathcal{K}^\lambda_L := L^{\otimes 2\lambda}$. In most cases we will drop the mentioning of $L$, but we have to keep the choice in mind. The fine-structure of the algebras we are about to define will depend on the choice. But the main properties will remain the same.

REMARK 3.2. A Riemann surface of genus $g$ has exactly $2^{2g}$ non-isomorphic square roots of $\mathcal{K}$. For $g = 0$ we have $\mathcal{K} = \mathcal{O}(-2)$, and $L = \mathcal{O}(-1)$, the tautological bundle, is the unique square root. Already for $g = 1$ we have four non-isomorphic ones. As in this case $\mathcal{K}=\mathcal{O}$ one solution is $L_0=\mathcal{O}$. But we have also other bundles $L_i$, $i=1,2,3$. Note that $L_0$ has a nonvanishing global holomorphic section, whereas this is not the case for $L_1, L_2$ and $L_3$. In general, depending on the parity of the dimension of the space of globally holomorphic sections, i.e. of $\dim \mathrm{H}^0(\Sigma, L)$, one distinguishes even and odd theta characteristics $L$. For $g=1$ the bundle $\mathcal{O}$ is an odd, the others are even theta characteristics. The choice of a theta characteristic is also called a spin structure on $\Sigma$ [**3**].

We set

$$\mathcal{F}^\lambda(A) := \{f \text{ is a global meromorphic section of } \mathcal{K}^\lambda \mid \\ f \text{ is holomorphic on } \Sigma\setminus A\}. \tag{3.8}$$

Obviously this is a $\mathbb{C}$-vector space. To avoid cumbersome notation, we will often drop the set $A$ in the notation if $A$ is fixed and/or clear from the context. Recall that in the case of half-integer $\lambda$ everything depends on the theta characteristic $L$.

DEFINITION 3.3. The elements of the space $\mathcal{F}^\lambda(A)$ are called *meromorphic forms of weight* $\lambda$ (with respect to the theta characteristic $L$).

REMARK 3.4. In the two extremal cases for the set $A$ we obtain $\mathcal{F}^\lambda(\emptyset)$ the global holomorphic forms, and $\mathcal{F}^\lambda(\Sigma)$ all meromorphic forms. By compactness each $f\in\mathcal{F}^\lambda(\Sigma)$ will have only finitely many poles. In the case that $f\not\equiv 0$ it will also have only finitely many zeros.

Let us assume that $z_i$ and $z_j$ are local coordinates for the same point $P\in\Sigma$. For the bundle $\mathcal{K}$ both $dz_i$ and $dz_j$ are frames. If we represent the same form $f$ locally by $f_i dz_i$ and $f_j dz_j$ then we conclude from (3.7) that $f_j = f_i\cdot c_1$ and that the transition function $c_1$ is given by

$$c_1 = \left(\frac{dz_j}{dz_i}\right)^{-1} = \frac{dz_i}{dz_j}. \tag{3.9}$$

For sections of $\mathcal{K}^\lambda$ with $\lambda\in\mathbb{Z}$ the transition functions are $c_\lambda=(c_1)^\lambda$. The corresponding is true also for half-integer $\lambda$. In this case the basic transition function of the chosen theta characteristics $L$ is given as $c_{1/2}$ and all others are integer powers of it. Symbolically, we write $\sqrt{dz_i}$ or $(dz)^{1/2}$ for the local frame, keeping in mind that the signs for the square root is not uniquely defined but depends on the bundle $L$.

If $f$ is a meromorphic $\lambda$-form it can be represented locally by meromorphic functions $f_i$. If $f\not\equiv 0$ the local representing functions have only finitely many zeros and poles. Whether a point $P$ is a zero or a pole of $f$ does not depend on the coordinate $z_i$ chosen, as the transition function $c_\lambda$ will be a nonvanishing function. Moreover, we can define for $P\in\Sigma$ the *order*

$$\mathrm{ord}_P(f) := \mathrm{ord}_P(f_i), \tag{3.10}$$

where $\mathrm{ord}_P(f_i)$ is the lowest nonvanishing order in the Laurent series expansion of $f_i$ in the variable $z_i$ around $P$. It will not depend on the coordinate $z_i$ chosen. The order $\mathrm{ord}_P(f)$ is (strictly) positive if and only if $P$ is a zero of $f$. It is negative if and only if $P$ is a pole of $f$. Moreover, its value gives the order of the zero and pole respectively. By compactness our Riemann surface $\Sigma$ can be covered by finitely many coordinate patches. Hence, $f$ can only have finitely many zeros and poles. We define the *(sectional) degree* of $f$ to be

$$\operatorname{sdeg}(f) := \sum_{P\in\Sigma} \operatorname{ord}_P(f). \tag{3.11}$$

PROPOSITION 3.5. *Let $f \in \mathcal{F}^\lambda$, $f \not\equiv 0$ then*

$$\operatorname{sdeg}(f) = 2\lambda(g-1). \tag{3.12}$$

For this and related results see [**85**].

## 4. Algebraic Structures

Next we introduce algebraic operations on the vector space of meromorphic forms of arbitrary weights. This space is obtained by summing over all weights

$$\mathcal{F} := \bigoplus_{\lambda\in\frac{1}{2}\mathbb{Z}} \mathcal{F}^\lambda. \tag{4.1}$$

The basic operations will allow us to introduce finally the algebras we are heading for.

**4.1. Associative structure.** In this section $A$ is still allowed to be an arbitrary subset of points in $\Sigma$. We will drop the subset $A$ in the notation. The natural map of the locally free sheaves of rang one

$$\mathcal{K}^\lambda \times \mathcal{K}^\nu \to \mathcal{K}^\lambda \otimes \mathcal{K}^\nu \cong \mathcal{K}^{\lambda+\nu}, \quad (s,t) \mapsto s\otimes t, \tag{4.2}$$

defines a bilinear map

$$\cdot : \mathcal{F}^\lambda \times \mathcal{F}^\nu \to \mathcal{F}^{\lambda+\nu}. \tag{4.3}$$

With respect to local trivialisations this corresponds to the multiplication of the local representing meromorphic functions

$$(s\,dz^\lambda, t\,dz^\nu) \mapsto s\,dz^\lambda \cdot t\,dz^\nu = s\cdot t\,dz^{\lambda+\nu}. \tag{4.4}$$

If there is no danger of confusion then we will mostly use the same symbol for the section and for the local representing function.

The following is obvious

PROPOSITION 4.1. *The space $\mathcal{F}$ is an associative and commutative graded (over $\frac{1}{2}\mathbb{Z}$) algebra. Moreover, $\mathcal{A} = \mathcal{F}^0$ is a subalgebra and the $\mathcal{F}^\lambda$ are modules over $\mathcal{A}$.*

Of course, $\mathcal{A}$ is the algebra of those meromorphic functions on $\Sigma$ which are holomorphic outside of $A$. In the case of $A = \emptyset$, it is the algebra of global holomorphic functions. By compactness, these are only the constants, hence $\mathcal{A}(\emptyset) = \mathbb{C}$. In the case of $A = \Sigma$ it is the field of all meromorphic functions $\mathcal{M}(\Sigma)$.

**4.2. Lie and Poisson algebra structure.** Next we define a Lie algebra structure on the space $\mathcal{F}$. The structure is induced by the map

$$\mathcal{F}^\lambda \times \mathcal{F}^\nu \to \mathcal{F}^{\lambda+\nu+1}, \qquad (e,f) \mapsto [e,f], \tag{4.5}$$

which is defined in local representatives of the sections by

$$(e\,dz^\lambda, f\,dz^\nu) \mapsto [e\,dz^\lambda, f\,dz^\nu] := \left((-\lambda)e\frac{df}{dz} + \nu f\frac{de}{dz}\right) dz^{\lambda+\nu+1}, \tag{4.6}$$

and bilinearly extended to $\mathcal{F}$. Of course, we have to show the following

PROPOSITION 4.2 ([**91**, Prop. 2.6 and 2.7]). *The prescription* $[.,.]$ *given by* (4.6) *is well-defined and defines a Lie algebra structure on the vector space* $\mathcal{F}$.

PROOF. It is a nice exercise to show that the expressions on the right hand side in (4.6) transform correctly as local representing functions for $\lambda+\nu+1$ forms. That the Jacobi identity is true follows from direct calculations, see the above reference. □

PROPOSITION 4.3 ([**91**, Prop. 2.8]). *The subspace* $\mathcal{L}=\mathcal{F}^{-1}$ *is a Lie subalgebra, and the* $\mathcal{F}^\lambda$*'s are Lie modules over* $\mathcal{L}$.

PROOF. For illustration we supply the proof. For $\lambda=\nu=-1$ we get as weight of the Lie product $\lambda+\nu+1=-1$, hence the subspace is closed under the bracket and a Lie subalgebra. For $e\in\mathcal{L}$ and $h\in\mathcal{F}^\lambda$ the Lie module structure is given by $e\,.\,h:=[e,h]\in\mathcal{F}^\lambda$. The Jacobi identity for $e,f\in\mathcal{L}$ and $h\in\mathcal{F}^\lambda$ reads as

$$0=[[e,f],h]+[[f,h],e]+[[h,e],f]=[e,f]\,.\,h-e\,.\,(f\,.\,h)+f\,.\,(e\,.\,h). \tag{4.7}$$

This is exactly the condition for $\mathcal{F}^\lambda$ being a Lie module. □

DEFINITION 4.4. An algebra $(\mathcal{B},\cdot,[.,.])$ such that $\cdot$ defines the structure of an associative algebra on $\mathcal{B}$ and $[.,.]$ defines the structure of a Lie algebra on $\mathcal{B}$ is called a *Poisson algebra* if and only if the Leibniz rule is true, i.e.

$$\forall e,f,g\in\mathcal{B}:\ [e,f\cdot g]=[e,f]\cdot g+f\cdot[e,g]. \tag{4.8}$$

In other words, via the Lie product $[.,.]$ the elements of the algebra act as derivations on the associative structure. The reader should be warned that $[.,.]$ is not necessarily the commutator of the algebra $(\mathcal{B},\cdot)$.

THEOREM 4.5 ([**91**, Thm. 2.10]). *The space* $\mathcal{F}$ *with respect to* $\cdot$ *and* $[.,.]$ *is a Poisson algebra.*

Next we consider important substructures. We already encountered the subalgebras $\mathcal{A}$ and $\mathcal{L}$. But there are more structures around.

**4.3. The vector field algebra and the Lie derivative.** First we look again on the Lie subalgebra $\mathcal{L}=\mathcal{F}^{-1}$. Here the Lie action respect the homogeneous subspaces $\mathcal{F}^\lambda$. As forms of weight $-1$ are vector fields, it could also be defined as the Lie algebra of those meromorphic vector fields on the Riemann surface $\Sigma$ which are holomorphic outside of $A$. For vector fields we have the usual Lie bracket and the usual Lie derivative for their actions on forms. For the vector fields we have (again naming the local functions with the same symbol as the section) for $e,f\in\mathcal{L}$

$$[e,f]_| = [e(z)\frac{d}{dz},f(z)\frac{d}{dz}]=\left(e(z)\frac{df}{dz}(z)-f(z)\frac{de}{dz}(z)\right)\frac{d}{dz}\,. \tag{4.9}$$

For the Lie derivative we get

$$\nabla_e(f)_| = L_e(g)_| = e\,.\,g_| = \left(e(z)\frac{df}{dz}(z)+\lambda f(z)\frac{de}{dz}(z)\right)\frac{d}{dz}\,. \tag{4.10}$$

Obviously, these definitions coincide with the definitions already given above. But now we obtained a geometric interpretation.

**4.4. The algebra of differential operators.** If we look at $\mathcal{F}$, considered as Lie algebra, more closely, we see that $\mathcal{F}^0$ is an abelian Lie subalgebra and the vector space sum $\mathcal{F}^0 \oplus \mathcal{F}^{-1} = \mathcal{A} \oplus \mathcal{L}$ is also a Lie subalgebra. In an equivalent way this can also be constructed as semidirect sum of $\mathcal{A}$ considered as abelian Lie algebra and $\mathcal{L}$ operating on $\mathcal{A}$ by taking the derivative.

DEFINITION 4.6. The *Lie algebra of differential operators of degree* $\leq 1$ is defined as the semidirect sum of $\mathcal{A}$ with $\mathcal{L}$ and is denoted by $\mathcal{D}^1$.

In terms of elements the Lie product is

$$[(g,e),(h,f)] = (e\,.\,h - f\,.\,g\,,\,[e,f]). \tag{4.11}$$

Using the fact, that $\mathcal{A}$ is an abelian subalgebra in $\mathcal{F}$ this is exactly the definition for the Lie product given for this algebra. Hence, $\mathcal{D}^1$ is a Lie algebra.

The projection on the second factor $(g,e) \mapsto e$ is a Lie homomorphism and we obtain a short exact sequences of Lie algebras

$$0 \longrightarrow \mathcal{A} \longrightarrow \mathcal{D}^1 \longrightarrow \mathcal{L} \longrightarrow 0\,. \tag{4.12}$$

Hence $\mathcal{A}$ is an (abelian) Lie ideal of $\mathcal{D}^1$ and $\mathcal{L}$ a quotient Lie algebra. Obviously $\mathcal{L}$ is also a subalgebra of $\mathcal{D}^1$.

PROPOSITION 4.7. *The vector space* $\mathcal{F}^\lambda$ *becomes a Lie module over* $\mathcal{D}^1$ *by the operation*

$$(g,e).f := g\cdot f + e.f, \quad (g,e) \in \mathcal{D}^1(A), f \in \mathcal{F}^\lambda(A). \tag{4.13}$$

**4.5. Differential operators of all degree.** We want to consider also differential operators of arbitrary degree acting on $\mathcal{F}^\lambda$. This is obtained via some universal constructions. First we consider the universal enveloping algebra $U(\mathcal{D}^1)$. We denote its multiplication by $\odot$ and its unit by $\mathbf{1}$.

The universal enveloping algebra contains many elements which act in the same manner on $\mathcal{F}^\lambda$. For example, if $h_1$ and $h_2$ are functions different from constants then $h_1 \cdot h_2$ and $h_1 \odot h_2$ are different elements of $U(\mathcal{D}^1)$. Nevertheless, they act in the same way on $\mathcal{F}^\lambda$.

Hence we will divide out further relations

$$\mathcal{D} = U(\mathcal{D}^1)/J, \quad \text{respectively} \quad \mathcal{D}_\lambda = U(\mathcal{D}^1)/J_\lambda \tag{4.14}$$

with the two-sided ideals

$$J := (\, a\odot b - a\cdot b,\ \mathbf{1} - 1 \mid a,b \in \mathcal{A}\,),$$
$$J_\lambda := (\, a\odot b - a\cdot b,\ \mathbf{1} - 1,\ a\odot e - a\cdot e + \lambda\, e\,.\,a \mid a,b \in \mathcal{A}, e \in L\,).$$

By universality the $\mathcal{F}^\lambda$ are modules over $U(\mathcal{D}^1)$. The relations in $J$ are fulfilled as $(a\odot b)\cdot f = a\cdot(b\cdot f) = (a\cdot b)\cdot f$. Hence for all $\lambda$ the $\mathcal{F}^\lambda$ are modules over $\mathcal{D}$.

If $\lambda$ is fixed then the additional relations in $J_\lambda$ are also true. For this we calculate

$$\begin{aligned}
&(a\odot e)\,.\,f = a\cdot(e\,.\,f) = ae\frac{df}{dz} + \lambda af\frac{de}{dz},\\
&(a\cdot e)\,.\,f = (ae)\frac{df}{dz} + \lambda f\frac{d(ae)}{dz} = (ae)\frac{df}{dz} + \lambda fe\frac{da}{dz} + \lambda fa\frac{de}{dz},\\
&\lambda(e\,.\,a)\cdot f = \lambda\, ef\frac{da}{dz}.
\end{aligned} \tag{4.15}$$

Hence,

$$(a \odot e - a \cdot e + \lambda(e\,.\,a))\,.\,f = 0. \tag{4.16}$$

Consequently, for a fixed $\lambda$ the space $\mathcal{F}^\lambda$ is a module over $\mathcal{D}_\lambda$.

DEFINITION 4.8 ([**36**, IV,16.8,16.11] and [**9**]). A linear map $D : \mathcal{F}^\lambda \to \mathcal{F}^\lambda$ is called an (algebraic) *differential operator* of degree $\le n$ with $n \ge 0$ if and only if

(1) If $n = 0$ then $D = b$, the multiplication with a function $b \in \mathcal{A}$.
(2) If $n > 0$, then for $a \in \mathcal{A}$ (considered as multiplication operator)

$$[D, a] \quad : \quad \mathcal{F}^\lambda \to \mathcal{F}^\lambda$$

is a differential operator of degree $\le (n-1)$.

Let $\mathrm{Diff}^{(n)}(\mathcal{F}^\lambda)$ be the subspace of all differential operators on $\mathcal{F}^\lambda$ of degree $\le n$. By composing the operators

$$\mathrm{Diff}(\mathcal{F}^\lambda) := \bigcup_{n \in \mathbb{N}_0} \mathrm{Diff}^{(n)}(\mathcal{F}^\lambda)$$

becomes an associative algebra which is a subalgebra of $\mathrm{End}(\mathcal{F}^\lambda)$.

Let $D \in \mathcal{D}$ and assume that $D$ is one of the generators

$$D = a_0 \odot e_1 \odot a_1 \odot e_2 \odot \cdots \odot a_{n-1} \odot e_n \odot a_n \tag{4.17}$$

with $e_i \in \mathcal{L}$ and $a_i \in \mathcal{A}$ (written as element in $U(\mathcal{D}^1)$).

PROPOSITION 4.9 ([**91**, Prop. 2.14]). *Every element $D \in \mathcal{D}$ respectively of $\mathcal{D}_\lambda$ of the form (4.17) operates as (algebraic) differential operator of degree $\le n$ on $\mathcal{F}^\lambda$.*

In fact, we get (associative) algebra homomorphisms

$$\mathcal{D} \to \mathrm{Diff}(\mathcal{F}^\lambda), \qquad \mathcal{D}_\lambda \to \mathrm{Diff}(\mathcal{F}^\lambda)\ . \tag{4.18}$$

In case the set $A$ of points where poles are allowed is finite and non-empty the complement $\Sigma \setminus A$ is affine [**39**, p.297]. Hence, as shown in [**36**] every differential operator can be obtained by successively applying first order operators, i.e. by applying elements from $U(\mathcal{D}^1)$. In other words the homomorphisms (4.18) are surjective.

**4.6. Lie superalgebras of half forms.** Recall from Remark 2.2 the definition of a Lie superalgebra.

With the help of our associative product (4.2) we will obtain examples of Lie superalgebras. First we consider

$$\cdot\ \mathcal{F}^{-1/2} \times \mathcal{F}^{-1/2} \to \mathcal{F}^{-1} = \mathcal{L}\,, \tag{4.19}$$

and introduce the vector space $\mathcal{S}$ with the product

$$\mathcal{S} := \mathcal{L} \oplus \mathcal{F}^{-1/2}, \quad [(e, \varphi), (f, \psi)] := ([e, f] + \varphi \cdot \psi, e\,.\,\varphi - f\,.\,\psi). \tag{4.20}$$

The elements of $\mathcal{L}$ are denoted by $e, f, \ldots$, and the elements of $\mathcal{F}^{-1/2}$ by $\varphi, \psi, \ldots$.

The definition (4.20) can be reformulated as an extension of $[.,.]$ on $\mathcal{L}$ to a super-bracket (denoted by the same symbol) on $\mathcal{S}$ by setting

$$[e, \varphi] := -[\varphi, e] := e\,.\,\varphi = (e\frac{d\varphi}{dz} - \frac{1}{2}\varphi\frac{de}{dz})(dz)^{-1/2} \tag{4.21}$$

and

$$[\varphi, \psi] := \varphi \cdot \psi\ . \tag{4.22}$$

We call the elements of $\mathcal{L}$ elements of even parity, and the elements of $\mathcal{F}^{-1/2}$ elements of odd parity. For such elements $x$ we denote by $\bar{x} \in \{\bar{0}, \bar{1}\}$ their parity.

The sum (4.20) can also be described as $\mathcal{S} = \mathcal{S}_{\bar{0}} \oplus \mathcal{S}_{\bar{1}}$, where $\mathcal{S}_{\bar{i}}$ is the subspace of elements of parity $\bar{i}$.

PROPOSITION 4.10 ([**91**, Prop. 2.15]). *The space $\mathcal{S}$ with the above introduced parity and product is a Lie superalgebra.*

REMARK 4.11. The choice of the theta characteristics corresponds to choosing a *spin structure* on $\Sigma$. For the relation of the Neveu-Schwarz superalgebra to the geometry of graded Riemann surfaces see Bryant [**17**].

**4.7. Jordan superalgebra.** Leidwanger and Morier-Genoux introduced in [**61**] a *Jordan superalgebra* in our geometric setting. They put

$$\mathcal{J} := \mathcal{F}^0 \oplus \mathcal{F}^{-1/2} = \mathcal{J}_{\bar{0}} \oplus \mathcal{J}_{\bar{1}}. \tag{4.23}$$

Recall that $\mathcal{A} = \mathcal{F}^0$ is the associative algebra of meromorphic functions. They define the (Jordan) product $\circ$ via the algebra structures for the spaces $\mathcal{F}^\lambda$ by

$$\begin{aligned} f \circ g &:= f \cdot g \quad && \in \mathcal{F}^0, \\ f \circ \varphi &:= f \cdot \varphi \quad && \in \mathcal{F}^{-1/2}, \\ \varphi \circ \psi &:= [\varphi, \psi] \quad && \in \mathcal{F}^0. \end{aligned} \tag{4.24}$$

By rescaling the second definition with the factor $1/2$ one obtains a *Lie anti-algebra* as introduced by Ovsienko [**72**]. See [**61**] for more details and additional results on representations.

**4.8. Higher genus current algebras.** We fix an arbitrary finite-dimensional complex Lie algebra $\mathfrak{g}$. Our goal is to generalize the classical current algebra to higher genus. For this let $(\Sigma, A)$ be the geometrical data consisting of the Riemann surface $\Sigma$ and the subset of points $A$ used to define $\mathcal{A}$, the algebra of meromorphic functions which are holomorphic outside of the set $A \subseteq \Sigma$.

DEFINITION 4.12. The *higher genus current algebra* associated to the Lie algebra $\mathfrak{g}$ and the geometric data $(\Sigma, A)$ is the Lie algebra $\overline{\mathfrak{g}} = \overline{\mathfrak{g}}(A) = \overline{\mathfrak{g}}(\Sigma, A)$ given as vector space by $\overline{\mathfrak{g}} = \mathfrak{g} \otimes_{\mathbb{C}} \mathcal{A}$ with the Lie product

$$[x \otimes f, y \otimes g] = [x, y] \otimes f \cdot g, \qquad x, y \in \mathfrak{g}, \quad f, g \in \mathcal{A}. \tag{4.25}$$

PROPOSITION 4.13. *$\overline{\mathfrak{g}}$ is a Lie algebra.*

PROOF. The antisymmetry is clear from the definition. Moreover

$$[[x \otimes f, y \otimes g], z \otimes h] = [[x, y], z] \otimes ((f \cdot g) \cdot h).$$

As $\mathcal{A}$ is associative and commutative summing up cyclically the Jacobi identity follows directly from the Jacobi identity for $\mathfrak{g}$. □

As usual we will suppress the mentioning of $(\Sigma, A)$ if not needed. The elements of $\overline{\mathfrak{g}}$ can be interpreted as meromorphic functions $\Sigma \to \mathfrak{g}$ from the Riemann surface $\Sigma$ to the Lie algebra $\mathfrak{g}$, which are holomorphic outside of $A$.

Later we will introduce central extensions for these current algebras. They will generalize affine Lie algebras, respectively affine Kac-Moody algebras of untwisted type.

For some applications it is useful to extend the definition by considering differential operators (of degree $\le 1$) associated to $\overline{\mathfrak{g}}$. We define $\mathcal{D}^1_{\mathfrak{g}} := \overline{\mathfrak{g}} \oplus \mathcal{L}$ and take in the summands the Lie product defined there and put additionally

$$[e, x \otimes g] := -[x \otimes g, e] := x \otimes (e.g). \tag{4.26}$$

This operation can be described as semidirect sum of $\overline{\mathfrak{g}}$ with $\mathcal{L}$ and we get

PROPOSITION 4.14 ([**91**, Prop. 2.15]). *$\mathcal{D}^1_{\mathfrak{g}}$ is a Lie algebra.*

**4.9. Krichever–Novikov type algebras.** Above the set $A$ of points where poles are allowed was arbitrary. In case that $A$ is finite and moreover $\#A \ge 2$ the constructed algebras are called Krichever–Novikov (KN) type algebras. In this way we get the KN vector field algebra, the function algebra, the current algebra, the differential operator algebra, the Lie superalgebra, etc. The reader might ask what is so special about this situation so that these algebras deserve special names. In fact in this case we can endow the algebra with a (strong) almost-graded structure. This will be discussed in the next section. The almost-grading is a crucial tool for extending the classical result to higher genus. Recall that in the classical case we have genus zero and $\#A = 2$.

Strictly speaking, a KN type algebra should be considered to be one of the above algebras for $2 \le \#A < \infty$ together with a fixed chosen almost-grading induced by the splitting $A = I \cup O$ into two disjoint non-empty subset, see Definition 5.1.

## 5. Almost-Graded Structure

**5.1. Definition of almost-gradedness.** In the classical situation discussed in Section 2 the algebras introduced in the last section are graded algebras. In the higher genus case and even in the genus zero case with more than two points where poles are allowed there is no non-trivial grading anymore. As realized by Krichever and Novikov [**56**] there is a weaker concept, an almost-grading, which to a large extend is a valuable replacement of a honest grading. Such an almost-grading is induced by a splitting of the set $A$ into two non-empty and disjoint sets $I$ and $O$. The (almost-)grading is fixed by exhibiting certain basis elements in the spaces $\mathcal{F}^\lambda$ as homogeneous.

DEFINITION 5.1. Let $\mathcal{L}$ be a Lie or an associative algebra such that $\mathcal{L} = \oplus_{n\in\mathbb{Z}}\mathcal{L}_n$ is a vector space direct sum, then $\mathcal{L}$ is called an *almost-graded* (Lie-) algebra if

(i) $\dim \mathcal{L}_n < \infty$,
(ii) There exists constants $L_1, L_2 \in \mathbb{Z}$ such that

$$\mathcal{L}_n \cdot \mathcal{L}_m \subseteq \bigoplus_{h=n+m-L_1}^{n+m+L_2} \mathcal{L}_h, \qquad \forall n, m \in \mathbb{Z}.$$

The elements in $\mathcal{L}_n$ are called *homogeneous* elements of degree $n$, and $\mathcal{L}_n$ is called *homogeneous subspace* of degree $n$.

If $\dim \mathcal{L}_n$ is bounded with a bound independent of $n$ we call $\mathcal{L}$ *strongly almost-graded.* If we drop the condition that $\dim \mathcal{L}_n$ is finite we call $\mathcal{L}$ *weakly almost-graded.*

In a similar manner almost-graded modules over almost-graded algebras are defined. We can extend in an obvious way the definition to superalgebras, respectively even to more general algebraic structures. This definition makes complete sense also for more general index sets $\mathbb{J}$. In fact we will consider the index set

$\mathbb{J} = (1/2)\mathbb{Z}$ in the case of superalgebras. The even elements (with respect to the super-grading) will have integer degree, the odd elements half-integer degree.

**5.2. Separating cycle and Krichever-Novikov pairing.** Before we give the almost-grading we introduce an important structure. Let $C_i$ be positively oriented (deformed) circles around the points $P_i$ in $I$, $i = 1, \ldots, K$ and $C_j^*$ positively oriented circles around the points $Q_j$ in $O$, $j = 1, \ldots, M$.

A cycle $C_S$ is called a separating cycle if it is smooth, positively oriented of multiplicity one and if it separates the in-points from the out-points. It might have more than one component. In the following we will integrate meromorphic differentials on $\Sigma$ without poles in $\Sigma \setminus A$ over closed curves $C$. Hence, we might consider $C$ and $C'$ as equivalent if $[C] = [C']$ in $\mathrm{H}_1(\Sigma \setminus A, \mathbb{Z})$. In this sense we write for every separating cycle

$$[C_S] = \sum_{i=1}^{K} [C_i] = -\sum_{j=1}^{M} [C_j^*]. \tag{5.1}$$

The minus sign appears due to the opposite orientation. Another way for giving such a $C_S$ is via level lines of a "proper time evolution", for which I refer to [**91**, Section 3.9].

Given such a separating cycle $C_S$ (respectively cycle class) we define a linear map

$$\mathcal{F}^1 \to \mathbb{C}, \qquad \omega \mapsto \frac{1}{2\pi \mathrm{i}} \int_{C_S} \omega. \tag{5.2}$$

The map will not depend on the separating line $C_S$ chosen, as two of such will be homologous and the poles of $\omega$ are only located in $I$ and $O$.

Consequently, the integration of $\omega$ over $C_S$ can also be described over the special cycles $C_i$ or equivalently over $C_j^*$. This integration corresponds to calculating residues

$$\omega \quad \mapsto \quad \frac{1}{2\pi \mathrm{i}} \int_{C_S} \omega \;=\; \sum_{i=1}^{K} \mathrm{res}_{P_i}(\omega) \;=\; -\sum_{l=1}^{M} \mathrm{res}_{Q_l}(\omega). \tag{5.3}$$

DEFINITION 5.2. The pairing

$$\mathcal{F}^\lambda \times \mathcal{F}^{1-\lambda} \to \mathbb{C}, \quad (f, g) \mapsto \langle f, g \rangle := \frac{1}{2\pi \mathrm{i}} \int_{C_S} f \cdot g, \tag{5.4}$$

between $\lambda$ and $1 - \lambda$ forms is called *Krichever-Novikov (KN) pairing.*

Note that the pairing depends not only on $A$ (as the $\mathcal{F}^\lambda$ depend on it) but also critically on the splitting of $A$ into $I$ and $O$ as the integration path will depend on it. Once the splitting is fixed the pairing will be fixed too.

In fact there exit dual basis elements (see (5.9)) hence the pairing is non-degenerate.

**5.3. The homogeneous subspaces.** Given the vector spaces $\mathcal{F}^\lambda$ of forms of weight $\mathcal{L}$ we will now single out subspaces $\mathcal{F}_m^\lambda$ of degree $m$ by giving a basis of these subspaces. This has been done in the 2-point case by Krichever and Novikov [**56**] and in the multi-point case by the author [**77**], [**78**], [**79**], [**80**], see also Sadov [**76**]. See in particular [**91**, Chapters 3,4,5] for a complete treatment. All proofs of the statements to come can be found there.

Depending on whether the weight $\lambda$ is integer or half-integer we set $\mathbb{J}_\lambda = \mathbb{Z}$ or $\mathbb{J}_\lambda = \mathbb{Z} + 1/2$. For $\mathcal{F}^\lambda$ we introduce for $m \in \mathbb{J}_\lambda$ subspaces $\mathcal{F}^\lambda_m$ of dimension $K$, where $K = \#I$, by exhibiting certain elements $f^\lambda_{m,p} \in \mathcal{F}^\lambda$, $p = 1, \ldots, K$ which constitute a basis of $\mathcal{F}^\lambda_m$. Recall that the spaces $\mathcal{F}^\lambda$ for $\lambda \in \mathbb{Z} + 1/2$ depend on the chosen square root $L$ (the theta characteristic) of the bundle chosen. The elements are the elements of degree $m$. As explained in the following, the degree is in an essential way related to the zero orders of the elements at the points in $I$.

Let $I = \{P_1, P_2, \ldots, P_K\}$ then we have for the zero-order at the point $P_i \in I$ of the element $f^\lambda_{n,p}$

$$\operatorname{ord}_{P_i}(f^\lambda_{n,p}) = (n + 1 - \lambda) - \delta^p_i, \quad i = 1, \ldots, K \,. \tag{5.5}$$

The prescription at the points in $O$ is made in such a way that the element $f^\lambda_{m,p}$ is essentially uniquely given. Essentially unique means up to multiplication with a constant[4]. After fixing as additional geometric data a system of coordinates $z_l$ centered at $P_l$ for $l = 1, \ldots, K$ and requiring that

$$f^\lambda_{n,p}(z_p) = z_p^{n-\lambda}(1 + O(z_p))(dz_p)^\lambda \tag{5.6}$$

the element $f_{n,p}$ is uniquely fixed. In fact, the element $f^\lambda_{n,p}$ only depends on the first jet of the coordinate $z_p$.

EXAMPLE. Here we will not give the general recipe for the prescription at the points in $O$. Just to give an example which is also an important special case, assume $O = \{Q\}$ is a one-element set. If either the genus $g = 0$, or $g \geq 2$, $\lambda \neq 0,\ 1/2,\ 1$ and the points in $A$ are in generic position then we require

$$\operatorname{ord}_Q(f^\lambda_{n,p}) = -K \cdot (n + 1 - \lambda) + (2\lambda - 1)(g - 1). \tag{5.7}$$

In the other cases (e.g. for $g = 1$) there are some modifications at the point in $O$ necessary for finitely many $m$.

THEOREM 5.3 ([**91**, Thm. 3.6]). *Set*

$$\mathcal{B}^\lambda := \{\, f^\lambda_{n,p} \mid n \in \mathbb{J}_\lambda,\ p = 1, \ldots, K \,\}. \tag{5.8}$$

*Then (a) $\mathcal{B}^\lambda$ is a basis of the vector space $\mathcal{F}^\lambda$.*

*(b) The introduced basis $\mathcal{B}^\lambda$ of $\mathcal{F}^\lambda$ and $\mathcal{B}^{1-\lambda}$ of $\mathcal{F}^{1-\lambda}$ are dual to each other with respect to the Krichever-Novikov pairing (5.4), i.e.*

$$\langle f^\lambda_{n,p}, f^{1-\lambda}_{-m,r} \rangle = \delta^r_p\, \delta^m_n, \quad \forall n, m \in \mathbb{J}_\lambda, \quad r, p = 1, \ldots, K. \tag{5.9}$$

In particular, from part (b) of the theorem it follows that the Krichever-Novikov pairing is non-degenerate. Moreover, any element $v \in \mathcal{F}^{1-\lambda}$ acts as linear form on $\mathcal{F}^\lambda$ via

$$\Phi_v : \mathcal{F}^\lambda \mapsto \mathbb{C}, \quad w \mapsto \Phi_v(w) := \langle v, w \rangle. \tag{5.10}$$

Via this pairing $\mathcal{F}^{1-\lambda}$ can be considered as restricted dual of $\mathcal{F}^\lambda$. The identification depends on the splitting of $A$ into $I$ and $O$ as the KN pairing depends on it. The full space $(\mathcal{F}^\lambda)^*$ can even be described with the help of the pairing in a "distributional interpretation" via the distribution $\Phi_{\hat{v}}$ associated to the formal series

$$\hat{v} := \sum_{m \in \mathbb{J}_\lambda} \sum_{p=1}^{K} a_{m,p} f^{1-\lambda}_{m,p}, \quad a_{m,p} \in \mathbb{C} \,. \tag{5.11}$$

---

[4]Strictly speaking, there are some special cases where some constants have to be added such that the Krichever-Novikov duality (5.9) is valid.

The dual elements of $\mathcal{L}$ will be given by the formal series (5.11) with basis elements from $\mathcal{F}^2$, the quadratic differentials, the dual elements of $\mathcal{A}$ correspondingly from $\mathcal{F}^1$, the differentials, and the dual elements of $\mathcal{F}^{-1/2}$ correspondingly from $\mathcal{F}^{3/2}$.

It is quite convenient to use special notations for elements of some important weights:

$$\begin{aligned} e_{n,p} := f^{-1}_{n,p}, \quad \varphi_{n,p} := f^{-1/2}_{n,p}, \quad A_{n,p} := f^{0}_{n,p}, \\ \omega^{n,p} := f^{1}_{-n,p}, \quad \Omega^{n,p} := f^{2}_{-n,p}. \end{aligned} \tag{5.12}$$

In view of (5.9) for the forms of weights 1 and 2 we invert the index $n$ and write it as a superscript.

REMARK 5.4. It is also possible (and for certain applications necessary) to write explicitely down the basis elements $f^{\lambda}_{n,p}$ in terms of "usual" objects defined on the Riemann surface $\Sigma$. For genus zero they can be given with the help of rational functions in the quasi-global variable $z$. For genus one (i.e. the torus case) representations with the help of Weierstraß $\sigma$ and Weierstraß $\wp$ functions exists. For genus $\geq 1$ there exists expressions in terms of theta functions (with characteristics) and prime forms. Here the Riemann surface has first to be embedded into its Jacobian via the Jacobi map. See [**91**, Chapter 5], [**78**], [**81**] for more details.

### 5.4. The algebras.

THEOREM 5.5 ([**91**, Thm. 3.8]). *There exists constants $R_1$ and $R_2$ (depending on the number and splitting of the points in $A$ and on the genus $g$) independent of $\lambda$ and $\nu$ and independent of $n, m \in \mathbb{J}$ such that for the basis elements*

$$\begin{aligned} f^{\lambda}_{n,p} \cdot f^{\nu}_{m,r} = \ & f^{\lambda+\nu}_{n+m,r}\delta^r_p \\ & + \sum_{h=n+m+1}^{n+m+R_1} \sum_{s=1}^{K} a^{(h,s)}_{(n,p)(m,r)} f^{\lambda+\nu}_{h,s}, \qquad a^{(h,s)}_{(n,p)(m,r)} \in \mathbb{C}, \\ [f^{\lambda}_{n,p}, f^{\nu}_{m,r}] = \ & (-\lambda m + \nu n)\, f^{\lambda+\nu+1}_{n+m,r}\delta^r_p \\ & + \sum_{h=n+m+1}^{n+m+R_2} \sum_{s=1}^{K} b^{(h,s)}_{(n,p)(m,r)} f^{\lambda+\nu+1}_{h,s}, \qquad b^{(h,s)}_{(n,p)(m,r)} \in \mathbb{C}. \end{aligned} \tag{5.13}$$

This says in particular that with respect to both the associative and Lie structure the algebra $\mathcal{F}$ is weakly almost-graded. In generic situations and for $N = 2$ points one obtains $R_1 = g$ and $R_2 = 3g$.

The reason why we only have weakly almost-gradedness is that

$$\mathcal{F}^{\lambda} = \bigoplus_{m \in \mathbb{J}_{\lambda}} \mathcal{F}^{\lambda}_m, \qquad \text{with} \quad \dim \mathcal{F}^{\lambda}_m = K, \tag{5.14}$$

and if we add up for a fixed $m$ all $\lambda$ we get that our homogeneous spaces are infinite dimensional.

In the definition of our KN type algebra only finitely many $\lambda$s are involved, hence the following is immediate

THEOREM 5.6. *The Krichever-Novikov type vector field algebras $\mathcal{L}$, function algebras $\mathcal{A}$, differential operator algebras $\mathcal{D}^1$, Lie superalgebras $\mathcal{S}$, and Jordan superalgebras $\mathcal{J}$ are all (strongly) almost-graded algebras and the corresponding modules $\mathcal{F}^\lambda$ are almost-graded modules.*

We obtain with $n \in \mathbb{J}_\lambda$

$$\begin{gathered} \dim \mathcal{L}_n = \dim \mathcal{A}_n = \dim \mathcal{F}^\lambda_n = K, \\ \dim \mathcal{S}_n = \dim \mathcal{J}_n = 2K, \quad \dim \mathcal{D}^1_n = 3K\,. \end{gathered} \tag{5.15}$$

If $\mathcal{U}$ is any of these algebras, with product denoted by $[\,,\,]$ then

$$[\mathcal{U}_n, \mathcal{U}_m] \subseteq \bigoplus_{h=n+m}^{n+m+R_i} \mathcal{U}_h, \tag{5.16}$$

with $R_i = R_1$ for $\mathcal{U} = \mathcal{A}$ and $R_i = R_2$ otherwise.

For further reference let us specialize the lowest degree term component in (5.13) for certain special cases.

$$\begin{aligned} A_{n,p} \cdot A_{m,r} &= A_{n+m,r}\, \delta^p_r \;+\; \text{h.d.t.}, \\ A_{n,p} \cdot f^\lambda_{m,r} &= f^\lambda_{n+m,r}\, \delta^p_r \;+\; \text{h.d.t.}, \\ [e_{n,p}, e_{m,r}] &= (m-n) \cdot e_{n+m,r}\, \delta^p_r \;+\; \text{h.d.t.}, \\ e_{n,p} \cdot f^\lambda_{m,r} &= (m+\lambda n) \;\cdot f^\lambda_{n+m,r}\, \delta^p_r \;+\; \text{h.d.t.} \end{aligned} \tag{5.17}$$

Here h.d.t. denote linear combinations of basis elements of degree between $n+m+1$ and $n+m+R_i$,

Finally, the almost-grading of $\mathcal{A}$ induces an almost-grading of the current algebra $\overline{\mathfrak{g}}$ by setting $\overline{\mathfrak{g}}_n = \mathfrak{g} \otimes \mathcal{A}_n$. We obtain

$$\overline{\mathfrak{g}} = \bigoplus_{n \in \mathbb{Z}} \overline{\mathfrak{g}}_n, \quad \dim \overline{\mathfrak{g}}_n = K \cdot \dim \mathfrak{g}. \tag{5.18}$$

**5.5. Triangular decomposition and filtrations.** Let $\mathcal{U}$ be one of the above introduced algebras (including the current algebra). On the basis of the almost-grading we obtain a triangular decomposition of the algebras

$$\mathcal{U} = \mathcal{U}_{[+]} \oplus \mathcal{U}_{[0]} \oplus \mathcal{U}_{[-]}, \tag{5.19}$$

where

$$\mathcal{U}_{[+]} := \bigoplus_{m>0} \mathcal{U}_m, \quad \mathcal{U}_{[0]} = \bigoplus_{m=-R_i}^{m=0} \mathcal{U}_m, \quad \mathcal{U}_{[-]} := \bigoplus_{m<-R_i} \mathcal{U}_m. \tag{5.20}$$

By the almost-gradedness the $[+]$ and $[-]$ subspaces are (infinite dimensional) subalgebras. The $[0]$ spaces, in general, are not. Sometimes we will use the *critical strip* for them.

With respect to the almost-grading of $\mathcal{F}^\lambda$ we introduce a filtration

$$\begin{gathered} \mathcal{F}^\lambda_{(n)} := \bigoplus_{m \geq n} \mathcal{F}^\lambda_m, \\ \dots \;\supseteq\; \mathcal{F}^\lambda_{(n-1)} \;\supseteq\; \mathcal{F}^\lambda_{(n)} \;\supseteq\; \mathcal{F}^\lambda_{(n+1)} \;\dots \end{gathered} \tag{5.21}$$

PROPOSITION 5.7 ([**91**, Prop. 3.15]).

$$\mathcal{F}^\lambda_{(n)} = \{\, f \in \mathcal{F}^\lambda \mid \operatorname{ord}_{P_i}(f) \geq n - \lambda, \forall i = 1, \dots, K \,\}. \tag{5.22}$$

This proposition is very important. In case that $O$ has more than one point there are certain choices, e.g. numbering of the points in $O$, different rules, etc. involved in defining the almost-grading. Hence, if the choices are made differently the subspaces $\mathcal{F}_n^\lambda$ might depend on them, and consequently also the almost-grading. But by this proposition the induced filtration is indeed canonically defined via the splitting of $A$ into $I$ and $O$.

Moreover, different choices will give equivalent almost-grading. We stress the fact, that under a KN algebra we will always understand one of the introduced algebras together with an almost-grading (respectively equivalence class of almost-grading, respectively filtration) introduced by the splitting $A = I \cup O$.

## 6. Central Extensions

Central extension of our algebras appear naturally in the context of quantization and regularization of actions. Of course they are also of independent mathematical interest.

**6.1. Central extensions and cocycles.** For the convenience of the reader let us repeat the relation between central extensions and the second Lie algebra cohomology with values in the trivial module. A central extension of a Lie algebra $W$ is a special Lie algebra structure on the vector space direct sum $\widehat{W} = \mathbb{C} \oplus W$. If we denote $\hat{x} := (0, x)$ and $t := (1, 0)$ then the Lie structure is given by

$$[\hat{x}, \hat{y}] = \widehat{[x, y]} + \psi(x, y) \cdot t, \quad [t, \widehat{W}] = 0, \quad x, y \in W, \tag{6.1}$$

with bilinear form $\psi$. The map $x \mapsto \hat{x} = (0, x)$ is a linear splitting map. $\widehat{W}$ will be a Lie algebra, e.g. will fulfill the Jacobi identity, if and only if $\psi$ is an antisymmetric bilinear form and fulfills the Lie algebra 2-cocycle condition

$$0 = d_2\psi(x, y, z) := \psi([x, y], z) + \psi([y, z], x) + \psi([z, x], y). \tag{6.2}$$

Two central extensions are equivalent if they essentially correspond only to the choice of different splitting maps. A 2-cochain $\psi$ is a coboundary if there exists a linear form $\varphi : W \to \mathbb{C}$ such that

$$\psi(x, y) = \varphi([x, y]). \tag{6.3}$$

Every coboundary is a cocycle. The second Lie algebra cohomology $\mathrm{H}^2(W, \mathbb{C})$ of $W$ with values in the trivial module $\mathbb{C}$ is defined as the quotient of the space of cocycles modulo coboundaries. Moreover, two central extensions are equivalent if and only if the difference of their defining 2-cocycles $\psi$ and $\psi'$ is a coboundary. In this way the second Lie algebra cohomology $\mathrm{H}^2(W, \mathbb{C})$ classifies equivalence classes of central extensions. The class $[0]$ corresponds to the trivial central extension. In this case the splitting map is a Lie homomorphism. To construct central extensions of our algebras we have to find such Lie algebra 2-cocycles.

Clearly, equivalent central extensions are isomorphic. The opposite is not true. In our case we can always rescale the central element by multiplying it with a nonzero scalar. This is an isomorphism but not an equivalence of central extensions. Nevertheless it is an irrelevant modification. Hence we will be mainly interested in central extensions modulo equivalence and rescaling. They are classified by $[0]$ and the elements of the projectivized cohomology space $\mathbb{P}(\mathrm{H}^2(W, \mathbb{C}))$.

In the classical case we have $\dim \mathrm{H}^2(\mathcal{W}, \mathbb{C}) = 1$, hence there are only two essentially different central extensions, the splitting one given by the direct sum

$\mathbb{C} \oplus \mathcal{W}$ of Lie algebras and the up to equivalence and rescaling unique non-trivial one, the Virasoro algebra $\mathcal{V}$.

**6.2. Geometric cocycles.** The defining cocycle

$$\frac{1}{12}(n^3 - n)\delta_n^{-m} \tag{6.4}$$

for the Virasoro algebra is very special. Obviously it does not make any sense in the higher genus and/or multi-point case. We need to find a geometric description. For this we have first to introduce connections.

6.2.1. *Projective and affine connections.* Let $(U_\alpha, z_\alpha)_{\alpha \in J}$ be a covering of the Riemann surface by holomorphic coordinates with transition functions $z_\beta = f_{\beta\alpha}(z_\alpha)$.

DEFINITION 6.1. (a) A system of local (holomorphic, meromorphic) functions $R = (R_\alpha(z_\alpha))$ is called a (holomorphic, meromorphic) *projective connection* if it transforms as

$$R_\beta(z_\beta) \cdot (f'_{\beta,\alpha})^2 = R_\alpha(z_\alpha) + S(f_{\beta,\alpha}), \qquad \text{with} \quad S(h) = \frac{h'''}{h'} - \frac{3}{2}\left(\frac{h''}{h'}\right)^2, \tag{6.5}$$

the Schwartzian derivative. Here $'$ denotes differentiation with respect to the coordinate $z_\alpha$.

(b) A system of local (holomorphic, meromorphic) functions $T = (T_\alpha(z_\alpha))$ is called a (holomorphic, meromorphic) *affine connection* if it transforms as

$$T_\beta(z_\beta) \cdot (f'_{\beta,\alpha}) = T_\alpha(z_\alpha) + \frac{f''_{\beta,\alpha}}{f'_{\beta,\alpha}}. \tag{6.6}$$

Every Riemann surface admits a holomorphic projective connection [**40**],[**38**]. Given a point $P$ then there exists always a meromorphic affine connection holomorphic outside of $P$ and having maximally a pole of order one there [**80**].

From their very definition it follows that the difference of two affine (projective) connections will be a (quadratic) differential. Hence, after fixing one affine (projective) connection all others are obtained by adding (quadratic) differentials.

6.2.2. *The function algebra $\mathcal{A}$.* We consider it as abelian Lie algebra. Let $C$ be an arbitrary smooth but not necessarily connected curve. We set

$$\psi^1_C(g,h) := \frac{1}{2\pi \mathrm{i}}\int_C g dh, \quad g,h \in \mathcal{A}. \tag{6.7}$$

6.2.3. *The current algebra $\overline{\mathfrak{g}}$.* For $\overline{\mathfrak{g}} = \mathfrak{g} \otimes \mathcal{A}$ we fix a symmetric, invariant, bilinear form $\beta$ on $\mathfrak{g}$ (not necessarily non-degenerate). Recall, that invariance means that we have $\beta([x,y],z) = \beta(x,[y,z])$ for all $x,y,z \in \mathfrak{g}$. Then a cocycle is given as

$$\psi^2_{C,\beta}(x \otimes g, y \otimes h) := \beta(x,y) \cdot \frac{1}{2\pi \mathrm{i}}\int_C g dh, \quad x,y \in \mathfrak{g}, \; g,h \in \mathcal{A}. \tag{6.8}$$

6.2.4. *The vector field algebra $\mathcal{L}$.* Here it is a little bit more delicate. First we have to choose a (holomorphic) projective connection $R$. We define

$$\psi^3_{C,R}(e,f) := \frac{1}{24\pi \mathrm{i}}\int_C \left(\frac{1}{2}(e'''f - ef''') - R \cdot (e'f - ef')\right) dz\,. \tag{6.9}$$

Only by the term coming with the projective connection it will be a well-defined differential, i.e. independent of the coordinate chosen. It is shown in [**80**] (and [**91**]) that it is a cocycle. Another choice of a projective connection will result in a cohomologous one. Hence, the equivalence class of the central extension will be the same.

6.2.5. *The differential operator algebra $\mathcal{D}^1$.* For the differential operator algebra the cocycles of type (6.7) for $\mathcal{A}$ can be extended by zero on the subspace $\mathcal{L}$. The cocycles for $\mathcal{L}$ can be pulled back. In addition there is a third type of cocycles mixing $\mathcal{A}$ and $\mathcal{L}$:

$$\psi^4_{C,T}(e,g) := \frac{1}{24\pi\mathrm{i}}\int_C (eg'' + Teg')dz, \qquad e \in \mathcal{L}, g \in \mathcal{A}, \tag{6.10}$$

with an affine connection $T$, with at most a pole of order one at a fixed point in $O$. Again, a different choice of the connection will not change the cohomology class. For more details on the cocycles see [**83**], [**91**].

6.2.6. *The Lie superalgebra $\mathcal{S}$.* Here we have to take into account that it is not a Lie algebra. Hence, the Jacobi identity has to be replaced by the super-Jacobi identity. The conditions for being a cocycle for the superalgebra cohomology will change too. Recall the definition of the algebra from Section 4.6, in particular that the even elements (parity 0) are the vector fields and the odd elements (parity 1) are the half-forms. A bilinear form $c$ is a cocycle if the following is true. The bilinear map $c$ will be symmetric if both $x$ and $y$ are odd, otherwise it will be antisymmetric:

$$c(x,y) = -(-1)^{\bar{x}\bar{y}}c(x,y). \tag{6.11}$$

The super-cocycle condition reads as

$$(-1)^{\bar{x}\bar{z}}c(x,[y,z]) + (-1)^{\bar{y}\bar{x}}c(y,[z,x]) + (-1)^{\bar{z}\bar{y}}c(z,[x,y]) \;=\; 0. \tag{6.12}$$

With the help of $c$ we can define central extensions in the Lie superalgebra sense. If we put the condition that the central element is even then the cocycle $c$ has to be an even map and $c$ vanishes for pairs of elements of different parity.

By convention we denote vector fields by $e, f, g, \ldots$ and -1/2-forms by $\varphi, \psi, \chi, ..$ and get

$$c(e,\varphi) = 0, \quad e \in \mathcal{L}, \;\; \varphi \in \mathcal{F}^{-1/2}. \tag{6.13}$$

The super-cocycle conditions for the even elements is just the cocycle condition for the Lie subalgebra $\mathcal{L}$. The only other nonvanishing super-cocycle condition is for the *(even,odd,odd)* elements and reads as

$$c(e,[\varphi,\psi]) - c(\varphi, e\,.\,\psi) - c(\psi, e\,.\,\varphi) = 0. \tag{6.14}$$

Here the definition of the product $[e,\psi] := e\,.\,\psi$ was used.

If we have a cocycle $c$ for the algebra $\mathcal{S}$ we obtain by restriction a cocycle for the algebra $\mathcal{L}$. For the mixing term we know that $c(e,\psi) = 0$. A naive try to put just anything for $c(\varphi,\psi)$ (for example 0) will not work as (6.14) relates the restriction of the cocycle on $\mathcal{L}$ with its values on $\mathcal{F}^{-1/2}$.

Proposition 6.2 ([**88**]). *Let $C$ be any closed (differentiable) curve on $\Sigma$ not meeting the points in $A$, and let $R$ be any (holomorphic) projective connection then the bilinear extension of*

$$\begin{aligned}
\Phi_{C,R}(e,f) &:= \frac{1}{24\pi\mathrm{i}}\int_C \left(\frac{1}{2}(e'''f - ef''') - R\cdot(e'f - ef')\right)dz,\\
\Phi_{C,R}(\varphi,\psi) &:= -\frac{1}{24\pi\mathrm{i}}\int_C (\varphi''\cdot\psi + \varphi\cdot\psi'' - R\cdot\varphi\cdot\psi)\,dz,\\
\Phi_{C,R}(e,\varphi) &:= 0,
\end{aligned} \tag{6.15}$$

*gives a Lie superalgebra cocycle for $\mathcal{S}$, hence defines a central extension of $\mathcal{S}$. A different projective connection will yield a cohomologous cocycle.*

A similar formula was given by Bryant in [**17**]. By adding the projective connection in the second part of (6.15) he corrected some formula appearing in [**12**]. He only considered the two-point case and only the integration over a separating cycle. See also [**54**] for the multi-point case, where still only the integration over a separating cycle is considered.

In contrast to the differential operator algebra case the two parts cannot be prescribed independently. Only with the same integration path (more precisely, homology class) and the given factors in front of the integral it will work. The reason for this is that (6.14) relates both.

**6.3. Uniqueness and classification of central extensions.** The above introduced cocycles depend on the choice of the connections $R$ and $T$. Different choices will not change the cohomology class. Hence, this ambiguity does not disturb us. What really matters is that they depend on the integration curve $C$ chosen.

In contrast to the classical situation, for the higher genus and/or multi-point situation there are many essentially different closed curves and hence many non-equivalent central extensions defined by the integration.

But we should take into account that we want to extend the almost-grading from our algebras to the centrally extended ones. This means we take $\deg \hat{x} := \deg x$ and assign a degree $deg(t)$ to the central element $t$, and still we want to obtain almost-gradedness.

This is possible if and only if our defining cocycle $\psi$ is "local" in the following sense (the name was introduced in the two point case by Krichever and Novikov in [**56**]). There exists $M_1, M_2 \in \mathbb{Z}$ such that

$$\forall n, m: \quad \psi(W_n, W_m) \neq 0 \implies M_1 \leq n + m \leq M_2. \tag{6.16}$$

Here $W$ stands for any of our algebras (including the supercase). Very important, "local" is defined in terms of the almost-grading, and the almost-grading itself depends on the splitting $A = I \cup O$. Hence what is "local" depends on the splitting too.

We will call a cocycle *bounded* (from above) if there exists $M \in \mathbb{Z}$ such that

$$\forall n, m: \quad \psi(W_n, W_m) \neq 0 \implies n + m \leq M. \tag{6.17}$$

Similarly bounded from below can be defined. Locality means bounded from above and from below.

Given a cocycle class we call it bounded (respectively local) if and only if it contains a representing cocycle which is bounded (respectively local). Not all cocycles in a bounded class have to be bounded. If we choose as integration path a separating cocycle $C_S$, or one of the $C_i$ then the above introduced geometric cocycles are local, respectively bounded. Recall that in this case integration can be done by calculating residues at the in-points or at the out-points. All these cocycles are cohomologically nontrivial. The theorems in the following concern the opposite direction. They were treated in my works [**83**], [**84**], [**88**]. See also [**91**] for a complete and common treatment.

The following result for the vector field algebra $\mathcal{L}$ gives the principal structure of the classification results.

THEOREM 6.3 ([**83**], [**91**, Thm. 6.41]). *Let $\mathcal{L}$ be the Krichever–Novikov vector field algebra with a given almost-grading induced by the splitting $A = I \cup O$.*
*(a) The space of bounded cohomology classes is $K$-dimensional ($K = \#I$). A basis is given by setting the integration path in (6.9) to $C_i$, $i = 1, \dots, K$ the little (deformed) circles around the points $P_i \in I$.*
*(b) The space of local cohomology classes is one-dimensional. A generator is given by integrating (6.9) over a separating cocycle $C_S$, i.e.*

$$\psi^3_{C_S,R}(e,f) = \frac{1}{24\pi \mathrm{i}} \int_{C_S} \left( \frac{1}{2}(e'''f - ef''') - R \cdot (e'f - ef') \right) dz \ . \tag{6.18}$$

*(c) Up to equivalence and rescaling there is only one non-trivial one-dimensional central extension $\widehat{\mathcal{L}}$ of the vector field algebra $\mathcal{L}$ which allows an extension of the almost-grading.*

REMARK 6.4. In the classical situation, Part (c) shows also that the Virasoro algebra is the unique non-trivial central extension of the Witt algebra (up to equivalence and rescaling). This result extends to the more general situation under the condition that one fixes the almost-grading, hence the splitting $A = I \cup O$. Here I like to repeat the fact that for $\mathcal{L}$ depending on the set $A$ and its possible splittings into two disjoint subsets there are different almost-gradings. Hence, the "unique" central extension finally obtained will also depend on the splitting. Only in the two point case there is only one splitting possible. In the case that the genus $g \geq 1$ there are even integration paths possible in the definition of (6.9) which are not homologous to a separating cycle of any splitting. Hence, there are other central extensions possible not corresponding to any almost-grading.

The above theorem is a model for all other classification results. We will always obtain a statement about the bounded (from above) cocycles and then for the local cocycles.

If we consider the function algebra $\mathcal{A}$ as an abelian Lie algebra then every skew-symmetric bilinear form will be a non-trivial cocycle. Hence, there is no hope of uniqueness. But if we add the condition of $\mathcal{L}$-invariance, which is given as

$$\psi(e.g, h) + \psi(g, e.h) = 0, \quad \forall e \in \mathcal{L},\ g, h \in \mathcal{A} \tag{6.19}$$

things will change.

Let us denote the subspace of local cohomology classes by $\mathrm{H}^2_{loc}$, and the subspace of local and $\mathcal{L}$-invariant cohomology classes by $\mathrm{H}^2_{\mathcal{L},loc}$. Note that the condition is only required for at least one representative in the cohomology class. We collect a part of the results for the cocycle classes of the other algebras in the following theorem.

THEOREM 6.5 ([**91**, Cor. 6.48]).
*(a)* $\dim \mathrm{H}^2_{\mathcal{L},loc}(\mathcal{A}, \mathbb{C}) = 1,$
*(b)* $\dim \mathrm{H}^2_{loc}(\mathcal{L}, \mathbb{C}) = 1,$
*(c)* $\dim \mathrm{H}^2_{loc}(\mathcal{D}^1, \mathbb{C}) = 3,$
*(d)* $\dim \mathrm{H}^2_{loc}(\overline{\mathfrak{g}}, \mathbb{C}) = 1$ *for $\mathfrak{g}$ a simple finite-dimensional Lie algebra,*
*(e)* $\dim \mathrm{H}^2_{loc}(\mathcal{S}, \mathbb{C}) = 1.$
*A basis of the cohomology spaces are given by taking the cohomology classes of the cocycles (6.7), (6.9), (6.10), (6.8), (6.15) obtained by integration over a separating cycle $C_S$.*

Consequently, we obtain also for these algebras the corresponding result about uniqueness of almost-graded central extensions. For the differential operator algebra we get three independent cocycles. This generalizes results of [**2**] for the classical case.

For result on the bounded cocycle classes we have to multiply the dimensions above by $K = \#I$. For the supercase with odd central element the bounded cohomology vanishes.

For $\mathfrak{g}$ a reductive Lie algebra and if the cocycle is $\mathcal{L}$-invariant if restricted to the abelian part, a complete classification of local cocycle classes for both $\overline{\mathfrak{g}}$ and $\mathcal{D}^1_{\mathfrak{g}}$ can be found in [**84**], [**91**, Chapter 9]. See also the examples below. Note that in the case of a simple Lie algebra every symmetric, invariant bilinear form $\beta$ is a multiple of the Cartan-Killing form.

I would like to mention that in all the applications I know of, the cocycles coming from representations, regularizations, etc., are local. Hence, the uniqueness or classification result above can be used.

## 7. Examples and Generalizations

**7.1. Examples: $\mathfrak{sl}(n)$ and $\mathfrak{gl}(n)$.** For illustration we give the cocycles for the important special cases $\mathfrak{sl}(n)$ (which is a simple algebra) and the case of $\mathfrak{gl}(n)$ (which is reductive but not semisimple). We will study the current and the affine algebras, the differential operator algebras, and their central extensions.

7.1.1. $\mathfrak{sl}(n)$. This is the Lie algebra of trace-less complex $n \times n$ matrices. Up to multiplication with a scalar the Cartan-Killing form $\beta(x, y) = \operatorname{tr}(xy)$ is the unique symmetric invariant bilinear form. It is non-degenerate.

PROPOSITION 7.1 ([**91**, Prop. 9.47]). *(a) Every local cocycle for the current algebra $\overline{\mathfrak{sl}}(n)$ is cohomologous to*

$$\gamma(x \otimes g, y \otimes h) = r \frac{\operatorname{tr}(xy)}{2\pi \mathrm{i}} \int_{C_S} g dh, \quad r \in \mathbb{C}. \tag{7.1}$$

*(b) Every $\mathcal{L}$-invariant local cocycle equals the cocycle (7.1) with a suitable $r$.*
*(c) Every local cocycle for the differential operator algebra $\mathcal{D}^1_{\mathfrak{sl}(n)}$ is cohomologous to a linear combination of (7.1) and the standard local cocycle (6.18) for the vector field algebra. In particular, no cocycles of pure mixing type exists.*

7.1.2. $\mathfrak{gl}(n)$. This is the Lie algebra of all complex $n \times n$-matrices. Recall that $\mathfrak{gl}(n)$ can be written as Lie algebra direct sum

$$\mathfrak{gl}(n) = \mathfrak{s}(n) \oplus \mathfrak{sl}(n) \cong \mathbb{C} \oplus \mathfrak{sl}(n). \tag{7.2}$$

Here $\mathfrak{s}(n)$ denotes the $n \times n$ scalar matrices. This decomposition is the decomposition as reductive Lie algebra into its abelian and semisimple summands.

After tensoring with $\mathcal{A}$ we obtain

$$\overline{\mathfrak{gl}}(n) = \overline{\mathfrak{s}}(n) \oplus \overline{\mathfrak{sl}}(n) \cong \mathcal{A} \oplus \overline{\mathfrak{sl}}(n). \tag{7.3}$$

PROPOSITION 7.2 ([**91**, Prop. 9.48]). *(a) A cocycle $\gamma$ for $\overline{\mathfrak{gl}}(n)$ is local and restricted to $\overline{\mathfrak{s}}(n)$ is $\mathcal{L}$-invariant if and only if it is cohomologous to a linear combination $\gamma'$ of the following two cocycles*

$$\gamma_1(x\otimes g, y\otimes h) = \frac{\operatorname{tr}(xy)}{2\pi\mathrm{i}}\int_{C_S} gdh,$$
$$\gamma_2(x\otimes g, y\otimes h) = \frac{\operatorname{tr}(x)\operatorname{tr}(y)}{2\pi\mathrm{i}}\int_{C_S} gdh. \tag{7.4}$$

*(b) If the cocycle $\gamma$ of (a) is $\mathcal{L}$-invariant then $\gamma$ is equal to a linear combination $r_1\gamma_1 + r_2\gamma_2$ of the cocycles (7.4)*
*(c)* $\dim \mathrm{H}^2_{loc}(\mathfrak{gl}(n), \mathbb{C}) = 2$.

PROPOSITION 7.3. [**91**, Prop. 9.49] *(a) Every local cocycle $\gamma$ for $\mathcal{D}^1_{\mathfrak{gl}(n)}$ is cohomologous to a linear combination of the cocycles $\gamma_1$ and $\gamma_2$ of (7.4), of the mixing cocycle*

$$\gamma^{(m)}_{S,tr,T}(e, x(g)) = \frac{\operatorname{tr}(x)}{2\pi\mathrm{i}}\int_{C_S}(\tilde{e}g'' + T\tilde{e}g')dz, \tag{7.5}$$

*and of the standard local cocycle $\psi^3_{C_S,R}$ for the vector field algebra, i.e.*

$$\gamma = r_1\gamma_1 + r_2\gamma_2 + r_3\gamma^{(m)}_{S,tr,T} + r_4\psi^3_{C_S,R} + \textit{coboundary}, \tag{7.6}$$

*with suitable $r_1, r_2, r_3, r_4 \in \mathbb{C}$.*
*(b) If the cocycle $\gamma$ is local and restricted to $\overline{\mathfrak{gl}}(n)$ is $\mathcal{L}$-invariant, and $r_3, r_4 \neq 0$ then there exist an affine connection $T'$ and a projective connection $R'$ holomorphic outside $A$ such that $\gamma = r_1\gamma_1 + r_2\gamma_2 + r_3\gamma_{S,tr,T'} + r_4\gamma^{\mathcal{L}}_{S,R'}$.*
*(c)* $\dim \mathrm{H}^2_{loc}(\mathcal{D}^1_{\mathfrak{gl}(n)}, \mathbb{C}) = 4$.

**7.2. The genus zero and three-point situation.** For illustration let us consider the three-point KN type algebras of genus zero. We consider the Riemann sphere $S^2 = \mathbb{P}^1$ and a set $A$ consisting of 3 points. Given any triple of 3 points there exists always an analytic automorphism of $\mathbb{P}^1$ mapping this triple to $\{a, -a, \infty\}$, with $a \neq 0$. In fact $a = 1$ would suffice. Without restriction we can take

$$I := \{\infty\}, \quad O := \{a, -a\}\ .$$

Due to the symmetry of the situation it is more convenient to take a symmetrized basis of $\mathcal{A}$ (with $k \in \mathbb{Z}$)

$$A_{2k} := (z-a)^k(z+a)^k, \qquad A_{2k+1} := z(z-a)^k(z+a)^k\,, \tag{7.7}$$

for $\mathcal{L}$ (with $k \in \mathbb{Z}$)

$$V_{2k} := z(z-a)^k(z+a)^k\frac{d}{dz}, \qquad V_{2k+1} := (z-a)^{k+1}(z+a)^{k+1}\frac{d}{dz}\,, \tag{7.8}$$

and for the $-1/2$-forms

$$\varphi_{2k-1/2} := (z-a)^k(z+a)^k(\frac{d}{dz})^{-1/2}, \quad \varphi_{2k+1/2} := z(z-a)^k(z+a)^k(\frac{d}{dz})^{-1/2}\,. \tag{7.9}$$

Also we inverted the grading. By straight-forward calculations we obtain for the algebras the following structures.

**The function algebra.**

$$A_n \cdot A_m = \begin{cases} A_{n+m}, & n \text{ or } m \text{ even}, \\ A_{n+m} + a^2 \otimes A_{n+m-2}, & n \text{ and } m \text{ odd}. \end{cases} \tag{7.10}$$

**The vector field algebra.**

$$[V_n, V_m] = \begin{cases} (m-n)V_{n+m}, & n, m \text{ odd}, \\ (m-n)(V_{n+m} + a^2 V_{n+m-2}), & n, m \text{ even}, \\ (m-n)V_{n+m} + (m-n-1)a^2 V_{n+m-2}, & n \text{ odd}, \ m \text{ even}. \end{cases} \tag{7.11}$$

**The current algebra.**

$$[x \otimes A_n, y \otimes A_m] = \begin{cases} [x,y] \otimes A_{n+m}, & n \text{ or } m \text{ even}, \\ [x,y] \otimes A_{n+m} + a^2 [x,y] \otimes A_{n+m-2}, & n \text{ and } m \text{ odd}. \end{cases} \tag{7.12}$$

The structure equations for the superalgebra look similar and can be easily calculated.

The central extensions can be given by determining the cocycle values by calculating the residues of the integrand at $\infty$. For example the local cocycle $\psi^1_{C_S}$ for the function algebra calculates as (see [**30**, A.13 and A.14])

$$\frac{1}{2\pi\mathrm{i}} \int_{C_S} A_n dA_m = \begin{cases} -n\delta_m^{-n}, & n, \ m \text{ even}, \\ 0, & n, \ m \text{ different parity}, \\ -n\delta_m^{-n} + a^2(-n+1)\delta_m^{-n+2}, & n, \ m \text{ odd}. \end{cases} \tag{7.13}$$

The **affine algebra** is now given as the almost-graded central extension $\widehat{\mathfrak{g}}_{\beta,S}$ of the current algebra given by the cocycle

$$\psi^2_{C_S,\beta}(x \otimes A_n, y \otimes A_m) = \beta(x,y) \cdot \frac{1}{2\pi\mathrm{i}} \int_{C_S} A_n dA_m = \beta(x,y) \cdot \psi^1_{C_S}(A_n, A_m). \tag{7.14}$$

**Three-point $\mathfrak{sl}(2,\mathbb{C})$-current algebra for genus 0.**
Given a simple Lie algebra $\mathfrak{g}$ with generators and structure equations the relations above can be written in these terms. An important example is $\mathfrak{sl}(2,\mathbb{C})$ with the standard generators

$$h := \begin{pmatrix} 1 & 0 \\ 0 & -1 \end{pmatrix}, \quad e := \begin{pmatrix} 0 & 1 \\ 0 & 0 \end{pmatrix}, \quad f := \begin{pmatrix} 0 & 0 \\ 1 & 0 \end{pmatrix}.$$

We set $e_n := e \otimes A_n$, $n \in \mathbb{Z}$ and in the same way $f_n$ and $h_n$. Recall that the invariant bilinear form $\beta(x,y) = \mathrm{tr}(x \cdot y)$. We calculate

$$[e_n, f_m] = \begin{cases} h_{n+m}, & n \text{ or } m \text{ even}, \\ h_{n+m} + a^2 h_{n+m-2}, & n \text{ and } m \text{ odd}, \end{cases} \tag{7.15}$$

$$[h_n, e_m] = \begin{cases} 2e_{n+m}, & n \text{ or } m \text{ even}, \\ 2e_{n+m} + 2a^2 e_{n+m-2}, & n \text{ and } m \text{ odd}, \end{cases} \tag{7.16}$$

$$[h_n, f_m] = \begin{cases} -2f_{n+m}, & n \text{ or } m \text{ even}, \\ -2f_{n+m} - 2a^2 f_{n+m-2}, & n \text{ and } m \text{ odd}. \end{cases} \tag{7.17}$$

For the central extension we obtain

$$[e_n, f_m] = \begin{cases} h_{n+m} - n\delta_m^{-n}, & n \text{ or } m \text{ even}, \\ h_{n+m} + a^2 h_{n+m-2} - n\delta_m^{-n} - a^2(n-1)\delta_m^{-n+2}, & n \text{ and } m \text{ odd}, \end{cases} \tag{7.18}$$

and

$$[h_n, h_m] = \begin{cases} -2n\delta_m^{-n}, & n,\ m \text{ even}, \\ 0, & n,\ m \text{ different parity}, \\ -2n\delta_m^{-n} + 2a^2(-n+1)\delta_m^{-n+2}, & n,\ m \text{ odd}. \end{cases} \tag{7.19}$$

For the other commutators we do not have contributions to the center. See also [**92**] for more information.

**7.3. Deformations.** As the second Lie algebra cohomology of the Witt and Virasoro algebra in their adjoint module vanishes [**87**], [**27**], [**28**] both are formally and infinitesimally rigid. This means that all formal (and infinitesimal) families with special fiber these algebras are equivalent to the trivial one. If we consider the examples of Section 7.2 parameterized by a variable $a$, then they are non-trivial (even locally non-trivial) families which have themselves as special elements for $a = 0$ the classical algebras. The geometric context is clear: the two points $a$ and $-a$ move together. By Fialowski and Schlichenmaier [**29**], [**30**], [**31**] the above algebras and similar families of algebras on tori, were used to exhibit the fact, that e.g. the Witt and Virasoro algebra despite their formal rigidity allow non-trivial algebraic-geometric deformations. This is an effect that cannot appear in the finite-dimensional algebra setting. For families on tori see the above quoted results, respectively [**91**, Chapter 12]. See also [**78**], [**13**], [**15**], [**19**] [**24**], [**74**].

**7.4. Genus zero multi-point algebras – integrable systems.** Already the Witt and Virasoro algebra in genus zero with two points where poles are allowed are mathematically highly interesting objects which have e.g. a non-trivial representation theory. If we remain on the Riemann sphere but now allow more than two poles we obtain an even more demanding mathematical theory. For the multi-point case the related systems are important. For example the classical Knizhnik-Zamolodchikov models of Conformal Field Theory (CFT) are of this type, see e.g. [**53**]. Integrable systems show up.

Due to the connection between CFT and statistical mechanics it is not a surprise that the genus zero multi-point Krichever–Novikov algebras turn out to be related to algebras appearing in statistical mechanics. For example the *Onsager algebra* appears as subalgebra of the three-point, $g = 0$, $\mathfrak{sl}(2,\mathbb{C})$-Krichever–Novikov algebra. In this context see e.g. the work of Terwilliger and collaborators [**41**], [**5**], [**46**].

For the genus zero multi-point situation quite a number of publications appeared. Some references are [**78**], [**29**], [**30**], [**31**], [**14**], [**16**], [**86**], [**1**], [**21**]. Recently, the author [**92**] gave a thorough and unified treatment of universal central extensions of the genus zero algebras.

From the point of view of symmetries of integrable systems the concept of *automorphic Lie algebras* shows up. It was e.g. developed by Lombardo, Mikailov, and Sanders in [**64**], [**65**], [**66**]. Invariant objects under finite subgroups of $PGL(2,\mathbb{C})$, the symmetry group of the Riemann sphere, are studied. Of course, there are relations to the $g = 0$, multi-point Krichever–Novikov type algebras. Chopp [**20**] obtained some results for the genus one multi-point setting.

**7.5. Toroidal Lie algebras.** The path of generalization starting from the classical picture as taken by Krichever–Novikov goes from genus zero to higher genus. There is another path by considering instead of the classical algebra of Laurent polynomial $\mathbb{C}[z^{-1}, z]$, the algebra generated by several "Laurent variables". This generalization is in the context of current algebras quite natural, as there the finite-dimensional Lie algebra can be tensorized by an arbitrary commutative algebra. The algebras obtained in this way are called *toroidal algebras*. They correspond to increasing the dimension instead of the genus. A short collection of references and names is given by Bermann, Billig, Buelk, Cox, Futorny, Hu, Jurisich, Kashuba, Penkov, Szmigielski, Xia, Yokunuma, [**8**], [**18**], [**35**], [**6**], [**7**], [**22**], [**10**], [**25**], [**69**], [**70**], [**108**].

**7.6. Other generalizations of KN type algebras.** We considered geometric vector fields, respectively differential operators. Donin and Khesin [**26**] showed that also pseudo-differential symbols could be treated via some Krichever–Novikov like objects.

Furthermore there are *discretized and q-deformed Krichever–Novikov type algebras.* In the definition of the KN vector field algebra the differential can be replaced by a difference operator assuming that we have a geometric situation appropriate to this discretization. See e.g. Meiler and Ruffing [**73**], [**67**].

Such discretizations are related to $q$-deformed versions of the Krichever–Novikov vector field algebra - again for special geometric cases. In some sense the structure equations are deformed by expressions depending on a formal parameter $q$. One does not obtain Lie algebras anymore, but $q$-Lie algebras. The $q$-Witt and $q$-Virasoro algebras are of certain importance in the context of integrable systems. One might guess that the same will be the case for the $q$-deformed Krichever–Novikov vector field algebra. See e.g. the references Kuang [**60**], Oh and Singh [**71**].

## 8. Lax Operator Algebras

Recently, a new class of current type algebras appeared, the Lax operator algebras. As the naming indicates, they are related to integrable systems [**102**]. The algebras were introduced by Krichever [**55**], and Krichever and Sheinman [**59**]. Here I will report on their definition. See the book [**103**] of Sheinman for more details.

Compared to the KN current type algebra we will allow additional singularities which will play a special role. The points where these singularities are allowed are called *weak singular points.* The set of such points is denoted by

$$W = \{\gamma_s \in \Sigma \setminus A \mid s = 1, \dots, R\}. \tag{8.1}$$

Here let $\mathfrak{g}$ be one of the classical matrix algebras $\mathfrak{gl}(n)$, $\mathfrak{sl}(n)$, $\mathfrak{so}(n)$, $\mathfrak{sp}(2n)$. We assign to every point $\gamma_s$ a vector $\alpha_s \in \mathbb{C}^n$ (respectively $\in \mathbb{C}^{2n}$ for $\mathfrak{sp}(2n)$). The system

$$\mathcal{T} := \{(\gamma_s, \alpha_s) \in \Sigma \times \mathbb{C}^n \mid s = 1, \dots, R\} \tag{8.2}$$

is called *Tyurin data.*

REMARK 8.1. In case that $R = n \cdot g$ and for generic values of $(\gamma_s, \alpha_s)$ with $\alpha_s \neq 0$ the tuples of pairs $(\gamma_s, [\alpha_s])$ with $[\alpha_s] \in \mathbb{P}^{n-1}(\mathbb{C})$ parameterize semi-stable rank $n$ and degree $n\,g$ framed holomorphic vector bundles as shown by Tyurin [**107**]. Hence, the name Tyurin data.

We consider $\mathfrak{g}$-valued meromorphic functions [5]

$$L:\ \Sigma\ \to\ \mathfrak{g}, \tag{8.3}$$

which are holomorphic outside $W\cup A$, have at most poles of order one (respectively of order two for $\mathfrak{sp}(2n)$) at the points in $W$, and fulfill certain conditions at $W$ depending on $\mathcal{T}$. To describe them let us fix local coordinates $w_s$ centered at $\gamma_s$, $s=1,\dots,R$. For $\mathfrak{gl}(n)$ the conditions are as follows. For $s=1,\dots,R$ we require that there exist $\beta_s\in\mathbb{C}^n$ and $\kappa_s\in\mathbb{C}$ such that the function $L$ has the following expansion at $\gamma_s\in W$

$$L(w_s)=\frac{L_{s,-1}}{w_s}+L_{s,0}+\sum_{k>0}L_{s,k}w_s^k, \tag{8.4}$$

with

$$L_{s,-1}=\alpha_s\beta_s^t,\quad \mathrm{tr}(L_{s,-1})=\beta_s^t\alpha_s=0,\quad L_{s,0}\,\alpha_s=\kappa_s\alpha_s. \tag{8.5}$$

In particular, if $L_{s,-1}$ is non-vanishing then it is a rank 1 matrix, and if $\alpha_s\neq 0$ then it is an eigenvector of $L_{s,0}$. The requirements (8.5) are independent of the chosen coordinates $w_s$.

It is not at all clear that the commutator of two such matrix functions fulfills again these conditions. But it is shown in [**59**] that they indeed close to a Lie algebra (in fact in the case of $\mathfrak{gl}(n)$ they constitute an associative algebra under the matrix product). If one of the $\alpha_s=0$ then the conditions at the point $\gamma_s$ correspond to the fact, that $L$ has to be holomorphic there. If all $\alpha_s$'s are zero or $W=\emptyset$ we obtain back the current algebra of KN type. For the algebra under consideration here, in some sense the Lax operator algebras generalize them. In the bundle interpretation of the Tyurin data the KN case corresponds to the trivial rank $n$ bundle.

For $\mathfrak{sl}(n)$ the only additional condition is that in (8.4) all matrices $L_{s,k}$ are trace-less. The conditions (8.5) remain unchanged.

In the case of $\mathfrak{so}(n)$ one requires that all $L_{s,k}$ in (8.4) are skew-symmetric. In particular, they are trace-less. Following [**59**] the set-up has to be slightly modified. First only those Tyurin parameters $\alpha_s$ are allowed which satisfy $\alpha_s^t\alpha_s=0$. Then, (8.5) is changed in the following way:

$$L_{s,-1}=\alpha_s\beta_s^t-\beta_s\alpha_s^t,\quad \mathrm{tr}(L_{s,-1})=\beta_s^t\alpha_s=0,\quad L_{s,0}\,\alpha_s=\kappa_s\alpha_s. \tag{8.6}$$

For $\mathfrak{sp}(2n)$ we consider a symplectic form $\hat\sigma$ for $\mathbb{C}^{2n}$ given by a non-degenerate skew-symmetric matrix $\sigma$. The Lie algebra $\mathfrak{sp}(2n)$ is the Lie algebra of matrices $X$ such that $X^t\sigma+\sigma X=0$. The condition $\mathrm{tr}(X)=0$ will be automatic. At the weak singularities we have the expansion

$$L(w_s)=\frac{L_{s,-2}}{w_s^2}+\frac{L_{s,-1}}{w_s}+L_{s,0}+L_{s,1}w_s+\sum_{k>1}L_{s,k}w_s^k. \tag{8.7}$$

---

[5] Strictly speaking, the interpretation as function is a little bit misleading, as they behave under differentiation like operators on trivialized sections of vector bundles.

The condition (8.5) is modified as follows (see [**59**]): there exist $\beta_s \in \mathbb{C}^{2n}$, $\nu_s, \kappa_s \in \mathbb{C}$ such that

(8.8)
$$L_{s,-2} = \nu_s \alpha_s \alpha_s^t \sigma, \quad L_{s,-1} = (\alpha_s \beta_s^t + \beta_s \alpha_s^t)\sigma, \quad {\beta_s}^t \sigma \alpha_s = 0, \quad L_{s,0}\, \alpha_s = \kappa_s \alpha_s.$$

Moreover, we require $\alpha_s^t \sigma L_{s,1} \alpha_s = 0$. Again under the point-wise matrix commutator the set of such maps constitute a Lie algebra.

It is possible to introduce an almost-graded structure for these Lax operator algebras induced by a splitting of the set $A = I \cup O$. This is done for the two-point case in [**59**] and for the multi-point case in [**90**]. From the applications there is again a need to classify almost-graded central extensions.

The author obtained this jointly with O. Sheinman in [**96**] for the two-point case. For the multi-point case see [**90**]. For the Lax operator algebras associated to the simple algebras $\mathfrak{sl}(n), \mathfrak{so}(n), \mathfrak{sp}(n)$ it will be unique (meaning: given a splitting of $A$ there is an almost-grading and with respect to this there is up to equivalence and rescaling only one non-trivial almost-graded central extension). For $\mathfrak{gl}(n)$ we obtain two independent local cocycle classes if we assume $\mathcal{L}$-invariance on the reductive part. Both in the definition of the cocycle and in the definition of $\mathcal{L}$-invariance a connection shows up.

REMARK 8.2. Recently, Sheinman extended the set-up to $G_2$ [**104**] and moreover gave a recipe for all semi-simple Lie algebras [**105**].

## 9. Fermionic Fock Space

**9.1. Semi-infinite forms and fermionic Fock space representations.** Our Krichever-Novikov vector field algebras $\mathcal{L}$ have as Lie modules the spaces $\mathcal{F}^\lambda$. These representations are not of the type physicists are usually interested in. There are neither annihilation nor creation operators which can be used to construct the full representation out of a vacuum state.

To obtain representation with the required properties the almost-grading again comes into play. First, using the grading of $\mathcal{F}^\lambda$ it is possible to construct starting from $\mathcal{F}^\lambda$, the forms of weight $\lambda \in 1/2\mathbb{Z}$, the *semi-infinite wedge forms* $\mathcal{H}^\lambda$s.

The vector space $\mathcal{H}^\lambda$ is generated by basis elements which are formal expressions of the type

$$\Phi = f^\lambda_{(i_1)} \wedge f^\lambda_{(i_2)} \wedge f^\lambda_{(i_3)} \wedge \cdots, \tag{9.1}$$

where $(i_1) = (m_1, p_1)$ is a double index indexing our basis elements. The indices are in strictly increasing lexicographical order. They are stabilizing in the sense that they will increase exactly by one starting from a certain index which depends on $\Phi$. The action of $\mathcal{L}$ should be extended by Leibniz rule from $\mathcal{F}^\lambda$ to $\mathcal{H}^\lambda$. But a problem arises. For elements of the critical strip $\mathcal{L}_{[0]}$ of the algebra $\mathcal{L}$ it might happen that they produce infinitely many contributions. The action has to be regularized (as physicists like to call it), which is a well-defined mathematical procedure.

Here the almost-grading has another appearance. By the (strong) almost-graded module structure of $\mathcal{F}^\lambda$ the algebra $\mathcal{L}$ can be embedded into the Lie algebra of both-sided infinite matrices

$$\overline{gl}(\infty) := \{A = (a_{ij})_{i,j \in \mathbb{Z}} \mid \exists r = r(A), \text{ such that } a_{ij} = 0 \text{ if } |i-j| > r\ \}, \tag{9.2}$$

with "finitely many diagonals". The embedding will depend on the weight $\lambda$. For $\overline{gl}(\infty)$ there exists a procedure for the regularization of the action on the semi-infinite wedge product [**23**], [**50**], see also [**51**]. In particular, there is a unique non-trivial central extension $\widehat{gl}(\infty)$. If we pull-back the defining cocycle for the extension we obtain a central extension $\widehat{\mathcal{L}}_\lambda$ of $\mathcal{L}$ and the required regularization of the action of $\widehat{\mathcal{L}}_\lambda$ on $\mathcal{H}^\lambda$. As the embedding of $\mathcal{L}$ depends on the weight $\lambda$ the cocycle will depend too. The pull-back cocycle will be local. Hence, by the classification results of Section 6.3 it is the unique central extension class defined by (6.9) integrated over $C_S$ (up to a rescaling).

In $\mathcal{H}^\lambda$ there are invariant subspaces, which are generated by a certain "vacuum vectors". The subalgebra $\mathcal{L}_{[+]}$ annihilates the vacuum, the central element and the other elements of degree zero act by multiplication with a constant and the whole representation space is generated by $\mathcal{L}_{[-]} \oplus \mathcal{L}_{[0]}$ from the vacuum.

As the function algebra $\mathcal{A}$ operates as multiplication operators on $\mathcal{F}^\lambda$ the above representation can be extended to the algebra $\mathcal{D}^1$ (see details in [**80**], [**91**]) after one passes to central extensions. The cocycle again is local and hence, up to coboundary, it will be a certain linear combination of the 3 generating cocycles for the differential operator algebra. In fact its class will be

$$c_\lambda \cdot [\psi^3_{C_S}] + \frac{2\lambda - 1}{2}[\psi^4_{C_S}] - [\psi^1_{C_S}], \qquad c_\lambda := -2(6\lambda^2 - 6\lambda + 1). \tag{9.3}$$

Recall that $\psi^3$ is the cocycle for the vector field algebra, $\psi^1$ the cocycle for the function algebra, and $\psi^4$ the mixing cocycle. Note that the expression for $c_\lambda$ appears also in Mumford's formula [**85**] relating divisors on the moduli space of curves.

For $\mathcal{L}$ we could rescale the central element. Hence essentially, the central extension $\widehat{\mathcal{L}}$ did not depend on the weight. Here this is different. The central extension $\widehat{\mathcal{D}^1}_\lambda$ depends on it. Furthermore, the representation on $\mathcal{H}^\lambda$ gives a projective representation of the algebra of $\mathcal{D}_\lambda$ of differential operators of all orders. It is exactly the combination (9.3) which lifts to a cocycle for $\mathcal{D}_\lambda$ and gives a central extension $\widehat{\mathcal{D}}_\lambda$.

For the centrally extended algebras $\widehat{\mathfrak{g}}$ in a similar way fermionic Fock space representations can be constructed, see [**101**], [**94**].

**9.2.** $b - c$ **systems.** Related to the above there are other quantum algebra systems which can be realized on $\mathcal{H}^\lambda$. On the space $\mathcal{H}^\lambda$ the forms $\mathcal{F}^\lambda$ act by wedging elements $f^\lambda \in \mathcal{F}^\lambda$ in front of the semi-infinite wedge form, i.e.

$$\Phi \mapsto f^\lambda \wedge \Phi. \tag{9.4}$$

Using the Krichever-Novikov duality pairing (5.4) and by contracting the elements in the semi-infinite wedge forms, the forms $f^{1-\lambda} \in \mathcal{F}^{1-\lambda}$ will act on them too. For $\Phi$ a basis element (9.1) of $\mathcal{H}^\lambda$ the contraction is defined via

$$i(f^{1-\lambda})\Phi = \sum_{l=1}^{\infty}(-1)^{l-1}\langle f^{1-\lambda}, f^\lambda_{i_l}\rangle \cdot f^\lambda_{(i_1)} \wedge f^\lambda_{(i_2)} \wedge \cdots \check{f}^\lambda_{(i_l)}. \tag{9.5}$$

Here $\check{f}^\lambda_{(i_l)}$ indicates as usual that this element will not be there anymore.

Both operations create a Clifford algebra like structure, which is sometimes called a $b - c$ system, see [**91**, Chapters 7 and 8].

**9.3. Vertex algebras.** From $b-c$ systems it is not far to describe the mathematical notion of a global operator field. Furthermore, it is possible to describe operator product expansions also in the Krichever–Novikov setting. Above we discussed *fermionic representations.* In physics also bosonic representations are needed. From the physicists' point of view vertex operators give a "boson-fermion correspondence".

For the mathematical background of vertex algebras in the classical genus zero setting see [**48**], [**52**], [**34**], [**49**]. We will not recall their definition here. Let me only say, that there is a *state-field correspondence* fulfilling certain axioms.

It has to be pointed out that vertex algebras do not only play a role in field theory. They were also crucial in understanding the *Monster and Moonshine* phenomena which refers to the fact that dimensions of irreducible representations of the largest sporadic finite group, the monster group, show up in the coefficients of the $q$-expansion of the elliptic modular function $j$. This was first seen experimentally and later explained with the help of representations of a certain vertex algebra which was related to the monster. The $j$-function appears as graded dimension of a representation of this vertex algebra. The details can be found in [**34**]. Also with the help of vertex algebras representations of Kac-Moody algebras can be constructed.

To construct vertex algebras in higher genus there are different strategies. One is by some kind of semi-local approach very much in the spirit of Tsuchiya, Ueno and Yamada [**106**]. An example is given by Zhu [**109**]. Another direction is based on an operadic approach. See for example Huang and Lepowsky [**42**], [**43**], [**44**], [**45**]. Also there is a sheaf theoretic approach due to Frenkel and Ben-Zvi [**32**], [**33**].

A mathematical treatment via the Krichever–Novikov objects which stays very close to the axiomatic treatment in genus zero is given by Linde [**62**], [**63**]. Strictly speaking, he does it only for the two-point case. His objects, as they are formulated in terms of the KN basis, should extend to the multi-point situation too. The details are not yet done.

A physicist's approach via Krichever–Novikov objects in the context of explicit types of field theories and their special properties is given by Bonora and collaborators [**11**], [**75**]. For a general use of KN type algebras in Quantum Field Theory by physicists see [**91**, Section 14.5]. There an extensive list of names and references can be found.

## 10. Sugawara Representation

In the classical set-up the Sugawara construction relates to a representation of the classical affine Lie algebra $\widehat{\mathfrak{g}}$ a representation of the Virasoro algebra, see e.g. [**48**], [**51**]. In joint work with O. Sheinman the author succeeded in extending it to arbitrary genus and the multi-point setting [**93**]. For an updated treatment, incorporating also the uniqueness results of central extensions, see [**91**, Chapter 10]. Here we will give a very rough sketch.

We start with an admissible representation $V$ of a centrally extended current algebra $\widehat{\mathfrak{g}}$. Admissible means, that the central element operates as constant $\times$ identity, and that every element $v$ in the representation space will be annihilated by the elements in $\widehat{\mathfrak{g}}$ of sufficiently high degree (the degree depends on the element $v$).

For simplicity let $\mathfrak{g}$ be either abelian or simple and $\beta$ the non-degenerate symmetric invariant bilinear form used to construct $\widehat{\mathfrak{g}}$ (now we need that it is non-degenerate). Let $\{u_i\}$, $\{u^j\}$ be a system of dual basis elements for $\mathfrak{g}$ with respect to $\beta$, i.e. $\beta(u_i, u^j) = \delta_i^j$. Note that the Casimir element of $\mathfrak{g}$ can be given by $\sum_i u^i$. For $x \in \mathfrak{g}$ we consider the family of operators $x(n,p)$ given by the operation of $x \otimes A_{n,p}$ on $V$. We group them together in a formal sum

$$\widehat{x}(Q) := \sum_{n\in\mathbb{Z}} \sum_{p=1}^{K} x(n,p)\omega^{n,p}(Q), \quad Q \in \Sigma. \tag{10.1}$$

Such a formal sum is called a field if applied to a vector $v \in V$ it gives again a formal sum (now of elements from $V$) which is bounded from above. By the condition of admissibility $\widehat{x}(Q)$ is a field. It is of conformal weight one, as the one-differentials $\omega^{n,p}$ show up.

The current operator fields are defined as [6]

$$J_i(Q) := \widehat{u_i}(Q) = \sum_{n,p} u_i(n,p)\omega^{n,p}(Q). \tag{10.2}$$

The Sugawara operator field $T(Q)$ is defined by

$$T(Q) := \frac{1}{2} \sum_i :J_i(Q)J^i(Q): \;. \tag{10.3}$$

Here :...: denotes some normal ordering, which is needed to make the product of two fields again a field. The *standard normal ordering* is defined as

$$:x(n,p)y(m,r): \;:= \begin{cases} x(n,p)y(m,r), & (n,p) \le (m,r) \\ y(m,r)x(n,p), & (n,p) > (m,r) \end{cases} \tag{10.4}$$

where the indices $(n,p)$ are lexicographically ordered. By this prescription the annihilation operator, i.e. the operators of positive degree, are brought as much as possible to the right so that they act first.

As the current operators are fields of conformal weights one the Sugawara operator field is a field of weight two. Hence we write it as

$$T(Q) = \sum_{k\in\mathbb{Z}} \sum_{p=1}^{K} L_{k,p} \cdot \Omega^{k,p}(Q) \tag{10.5}$$

with certain operators $L_{k,p}$. The $L_{k,p}$ are called *modes of the Sugawara field $T$* or just Sugawara operators.

Let $2\kappa$ be the eigenvalue of the Casimir operator in the adjoint representation. For $\mathfrak{g}$ abelian $\kappa = 0$. For $\mathfrak{g}$ simple and $\beta$ normalized such that the longest roots have square length 2 then $\kappa$ is the dual Coxeter number. Recall that the central element $t$ acts on the representation space $V$ as $c \cdot id$ with a scalar $c$. This scalar is called the *level* of the representation. The key result is (where $x(g)$ denotes the operator corresponding to the element $x \otimes g$)

PROPOSITION 10.1 ([**91**, Prop. 10.8]). *Let $\mathfrak{g}$ be either an abelian or a simple Lie algebra. Then*

$$[L_{k,p}, x(g)] = -(c+\kappa) \cdot x(e_{k,p} \cdot g), \tag{10.6}$$

[6] For simplicity we drop mentioning the range of summation here and in the following when it is clear.

$$[L_{k,p}, \widehat{x}(Q)] = (c+\kappa)\cdot(e_{k,p}\,.\,\widehat{x}(Q))\,. \tag{10.7}$$

Recall that $e_{k,p}$ are the KN basis elements for the vector field algebra $\mathcal{L}$.

In the next step the commutators of the operators $L_{k,p}$ can be calculated. In the case the $c+\kappa=0$, called the *critical level*, these operators generate a subalgebra of the center of $\mathfrak{gl}(V)$. If $c+\kappa\neq 0$ (i.e. at a non-critical level) the $L_{k,p}$ can be replaced by rescaled elements $L^*_{k,p}=\frac{-1}{c+\kappa}L_{k,p}$ and we we denote by $T[..]$ the linear representation of $\mathcal{L}$ induced by

$$T[e_{k,p}] = \mathcal{L}^*_{k,p}. \tag{10.8}$$

The result is that $T$ defines a projective representation of $\mathcal{L}$ with a local cocycle. This cocycle is up to rescaling our geometric cocycle $\psi^3_{C_S,R}$ with a projective connection [7] $R$. In detail,

$$T[[e,f]] = [T[e],T[f]] + \frac{c\dim\mathfrak{g}}{c+\kappa}\,\psi^3_{C_S,R}(e,f)id. \tag{10.9}$$

Consequently, by setting

$$T[\hat{e}] := T[e],\quad T[t] := \frac{c\dim\mathfrak{g}}{c+\kappa}\,id \tag{10.10}$$

we obtain a honest Lie representation of the centrally extended vector field algebra $\widehat{\mathcal{L}}$ given by this local cocycle. For the general reductive case, see [**91**, Section 10.2.1].

## 11. Application to Moduli Space

This application deals with *Wess-Zumino-Novikov-Witten models* and *Knizhnik-Zamolodchikov Connection.* Despite the fact, that it is a very important application, the following description is very condensed. More can be found in [**94**], [**95**]. See also [**91**], [**103**]. Wess-Zumino-Novikov-Witten (WZNW) models are defined on the basis of a fixed finite-dimensional simple (or semi-simple) Lie algebra $\mathfrak{g}$. One considers families of representations of the affine algebras $\widehat{\mathfrak{g}}$ (which is an almost-graded central extension of $\overline{\mathfrak{g}}$) defined over the moduli space of Riemann surfaces of genus $g$ with $K+1$ marked points and splitting of type $(K,1)$. The single point in $O$ will be a reference point. The data of the moduli of the Riemann surface and the marked points enter the definition of the algebra $\widehat{\mathfrak{g}}$ and the representation. The construction of certain co-invariants yields a special vector bundle of finite rank over moduli space, called the vector bundle of conformal blocks, or Verlinde bundle. With the help of the Krichever Novikov vector field algebra, and using the Sugawara construction, the *Knizhnik-Zamolodchikov (KZ) connection* is given. It is projectively flat. An essential fact is that certain elements in the critical strip $\mathcal{L}_{[0]}$ correspond to infinitesimal deformations of the moduli and to moving the marked points. This gives a global operator approach in contrast to the semi-local approach of Tsuchia, Ueno, and Yamada [**106**].

## References

[1] A. Anzaldo-Meneses, *Krichever-Novikov algebras on Riemann surfaces of genus zero and one with N punctures*, J. Math. Phys. **33** (1992), no. 12, 4155–4163, DOI 10.1063/1.529814. MR1191775 (93j:81025)

[2] E. Arbarello, C. De Concini, V. G. Kac, and C. Procesi, *Moduli spaces of curves and representation theory*, Comm. Math. Phys. **117** (1988), no. 1, 1–36. MR946992 (89i:14019)

[7] The projective connection takes care of the "up to coboundary".

[3] Michael F. Atiyah, *Riemann surfaces and spin structures*, Ann. Sci. École Norm. Sup. (4) **4** (1971), 47–62. MR0286136 (44 #3350)

[4] A. A. Belavin, A. M. Polyakov, and A. B. Zamolodchikov, *Infinite conformal symmetry in two-dimensional quantum field theory*, Nuclear Phys. B **241** (1984), no. 2, 333–380, DOI 10.1016/0550-3213(84)90052-X. MR757857 (86m:81097)

[5] Georgia Benkart and Paul Terwilliger, *The universal central extension of the three-point $\mathfrak{sl}_2$ loop algebra*, Proc. Amer. Math. Soc. **135** (2007), no. 6, 1659–1668, DOI 10.1090/S0002-9939-07-08765-5. MR2286073 (2007m:17031)

[6] Stephen Berman and Yuly Billig, *Irreducible representations for toroidal Lie algebras*, J. Algebra **221** (1999), no. 1, 188–231, DOI 10.1006/jabr.1999.7961. MR1722910 (2000k:17004)

[7] Stephen Berman, Yuly Billig, and Jacek Szmigielski, *Vertex operator algebras and the representation theory of toroidal algebras*, Recent developments in infinite-dimensional Lie algebras and conformal field theory (Charlottesville, VA, 2000), Contemp. Math., vol. 297, Amer. Math. Soc., Providence, RI, 2002, pp. 1–26, DOI 10.1090/conm/297/05090. MR1919810 (2003j:17037)

[8] Stephen Berman and Ben Cox, *Enveloping algebras and representations of toroidal Lie algebras*, Pacific J. Math. **165** (1994), no. 2, 239–267. MR1300833 (95i:17007)

[9] I. N. Bernšteĭn, I. M. Gel′fand, and S. I. Gel′fand, *Differential operators on the base affine space and a study of $\mathfrak{g}$-modules*, Lie groups and their representations (Proc. Summer School, Bolyai János Math. Soc., Budapest, 1971), Halsted, New York, 1975, pp. 21–64. MR0578996 (58 #28285)

[10] Yuly Billig, *Representations of toroidal Lie algebras and Lie algebras of vector fields*, Resenhas **6** (2004), no. 2-3, 111–119. MR2215973 (2007h:17025)

[11] Loriano Bonora, Adrian Lugo, Marco Matone, and Jorge Russo, *A global operator formalism on higher genus Riemann surfaces: b-c systems*, Comm. Math. Phys. **123** (1989), no. 2, 329–352. MR1002042 (91a:81184)

[12] L. Bonora, M. Martellini, M. Rinaldi, and J. Russo, *Neveu-Schwarz- and Ramond-type superalgebras on genus-g Riemann surfaces*, Phys. Lett. B **206** (1988), no. 3, 444–450, DOI 10.1016/0370-2693(88)91607-3. MR944268 (90d:81106)

[13] Murray R. Bremner, *On a Lie algebra of vector fields on a complex torus*, J. Math. Phys. **31** (1990), no. 8, 2033–2034, DOI 10.1063/1.528652. MR1067650 (91j:17040)

[14] Murray R. Bremner, *Structure of the Lie algebra of polynomial vector fields on the Riemann sphere with three punctures*, J. Math. Phys. **32** (1991), no. 6, 1607–1608, DOI 10.1063/1.529499. MR1109216 (92i:17030)

[15] Murray Bremner, *Universal central extensions of elliptic affine Lie algebras*, J. Math. Phys. **35** (1994), no. 12, 6685–6692, DOI 10.1063/1.530700. MR1303073 (95i:17024)

[16] Murray Bremner, *Four-point affine Lie algebras*, Proc. Amer. Math. Soc. **123** (1995), no. 7, 1981–1989, DOI 10.2307/2160931. MR1249871 (95i:17025)

[17] Peter Bryant, *Graded Riemann surfaces and Krichever-Novikov algebras*, Lett. Math. Phys. **19** (1990), no. 2, 97–108, DOI 10.1007/BF01045879. MR1039517 (91f:17027)

[18] Samuel Buelk, Ben Cox, and Elizabeth Jurisich, *A Wakimoto type realization of toroidal $\mathfrak{sl}_{n+1}$*, Algebra Colloq. **19** (2012), no. Special Issue No.1, 841–866, DOI 10.1142/S1005386712000727. MR2999239

[19] André Bueno, Ben Cox, and Vyacheslav Futorny, *Free field realizations of the elliptic affine Lie algebra $\mathfrak{sl}(2,\mathbf{R}) \oplus (\Omega_R/d\mathrm{R})$*, J. Geom. Phys. **59** (2009), no. 9, 1258–1270, DOI 10.1016/j.geomphys.2009.06.007. MR2541818 (2010k:17035)

[20] Chopp, M., *Lie-admissible structures on Witt type algebras and automorphic algebras*, Phd thesis 2011, University of Luxembourg and University of Metz.

[21] Ben Cox, *Realizations of the four point affine Lie algebra $\mathfrak{sl}(2,R)\oplus(\Omega_R/dR)$*, Pacific J. Math. **234** (2008), no. 2, 261–289, DOI 10.2140/pjm.2008.234.261. MR2373448 (2008k:17030)

[22] B. Cox and V. Futorny, *Borel subalgebras and categories of highest weight modules for toroidal Lie algebras*, J. Algebra **236** (2001), no. 1, 1–28, DOI 10.1006/jabr.2000.8509. MR1808344 (2001k:17011)

[23] Etsurō Date, Masaki Kashiwara, Michio Jimbo, and Tetsuji Miwa, *Transformation groups for soliton equations*, Nonlinear integrable systems—classical theory and quantum theory (Kyoto, 1981), World Sci. Publishing, Singapore, 1983, pp. 39–119. MR725700 (86a:58093)

[24] Thomas Deck, *Deformations from Virasoro to Krichever-Novikov algebras*, Phys. Lett. B **251** (1990), no. 4, 535–540, DOI 10.1016/0370-2693(90)90793-6. MR1084542 (92d:17027)

[25] Ivan Dimitrov, Vyacheslav Futorny, and Ivan Penkov, *A reduction theorem for highest weight modules over toroidal Lie algebras*, Comm. Math. Phys. **250** (2004), no. 1, 47–63, DOI 10.1007/s00220-004-1142-3. MR2092028 (2006g:17010)

[26] Dmitry Donin and Boris Khesin, *Pseudodifferential symbols on Riemann surfaces and Krichever-Novikov algebras*, Comm. Math. Phys. **272** (2007), no. 2, 507–527, DOI 10.1007/s00220-007-0234-2. MR2300251 (2008g:17027)

[27] Alice Fialowski, *Formal rigidity of the Witt and Virasoro algebra*, J. Math. Phys. **53** (2012), no. 7, 073501, 5, DOI 10.1063/1.4731220. MR2985241

[28] Alice Fialowski, *Deformations of some infinite-dimensional Lie algebras*, J. Math. Phys. **31** (1990), no. 6, 1340–1343, DOI 10.1063/1.528720. MR1054321 (91i:17028)

[29] Alice Fialowski and Martin Schlichenmaier, *Global deformations of the Witt algebra of Krichever-Novikov type*, Commun. Contemp. Math. **5** (2003), no. 6, 921–945, DOI 10.1142/S0219199703001208. MR2030563 (2005b:17048)

[30] Alice Fialowski and Martin Schlichenmaier, *Global geometric deformations of current algebras as Krichever-Novikov type algebras*, Comm. Math. Phys. **260** (2005), no. 3, 579–612, DOI 10.1007/s00220-005-1423-5. MR2183958 (2006k:17039)

[31] A. Fialowski and M. Schlichenmaier, *Global geometric deformations of the Virasoro algebra, current and affine algebras by Krichever-Novikov type algebras*, Internat. J. Theoret. Phys. **46** (2007), no. 11, 2708–2724, DOI 10.1007/s10773-007-9383-5. MR2363750 (2009b:17063)

[32] Edward Frenkel, *Vertex algebras and algebraic curves*, Astérisque **276** (2002), 299–339. Séminaire Bourbaki, Vol. 1999/2000. MR1886764 (2003f:17037)

[33] Edward Frenkel and David Ben-Zvi, *Vertex algebras and algebraic curves*, Mathematical Surveys and Monographs, vol. 88, American Mathematical Society, Providence, RI, 2001. MR1849359 (2003f:17036)

[34] Igor Frenkel, James Lepowsky, and Arne Meurman, *Vertex operator algebras and the Monster*, Pure and Applied Mathematics, vol. 134, Academic Press, Inc., Boston, MA, 1988. MR996026 (90h:17026)

[35] Vyacheslav Futorny and Iryna Kashuba, *Verma type modules for toroidal Lie algebras*, Comm. Algebra **27** (1999), no. 8, 3979–3991, DOI 10.1080/00927879908826677. MR1700209 (2000c:17040)

[36] A. Grothendieck and J. A. Dieudonné, *Eléments de géométrie algébrique. I* (French), Grundlehren der Mathematischen Wissenschaften [Fundamental Principles of Mathematical Sciences], vol. 166, Springer-Verlag, Berlin, 1971. MR3075000

[37] Laurent Guieu and Claude Roger, *L'algèbre et le groupe de Virasoro* (French), Les Publications CRM, Montreal, QC, 2007. Aspects géométriques et algébriques, généralisations. [Geometric and algebraic aspects, generalizations]; With an appendix by Vlad Sergiescu. MR2362888 (2009b:17064)

[38] R. C. Gunning, *Lectures on Riemann surfaces*, Princeton Mathematical Notes, Princeton University Press, Princeton, N.J., 1966. MR0207977 (34 #7789)

[39] Robin Hartshorne, *Algebraic geometry*, Springer-Verlag, New York-Heidelberg, 1977. Graduate Texts in Mathematics, No. 52. MR0463157 (57 #3116)

[40] N. S. Hawley and M. Schiffer, *Half-order differentials on Riemann surfaces*, Acta Math. **115** (1966), 199–236. MR0190326 (32 #7739)

[41] Brian Hartwig and Paul Terwilliger, *The tetrahedron algebra, the Onsager algebra, and the $\mathfrak{sl}_2$ loop algebra*, J. Algebra **308** (2007), no. 2, 840–863, DOI 10.1016/j.jalgebra.2006.09.011. MR2295093 (2007m:17034)

[42] Yi-Zhi Huang, *Operadic formulation of topological vertex algebras and Gerstenhaber or Batalin-Vilkovisky algebras*, Comm. Math. Phys. **164** (1994), no. 1, 105–144. MR1288155 (95g:17029)

[43] Yi-Zhi Huang, *Meromorphic open-string vertex algebras*, J. Math. Phys. **54** (2013), no. 5, 051702, 33, DOI 10.1063/1.4806686. MR3098909

[44] Yi-Zhi Huang and Liang Kong, *Open-string vertex algebras, tensor categories and operads*, Comm. Math. Phys. **250** (2004), no. 3, 433–471, DOI 10.1007/s00220-004-1059-x. MR2094470 (2005h:17049)

[45] Yi-Zhi Huang and James Lepowsky, *On the $\mathcal{D}$-module and formal-variable approaches to vertex algebras*, Topics in geometry, Progr. Nonlinear Differential Equations Appl., vol. 20, Birkhäuser Boston, Boston, MA, 1996, pp. 175–202. MR1390314 (97f:17032)

[46] Tatsuro Ito and Paul Terwilliger, *Finite-dimensional irreducible modules for the three-point $\mathfrak{sl}_2$ loop algebra*, Comm. Algebra **36** (2008), no. 12, 4557–4598, DOI 10.1080/00927870802185963. MR2473348 (2009j:17019)

[47] V. G. Kac, *Simple irreducible graded Lie algebras of finite growth* (Russian), Izv. Akad. Nauk SSSR Ser. Mat. **32** (1968), 1323–1367. MR0259961 (41 #4590)

[48] Victor G. Kac, *Infinite-dimensional Lie algebras*, 3rd ed., Cambridge University Press, Cambridge, 1990. MR1104219 (92k:17038)

[49] Victor Kac, *Vertex algebras for beginners*, 2nd ed., University Lecture Series, vol. 10, American Mathematical Society, Providence, RI, 1998. MR1651389 (99f:17033)

[50] Victor G. Kac and Dale H. Peterson, *Spin and wedge representations of infinite-dimensional Lie algebras and groups*, Proc. Nat. Acad. Sci. U.S.A. **78** (1981), no. 6, 3308–3312. MR619827 (82j:17019)

[51] V. G. Kac and A. K. Raina, *Bombay lectures on highest weight representations of infinite-dimensional Lie algebras*, Advanced Series in Mathematical Physics, vol. 2, World Scientific Publishing Co., Inc., Teaneck, NJ, 1987. MR1021978 (90k:17013)

[52] Kac, V., Raina, A., Rozhkovskaya, N., *Highest weight representations of infinite dimensional Lie algebras, 2. Edition*, World Scientific, 2013.

[53] V. G. Knizhnik and A. B. Zamolodchikov, *Current algebra and Wess-Zumino model in two dimensions*, Nuclear Phys. B **247** (1984), no. 1, 83–103, DOI 10.1016/0550-3213(84)90374-2. MR853258 (87h:81129)

[54] Marie Kreusch, *Extensions of superalgebras of Krichever-Novikov type*, Lett. Math. Phys. **103** (2013), no. 11, 1171–1189, DOI 10.1007/s11005-013-0628-3. MR3095141

[55] I. Krichever, *Vector bundles and Lax equations on algebraic curves*, Comm. Math. Phys. **229** (2002), no. 2, 229–269, DOI 10.1007/s002200200659. MR1923174 (2003m:14054)

[56] Igor Moiseevich Krichever and S. P. Novikov, *Algebras of Virasoro type, Riemann surfaces and the structures of soliton theory* (Russian), Funktsional. Anal. i Prilozhen. **21** (1987), no. 2, 46–63. MR902293 (88i:17016)

[57] Igor Moiseevich Krichever and S. P. Novikov, *Algebras of Virasoro type, Riemann surfaces and strings in Minkowski space* (Russian), Funktsional. Anal. i Prilozhen. **21** (1987), no. 4, 47–61, 96. MR925072 (89f:17020)

[58] Igor Moiseevich Krichever and S. P. Novikov, *Algebras of Virasoro type, the energy-momentum tensor, and operator expansions on Riemann surfaces* (Russian), Funktsional. Anal. i Prilozhen. **23** (1989), no. 1, 24–40, DOI 10.1007/BF01078570; English transl., Funct. Anal. Appl. **23** (1989), no. 1, 19–33. MR998426 (90k:17049)

[59] I. M. Krichever and O. K. Sheĭnman, *Lax operator algebras* (Russian, with Russian summary), Funktsional. Anal. i Prilozhen. **41** (2007), no. 4, 46–59, 96, DOI 10.1007/s10688-007-0026-7; English transl., Funct. Anal. Appl. **41** (2007), no. 4, 284–294. MR2411605 (2009i:37177)

[60] Le Man Kuang, *The q-deformed Krichever-Novikov algebra and Ward q-identities for correlators on higher genus Riemann surfaces*, J. Phys. A **26** (1993), no. 20, 5435–5441. MR1248725 (94j:81084)

[61] Séverine Leidwanger and Sophie Morier-Genoud, *Superalgebras associated to Riemann surfaces: Jordan algebras of Krichever-Novikov type*, Int. Math. Res. Not. IMRN **19** (2012), 4449–4474, DOI 10.1093/imrn/rnr196. MR2981716

[62] K. J. Linde, *Global vertex algebras on Riemann surfaces*. München, Dissertation, 200 pp., 2004.

[63] K. J. Linde, *Towards vertex algebras of Krichever–Novikov type, Part I.* arXiv: math/0305428 (2003), 21 pages.

[64] S. Lombardo and A. V. Mikhailov, *Reduction groups and automorphic Lie algebras*, Comm. Math. Phys. **258** (2005), no. 1, 179–202, DOI 10.1007/s00220-005-1334-5. MR2166845 (2006g:17026)

[65] S. Lombardo and A. V. Mikhailov, *Reductions of integrable equations and automorphic Lie algebras*, SPT 2004—Symmetry and perturbation theory, World Sci. Publ., Hackensack, NJ, 2005, pp. 183–192, DOI 10.1142/9789812702142_0022. MR2331220 (2008d:37126)

[66] Sara Lombardo and Jan A. Sanders, *On the classification of automorphic Lie algebras*, Comm. Math. Phys. **299** (2010), no. 3, 793–824, DOI 10.1007/s00220-010-1092-x. MR2718933 (2011h:17042)

[67] Maria Meiler and Andreas Ruffing, *Constructing similarity solutions to discrete basic diffusion equations*, Adv. Dyn. Syst. Appl. **3** (2008), no. 1, 41–51. MR2547660 (2010g:39007)

[68] Robert V. Moody, *Euclidean Lie algebras*, Canad. J. Math. **21** (1969), 1432–1454. MR0255627 (41 #287)

[69] S. Eswara Rao and R. V. Moody, *Vertex representations for n-toroidal Lie algebras and a generalization of the Virasoro algebra*, Comm. Math. Phys. **159** (1994), no. 2, 239–264. MR1256988 (94m:17028)

[70] Robert V. Moody, Senapathi Eswara Rao, and Takeo Yokonuma, *Toroidal Lie algebras and vertex representations*, Geom. Dedicata **35** (1990), no. 1-3, 283–307, DOI 10.1007/BF00147350. MR1066569 (91i:17032)

[71] C. H. Oh and K. Singh, *q-deformation of the Krichever-Novikov algebra*, Lett. Math. Phys. **42** (1997), no. 2, 115–122, DOI 10.1023/A:1007494105829. MR1479355 (98f:17013)

[72] V. Ovsienko, *Lie antialgebras: prémices*, J. Algebra **325** (2011), 216–247, DOI 10.1016/j.jalgebra.2010.10.003. MR2745538 (2011k:17054)

[73] Andreas Ruffing, *On discretized generalizations of Krichever-Novikov algebras*, Open Syst. Inf. Dyn. **7** (2000), no. 3, 237–254, DOI 10.1023/A:1009636717273. MR1826132 (2002b:81052)

[74] Andreas Ruffing, Thomas Deck, and Martin Schlichenmaier, *String branchings on complex tori and algebraic representations of generalized Krichever-Novikov algebras*, Lett. Math. Phys. **26** (1992), no. 1, 23–32, DOI 10.1007/BF00420515. MR1193623 (93j:17054)

[75] Jorge Russo, *Bosonization and fermion vertices on an arbitrary genus Riemann surface by using a global operator formalism*, Modern Phys. Lett. A **4** (1989), no. 24, 2349–2362, DOI 10.1142/S0217732389002641. MR1049109 (91j:17049)

[76] V. A. Sadov, *Bases on multipunctured Riemann surfaces and interacting strings amplitudes*, Comm. Math. Phys. **136** (1991), no. 3, 585–597. MR1099697 (92e:81069)

[77] Martin Schlichenmaier, *Krichever-Novikov algebras for more than two points*, Lett. Math. Phys. **19** (1990), no. 2, 151–165, DOI 10.1007/BF01045886. MR1039524 (91a:17039)

[78] Martin Schlichenmaier, *Krichever-Novikov algebras for more than two points: explicit generators*, Lett. Math. Phys. **19** (1990), no. 4, 327–336, DOI 10.1007/BF00429952. MR1051815 (91i:17034)

[79] Martin Schlichenmaier, *Central extensions and semi-infinite wedge representations of Krichever-Novikov algebras for more than two points*, Lett. Math. Phys. **20** (1990), no. 1, 33–46, DOI 10.1007/BF00417227. MR1058993 (92c:17035)

[80] Martin Schlichenmaier, *Verallgemeinerte Krichever–Novikov Algebren und deren Darstellungen.* Ph.D. thesis, Universität Mannheim, 1990.

[81] Martin Schlichenmaier, *Degenerations of generalized Krichever-Novikov algebras on tori*, J. Math. Phys. **34** (1993), no. 8, 3809–3824, DOI 10.1063/1.530008. MR1230553 (94e:17040)

[82] Martin Schlichenmaier, *Zwei Anwendungen algebraisch-geometrischer Methoden in der theoretischen Physik: Berezin-Toeplitz-Quantisierung und globale Algebren der zweidimensionalen konformen Feldtheorie"*, Habilitation Thesis, University of Mannheim, June 1996.

[83] Martin Schlichenmaier, *Local cocycles and central extensions for multipoint algebras of Krichever-Novikov type*, J. Reine Angew. Math. **559** (2003), 53–94, DOI 10.1515/crll.2003.052. MR1989644 (2004c:17056)

[84] Martin Schlichenmaier, *Higher genus affine algebras of Krichever-Novikov type* (English, with English and Russian summaries), Mosc. Math. J. **3** (2003), no. 4, 1395–1427. MR2058804 (2005f:17025)

[85] Martin Schlichenmaier, *An introduction to Riemann surfaces, algebraic curves and moduli spaces*, Lecture Notes in Physics, vol. 322, Springer-Verlag, Berlin, 1989. With an introduction by Ian McArthur. MR981595 (90c:32033)

[86] Martin Schlichenmaier, *Higher genus affine Lie algebras of Krichever-Novikov type*, Difference equations, special functions and orthogonal polynomials, World Sci. Publ., Hackensack, NJ, 2007, pp. 589–599, DOI 10.1142/9789812770752_0051. MR2451204 (2009i:17038)

[87] Martin Schlichenmaier, *An elementary proof of the vanishing of the second cohomology of the Witt and Virasoro algebra with values in the adjoint module*, arXiv:1111.6624v1, 2011, DOI: 10.1515/forum-2011-0143, February 2012.

[88] Martin Schlichenmaier, *Lie superalgebras of Krichever-Novikov type and their central extensions*, Anal. Math. Phys. **3** (2013), no. 3, 235–261, DOI 10.1007/s13324-013-0056-7. MR3084396

[89] Martin Schlichenmaier, *From the Virasoro algebra to Krichever-Novikov type algebras and beyond*, Harmonic and complex analysis and its applications, Trends Math., Birkhäuser/Springer, Cham, 2014, pp. 325–358, DOI 10.1007/978-3-319-01806-5_7. MR3203105

[90] M. Schlichenmaĭer, *Multipoint Lax operator algebras. Almost-graded structure and central extensions* (Russian, with Russian summary), Mat. Sb. **205** (2014), no. 5, 117–160; English transl., Sb. Math. **205** (2014), no. 5-6, 722–762. MR3242635

[91] Martin Schlichenmaier, *Krichever-Novikov type algebras*, De Gruyter Studies in Mathematics, vol. 53, De Gruyter, Berlin, 2014. Theory and applications. MR3237422

[92] Martin Schlichenmaier, *$N$-point Virasoro algebras are multi-point Krichever–Novikov type algebras* , arXiv:1505.00736.

[93] M. Schlichenmaier and O. K. Scheinman, *The Sugawara construction and Casimir operators for Krichever-Novikov algebras*, J. Math. Sci. (New York) **92** (1998), no. 2, 3807–3834, DOI 10.1007/BF02434007. Complex analysis and representation theory, 1. MR1666274 (2000g:17036)

[94] M. Schlichenmaier and O. K. Scheinman, *Wess-Zumino-Witten-Novikov theory, Knizhnik-Zamolodchikov equations, and Krichever-Novikov algebras, I.*. Russian Math. Surv. (Uspekhi Math. Nauk.) **54** (1999), 213–250, math.QA/9812083.

[95] M. Schlichenmaĭer and O. K. Sheĭnman, *The Knizhnik-Zamolodchikov equations for positive genus, and Krichever-Novikov algebras* (Russian, with Russian summary), Uspekhi Mat. Nauk **59** (2004), no. 4(358), 147–180, DOI 10.1070/RM2004v059n04ABEH000760; English transl., Russian Math. Surveys **59** (2004), no. 4, 737–770. MR2106647 (2005k:32016)

[96] M. Schlichenmaĭer and O. K. Sheĭnman, *Central extensions of Lax operator algebras* (Russian, with Russian summary), Uspekhi Mat. Nauk **63** (2008), no. 4(382), 131–172, DOI 10.1070/RM2008v063n04ABEH004550; English transl., Russian Math. Surveys **63** (2008), no. 4, 727–766. MR2483201 (2009k:37161)

[97] O. K. Sheĭnman, *Elliptic affine Lie algebras* (Russian), Funktsional. Anal. i Prilozhen. **24** (1990), no. 3, 51–61, 96, DOI 10.1007/BF01077962; English transl., Funct. Anal. Appl. **24** (1990), no. 3, 210–219 (1991). MR1082031 (93i:17027)

[98] O. K. Sheĭnman, *Highest Weight modules over certain quasi-graded Lie algebras on elliptic curves*, Funktional Anal. i Prilozhen. **26**, No.3 (1992), 203–208.

[99] O. K. Sheĭnman, *Affine Lie algebras on Riemann surfaces* (Russian, with Russian summary), Funktsional. Anal. i Prilozhen. **27** (1993), no. 4, 54–62, 96, DOI 10.1007/BF01078844; English transl., Funct. Anal. Appl. **27** (1993), no. 4, 266–272 (1994). MR1264318 (95c:17033)

[100] O. K. Sheĭnman, *Highest-weight modules for affine Lie algebras on Riemann surfaces* (Russian, with Russian summary), Funktsional. Anal. i Prilozhen. **29** (1995), no. 1, 56–71, 96, DOI 10.1007/BF01077040; English transl., Funct. Anal. Appl. **29** (1995), no. 1, 44–55. MR1328538 (96b:17005)

[101] O. K. Sheĭnman, *A fermionic model of representations of affine Krichever-Novikov algebras* (Russian, with Russian summary), Funktsional. Anal. i Prilozhen. **35** (2001), no. 3, 60–72, 96, DOI 10.1023/A:1012378912526; English transl., Funct. Anal. Appl. **35** (2001), no. 3, 209–219. MR1864989 (2003h:17011)

[102] O. K. Sheĭnman, *Lax operator algebras and Hamiltonian integrable hierarchies* (Russian, with Russian summary), Uspekhi Mat. Nauk **66** (2011), no. 1(397), 151–178, DOI 10.1070/RM2011v066n01ABEH004730; English transl., Russian Math. Surveys **66** (2011), no. 1, 145–171. MR2841688 (2012k:37146)

[103] O. K. Sheĭnman, *Current algebras on Riemann surfaces*, Expositions in Mathematics, Vol. 58, De Gruyter, 2012.

[104] Oleg K. Sheinman, *Lax operator algebras of type $G_2$*, Topology, geometry, integrable systems, and mathematical physics, Amer. Math. Soc. Transl. Ser. 2, vol. 234, Amer. Math. Soc., Providence, RI, 2014, pp. 373–392. MR3307157

[105] Oleg K. Sheinman, *Lax operator algebras and gradings on semi-simple Lie algebras*, 18 pp., arXiv:1406.5017.

[106] Akihiro Tsuchiya, Kenji Ueno, and Yasuhiko Yamada, *Conformal field theory on universal family of stable curves with gauge symmetries*, Integrable systems in quantum field theory and statistical mechanics, Adv. Stud. Pure Math., vol. 19, Academic Press, Boston, MA, 1989, pp. 459–566. MR1048605 (92a:81191)

[107] A. N. Tjurin, *The classification of vector bundles over an algebraic curve of arbitrary genus* (Russian), Izv. Akad. Nauk SSSR Ser. Mat. **29** (1965), 657–688. MR0217071 (36 #166a)

[108] Li-Meng Xia and Nai-Hong Hu, *Irreducible representations for Virasoro-toroidal Lie algebras*, J. Pure Appl. Algebra **194** (2004), no. 1-2, 213–237, DOI 10.1016/j.jpaa.2004.05.001. MR2086083 (2006a:17028)

[109] Yongchang Zhu, *Global vertex operators on Riemann surfaces*, Comm. Math. Phys. **165** (1994), no. 3, 485–531. MR1301621 (96i:17025)

University of Luxembourg, Mathematics Research Unit, FSTC, Campus Kirchberg, 6, rue Coudenhove-Kalergi, L-1359 Luxembourg-Kirchberg, Luxembourg

*E-mail address*: martin.schlichenmaier@uni.lu

Proceedings of Symposia in Pure Mathematics
Volume **92**, 2016
http://dx.doi.org/10.1090/pspum/092/01586

# Lax operator algebras and Lax equations

Oleg K. Sheinman

Abstract. A review of the current state of the theory, and applications of Lax operator algebras is given. It reflects crucial changes in the approach to the Lax operator algebras that happened in the beginning of 2014. The presentation is being carried on from the point of view of gradings on semi-simple Lie algebras. The Tyurin parametrization is retrieved in the framework of this approach. A big part of the article is devoted to new approaches to the commuting hierarchies of finite-dimensional Lax equations, including their Hamiltonianity. The last section treats the relationship of finite-dimensional Lax equations to CFT.

Contents



## 1. Introduction

The present paper provides a review of the current state of the theory of Lax operator algebras, their applications to integrable systems, and their quantization. It follows the lines of the series of authors' talks at the Southeastern Workshop on Lie Theory, 2012, Charleston, SC, USA. However, the approach to the Lax operator algebras has been crucially changed since that time: the older definition based on the Tyurin parametrization of holomorphic vector bundles on Riemann surfaces has given place to the new one, based on gradings of semi-simple Lie algebras. We will speak of the old, and new approaches to Lax operator algebras in this context, despite the fact that the first of them is only seven years old. We would like stress that the two approaches to Lax operator algebras do not cancel each other. Vice versa, they reflect relations of Lax operator algebras to holomorphic vector bundles

The work is supported in part by the Russian Foundation for Basic Research under the grant 13-01-12469-ofi-m2, and by the research project "Geometry and Mathematical Quantization" of the University of Luxembourg.

on Riemann surfaces on the one hand, and to semi-simple Lie algebras on the other which are a subject for a further investigation.

Lax operator algebras are Lie algebras of current type on complex compact Riemann surfaces. The history of their emergence is as follows.

In 2001, in his programmatic work [**14**], I.M. Krichever proposed his theory of Lax operators with the spectral parameter on a Riemann surface. From a general and unifying point of view, related to the Tyurin parameters of holomorphic vector bundles, he has given a definition of Lax operators, proved existence and Hamiltonianity of their commuting hierarchies. His theory provides an effective approach to many classical and recently discovered finite-dimensional integrable systems including Calogero–Moser and Hitchin systems corresponding to the root system $A_n$. It provides explicit expressions for Hamiltonians of the corresponding hierarchies, description of action-angle coordinates, and symplectic geometry of the corresponding integrable systems. I.M. Krichever proposed also a similar approach to the zero-curvature, and Schlesinger-type equations which we don't discuss here. His approach goes back to his joint with S.P. Novikov technique of finding high rank finite-zone solutions to Kadomtsev–Petviashvili and Shrödinger equations [**8**, **9**].

In 2006, I.M. Krichever and the author discovered that the Lax operators with spectral parameter on a Riemann surface form an associative algebra [**15**], and constructed their orthogonal and symplectic analogs which form Lie algebras. They were called Lax operator algebras. These algebras possess an important property of almost grading which is weaker than grading but stronger than filtration. It leads to many consequences, in particular, enables one to construct analogs of highest weight representations. Lax operator algebras possess also a unique (in the class of almost graded Lie algebras) central extension. Both almost grading structure, and the central extension of the Lax operator algebras are quite similar to those of loop algebras (where the almost grading is really a grading). These structures have been introduced in [**15**]. A more general approach to the almost gradings has been given in [**21**], a proof of uniqueness of the almost graded central extension – in [**23**]. A new approach to the Lax operator algebras is given in [**32**, **33**]. We reproduce it here, in the beginning of the present paper, because it provides simple and unifying treatment of Lax operator algebras and Lax operators with spectral parameter on an arbitrary Riemann surface, and values in an arbitrary semi-simple complex Lie algebra. We derive the Tyurin parametrization of Lax operator algebras in the framework of the new approach, and use it below for the purposes of the theory of integrable systems.

In the case of classical Lie algebras, the theory of Lax equations with Lax operators of the above type has been quite far developed in [**14**, **26**, **27**] where the existence and Hamiltonianity of a hierarchy of commuting flows has been proven for any Lax operator of that type. The treatment was given in terms of Tyurin parametrization. The proofs there are specific for Lax operators corresponding to different classical Lie algebras, though, obviously, they follow similar lines. A general a posteriori observation is as follows: the constraints on Tyurin parameters making the space of Lax operators closed with respect to their Lie bracket, simultaneously provide sufficient conditions of Hamiltonianity of the corresponding hierarchies.

The existence of commuting hierarchies, and their Hamiltonian theory recently have been reworked in terms of the new approach [**34**–**36**]. Here we reproduce the main results, and sketch proofs of the main theorems referring to the original papers for details. In particular, we outline here a unified approach to existence of commuting hierarchies for all semisimple Lie algebras [**34**]. The Hamiltonian theory is presented, basically, following the lines of [**29**], for the reason that more recent investigations [**35**] were not on time to get into this survey. However, we would like stress that using the *ansatz* for the operators of Lax pairs described here the remainder of the theory (hierarchies and Hamiltonian theory) has been constructed in general terms of semisimple Lie algebras and their invariants.

A Lax operator algebra has a canonical representation, and an analog of Cartan subalgebra related to a generic element. This enables us to assign a kind of CFT to every Lax equation of the type considered here. Mathematically, this is a projective unitary representation of the Poisson algebra of classical observables of the Lax system by Knizhnik–Zamolodchikov-type operators. Physically, this is a prequantization of the Lax integrable system. We treat these questions in Section 5 of the present paper, following [**28**].

**Acknowledgments.** I would like to thank I.M. Krichever, M. Schlichenmaier, and E.B. Vinberg for fruitful discussions. I also would like to thank the organizers of the Southeastern Workshop on Lie Theory, 2012, in Charleston, and the participants for their interest. I am also thankful to the referee for the careful reading of the manuscript.

## 2. Lax operator algebras

**2.1. Basic definition.** In this section, following [**33**], we introduce the Lax operator algebras via gradings on the semisimple Lie algebras, and state their basic properties like almost graded structure, existence and uniqueness of almost graded central extensions. We omit all proofs referring to [**33**] for them. Some important examples follow in the next section.

Let $\mathfrak{g}$ be a semisimple finite dimensional Lie algebra over $\mathbb{C}$, $\mathfrak{h}$ its Cartan subalgebra, $R$ the root system of $\mathfrak{g}$ with respect to $\mathfrak{h}$, $h \in \mathfrak{h}$ such that $\alpha_i(h) \in \mathbb{Z}_+$ for any simple root $\alpha_i$. Given a pair $\mathfrak{h}$, $h$ satisfying the above conditions, and $p \in \mathbb{Z}$, let $\mathfrak{g}_p = \{X \in \mathfrak{g} \mid (\operatorname{ad} h)X = pX\}$, and $k = \max\{p \mid \mathfrak{g}_p \neq 0\}$. Then $\mathfrak{g} = \bigoplus\limits_{i=-k}^{k} \mathfrak{g}_p$ gives a $\mathbb{Z}$-grading on $\mathfrak{g}$. For the theory and classification results on such kind of gradings we refer to [**38**]. We call $k$ the depth of the grading. Obviously, $\mathfrak{g}_p = \bigoplus\limits_{\substack{\alpha\in R \\ \alpha(h)=p}} \mathfrak{g}_\alpha$.

Define also the following filtration of $\mathfrak{g}$: $\tilde{\mathfrak{g}}_p = \bigoplus\limits_{q=-k}^{p} \mathfrak{g}_q$. Then, $\tilde{\mathfrak{g}}_p \subset \tilde{\mathfrak{g}}_{p+1}$ $(p \geq -k)$, $\tilde{\mathfrak{g}}_{-k} = \mathfrak{g}_{-k}, \ldots, \tilde{\mathfrak{g}}_k = \mathfrak{g}$, $\tilde{\mathfrak{g}}_p = \mathfrak{g}$, $p > k$.

Let $\Sigma$ be a complex compact Riemann surface with two disjoint fixed finite sets of marked points: $\Pi$ and $\Gamma$. Let $L$ be a meromorphic map $\Sigma \to \mathfrak{g}$ which is holomorphic outside the marked points, may have poles of arbitrary orders at the points in $\Pi$, and has expansions of the following form at the points in $\Gamma$:

$$L(z) = \sum_{p=-k}^{\infty} L_p z^p, \quad L_p \in \tilde{\mathfrak{g}}_p \tag{2.1}$$

where $z$ is a local coordinate in a neighborhood of $\gamma \in \Gamma$ such that $z(\gamma) = 0$. The grading element $h$ may vary from one $\gamma$ to another. For simplicity we assume that

$k$ is constant at all $\gamma \in \Gamma$ though nothing would change below if we did not assume that.

Let us denote the linear space of all such maps by $\mathcal{L}$. Since the relation (2.1) holds true under the commutator, $\mathcal{L}$ is a Lie algebra (see assertion 1° of Theorem 2.2 below). The Lie algebra $\mathcal{L}$, its properties and applications are the main subjects of the present paper. We keep the name *Lax operator algebras* for this class of current algebras in order to emphasize their succession to those in [**15**, **29**].

**2.2. Representation via Tyurin parameters** [**33**]. It is instructive to derive the initial definition of Lax operator algebras [**15**, **29**] from the general one given above. For this purpose we consider particular examples of gradings. These are gradings of depth 1 or 2 of the classical Lie algebras, and $G_2$ given by simple roots. The last means that given a simple root $\alpha_i$ we put $h \in \mathfrak{h}$ to be the dual element: $\alpha_i(h) = 1$, $\alpha_j(h) = 0$ $(1 \le j \le n,\ j \ne i)$, and proceed as described in the previous section. By [**38**, Sect.2, §3.5], the grading subspace $\mathfrak{g}_p$ is then a direct sum of the root subspaces $\mathfrak{g}_\alpha$ such that $\alpha_i$ is contained in the expansion of $\alpha$ over simple roots with multiplicity $p$, and the depth of the grading given by a simple root is equal to the multiplicity of this simple root in the expansion of the highest root. Below, we keep the following conventions: $\mathfrak{g}_i \subset \mathfrak{g}$ is a subspace with the eigenvalue $(-i)$; in the pictures below the longest lines correspond to the medians of the corresponding matrices.

2.2.1. *Type $A_n$*. In this case $\mathfrak{g}$ has $\left[\frac{n}{2}\right]$ gradings of depth 1 (and no gradings of depth 2 given by simple roots). Let the grading number $r$ $(1 \le r \le \left[\frac{n}{2}\right])$ correspond to the simple root $\alpha_r$. For the grading number 1 the block structure

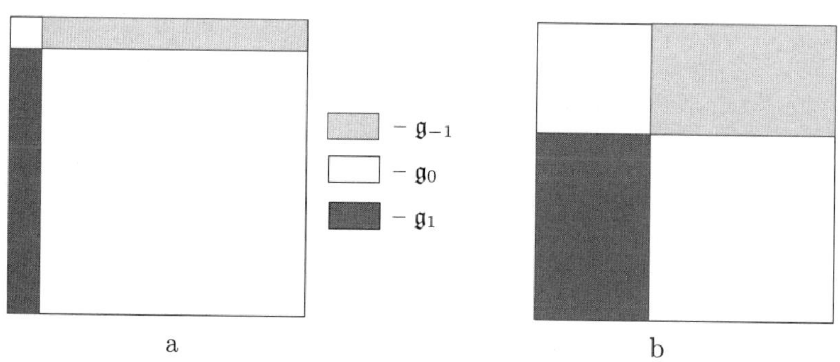


FIGURE 1. Case of $A_n$

of the grading subspaces is given in the figure 1,a. Observe that the matrices corresponding to the subspace $\mathfrak{g}_{-1} = \tilde{\mathfrak{g}}_{-1}$ can be represented as $\alpha\beta^t$ where $\alpha$ is given by $\alpha^t = (1, 0, \ldots, 0)$, and $\beta \in \mathbb{C}^n$ is such that $\beta^t\alpha = 0$. An element $L_0 \in \mathfrak{g}$ belongs to the filtration subspace $\tilde{\mathfrak{g}}_0 = \mathfrak{g}_{-1} \oplus \mathfrak{g}_0$ if, and only if $\alpha$ is its eigenvector. Thus we arrive to the following expansion of a Lax operator for $\mathfrak{sl}(n)$ at a $\gamma$-point [**14**] (see also [**15**, **29**]):

$$L(z) = \alpha\beta^t z^{-1} + L_0 + \ldots$$

where $\beta^t\alpha = 0$, and $L_0\alpha = \varkappa\alpha$ (for some $\varkappa \in \mathbb{C}$). It was this expansion followed by the two conditions what had been used in the initial definitions of Lax operators with spectral parameter on a Riemann surface [**14**], and of Lax operator algebras of type $A_n$ [**15**, **29**]. Since the whole picture is considered up to an inner automorphism, $\alpha \in \mathbb{C}^n$ can be arbitrary here (it is true for all examples in this section).

Here, and below, the (sets of) pairs $\{(\alpha, \gamma) \mid \gamma \in \Gamma\}$ are called *Tyurin parameters* of Lax operators (Lax operator algebras). They appeared first in the theory of holomorphic vector bundles on Riemann surfaces [**37**].

The grading number $r$ is given by the simple root $\alpha_r$. The corresponding matrix realization is presented in the figure 1,b.

2.2.2. *Type $B_n$.* To fix notation, consider the Dynkin diagram $B_n$:

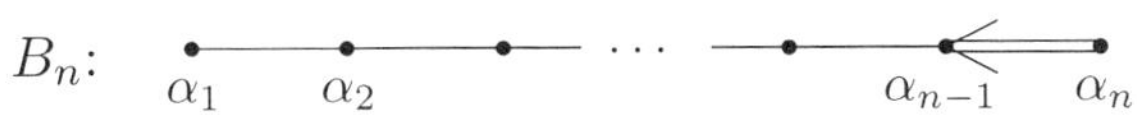


The highest root is $\theta = e_1 + e_2 = \alpha_1 + 2\alpha_2 + \ldots + 2\alpha_n$. We will consider here the grading of depth 1 given by $\alpha_1$, and the grading of depth 2 given by $\alpha_n$. The Lax operator algebra corresponding to the first of them has been found in [**15**], see also [**29**]. The algebra corresponding to the second grading has been found in [**33**].

The blocks corresponding to the grading subspaces in the matrix realization of $\mathfrak{g} = \mathfrak{so}(2n+1)$ with respect to the quadratic form given by the matrix

$$\sigma = \begin{pmatrix} 0 & 0 & E \\ 0 & 1 & 0 \\ E & 0 & 0 \end{pmatrix}$$

are represented in the figure 2,a for $\alpha_1$, and in the figure 2,b for $\alpha_n$.

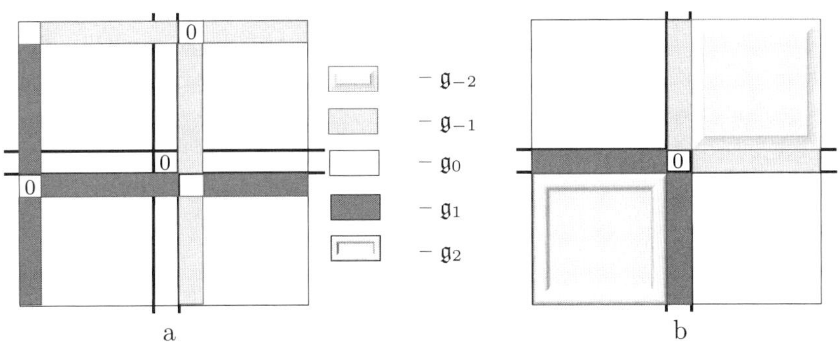


FIGURE 2. Case of $B_n$

In figure 2(a), the subspace $\mathfrak{g}_{-1}$ consists of matrices of the form $(\alpha\beta^t - \beta\alpha^t)\sigma$ where $\alpha, \beta \in \mathbb{C}^{2n+1}$, $\alpha^t = (1, 0, \ldots, 0)$, and $\beta^t\sigma\alpha = 0$ (observe that also $\alpha^t\sigma\alpha = 0$). Again, $\alpha$ is an eigenvector with respect to $\mathfrak{g}_0$. Hence we arrive at the following form of the expansion (2.1) in this case:

$$L(z) = (\alpha\beta^t - \beta\alpha^t)\sigma z^{-1} + L_0 + \ldots$$

where $\alpha$, $\beta$, $L_0$ satisfy the just listed relations (see also [**15, 29**]).

The matrix realization of the grading given by a simple root $\alpha_n$ is represented in the figure 2,b.

2.2.3. *Type $C_n$.* The Dynkin diagram $C_n$ is as follows:

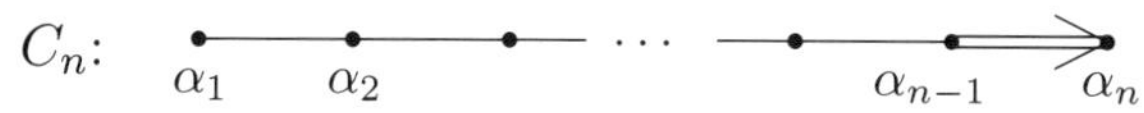


The highest root is $\theta = 2\alpha_1 + 2\alpha_2 + \ldots + 2\alpha_{n-1} + \alpha_n$. We will consider here the grading of depth 2 given by $\alpha_1$, and the grading of depth 1 given by $\alpha_n$. The Lax operator algebra corresponding to the first of them has been found in [**15**], see also [**29**]. The algebra corresponding to the second grading has been found in [**33**].

The block structures of the grading subspaces in the matrix realization of $\mathfrak{g} = \mathfrak{sp}(2n)$ with respect to the symplectic form $\sigma = \begin{pmatrix} 0 & E \\ -E & 0 \end{pmatrix}$ are represented at figure 3,a (for $\alpha_1$), and at figure 3,b (for $\alpha_n$).

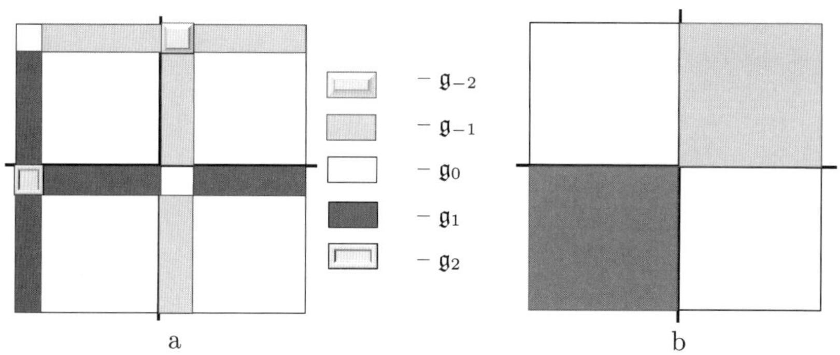


FIGURE 3. Case of $C_n$

In the figure 3,a (the case of $\alpha_1$), the subspace $\mathfrak{g}_{-1}$ consists of matrices of the form $(\alpha\beta^t + \beta\alpha^t)\sigma$ where $\alpha, \beta \in \mathbb{C}^{2n}$, $\alpha^t = (1, 0, \ldots, 0)$, $\beta^t\sigma\alpha = 0$. The subspace $\mathfrak{g}_{-2}$ is one-dimensional, and corresponds to the highest root. It consists of matrices of the form $\nu\alpha\alpha^t\sigma$ where $\alpha \in \mathbb{C}^{2n}$, $\nu \in \mathbb{C}$. Again, $\alpha$ is an eigenvector with respect to $\mathfrak{g}_0$. Finally, observe that for every $L_1 \in \tilde{\mathfrak{g}}_1$ we have $\alpha^t\sigma L_1\alpha = 0$. Hence the relation (2.1) reads as follows in this case:

$$L(z) = \nu\alpha\alpha^t\sigma z^{-2} + (\alpha\beta^t + \beta\alpha^t)\sigma z^{-1} + L_0 + L_1 z + \ldots$$

where $\alpha$, $\beta$, $L_0$, $L_1$ satisfy the just listed relations (see also [**15, 29**]).

The matrix realization of the grading given by a simple root $\alpha_n$ is represented in the figure 3,b.

2.2.4. *Type $D_n$.* The highest root is $\theta = \alpha_1 + 2\alpha_2 + \ldots + 2\alpha_{n-2} + \alpha_{n-1} + \alpha_n$, hence the simple roots $\alpha_1$, $\alpha_{n-1}$, and $\alpha_n$ give gradings of depth 1, but the last two are equivalent under an outer automorphism.

Consider the gradings corresponding to the simple roots $\alpha_1$, and $\alpha_n$. The blocks corresponding to the grading subspaces in the matrix realization of $\mathfrak{g} = \mathfrak{so}(2n)$ with respect to the quadratic form $\sigma = \begin{pmatrix} 0 & E \\ E & 0 \end{pmatrix}$ are represented at figures 4,a (for $\alpha_1$), and 4,b (for $\alpha_n$):

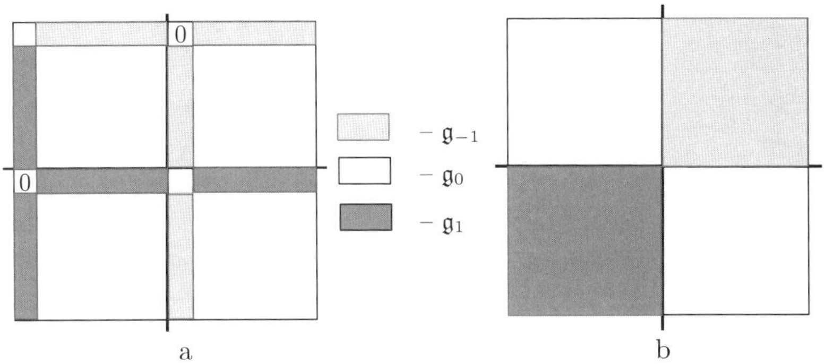


FIGURE 4. Case of $D_n$

Observe that in figure 4(a), $\mathfrak{g}_{-1}$ can be represented as the subspace of rank 2 matrices of the form $(\alpha\beta^t - \beta\alpha^t)\sigma$ where $\alpha, \beta \in \mathbb{C}^{2n}$, $\alpha^t = (1, 0, \ldots, 0)$, and $\beta^t\sigma\alpha = 0$. Observe also that $\alpha$ is an eigenvector with respect to $\mathfrak{g}_0$, and $\alpha^t\sigma\alpha = 0$. Hence we obtain the following expansion (first found in [**15**]) for $L$ at a $\gamma$-point:

$$L(z) = (\alpha\beta^t - \beta\alpha^t)\sigma z^{-1} + L_0 + \ldots$$

where $\alpha$, $\beta$ and $L_0$ satisfy the above relations.

The Lax operator algebra corresponding to the simple root $\alpha_n$ appeared in [**33**] for the first time.

2.2.5. *Type $G_2$*. The Dynkin diagram $G_2$ is as follows:

$G_2$: 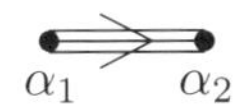


The highest root is $\theta = 3\alpha_1 + 2\alpha_2$.

Below, the block structure of the grading subspaces is described with respect to the exact 7-dimensional representation of the Lie algebra $G_2$ ginen by matrices of the form presented in the figure 5,b. In the figure, dependent blocks are of the same color (bright gray, dark gray, or white). By $[x]$ (where $x \in \mathbb{C}^3$, $x^T = (x_1, x_2, x_3)$) we denote a skew-symmetric matrix given by $[x] = \begin{pmatrix} 0 & x_3 & -x_2 \\ -x_3 & 0 & x_1 \\ x_2 & -x_1 & 0 \end{pmatrix}$. Let us consider first the grading of depth 2 corresponding to $\alpha_2$ (figure 5,a).

It is easy to check that the subspace $\mathfrak{g}_{-2}$ consists of matrices of the form

$$L_{-2} = \mu \begin{pmatrix} 0 & 0 & 0 \\ 0 & \tilde{\alpha}_1\tilde{\alpha}_2^t & 0 \\ 0 & 0 & -\tilde{\alpha}_2\tilde{\alpha}_1^t \end{pmatrix}, \quad \mu \in \mathbb{C} \tag{2.2}$$

where $\tilde{\alpha}_1 = (0, 1, 0)$, $\tilde{\alpha}_2 = (0, 0, 1)$ while the subspace $\mathfrak{g}_{-1}$ consists of matrices of the form

$$L_{-1} = \begin{pmatrix} 0 & -\sqrt{2}\beta_{02}\tilde{\alpha}_2^t & -\sqrt{2}\beta_{01}\tilde{\alpha}_1^t \\ \sqrt{2}\beta_{01}\tilde{\alpha}_1 & \tilde{\alpha}_1\beta_2^t - \beta_1\tilde{\alpha}_2^t & \beta_{02}[\tilde{\alpha}_2] \\ \sqrt{2}\beta_{02}\tilde{\alpha}_2 & \beta_{01}[\tilde{\alpha}_1] & \tilde{\alpha}_2\beta_1^t - \beta_2\tilde{\alpha}_1^t \end{pmatrix} \tag{2.3}$$

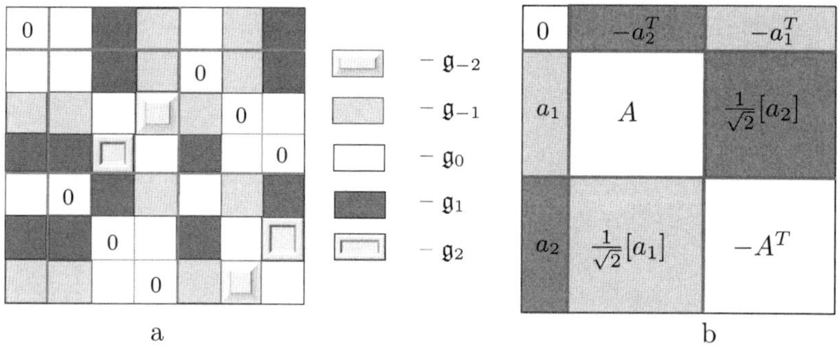


FIGURE 5. Case of $G_2$: depth 2

where $\beta_{01}, \beta_{02} \in \mathbb{C}$ are arbitrary, $\beta_1, \beta_2 \in \mathbb{C}^3$ satisfy the following orthogonality relations: $\tilde{\alpha}_1^t \beta_2 = 0$, $\tilde{\alpha}_2^t \beta_1 = 0$. Observe also that $\tilde{\alpha}_1^t \tilde{\alpha}_2 = 0$ and if $L_0 \in \tilde{\mathfrak{g}}_0$ is given as in the figure 5,b then

$$\tilde{\alpha}_1^t a_2 = 0, \quad \tilde{\alpha}_2^t a_1 = 0, \quad A\tilde{\alpha}_1 = \varkappa_1 \tilde{\alpha}_1, \quad -A^t \tilde{\alpha}_2 = \varkappa_2 \tilde{\alpha}_2 \tag{2.4}$$

where $\varkappa_1, \varkappa_2 \in \mathbb{C}$.

As a result we have obtained the Lax operator algebra recently found in [**30**, **31**]. Hence we may claim that this Lax operator algebra corresponds to the depth 2 grading of $G_2$ given by the simple root $\alpha_2$.

Besides, the Lie algebra $G_2$ has a grading of depth 3 given by the simple root $\alpha_1$. The matrix realization of this grading is given in the figure 6.

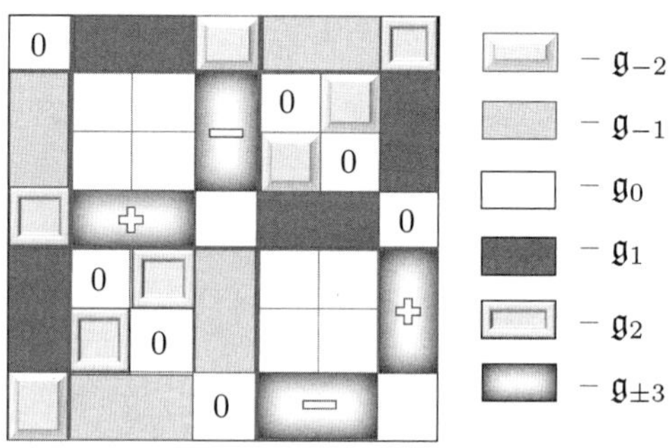

FIGURE 6. Case of $G_2$: depth 3

## 2.3. Almost graded structures [18, 19, 21, 33].

DEFINITION 2.1. Given a Lie algebra $\mathcal{L}$, by an *almost graded structure* on it we mean a system of finite dimensional subspaces $\mathcal{L}_m$, and two non-negative integers $R$, $S$ such that $\mathcal{L} = \bigoplus_{m=-\infty}^{\infty} \mathcal{L}_m$, and $[\mathcal{L}_m, \mathcal{L}_n] \subseteq \bigoplus_{r=m+n-R}^{m+n+S} \mathcal{L}_r$ ($R$, $S$ are independent of $m$, $n$).

The almost graded structure on associative and Lie algebras, and modules over them, was introduced by I.M. Krichever and S.P. Novikov in [**10**]. For the Lax operator algebras and the two point case ($card(\Pi) = 2$) it has been investigated in [**15**]. The most general setup for both Krichever–Novikov and Lax operator algebras has been considered by M. Schlichenmaier [**18**, **19**, **21**].

The above introduced $\mathcal{L}$ possesses a number of almost graded structures. To define one, represent $\Pi$ as a disjoint union of two subsets: $\Pi = \{P_i \mid i = 1, \dots, N\} \cup \{Q_j \mid j = 1, \dots, M\}$. For every $m \in \mathbb{Z}$ consider three divisors:

$$D_m^P = -m \sum_{i=1}^{N} P_i,\ D_m^Q = \sum_{j=1}^{M} (a_j m + b_{m,j})\, Q_j,\ D^\Gamma = k \sum_{\gamma\in\Gamma} \gamma \tag{2.5}$$

where $a_j, b_{m,j} \in \mathbb{Q}$, $a_j > 0$, $a_j m + b_{m,j}$ is an ascending $\mathbb{Z}$ -valued function of $m$, and there exists a $B \in \mathbb{R}_+$ such that $|b_{m,j}| \le B, \forall m \in \mathbb{Z}, j = 1, \dots, M$. We require that

$$\sum_{j=1}^{M} a_j = N, \qquad \sum_{j=1}^{M} b_{m,j} = N + g - 1. \tag{2.6}$$

Let

$$D_m = D_m^P + D_m^Q + D^\Gamma \tag{2.7}$$

and

$$\mathcal{L}_m = \{ L \in \mathcal{L} \mid (L) + D_m \ge 0 \}, \tag{2.8}$$

where $(L)$ is the divisor of a $\mathfrak{g}$-valued function $L$. To be more specific about $(L)$, let us notice that by the order of a meromorphic vector-valued function we mean the minimal order of its entries.

We call $\mathcal{L}_m$ *the (homogeneous, grading) subspace of degree* $m$ of the Lie algebra $\mathcal{L}$.

THEOREM 2.2 ([**33**]).

1° *$\mathcal{L}$ is closed with respect to the point-wise commutator* $[L, L'](P) = [L(P), L'(P)]$ $(P \in \Sigma)$.

2° $\dim \mathcal{L}_m = N \dim \mathfrak{g}$*;*

3° $\mathcal{L} = \bigoplus_{m=-\infty}^{\infty} \mathcal{L}_m$ *;*

4° $[\mathcal{L}_m, \mathcal{L}_n] \subseteq \bigoplus_{r=m+n}^{m+n+S} \mathcal{L}_r$ *where $S$ is a positive integer depending on $N$, $M$, and $g$, and independent of $m$, $n$.*

As it was mentioned in Section 2.1, a proof of the assertion 1° easily follows from the fact that the subspaces $\tilde{\mathfrak{g}}_p$ give a filtration of $\mathfrak{g}$, and for this reason the relation (2.1) holds true under the commutator. For the proof of the assertions 3°, 4° we refer to [**21**] (where they are given for classical Lie algebras but actually are true in our present set-up). The proof of the assertion 2° demonstrates a crucial interaction between the structure of semi-simple Lie algebras, and the Riemann-Roch theorem. We refer to [**32**, **33**] for this proof. Here we only mention that it is based on the dimension formula (3.2) below.

**2.4. Central extensions [10, 15, 21, 23, 29, 33].** In this section we construct the almost graded central extensions of $\mathcal{L}$. We call a central extension almost graded if it inherits the almost graded structure from the original Lie algebra while central elements are relegated to the degree 0 subspace.

Almost graded central extensions are given by local cocycles. Let us recall from [**10**, **15**, **21**, **23**, **29**] that a two-cocycle $\eta$ on $\mathcal{L}$ is called local if $\exists M \in \mathbb{Z}_+$ such that for any $m, n \in \mathbb{Z}$, $|m+n| > M$, and any $L \in \mathcal{L}_m$, $L' \in \mathcal{L}_n$ we have $\eta(L, L') = 0$.

Below, let $\langle \cdot, \cdot \rangle$ denote an invariant symmetric bilinear form on $\mathfrak{g}$. In abuse of notation, we denote by the same symbol the natural pointwise continuation of this form to $\mathfrak{g}$-valued functions, and 1-forms on $\Sigma$. For example, for $L, L' \in \mathcal{L}$ by $\langle L, L' \rangle$ we denote the scalar function on $\Sigma$ taking the value $\langle L(P), L'(P) \rangle$ at every $P \in \Sigma$.

Finally, let $\omega$ be a $\mathfrak{g}_0$-valued 1-form on $\Sigma$ having the expansion of the form

$$\omega(z) = \left( \frac{h}{z} + \omega_0 + \dots \right) dz$$

at any $\gamma \in \Gamma$, where $h \in \mathfrak{h}$ is the element giving the grading on $\mathfrak{g}$ at the point $\gamma$.

THEOREM 2.3 ([**33**]).

1° *For any $L, L' \in \mathcal{L}$ the 1-form $\langle L, (d - \operatorname{ad}\omega) L' \rangle$ is holomorphic except at the points $P_i, Q_j \in \Pi$.*

2° *For any invariant symmetric bilinear form $\langle \cdot, \cdot \rangle$ on $\mathfrak{g}$*

$$\eta(L, L') = \sum_{i=1}^{N} \operatorname{res}_{P_i} \langle L, (d - \operatorname{ad}\omega) L' \rangle$$

*gives a local cocycle on $\mathcal{L}$.*

3° *Up to equivalence, the almost-graded central extensions of $\mathcal{L}$ are in a one-to-one correspondence with the invariant symmetric bilinear forms on $\mathfrak{g}$. In particular, if $\mathfrak{g}$ is simple then the central extension given by the cocycle $\eta$ is unique (in the class of the almost graded central extensions) up to equivalence and rescaling the central element.*

## 3. Lax equations with spectral parameter on Riemann surfaces, and their hierarchies

As it was mentioned in the Introduction, the theory of Lax equations with Lax operators given by Tyurin parameters has been quite far developed in [**14**, **26**, **27**, **29**].

Here, we present an approach to finite-dimensional Lax equations in the more general set-up corresponding to Section 2. Following the multiply quoted here Krichever's program [**14**], such an approach should at least include a proof of existence and Hamiltonianity of hierarchies of Lax equations given by a Lax operator. With this aim, we treat $M$-operators, the counterparts of $L$-operators in Lax equations, in general terms of semi-simple Lie algebras, and their $\mathbb{Z}$-gradings, as it is being done in Section 2. In this generality, we are able to prove a criterium for the Lax equation to be well defined (Theorem 3.1 below), and certain properties of $M$-operators. The theorem of existence of the commuting hierarchy given here (Theorem 3.3) relies on the dimension formula for $M$-operators (3.3). The results are applicable to a broad class of finite-dimensional integrable systems, including Hitchin systems, integrable gyroscopes, integrable cases of flow around a solid, and more.

From now on, we regard the Lax operators as follows. Consider data consisting of the sets $\Pi$ and $\Gamma$, corresponding sets of grading elements, and corresponding sets of spaces of $\mathfrak{g}$-valued Laurent expansions satisfying the restrictions defined in Section 2 with respect to the gradings. In Section 2 we have assigned a Lax operator algebra to such data. Supplying the space of the data with an appropriate topology (and even a complex structure) we can consider a sheaf of Lax operator algebras over the space of just specified data. By *Lax operator* we mean any section of this sheaf.

In Section 3.1 following [**34**] we introduce *M-operators* and study their analytic and algebraic properties. The main problem here is to formulate requirements to the main parts of $M$-operators at the points in $\Gamma$. We quote from [**34**] the dimension formula for $M$-operators. This formula goes back to Krichever [**14**], and plays a fundamental role in the proof of existence of commuting hierarchies.

In Section 3.2 we investigate the equations of motion for main parts of $L$-operators following from Lax equations. We prove a criterion of the dynamical systems given by a Lax equation on certain finite-dimensional subspaces $\mathcal{L}^D \subset \mathcal{L}$ corresponding to non-negative divisors $D$ to be well defined. In the case of existence of the Tyurin parametrization of the system, the just mentioned equations of motion correspond to the dynamics of the Tyurin parameters. The last play a fundamental role in the theory of Kadomtsev–Petviashvili equation [**9**], and is heavily used in [**14**].

In Section 3.3 we formulate, and sketch the proof of the existence of commuting hierarchies of Lax equations following [**29**]. We make a remark on how to generalize this approach to the new setup, and refer to [**34**] for details.

**3.1. $M$-operators and dimension formulas [34].** We call an $M$-operator any meromorphic mapping $M : \Sigma \to \mathfrak{g}$ holomorphic outside the sets $\Pi$, $\Gamma$, such that at every point $\gamma \in \Gamma$,

$$M(z) = \frac{\nu h}{z} + \sum_{i=-k}^{\infty} M_i z^i, \tag{3.1}$$

where $M_i \in \tilde{\mathfrak{g}}_i$ for $i < 0$, $M_i \in \mathfrak{g}$ for $i \geq 0$, $h \in \mathfrak{h}$ is the grading element at $\gamma$, and $\nu \in \mathbb{C}$. As in the relation (2.1), $z$ is a local coordinate in a neighborhood of $\gamma$ such that $z(\gamma) = 0$. Let us denote the set of $M$-operators by $\mathcal{M}$. Obviously, $\mathcal{L} \subset \mathcal{M}$ for $\nu = 0$.

Let us choose an arbitrary non-negative divisor

$$D = \sum_{i=1}^{N} m_i P_i, \quad m_i \geq 0 \ (i = 1, \ldots, N),$$

We now compute the dimension of the spaces $\mathcal{M}^D = \{M \in \mathcal{M} \mid (M) + D + k \sum_{\gamma \in \Gamma} \gamma \geq 0\}$, and $\mathcal{L}^D = \mathcal{M}^D \cap \mathcal{L}$. By the Riemann–Roch theorem, taking into account additional relations, we obtain

$$\dim \mathcal{L}^D = (\dim \mathfrak{g})(\deg D - g + 1) \tag{3.2}$$

independently of $\Gamma$.

If we choose $|\Gamma|$ so that $\left( \sum_{i=-k}^{-1} \dim \tilde{\mathfrak{g}}_i^{\gamma} + 1 \right) |\Gamma| = (\dim \mathfrak{g}) l$ where $l \in \mathbb{Z}_+$ (which is always possible) then

$$\dim \mathcal{M}^D = (\dim \mathfrak{g})(\deg D + l - g + 1). \tag{3.3}$$

Below we always assume that

$$l - g + 1 \geq 0.$$

For example, for the Lie algebras $\mathfrak{gl}(n)$, $\mathfrak{so}(2n)$, $\mathfrak{sp}(2n)$, $G_2$ we can take $l = g$, $|\Gamma| = ng$ where $n = \operatorname{rank}\mathfrak{g}$ (see Examples 1-4 below), and obtain

$$\dim \mathcal{M}^D = (\dim \mathfrak{g})(\deg D + 1). \tag{3.4}$$

In this form, the dimension formula for $\mathcal{M}^D$ has been obtained by I.M. Krichever in [**14**].

EXAMPLE 1. $\mathfrak{g} = \mathfrak{gl}(n)$, the grading of depth 1 is given by means of the simple root $\alpha_1$. Then $k = 1$, $\dim \tilde{\mathfrak{g}}_{-1} = n - 1$, $\dim \mathfrak{g} - (\dim \tilde{\mathfrak{g}}_{-1} + 1)n = \dim \mathfrak{g} - n^2 = 0$.

EXAMPLE 2. $\mathfrak{g} = \mathfrak{so}(2n)$, the grading of depth 1 is given by means of the simple root $\alpha_1$, $k = 1$, $\dim \tilde{\mathfrak{g}}_{-1} = 2n - 2$, $\dim \mathfrak{g} - (\dim \tilde{\mathfrak{g}}_{-1} + 1)n = \dim \mathfrak{g} - (2n - 1)n = 0$.

EXAMPLE 3. $\mathfrak{g} = \mathfrak{sp}(2n)$, the grading of depth 2 is given by means of the simple root $\alpha_1$, $k = 2$, $\dim \mathfrak{g}_{-2} = 1$, $\dim \mathfrak{g}_{-1} = 2n - 2$, $\dim \tilde{\mathfrak{g}}_{-1} = \dim \mathfrak{g}_{-2} + \dim \mathfrak{g}_{-1} = 2n - 1$, $\dim \mathfrak{g} - (\dim \tilde{\mathfrak{g}}_{-2} + \dim \tilde{\mathfrak{g}}_{-1} + 1)n = \dim \mathfrak{g} - (2n + 1)n = 0$.

EXAMPLE 4. $\mathfrak{g} = G_2$, the grading of depth 2 is given by means of the simple root $\alpha_1$, $k = 2$, $n = 2$, $\dim \mathfrak{g}_{-2} = 1$, $\dim \mathfrak{g}_{-1} = 4$, $\dim \tilde{\mathfrak{g}}_{-1} = \dim \mathfrak{g}_{-2} + \dim \mathfrak{g}_{-1} = 5$, $\dim \mathfrak{g} - (\dim \tilde{\mathfrak{g}}_{-2} + \dim \tilde{\mathfrak{g}}_{-1} + 1)n = \dim \mathfrak{g} - 7 \cdot 2 = 0$.

Consider the case $g = 1$. Then the Example 1 corresponds to the Calogero–Moser system for the root system $A_n$.

The following two examples give the Calogero–Moser systems for the root systems $D_n$, $C_n$ and $B_n$.

EXAMPLE 5. $\mathfrak{g} = \mathfrak{so}(2n)$ and $\mathfrak{g} = \mathfrak{sp}(2n)$, $|\Gamma| = 2n$. The gradings on $\mathfrak{g}$ are the same as in the Examples 2 and 3, respectively. We take $l = 2$, and obtain

$$\dim \mathcal{M}^D = (\dim \mathfrak{g})(\deg D + 2). \tag{3.5}$$

EXAMPLE 6. $\mathfrak{g} = \mathfrak{so}(2n + 1)$, the grading of depth 1 is given by means the simple root $\alpha_1$, $k = 1$, $\dim \tilde{\mathfrak{g}}_{-1} = 2n - 1$, $\dim \mathfrak{g} = n(2n + 1)$. Thus $2 \dim \mathfrak{g} = (\dim \tilde{\mathfrak{g}}_{-1} + 1)(2n + 1)$, and we can take $|\Gamma| = 2n + 1$, $l = 2$. In this case the dimension formula has the form (3.5) again.

Observe that the dimension formula (3.4) is not true for $\mathfrak{g} = \mathfrak{so}(2n + 1)$ and, moreover, for $\mathfrak{g} = \mathfrak{sl}(n)$.

**3.2. Lax equations.** Let us give a smooth dependence of the set $\Gamma$ on time, i.e., consider a smooth curve $\Gamma(t)$ in $\Sigma \times \ldots \times \Sigma$ (where the number of co-factors is equal to $|\Gamma|$) avoiding non-generic points. Since $\mathcal{L}$ and $\mathcal{M}$ depend on $\Gamma$, we shall denote them $\mathcal{L}_t$ and $\mathcal{M}_t$ over $\Gamma(t)$ when necessary. These are the fibers of the above mentioned sheaf of Lax operator algebras, and the similar sheaf of spaces of $\mathcal{M}$-operators. In abuse of notation, we will denote these sheaves by the same letters $\mathcal{L}$ and $\mathcal{M}$, respectively.

Let $L = L(t)$, $M = M(t)$ be two smooth curves, such that $L(t) \in \mathcal{L}_t$, and $M(t) \in \mathcal{M}_t$. Consider a *Lax equation*

$$\dot{L} = [L, M] \tag{3.6}$$

where $\dot{L} = \frac{dL}{dt}$. In order that a pair $L$, $M$ would be a solution to (3.6), it is necessary that at every point $\gamma \in \Gamma$ the following relations hold:

$$\dot{z} = -\nu, \quad \dot{L}_p = \sum_{i+j=p} [L_i, M_j] + \nu \sum_{s=-k}^{p} (p + 1 - s) L_{p+1}^s \quad (p = -k, \ldots, 0), \tag{3.7}$$

where $L^s_{p+1}$ is the projection of $L_{p+1}$ onto $\mathfrak{g}_s$.

Here, the local coordinate $z$ is assumed to be dependent on $t$ (see the remark below). To derive (3.7), it is sufficient to write down the equation (3.6) for the expansions at the points $\gamma \in \Gamma$ assuming $\gamma$ to be dependent on $t$. These are the expansions in $z - z_{\gamma(t)}$ where $\frac{d}{dt}(z - z_\gamma) = -\dot{z}_\gamma$.

REMARK. In case of existence of the Tyurin parametrization (the examples of the previous section, except for the last one where it is not verified) the relations (3.7) reduce to the equations of motion of Tyurin parameters [**14**, **29**].

Let $T_L\mathcal{L}^D$ be a tangent space to $\mathcal{L}^D$ at a point $L \in \mathcal{L}^D$, and $M \in \mathcal{M}$. The following theorem provides a criterium for the equation (3.6) to be well defined on the total space of the subsheaf $\mathcal{L}^D$ of the sheaf $\mathcal{L}$. Simply speaking, it reduces the question to certain local conditions on the sets $\Pi$, and $\Gamma$, separately.

THEOREM 3.1 ([**14**, **29**, **34**]). *Assume that $\dot{L}$, $M$ satisfy the relations (3.7) for every $\gamma \in \Gamma$. Then $[L, M] \in T_L\mathcal{L}^D$ if, and only if $([L, M]) + D \geq 0$ outside $\Gamma$.*

**3.3. Hierarchies of commuting flows given by Lax equations.** The treatment of hierarchies is given only for classical Lie algebras here, and follows the lines of [**29**]. Recently, a general treatment is obtained [**34**] in terms of invariant polynomials of Lie algebras. Namely, $L^p(w)$ in the Lemma 3.2 below should be replaced with $\delta\chi(L(w))$ where $\chi$ is an invariant polynomial on $\mathfrak{g}$, $\delta\chi$ is its gradient with respect to the Cartan–Killing form.

Let $\mathfrak{g}$ be a Lie algebra of one of the types $A_n$, $B_n$, $C_n$, $D_n$, $G_2$, where $B_n$ is considered here only in combination with elliptic curves ($g = 1$, see the remark at the end of section 3.1) [1]. The indices $a$, $b$ introduced in the previous section, denote the triples $\{P \in \Pi, p \in \mathbb{Z}^+, m > -m_P\}$ here, where $m_P$ is a multiplicity of $P$ in the divisor $D$. For $\mathfrak{g} = \mathfrak{so}(n)$, $\mathfrak{g} = \mathfrak{sp}(2n)$ we assume that $p \equiv 1(\text{mod } 2)$.

LEMMA 3.2 ([**29**]). *For every $a = \{P, p, m\}$, and almost every $L \in \mathcal{L}$, there exists a unique $M$-operator $M_a$ such that*

$$M_a(w) = w^{-m}L^p(w) + O(1) \tag{3.8}$$

*where $w$ is a local parameter in a neighborhood of $P$, $P$ is a unique pole of $M_a$ outside $\Gamma$, and $M_a(P_0) = 0$ at a fixed point $P_0 \notin (\Pi \cup \Gamma)$. If $L \in \mathcal{L}^D$, then $([L, M_a]) + D \geq 0$ outside $\Gamma$.*

The proof of the Lemma basically relies on the dimension formula (3.4). By Lemma 3.2, $M_a$ is determined by $L$. We express this correspondence by means of the notation $M_a = M_a(L)$.

THEOREM 3.3 ([**34**]). *The relations*

$$\partial_a L = [L, M_a] \tag{3.9}$$

*where $M_a = M_a(L)$, define the set of commuting vector fields on an open set in $\mathcal{L}^D$.*

We give a sketch of the proof here (see [**29**, **34**] for details). By Lemma 3.2, and Theorem 3.1 the $\partial_a$ is a tangent vector field on $\mathcal{L}^D$.

First, we prove that by (3.9), $\partial_a M_b - \partial_b M_a + [M_a, M_b]$ is an $M$-operator. It follows that this $M$-operator has a zero degree divisor outside the set $\Gamma$. By (3.4), the dimension of the space of such $M$-operators is equal to 1. $M_a$, $M_b$ are assumed

[1] It is no longer important in the general approach [**34**]

to satisfy a normalizing condition $M_a(P_0) = M_b(P_0) = 0$, hence $\partial_a M_b - \partial_b M_a + [M_a, M_b]$ also satisfies it, hence it is equal to 0. The last is equivalent to commuting of $\partial_a, \partial_b$.

The following statement is the main point of the section.

COROLLARY 3.4. *The Lax equations $\dot{L} = [L, M_a]$ give commuting flows on $\mathcal{L}^D$.*

Indeed, these equations can be represented in the form $\dot{L} = \partial_a L$ where $\partial_a$ are commuting vector fields on $\mathcal{L}^D$.

## 4. Hamiltonian theory

A Hamiltonian theory for Lax equations with the spectral parameter on a Riemann surface has been constructed in [**14**] before emergence of Lax operator algebras. In the last context it corresponds to the Lax operator algebras for $\mathfrak{g} = \mathfrak{gl}(n)$. The main points of the theory are the definition of a universal symplectic structure of the soliton theory (going back to [**12**] and called here the *Krichever–Phong symplectic structure*), and construction of Hamiltonians for the commuting flows considered in the previous section of the present paper, in the case $\mathfrak{g} = \mathfrak{gl}(n)$. The further development [**27**, **29**] consisted in generalization of the theory to the Lax operators with spectral parameter on a Riemann surface, and taking values in the classical Lie algebras. Despite the new approach to Lax operator algebras, which already involves the Hamiltonian theory [**35**], we give here the older version of the last, formulated in terms of Tyurin parameters, for the reason this approach has an independent significance. We give here only a brief review of it following [**29**]. All proofs are omitted, except for a brief outline of the proof of the main theorem.

In Section 4.1 we define the Krichever–Phong symplectic structure. In Section 4.2 we formulate, and briefly prove the main theorem. In Section 4.3 we consider the elliptic Calogero–Moser systems corresponding to classical Lie algebras as examples of Hamiltonian Lax equations with spectral parameter on an elliptic curve. The presentation here is improved compared to [**29**] in a sense that we renounce the difference between ranges of values of $L$- and $M$-operators.

Observe that in [**5**] Lax representations of the Calogero–Moser systems with spectral parameters on elliptic curves are obtained for all irreducible reduced root systems. They are of Baker–Akhiezer type, and take values in Lie algebras of endomorphisms of different $\mathfrak{g}$-modules, while our Lax operators are meromorphic, and take values in $\mathfrak{g}$.

**4.1. Symplectic structure.** Following the lines of [**14**] we introduce here a symplectic structure on a certain subspace $\mathcal{P}^D \subset \mathcal{L}^D/G$ (where $G$ is a classical Lie group with the Lie algebra $\mathfrak{g}$). We call it *Krichever–Phong symplectic structure.*

Let $\Psi$ be the matrix formed by the canonically normalized left eigenvectors of $L$. In the case $\mathfrak{g} = \mathfrak{gl}(n)$ we consider a vector $\psi$ to be canonically normalized if $\sum \psi_i = 1$. In the other cases we require that $\Psi^t = -\varepsilon\sigma\Psi^{-1}\sigma^{-1}$, and $\varepsilon$ satisfies $\sigma^t = -\varepsilon\sigma$, i.e. $\varepsilon = -1$ if $\sigma$ is symmetric (the case $\mathfrak{g} = \mathfrak{so}(n)$), $\varepsilon = 1$ if $\sigma$ is skew-symmetric (the case $\mathfrak{g} = \mathfrak{sp}(2n)$). At the non-singular points it is equivalent to the requirement $\Psi \in G$ point-wise where $G = SO(n), Sp(2n)$ depending on $\mathfrak{g}$.

The $\Psi$ is defined modulo permutations of its rows. We consider $L$ and $\Psi$ as matrix-valued functions on $\mathcal{L}^D$. Let $\delta L$ and $\delta\Psi$ denote their exterior differentials which are 1-forms on $\mathcal{L}^D$. In the same way we consider the diagonal matrix $K$ defined by

$$\Psi L = K\Psi,$$

i.e. formed by the eigenvalues of $L$, and the matrix-valued 1-form $\delta K$. Let $\Omega$ be a 2-form on $\mathcal{L}^D$ with values in the space of meromorphic functions on $\Sigma$ defined by the relation

$$\Omega = \operatorname{tr}(\delta\Psi \wedge \delta L \cdot \Psi^{-1} - \delta K \wedge \delta\Psi \cdot \Psi^{-1}).$$

$\Omega$ does not depend on the order of the eigenvalues, hence it is well defined on $\mathcal{L}$.

Fix a holomorphic differential $\varpi$ on $\Sigma$ and define a scalar-valued 2-form $\omega$ on $\mathcal{L}^D$ by the relation

$$\omega = -\frac{1}{2}\left(\sum_{\gamma\in\Gamma} \operatorname{res}_\gamma \Omega\varpi + \sum_{P\in\Pi} \Omega\varpi\right).$$

There is another representation for $\Omega$:

$$\Omega = 2\delta \operatorname{tr}\left(\delta\Psi \cdot \Psi^{-1}K\right)$$

which implies that $\omega$ is apparently closed. First we want prove that it is nondegenerate when restricted to the space of Tyurin parameters, i.e., $\omega$ yields a symplectic form on this space. We will point out a canonical form of that restriction.

LEMMA 4.1 ([**29**]). *The contribution of Tyurin parameters to $\omega$ is of the form*

$$\omega_0 = \sum_{\gamma\in\Gamma} (a\,\delta z_\gamma \wedge \delta\varkappa_\gamma + \delta\beta_\gamma^t \wedge \delta\alpha_\gamma)$$

*where $a = 1$ for $\mathfrak{g} = \mathfrak{gl}(n)$, $a = 2$ for $\mathfrak{g} = \mathfrak{so}(n)$ and $\mathfrak{g} = \mathfrak{sp}(2n)$.*

Let us consider now the contribution of the points $P\in\Pi$. Let $\omega_P = -\frac{1}{2}\operatorname{res}_P \Omega\varpi$. Define $\mathcal{P}_0^D$ as a subspace in $\mathcal{L}^D$ where the 1-form $\delta\varkappa\,\varpi$ is holomorphic. It is the same as the set of common zeroes of the following functions on $\mathcal{L}^D$:

$$T_{P,j,l} = \operatorname{res}_{P^l}\left((z-z(P))^j \varkappa\,\varpi\right),\ j = 0,\dots,(m_P - d_P),$$

where $l$ labels sheets of the spectral curve (as a branch cover of $\Sigma$), $d_P = \operatorname{ord}_P \varpi$, and $z$ is a local coordinate in the neighborhood of $P$.

LEMMA 4.2. *On the space $\mathcal{P}_0^D$*

$$\omega_P = \operatorname{res}_P \operatorname{tr}(L\,\Psi^{-1}\delta\Psi \wedge \Psi^{-1}\delta\Psi)\varpi. \tag{4.1}$$

Let $\xi$ be a tangent vector to $\mathcal{P}_0^D$ at a point $L$. It is a $\mathfrak{g}$-valued meromorphic function on $\Sigma$, and $\delta\Psi(\xi)\cdot\Psi^{-1} = \xi$. Hence for any pair $\xi$, $\eta$ of tangent vectors we have $\omega_P(\xi,\eta) = \operatorname{res}_P \operatorname{tr}(L[\xi,\eta])$. This is a 2-form of Kirillov type. In general it is degenerate and has symplectic leaves. In particular in the case when $L\varpi$ has a simple pole at $P$ with the residue $L_P$ the $\omega_P$ descends to the canonical Kirillov form on the orbit $O_P$ of the element $L_P\in\mathfrak{g}$. It is obvious that $(L\varpi) + D - (\varpi) \geq 0$ for $L\in\mathcal{L}^D$ where $(\varpi)$ is a zero divisor of the form $\varpi$. On the contrary to the previous case, if some components of the divisor $D - (\varpi)$ are negative the corresponding $\omega_P$ vanish and $\omega$ obviously degenerates. We omit details and refer to the complete analogy with [**14**] with this respect.

Since $\omega$ is $G$-invariant it is actually defined on $\mathcal{P}^D = \mathcal{P}_0^D/G$. Together with what has been already proven in this section we have the following statement which is quite similar to the one in [**14**, Theorem 4.1].

THEOREM 4.3. *If $D-(\varpi)\geq 0$ then the form*

$$\omega=\sum_{\gamma\in\Gamma}(a\,\delta\varkappa_\gamma\wedge\delta z_\gamma+\delta\alpha_\gamma^t\wedge\delta\beta_\gamma)+\sum_{P\in\Pi}\omega_P, \tag{4.2}$$

*is nondegenerate on $\mathcal{P}^D$. Thus it gives a symplectic structure on $\mathcal{P}^D$.*

REMARK. In the case $D=(\varpi)$ the form $\omega$ still is non-degenerate but the contribution of the points $P\in\Pi$ vanishes. This corresponds to the case of Riemann surfaces without marked points, i.e. to the Hitchin case.

**4.2. Hamiltonians for commuting flows.** Following the lines of [**14**, **29**] we state here that hierarchies in Theorem 3.3 are Hamiltonian for all classical Lie algebras in question, and compute the corresponding Hamiltonians.

Let us note that the foliation given by the common level sets of the functions $T_{P,j,l}$ is invariant with respect to the flows of our commuting hierarchy since those preserve the spectrum $\varkappa$ as it can be proved. In particular, $\mathcal{P}_0$ is invariant for this reason.

For a vector field $e$ on $\mathcal{L}^D$ let $i_e\omega$ be the 1-form defined by $i_e\omega(X)=\omega(e,X)$ (where $X$ is an arbitrary vector field). By definition, $e$ is Hamiltonian if $i_e\omega=\delta H$ where $H$ is some function called the Hamiltonian of $e$. The following theorem is the main purpose of this section.

THEOREM 4.4. *Let $\partial_a$ be a vector field defined by (3.9). Then*

$$i_{\partial_a}\omega=\delta H_a$$

*where*

$$H_a=-\frac{1}{k+1}\operatorname{res}_P\operatorname{tr}(w^{-m}L^{k+1})\varpi,\ a=(P,k,m).$$

Let us give a brief outline of the proof. The operators $L$, $\partial_a+M_a$ commute due to the Lax equation, hence the rows of $\Psi$ are eigenvectors for both of them. The diagonal forms of those two operators are as follows:

$$K=\Psi L\Psi^{-1},\ F_a=\Psi(\partial_a+M_a)\Psi^{-1}. \tag{4.3}$$

It is not difficult to show that $L$ is conjugated to a function holomorphic at the points $\gamma\in\Gamma$, hence its spectrum $K$ is also holomorphic there. A key, and most difficult point of the proof is that $F_a$ is also holomorphic there (except, perhaps, for certain cases of degeneration of the Lax operators explicitly listed in [**29**, Lemma 5.13]).

The main steps of the proof are as follows. First, by direct computations combined with the statement of holomorphy of the the spectra of $K$, and $F_a$ at the points $\gamma\in\Gamma$ we obtain

$$i_{\partial_a}\omega=\sum_{P\in\Pi}\operatorname{res}_P\operatorname{tr}(\delta K\,F_a)\varpi. \tag{4.4}$$

Observe that the function $F_a$ has singularities other than $\gamma$-points and the points $P\in\Pi$. Indeed $F_a=-\partial_a\Psi\cdot\Psi^{-1}-\Psi M_a\Psi^{-1}$, and $\Psi^{-1}$ has poles at the branching points of eigenvalues of $L$.

For $a=(P,k,m)$ we will write $P=P_a$, etc. By Lemma 3.2 the $M_a$ is holomorphic at $P$, $P\neq P_a$, hence the spectrum $F_a$ of $\partial_a+M_a$ is too, and (4.4) results in

$$i_{\partial_a}\omega=\operatorname{res}_{P_a}\operatorname{tr}(\delta K\,F_a)\varpi.$$

By (4.3) $F_a = -\partial_a\Psi\cdot\Psi^{-1} - \Psi M_a\Psi^{-1}$, and $\partial_a\Psi\cdot\Psi^{-1}$ is holomorphic at $P_a$ because the coordinate of $P_a$ is independent of any time, hence $\partial_a\Psi$ and $\Psi$ have the same order at $P_a$. Hence $F_a = -\Psi M_a\Psi^{-1} + O(1)$. Let us plug in $M_a = w^{-m}L^k + O(1)$ (at $P_a$) from Lemma 3.2. This implies $\Psi M_a\Psi^{-1} = w^{-m}\Psi L^k\Psi^{-1} + O(1)$ since $\Psi$ is holomorphically invertible at $P_a$ (a genericity asumption). Here, we plug in $\Psi L^k\Psi^{-1} = K^k$ from (4.3), and obtain $F_a = w^{-m}K^k + O(1)$ at $P_a$. Since $\delta K\varpi$ is holomorphic at $P_a$ we obtain

$$\begin{aligned} i_{\partial_a}\omega &= -\operatorname{res}_{P_a}\operatorname{tr}(w^{-m}\delta K\,K^k)\varpi = -\frac{1}{k+1}\operatorname{res}_{P_a}\delta\operatorname{tr}(w^{-m}K^{k+1})\varpi \\ &= -\frac{1}{k+1}\delta\operatorname{res}_{P_a}\operatorname{tr}(w^{-m}L^{k+1})\varpi = \delta H_a. \end{aligned}$$

The Hamiltonians $H_a$ are in involution since they depend only on the spectra. We refer to [**14**, **29**] for more details.

**4.3. Examples: Calogero–Moser systems.** Let us start with the example considered in [**14**] — the elliptic Calogero–Moser model corresponding to the root system $A_n$. Define a $\mathfrak{gl}(n)$-valued Lax operator by

$$L_{ij} = f_{ij}\frac{\sigma(z+q_j-q_i)\sigma(z-q_j)\sigma(q_i)}{\sigma(z)\sigma(z-q_i)\sigma(q_i-q_j)\sigma(q_j)}\ (i\neq j),\quad L_{jj}=p_j \tag{4.5}$$

where $\sigma$ (and $\wp$ below) are Weierstraß functions, $f_{ij}\in\mathbb{C}$ are constant. Up to the constants $f_{ij}$ this form of $L$ is determined by the requirements that $L$ is elliptic and that it has simple poles at the points $z=q_i$ $(i=1,\dots,n)$ and $z=0$. The last is the only element of $\Pi$. By reduction of the remaining gauge freedom it is obtained in [**14**] that $f_{ij}f_{ji}=1$. For the second order Hamiltonian corresponding to that pole we have according to Theorem 4.4, up to normalization,

$$H = \operatorname{res}_{z=0}\ z^{-1}\left(-\frac{1}{2}\sum_{j=1}^n p_j^2 - \sum_{i<j}L_{ij}L_{ji}\right).$$

By the addition theorem for Weierstrass functions

$$-L_{ij}L_{ji} = \frac{\sigma(z+q_i-q_j)\sigma(z+q_j-q_i)}{\sigma(z)^2\sigma(q_i-q_j)^2} = \wp(q_i-q_j)-\wp(z),$$

hence

$$H = -\frac{1}{2}\sum_{j=1}^n p_j^2 + \sum_{i<j}\wp(q_i-q_j).$$

Let us consider now the cases $\mathfrak{g}=\mathfrak{so}(2n)$, $\mathfrak{g}=\mathfrak{sp}(2n)$. We assume that the Tyurin divisor is supported at $|\Gamma|=2n$ points $\pm q_1,\dots,\pm q_n$, and make use of Example 5 of Section 3.1.

To obtain the Calogero–Moser system corresponding to $D_n$ let us take the Lax operator in the form

$$L = \begin{pmatrix} A & B \\ C & -A^t\end{pmatrix} \tag{4.6}$$

where $A$, $B$, $C$ are $n\times n$ matrices, $B=-B^t$, $C=-C^t$. It is a standard form of an element of $\mathfrak{so}(2n)$.

Assume $A$ to be given by (4.5). For $i<j$ take

$$B_{ij}=f^B_{ij}\frac{\sigma(z+q_j+q_i)\sigma(z-q_j)}{\sigma(z)\sigma(z+q_i)\sigma(q_i+q_j)},\quad C_{ji}=f^C_{ji}\frac{\sigma(z-q_j-q_i)\sigma(z+q_i)}{\sigma(z)\sigma(z-q_j)\sigma(q_i+q_j)} \tag{4.7}$$

where $f^B_{ij}, f^C_{ij}\in\mathbb{C}$ are constant. These relations determine the matrices $B$ and $C$ due to skew-symmetry. Similarly to the case of $\mathfrak{gl}(n)$ we obtain $f^B_{ij}f^C_{ji}=-1$ by reduction of the remaining gauge freedom (taking account of the relation $\alpha^t\sigma\alpha=0$ which descends to $\alpha'^t\alpha''=0$ in this case). For the Hamiltonian we have

$$\begin{aligned}H&=-\operatorname{res}_{z=0}\ z^{-1}\left(\sum_{i=1}^n p_i^2+2\sum_{i<j}A_{ij}A_{ji}+2\sum_{i<j}B_{ij}C_{ji}\right)\\&=-\sum_{i=1}^n p_i^2+2\sum_{i<j}\wp(q_i-q_j)+2\sum_{i<j}\wp(q_i+q_j)\end{aligned}$$

which is a conventional form of the second order Hamiltonian of the elliptic $D_n$ Calogero–Moser model.

Now, we come to the Calogero–Moser systems corresponding to the root system $C_n$ and take $L$ in the form (4.6) where $B=B^t$, $C=C^t$. The corresponding elements $A_{ij}$ are defined by the right hand sides of (4.5), $B_{ij}$ and $C_{ij}$ are defined by (4.7). The relations (4.7) make sense for $i=j$ as well, and in this case their contribution to the second order Hamiltonian is equal to

$$B_{ii}C_{ii}=f^B_{ii}f^C_{ii}(\wp(2q_i)-\wp(z))$$

where we can set $f^B_{ii}f^C_{ii}=2$. Hence

$$H=-\sum_{i=1}^n p_i^2+2\sum_{i<j}\wp(q_i-q_j)+2\sum_{i<j}\wp(q_i+q_j)+2\sum_{i=1}^n\wp(2q_i)$$

which is a conventional form of the second order Hamiltonian of the elliptic Calogero–Moser model in the symplectic case.

Finally, for the root system $B_n$ let us take $\Gamma=\{\pm q_1,\dots,\pm q_n,q_0\}$ (see Example 6 of Section 3.1), and the Lax operator in the form

$$L=\begin{pmatrix}A & a & B\\ -b^t & 0 & -a^t\\ C & b & -A^t\end{pmatrix} \tag{4.8}$$

where $A$, $B$, $C$ are $n\times n$ matrices, $B=-B^t$, $C=-C^t$, $a,b\in\mathbb{C}^n$. It is a standard form of an element of the Lie algebra $\mathfrak{so}(2n+1)$. Let $A$, $B$, $C$ be given as for $D_n$, and $a$, $b$ are given by means of their coordinates as follows:

$$a_i=f^a_i\frac{\sigma(z-q_0-q_i)\sigma(z)}{\sigma(z-q_0)\sigma(z-q_i)\sigma(q_i)},\quad b_i=f^b_i\frac{\sigma(z-q_0+q_i)\sigma(z-q_i)}{\sigma(z)\sigma(z-q_0)\sigma(q_i)}. \tag{4.9}$$

By the addition formulae, $a_i b_i = f_i^a f_i^b(\wp(q_i) - \wp(z - q_0))$. For the Hamiltonian we have

$$H = -\operatorname{res}_{z=0}\; z^{-1}\left(\sum_{i=1}^{n} p_i^2 + 2\sum_{i<j} A_{ij}A_{ji} + 2\sum_{i<j} B_{ij}C_{ji} + 2\sum_{i=1}^{n} a_i b_i\right)$$

$$= -\sum_{i=1}^{n} p_i^2 + 2\sum_{i<j}\wp(q_i - q_j) + 2\sum_{i<j}\wp(q_i + q_j) + 2\sum_{i=1}^{n}(\wp(q_i) - \wp(q_0)).$$

Observe that $p_0 = 0$, hence $q_0 = const$ gives an invariant submanifold of the hierarchy of commuting flows (in the Lax form of the equation the corresponding fact is given by the relation $\nu_0 = 0$, following from the fact that $M$ is an element of $\mathfrak{so}(2n+1)$). Taking a restriction of the system to this invariant submanifold, and omitting $\wp(q_0)$ as a non-essential constant, we obtain the conventional form of the second order Hamiltonian of the elliptic Calogero–Moser system $B_n$.

## 5. Lax equations and Conformal Field Theory

We address here the following problem: given a Lax equation of the type discussed above, to construct a unitary projective representation of the corresponding Lie algebra of Hamiltonian vector fields. For the Lax equations in question, we propose a way to represent Hamiltonian vector fields by covariant derivatives with respect to the Knizhnik–Zamolodchikov connection. This is a Dirac-type prequantization from the physical point of view. There is a lot of activity on quantization of the Hitchin-type systems. Above all, the work [**4**] must be mentioned in this context. An early, and well-known work is [**1**]. In particular, it has run a theory of quantum integrable systems. In particular, detailed information is obtained on quantum Calogero–Moser systems due to the works by A.P. Veselov, A.N. Sergeev, G. Felder, M.V. Feigin. Another idea is to quantize Hitchin systems by means the Knizhnik–Zamolodchikov connection. It was addressed, or at least mentioned, many times in the theoretical physics literature (D. Ivanov [**6**], G. Felder and Ch. Wieczerkowski [**2**], M.A. Olshanetsky and A.M. Levin [**16, 17**]) but only the second order Hamiltonians have been involved. Unexpectedly, in [**28**], we observed such relations for all Hamiltonians and, moreover, for all observables of the Hamiltonian system given by the Lax equations in question. Our presentation here follows [**29**].

In Section 5.1 we briefly outline the main ideas and results of the section. Details are given in Sections 5.2–5.6. In particular, in Sections 5.2–5.3 the maximal Lie subalgebra in $\mathcal{L}$ commuting with a generic $L \in \mathcal{L}$ is described in terms of the spectral curve of $L$. Section 5.7 is devoted to a discussion of the relation to the Seiberg–Witten theory. For all details related to Krichever–Novikov algebras, functions, vector fields etc., the Sugawara representation, we refer to the contribution by M. Schlichenmaier in this volume, and to [**20, 29**].

**5.1. Conformal field theory related to a Lax equation.** By conformal field theory we mean a family of Riemann surfaces, a finite rank bundle (of *conformal blocks*) on this family, and a flat connection (*Knizhnik–Zamolodchikov connection*) on this bundle.

Below, we take the universal bundle of spectral curves over the phase space $\mathcal{P}^D$ of an integrable system (figure 7) as such a family of Riemann surfaces. For every $L \in \mathcal{P}^D$ (all arguments of $L$ are fixed except for $z$) the curve $\Sigma_L$ given by the

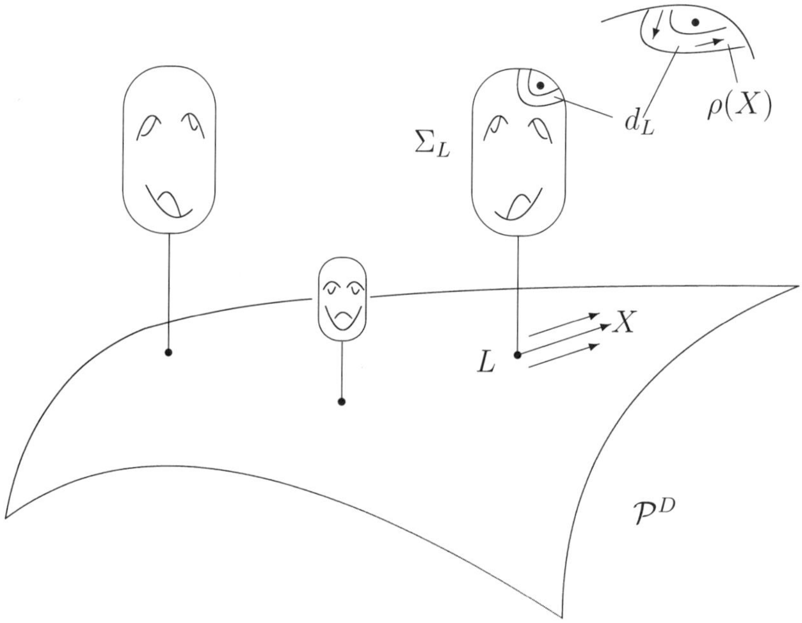


FIGURE 7. Family of spectral curves over the phase space

equation $\det(L(z)-\lambda)=0$ is called a *spectral curve* of $L$. It is an $n$-fold branch covering of $\Sigma$.

Following along the lines of [**22**, **29**] we construct an analog of the Knizhnik–Zamolodchikov connection for this family. For this end, we construct a sheaf of admissible representations of the Lax operator algebra in the space of semi-infinite wedge forms on a certain space $\mathcal{F}$ canonically related to the integrable system.

For any tangent vector $X$ on $\mathcal{P}^D$ let $\rho(X)$ denote its image under the Kodaira–Spencer mapping. It is defined as follows. Cut off a hat from the Riemann surface $\Sigma_L$ around a marked point which we denote by $P_\infty$ (figure 7). Let $d_L$ be a gluing function of the hat with the remainder of the surface (defined on the annulus – see the figure). Then $d_L$ is a local diffeomorphism. Let $\rho(X)=d_L^{-1}(\partial_X d_L)$. Then $\rho(X)$ is a vector field on the annulus. It is shown in [**22**] that $\rho(X)$ is a restriction of a Krichever–Novikov vector field, and can be considered as an element of $\mathcal{V}_L^{\rm reg}\backslash\mathcal{V}_L/\mathcal{V}_L^{(1)}$ where $\mathcal{V}_L$ is the Lie algebra of Krichever–Novikov vector fields on $\Sigma_L$, $\mathcal{V}_L^{(1)}$ is the direct sum of homogeneous subspaces of degree $\ge 0$ in $\mathcal{V}_L$, and $\mathcal{V}_L^{\rm reg}\subset\mathcal{V}_L$ is the subspace of vector fields having zero at $P_\infty$. Both $\mathcal{V}_L^{(1)}$ and $\mathcal{V}_L^{\rm reg}$ are Lie subalgebras in $\mathcal{V}_L$.

Having represented $\rho(X)$ by a Krichever–Novikov vector field, we may consider $T(\rho(X))$ where $T$ is the Sugawara representation, and, further on, define the high genus Knizhnik–Zamolodchikov connection

$$\nabla_X=\partial_X+T(\rho(X)).$$

The following property called *projective flatness* is the main property of the Knizhnik–Zamolodchikov connection [**22**, **29**]:

$$[\nabla_X, \nabla_Y] = \nabla_{[X,Y]} + \lambda(X,Y)\cdot id.$$

Another reading of this property is as follows: $X \to \nabla_X$ *is a projective representation of the Lie algebra of vector fields on* $\mathcal{P}^D$.

Now observe that $\mathcal{P}^D$ is a symplectic manifold with respect to the Krichever–Phong symplectic structure $\omega$ [**12**, **14**]. For any $f \in C^\infty(\mathcal{P}^D)$ we can consider the corresponding Hamiltonian vector field $X_f$. The correspondence $f \to \nabla_{X_f}$ is a representation of the Poisson algebra of observables of the classical system described by the Lax equation. If $\{H_a\}$ is a Poisson commuting family of Hamiltonians then the corresponding Knizhnik–Zamolodchikov operators projectively commute. Moreover, for Hamiltonians depending only on the action variables the representation operators commute without adding any *projectively.*

The restriction of the just introduced representation onto the Lie algebra of Hamiltonian vector fields is unitary. This claim relies on two facts: on Poincaré theorem on absolute integral invariants, and on unitarity of the Sugawara representation. For the Virasoro and loop algebras the last is proved in [**7**], and for Krichever–Novikov vector field and commutative current algebras in [**11**], see [**28**, **29**] for a general presentation.

Below, we comment on the most essential points of the brief outline above.

**5.2. From Lax operator algebra to commutative Krichever–Novikov algebra.** In this section, we canonically associate a commutative Krichever–Novikov algebra to a generic element $L \in \mathcal{L}$. We need it for the Sugawara construction below.

Let $\Psi$, $K$ be as at page 234. Then

$$L = \Psi^{-1} K \Psi. \tag{5.1}$$

As stated in Section 4.2, the matrix-valued function $K$ is holomorphic at all $\gamma \in \Gamma$ except for rare degenerations of $L$ listed there. Hence $K$ is a meromorphic diagonal matrix-valued function on $\Sigma$ holomorphic outside $P \in \Pi$. If we denote by $\mathcal{A}$ the algebra of scalar-valued functions on $\Sigma$ possessing this property then $K \in \mathfrak{h} \otimes \mathcal{A}$ where $\mathfrak{h} \subset \mathfrak{g}$ is a diagonal (Cartan) subalgebra. $\mathcal{A}$ is a Krichever–Novikov function algebra on $\Sigma$, and $\mathfrak{h} \otimes \mathcal{A}$ is the corresponding Krichever–Novikov current algebra. We obtain only commutative current algebras here. Let $\overline{\mathfrak{h}} = \mathfrak{h} \otimes \mathcal{A}$.

Let $\mathcal{A}_L$ be the Krichever–Novikov function algebra on $\Sigma_L$ having pre-images of the points $P \in \Pi$ (including $P_\infty$) as the collection of poles. Let us take an arbitrary element of $\mathcal{A}_L$ and push it down to $\Sigma$ as a diagonal matrix $h$. Every sheet of $\Sigma_L$ is associated with a certain row of $h$. At the branching points we may obtain a coincidence of eigenvalues of $h$. The order of diagonal elements of the matrix $h$ will depend on the order of the sheets. This ambiguity is the same as for $\Psi$. The permutation of the rows corresponding to an element $w$ of the Weil group descends to $\Psi \to w\Psi$ (which is easily verified for $w$ to be a transposition), and $h$ transforms as $h \to whw^{-1}$. Thus $L_h = \Psi^{-1} h \Psi$ does not depend on $w$. The mapping $\mathcal{A}_L \to \overline{\mathfrak{h}}$ is just the direct image of functions from $\mathcal{A}_L$, and $L_h$ is independent of the order of the sheets.

The commutative algebra $\mathcal{A}_L$ is canonically associated with $L$.

**5.3. The representation of $\mathcal{A}_L$, and Sugawara construction.** We need a representation of $\mathcal{A}_L$ in order to construct the bundle of conformal blocks, and Knizhnik–Zamolodchikov connection in the next section.

LEMMA 5.1 ([**28**, **29**]). *For any $h \in \overline{\mathfrak{h}}$ we have $\Psi^{-1}h\Psi \in \mathcal{L}$.*

The elements $\Psi^{-1}h\Psi \in \mathcal{L}$, $h \in \overline{\mathfrak{h}}$ obviously form a Lie algebra. For a generic $L$ it will be the maximal Lie subalgebra in $\mathcal{L}$ commuting with $L$ (where $\Psi$ is assumed to be the same).

By Lemma 5.1 any representation of $\mathcal{L}$ induces the corresponding representation of $\overline{\mathfrak{h}}$. Since $\Psi$ is meromorphic at $P_i$'s, the induction preserves almost grading of modules. Let us note that $\Psi$ is defined up to left multiplication by a diagonal matrix. For this reason $\overline{\mathfrak{h}}$ needs to be diagonal, otherwise the mapping $\overline{\mathfrak{h}} \to \mathcal{L}$ is not well defined.

Consider the following natural representation of $\mathcal{L}$. Let $\mathcal{F}$ be the space of meromorphic vector-valued functions $\psi$ holomorphic except at $P \in \Pi$, and $\gamma \in \Gamma$, such that

$$\psi(z) = \nu\frac{\alpha}{z} + \psi_0 + \dots$$

at any point $\gamma \in \Gamma$. $\mathcal{F}$ is an almost graded $\mathcal{L}$-module with respect to the Krichever–Novikov base introduced in [**25**, **29**]. Consider the semi-infinite degree of this module. Denote it by $\mathcal{F}^{\infty/2}$. The induced $\overline{\mathfrak{h}}$-module is what we needed. This is an *admissible* module in the sense that every one of its element annihilates having been multiply operated by an element of $\overline{\mathfrak{h}}$ of a positive degree. Moreover, it is generated by a vacuum vector. By the above constructed mapping $\mathcal{A}_L \to \overline{\mathfrak{h}}$ we also consider $\mathcal{F}^{\infty/2}$ as an $\mathcal{A}_L$-module.

Given an admissible representation of a Krichever–Novikov current algebra, the Sugawara construction yields the corresponding representation of the algebra of Krichever–Novikov vector fields on the same Riemann surface.

Here, we apply the Sugawara construction to the $\mathcal{A}_L$-module $\mathcal{F}^{\infty/2}$, and obtain a representation (denoted by $T$ above) of the Lie algebra of Krichever–Novikov vector fields on the spectral curve over $L$. We use here a "commutative" version of the Sugawara construction, developed in [**11**]. For general information on Sugawara construction see the contribution of M. Schlichenmaier in this volume, for more details in the concrete situation see [**28**, **29**].

**5.4. Conformal blocks and Knizhnik–Zamolodchikov connection.** Let us consider the sheaf of $\mathcal{A}_L$-modules $\mathcal{F}^{\infty/2}$ on $\mathcal{P}^D$. Let $\mathcal{A}_L^{reg} \subset \mathcal{A}_L$ be the subalgebra of functions regular at the preimages (or at a locally fixed single pre-image) of $P_\infty$. The sheaf of quotients $C = \mathcal{F}^{\infty/2}/\mathcal{A}_L^{reg}\mathcal{F}^{\infty/2}$ on $\mathcal{P}^D$ is called the sheaf of *coinvariants*.

Let $X$ be a vector field on $\mathcal{P}^D$. By definition

$$\nabla_X = \partial_X + T(\rho(X)) \tag{5.2}$$

where $\rho$ is the Kodaira–Spencer mapping, $T$ is the Sugawara representation in $\mathcal{F}^{\infty/2}$.

THEOREM 5.2 ([**28**, **29**]). *$\nabla$ is a projective flat connection on the sheaf of coinvariants:*

$$[\nabla_X, \nabla_Y] = \nabla_{[X,Y]} + \lambda(X,Y)\cdot id$$

*where $\lambda$ is a certain cocycle, $id$ is the identity operator.*

The horizontal sections of the sheaf of covariants are called *conformal blocks.*

**5.5. The representation of the algebra of Hamiltonian vector fields and commuting Hamiltonians.** By Theorem 5.2 $X \to \nabla_X$ is a projective representation of the Lie algebra of vector fields on $\mathcal{P}^D$ in the space of sections of the sheaf of covariants. Denote this representation by $\nabla$. The restriction of $\nabla$ to the subalgebra of Hamiltonian vector fields gives the projective representation of the last.

THEOREM 5.3. *If $X$, $Y$ are Hamiltonian vector fields such that their Hamiltonians Poisson commute, then $[\nabla_X, \nabla_Y] = \lambda(X,Y)\cdot id$. If the Hamiltonians depend only on action variables, then $[\nabla_X, \nabla_Y] = 0$.*

PROOF. The projective commutativity immediately follows from Theorem 5.2 since $[X,Y] = 0$.

For a Lax equation the spectral curve is an integral of motion (more precisely, the coefficients of its equation are integrals). This means that if $X$ is a Hamiltonian vector field such that its Hamiltonian depends only on the action variables then the complex structure, hence the transition functions, are invariant along the phase trajectories of $X$. Denote by $d_L$ the transition function over $L$. By its invariance $\partial_X d_L = 0$, hence $\rho(X) = d_L^{-1}\partial_X d_L = 0$, and $\nabla_X = \partial_X$. Let $H_X$, $H_Y$ be Hamiltonians depending only on the action variables. Then $[\nabla_X, \nabla_Y] = [\partial_X, \partial_Y] = \partial_{[X,Y]}$, and $[X,Y] = 0$. This implies the commutativity of $\nabla_X$, $\nabla_Y$. □

**5.6. Unitarity.** Here we specify a subspace of the representation space and introduce a scalar product there such that the above representation $\nabla$ becomes unitary.

Let $\mathcal{G}$ be a Lie algebra and $T$ its representation in the space $V$. Consider an antilinear antiinvolution $\dagger$ on $\mathcal{G}$ ($X \to X^\dagger$, $X \in \mathcal{G}$). An Hermitian scalar product in $V$ is called contravariant if $T(X)^\dagger = T(X^\dagger)$ where the $\dagger$ on the left hand side means the Hermitian conjugation. A pair consisting of $T$ and a contravariant scalar product is called a unitary representation of $\mathcal{G}$ [**7**]. The restriction of $T$ to the Lie subalgebra of the elements such that $X^\dagger = -X$ is unitary in the classical sense.

To construct a contravariant Hermitian scalar product in the space of the representation $\nabla$ we first introduce a point-wise scalar product in the sheaf of covariants, and then integrate it over the phase space $\mathcal{P}^D$ by an invariant volume form.

Over every point in $\mathcal{P}^D$ we introduce an Hermitian scalar product in $\mathcal{F}^{\infty/2}$ declaring semi-infinite monomials with basis entries to be orthonormal ([**7**, p.39], [**11**]). Theorem 5.4 below states that it gives a well-defined point-wise scalar product on coinvariants.

By Poincaré theorem the symplectic form and its degrees are absolute integral invariants of Hamiltonian phase flows. Hence $\omega^p/p!$ defines an invariant measure with respect to Hamiltonian phase flows. Let $\mathcal{L}^2(C, \omega^p/p!)$ be the space of quadratically integrable sections of the sheaf $C$ with respect to that measure where the square over a point is given by the above point-wise scalar product.

THEOREM 5.4. *The representation $\nabla : X \to \nabla_X$ of the Lie algebra of Hamiltonian vector fields on $\mathcal{P}^D$ in the subspace of smooth sections in $\mathcal{L}^2(C, \omega^p/p!)$ is unitary.*

Indeed, for every Hamiltonian vector field $X$ on $\mathcal{P}_D$ the operator $\partial_X$ is skew-Hermitian by invariance of the measure with respect to the corresponding phase flow. As for $T(\rho(X))$, it is skew-Hermitian fiber-wise, hence this property holds after invariant averaging.

**5.7. Remark on Seiberg–Witten theory.** The Seiberg–Witten theory deals with solutions to $\mathcal{N}=2$ supersymmetric Yang–Mills theories. A complete description of a $\mathcal{N}=2$ supersymmetric Yang–Mills theory is given by the *partition function* of the form $\widetilde{\mathcal{F}}=\int e^{i\hbar S(\psi)}D\psi$ where $S$ is the Yang–Mills action, $\psi$ runs over gauge fields. A certain low energy limit of $\widetilde{\mathcal{F}}$ is given by the *Seiberg–Witten prepotential* $\mathcal{F}$.

In [**24**] the following ansatz has been proposed for $\mathcal{F}$. There is a family of pairs $(\Sigma,\lambda)$ where $\Sigma$ is a Riemann surface, $\lambda$ is a 1-form on it. Let $a_i$, $b_i$ be the periods of $\lambda$ with respect to some canonical base of cycles $\{A_i,B_i\}$. Then $\mathcal{F}$ is a function of $\{a_i,b_i\}$ satisfying the following *Seiberg–Witten equations*:

$$b_i=\frac{\partial\mathcal{F}}{\partial a_i}.$$

The following observation [**3**] relates prepotentials and integrable systems: in all known cases the above family of Riemann surfaces is just the family of the spectral curves of an integrable system, fibered over the phase space of the latter. The corresponding family of 1-forms in our terms is given by[2]

$$\lambda=\varkappa\cdot\varpi$$

where $\varkappa$ is the spectral function of $L$, $\varpi$ is the same as in Section 4.1. Then

$$a_i=\oint_{A_i}\varkappa\cdot\varpi,\quad b_i=\oint_{B_i}\varkappa\cdot\varpi$$

are integrals of motion (depend only on the spectrum of Lax operator). They are dependent, so that

$$\sum_i da_i\wedge db_i=0.$$

In other words, $d(\sum_i b_i da_i)=0$, hence the 1-form $\sum_i b_i da_i$ has potential $\mathcal{F}$ obviously satisfying the Seiberg–Witten equations.

## References

[1] A. Beilinson, V. Drinfeld, *Quantization of Hitchin's integrable system and Hecke eigensheaves.* Preprint.

[2] Giovanni Felder and Christian Wieczerkowski, *Conformal blocks on elliptic curves and the Knizhnik-Zamolodchikov-Bernard equations*, Comm. Math. Phys. **176** (1996), no. 1, 133–161. MR1372821 (97g:32019)

[3] A. Gorsky, I. Krichever, A. Marshakov, A. Mironov, A. Morozov, *Integrability and Seiberg–Witten exact solution*, Arx.math. 9505035.

[4] N. J. Hitchin, *Flat connections and geometric quantization*, Comm. Math. Phys. **131** (1990), no. 2, 347–380. MR1065677 (91g:32022)

[5] E. D'Hocker, D. H. Phong, *Calogero-Moser Lax pairs with spectral parameter for general Lie algebras.* Hep-th/9804124.

[6] D. Ivanov, *Knizhnik-Zamolodchikov-Bernard equations as a quantization of nostationary Hitchin systems.* Hep-th/9610207.

[7] V. G. Kac and A. K. Raina, *Bombay lectures on highest weight representations of infinite-dimensional Lie algebras*, Advanced Series in Mathematical Physics, vol. 2, World Scientific Publishing Co., Inc., Teaneck, NJ, 1987. MR1021978 (90k:17013)

[2] I am obliged to I.M. Krichever for this remark.

[8] Krichever, I.M. *Elliptic solutions to the Kadomtsev-Petviashvili equation and integrable systems of particles.* Russian, Funktsional. Anal. i Prilozhen. **14** 1980, no. 4, 45–54, 95, MR595728 (82e:58046), English transl., Funct. Anal. and Appl., 1980, **14**:4 (1980), 282–290.

[9] I. M. Krichever, S. P. Novikov, *Holomorphic bundles over algebraic curves and non-linear equations.* Russ. Math. Surv., 6, **35** (1980), 53–79.

[10] I. M. Krichever, S. P. Novikov, *Algebras of Virasoro type, Riemann surfaces and structures of the theory of solitons* , Funct. Anal. and Appl. 2, **21** (1987), 126–142.

[11] I. M. Krichever, S. P. Novikov, *Algebras of Virasoro type, Riemann surfaces and strings in Minkowsky space*, Funct. Anal. and Appl., 4, **21** (1987), 294-307.

[12] I. M. Krichever and D. H. Phong, *Symplectic forms in the theory of solitons*, Surveys in differential geometry: integral systems [integrable systems], Surv. Differ. Geom., IV, Int. Press, Boston, MA, 1998, pp. 239–313, DOI 10.4310/SDG.1998.v4.n1.a6. MR1726930 (2001k:37114)

[13] I. M. Krichever and S. P. Novikov, *Holomorphic bundles and commuting difference operators. Two-term constructions* (Russian), Uspekhi Mat. Nauk **55** (2000), no. 3(333), 181–182, DOI 10.1070/rm2000v055n03ABEH000302; English transl., Russian Math. Surveys **55** (2000), no. 3, 586–588. MR1777362 (2001j:37127)

[14] I. Krichever, *Vector bundles and Lax equations on algebraic curves*, Comm. Math. Phys. **229** (2002), no. 2, 229–269, DOI 10.1007/s002200200659. MR1923174 (2003m:14054)

[15] I. M. Krichever and O. K. Sheinman, *Lax operator algebras* (Russian, with Russian summary), Funktsional. Anal. i Prilozhen. **41** (2007), no. 4, 46–59, 96, DOI 10.1007/s10688-007-0026-7; English transl., Funct. Anal. Appl. **41** (2007), no. 4, 284–294. MR2411605 (2009i:37177)

[16] A. Levin, M. Olshanetsky. *Non-autonomous Hamiltonian systems related to highest Hitchin integrals.* Proceedings of the "International Seminar on Integrable systems". In memoriam Mikail V. Saveliev. Bonn, February, 1999. ArXiv:math-ph/9904023.

[17] M. A. Olshanetsky, *Generalized Hitchin systems and the Knizhnik-Zamolodchikov-Bernard equation on elliptic curves*, Lett. Math. Phys. **42** (1997), no. 1, 59–71, DOI 10.1023/A:1007308105086. MR1473360 (98j:14036)

[18] Martin Schlichenmaier, *Krichever-Novikov algebras for more than two points*, Lett. Math. Phys. **19** (1990), no. 2, 151–165, DOI 10.1007/BF01045886. MR1039524 (91a:17039)

[19] Martin Schlichenmaier, *Central extensions and semi-infinite wedge representations of Krichever-Novikov algebras for more than two points*, Lett. Math. Phys. **20** (1990), no. 1, 33–46, DOI 10.1007/BF00417227. MR1058993 (92c:17035)

[20] Martin Schlichenmaier, *Krichever-Novikov type algebras: Theory and applications*, De Gruyter Studies in Mathematics, vol. 53, De Gruyter, Berlin, 2014. MR3237422

[21] M. Shlichenmaier, *Multipoint Lax operator algebras. Almost-graded structure and central extensions* (Russian, with Russian summary), Mat. Sb. **205** (2014), no. 5, 117–160; English transl., Sb. Math. **205** (2014), no. 5-6, 722–762. MR3242635

[22] M. Shlichenmaier and O. K. Sheinman, *The Knizhnik-Zamolodchikov equations for positive genus, and Krichever-Novikov algebras* (Russian, with Russian summary), Uspekhi Mat. Nauk **59** (2004), no. 4(358), 147–180, DOI 10.1070/RM2004v059n04ABEH000760; English transl., Russian Math. Surveys **59** (2004), no. 4, 737–770. MR2106647 (2005k:32016)

[23] M. Shlichenmaier and O. K. Sheinman, *Central extensions of Lax operator algebras* (Russian, with Russian summary), Uspekhi Mat. Nauk **63** (2008), no. 4(382), 131–172, DOI 10.1070/RM2008v063n04ABEH004550; English transl., Russian Math. Surveys **63** (2008), no. 4, 727–766. MR2483201 (2009k:37161)

[24] N. Seiberg and E. Witten, *Electric-magnetic duality, monopole condensation, and confinement in $N = 2$ supersymmetric Yang-Mills theory*, Nuclear Phys. B **426** (1994), no. 1, 19–52, DOI 10.1016/0550-3213(94)90124-4. MR1293681 (95m:81202a)

[25] O. K. Sheinman, *A fermionic model of representations of affine Krichever-Novikov algebras* (Russian, with Russian summary), Funktsional. Anal. i Prilozhen. **35** (2001), no. 3, 60–72, 96, DOI 10.1023/A:1012378912526; English transl., Funct. Anal. Appl. **35** (2001), no. 3, 209–219. MR1864989 (2003h:17011)

[26] O. K. Sheinman, *Lax operator algebras and integrable hierarchies* (Russian, with Russian summary), Tr. Mat. Inst. Steklova **263** (2008), no. Geometriya, Topologiya i Matematicheskaya Fizika. I, 216–226, DOI 10.1134/S0081543808040159; English transl., Proc. Steklov Inst. Math. **263** (2008), no. 1, 204–213. MR2599381 (2011e:17047)

[27] O. K. Sheinman, *Lax operator algebras and Hamiltonian integrable hierarchies* (Russian, with Russian summary), Uspekhi Mat. Nauk **66** (2011), no. 1(397), 151–178, DOI

10.1070/RM2011v066n01ABEH004730; English transl., Russian Math. Surveys **66** (2011), no. 1, 145–171. MR2841688 (2012k:37146)

[28] O. K. Sheinman, *Lax equations and Knizhnik–Zamolodchikov connection*, Geometric Methods in Physics, XXX Workshop, Bialowieza, Poland, 2011, Trends in Mathematics, Springer, Basel, 2013, 405–413 arXiv: 1009.4706.

[29] O. K. Sheinman, *Current algebras on Riemann surfaces: New results and applications*, de Gruyter Expositions in Mathematics, vol. 58, Walter de Gruyter GmbH & Co. KG, Berlin, 2012. MR2985911

[30] O. K. Sheinman, *Lax operator algebras of type $G_2$* (Russian), Dokl. Akad. Nauk **455** (2014), no. 1, 23–25; English transl., Dokl. Math. **89** (2014), no. 2, 151–153. MR3237616

[31] O. K. Sheinman, *Lax operator algebras of type $G_2$*, Topology, geometry, integrable systems, and mathematical physics, Amer. Math. Soc. Transl. Ser. 2, vol. 234, Amer. Math. Soc., Providence, RI, 2014, pp. 373–392. MR3307157

[32] O. K. Sheinman, *Lax operator algebras and gradings on semi-simple Lie algebras.* Doklady Mathematics, Vol. 91, No. 2, 2015, pp. 160–162. DOI: 10.1134/S106456241502012X.

[33] O. K. Sheinman, *Lax operator algebras and gradings on semi-simple Lie algebras*, Transformation Groups **21** (1), 181–196, DOI 10.1007/s00031-015-9340-y.

[34] O. K. Sheinman, *Hierarchies of finite-dimensional Lax equations with spectral parameter on a Riemann surface, and semi-simple Lie algebras*, Teoret. Mat. Fiz. **185** (2015), no. 3, 527–544 (Russian). MR3438634; English transl., Theoretical and Mathematical Physics **185** (3), 1816–1831 (2015).

[35] O. K. Sheinman, *Semisimple Lie algebras, and Hamiltonian theory of Lax equations with spectral parameter on a Riemann surface*, Proceedings of Steklov Mathematical Institute, 2015, V. 290, 191–201. DOI: 10.1134/S03719685..... (in Russian). English transl., Proc. of the Steklov Institute of Mathematics, 2015, Vol. 290, pp. 178-188.

[36] O. K. Sheinmanm *Lax operator algebras and finite-dimensional integrable systems*, (in Russian), Uspekhi Mat. Nauk 71:1 (427) (2016), 117-168, arXiv 1602.04320

[37] A. N. Tjurin, *The classification of vector bundles over an algebraic curve of arbitrary genus* (Russian), Izv. Akad. Nauk SSSR Ser. Mat. **29** (1965), 657–688. MR0217071 (36 #166a)

[38] È. B. Vinberg, V. V. Gorbatsevich, and A. L. Onishchik, *Structure of Lie groups and Lie algebras* (Russian), Current problems in mathematics. Fundamental directions, Vol. 41 (Russian), Itogi Nauki i Tekhniki, Akad. Nauk SSSR, Vsesoyuz. Inst. Nauchn. i Tekhn. Inform., Moscow, 1990, pp. 5–259, DOI 10.1007/978-1-4614-1068-3_2. MR1056486 (91j:17004)

Department of Geometry and Topology, Steklov Mathematical Institute, Moscow

Proceedings of Symposia in Pure Mathematics
Volume **92**, 2016
http://dx.doi.org/10.1090/pspum/092/01591

# From forced gradings to Q-Koszul algebras

Brian J. Parshall and Leonard L. Scott

*We dedicate this paper to the memory of Julie Riddleberger.*

ABSTRACT. This paper has two parts. Part I surveys some recent work by the authors in which "forced" grading constructions have played a significant role in the representation theory of semisimple algebraic groups $G$ in positive characteristic. The constructions begin with natural finite dimensional quotients of the distribution algebra $\mathrm{Dist}(G)$, but then "force" gradings into the picture by passing to positively graded algebras constructed from certain ideal filtrations of these quotients. This process first guaranteed a place for itself by proving, for large primes, that all Weyl modules have $p$-Weyl filtrations. Later it led, under similar circumstances, to a new "good filtration" result for restricted Lie algebra Ext groups between $p$-restricted irreducible $G$-modules. In the process of proving these results, a new kind of graded algebra was invented, called a Q-Koszul algebra. Recent conjectures suggest these algebras arise in forced grading constructions as above, from quotients of $\mathrm{Dist}(G)$, even for small primes and even in settings involving singular weights. Related conjectures point to a promising future for using Kazhdan-Lusztig theory to relate quantum and algebraic group cohomology and Ext groups in these same small prime and singular weight settings. One of these conjectures is partly proved in Part II of this paper. The proof is introduced by remarks of general interest on positively graded algebras and Morita equivalence, followed by a discussion of recent Koszulity results of Shan-Varagnolo-Vasserot, observing some extensions of their work.

## Introduction

Many questions in the modular representation theory of a semisimple algebraic group $G$ reduce to questions involving certain families of finite dimensional algebras $A$ which capture part of the story for $G$. This applies both to structural questions of specific classes of modules, but also to the cohomology theory of $G$. In addition, the existence of a suitable positive grading on $A$ would provide a powerful tool to attacking many natural questions. Unlike the case of the BGG category $\mathcal{O}$, such positive gradings are difficult (and maybe impossible) to obtain. Several years ago the authors discovered a way to construct (or "force") from $A$ an associated positively graded algebra $\widetilde{\mathrm{gr}}A$ which, remarkably, enjoys many nice properties. For example, it can often be shown that $\widetilde{\mathrm{gr}}A$ is a quasi-hereditary algebra. In addition,

2010 *Mathematics Subject Classification.* Primary 17B55, 20Gxx; Secondary 17B50.
This research was supported in part by the National Science Foundation.

the authors conjecture that $\widetilde{\mathrm{gr}}A$ is a standard Q-Koszul algebra, a new class of algebras which stand between quasi-hereditary and Koszul algebras. This conjecture is made for almost all characteristics $p$ of the base field of $G$. To date, we have verified the conjecture for $p$ very large, but also in at least one highly non-trivial case when $p = 2$. Taken together with several companion conjectures, these results suggest a new global picture for the representation/cohomology theory of $G$ which is largely independent of the validity of the Lusztig character formula. It is "global" for at least two reasons: first, it should hold for "small" characteristics, and, second, it is not bound by any Jantzen region, applying instead for any arbitrarily large finite saturated set of dominant weights.

This paper is organized as follows. Part I surveys the authors' work over the past five years on forced graded methods and results. These lead up to the definition of Q-Koszul and standard Q-Koszul algebras in §4, where the main conjecture mentioned above is stated. (The notions of Q-Koszul and standard Q-Koszul algebras first appear in [**PS13b**], [**PS14b**], and then in [**PS14a**], which first states the conjecture and several others.) Part II contains a number of new results. For example, in §5, it is proved that the Q-Koszul property for a finite dimensional algebra is invariant under Morita equivalence. In §6, we discuss a recent result of Shan-Varagnolo-Vasserot [**SVV14**] on the Koszulity of $q$-Schur algebras, focusing on their general "parabolic-singular duality" theory for affine Lie algebras. Then we indicate how the theory may be used to treat Koszulity, with some restrictions, for other quantum types (i.e., other than type $A$). With this result in hand, §7 returns to the conjectures given in [**PS14a**], proving part of one of them.

The authors dedicate this paper to the memory of Julie Riddleberger. For decades, she held an iconic position on our departmental staff. Without visible complaint, she typed or assisted in the typing, of many mathematical papers, often catching mathematical errors and, sometimes, even correcting them as she worked.

## Notation

We generally follow Jantzen's book [**Jan03**] for basic material on the representation theory of a reductive algebraic group $G$. A notable exception: the root system of $G$ is denoted $\Phi$, and $\Pi$ is a fixed set of simple roots (rather than $R$ and $S$ as in [**Jan03**]). Nevertheless, here are some standard notations to be used:

(1) Given a finite root system $\Phi$, let $m$ be (temporarily) the maximum of all heights of roots $\alpha \in \Phi$ (taken with respect to some fixed set of simple roots). Then $h := m + 1$ is the Coxeter number of $\Phi$. The Coxeter number of a reductive group $G$ is defined to be the Coxeter number of its root system $\Phi$..

(2) Let $G$ be a simply connected reductive algebraic group over an algebraically closed field $k$. Let $T$ be a fixed maximal torus and let $X = X(T)$ be the weight lattice of $T$. Fix a positive Borel subgroup, and let $X_+ = X(T)_+$ be the corresponding set of dominant weights. Recall $\lambda \in X$ is $p$-regular if $(\lambda + \rho, \alpha^\vee) \not\equiv 0 \bmod p$, for all roots $\alpha$ (letting $\rho$ be the half-sum of the positive roots). Each $p$-alcove contains a $p$-regular (integral) weight if and only if $p \geq h$. The set $X_+$ is a poset, setting $\lambda \leq \mu$ if and only if $\mu - \lambda$ is a sum of positive roots. (This is the usual "dominance order.") Finite subposets of $X_+$ play an important role. For example,

$\Gamma_{\rm res,reg}$ is the poset ideal in the poset of all $p$-regular dominant weights generated by the set of $p$-restricted $p$-regular weights.

(3) The Jantzen region is defined by

$$\Gamma_{\rm Jan} := \{\lambda \in X_+ \mid (\lambda + \rho, \alpha_0^\vee) \leq p(p - h + 2)$$
$$\text{whenever } \alpha_0 \text{ is a maximal short root for a component of } \Phi\}.$$

(4) The Lusztig character formula (LCF) is a conjectured formula giving the characters of irreducible modules $L(\gamma)$ in terms of characters of standard (Weyl) modules $\Delta(\lambda)$, for $\gamma, \lambda, \in \Gamma_{\rm Jan}$. We need not describe this expression explicitly here; see [**Jan03**, Appendix C]. We note that the LCF holds if $p$ is sufficiently large, depending on the root system $\Phi$. (See [**Fi12**] for a specific bound.) Also, (2.1.1) below provides an equivalent formulation—convenient for this paper—for the LCF (or, more accurately, for its validity for $\Gamma_{\rm Jan}$), stated in terms of all dominant weights.

# Part I: A survey

## 1. First forced grading results

We assume the reader has some familiarity with the theory of rational representations of a reductive group $G$ over an algebraically closed field $k$ of characteristic $p > 0$. Nevertheless, the results we discuss here are already meaningful and nontrivial even for $G = GL_n$, a general linear group, or the corresponding "simple" (and simply connected) algebraic group $SL_n$, the special linear group of all $n \times n$ determinant 1 matrices.

The finite dimensional modules with composition factors having highest weights in some fixed finite saturated set (poset ideal) $\Gamma$ of dominant weights identify with the finite dimensional modules for a finite dimensional algebra $A$. The algebra $A$ may be constructed as the quotient $\text{Dist}(G)_\Gamma$ of the distribution algebra $\text{Dist}(G)$ of $G$ by the annihilator in $\text{Dist}(G)$ of all finite dimensional modules with composition factors having highest weight in $\Gamma$. (It is not entirely obvious that such a quotient is finite dimensional, but it is.) The homological algebra of $A$ closely parallels that of $G$; for example, given $M, N \in A$–mod, $\text{Ext}^\bullet_A(M, N) \cong \text{Ext}^\bullet_G(M, N)$, where, on the right-hand side, $M, N$ are identified with rational $G$-modules. By varying $\Gamma$, the representation theory of $G$ can be largely recaptured from that of the algebras $A$. Sometimes it is useful to study smaller categories in this way, such as blocks or unions of blocks of $G$-modules, in which case corresponding statements hold for evident modifications of the posets $\Gamma$ and algebras $A = \text{Dist}(G)_\Gamma$.

All these algebras $A$ are examples of quasi-hereditary algebras, which have been extensively studied in their own right, but often with an eye toward applications to representation theory. When $G = GL_n$, the famous Schur algebras arise this way [**GES10**]. The representation/cohomology theory of a (Lusztig) quantum enveloping algebra $U_\zeta$ at a root of unity also can be studied by means of similarly defined finite dimensional algebras, using $U_\zeta$ in the constructions above. The $q$-Schur algebras of Dipper-James [**DJ89**] (see also [**Do99**], [**DDPW08**] for exhaustive treatments) are the most well-known examples.

A central theme of this article is the suggestion that many (perhaps all) of these algebras $A$ have positively graded companion algebras $\widetilde{\rm gr}A$ with many nice properties. Our interest in graded structures goes back the mid-nineties, especially to [**CPS97**, Remark (2.3.5)], which observed that the Lusztig conjecture would hold for all the irreducible $A$-modules provided $A$ itself had a sufficiently nice positive grading. In particular it was required that $A$ be isomorphic to $\mathrm{gr}A$, the graded algebra obtained from $A$ by adding together all quotients of successive terms in its radical filtration. (A formal definition is given below.) Unfortunately, the same paper gave an example [**CPS97**, 3.2] that showed an abstract finite dimensional algebra $A$ could have many standard Lie-theoretic properties, including known ungraded homological consequences of Koszulity, yet fail to have a Koszul grading, or even have $A \cong \mathrm{gr}A$. That depressing fact considerably slowed the search for good positive gradings for well over a decade. Finally, we came to the idea of not requiring $A$ be isomorphic to $\mathrm{gr}A$ but only that it share major properties with the latter, such as quasi-heredity. Anyone who has worked with $\mathrm{gr}A$ knows that this requirement sets a very high bar, likely involving difficult proofs, if possible at all. Nevertheless, with a great deal of effort and use of a highly nontrivial result in [**CPS97**][1] we were able to obtain a result [**PS13a**, Thm.10.6] of this type. The following theorem gives a slightly abbreviated version. We assume the reader is familiar with the quasi-hereditary algebra notion, but we will give a formal definition of "standard Koszul" later. (See particularly Definition 4.3 below.)

THEOREM 1.1. *Assume that $G$ is a semisimple, simply connected algebraic group over a field of characteristic $p \geq 2h-2$. Also, assume that $p$ is large enough so that the LCF holds for all $p$-regular weights in $\Gamma_{\mathrm{Jan}}$. Let $\Gamma$ be a poset ideal in $\Gamma_{\mathrm{res,reg}}$. [Let $N$ be the global dimension of $A := \mathrm{Dist}(G)_\Gamma$ or $N = 2$ in case this global dimension is 1. Assume also that $p \geq 2N(h-1)-1$.] Then the algebra $\mathrm{gr}A$ is a quasi-hereditary algebra with a standard Koszul grading. Its standard modules have the form $\mathrm{gr}\Delta(\lambda)$. $\lambda \in \Gamma$.*

The condition on $N$ in the statement of the theorem was removed in [**PS13b**, Cor. 3.8], which also implies the ideal $\Gamma_{\text{res, reg}}$ can be replaced by the (po)set $\Gamma_{\text{Jan, reg}}$ of all $p$-regular weights in the the Jantzen region $\Gamma_{\mathrm{Jan}}$. The hypothesis on $p$ in any of these formulations is satisfied for $p$ sufficiently large, depending on the root system, as noted in (4) above in the Notation. The algebra $\mathrm{gr}A$ is explicitly given by

$$\mathrm{gr}A := \bigoplus_{n\geq 0} \mathrm{rad}^n A/\mathrm{rad}^{n+1} A.$$

A similar definition may be made for $A$-modules, such as $\Delta(\gamma)$, resulting in $\mathrm{gr}A$-modules, such as $\mathrm{gr}\Delta(\gamma)$.

The algebras $\mathrm{gr}A$ provide the first examples in this paper of the "forced grading" process. In general terms, this just means the building of a graded algebra from a filtration by ideals. In the present context, the process itself is quite familiar.

[1]This result is [**CPS97**, Thm.2.2.1]. Its hypothesis is a very strong version of the parity conditions on Ext groups common in Kazhdan-Lusztig theory; this strong version holds by [**CPS94**, Thm.3.8] if a Koszul grading is also present. The usual parity conditions involving Ext between standard and irreducible modules have to be extended, in the hypotheses of (a dual version of) this theorem, to involve radical series of standard modules. These hypotheses are sufficiently involved that it was not clear it was even practical to verify them without knowing *a priori* a Koszul grading was present. Independently of [**PS13a**], such a verification was first done in a Lie-theoretic setting (for which a Koszul grading on $A$ itself was—and still is—unknown) in the case of (truncated versions of) Virasoro algebras in characteristic 0 [**BNW08**].

The novelty, however, is that it is possible to prove serious theorems about these algebras, either that gr$A$ inherits an important property from $A$, or has some other important new property. Later, the "forced grading" process itself will involve a more sophisticated construction.

A characteristic 0 quantum version of the above theorem was also proved [**PS13a**, Thm. 8.4]. We state a version below, and then describe some recent improvements. Let $U_\zeta$ be a (Lusztig) quantum enveloping algebra over $\mathbb{Q}(\zeta)$ with $\zeta^2$ a primitive $e$th root of unity. Assume $e$ is odd,[2] and not divisible by three in case the underlying root system $\Phi$ has a component of type $G_2$. Assume also $e > h$. The notation $\Delta_\zeta(\lambda)$ refers to a standard module in the quantum context.

THEOREM 1.2. *Let $\Gamma$ be a finite non-empty poset ideal in the poset of $e$-regular dominant weights and let $B = U_{\zeta,\Gamma}$. Then the algebra* gr$B$ *is a quasi-hereditary algebra with a standard Koszul grading. Its standard modules have the form* gr$\Delta_\zeta(\lambda)$, $\lambda \in \Gamma$.

It is a deeper fact that, under similar, often weaker assumptions on $e$ (none for type $A$), the use of $e$-regular (rather than all) dominant weights is unnecessary, and that $B$ and gr$B$ become isomorphic upon base change to $\mathbb{C}$. This was essentially conjectured by us in a remark following [**PS14a**, Conj.IIb]. We noted there that such a result for type $A$ had been proved by Shan-Varagnolo-Vasserot in [**SVV14**], and we speculated that their methods should give similar results in the other types. This is indeed the case, and we state and prove such a general result as Corollary 6.2, after giving an exposition of its main ingredient, the affine Lie algebra result [**SVV14**, Thm.3.6].

Before moving on, we mention that the restricted enveloping algebra $u \subset \mathrm{Dist}(G)$ and its quantum analogue $u_\zeta \subset U_\zeta$ (usually referred to as the small quantum group) play a key role in this work (though it is not always mentioned in this survey). Let $u'_\zeta$ denote the sum of the $e$-regular blocks of $u_\zeta$. When $e > h$, the Lusztig character formula holds for $U_\zeta$ and it is known from [**AJS94**] that $u'_\zeta$ is a Koszul algebra. The sum of the $p$-regular blocks for $u$ is denoted $u'$ or $\mathfrak{a}$. When $p > h$ and the LCF holds for $G$, [**AJS94**] also proves that the algebra $u'$ is Koszul. These graded algebras $u'$ (usually denoted by $\mathfrak{a}$) and $u'_\zeta$ (with $e = p$) play a central role in the proofs of the theorems above. At the end of §2, we will discuss another compatible (not necessarily Koszul) grading on $\mathfrak{a}$ which does not require that the LCF holds, only that $p > h$. The grading on $\mathfrak{a}$ under the hypothesis of $p > h$ arises from a grading at the integral level via base change, as will be discussed.

## 2. $p$-filtrations

Continuing now with the time-line initiated by the above theorems, we have much more to say about forced gradings. Actually, the next result [**PS15**, Cor.5.2] doesn't mention forced gradings at all, but they are there—in its proof! The theorem answers, in the large $p$ case, a question raised by Jantzen [**Jan80**] more than 35 years ago. As previously remarked, the hypotheses hold for any sufficiently large $p$, depending on the root system.

[2]A slightly different notation was used in [**PS13a**], effectively requiring that $\zeta$ itself, rather than $\zeta^2$, be an odd root of unity. However, this was unnecessary, as follows, for example, from [**PS12a**, Thm. 7.3]. The current notation fits with later usage of $e$, as in [**PS12a**] and our discussion in §6 below of [**SVV14**]. As discussed there, the assumptions on $e$ and $\zeta$ can be further relaxed.

THEOREM 2.1. *Assume that $p \geq 2h-2$ is an odd prime and that the LCF holds for $G$. Then for any $\gamma \in X_+$, $\Delta(\gamma)$ has a $\Delta^{\rm red} = \Delta^p$-filtration.*

The modules $\Delta^p(\gamma)$ date back to Jantzen's paper. If a dominant weight $\gamma$ is written $\gamma = \gamma_0 + p\gamma_1$ with $\gamma_0$ $p$-restricted and $\gamma_1$ dominant, then $\Delta^p(\gamma) := L(\gamma_0) \otimes \Delta(\gamma_1)^{[1]}$, the tensor product of a restricted irreducible module with the Frobenius twist of a standard module. The modules $\Delta^{\rm red}(\gamma)$ are obtained by "reduction mod $p$", using a lattice generated by a highest weight vector (of weight $\gamma$), from an irreducible module for the quantum enveloping algebra $U_\zeta$. They were first studied substantively by Lin [**Lin98**] and later in [**CPS09**] (with $e = p$), which gave them the name used here. Lin's paper shows $\Delta^{\rm red}(\gamma) \cong \Delta^{\rm red}(\gamma_0) \otimes \Delta(\gamma_1)^{[1]}$. The LCF, as a traditional character formula for irreducible modules, follows in the quantum (root of unity, dominant weight) case from an analog for negative level affine Lie algebras, whenever the Kazhdan-Lusztig functor (see [**T04**] and the discussion in §6 below) is an equivalence. This is true, even in the case of singular highest weights. Thus, whenever the LCF is known[3] in the quantum case (e.g, the cases where the Kazhdan-Lusztig functor is an equivalence of categories), the validity of the LCF in a corresponding algebraic groups case, for a given prime characteristic $p > 0$, is equivalent to the assertion

$$\Delta^{\rm red}(\gamma) \cong \Delta^p(\gamma), \tag{2.1.1}$$

for all dominant weights $\gamma$. This frees the LCF from the Jantzen region, and is a formulation that can sometimes be used in dominant weight poset ideals, without requiring the above isomorphism to hold for all dominant weights. See [**PS15**, Thm.B.1] for a modest application of this kind, a variation on the theorem above in which the LCF is not assumed for the weight $\gamma$ itself. For other illustrations, see [**PS14b**].

For our purposes, here, however, the main advantage of the above equation is that it shows, for $p \gg 0$, that finding $\Delta^p$-filtrations is equivalent to finding $\Delta^{\rm red}$-filtrations. Since standard modules in all cases can be obtained by "reduction mod $p$" from analogous quantum modules, this allows Jantzen's problem to be considered from the quantum group point of view.[4]

Actually, thinking entirely in terms of the modules $\Delta^{\rm red}$, a much stronger analog of the Jantzen question had been conjectured in [**CPS09**, Conj.6.11] for these primes $p > h$ and $p$-regular dominant weights $\gamma$, namely, that $\Delta(\gamma)$ has a $\Delta^{\rm red}$-filtration compatible with the filtration induced on it through the quantum radical series. The result [**CPS09**, Thm. 6.9] established that the top two sections of this filtration had such a $\Delta^{\rm red}$-filtration (the topmost section just being $\Delta^{\rm red}(\gamma)$ itself) assuming $p$ was also large enough that the LCF held. In essence, it is this conjecture that is proved (in the large $p$ case) in [**PS15**] to obtain the theorem above. First, the proof reduces to the case where the weight $\gamma$ in Theorem 2.1 is $p$-regular. Then

[3]There are modest standing assumptions on the root of unity $\zeta$ used in both the cited papers, but these can surely be weakened with more modern references. In particular, it is likely that Lin's result holds in all the known cases where the Kazhdan-Lusztig functor is an equivalence (as discussed briefly in §6), though we have not checked it.

[4]Moreover, even for small primes $\Delta^{\rm red}$-filtrations give $\Delta^p$-filtrations, since the modules $\Delta^{\rm red}(\gamma)$ all have $\Delta^p$-filtrations. However, there is no complete equivalence of the two filtration types for small primes. In fact, it is known, through an example of Will Turner (see [**PS14a**, Prop.5.1]) for $p = 2$, that there are standard modules that do not have $\Delta^{\rm red}$-filtrations, though they do have $\Delta^p$-filtrations.

the standard module $\Delta(\gamma)$ is lifted to an integral form for the quantum group, and a "radically new" version of forced grading is used.

An explanation of this new version requires the notion, first given in [**CPS90**], of an *integral* quasi-hereditary algebra $\widetilde{A}$ over a commutative ring $\mathcal{O}$—in our case, $\mathcal{O}$ is a DVR. Denote its fraction field by $K$ and its residue field by $k$. For simplicity, we also assume that $\widetilde{A}$ is *split* in the sense that all irreducible modules of $\widetilde{A}_k$ or $\widetilde{A}_K$ are absolutely irreducible. Then $\widetilde{A}$ is a (split) integral quasi-hereditary algebra if it is, first of all, an $\mathcal{O}$-algebra which is finitely generated and free over $\mathcal{O}$(i.e., $\widetilde{A}$ is an $\mathcal{O}$-order). Second, $\widetilde{A}_k$, $\widetilde{A}_K$ must be both quasi-hereditary with respect to the same (finite) poset $\Gamma$. Finally, for any $\gamma \in \Gamma$ the standard modules $\Delta_k(\gamma)$ and $\Delta_K(\gamma)$ for both of these algebras are required to be obtained by base change from the same $\widetilde{A}$-lattice $\widetilde{\Delta}(\gamma)$.[5] The new version of forced grading is the construction

$$\mathrm{gr}\widetilde{A} := \bigoplus_{i\geq 0} \frac{\widetilde{A}\cap \mathrm{rad}^i\,\widetilde{A}_K}{\widetilde{A}\cap \mathrm{rad}^{i+1}\,\widetilde{A}_K}$$

which gives a positively graded $\mathcal{O}$-order. A similar construction may be applied to $\widetilde{A}$-lattices $\widetilde{M}$, yielding graded lattices $\mathrm{gr}\widetilde{M}$ for $\mathrm{gr}\widetilde{A}$. Each of these constructions can be base-changed to $k$. We sometimes write $A{=}k\otimes\widetilde{A}$ and $\widetilde{\mathrm{gr}}A{=}k\otimes \mathrm{gr}\widetilde{A}$, and use similar conventions for lattices. The graded $k$-algebra $\widetilde{\mathrm{gr}}A$ can be quite different from $\mathrm{gr}A$, and, generally speaking, we show it has better properties than the latter. A main result in this direction is [**PS12b**, Thm 6.3], below, stated at the integral level. Here the algebra $\mathcal{O}$ is the localization of $\mathbb{Z}[\zeta]$ at the ideal generated by $\zeta - 1$, with $\zeta$ a primitive $p$th root of 1. The algebra $\widetilde{A} = \widetilde{U}_{\zeta,\Gamma}$ is the natural $\mathcal{O}$-order in the truncation $U_{\zeta,\Gamma}$ of the Lusztig quantum group over $\mathbb{Q}(\zeta)$.

THEOREM 2.2. *Assume that $e = p > 2h - 2$ is an odd prime and consider the algebra $\widetilde{A} = \widetilde{U}_{\zeta,\Gamma}$, where $\Gamma$ is an non-empty (finite) ideal of $p$-regular dominant weights. Then $\mathrm{gr}\widetilde{A}$ is a quasi-hereditary algebra over $\mathcal{O}$, with standard modules $\mathrm{gr}\,\widetilde{\Delta}(\lambda)$, $\lambda \in \Gamma$.*

The condition on $p$ can likely be improved, and $p$-regularity may not be necessary. Nevertheless, the theorem as stated played a key role in the proof of Theorem 2.1. Very briefly, an $\mathrm{Ext}^1$-vanishing criterion for grade-by-grade $\Delta^{\mathrm{red}}$-filtrations of graded lattices $\mathrm{gr}\widetilde{M}$ is obtained in [**PS15**, Thm.4.4]. The criterion is that $\mathrm{ext}^1_{\mathrm{gr}\widetilde{A}}(\mathrm{gr}\widetilde{M}, \widetilde{X}) = 0$ for certain graded $\mathrm{gr}\widetilde{A}$-lattices $\widetilde{X}$. (Here ext denotes Ext computed in the category of graded modules, a notation that will reemerge again in §4.) The graded modules $\widetilde{X}$ used in the second variable are graded $\mathrm{gr}\widetilde{A}$-lattices arising from injective $\widetilde{A}_K$-modules.[6] The construction of these modules $\widetilde{X}$ gives them, in the language of the graded integral quasi-hereditary algebra $\mathrm{gr}\widetilde{A}$, a costandard filtration, so that the required $\mathrm{Ext}^1$-vanishing is satisfied when $\mathrm{gr}\widetilde{M} = \mathrm{gr}\widetilde{\Delta}(\gamma)$.

To return to the end of §1, the small quantum group $u_\zeta$ admits an integral form, as does $u'_\zeta$. The latter is denoted $\widetilde{\mathfrak{a}}$. We have $\widetilde{\mathfrak{a}}_k \cong \mathfrak{a} = u'$. It is proved in

[5]To obtain agreement with the definition in [**CPS90**] it is necessary to build some projective $\widetilde{A}$ modules with appropriate $\widetilde{\Delta}$-filtrations. These can be obtained by an iterative construction, adapting the arguments of [**DS94**, 2A.3]. Alternately, [**Rou08**, Thm.4.15 ] may be quoted.

[6]The proof of the criterion requires control on the grade 0 part of the base change of these lattice to $K$, most easily obtained if the base change gives an indecomposable module at the $\widetilde{A}_K$ level. This requires the LCF.

[**PS12b**, Thm. 8.1] that, when $p > h$, the algebra $\widetilde{\mathfrak{a}}$ has a positive grading which base changes to the Koszul grading on $u'_\zeta$. In turn, the grading on $\widetilde{\mathfrak{a}}$ induces a positive grading on $\mathfrak{a}$. Also, when the LCF holds, the grading on $\mathfrak{a}$ agrees with its Koszul grading (cf. the remarks at the end of §1).

## 3. Applications to good filtrations

The forced grading methods described in the previous section form a basis for extending their scope from module structure theory to the study of homological resolutions. In particular, this leads to some striking results concerning good filtrations of cohomology spaces. The algebra $\mathfrak{a}$ (which captures part of the representation theory of the infinitesimal group $G_1$) is used as both a tool and a target. For example, the following result is proved in [**PS13b**, §§5,6].

THEOREM 3.1. *Let $G$ be a semsimple, simply connected algebraic group, and assume that $p \geq 2h-2$ is odd and sufficiently large that the LCF holds. Let $\lambda, \mu$ be arbitrary $p$-regular dominant weights. Then the following statements hold:*

*(a) For any integer $n \geq 0$, the rational $G$-module $\mathrm{Ext}^n_{G_1}(\Delta^{\mathrm{red}}(\lambda), \nabla_{\mathrm{red}}(\mu))^{[-1]}$ has a $\nabla$-filtration.*

*(b) For any integer $n \geq 0$, the rational $G$-module $\mathrm{Ext}^n_{G_1}(\Delta(\lambda), \nabla_{\mathrm{red}}(\mu))^{[-1]}$ has a $\nabla$-filtration. Also, the natural map*

$$\mathrm{Ext}^n_{G_1}(\Delta^{\mathrm{red}}(\nu), \nabla_{\mathrm{red}}(\mu)) \to \mathrm{Ext}^n_{G_1}(\Delta(\nu), \nabla_{\mathrm{red}}(\mu))$$

*induced by the quotient map $\Delta(\nu) \twoheadrightarrow \Delta^{\mathrm{red}}(\nu)$ is surjective.*

*(c) Dually, for any integer $n \geq 0$, the rational $G$-module $\mathrm{Ext}^n_{G_1}(\Delta^{\mathrm{red}}(\lambda), \nabla(\mu))^{[-1]}$ has a $\nabla$-filtration. Also, the natural map*

$$\mathrm{Ext}^n_{G_1}(\Delta^{\mathrm{red}}(\mu), \nabla_{\mathrm{red}}(\nu)) \to \mathrm{Ext}^n_{G_1}(\Delta^{\mathrm{red}}(\mu), \nabla(\nu))$$

*induced by the inclusion $\nabla_{\mathrm{red}}(\nu) \hookrightarrow \nabla(\nu)$ is surjective.*

In addition, [**PS13b**, §7] gives explicit formulas for the multiplicities

$$\begin{cases} [\mathrm{Ext}^n_{G_1}(\Delta^{\mathrm{red}}(\lambda), \nabla_{\mathrm{red}}(\mu))^{[-1]} : \nabla(\omega)], \\ [\mathrm{Ext}^n_{G_1}(\Delta(\lambda), \nabla_{\mathrm{red}}(\mu))^{[-1]} : \nabla(\omega)], \\ [\mathrm{Ext}^n_{G_1}(\Delta^{\mathrm{red}}(\lambda), \nabla(\mu))^{[-1]} : \nabla(\omega)], \end{cases}$$

for any dominant weight $\omega$. These multiplicities are expressed in terms of Kazhdan-Lusztig polynomial coefficients for the affine Weyl group $W_p$ attached to $G$.

REMARKS 3.2. (a) Theorem 3.1 is suggested by the work [**KLT99**] of Kumar, Lauritzen, and Thomsen, showing that, if $p > h$, then, for any $n \geq 0$, $H^n(G_1, \nabla(\lambda))^{[-1]} = \mathrm{Ext}^n_{G_1}(k, \nabla(\tau))^{[-1]}$ always has a $\nabla$-filtration.

(b) When the conclusion of Theorem 3.1(a) holds, observe that

$$\mathrm{Ext}^n_G(\Delta^{\mathrm{red}}(\lambda), \nabla_{\mathrm{red}}(\mu)) \cong (\mathrm{Ext}^n_{G_1}(\Delta^{\mathrm{red}}(\lambda), \nabla_{\mathrm{red}}(\mu))^{[-1]})^G$$

since the corresponding Hochschild-Serre spectral sequence

$$E_2^{s,t} = H^s(G, \mathrm{Ext}^t_{G_1}(\Delta^{\mathrm{red}}(\lambda), \nabla_{\mathrm{red}}(\mu))^{[-1]}) \implies \mathrm{Ext}^{s+t}_G(\Delta^{\mathrm{red}}(\lambda), \nabla_{\mathrm{red}}(\mu))$$

collapses in the sense that $E_2^{s,t} = 0$ for $s > 0$ and $E_2^{0,s+t} \cong E_\infty^{0,s+t} \cong \mathrm{Ext}^{s+t}_G(\Delta^{\mathrm{red}}(\lambda), \nabla_{\mathrm{red}}(\mu)) = 0$. This is true because $H^n(G, \nabla(\tau))$ vanishes for $n > 0$ and all dominant weights $\tau$. In particular, $\mathrm{Ext}^n_G(\Delta^{\mathrm{red}}(\lambda), \nabla_{\mathrm{red}}(\mu)) \cong (\mathrm{Ext}^n_{G_1}(\Delta^{\mathrm{red}}(\lambda), \nabla_{\mathrm{red}}(\mu))^{[-1]})^G$, which can be determined from the multiplicity results mentioned above. Similar comments apply to parts (b), (c) of the theorem. See [**PSt14**, Conj.] for a general conjecture related to the above observations.

(c) When $p > h$, the dimensions of the spaces $\mathrm{Ext}^n_G(\Delta(\lambda)^{[1]}, \Delta(\mu))$, $\mathrm{Ext}^n_G(\Delta(\lambda), \nabla(\mu)^{[1]})$, and $\mathrm{Ext}^n_G(\Delta(\lambda)^{[1]}, \nabla(\mu))$ are determined in [**CPS09**, Thm. 5.4]. In this case, $\Delta^{\mathrm{red}}(p\lambda) \cong \Delta(\lambda)^{[1]}$ and $\nabla_{\mathrm{red}}(p\lambda) \cong \nabla(p)^{[1]}$.

An essential ingredient in the proof of Theorem 3.1 involves the construction of certain explicit exact complexes $\Xi_\bullet \twoheadrightarrow M$. Here $M$ is a graded $\widetilde{\mathrm{gr}}A$-module, where $A = A_\Gamma$ for a finite ideal $\Gamma$ of $p$-regular dominant weights. There is an increasing sequence $\Gamma \subseteq \Gamma_0 \subseteq \Gamma_1 \subseteq \cdots$ of finite ideals of $p$-regular weights such that each term $\Xi_i$, $i \geq 0$, is a graded $\widetilde{\mathrm{gr}}A_{\Gamma_i}$-module, and the map $\Xi_i \to \Xi_{i-1}$ is a morphism in $\widetilde{\mathrm{gr}}A_{\Gamma_i}$-grmod. This makes sense because $\Xi_{i-1}$ can be regarded as a graded $\widetilde{\mathrm{gr}}A_{\Gamma_i}$-module through the surjective algebra homomorphism $\widetilde{\mathrm{gr}}A_{\Gamma_i} \twoheadrightarrow \widetilde{\mathrm{gr}}A_{\Gamma_{i-1}}$. In addition, when $M$ is regarded as a graded $\widetilde{\mathfrak{a}}$-module, it is required that it be $\mathfrak{a}$-linear.[7] In other words, if $P_\bullet \twoheadrightarrow M$ is a minimal graded $\mathfrak{a}$-projective resolution, then $\ker(P_{i+1} \to P_i)$ is generated by its terms of grade $i+2$. When $p$ is sufficiently large that the LCF holds, then linearity holds for the modules $\Delta^{\mathrm{red}}(\lambda)$ (since $\mathfrak{a}$ is Koszul). In addition, the standard modules $\Delta(\lambda)$ for $p$-regular dominant weights $\lambda$ are also linear (see Theorem 3.3 below). Further, the resolution can be chosen so that each $\Xi_i$, when viewed as an $\mathfrak{a}$-module through the map $\mathfrak{a} \to \widetilde{\mathrm{gr}}A_{\Gamma_i}$, is projective. Also, the $\Xi_i$ and syzygy modules $\Omega_{i+1} = \ker(\Xi_i \to \Xi_{i-1})$ have $\Delta^{\mathrm{red}}$-filtrations.

We conclude this section with two final results which will be important later. Observe that part (a) of the first theorem and all the second theorem do not require any assumptions about the LCF. The proof of the first theorem is given in [**PS13b**, Thm. 6.3] and the proof of the second is found in [**PS13b**, Thm. 5.3(b), Thm. 6.5].

THEOREM 3.3. *Assume that $p \geq 2h-2$ is odd.*

*(a) For a $p$-regular dominant weight $\lambda$, the standard module $\Delta(\lambda)$ has a graded $\mathfrak{a}$-module structure, isomorphic to $\widetilde{\mathrm{gr}}\Delta(\lambda)$ over $\widetilde{\mathrm{gr}}\mathfrak{a} \cong \mathfrak{a}$.*

*(b) Assume that the Lusztig character formula holds. With the graded structure given in (a), $\Delta(\lambda)$ is linear over the graded algebra $\mathfrak{a}$.*

THEOREM 3.4. *Assume that $p \geq 2h-2$ is odd. Let $\lambda, \mu$ be $p$-regular dominant weights contained in a finite ideal $\Gamma$ of $p$-regular weights. Then there are graded isomorphisms*

$$\begin{cases} \mathrm{Ext}^\bullet_G(\Delta(\lambda), \nabla_{\mathrm{red}}(\mu)) \cong \mathrm{Ext}^\bullet_A(\Delta(\lambda), \nabla_{\mathrm{red}}(\mu)) \cong \mathrm{Ext}^\bullet_{\widetilde{\mathrm{gr}}A}(\widetilde{\mathrm{gr}}\Delta(\lambda), \nabla_{\mathrm{red}}(\mu)), \\ \mathrm{Ext}^\bullet_G(\Delta^{\mathrm{red}}(\lambda), \nabla_{\mathrm{red}}(\mu)) \cong \mathrm{Ext}^\bullet_A(\Delta^{\mathrm{red}}(\lambda), \nabla_{\mathrm{red}}(\mu)) \cong \mathrm{Ext}^\bullet_{\widetilde{\mathrm{gr}}A}(\Delta^{\mathrm{red}}(\lambda), \nabla_{\mathrm{red}}(\mu)), \\ \mathrm{Ext}^\bullet_G(\Delta^{\mathrm{red}}(\lambda), \nabla(\mu)) \cong \mathrm{Ext}^\bullet_A(\Delta^{\mathrm{red}}(\lambda), \nabla(\mu)) \cong \mathrm{Ext}^\bullet_{\widetilde{\mathrm{gr}}A}(\Delta^{\mathrm{red}}(\lambda), \widetilde{\mathrm{gr}}^\diamond\nabla(\mu)), \end{cases}$$

*where $\widetilde{\mathrm{gr}}^\diamond\nabla(\mu)$ denotes the dual Weyl module for $\widetilde{\mathrm{gr}}A$ of highest weight $\mu$.*

---

[7]The term "linear" generally refers to a certain kind of graded complex , though we often use it to refer to graded objects which have a resolution by such a complex, the nature of which depends on context. Typically, the objects in the $i^th$ term of any linear complex are shifts $X_i = X\langle i\rangle$ of graded objects $X$ which have some preferred or standard form for their grading–e.g., a projective object generated in grade 0, or an injective object with socle of grade 0. Mazorchuk [Mazor10] even defines linear complexes of tilting modules. One useful aspect of having a terminology "linear" for objects, and not just complexes, is that it is often useful to consider resolutions of linear objects which may not be linear, as we are doing here.

In the process of proving these results, new graded homological properties of reduced standard and costandard modules emerged. These are of a graded nature, and their description is best described in the context of a new class of graded algebras, which we now introduce.

## 4. Q-Koszul algebras

This section introduces the notion of a Q-Koszul algebra and a standard Q-Koszul algebra [**PS13b**], [**PS14b**], [**PS14a**]. We work with finite dimensional algebras $A$ over a field $k$ of characteristic $p \geq 0$. Generally, we assume that $A$ is split over $k$ in the sense that the irreducible $A$-modules remain irreducible upon extension to $\overline{k}$. Also, assume that $A$ has an $\mathbb{N}$-grading

$$A = \bigoplus_{i=0}^{\infty} A_j, \quad A_i A_j \subseteq A_{i+j}.$$

The subspace $A_i$ is called the term in grade $i$. The nilpotent ideal $A_{>0} := \bigoplus_{i>0} A_i$ plays an important role. The quotient algebra $A/A_{>0}$ is isomorphic to the subalgebra $A_0$ of $A$. In general, $A_0$ is *not* assumed to be semisimple.

Let $A$-grmod be the category of finite dimensional $\mathbb{Z}$-graded $A$-modules $M$. Thus,

$$M = \bigoplus_{i\in\mathbb{Z}} M_i, \quad A_j M_i \subseteq M_{i+j}.$$

Also, $A$–mod is the category of finite dimensional $A$-modules (no grading). There is a natural forgetful functor $A$-grmod $\to$ $A$–mod. Let $\Lambda = \Lambda_A$ be a fixed finite set indexing the distinct isomorphism classes of irreducible $A$-modules:

$$\lambda \in \Lambda \longleftrightarrow L(\lambda) \in A\text{–mod}.$$

If $M \in A$-grmod and $r \in \mathbb{Z}$, then $M\langle r\rangle$ is the graded $A$-module obtained from $M$ by "shifting" the grading $r$ steps to the right: $M\langle r\rangle_i := M_{i-r}$. Clearly, the set $\Lambda \times \mathbb{Z}$ indexes the isomorphism class of irreducible objects in $A$-grmod: if $r \in \mathbb{Z}$ and $\lambda \in \Lambda$, the pair $(\lambda, r)$ corresponds to the graded irreducible module $L(\lambda)\langle r\rangle$. Up to isomorphism every graded irreducible module is isomorphic to a unique $L(\lambda)\langle i\rangle$. The set $\Lambda$ also indexes the isomorphism classes of irreducible $A_0$-modules.

The categories $A$–mod and $A$-grmod both have enough projective and injective modules. If $P(\lambda)$ denotes the PIM of $L(\lambda)$ in $A$–mod, then $P(\lambda)$ has a unique $\mathbb{N}$-grading making it an object in $A$-grmod such that $P(\lambda)_0 \cong L(\lambda)$ as a $A_0$- or a $A/A_{>0}$-module. Given $M, N \in A$-grmod, the Ext-groups in $A$-grmod, are denoted $\text{ext}^n_A(M, N)$, $n = 0, 1, \cdots$. These are related to the Ext-groups in $A$–mod by means of the identity:

$$\text{Ext}^n_A(M, N) \cong \bigoplus_{r\in\mathbb{Z}} \text{ext}^n(M, N\langle r\rangle).$$

DEFINITION 4.1. Let $A$ be an $\mathbb{N}$-graded algebra, but assume that there is a poset structure on $\Lambda$ with respect to which $A_0$ is a quasi-hereditary algebra. For $\lambda \in \Lambda$, let $\Delta_0(\lambda)$ (resp., $\nabla^0(\lambda)$) denote the corresponding standard (resp., costandard) $A_0$-module, regarded as a graded $A$-module. Then $A$ is called Q-Koszul provided, for all integers $i > 0$,

$$\text{(4.1.1)} \qquad \text{ext}^i_A(\Delta_0(\lambda), \nabla^0(\mu)\langle j\rangle) \neq 0 \implies i = j, \quad \forall j \in \mathbb{Z}, \lambda, \mu \in \Lambda.$$

If $n \geq 0$ and (4.1.1) holds for $0 < i \leq n$, then $A$ is called $n$-Q-Koszul.

Thus, $A$ is Q-Koszul if it is $n$-Q-Koszul for all $n > 0$. The following result, found in [**PS14a**, Thm. 2.3], shows that the properties of being 1- or 2-Koszul are already quite strong.

THEOREM 4.2. *(a) Assume that $A$ is 1-Q-Koszul as explained in Definition 4.1. Then $A$ is tight, in the sense that for any $n \geq 1$, $A_n$ is the product $\underbrace{A_1 \cdots A_1}_{n}$.*

*(b) Assume that $A$ is 2-Q-Koszul. Then $A$ is a quadratic algebra, in the following sense. Let*

$$T_{A_0}(A_1) = \bigoplus_{n=0}^{\infty} T^n_{A_0}(A_1), \quad \text{where } T^n_{A_0}(A_1) := A_1 \otimes_{A_0} \otimes_{A_0} \cdots \otimes_{A_0} A_1$$

*be the tensor algebra of the $(A_0, A_0)$-bimodule $A_1$. Then the mapping $T_{A_0}(A_1) \to A$, $a_1 \otimes \cdots \otimes a_n \mapsto a_1 \cdots a_n$, is surjective with kernel generated by its terms in grade 2.*

Now let $A = \bigoplus_{n\geq 0} A_n$ be a (positively) graded quasi-hereditary algebra with weight poset $\lambda$. It is elementary to show that $A_0$ is also quasi-hereditary with weight poset $\lambda$. The standard (resp., costandard) modules for $A_0$ are just the grade 0 components $\Delta^0(\lambda)$ (resp., $\nabla_0(\lambda)$) of the standard (resp., costandard) modules of $A$.

DEFINITION 4.3. The positively graded algebra $A$ is a *standard Q-Koszul algebra* provided that, for all $\lambda, \mu \in \Lambda$,

$$\text{(4.3.1)} \qquad \begin{cases} (a) & \text{ext}^n_A(\Delta(\lambda), \nabla_0(\mu)\langle r\rangle) \neq 0 \implies n = r; \\ (b) & \text{ext}^n_A(\Delta^0(\mu), \nabla(\lambda)\langle r\rangle) \neq 0 \implies n = r \end{cases}$$

for all integers $n, r$.

The grading on $A$ matters: Observe that any quasi-hereditary algebra $A$, given the trivial grading $A = A_0$ is Q-Koszul and even standard Q-Koszul. A Koszul algebra is Q-Koszul, a more nontrivial example. Following Mazorchuk [**Maz08**], a Koszul algebra $A$ is called standard Koszul provided that $A$ is quasi-hereditary, and if each standard (costandard) module $\Delta(\lambda)$ (resp., $\nabla(\lambda)$) is linear. Thus, if $\Delta(\lambda)$ is given the unique grading with its head in grade 0, then it has a (graded) projective resolution $P_\bullet \twoheadrightarrow \Delta(\lambda)$ in which the head of $P_i$ has grade $i$. A dual property is required for the costandard modules. In particular, a standard Koszul algebra is standard Q-Koszul.

The following theorem, proved in [**PS14a**, Cor. 3.4], improves on the definition in [**PS13b**], showing that standard Q-Koszul algebras as defined above are automatically Q-Koszul.

THEOREM 4.4. *Let $A$ be a standard Q-Koszul algebra with weight poset $\lambda$. For $\lambda, \mu \in \Lambda$,*

$$\text{ext}^n_A(\Delta^0(\lambda), \nabla_0(\mu)\langle r\rangle) \neq 0 \implies n = r$$

*for all $n \geq 0$ and all $r \in \mathbb{Z}$. In particular, $A$ is Q-Koszul.*

The results and methods described in §3 now come together to prove the following important theorem.

THEOREM 4.5 ([**PS13b**, Thm. 3.7]). *Assume that $p \geq 2h-2$ is odd and that the LCF holds for the semisimple, simply connected algebraic group $G$. Let $\Lambda$ be a finite ideal of $p$-regular dominant weights, and put $A := A_\Lambda$. The graded algebra $\widetilde{\mathrm{gr}}A$ is standard Q-Koszul with poset $\Lambda$. In addition, the standard modules for $\widetilde{\mathrm{gr}}A$ are the modules $\widetilde{\mathrm{gr}}\Delta(\lambda)$, $\lambda \in \Lambda$. Also, the standard modules for $(\widetilde{\mathrm{gr}}A)_0$ are the modules $\Delta^{\mathrm{red}}(\lambda)$, $\lambda \in \Lambda$.*

The costandard modules for $\widetilde{\mathrm{gr}}A$ arise as certain linearly dual modules of right standard modules. The costandard modules for $(\widetilde{\mathrm{gr}}A)_0$ are just the $\nabla_{\mathrm{red}}(\lambda)$, $\lambda \in \Lambda$.

To prove Theorem 4.5, it is necessary to prove (4.3.1). As discussed earlier, the algebra $\widetilde{\mathrm{gr}}A$ is known to be quasi-hereditary with weight poset $\Lambda$ and standard objects $\widetilde{\mathrm{gr}}\Delta(\lambda)$, $\lambda \in \Lambda$. By [**PS15**], each section of $\widetilde{\mathrm{gr}}\Delta(\lambda)$ has a $\Delta^{\mathrm{red}}$-filtration. In addition, Theorem 3.3 implies that $\widetilde{\mathrm{gr}}\Delta(\lambda)$ is linear as an $\mathfrak{a}$-module. The required vanishing in (4.3.1) can then be obtained by using properties of the complex $\Xi \twoheadrightarrow \widetilde{\mathrm{gr}}\Delta(\lambda)$ briefly described in the previous section. A similar argument works for the costandard modules.

In [**PS14a**] the authors made several conjectures. A prime $p$ is called KL-good (for a given root system $X$) provided the Kazhdan-Lusztig functors $F_\ell$ associated to $X$ (see [**T04**] and [**KL93**] for the definition; a few details are given below above Corollary 6.2) are category equivalences for

$$\ell = \ell(p) := \begin{cases} 4 & p = 2; \\ p & p \neq 2. \end{cases} \tag{4.5.1}$$

So, in all cases, if $\zeta$ is a primitive $\ell(p)$th root of unity, $\zeta^2$ is a primitive $p$th root of unity. As discussed further in §6, a list of known KL-good primes is given in [**T04**] for each indecomposable type. (More precisely, values of $\ell$ are described for which the Kazhdan-Lusztig functor $F_\ell$ is known to be an equivalence.) If the root system $X$ is not indecomposable, call a prime KL-good for $X$ provided it is KL-good for each component.

The following conjecture is Conjecture I in [**PS14a**]. Evidence for this conjecture is given, for very large primes, by Theorem 4.5 above and, for $p = 2$, in the theorem below the conjecture.

CONJECTURE 4.6. Let $G$ be a semisimple, simply connected algebraic group defined and split over $\mathbb{F}_p$ for a KL-good prime $p$. Let $\Gamma$ be a finite ideal of dominant weights and form the quasi-hereditary algebra $A = A_\Gamma$. Then $\widetilde{\mathrm{gr}}A$ is standard Q-Koszul.

THEOREM 4.7. [**PS14a**, Thm. 6.2] *Let $S(5,5)$ be the Schur algebra for $GL_5(\overline{\mathbb{F}}_2)$ in characteristic 2. Let $A$ be the principal block of $S(5,5)$. Then $\widetilde{\mathrm{gr}}A$ is standard Q-Koszul.*

It is also stated in [**PS14a**], without additional details, that the same result holds for the full algebra $S(5,5)$.

# Part II: New results

## 5. Morita equivalence and positive gradings

Finite dimensional algebras $A$ behave quite well with respect to the interaction of Morita equivalences and positive gradings. For instance, [**AJS94**, Lem.F.3], stated for Artinian rings, shows that a finite dimensional algebra $A$ has a Koszul grading if it is Morita equivalent to an algebra $B$ with a Koszul grading.[8] The algebra $A$ is not assumed a priori to have any grading at all. It is the aim of this section to show that many more properties with a positive grading underpinning carry over under Morita equivalence. An especially fundamental one is the existence itself of a comparable positive grading. The proposition below makes this property precise and provides a framework for handling many others.

Recall that $A$–mod denotes the category of finite dimensional left $A$-modules, and, if $A$ is graded, $A$–grmod denotes the category of finite dimensional graded $A$-modules. If both $A$ and $B$ are finite dimensional graded algebras over the same field, and $F : B\text{–grmod} \to A\text{–grmod}$ is an (additive) functor, we say (following [**AJS94**, app. E]) that $F$ is *graded* if it commutes (up to natural isomorphism) with the grading shift functors, i.e., $F(M\langle r\rangle) \cong F(M)\langle r\rangle$, for $M \in B\text{–grmod}$, $r \in \mathbb{Z}$. We'll also say a graded functor $F$ is a *graded version* of an (additive) functor $E : B\text{–mod} \to A\text{–mod}$ if there is a functor composition diagram

$$\begin{array}{ccc} B\text{–grmod} & \xrightarrow{F} & A\text{–grmod} \\ {\scriptstyle v}\downarrow & & {\scriptstyle v}\downarrow \\ B\text{–mod} & \xrightarrow{E} & A\text{–mod} \end{array} \tag{5.0.1}$$

commutative up to a natural isomorphism, in which the vertical maps are forgetful functors (both denoted $v$, by abuse of notation). It is also useful to have the notion of a *grade-preserving* functor. This is a graded functor $F$ which takes any graded object $M$ whose nonzero grades $M_n$ all satisfy any given inequality $a \leq n \leq b$ to an object $F(M)$ with the same property. In the case of an exact graded functor $F$, this just reduces to the condition that $F(M)$ is pure of grade $n$ whenever $M$ is pure of grade $n$.

In any case, we only use the term grade-preserving for graded functors. If a grade-preserving functor $F$ is a graded version of a functor $E$, as above, we will sometimes simply say that $F$ is a grade-preserving version of $E$.

PROPOSITION 5.1. *Suppose $A$ is a finite dimensional algebra Morita equivalent to an algebra $B$ which has a given positive grading. Then there is positive grading on $A$ and a grade-preserving functor $F : B\text{–grmod} \to A\text{–grmod}$ which is a graded version of a functor $E : B\text{–mod} \to A\text{–mod}$ with both $F$ and $E$ equivalences of categories. Moreover, $F, E$ and inverse equivalences may be chosen so that the inverse of $F$ is a grade-preserving version of the inverse of $E$.*

PROOF. First, recall the well-known result from Morita theory that $A \cong eM_n(B)e$ for some positive integer $n$ and "full" idempotent $e \in M_n(B)$. (That is, $M_n(B)e$ is a progenerator for $M_n(B)$.) Here $n$ is some positive integer and

[8]Note that, if $A$ is a finite dimensional algebra over a field $k$, and $B$ is any ring Morita equivalent to $A$, then $B$ is also a finite dimensional algebra over $k$.

$M_n(B)$ is the algebra of $n \times n$ matrices over $B$. We will also write $M_{n,m}(X)$ for the set of all $n \times m$ matrices with entries from a set $X$, for any positive integers $n, m$. Generally, $X$ will have some kind of left or right $B$-module structure, leading to a corresponding structure for $M_{n,m}(X)$ over $M_n(B)$ or $M_m(B)$, respectively.

Next observe that $M_n(B)$ transparently inherits $B$'s given positive grading. There is a standard Morita equivalence from $B$-mod to $M_n(B)$-mod, given by tensoring over $B$ with $M_{n,1}(B)$. It sends a left $B$-module $N$ to the left $M_n(B)$-module $M_{n,1}(N)$. If $N$ is graded, then its graded structure is obviously inherited, so that the same recipe defines a grade-preserving graded analog of the original functor. Both graded and ungraded versions are equivalences of categories. (As is well-known in the ungraded case, an inverse functor is given by tensoring over $M_n(B)$ with $M_{1,n}(B)$. Identifying $B$ with $eM_n(B)e$, where $e$ is the matrix unit $e_{1,1}$, this inverse functor is naturally equivalent to mutlipication by $e$. The same recipe gives an inverse functor at the graded level, a grade-preserving version of the same multiplication functor, essentially identical to it.)

The proposition will now follow by taking compositions of functors, if we can prove it for the case where $A = eBe$ for an idempotent $e$ with $Be$ a progenerator for the category of $B$-modules. Henceforth, we consider that case. Another reduction we can make is to replace $A$ with any isomorphic algebra, and we proceed to construct a very useful one. Note that $B_0$ is isomorphic to the factor algebra $B/B_{\geq 1}$ by projection, so that there is an idempotent $e_0 \in B_0$ which has the same projection as $e$. Since the ideal $B_{\geq 1}$ is clearly nilpotent, the projective modules $Be_0$ has the same head as $Be$, and so is isomorphic to it. Therefore, the endomorphism algebra $\mathrm{End}_B(Be_0) \cong (e_0Be_0)^{\mathrm{op}}$ is isomorphic to the endomorphism algebra $\mathrm{End}_B(Be) \cong (eBe)^{\mathrm{op}} = A^{\mathrm{op}}$. So we may assume $A = e_0Be_0$, or, equivalently, we may assume $e = e_0$.

In particular, $A$ inherits $B$'s given positive grading. Also, if $N \in B$–grmod, then $eN$ inherits a graded $A$-module structure. The resulting functor $F : B\text{–grmod} \to A\text{–grmod}$, given by $N \mapsto eN$, is obviously a grade-preserving version of its ungraded analog $E$. The ungraded functor $E$ is well-known to be a Morita equivalence, with inverse given by $E^\dagger := Be \otimes_{eBe} (-)$.

We next construct a graded version $F^\dagger$ of this inverse. Given a graded $eBe$-module $Y$, let $Z = Be \otimes_{eBe} Y$. Let $k$ denote the ground field, and regard $Z$ as the quotient of $V := Be \otimes_k Y$; we will henceforth omit the $k$ subscript. The space $V$ becomes a graded $B$-module in an obvious way, if, for any integer $n$, we let $V_n$ denote the sum of all terms $(Be)_i \otimes Y_j$ with $i, j$ integers such that $i + j = n$. Let $R_n$ denote the $k$-span in $V_n$ of all expressions $st \otimes y - s \otimes ty$ with $s \in (Be)_{i'}, t \in (eBe)_m$ and $y \in Y_{j'}$ for some integers $i', m, j'$ with $i' + m + j' = n$. Then $R = \oplus_n R_n$ is a graded $B$-submodule of $V$. Also, $R$ is precisely the $k$-span of all elements $st \otimes y - s \otimes ty$ with $s \in Be, t \in eBe$ and $y \in Y$. It follows that the natural map of $B$-modules $Y \to Z$ has kernel $R$. This gives $Z = Be \otimes_{eBe} Y$ the structure of a graded $B$-module with $Z_n = Y_n/R_n$ for any integer $n$. Finally, if $f : Y \to Y'$ is a map of graded $eBe$-modules, then $1_{Be} \otimes f$ gives a graded map $V \to V'$, where $V' := Be \otimes Y'$. The image of $R$ is contained, grade by grade, in $R'$, defined by analogy with $R$. Thus, there is an induced graded map $V/R \to V'/R'$. As before, we identify $Z' = V'/R'$ with $Be \otimes_{eBe} Y'$. The map $V/R \to V'/R'$ induced by $1_{Be} \otimes f$ now agrees, with these identifications, with the map $1_{Be} \otimes_{eBe} f$. This gives the desired graded version $F^\dagger$, now defined on both objects and maps, of the inverse functor $E^\dagger = Be \otimes_{eBe} (-)$.

It remains to check that this graded version $F^\dagger$ is grade preserving. It is exact, since the "fullness" of $e$ (in a sense analogous to that explained at the start of this proof) can be stated as the symmetric condition $B = BeB$. This implies that multiplication on the right gives an equivalence from mod-$B$ to mod-$eBe$, so that, in particular $Be$ is a projective right $eBe$-module. It suffices, now, to apply the graded inverse functor to a graded irreducible $eBe$-module $L$ of pure grade 0, and determine that the result is again pure of grade 0. In view of the graded structure defined above on $Be \otimes_{eBe} Y$ in the case $Y = L$, it is enough to show that $B_{\geq 1}(Be \otimes_{eBe} L) = 0$. However, $B_{\geq 1}Be = (BeB)(B_{\geq 1})e \subseteq B(eB_{\geq 1}e)$, and, for many reasons, $eB_{\geq 1}eL = 0$. (One can use grade considerations, or the very general fact that $e\,\mathrm{rad}(Be) \subseteq \mathrm{rad}(eBe)$.) Thus, the graded version $F^\dagger$ that we have constructed of the inverse functor $E^\dagger$ is also grade-preserving. This completes the proof of the proposition. □

It is also known from [**CPS90**] that, if a finite dimensional algebra $B$ is quasi-hereditary and has a positive grading, then irreducible, projective and standard modules can be chosen to have a graded $B$-module structure, with heads of grade 0. Left-right symmetry of the quasi-hereditary property implies costandard modules and injective modules also have gradings, with socles of grade 0. It is pointed out in [**SVV14**] that such gradings, with the grade 0 properties given, are unique, and similar unique gradings are noted for indecomposable tilting modules, cf. [**SVV14**, Prop.2.7].[9] As a consequence, if $B$ is quasi-hereditary and positively graded, and $A$ is Morita equivalent of $B$, then all these graded modules get carried by the functor $F$ above into corresponding graded modules (irreducible, standard, projective, constandard, injective, tilting) for $A$, with the characteristic grade 0 property carrying over when present.

COROLLARY 5.2. *Suppose $A$, $B$ are finite dimensional Morita equivalent algebras, and that $B$ has one or more of the properties below (some of which require a positive grading on $B$). Then $A$ has the corresponding property or properties:*[10]

*quasi-hereditary, positively graded, Koszul, standard Koszul,*
*Q-Koszul, standard Q-Koszul.*

*Moreover, after giving $A$ an appropriate positive grading, the categories $A$-grmod and $B$-grmod are equivalent by a graded functor.*

PROOF. Assume the hypothesis on $A$, $B$ and apply Proposition 5.1. Most of the corollary is immediate from the proposition and remarks after it. We treat the slightly more involved Q-Koszul cases: First, just assuming $B$ is positively graded, note the category of $B_0$-modules identifies with the category of graded $B$-modules which are pure of grade 0. It follows that $A_0$ is Morita equivalent to $B_0$, with $F$ furnishing the required category equivalence. In particular, if $B_0$ is

[9] One way to phrase the uniqueness condition in the case of an indecomposable tilting module is to say its irreducible section of highest weight can be found in grade 0.

[10] The Mazorchuk property of a "balanced" quasi-hereditary algebra $B$ is heavily used in [**SVV14**]. It means that $B$ is a positively graded quasi-hereditary algebra. Additionally, it is assumed that all standard modules $\Delta(\lambda)$ have linear tilting resolutions $\Delta(\lambda) \hookrightarrow T^\bullet$ and all costandard modules $\nabla(\lambda)$ have linear tilting coresolutions $T'_\bullet \twoheadrightarrow \nabla(\lambda)$. See Mazorchuk [**Maz10**, p. 3] for a precise definition. Note that the shift functors $\langle r\rangle$ in [**Maz10**] are what we denote by $\langle -r\rangle$. (Our notation agrees with that in [**BGS96**]. Consequently, Proposition 5.1 shows "balanced" is also preserved by Morita equivalence.)

quasi-hereditary, then so is $A_0$, with standard and costandard modules, viewed as pure grade 0 modules for $B$ and $A$, respectively, corresponding under $F$. If also $B$ is Q-Koszul, these modules satisfy the defining ext properties (4.1.1) in $B$-grmod, which carry over to precisely to the same properties in $A$-grmod. Thus, $A$ is also Q-Koszul. A similar argument shows that the standard Q-Koszul property also carries over.

The other cases are much easier. The rest of the proof is left to the reader. □

## 6. Some complements to [SVV14]

We begin this section with a detailed exposition of a main theorem [**SVV14**, Thm. 3.12] of Shan-Varagnolo-Vasserot. In it these authors study, for affine Lie algebras $\mathbf{g}$, certain categories

$$ {}^{w}\mathbb{O}^{\nu}_{\mu,-} \text{ and } {}^{v}\mathbb{O}^{\mu}_{\nu,+} \tag{6.0.1} $$

associated to, and collectively determining, blocks at non-critical levels of associated integral weight parabolic categories $\mathbb{O}$. They show these categories are equivalent to finite dimensional module categories of finite dimensional $\mathbb{C}$-algebras which are standard Koszul, even "balanced" (see the ftn. 10 in §5 above). We will explain below the notation, but should say right away that, with an appropriate relation between the parameters $w$ and $v$ (elements of the ambient affine Weyl group), the two categories are Koszul dual to each other. (This means, if the respective ungraded module categories are equivalent to $A$–mod and $B$-mod respectively, then $A$ is Morita equivalent to $B^! = \operatorname{Ext}^{\bullet}_B(B_0, B_0)^{\rm op}$. It seems harmless, in view of §5, to say the category $A$–mod is Koszul or standard Koszul if the algebra $A$ has the corresponding property.) On both sides of (6.0.1) the symbols $\mu$ and $\nu$ denote proper subsets, possibly empty, of fundamental roots. (Neither symbol $\mu$ or $\nu$ has precedence over the other, and our usage tends to be the reverse of that in [**SVV14**].) These are used to parameterize the "parabolic" and "singular" features of the ambient parabolic blocks, denoted $\mathbb{O}^{\nu}_{\mu,-}$ and $\mathbb{O}^{\mu}_{\nu,+}$, respectively. The subset symbols $\mu$ and $\nu$ used for a superscript parameterize the "parabolic" nature of the block, and parameterize its "singular" aspect when used as a subscript. The signs $-$ and $+$ stand for two cases for the level of the block, whether it is $-e-g$ or $+e-g$ where $e$ is a positive integer and $g$ denotes the dual Coxeter number (called $N$ in [**SVV14**]). The critical level $-g$ is not allowed. The level itself is suppressed in the notation, since the equivalence class of the categories considered depends only on whether the level is below or above $-g$ (as given by the signs $-$ and $+$, respectively) together with other parameters that do not depend on the level. This is a general feature of the notation for blocks here, which is used in a generic way to describe any one of a family of blocks which are all equivalent as categories.

This is made more concrete in [**SVV14**] by choosing weights $o_{\mu,-}$ and $o_{\nu,+}$ in the anti-dominant and dominant cones (with shifted origin), respectively, so that the irreducible modules in $\mathbb{O}^{\nu}_{\mu,-}$ and $\mathbb{O}^{\mu}_{\nu,+}$, respectively, all have highest weights in the orbit of $o_{\mu,-}$ or $o_{\nu,+}$, respectively, under the dot action of the affine Weyl group. The notation is chosen so that the respective stabilizers of these two weights are $W_\mu$ and $W_\nu$, the (finite) Weyl groups associated to the respective root systems generated by $\mu$ and $\nu$. This captures the singularity of the blocks under consideration. The weights in the first orbit specific to $\mathbb{O}^{\nu}_{\mu,-}$ are those which are "$\nu$-dominant", in the sense that their coefficients are nonnegative at any fundamental weight associated to $\nu$. This describes the "parabolic" aspect of this block. Similar considerations apply to the second orbit, with the roles of $\mu$ and $\nu$ reversed. It is now fairly clear from translation arguments that the underlying categories of these two blocks are determined up to equivalence by $\mu, \nu$ and the choice of sign. (The arguments in [**PS12a**, Sec.6] show how translations available only "in one direction" can give equivalences using highest weight category theory. Alternately, instead of quoting translation arguments, one can argue from Fiebig's excellent characterizations of blocks at non-critical levels for the full category $\mathbb{O}$, [**Fi06**, Thm.11]. See Proposition 6.1 below and the discussion following it.)

Next, we discuss the categories on each side of (6.0.1). The left-hand side is a full subcategory of its ambient parabolic block, closed under extension, a highest weight subcategory associated to a finite poset ideal, in the original sense of [**CPS88**]. The right-hand side of the display is a quotient category, associated with a finite poset coideal. (It is a highest weight category, though the block itself does not exactly fit this formalism, failing to have enough injectives. However, the arguments in [**CPS88**, Thm.3.5(b)] can be used to construct the quotient, using projectives, rather than injectives.) The paper [**SVV14**] does not generally require their various categories $\mathbb{O}$ to consist of finitely generated objects, but this adjustment is needed (and used) here to make sure all objects in the quotient have finite length. (As noted in [**SVV14**] the adjustment is not entirely necessary, with its omission just giving a quotient identifying with the category of all modules for a finite dimensional algebra, rather than the category of finite dimensional modules.)

Finally, we discuss the parameters $w$ and $v$ and the poset ideals and coideals they control. Again following [**SVV14**], let $I_\mu^{\max}$ denote the set of maximal length left coset representatives (in the affine Weyl group, denoted $\widehat{W}$ in [**SVV14**]) for $W_\mu$, and define $I_\nu^{\min}$ as the set of minimal length left coset representatives for $W_\nu$. Define $(I_{\mu,-}, \preceq)$ to be the poset $I_\mu^{\max}$ with order relation $\preceq$ the Bruhat order, and define $(I_{\nu,+}, \preceq)$ to be the poset $I_\nu^{\min}$ with order relation $\preceq$ the opposite Bruhat order. Next, define $I_{\mu,-}^\nu$ to be the subposet consisting its elements $y$ for which $y \cdot o_{\mu,-}$ is $\nu$-dominant, and define $I_{\mu,+}^\nu$ analogously, interchanging the roles of $\mu$ and $\nu$, and of $+$ and $-$. Finally, for $w \in I_{\mu,-}^\nu$ let ${}^wI_{\mu,-}^\nu$ denote the (finite) poset ideal it generates, and, for $v \in I_{\nu,+}^\mu$ let ${}^vI_{\nu,+}^\mu$ denote the (finite) poset coideal it generates. At this point, it is useful to pause, and note that these posets can all be used to index irreducible modules, taking dot products with $o_{\mu,-}$ or $o_{\nu,+}$, as appropriate, to get highest weights for them. We can now describe the left hand side of (6.0.1) as the full subcategory, closed under extension, of $\mathbb{O}_{\mu,-}^\nu$ generated by the irreducible modules indexed by ${}^wI_{\mu,-}^\nu$. Similarly, the right-hand side is the quotient category of $\mathbb{O}_{\nu,+}^\mu$ obtained, for instance, by factoring out the Serre subcategory of all modules with no section an irreducible module indexed by an element of ${}^vI_{\nu,+}^\mu$. (As noted above, the quotient can also be taken using (Hom from) a suitable projective module. Also, as noted in the previous paragraph, a finitely generated module version of the parabolic block needs to be used, if only finite length objects are desired in the quotient.)

We can now give [**SVV14**]'s sufficient condition for the categories on both sides of (6.0.1) to be Koszul dual, and, at the same time, we give the quite elegant correspondence of labels for irreducibles that achieves this. In fact, the required correspondence

$$\Psi^\nu_\mu : I^\nu_{\mu,-} \xrightarrow{\sim} (I^\mu_{\nu,+})^{\mathrm{op}} \tag{6.0.2}$$

is given[11] uniformly for fixed $\mu$ and $\nu$ in the form of an anti-isomorphism from the posets $I^\nu_{\mu,-}$ to the poset $I^\mu_{\nu,+}$. Let $w_\mu$ and $w_\nu$ denote the long words in $W_\mu$ and $W_\nu$, respectively. Then the anti-isomorphism is given by letting letting $x \in I^\nu_{\mu,-}$ correspond to the affine Weyl group element $y = w_\mu x^{-1} w_\nu$. For any $x$, the element $y$ belongs to $I^\mu_{\nu,+}$ if and only if $x$ belongs to $I^\nu_{\mu,-}$ [**SVV14**, Lem.3.2]. The parameter $w$ in the left side of (6.0.1) can be any such $x$, in which case $v$ on the right is taken to be the corresponding element $y$. The anti-isomorphism $\Psi^\nu_\mu$ then restricts to an anti-isomorphism ${}^wI^\nu_{\mu,-} \xrightarrow{\sim} ({}^vI^\mu_{\nu,+})^{\mathrm{op}}$. With $w$ and $v$ chosen in this way, part of the assertion of [**SVV14**, Thm.3.6] is that the two sides of (6.0.1) are Koszul dual to each other, after relabeling the irreducibles (of one side or the other) using this anti-isomorphism. This also makes the underlying poset on one side of (6.0.1) the same as the opposite of the poset on the other side—all as expected for Koszul duality.

In some sense this completes our exposition of [**SVV14**, Thm.3.6] *per se*, but we have several more remarks to make which clarify the result and extend its scope. First, the observations of §5 show that *any* finite dimensional algebra whose module category is equivalent to one of the categories in the display inherits the standard Koszulity property, and is even balanced. Second, many more categories for affine Kac-Moody Lie algebras, allowing irreducible modules to have non-integral highest weights, also have these properties, as follows by applying the work of Fiebig [**Fi06**, Thm.11] (cited above). Since Fiebig works with the full (finitely generated version of) the category $\mathbb{O}$, one needs characterizations of the categories in (6.0.1) that intrinsically fit his framework. The "singular" labels of weight orbits carry over with no difficulty, but the "parabolic" labels need to be treated more carefully. Part (a) of the following proposition, which is a restatement of [**SVV14**, Cor.3.3], gives one way to do this. A somewhat more transparent "double coset" version is given in part (b). For the latter, we introduce the notion of a *regular* double coset $XzY$ for two subgroups $X, Y$ of a given group $Z$. This is a double coset for which the intersection $z^{-1}Xz \cap Y$ is trivial. The definition is independent of the representative element $z$ of the double coset. Note that the "inverse" double coset $Yz^{-1}X$ of a regular $(X, Y)$ double coset is a regular $(Y, X)$ double coset.

PROPOSITION 6.1. *(a) The set $I^\nu_{\mu,-}$ consists precisely of the elements $x$ with $xw_\mu$ in $(I^{\max}_\nu)^{-1} \cap I^{\min}_\mu$. Similarly, $I^\mu_{\nu,+}$ consists precisely of the elements $y$ with $yw_\nu$ in $(I^{\min}_\mu)^{-1} \cap I^{\max}_\nu$.*

*(b) Also, $I^\nu_{\mu,-}$ consists precisely of maximal length representatives of regular $(W_\nu, W_\mu)$ double cosets. Similarly, $I^\mu_{\nu,+}$ consists precisely of minimal length representatives of regular $(W_\mu, W_\nu)$ double cosets.*

PROOF. Part (a) is, as noted, an almost verbatim restatement of [**SVV14**, Cor.3.3]. We just reduce part (b) to it. Clearly the elements given in part (b) belong to the counterpart sets described in part (a). Suppose next $xw_\mu \in (I^{\max}_\nu)^{-1} \cap I^{\min}_\mu$. Then $w_\mu x^{-1}$ is a maximal length element of its left $W_\nu$-coset. Equivalently, $xw_\mu$ is a maximal length element of its right $W_\nu$ coset. This means it has the form $w_\nu d$ where the length of $xw_\mu$ is the sum of the lengths of $w_\nu$ and of $d$. Since also

[11] The correspondence is not given a name in [**SVV14**].

$xw_\mu \in I^{\min}_\mu$ the length of the element $x = w_\nu d w_\mu$ is the sum of the lengths of its factors $w_\nu$, $d$, and $w_\mu$. This is obviously the maximal possible length for an element of the double coset $W_\nu d W_\mu$. Applying a well-known theory of Howlett and Kilmoyer (see, e.g., [**DDPW08**, §4.3]), this can only occur if $d$ is a distinguished (minimal length) double coset representative, and the intersection $d^{-1}W_\nu d \cap W_\mu$ is trivial. Thus, $x$ is the maximal length element in a regular $(W_\nu, W_\mu)$ double coset. Hence, the two characterizations of $I^\nu_{\mu,-}$ in parts (a) and (b) agree. Applying the (inverse of) $\Psi^\nu_\mu$ in (6.0.2), if $y \in I^\mu_{\nu,+}$, then $w_\nu y^{-1} w_\mu$ belongs to $I^\nu_{\mu,-}$. So, we now know it is of maximal length in its $(W_\nu, W_\mu)$ double coset, which we also know is regular. The maximal length element of any such regular double coset has the form $w_\nu z w_\mu$, where $z$ is its distinguished element of minimal length. Comparing the two expressions we have for the maximal length element gives that $y = z^{-1}$ has minimal length in its $W_\mu, W_\nu$ double coset. The latter double coset is obviously regular, since its inverse is regular. This completes the proof. □

Now, in the framework of [**SVV14**], the parabolic blocks $\mathbb{O}^\nu_{\mu,-}$ and $\mathbb{O}^\mu_{\nu,+}$ are precisely the full subcategories, of their ambient category $\mathbb{O}$ block, of objects whose irreducible sections have highest weights indexed by the affine Weyl group elements in $I^\nu_{\mu,-}$ or $I^\mu_{\nu,+}$, respectively. Proposition 6.1 shows these indexing elements are characterized inside the affine Weyl group in terms of the subgroups $W_\mu, W_\nu$. All of this Coxeter group information, together with the signs $-$ or $+$ associated to the level, carry over to the context studied by Fiebig in [**Fi06**, p.34 bottom, Thm.11] for non-integral weights.

Specifically, working with symmetrizable Kac-Moody Lie algebras, [**Fi06**] considers the block $\Lambda$ of the (non-integral, finitely-generated) category $\mathbb{O}$ corresponding to a non-integral "dominant" or anti-dominant weight $\lambda$, not at the critical level. (Weights or blocks below the critical level are called "negative", and above are called "positive," just as with the signs $-$ and $+$ we have been using here.) Associated to $\Lambda$ is an "integral Weyl group" $\mathcal{W}(\Lambda)$ generated by the real root reflections in the ambient Kac-Moody Weyl group that move $\lambda$ by an integral multiple of the reflection's underlying root. This integral Weyl group is a Coxeter group with generators $\mathcal{S}(\Lambda)$, and $\Lambda = \mathcal{W}(\Lambda)\cdot\lambda$. The stabilizer stab$(\lambda)$ of $\lambda$ under the dot action is generated by a subset of $\mathcal{S}(\Lambda)$. Then, [**Fi06**, Thm.11] says, briefly, that, together with the "negative" or "positive" nature of $\lambda$, the Coxeter group $\mathcal{W}(\Lambda)$ with its generating set $\mathcal{S}(\Lambda)$ and subgroup stab$(\lambda)$, is sufficient to determine the block up to a category equivalence. The proof shows that any two blocks, possibly of different symmetrizable Kac-Moody Lie algebras, but with the same sign and Coxeter group information, are equivalent as $\mathbb{C}$-categories by an equivalence preserving the Coxeter group indexing of irreducibles.

In particular, it makes sense to define categories ${}^w\mathbb{O}^\nu_{\mu,-}$ and ${}^v\mathbb{O}^\mu_{\nu,+}$ in Fiebig's context, provided $\mathcal{W}(\Lambda)$ is an affine Weyl group $\neq$ stab$(\lambda)$. In that case we can think of $\nu$ and $\mu$ as proper subsets of $\mathcal{S}(\Lambda)$. (In the original [**SVV14**] set-up, they are sets of fundamental roots, but these can be identified with their corresponding sets of fundamental reflections.) So, for example, suppose $\lambda$ above is anti-dominant, and $\mu$ is defined by the equality $W_\mu = \text{stab}(\lambda)$. Let $\nu$ be any proper subset of $\mathcal{S}(\Lambda)$ and $W_\nu u W_\mu$ any regular double coset, with $w$ its element of maximal length. Then the category ${}^w\mathbb{O}^\nu_{\mu,-}$ is defined to be the full subcategory of $\Lambda$ formed by those of its objects for which all irreducible sections have highest weight $x\cdot\lambda$ with $x$ the longest element in a regular $(W_\nu, W_\mu)$ double coset and $x \leq w$. Of course, in many common situations (such as the original [**SVV14**] set-up) the condition on $x$ will imply some version of $\nu$-dominance, but we need not insist upon it. We still have the conclusion that the category just defined is standard Koszul and balanced, by combining [**SVV14**, Thm.3.6] and [**Fi06**, Thm. 11]. A similar definition and conclusion can be made for ${}^{v}\mathbb{O}^{\mu}_{\nu,+}$, though the construction must proceed in two steps, first to get a full subcategory $\mathbb{O}^{\mu}_{\nu,+}$ of $\Lambda$ (now a "positive" level block), then passing to a quotient category to get ${}^{v}\mathbb{O}^{\mu}_{\nu,+}$.

One important case where we can be sure that $\mathcal{W}(\Lambda)$ is an affine Weyl group occurs when $\lambda$ is an anti-dominant rational weight of a certain form for an affine Lie algebra, still called here $\mathbf{g}$ in keeping with earlier notation in this section. The underlying finite root system is assumed indecomposable, and $\lambda$ is required to have integer coefficients at fundamental weights corresponding to these roots, but the level $k < -g$ of $\lambda$ may be a rational number, not necessarily an integer. The set of all such weights $\lambda$ is called $\mathcal{C}^{-}_{\rm rat}$ in [**PS12a**], which develops further a theory discussed in [**T04**, §6] without naming the set of weights involved. While [**T04**, §6] works primarily with the commutator algebra $\tilde{\mathbf{g}} = [\mathbf{g}, \mathbf{g}]$, it nevertheless follows from their results that $\mathcal{W}(\Lambda)$, as defined above, is an identifiable affine Weyl group (not always of the same type as $\mathbf{g}$).

In [**T04**, §6], in preparation for discussing the Kazhdan-Lusztig functor, Tanisaki discusses a category $\mathcal{O}_k$ of certain $\tilde{\mathbf{g}}$ modules with level $k$ as above. Its blocks are naturally equivalent to categories of $\mathbf{g}$-modules, as discussed [**PS12a**], with the restriction functor providing the equivalence. More precisely, each block of $\mathcal{O}_k$ is equivalent to a category $\mathbb{O}^{+}(\lambda)$ discussed in [**PS12a**, §4, §5], with $\lambda \in \mathcal{C}^{-}_{\rm rat}$ of level $k$. When $\lambda$ is integral, we can take it as $o_{\mu,-}$ for a block $\mathbb{O}^{\nu}_{\mu,-}$ with $\nu$ the set of fundamental roots in the finite root system, and $\mu$ the set of fundamental roots corresponding to the fundamental reflections in $\mathrm{stab}(\lambda)$. If $\lambda$ isn't integral, we can use essentially the same notation, as discussed above. In any case, each resulting block is, by the discussion above, the union of full subcategories, corresponding to the finite poset ideals ${}^{w}I^{\nu}_{\mu,-}$, each of which is standard Koszul and balanced. (That is, these full subcategories are each equivalent to finite dimensional module categories for finite dimensional algebras with the standard Koszul and balanced properties.) We remark that, by general highest weight category theory, these properties are inherited by the full subcategories corresponding to any finite poset ideal in $I^{\nu}_{\mu,-}$.

We are now ready to deduce the same properties for quantum group blocks, whenever the Kazhdan-Lusztig functor is an equivalence. The latter, as discussed in [**T04**], is a functor $F_\ell : \mathcal{O}_k \to \mathcal{Q}_\ell$. The target on the right-hand side is, in effect, the category of all finite dimensional type 1 modules[12] for the Lusztig quantum group over $\mathbb{C}$ at a primitive $\ell$th root of unity $\zeta$. The root system associated to the quantum group is the finite root system whose affine version is that of $\mathbf{g}$. The relation between $\ell$ and $k$ is that $k = -(\ell/2D) - g$, where $D \in \{1, 2, 3\}$ is 1 when the associated finite root system is simply-laced, 2 for types $B, C$, or $F_4$ and 3 for type $G_2$. For any fixed type of root system, the functor $F_\ell$ is an equivalence, if $\ell$ is sufficiently large. In type $A$ $F_\ell$ is always an equivalence. For further details, see [**T04**].

[12]This terminology is not used in [**T04**], but instead the modules are described as having classical weight space decompositions with appropriate actions of standard generators.

When $F_\ell$ is an equivalence, it takes irreducible objects to irreducible objects. The statement of [**T04**, Thm.7.1] shows that the highest weights of these two corresponding irreducibles have the same coefficients at fundamental weights associated to the underlying finite root system. This makes it tempting to use the dominance order for the highest weight order in both domain and range of $F_\ell$, but this actually leads to complications on the left. Another "natural" choice is to use the intrinsic order arising from standard modules (which, in both domain and range, have the same composition factors as costandard modules with the same indexing). This works, but is a little abstract, especially on the left. There, we have been using, in our discussion above, block-by-block Bruhat orders. All of these issues are thoroughly discussed in [**PS12a**]. See especially [**PS12a**, Appendix I], which proves, among other ordering results, that the traditional "up arrow" order $\uparrow$ on the right is, within any given block, equivalent to a Bruhat order on the right–actually two of them, one using a base weight in the lowest dominant closed alcove, and the other starting from its anti-dominant counterpart. The latter gives an easy way to match up weights with and partial orders on both domain and range, working block-by-block. Then, as required, we can translate back to the order $\uparrow$ as needed.

As a corollary of all of the above discussion, we have the following result, ultimately a corollary of [**SVV14**, Thm.3.6], though also depending, as indicated, on work of other authors and our synthesis here. For an abelian category $\mathcal{C}$ and an indexing by a set $\Gamma$ of some of its simple objects, let $\mathcal{C}[\Gamma]$ denote the full subcategory of $\mathcal{C}$ consisting of all objects whose simple sections may all be found among those indexed by $\Gamma$.

COROLLARY 6.2. *Assume the Kazhdan-Lusztig functor $F_\ell$ above is an equivalence, for a given* $\mathbf{g}$ *and* $\ell$. *Let $\Gamma$ be any finite poset, using either the dominance or $\uparrow$ order, in the dominant weights for the finite root system associated to the quantum group module category $\mathcal{Q}_\ell$ above. Then $\mathcal{Q}_\ell[\Gamma]$ is a highest weight category with respect to the given partial order on $\Gamma$, equivalent to the module category of a finite dimensional quasi-hereditary algebra $B'$ over $\mathbb{C}$. The algebra $B'$, or any algebra Morita equivalent to it, is standard Koszul and balanced. In particular, $B' \cong \mathrm{gr}B'$.*

The proof has already been given in the preceding discussion. Note that any poset ideal of dominant weights with respect to the dominance order is also a poset ideal with respect to the order $\uparrow$.

REMARK 6.3. (a) Note that we work over $\mathbb{C}$ in the above corollary and throughout this section, whereas in similar situations in §1 we were working over the field $\mathbb{Q}(\zeta)$. The algebra $B'$ above may be obtained by base change from a corresponding algebra $B$ over $\mathbb{Q}(\zeta)$. We don't know if $B$ shares the standard Koszul and balanced properties that $B'$ has, but $\mathrm{gr}B$ does share them, from the last assertion of the corollary.

(b) In [**SVV14**, Cor.6.6] it is proved that the $q$-Schur algebra is Morita equivalent to an algebra that is (standard Koszul and) balanced. Context indicates $q$ is a root of unity (arbitrary), and the underlying field is $\mathbb{C}$. The previous section shows, then, that the $q$-Schur algebra itself, over $\mathbb{C}$, is standard Koszul and balanced. All the conclusions of the corollary apply to it, or to any Morita equivalent algebra. The $q$-Schur algebra may, of course, be obtained by base-change from an algebra over $\mathbb{Q}(\zeta)$, so the remark above applies, as well.

## 7. Some complements to [PS14a, Conj.II]

We have already mentioned one of the three main conjectures in [**PS14a**], namely Conjecture 4.6 concerning the ubiquity of standard Q-Koszul algebras in modular representation theory. The next conjecture appears not to involve Q-Koszul algebras or forced gradings at all, but, as we shall see, their roles are hidden. The conjecture says simply that certain Ext group dimensions (for the algebraic group $G$) can be computed from corresponding dimensions for a quantum group at a $\ell(p)$th root of unity $\zeta$. (Recall that $\ell(p) = p$ for each odd prime, and $= 4$ when $p = 2$.) The Ext groups for $A$ below are the same as the corresponding Ext groups for $G$, and the Ext groups for $B$ are the same as for the quantum enveloping algebra $U_\zeta$. (The algebraic groups case is well-known, and the quantum case may be found in [**DS94**].)

CONJECTURE 7.1. Continue to assume the hypotheses and notation of Conjecture 4.6. (So, in particular, $p$ is KL-good, $\Gamma$ is a finite poset ideal of dominant weights associated to the root system of a semisimple algebraic group $G$, and $A := \mathrm{Dist}(G)_\Gamma$ is a finite-dimensional quasi-hereditary algebra.) Using the same poset $\Gamma$, let $B = U_{\zeta,\Gamma}$ be the corresponding algebra for the quantum enveloping algebra $U_\zeta$ associated to that root system at a primitive $\ell(p)$th root of unity $\zeta$ (as in Theorem 1.2). Then

$$\begin{cases}(1)\quad \dim \mathrm{Ext}^n_B(\Delta_\zeta(\lambda), L_\zeta(\mu)) &= \dim \mathrm{Ext}^n_A(\Delta(\lambda), \nabla_{\mathrm{red}}(\mu)),\\ (2)\quad \dim \mathrm{Ext}^n_B(L_\zeta(\lambda), \nabla_\zeta(\mu)) &= \dim \mathrm{Ext}^n_A(\Delta^{\mathrm{red}}(\lambda), \nabla(\mu)), \qquad \forall \lambda,\mu\in\Lambda, \forall n\geq 0.\\ (3)\quad \dim \mathrm{Ext}^n_B(L_\zeta(\lambda), L_\zeta(\mu)) &= \dim \mathrm{Ext}^n_A(\Delta^{\mathrm{red}}(\lambda), \nabla_{\mathrm{red}}(\mu)),\end{cases}$$

When $p > h$ and the Lusztig character formula holds for $G$ and $p$-restricted dominant weights, this conjecture is proved for $p$-regular weights in [**CPS09**, Thm. 5.4]. Some interesting cases can be proved assuming only $p > h$, along the lines of Theorem 3.1(c); see the discussion in [**PS14a**]. In the latter paper, we showed that, in cases where Conjecture 4.6 can be shown to be true, Conjecture 7.1 above reduces to two further conjectures, one of these is stated below as a theorem. It follows from Corollary 6.2, proved in the previous section. (Note that the Ext group dimensions in the theorem remain the same after base change to $\mathbb{C}$.)

THEOREM 7.2. *Under the hypothesis of Conjecture* 7.1, *we have*

$$\begin{cases}(1)\quad \dim \mathrm{Ext}^n_B(\Delta_\zeta(\lambda), L_\zeta(\mu)) &= \dim \mathrm{Ext}^n_{\mathrm{gr}B}(\mathrm{gr}\Delta_\zeta(\lambda), L_\zeta(\mu)),\\ (2)\quad \dim \mathrm{Ext}^n_B(L_\zeta(\lambda), \nabla_\zeta(\mu)) &= \dim \mathrm{Ext}^n_{\mathrm{gr}B}(L_\zeta(\lambda), \mathrm{gr}^\diamond\nabla_\zeta(\mu)),\\ (3)\quad \dim \mathrm{Ext}^n_B(L_\zeta(\lambda), L_\zeta(\mu)) &= \dim \mathrm{Ext}^n_{\mathrm{gr}B}(L_\zeta(\lambda), L_\zeta(\mu)),\end{cases}$$

*for all* $\lambda, \mu \in \Gamma$ *and all* $n \in \mathbb{N}$. *Also,* $\mathrm{gr}B$ *is a standard Koszul algebra.*

Since this result holds, the argument in [**PS14a**] shows Conjecture 7.1 reduces, whenever Conjecture 4.6 is true, to the conjecture below (called Conjecture IIa in [**PS14a**]). The argument for the reduction is just a base-change carried out in the graded case, made extremely easy to handle because of the Q-Koszul property.

CONJECTURE 7.3. Under the hypothesis of Conjecture 7.1, we have

$$\begin{cases}(1)\quad \dim \mathrm{Ext}^n_A(\Delta(\lambda), \nabla_{\mathrm{red}}(\mu)) &= \dim \mathrm{Ext}^n_{\widetilde{\mathrm{gr}}A}(\widetilde{\mathrm{gr}}\Delta(\lambda), \nabla_{\mathrm{red}}(\mu)),\\ (2)\quad \dim \mathrm{Ext}^n_A(\Delta^{\mathrm{red}}(\mu), \nabla(\lambda)) &= \dim \mathrm{Ext}^n_{\widetilde{\mathrm{gr}}A}(\Delta^{\mathrm{red}}(\mu), \widetilde{\mathrm{gr}}^\diamond\nabla(\lambda)),\\ (3)\quad \dim \mathrm{Ext}^n_A(\Delta^{\mathrm{red}}(\mu), \nabla_{\mathrm{red}}(\lambda)) &= \dim \mathrm{Ext}^n_{\widetilde{\mathrm{gr}}A}(\Delta^{\mathrm{red}}(\mu), \nabla_{\mathrm{red}}(\lambda)).\end{cases}$$

for all $\lambda, \mu \in \Gamma$.

In part (2) above, $\widetilde{\text{gr}}^{\diamond}\nabla(\lambda)$ denotes the costandard module corresponding to $\lambda$ in the highest weight category $\widetilde{\text{gr}}A$-mod. It has a natural graded structure, concentrated in non-positive grades, with $\nabla_{\text{red}}(\lambda)$ its grade 0 term. Under the assumptions that $\lambda, \mu$ are $p$-regular and $p \geq 2h-2$ is an odd prime, the conjecture follows from Theorem 3.4. In particular, assuming Conjecture 4.6, both Conjecture 7.3 and Conjecture 7.1 are true for $p$ sufficiently large without any LCF assumptions. There is a third conjecture, called Conjecture III in [**PS14a**], which provides explicit formulas in terms of Kazhdan-Lusztig polynomials for the dimensions of the quantum Ext groups appearing in Conjecture 7.1. Although Conjecture III only gives a formula for the quantum Ext-dimension in part (1) of Conjecture 7.3, the dimension in part (2) may be obtained by a duality. Then part (3) may be computed as in [**PS09**, §4], using Theorem 7.2. In the $p$-regular weight case with $p > h$ such formulas are already known in all three cases.

Finally, it is an interesting question as to when $A$ itself has a positive grading such that $A \cong \widetilde{\text{gr}}A$. Geometric conjectures of Achar-Riche [**AR14**] suggest this could be true at least for modestly large $p$. Of course, in that case, Conjucture 7.3 is an easy consequence.

## References


[AR14] P. Achar and S. Riche, Modular perverse sheaves on flag varieties, III: positivity conditions, arXiv:1408.4189v.1 (2014).

[AJS94] H. H. Andersen, J. C. Jantzen, and W. Soergel, *Representations of quantum groups at a pth root of unity and of semisimple groups in characteristic p: independence of p* (English, with English and French summaries), Astérisque **220** (1994), 321. MR1272539 (95j:20036)

[BGS96] A. Beilinson, V. Ginzburg, and W. Soergel, *Koszul duality patterns in representation theory*, J. Amer. Math. Soc. **9** (1996), no. 2, 473–527, DOI 10.1090/S0894-0347-96-00192-0. MR1322847 (96k:17010)

[BNW08] B. D. Boe, D. K. Nakano, and E. Wiesner, *Category $\mathcal{O}$ for the Virasoro algebra: cohomology and Koszulity*, Pacific J. Math. **234** (2008), no. 1, 1–21, DOI 10.2140/pjm.2008.234.1. MR2375311 (2009j:17022)

[CPS88] E. Cline, B. Parshall, and L. Scott, *Finite-dimensional algebras and highest weight categories*, J. Reine Angew. Math. **391** (1988), 85–99. MR961165 (90d:18005)

[CPS90] E. Cline, B. Parshall, and L. Scott, *Integral and graded quasi-hereditary algebras. I*, J. Algebra **131** (1990), no. 1, 126–160, DOI 10.1016/0021-8693(90)90169-O. MR1055002 (91c:16009)

[CPS94] E. Cline, B. Parshall, and L. Scott, *The homological dual of a highest weight category*, Proc. London Math. Soc. (3) **68** (1994), no. 2, 294–316, DOI 10.1112/plms/s3-68.2.294. MR1253506 (94m:20093)

[CPS97] E. Cline, B. Parshall, and L. Scott, *Graded and non-graded Kazhdan-Lusztig theories*, Algebraic groups and Lie groups, Austral. Math. Soc. Lect. Ser., vol. 9, Cambridge Univ. Press, Cambridge, 1997, pp. 105–125. MR1635677 (99h:16015)

[CPS09] E. Cline, B. Parshall, and L. Scott, *Reduced standard modules and cohomology*, Trans. Amer. Math. Soc. **361** (2009), no. 10, 5223–5261, DOI 10.1090/S0002-9947-09-04633-9. MR2515810 (2010g:20092)

[DDPW08] B. Deng, J. Du, B. Parshall, and J. Wang, *Finite dimensional algebras and quantum groups*, Mathematical Surveys and Monographs, vol. 150, American Mathematical Society, Providence, RI, 2008. MR2457938 (2009i:17023)

[DJ89] R. Dipper and G. James, *The q-Schur algebra*, Proc. London Math. Soc. (3) **59** (1989), no. 1, 23–50, DOI 10.1112/plms/s3-59.1.23. MR997250 (90g:16026)

[Do99] S. Donkin, *The q-Schur algebra*, London Mathematical Society Lecture Note Series, vol. 253, Cambridge University Press, Cambridge, 1998. MR1707336 (2001h:20072)

[DS94] J. Du and L. Scott, *Lusztig conjectures, old and new. I*, J. Reine Angew. Math. **455** (1994), 141–182, DOI 10.1515/crll.1994.455.141. MR1293877 (95i:20062)

[DPS98] J. Du, B. Parshall, and L. Scott, *Quantum Weyl reciprocity and tilting modules*, Comm. Math. Phys. **195** (1998), no. 2, 321–352, DOI 10.1007/s002200050392. MR1637785 (99k:17026)

[Fi06] P. Fiebig, *The combinatorics of category $\mathcal{O}$ over symmetrizable Kac-Moody algebras*, Transform. Groups **11** (2006), no. 1, 29–49, DOI 10.1007/s00031-005-1103-8. MR2205072 (2006k:17040)

[Fi12] P. Fiebig, *An upper bound on the exceptional characteristics for Lusztig's character formula*, J. Reine Angew. Math. **673** (2012), 1–31, DOI 10.1515/CRELLE.2011.170. MR2999126

[GES10] J. A. Green, *Polynomial representations of* $\mathrm{GL}_n$, Second corrected and augmented edition, Lecture Notes in Mathematics, vol. 830, Springer, Berlin, 2007. With an appendix on Schensted correspondence and Littelmann paths by K. Erdmann, Green and M. Schocker. MR2349209 (2009b:20084)

[Jan80] J. C. Jantzen, *Darstellungen halbeinfacher Gruppen und ihrer Frobenius-Kerne* (German), J. Reine Angew. Math. **317** (1980), 157–199, DOI 10.1515/crll.1980.317.157. MR581341 (82b:20057)

[Jan03] J. C. Jantzen, *Representations of algebraic groups*, 2nd ed., Mathematical Surveys and Monographs, vol. 107, American Mathematical Society, Providence, RI, 2003. MR2015057 (2004h:20061)

[KL93] D. Kazhdan and G. Lusztig, *Tensor structures arising from affine Lie algebras. I, II*, J. Amer. Math. Soc. **6** (1993), no. 4, 905–947, 949–1011, DOI 10.2307/2152745. MR1186962 (93m:17014)

[KLT99] S. Kumar, N. Lauritzen, and J. F. Thomsen, *Frobenius splitting of cotangent bundles of flag varieties*, Invent. Math. **136** (1999), no. 3, 603–621, DOI 10.1007/s002220050320. MR1695207 (2000g:20088)

[Lin98] Z. Lin, *Highest weight modules for algebraic groups arising from quantum groups*, J. Algebra **208** (1998), no. 1, 276–303, DOI 10.1006/jabr.1998.7488. MR1644011 (99k:20087)

[Maz08] V. Mazorchuk, *Some homological properties of the category $\mathcal{O}$*, Pacific J. Math. **232** (2007), no. 2, 313–341, DOI 10.2140/pjm.2007.232.313. MR2366357 (2008m:17013)

[Maz10] V. Mazorchuk, *Koszul duality for stratified algebras. I. Balanced quasi-hereditary algebras*, Manuscripta Math. **131** (2010), no. 1-2, 1–10, DOI 10.1007/s00229-009-0313-0. MR2574987 (2011e:16053)

[PSt14] A. Parker and D. Stewart, Stabilisation of the LHS spectral sequence for algebraic groups, arXiv:1402.4625 (2014).

[PS95] B. Parshall and L. Scott, *Koszul algebras and the Frobenius automorphism*, Quart. J. Math. Oxford Ser. (2) **46** (1995), no. 183, 345–384, DOI 10.1093/qmath/46.3.345. MR1348822 (96k:20084)

[PS09] B. Parshall and L. Scott, *Some $\mathbb{Z}/2$-graded representation theory*, Q. J. Math. **60** (2009), no. 3, 327–351, DOI 10.1093/qmath/han014. MR2533662 (2011b:20132)

[PS12a] B. Parshall and L. Scott, *A semisimple series for q-Weyl and q-Specht modules*, Recent developments in Lie algebras, groups and representation theory, Proc. Sympos. Pure Math., vol. 86, Amer. Math. Soc., Providence, RI, 2012, pp. 277–310, DOI 10.1090/pspum/086/1423. MR2977009

[PS12b] B. Parshall and L. Scott, *Forced gradings in integral quasi-hereditary algebras with applications to quantum groups*, Recent developments in Lie algebras, groups and representation theory, Proc. Sympos. Pure Math., vol. 86, Amer. Math. Soc., Providence, RI, 2012, pp. 247–276, DOI 10.1090/pspum/086/1422. MR2977008

[PS13a] B. Parshall and L. Scott, *A new approach to the Koszul property in representation theory using graded subalgebras*, J. Inst. Math. Jussieu **12** (2013), no. 1, 153–197, DOI 10.1017/S1474748012000679. MR3001737

[PS13b] B. Parshall and L. Scott, New graded methods in the homological algebra of semisimple algebraic groups, arXiv:1304.1461v3 (2013).

[PS14a] B. Parshall and L. Scott, Q-Koszul algebras and three conjectures, arXiv:1405.4419 (2014).

[PS14b] B. Parshall and L. Scott, *Some Koszul properties of standard and irreducible modules*, Recent advances in representation theory, quantum groups, algebraic geometry, and related topics, Contemp. Math., vol. 623, Amer. Math. Soc., Providence, RI, 2014, pp. 243–265, DOI 10.1090/conm/623/12457. MR3288631

[PS15] B. Parshall and L. Scott, *On p-filtrations of Weyl modules*, J. Lond. Math. Soc. (2) **91** (2015), no. 1, 127–158, DOI 10.1112/jlms/jdu067. MR3338612

[Rou08] R. Rouquier, *q-Schur algebras and complex reflection groups* (English, with English and Russian summaries), Mosc. Math. J. **8** (2008), no. 1, 119–158, 184. MR2422270 (2010b:20081)

[SVV14] P. Shan, M. Varagnolo, and E. Vasserot, *Koszul duality of affine Kac-Moody algebras and cyclotomic rational double affine Hecke algebras*, Adv. Math. **262** (2014), 370–435, DOI 10.1016/j.aim.2014.05.012. MR3228432

[T04] T. Tanisaki, *Character formulas of Kazhdan-Lusztig type*, Representations of finite dimensional algebras and related topics in Lie theory and geometry, Fields Inst. Commun., vol. 40, Amer. Math. Soc., Providence, RI, 2004, pp. 261–276. MR2057399 (2005h:17046)

Department of Mathematics, University of Virginia, Charlottesville, Virginia 22903
*E-mail address*: bjp8w@virginia.edu

Department of Mathematics, University of Virginia, Charlottesville, Virginia 22903
*E-mail address*: lls2l@virginia.edu

Proceedings of Symposia in Pure Mathematics
Volume **92**, 2016
http://dx.doi.org/10.1090/pspum/092/01592

# Perverse sheaves on the nilpotent cone and Lusztig's generalized Springer correspondence

Laura Rider and Amber Russell

Abstract. In this note, we consider perverse sheaves on the nilpotent cone. We prove orthogonality relations for the equivariant category of sheaves on the nilpotent cone in a method similar to Lusztig's for character sheaves. We also consider cleanness for cuspidal perverse sheaves and the (generalized) Lusztig–Shoji algorithm.

## 1. Introduction

Let $G$ be a connected, reductive algebraic group defined over an algebraically closed field of good characteristic, and let $\mathcal{N}$ be its nilpotent cone. We consider $\mathrm{D}^{\mathrm{b}}_G(\mathcal{N})$, the $G$-equivariant derived category of constructible sheaves on $\mathcal{N}$. This category encodes the representation theory of the Weyl group $W$ of $G$ via a perverse sheaf called the Springer sheaf $\mathbb{A}$ (see Section 2 for a definition) and Springer's correspondence. Of course, the category $\mathcal{P}\mathrm{erv}_G(\mathcal{N})$ of $G$-equivariant perverse sheaves on $\mathcal{N}$ contains much more information—there are $G$-equivariant perverse sheaves on $\mathcal{N}$ that are not part of the Springer correspondence. In [**L1**], Lusztig accounted for the extra information and related it to the representation theory of relative Weyl groups in his generalized Springer correspondence. His classification relies on understanding the *cuspidal data* associated to $G$ (see Definition 2.3).

Lusztig's work also reveals that understanding the stalks of these perverse sheaves on $\mathcal{N}$ (given by generalized Green functions) is an important part of the computation of characters of finite groups of Lie type. Using the Lusztig–Shoji algorithm [**S**, **L5**], these stalks can be computed using only knowledge of Weyl (or relative Weyl) group representations.

Our primary goal is to understand Lusztig's work on the generalized Springer correspondence using methods inspired by [**A**] in the Springer setting. One of our main results is Theorem 3.5, which gives an orthogonal decomposition:

$$\mathrm{D}^{\mathrm{b}}_G(\mathcal{N}) \cong \bigoplus_{\mathbf{c}/\sim} \mathrm{D}^{\mathrm{b}}_G(\mathcal{N}, \mathbb{A}_{\mathbf{c}}).$$

2010 *Mathematics Subject Classification.* Primary 17B08, Secondary 20G05, 14F05.

The first author was supported by an NSF Postdoctoral Research Fellowship. The second author was supported in part by the UGA VIGRE II grant DMS-07-38586 and the NSF RTG grant DMS-1344994 of the AGANT research group at UGA.

Here, $\mathbf{c}$ denotes a cuspidal datum and $\mathrm{D}^{\mathrm{b}}_G(\mathcal{N}, \mathbb{A}_{\mathbf{c}})$ is the triangulated category generated by the simple summands of the perverse sheaf $\mathbb{A}_{\mathbf{c}}$ induced from the cuspidal datum $\mathbf{c}$. This result relies on a key property enjoyed by cuspidal local systems on the nilpotent cone in good characteristic proven by Lusztig. That is, distinct cuspidals have distinct central characters.

In Section 5, we interpret the definitions of generalized Green functions directly in terms of perverse sheaves on the nilpotent cone avoiding characteristic functions on the corresponding group. We reprove the Lusztig–Shoji algorithm in Theorem 5.6 using the re-envisioned Green functions. Our proofs are heavily influenced by those of Lusztig in [**L5**]. From this point of view, Theorem 3.5 may be viewed as a categorical version of the orthogonality relations among generalized Green functions.

In Section 4, we prove cleanness for cuspidal local systems on the nilpotent cone (see Proposition 4.2). In [**L5**], Lusztig proves the much stronger result that cuspidal character sheaves are clean. Logically, our proof is similar in spirit to Lusztig's. However, a key role in our argument is played by the orthogonal decomposition in Theorem 3.5 considerably simplifying the exposition. In particular, from this point of view, it is easy to see that cuspidality for local systems is implied by the fact that the triangulated categories generated by the corresponding simple perverse sheaf and its Verdier dual are orthogonal to the 'rest' of the category.

Proposition 4.7 provides a computation of the Ext-groups between perverse sheaves on the nilpotent cone in terms of relative Weyl group invariants. This proposition is a first step towards proving formality for non-Springer blocks of sheaves on the nilpotent cone. Formality for the other blocks would complete the description of the equivariant derived category on the nilpotent cone, as was initiated by the first author [**R**] in the Springer case.

**Organization of the paper.** We review perverse sheaves on the nilpotent cone and the generalized Springer correspondence in Section 2. In Section 3, we prove our orthogonal decomposition (Theorem 3.5). In Section 4, we prove cleanness for cuspidal local systems (4.2), give some Ext computations (Proposition 4.7), and discuss purity for these Ext-groups (Corollary 4.8). In Section 5, we discuss generalized Green functions and the Lusztig–Shoji algorithm. Finally, in the appendix, we review some properties of central characters for equivariant perverse sheaves. In particular, we prove Proposition A.8, that perverse sheaves with distinct central characters have no extensions betweem them.

## 2. Preliminaries

We fix an algebraically closed field $k$ of positive characteristic. All varieties we consider will be defined over $k$ except in Section 5 when we need to employ mixed sheaves as developed in [**De**] and [**BBD**, Section 5]. For an action of an algebraic group $G$ on a variety $X$, we consider the categories, denoted $Perv_G(X) \subset D_G(X)$, of $G$-equivariant perverse sheaves on $X$ and the $G$-equivariant (bounded) derived category of sheaves on $X$. See [**BL**] for background and definitions related to these categories. All of our sheaves will have $\bar{\mathbb{Q}}_\ell$-coefficients. For $\mathcal{F}, \mathcal{G} \in D_G(X)$, we let $\mathrm{Hom}^i(\mathcal{F}, \mathcal{G}) := \mathrm{Hom}_{D_G(X)}(\mathcal{F}, \mathcal{G}[i])$. All sheaf functors are understood to be derived. We denote the constant sheaf on $X$ by $\underline{\mathbb{C}}_X$ or just $\underline{\mathbb{C}}$ when there is no ambiguity.

**2.1. The Springer correspondence.** From now on, we fix a connected, reductive algebraic group $G$ defined over $k$, and we let $\mathcal{N}$ denote its nilpotent cone. We also assume that $k$ has good characteristic for $G$. We consider $G$-equivariant sheaves on $\mathcal{N}$ with respect to the adjoint $G$-action.

The nilpotent cone $\mathcal{N}$ is a singular variety and has a well-studied desingularization called the Springer resolution, denoted $\mu : \widetilde{\mathcal{N}} \to \mathcal{N}$. Let $B$ be a Borel subgroup of $G$ with Levi decomposition $B = TU$, where $T$ is a maximal torus in $G$ and $U$ is the unipotent radical of $B$. Then, $\widetilde{\mathcal{N}} = G \times^B \mathfrak{u}$, where $\mathfrak{u} = \mathrm{Lie}(U)$. We also note that $\widetilde{\mathcal{N}}$ can be identified with the cotangent bundle of the flag variety $T^*G/B$. Hence, we have maps

$$\mathcal{N} \xleftarrow{\mu} \widetilde{\mathcal{N}} \xrightarrow{\pi} G/B. \tag{2.1}$$

Definition 2.1. The *Springer sheaf*, denoted $\mathbb{A}$, is defined by

$$\mathbb{A} := \mu_! \pi^* \, \underline{\mathbb{C}}[2d],$$

where $\underline{\mathbb{C}}$ is the constant sheaf on $G/B$ and $d = \dim \mathfrak{u}$.

It is well known that the Springer sheaf $\mathbb{A}$ is a semisimple perverse sheaf (see [**BM1**]), and it is $G$-equivariant since the Springer resolution is a $G$-map. Furthermore, its endomorphism ring is isomorphic as an algebra to the group algebra of the Weyl group $W$ of $G$,

$$\mathrm{End}(\mathbb{A}) \cong \bar{\mathbb{Q}}_\ell[W].$$

See [**BM1**]. This ring isomorphism allows us to link the simple summands of $\mathbb{A}$ with irreducible $W$ representations. This is known as the Springer correspondence.

**2.2. The generalized Springer correspondence and cuspidal data.** In general, not all simple ($G$-equivariant) perverse sheaves on $\mathcal{N}$ occur as part of the above correspondence. The goal of Lusztig's generalized Springer correspondence [**L1**] is to systematically identify the missing pieces and to assign some representation theoretic meaning to them. To this end, we define analogues of the maps and varieties in (2.1).

Let $P$ be a parabolic subgroup of $G$ with Levi decomposition $P = LU$. We denote by $\mathcal{N}_L$ the nilpotent cone for $L$ and $\mathfrak{u} = \mathrm{Lie}(U)$. We consider the following $G$-varieties and $G$-maps

$$\widetilde{\mathcal{N}}^P := G \times^P (\mathfrak{u} + \mathcal{N}_L) \quad \text{and} \quad \mathcal{C}_P := G \times^P \mathcal{N}_L,$$

$$\mathcal{N} \xleftarrow{\mu_P} \widetilde{\mathcal{N}}^P \xrightarrow{\pi_P} \mathcal{C}_P.$$

The variety $\widetilde{\mathcal{N}}^P$ is called a partial resolution of $\mathcal{N}$ and is studied in [**BM2**], where it appears with the same notation. When there is no ambiguity, we omit the subscript $P$ on the maps for simplicity of notation. Note that $\mu$ is proper and $\pi$ is smooth of relative dimension $d$, so we have $\mu_! = \mu_*$ and $\pi^! = \pi^*[2d]$, where $d$ is the dimension of $\mathfrak{u}$. We consider the following functors

$$\mathcal{I}_P^G = \mu_! \pi^*[d] \cong \mu_* \pi^![-d], \qquad \mathcal{R}_P^G = \pi_* \mu^![-d], \qquad \text{and} \qquad \widetilde{\mathcal{R}}_P^G = \pi_! \mu^*[d], \tag{2.2}$$

which we will refer to as induction and restriction functors. We have adjoint pairs $(\mathcal{I}_P^G, \mathcal{R}_P^G)$ and $(\widetilde{\mathcal{R}}_P^G, \mathcal{I}_P^G)$. In the case when our parabolic is a Borel $B$, we get the usual Springer resolution diagram from (2.1). By (equivariant) induction equivalence [**BL**], we have an equivalence of categories $\mathrm{D}^{\mathrm{b}}_G(\mathcal{C}_P) \cong \mathrm{D}^{\mathrm{b}}_L(\mathcal{N}_L)$ that preserves the perverse $t$-structure. Thus, often we will think of the induction (respectively, restriction) functor as having domain (respectively, codomain) $\mathrm{D}^{\mathrm{b}}_L(\mathcal{N}_L)$:

$$\mathcal{I}_P^G : \mathrm{D}^{\mathrm{b}}_L(\mathcal{N}_L) \to \mathrm{D}^{\mathrm{b}}_G(\mathcal{N}) \text{ and } \mathcal{R}_P^G, \widetilde{\mathcal{R}}_P^G : \mathrm{D}^{\mathrm{b}}_G(\mathcal{N}) \to \mathrm{D}^{\mathrm{b}}_L(\mathcal{N}_L).$$

The following theorem is due to Lusztig; see [**L2**, Theorem 4.4] for part of the proof in the setting of character sheaves. For a proof in the setting of more general coefficient rings (Noetherian commutative ring of finite global dimension), see [**AHR**, Proposition 4.7].

THEOREM 2.2. *The functors $\mathcal{I}_P^G$, $\mathcal{R}_P^G$, and $\widetilde{\mathcal{R}}_P^G$ are exact with respect to the perverse $t$-structure.*

DEFINITION 2.3. A simple perverse sheaf $\mathcal{F} \in \mathrm{D}^{\mathrm{b}}_G(\mathcal{N})$ is called *cuspidal* if $\mathcal{R}_P^G(\mathcal{F}) = 0$ for all proper parabolics $P$. Let $\mathcal{L}$ be a local system on a nilpotent orbit $\mathcal{O} \subset \mathcal{N}$. Then $\mathcal{L}$ is called a *cuspidal local system* if $\mathrm{IC}(\mathcal{O}, \mathcal{L})$ is cuspidal.

A *cuspidal datum* is a tuple $\mathbf{c} = (L, \mathcal{O}_L, \mathcal{L})$ where $L$ is a Levi subgroup of $G$, $\mathcal{O}_L$ is an $L$-orbit in $\mathcal{N}_L$, and $\mathcal{L}$ is a cuspidal local system on $\mathcal{O}_L$. For brevity, we will often denote the simple perverse sheaf which corresponds to the cuspidal datum $\mathbf{c}$ as $\mathrm{IC}_{\mathbf{c}}$.

Two cuspidal data $\mathbf{c}$ and $\mathbf{c}'$ are equivalent if they are conjugate in $G$, and in this case, we write $\mathbf{c} \sim \mathbf{c}'$.

DEFINITION 2.4. For the cuspidal datum $\mathbf{c} = (G, \mathcal{O}_G, \mathcal{L})$, we set $\mathbb{A}_{\mathbf{c}} = \mathrm{IC}_{\mathbf{c}}$. Let $\mathbf{c} = (L, \mathcal{O}_L, \mathcal{L})$ be a cuspidal datum for $G$. We define the perverse sheaf $\mathbb{A}_{\mathbf{c}} = \mathcal{I}_P^G(\mathrm{IC}_{\mathbf{c}})$, and call it a *Lusztig sheaf*. Let $\mathrm{D}^{\mathrm{b}}_G(\mathcal{N}, \mathbb{A}_{\mathbf{c}})$ be the triangulated subcategory of $\mathrm{D}^{\mathrm{b}}_G(\mathcal{N})$ generated by the simple summands of $\mathbb{A}_{\mathbf{c}}$.

REMARK 2.5. The only cuspidal datum when the Levi is a torus $T$ is $(T, \mathrm{pt}, \underline{\mathbb{C}})$. This datum gives rise to the Springer sheaf $\mathbb{A}$. Also, as Lusztig noted in [**L1**, Section 2.5], if $\mathrm{IC}(\mathcal{O}, \mathcal{L})$ is cuspidal, then so is $\mathbb{D}\,\mathrm{IC}(\mathcal{O}, \mathcal{L}) = \mathrm{IC}(\mathcal{O}, \mathcal{L}^\vee)$, where $\mathbb{D}$ is the Verdier duality functor and $\mathcal{L}^\vee$ is the local system dual to $\mathcal{L}$.

Fix a cuspidal datum $\mathbf{c} = (L, \mathcal{O}, \mathcal{L})$ so that $\mathbb{A}_{\mathbf{c}} = \mathcal{I}_P^G(\mathrm{IC}_{\mathbf{c}})$, where $\mathrm{IC}_{\mathbf{c}}$ is cuspidal in $\mathrm{D}^{\mathrm{b}}_L(\mathcal{N}_L)$. Let $W(L) = N_G(L)/L$. This group is referred to as a relative Weyl group. Let $\mathrm{Irr}(W(L))$ denote the set of isomorphism classes of irreducible $W(L)$-representations. For each $\psi \in \mathrm{Irr}(W(L))$, we fix a representative $V_\psi$. Lusztig proves in [**L1**, Theorem 9.2] that we have an algebra isomorphism:

$$\mathrm{End}(\mathbb{A}_{\mathbf{c}}) \cong \bar{\mathbb{Q}}_\ell[W(L)]. \tag{2.3}$$

This, together with the fact that $\mathcal{P}\mathrm{erv}_G(\mathcal{N})$ is a semisimple abelian category imply that we have a decomposition of $\mathbb{A}_{\mathbf{c}}$ into isotypic components indexed by the irreducible representations of $W(L)$

$$\mathbb{A}_{\mathbf{c}} \cong \bigoplus_{\psi \in \mathrm{Irr}(W(L))} \mathrm{IC}_\psi \otimes V_\psi.$$

The assignment $\mathrm{IC}_\psi \mapsto V_\psi^*$ is known as the generalized Springer correspondence [**L1**], where $V_\psi^*$ is the contragredient representation of $V_\psi$. Hence, we have an equivalence

$$\mathcal{P}\mathrm{erv}_G(\mathcal{N}) \cong \prod_{\mathbf{c}/\sim} \mathrm{Rep}\, W(L_{\mathbf{c}}).$$

A simple $G$-equivariant perverse sheaf on $\mathcal{N}$ is either cuspidal or induced from a cuspidal perverse sheaf for a proper Levi subgroup of $G$. In [**L2**, Theorem 4.4] Lusztig proves the statement for character sheaves, but it is easy to see that it also applies in this setting. Thus, it is sufficient to classify the cuspidal data (up to equivalence) for each group, which Lusztig does in [**L1**]. We use this classification to prove the following proposition.

PROPOSITION 2.6. *Let* $\mathbf{c}=(L,\mathcal{O},\mathcal{L})$ *be a cuspidal datum for* $G$. *Then* $\mathcal{R}_P^G\mathcal{I}_P^G \operatorname{IC}_{\mathbf{c}}$ *is a direct sum of copies of* $\operatorname{IC}_{\mathbf{c}}$. *Moreover, the same statement holds if* $\mathcal{R}_P^G$ *is replaced by* $\widetilde{\mathcal{R}}_P^G$.

PROOF. The result relies on Lusztig's generalized Springer correspondence as a classification of the simple perverse sheaves on $\mathcal{N}$.

Let $S$ be a simple summand of $\mathcal{R}_P^G\mathcal{I}_P^G \operatorname{IC}_{\mathbf{c}}$ in $\mathcal{P}\mathrm{erv}_L(\mathcal{N}_L)$. We have that $S$ is a direct summand of a Lusztig sheaf $\mathcal{I}_Q^L \operatorname{IC}_{\mathbf{c}'}$ on $\mathcal{N}_L$, where $\mathbf{c}'$ is a cuspidal datum for $L$. Note that we allow the case where $Q = L$. Hence, $\operatorname{Hom}(S, \mathcal{R}_P^G\mathcal{I}_P^G \operatorname{IC}_{\mathbf{c}}) \neq 0$ implies

$$\operatorname{Hom}(\mathcal{I}_Q^L \operatorname{IC}_{\mathbf{c}'}, \mathcal{R}_P^G\mathcal{I}_P^G \operatorname{IC}_{\mathbf{c}}) \neq 0.$$

Let $\tilde{Q} = QU_P$, where $U_P$ is the unipotent radical of $P$. Then $\tilde{Q}$ is a parabolic subgroup of $G$. By adjunction and the transitivity of induction $\mathcal{I}_{\tilde{Q}}^G \cong \mathcal{I}_P^G\mathcal{I}_Q^L$ [**L2**, Proposition 4.2],

$$\operatorname{Hom}(\mathcal{I}_{\tilde{Q}}^G \operatorname{IC}_{\mathbf{c}'}, \mathcal{I}_P^G \operatorname{IC}_{\mathbf{c}}) \neq 0.$$

This implies that $\mathcal{I}_{\tilde{Q}}^G \operatorname{IC}_{\mathbf{c}'}$ and $\mathcal{I}_P^G \operatorname{IC}_{\mathbf{c}}$ have a simple summand in common. However, Lusztig's generalized Springer correspondence partitions the set of simple $G$-equivariant perverse sheaves on $\mathcal{N}$ into distinct classes, each labeled by a unique cuspidal datum up to $G$-conjugation. Hence, if $\mathcal{I}_{\tilde{Q}}^G \operatorname{IC}_{\mathbf{c}'}$ and $\mathcal{I}_P^G \operatorname{IC}_{\mathbf{c}}$ have a simple summand in common, then it must be that

$$\mathcal{I}_{\tilde{Q}}^G \operatorname{IC}_{\mathbf{c}'} \cong \mathcal{I}_P^G \operatorname{IC}_{\mathbf{c}},$$

each labeled by cuspidal data that are $G$-conjugate. Since $Q \subset L$, we have $L = Q$. Thus, $S$ is a summand of (and so, must be equal to) $\mathcal{I}_Q^L(\operatorname{IC}_{\mathbf{c}'}) \cong \operatorname{IC}_{\mathbf{c}}$.

The case for $\widetilde{\mathcal{R}}_P^G$ follows in a similar manner. □

## 3. Orthogonal decomposition

In this section, we prove our main result: an orthogonal decomposition of the category $\mathrm{D}_G^{\mathrm{b}}(\mathcal{N})$ into blocks, each corresponding to a cuspidal datum for $G$.

DEFINITION 3.1. Let $\mathcal{T}$ be a triangulated category. We say that $\mathcal{T}_1$ and $\mathcal{T}_2$, two triangulated subcategories of $\mathcal{T}$, are *orthogonal* if for all objects $\mathcal{F} \in \mathcal{T}_1$ and $\mathcal{G} \in \mathcal{T}_2$, we have that

$$\operatorname{Hom}_{\mathcal{T}}(\mathcal{F}, \mathcal{G}) = \operatorname{Hom}_{\mathcal{T}}(\mathcal{G}, \mathcal{F}) = 0.$$

If we have a (finite) collection $\mathcal{T}_i$ of triangulated subcategories of $\mathcal{T}$ that are pairwise orthogonal and that generate $\mathcal{T}$ such that each $\mathcal{T}_i$ cannot be split further, then we call them *blocks*. The equivalence

$$\mathcal{T} \cong \bigoplus_i \mathcal{T}_i$$

is called a *block decomposition* of $\mathcal{T}$.

Suppose we have cuspidal perverse sheaves $\mathrm{IC}_{\mathbf{c}}$ on $\mathcal{N}_L$ and $\mathrm{IC}_{\mathbf{c}'}$ on $\mathcal{N}_{L'}$, where $L$ and $L'$ are the Levi factors of parabolics $P$ and $P'$ in $G$. Let $E = \mathcal{N}_{P'\cap L} \times_{\mathcal{N}_{L\cap L'}} \mathcal{N}_{P\cap L'}$.

PROPOSITION 3.2. *If $P' \cap L \times P \cap L'$ is properly contained within $L \times L'$, then $\mathrm{H}^i_c(E, \mathrm{IC}_{\mathbf{c}} \boxtimes \mathrm{IC}_{\mathbf{c}'}) = 0$ for all $i$.*

PROOF. Consider the following diagram.

$$\begin{array}{ccc}
\mathcal{N}_L \times \mathcal{N}_{L'} & & \\
\uparrow{\scriptstyle i} & & \\
\mathcal{N}_{P'\cap L} \times \mathcal{N}_{P\cap L'} & \xrightarrow{\varpi} & \mathcal{N}_{L\cap L'} \times \mathcal{N}_{L\cap L'} \\
\uparrow{\scriptstyle s} & & \uparrow{\scriptstyle \Delta} \\
E = \mathcal{N}_{P'\cap L} \times_{\mathcal{N}_{L\cap L'}} \mathcal{N}_{P\cap L'} & \xrightarrow{\tilde{\varpi}} & \mathcal{N}_{L\cap L'}
\end{array}$$

The maps $i$ and $s$ are inclusions, and $\Delta$ is diagonal inclusion. Note that $P' \cap L$ (resp. $P \cap L'$) is a parabolic subgroup of $L$ (resp. $L'$) with Levi factor $L \cap L'$. Thus, $\varpi$ is induced by the quotient $P' \cap L \times P \cap L' \to L \cap L' \times L \cap L'$ with unipotent kernel, $\tilde{\varpi}$ is its restriction, and the square commutes.

By base-change, we have an isomorphism $\tilde{\varpi}_! s^* \cong \Delta^* \varpi_!$. The functor $\varpi_! i^*$ is easily seen to be the non-equivariant version of restriction $\widetilde{\mathcal{R}}$ as in (2.2). Hence, $\varpi_! i^*(\mathrm{IC}_{\mathbf{c}} \boxtimes \mathrm{IC}_{\mathbf{c}'}) = 0$ since $\mathrm{IC}_{\mathbf{c}} \boxtimes \mathrm{IC}_{\mathbf{c}'}$ is cuspidal on $\mathcal{N}_L \times \mathcal{N}_{L'}$ and $P' \cap L \times P \cap L'$ is a proper parabolic in $L \times L'$. Hence, we have that $0 = \Delta^* \varpi_! i^*(\mathrm{IC}_{\mathbf{c}} \boxtimes \mathrm{IC}_{\mathbf{c}'}) = \tilde{\varpi}_! s^* i^*(\mathrm{IC}_{\mathbf{c}} \boxtimes \mathrm{IC}_{\mathbf{c}'})$. Therefore, $\mathrm{H}^i_c(E, \mathrm{IC}_{\mathbf{c}} \boxtimes \mathrm{IC}_{\mathbf{c}'}) = 0$ for all $i$. $\square$

Let $Z = \widetilde{\mathcal{N}}^P \times_{\mathcal{N}} \widetilde{\mathcal{N}}^{P'}$ be a generalized Steinberg variety. We use $\mathrm{IC}_{\mathbf{c}}$ and $\mathrm{IC}_{\mathbf{c}'}$ to also denote the $G$-equivariant perverse sheaves on $\mathcal{C}_P$ and $\mathcal{C}_{P'}$ which correspond to the cuspidal perverse sheaves on $\mathcal{N}_L$ and $\mathcal{N}_{L'}$ under induction equivalence. Consider the composition

$$\tau : \widetilde{\mathcal{N}}^P \times_{\mathcal{N}} \widetilde{\mathcal{N}}^{P'} \overset{i}{\hookrightarrow} \widetilde{\mathcal{N}}^P \times \widetilde{\mathcal{N}}^{P'} \overset{\Pi := \pi \times \pi'}{\longrightarrow} \mathcal{C}_P \times \mathcal{C}_{P'}. \tag{3.1}$$

Our goal in the following proposition is to prove that the cohomology groups $\mathrm{H}^i_G(Z, \tau^! \mathrm{IC}_{\mathbf{c}} \boxtimes \mathrm{IC}_{\mathbf{c}'})$ vanish for all $i$ when $\mathrm{IC}_{\mathbf{c}}$ and $\mathrm{IC}_{\mathbf{c}'}$ arise from cuspidals with non-conjugate Levis. The outline of our proof follows that of a similar result due to Lusztig [**L3**, Prop. 7.2] in the setting of character sheaves.

PROPOSITION 3.3. *Suppose $P$ and $P'$ are parabolics in $G$ such that their Levi factors $L$ and $L'$ are non-conjugate. Let $Z$ be as above. Then, $\mathrm{H}^i(Z, \tau^! \mathrm{IC}_{\mathbf{c}} \boxtimes \mathrm{IC}_{\mathbf{c}'}) = 0$ for all $i$.*

PROOF. We will first show an analogous statement for cohomology with compact support. Let $x \in G/P \times G/P'$, and consider the following diagram.

$$\begin{CD}
\phi^{-1}(x) @>{\iota}>> Z \\
@V{\phi_x}VV @VV{\phi}V \\
\{x\} @>>> G/P \times G/P' \\
@. @VVV \\
@. \mathrm{pt}
\end{CD}$$

Clearly, it is enough to show that $\phi_! \tau^*(\mathrm{IC}_{\mathbf{c}} \boxtimes \mathrm{IC}_{\mathbf{c}'}) = 0$. This holds if all the stalks of $\phi_! \tau^*(\mathrm{IC}_{\mathbf{c}} \boxtimes \mathrm{IC}_{\mathbf{c}'})$ vanish. This is equivalent to $\phi_{x\,!}(\iota^* \tau^*(\mathrm{IC}_{\mathbf{c}} \boxtimes \mathrm{IC}_{\mathbf{c}'})) = 0$ for all $x \in G/P \times G/P'$ by base change. In other words, it is enough to show that $\mathrm{H}_c^\bullet(\phi^{-1}(x), \tau^*(\mathrm{IC}_{\mathbf{c}} \boxtimes \mathrm{IC}_{\mathbf{c}'})) = 0$ for all $x$. Since $\phi_! \tau^*(\mathrm{IC}_{\mathbf{c}} \boxtimes \mathrm{IC}_{\mathbf{c}'})$ is $G$-equivariant, it is enough to check

$$\phi_{x\,!}(\iota^* \tau^*(\mathrm{IC}_{\mathbf{c}} \boxtimes \mathrm{IC}_{\mathbf{c}'})) = 0$$

for a single $x$ in each $G$-orbit of $G/P \times G/P'$. Let $n \in G$ be such that $L$ and $nL'n^{-1}$ share a maximal torus and consider $Z(n) = \phi^{-1}((P, nP'))$, a subvariety of $Z$. We must show that $\mathrm{H}_c^i(Z(n), \tau^*(\mathrm{IC}_{\mathbf{c}} \boxtimes \mathrm{IC}_{\mathbf{c}'})) = 0$ for all $i$. Consider the following commutative diagram where $L'' = nL'n^{-1}, P'' = nP'n^{-1}, f : \mathcal{N}_{L''} \xrightarrow{\sim} \mathcal{N}_{L'}$, and $E = \mathcal{N}_{P'' \cap L} \times_{\mathcal{N}_{L \cap L''}} \mathcal{N}_{P \cap L''}$.

$$\begin{CD}
Z @>{\tau}>> \mathcal{C}_P \times \mathcal{C}_{P'} @>{\sim}>> \mathcal{C}_P \times \mathcal{C}_{P''} \\
@AAA @. @AAA \\
@. @. \mathcal{N}_L \times \mathcal{N}_{L''} \\
@. @. @AAA \\
@. @. \mathcal{N}_{P'' \cap L} \times \mathcal{N}_{P \cap L''} \\
@. @. @AAA \\
Z(n) @>{\sim}>> \mathcal{N}_P \cap \mathcal{N}_{P''} @>{\alpha}>> E
\end{CD}$$

Let $\mathrm{IC}_{\mathbf{c}''} = f^* \mathrm{IC}_{\mathbf{c}'}$. Hence, $\mathrm{H}_c^i(Z(n), \tau^*(\mathrm{IC}_{\mathbf{c}} \boxtimes \mathrm{IC}_{\mathbf{c}'})) \cong \mathrm{H}_c^i(Z(n), \alpha^*(\mathrm{IC}_{\mathbf{c}} \boxtimes \mathrm{IC}_{\mathbf{c}''}))$. The map $\alpha$ is smooth of relative dimension $d$ where $d = \dim U_{P \cap P''}$, hence $\alpha^*[2d] \cong \alpha^!$.

Finally, we apply a special case of Braden's hyperbolic localization: equation (1) in [**Br**, Section 3]. In the diagram below, we let the multiplicative group $\mathbb{G}_m$ act on all varieties with compatible positive weights. Let $e : \{(0,0)\} \hookrightarrow E$ be the inclusion of the fixed point of this action, and let $\ell : E \to \{(0,0)\}$ be the map that sends every point to its limit. Note that Proposition 3.2 implies that $\ell_! \mathrm{IC}_{\mathbf{c}} \boxtimes \mathrm{IC}_{\mathbf{c}''} = 0$.

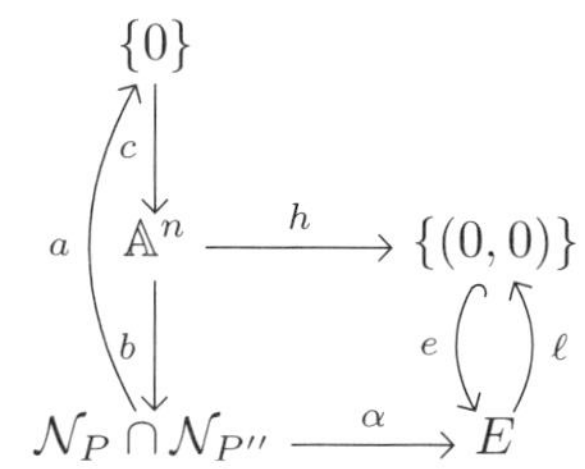

Then hyperbolic localization implies that we have isomorphisms $e^! \cong \ell_!$ and $a_! \cong (b \circ c)^!$. Furthermore, the diagram commutes. Combining these, we see that

$$a_!\alpha^! \cong c^!b^!\alpha^! \cong c^!h^!e^! \cong c^!h^!\ell_!$$

Now, we apply this to the cuspidal perverse sheaf $\mathrm{IC}_{\mathbf{c}} \boxtimes \mathrm{IC}_{\mathbf{c}''}$ to see that

$$a_!\alpha^! \, \mathrm{IC}_{\mathbf{c}} \boxtimes \mathrm{IC}_{\mathbf{c}''} \cong c^!h^!\ell_! \, \mathrm{IC}_{\mathbf{c}} \boxtimes \mathrm{IC}_{\mathbf{c}''} = 0.$$

Thus, we've shown $\mathrm{H}^i_c(Z(n), \tau^*(\mathrm{IC}_{\mathbf{c}} \boxtimes \mathrm{IC}_{\mathbf{c}'})) = 0$ for all $i$ and $n$, which implies $\mathrm{H}^i_c(Z, \tau^*(\mathrm{IC}_{\mathbf{c}} \boxtimes \mathrm{IC}_{\mathbf{c}'})) = 0$. Now we show that this implies $\mathrm{H}^i(Z, \tau^*(\mathrm{IC}_{\mathbf{c}} \boxtimes \mathrm{IC}_{\mathbf{c}'})) = 0$ for all $i$. If $\mathrm{IC}_{\mathbf{c}}$ and $\mathrm{IC}_{\mathbf{c}'}$ are distinct cuspidals, then $\mathbb{D}\,\mathrm{IC}_{\mathbf{c}}$ and $\mathbb{D}\,\mathrm{IC}_{\mathbf{c}'}$ are also distinct cuspidals by Remark 2.5. By the above argument, $p_!\tau^* \,\mathbb{D}\,\mathrm{IC}_{\mathbf{c}} \boxtimes \mathbb{D}\,\mathrm{IC}_{\mathbf{c}'} = 0$ where $p : Z \to \mathrm{pt}$. Hence, $p_!\tau^* \,\mathbb{D}(\mathrm{IC}_{\mathbf{c}} \boxtimes \mathrm{IC}_{\mathbf{c}'}) = \mathbb{D}\, p_*\tau^! \,\mathrm{IC}_{\mathbf{c}} \boxtimes \mathrm{IC}_{\mathbf{c}'} = 0$, which implies that $p_*\tau^! \,\mathrm{IC}_{\mathbf{c}} \boxtimes \mathrm{IC}_{\mathbf{c}'} = 0$. □

Since forgetting equivariance commutes with sheaf functors for equivariant maps, the following corollary also holds.

COROLLARY 3.4. *Suppose $P$ and $P'$ are parabolics in $G$ such that their Levi factors $L$ and $L'$ are non-conjugate. Then,* $\mathrm{H}^i_G(Z, \tau^! \,\mathrm{IC}_{\mathbf{c}} \boxtimes \mathrm{IC}_{\mathbf{c}'}) = 0$ *for all $i$.*

Now we come to the main theorem of this section: an orthogonal decomposition of the category $\mathrm{D}^{\mathrm{b}}_G(\mathcal{N})$. The proof of this decomposition mostly follows from arguments found in [**L1**] and [**L2**].

THEOREM 3.5. *We have an orthogonal decomposition indexed by cuspidal data up to equivalence*

$$\mathrm{D}^{\mathrm{b}}_G(\mathcal{N}) \cong \bigoplus_{\mathbf{c}/\sim} \mathrm{D}^{\mathrm{b}}_G(\mathcal{N}, \mathbb{A}_{\mathbf{c}}).$$

PROOF. Let $\mathbf{c} = (L, \mathcal{O}, \mathcal{L})$ and $\mathbf{c}' = (L', \mathcal{O}', \mathcal{L}')$ be two cuspidal data such that $\mathbf{c} \not\sim \mathbf{c}'$ with Lusztig sheaves $\mathbb{A}_{\mathbf{c}}$ and $\mathbb{A}_{\mathbf{c}'}$. We want to show that the two triangulated categories $\mathrm{D}^{\mathrm{b}}_G(\mathcal{N}, \mathbb{A}_{\mathbf{c}})$ and $\mathrm{D}^{\mathrm{b}}_G(\mathcal{N}, \mathbb{A}_{\mathbf{c}'})$ are orthogonal in $\mathrm{D}^{\mathrm{b}}_G(\mathcal{N})$. It is sufficient to show that $\mathrm{Hom}^i(S, S') = \mathrm{Hom}^i(S', S) = 0$ for all $i \in \mathbb{Z}$ and all simple summands $S$ of $\mathbb{A}_{\mathbf{c}}$ and $S'$ of $\mathbb{A}_{\mathbf{c}'}$.

We have two cases to consider: the Levis $L$ and $L'$ are either conjugate or not. In the second case, we apply equation (8.6.4) from [**CG**, Lemma 8.6.1] to yield

$$\mathrm{Hom}^i(\mathbb{A}_{\mathbf{c}}, \mathbb{A}_{\mathbf{c}'}) \cong \mathrm{H}^i_G(Z, \tau^!(\mathbb{D}\,\mathrm{IC}_{\mathbf{c}} \boxtimes \mathrm{IC}_{\mathbf{c}'})),$$

which vanishes by Corollary 3.4.

In the first case,

$$\mathrm{Hom}^i(\mathbb{A}_{\mathbf{c}'}, \mathbb{A}_{\mathbf{c}}) = \mathrm{Hom}^i(\mathrm{IC}_{\mathbf{c}'}, \mathcal{R}^G_P \mathcal{I}^G_P \,\mathrm{IC}_{\mathbf{c}}).$$

Here, we apply Proposition 2.6 to see that $\mathcal{R}_P^G \mathcal{I}_P^G \mathrm{IC}_{\mathbf{c}}$ is a finite direct sum of copies of $\mathrm{IC}_{\mathbf{c}}$. Hence, it is sufficient to see that $\mathrm{Hom}^i(\mathrm{IC}_{\mathbf{c}'}, \mathrm{IC}_{\mathbf{c}}) = 0$ for non-isomorphic cuspidals $\mathrm{IC}_{\mathbf{c}'}$ and $\mathrm{IC}_{\mathbf{c}}$. Lusztig proves in [**L1**] that non-isomorphic cuspidal perverse sheaves have distinct central characters. Therefore we may apply Proposition A.8 to complete the proof. □

Let $u : Y \hookrightarrow \mathcal{N}$ be a $G$-stable locally closed subvariety. Fix a Lusztig sheaf $\mathbb{A}_{\mathbf{c}}$. We define $\mathrm{D}^{\mathrm{b}}_G(Y, \mathbb{A}_{\mathbf{c}})$ as the triangulated subcategory of $\mathrm{D}^{\mathrm{b}}_G(Y)$ generated by restrictions $u^*S$ for all simple summands $S$ of $\mathbb{A}_{\mathbf{c}}$. (Note that $\mathrm{D}^{\mathrm{b}}_G(Y, \mathbb{A}_{\mathbf{c}})$ may become trivial.)

COROLLARY 3.6. *Let $Y$ be a $G$-stable locally closed subvariety of $\mathcal{N}$. Then restriction to $Y$ preserves orthogonality. That is, we have an equivalence*

$$\mathrm{D}^{\mathrm{b}}_G(Y) \cong \bigoplus_{\mathbf{c}/\sim} \mathrm{D}^{\mathrm{b}}_G(Y, \mathbb{A}_{\mathbf{c}}).$$

PROOF. The proof is by induction on the number of $G$-orbits in $\bar{Y}/Y$. Theorem 3.5 implies the base case. The argument then follows without modification from the proof in [**A**, Theorem 5.1]. □

COROLLARY 3.7. *Suppose $\mathcal{L}$ is a local system on the orbit $\mathcal{O}$ of $\mathcal{N}$ which appears as a composition factor of $\mathrm{H}^i(\mathrm{IC}_\chi|_{\mathcal{O}})$ for $\chi \in \mathrm{Irr}\, W(L)$ for some Levi $L$. Then $\mathrm{IC}(\mathcal{O}, \mathcal{L}) \cong \mathrm{IC}_\psi$ for some $\psi \in \mathrm{Irr}\, W(L)$ where $\mathrm{IC}_\chi$ and $\mathrm{IC}_\psi$ are in the same block.*

## 4. Applications: cleanness and Ext computations

DEFINITION 4.1. For a nilpotent orbit $\mathcal{O}$ of $\mathcal{N}$ consider the inclusion $j : \mathcal{O} \hookrightarrow \mathcal{N}$. A local system $\mathcal{L}$ on $\mathcal{O}$ is called *clean* if $j_!\mathcal{L}[\dim \mathcal{O}] = j_*\mathcal{L}[\dim \mathcal{O}] = \mathrm{IC}(\mathcal{O}, \mathcal{L})$.

The following statement is well known in the setting of perverse sheaves on the nilpotent cone and character sheaves. In good characteristic, this follows from work of Lusztig [**L1, L3, L4, L5**] and in particular, [**L5**, Theorem 23.1]. His argument uses the fact that all cuspidal local systems in good characteristic must have distinct central characters, and he goes on to show that this implies an orthogonality relationship which gives the result. See [**L2**, Proposition 7.9] for character sheaves on a semisimple group, for instance. The proof given here relies on these facts as well since they were needed to prove the orthogonal decomposition of the previous section. However, this proof also applies whenever a simple perverse sheaf and its Verdier dual are orthogonal to all other simples assuming some finiteness conditions on the variety.

PROPOSITION 4.2. *Cuspidal local systems on the nilpotent cone are clean.*

PROOF. Let $\mathcal{L}$ be a cuspidal local system on the orbit $\mathcal{O}$, and here, let $j : \mathcal{O} \hookrightarrow \mathcal{N}$. By the orthogonal decomposition, $\mathrm{IC}(\mathcal{O}, \mathcal{L})$ is orthogonal to all other simple perverse sheaves on $\mathcal{N}$. Let $\iota : \mathcal{O}' \hookrightarrow \mathcal{N}$ be the inclusion of an orbit $\mathcal{O}' \subset \overline{\mathcal{O}}$ with $\mathcal{O} \neq \mathcal{O}'$.

Suppose that $\iota^* \mathrm{IC}(\mathcal{O}, \mathcal{L}) \neq 0$. We assume without loss of generality that $\mathcal{O}'$ is minimal among orbits in $\overline{\mathcal{O}}$ with this property, i.e. we assume that no orbit in $\overline{\mathcal{O}'}/\mathcal{O}'$ has this property. Then, there exists $i \in \mathbb{Z}$ and a simple perverse sheaf $S$

(a shifted local system) on $\mathcal{O}'$ such that $\mathrm{Hom}^i(\iota^* \mathrm{IC}(\mathcal{O},\mathcal{L}), S) \neq 0$. By adjunction, we have $\mathrm{Hom}^i(\mathrm{IC}(\mathcal{O},\mathcal{L}), \iota_* S) \neq 0$. Consider the distinguished triangle

$$ {}^p\iota_* S \to \iota_* S \to A \to {}^p\iota_* S[1] $$

where $A \in {}^pD^{>0}$. This gives an exact sequence

$$ \mathrm{Hom}^i(\mathrm{IC}(\mathcal{O},\mathcal{L}), {}^p\iota_* S) \to \mathrm{Hom}^i(\mathrm{IC}(\mathcal{O},\mathcal{L}), \iota_* S) \to \mathrm{Hom}^i(\mathrm{IC}(\mathcal{O},\mathcal{L}), A). $$

Note that $\mathrm{Hom}^i(\mathrm{IC}(\mathcal{O},\mathcal{L}), A) = 0$ since we assumed minimality of $\mathcal{O}'$ and the support of $A$ is contained within $\overline{\mathcal{O}'}/\mathcal{O}'$. In particular, this implies that ${}^p\iota_* S$ has $\mathrm{IC}(\mathcal{O},\mathcal{L})$ as a direct summand since $\mathrm{IC}(\mathcal{O},\mathcal{L})$ is orthogonal to all other simple perverse sheaves on $\mathcal{N}$. However, the support of ${}^p\iota_* S$ is contained within $\overline{\mathcal{O}'}$, which is a contradiction. Therefore, it must be that the stalk of $\mathrm{IC}(\mathcal{O},\mathcal{L})$ at any $x \in \mathcal{O}'$ is 0. Thus, $j_!\mathcal{L}[d] = \mathrm{IC}(\mathcal{O},\mathcal{L})$, where $d = \dim \mathcal{O}$.

Recall that $\mathbb{D}\mathcal{L}$ is also cuspidal on $\mathcal{O}$ (Remark 2.5). The above argument proves that $j_!(\mathbb{D}\mathcal{L})[d] = \mathrm{IC}(\mathcal{O},\mathcal{L}^\vee)$, where $\mathcal{L}^\vee$ is the dual local system to $\mathcal{L}$. Thus, $\mathbb{D}j_*\mathcal{L}[d] = \mathbb{D}\,\mathrm{IC}(\mathcal{O},\mathcal{L})$ which implies $j_*\mathcal{L}[d] \cong \mathrm{IC}(\mathcal{O},\mathcal{L})$. □

REMARK 4.3. This result is now known to hold for cuspidal character $\bar{\mathbb{Q}}_\ell$- sheaves on $G$ defined over a field of any characteristic. The final cases were considered in [**O**] and [**L7**].

**4.1. Some Ext computations.** For the remainder of this section, let us assume we are in a particular block of the decomposition corresponding to the cuspidal datum $\mathbf{c} = (L, \mathcal{O}, \mathcal{L})$. This block will have cuspidal simple perverse sheaf $\mathrm{IC}_\mathbf{c}$ on $\mathcal{N}_L$. The simple summand of $\mathbb{A}_\mathbf{c}$ corresponding to the representation $V_\psi^*$ of $W(L)$ will be denoted $\mathrm{IC}_\psi$.

LEMMA 4.4. *We have that*

$$ \mathrm{Hom}^i_{\mathrm{D}^\mathrm{b}_L(\mathcal{N}_L)}(\mathrm{IC}_\mathbf{c}, \mathrm{IC}_\mathbf{c}) \cong \mathrm{H}^i_L(\mathcal{O}) \cong \mathrm{H}^i_G(G \times^P \mathcal{O}). $$

*In particular,* $\mathrm{Hom}^i_{\mathrm{D}^\mathrm{b}_L(\mathcal{N}_L)}(\mathrm{IC}_\mathbf{c}, \mathrm{IC}_\mathbf{c})$ *is pure of weight $i$ where Frobenius acts by $q^{i/2}$ and vanishes for $i$ odd.*

PROOF. By Proposition 4.2, cuspidal local systems are clean, so we have that

$$ \mathrm{Hom}^i(\mathrm{IC}_\mathbf{c}, \mathrm{IC}_\mathbf{c}) = \mathrm{Hom}^i(j_!\mathcal{L}[\dim \mathcal{O}], j_!\mathcal{L}[\dim \mathcal{O}]). $$

This reduces to a computation of local systems: $\mathrm{Hom}^i(j_!\mathcal{L}, j_!\mathcal{L}) = \mathrm{Hom}^i(\mathcal{L}, j^! j_!\mathcal{L}) = \mathrm{Hom}^i(\mathcal{L},\mathcal{L})$. On the other hand, $\mathrm{Hom}^i(\mathcal{L},\mathcal{L}) = \mathrm{H}^i(\mathcal{O}, R\mathcal{H}om(\mathcal{L},\mathcal{L}))$, and for rank one local systems, $R\mathcal{H}om(\mathcal{L},\mathcal{L}) = \underline{\mathbb{C}}_\mathcal{O}$. Thus, $\mathrm{Hom}^i(\mathrm{IC}_\mathbf{c}, \mathrm{IC}_\mathbf{c}) \cong \mathrm{H}^i_L(\mathcal{O})$, as desired.

We assume that all algebraic groups are split over $\mathbb{F}_q$. Let $x$ be a closed point in $\mathcal{O}$ fixed by Frobenius, so $\mathcal{O} \cong L \cdot x$. Let $S = \mathrm{Stab}_L(x)$ be the stabilizer of $x$ in $L$, $S^\circ = \mathrm{Stab}^\circ_L(x)$ its identity component, and $Z(L)^\circ$ be the identity component of the center of $L$.

First, we prove that $\mathrm{H}^\bullet_S(x) \cong \mathrm{H}^\bullet_{Z(L)^\circ}(x)$. By [**L1**, Proposition 2.8], the group $S^\circ/Z(L)^\circ$ is unipotent, which implies $\mathrm{H}^\bullet_{S^\circ}(x) \cong \mathrm{H}^\bullet_{Z(L)^\circ}(x)$. Thus, we have a homomorphism

$$ \mathrm{H}^\bullet_{S^\circ}(x)^{S/S^\circ} \hookrightarrow \mathrm{H}^\bullet_{S^\circ}(x) \cong \mathrm{H}^\bullet_{Z(L)^\circ}(x). $$

Moreover, $\pi_1^L(\mathcal{O}) \cong S/S^\circ$ acts trivially on $Z(L)^\circ$. Hence, the above injection is an isomorphism. Finally, we have an isomorphism $\mathrm{H}^\bullet_S(x) \cong \mathrm{H}^\bullet_{S^\circ}(x)^{S/S^\circ}$ (see, for instance, [**L6**, 1.9(a)]).

By equivariant induction, we have $\mathrm{H}^\bullet_L(\mathcal{O}) \cong \mathrm{H}^\bullet_S(x)$, which is isomorphic to $\mathrm{H}^\bullet_{Z(L)^\circ}(x)$ by the above argument. Now, it is well known that

$$\mathrm{H}^\bullet_{Z(L)^\circ}(x) \cong \mathrm{H}^\bullet((\mathbb{P}^\infty)^r)$$

where $r$ is the rank of the torus $Z(L)^\circ$ and that Frobenius acts by multiplication by $q$ in degree 2. □

LEMMA 4.5. *Let* $\mathrm{IC}_\psi$ *be a simple summand of* $\mathbb{A}_{\mathbf{c}}$ *as described above. We have an isomorphism* $\mathcal{R}^G_P(\mathrm{IC}_\psi) \cong \widetilde{\mathcal{R}}^G_P(\mathrm{IC}_\psi) \cong \mathrm{IC}_{\mathbf{c}} \otimes V^*_\psi$.

PROOF. By Proposition 2.6, we know $\mathcal{R}^G_P(\mathrm{IC}_\psi)$ is contained in the block $\mathrm{D}^{\mathrm{b}}_L(\mathcal{N}_L, \mathrm{IC}_{\mathbf{c}})$. Since our restriction functor $\mathcal{R}^G_P$ is $t$-exact, we know $\mathcal{R}^G_P(\mathrm{IC}_\psi)$ must be perverse. As $\mathrm{IC}_{\mathbf{c}}$ is the only simple perverse sheaf in $\mathrm{D}^{\mathrm{b}}_L(\mathcal{N}_L, \mathrm{IC}_{\mathbf{c}})$, we have $\mathcal{R}^G_P(\mathrm{IC}_\psi) \cong \mathrm{IC}_{\mathbf{c}} \otimes \mathrm{Hom}(\mathrm{IC}_{\mathbf{c}}, \mathcal{R}^G_P(\mathrm{IC}_\psi))$. Furthermore, we have

$$\mathrm{Hom}(\mathrm{IC}_{\mathbf{c}}, \mathcal{R}^G_P(\mathrm{IC}_\psi)) \cong \mathrm{Hom}(\mathbb{A}_{\mathbf{c}}, \mathrm{IC}_\psi) \cong \mathrm{Hom}(\mathrm{IC}_\psi \otimes V_\psi, \mathrm{IC}_\psi) \cong V^*_\psi.$$

Now, we need only to show that $\mathcal{R}$ and $\widetilde{\mathcal{R}}$ give isomorphic objects when restricted to the category of perverse sheaves. Following the same reasoning as above, we have $\widetilde{\mathcal{R}}^G_P(\mathrm{IC}_\psi) \cong \mathrm{IC}_{\mathbf{c}} \otimes V_\psi$. Since $W(L)$ is a Weyl group, which can be deduced from [**L0**, Theorem 5.9], we have $V_\psi \cong V^*_\psi$, and the result follows. □

PROPOSITION 4.6. *We have an isomorphism*

$$\mathrm{Hom}^i_{\mathrm{D}^{\mathrm{b}}_G(\mathcal{N})}(\mathbb{A}_{\mathbf{c}}, \mathbb{A}_{\mathbf{c}}) \cong \mathrm{H}^i_L(\mathcal{O}) \otimes \bar{\mathbb{Q}}_\ell[W(L)].$$

PROOF. First, using a similar argument as in the proof of Lemma 4.5 and Lusztig's isomorphism (2.3), we have that $\mathcal{R}^G_P\mathbb{A}_{\mathbf{c}} \cong \mathrm{IC}_{\mathbf{c}} \otimes \mathrm{Hom}(\mathrm{IC}_{\mathbf{c}}, \mathcal{R}^G_P\mathbb{A}_{\mathbf{c}}) \cong \mathrm{IC}_{\mathbf{c}} \otimes \bar{\mathbb{Q}}_\ell[W(L)]$,

$$\begin{aligned}\mathrm{Hom}^i_{\mathrm{D}^{\mathrm{b}}_G(\mathcal{N})}(\mathbb{A}_{\mathbf{c}}, \mathbb{A}_{\mathbf{c}}) &\cong \mathrm{Hom}^i_{\mathrm{D}^{\mathrm{b}}_L(\mathcal{N}_L)}(\mathrm{IC}_{\mathbf{c}}, \mathcal{R}^G_P\mathbb{A}_{\mathbf{c}}) \\ &\cong \mathrm{Hom}^i_{\mathrm{D}^{\mathrm{b}}_L(\mathcal{N}_L)}(\mathrm{IC}_{\mathbf{c}}, \mathrm{IC}_{\mathbf{c}} \otimes \bar{\mathbb{Q}}_\ell[W(L)]) \\ &\cong \mathrm{Hom}^i_{\mathrm{D}^{\mathrm{b}}_L(\mathcal{N}_L)}(\mathrm{IC}_{\mathbf{c}}, \mathrm{IC}_{\mathbf{c}}) \otimes \bar{\mathbb{Q}}_\ell[W(L)].\end{aligned}$$

Finally, we apply Lemma 4.4, and the result follows. □

PROPOSITION 4.7. *For simple perverse sheaves* $\mathrm{IC}_\psi$ *and* $\mathrm{IC}_\chi$ *in the same block corresponding to* $\psi, \chi \in \mathrm{Irr}\, W(L)$, *we have an isomorphism*

$$\mathrm{Hom}^i_{\mathrm{D}^{\mathrm{b}}_L(\mathcal{N}_L)}(\mathcal{R}^G_P(\mathrm{IC}_\psi), \mathcal{R}^G_P(\mathrm{IC}_\chi)) \cong V_\psi \otimes \mathrm{H}^i_L(\mathcal{O}) \otimes V^*_\chi.$$

*Moreover, we have that*

$$\begin{aligned}\mathrm{Hom}^i_{\mathrm{D}^{\mathrm{b}}_G(\mathcal{N})}(\mathrm{IC}_\psi, \mathrm{IC}_\chi) &\cong \mathrm{Hom}^i_{\mathrm{D}^{\mathrm{b}}_L(\mathcal{N}_L)}(\mathcal{R}^G_P(\mathrm{IC}_\psi), \mathcal{R}^G_P(\mathrm{IC}_\chi))^{W(L)} \\ &\cong (V_\psi \otimes \mathrm{H}^i_L(\mathcal{O}) \otimes V^*_\chi)^{W(L)}.\end{aligned}$$

PROOF. The first statement follows quickly from combining Lemmas 4.4 and 4.5. For the second statement, the proof follows in the same manner as in [**A**, Theorem 4.6]. □

COROLLARY 4.8. *Let* $P_1, P_2 \in \mathcal{P}\mathrm{erv}_G(\mathcal{N})$. *Then* $\mathrm{Hom}^i_{\mathrm{D}^{\mathrm{b}}_G(\mathcal{N})}(P_1, P_2)$ *is pure of weight* $i$ *for all even* $i \in \mathbb{Z}$ *and vanishes for all odd* $i$.

PROOF. First, suppose that $P_1$ and $P_2$ are simple. If they are not in the same block, then $\mathrm{Hom}^i(P_1, P_2)$ vanishes by the orthogonal decomposition, Theorem 3.5. If they are in the same block, then the result follows from Proposition 4.7 and the fact that $\mathrm{H}^i_L(\mathcal{O})$ is pure of weight $i$ for $i$ even and vanishes for $i$ odd. For general ($G$-equivariant) perverse sheaves on $\mathcal{N}$, we use the facts that $\mathcal{P}\mathrm{erv}_G(\mathcal{N})$ is a semisimple category and that $\mathrm{Hom}^i(-,-)$ commutes with direct sum. □

## 5. Generalized Green functions

In this section, we talk about generalized Green functions. For this, we need to work in the mixed category. For $X$ an algebraic variety defined over $\mathbb{F}_q$, we consider the category of mixed $\ell$-adic complexes $\mathrm{D}^{\mathrm{b}}_{\mathrm{m}}(X)$. There is a natural functor forgetting the Weil structure $\xi : \mathrm{D}^{\mathrm{b}}_{\mathrm{m}}(X) \to \mathrm{D}^{\mathrm{b}}_{\mathrm{c}}(X \times_{\mathrm{Spec}(\mathbb{F}_q)} \mathrm{Spec}(\bar{\mathbb{F}}_q))$. The standard reference for the definition and properties of $\mathrm{D}^{\mathrm{b}}_{\mathrm{m}}(X)$ is [**BBD**, Section 5].

We define $K(X)$ as the quotient of the Grothendieck group $K(\mathrm{D}^{\mathrm{b}}_{\mathrm{m}}(X))/\sim$ where we identify isomorphism classes of simple perverse sheaves $[S_1] \sim [S_2]$ if $S_1$ and $S_2$ have the same weight and $\xi(S_1) \cong \xi(S_2)$. We fix a square root of the Tate sheaf. Then $K(X)$ has the structure of a $\mathbb{Z}[t^{1/2}, t^{-1/2}]$-module so that the action of $t$ corresponds to Tate twist: $[\mathcal{F}(i/2)] = t^{-i/2}[\mathcal{F}]$. For the rest of this section, we assume that $G$ is a split connected, reductive algebraic group defined over $\mathbb{F}_q$ and let $\mathcal{N}$ denote its nilpotent cone. We also assume that $\mathbb{F}_q$ is sufficiently large so that all nilpotent orbits are non-empty. If $\mathcal{O}$ is a nilpotent orbit, then $K(\mathcal{O})$ is the free $\mathbb{Z}[t^{1/2}, t^{-1/2}]$-module generated by classes of (weight 0) irreducible local systems on $\mathcal{O}$.

In what follows, we define four sets of polynomials: $p_{S,S'}, \tilde{p}_{S,S'}, \lambda_{S,S'}$, and $\omega_{S,S'}$. For simplicity of notation, we assume throughout the section that $S$ and $S'$ are in the same block. For $\chi, \psi \in \mathrm{Irr}\, W(L)$, let $\mathrm{IC}_\chi$ and $\mathrm{IC}_\psi$ be the simple perverse sheaves that are pure of weight 0 corresponding to $V^*_\chi$ and $V^*_\psi$. In this case, we will instead use the notation: $p_{\chi,\psi}(t), \lambda_{\chi,\psi}(t), \tilde{p}_{\chi,\psi}(t)$, and $\omega_{\chi,\psi}(t)$. Also, we denote by $\mathcal{L}_\psi$ the local system such that $\mathrm{IC}_\psi|_{\mathcal{O}_\psi} = \mathcal{L}_\psi[\dim \mathcal{O}_\psi](\frac{1}{2}\dim \mathcal{O}_\psi)$, where $\mathcal{O}_\psi$ is the orbit open in the support of $\mathrm{IC}_\psi$.

We define $p_{\chi,\psi}(t) \in \mathbb{Z}[t^{1/2}, t^{-1/2}]$ as

$$[\mathrm{IC}_\chi|_{\mathcal{O}}] = \sum_{\{\psi | \mathcal{O}_\psi = \mathcal{O}\}} p_{\chi,\psi}(t)[\mathcal{L}_\psi]. \tag{5.1}$$

We also define 'dual' polynomials. Let $j_{\mathcal{O}} : \mathcal{O} \to \mathcal{N}$ be the inclusion of an orbit. Let $\tilde{p}_{\chi,\psi}(t) \in \mathbb{Z}[t^{1/2}, t^{-1/2}]$ be given by

$$[j^!_{\mathcal{O}}\, \mathrm{IC}_\chi] = \sum_{\{\psi | \mathcal{O}_\psi = \mathcal{O}\}} \tilde{p}_{\chi,\psi}(t)[\mathcal{L}_\psi]. \tag{5.2}$$

LEMMA 5.1. *The polynomials $p$ and $\tilde{p}$ satisfy the following relation: $\tilde{p}_{\chi,\psi}(t^{-1}) = t^{\dim \mathcal{O}} p_{\chi^*,\psi^*}(t)$.*

PROOF. Let $\mathrm{IC}_\chi = \mathrm{IC}(\mathcal{O}_\chi, \mathcal{L}_\chi)$. First, we have that $j^!_{\mathcal{O}}(\mathrm{IC}_\chi) \cong \mathbb{D}\, j^*_{\mathcal{O}}\, \mathbb{D}\, \mathrm{IC}_\chi = \mathbb{D}(\mathrm{IC}(\mathcal{O}_\chi, \mathcal{L}^\vee_\chi)|_{\mathcal{O}})$, where $\mathcal{L}^\vee$ denotes the dual local system to $\mathcal{L}$. Furthermore, Verdier dual transforms local systems in the following way:

$$\mathbb{D}(\mathcal{L}(-i)) = \mathcal{L}^\vee[2\dim \mathcal{O}](\dim \mathcal{O} + i). \tag{5.3}$$

Hence, Verdier dual induces a morphism $\mathbb{D} : K(\mathcal{O}) \to K(\mathcal{O})$ given by $t^i[\mathcal{L}] \mapsto t^{-\dim\mathcal{O}-i}[\mathcal{L}^\vee]$. We apply $\mathbb{D}$ to equation (5.1) to get

$$\begin{aligned}[j_{\mathcal{O}}^! \operatorname{IC}_\chi] &= [\mathbb{D}\, j_{\mathcal{O}}^* \operatorname{IC}(\mathcal{O}_\chi, \mathcal{L}_\chi^*)] \\ &= \mathbb{D} \sum_{\{\psi|\mathcal{O}_\psi=\mathcal{O}\}} p_{\chi^*,\psi}(t)[\mathcal{L}_\psi] \\ &= \sum_{\{\psi|\mathcal{O}_\psi=\mathcal{O}\}} t^{-\dim\mathcal{O}} p_{\chi^*,\psi}(t^{-1})[\mathcal{L}_\psi^*] \\ &= \sum_{\{\psi^*|\mathcal{O}_{\psi^*}=\mathcal{O}\}} t^{-\dim\mathcal{O}} p_{\chi^*,\psi^*}(t^{-1})[\mathcal{L}_\psi].\end{aligned}$$

Since the irreducible local systems (of weight 0) $[\mathcal{L}_\psi]$ are linearly independent in $K(\mathcal{O})$, we have that

$$\tilde{p}_{\chi,\psi}(t) = t^{-\dim\mathcal{O}} p_{\chi^*,\psi^*}(t^{-1}).$$

The result follows. □

Since we know $\operatorname{IC}_\chi|_{\mathcal{O}_\chi} = \mathcal{L}_\chi[\dim\mathcal{O}_\chi](\frac{1}{2}\dim\mathcal{O}_\chi)$ and $\operatorname{IC}_\chi$ vanishes off $\overline{\mathcal{O}_\chi}$, we have

$$p_{\chi,\psi}(t) = \begin{cases} t^{-\dim\mathcal{O}_\chi/2} & \text{if } \chi=\psi \\ 0 & \text{if } \mathcal{O}_\psi \not\subset \overline{\mathcal{O}_\chi} \text{ or if } \mathcal{O}_\psi = \mathcal{O}_\chi \text{ with } \chi \neq \psi. \end{cases} \tag{5.4}$$

We also define $\lambda_{\chi,\psi}(t) \in \mathbb{Z}[t^{1/2}, t^{-1/2}]$ by

$$\begin{aligned} [R\Gamma_c(\mathcal{O}_\chi, \mathcal{L}_\chi \otimes \mathcal{L}_\psi^\vee)] &= \lambda_{\chi,\psi}(t)[\underline{\mathbb{C}}_{\text{pt}}] && \text{if } \mathcal{O}_\chi = \mathcal{O}_\psi \\ \lambda_{\chi,\psi}(t) &= 0 && \text{if } \mathcal{O}_\chi \neq \mathcal{O}_\psi \end{aligned} \tag{5.5}$$

and $\omega_{\chi,\psi}(t) \in \mathbb{Z}[t^{1/2}, t^{-1/2}]$ by

$$[\mathbb{D}\, R\operatorname{Hom}(\operatorname{IC}_\chi, \operatorname{IC}_\psi)] = \omega_{\chi,\psi}(t)[\underline{\mathbb{C}}_{\text{pt}}]. \tag{5.6}$$

Using Corollary 4.8, we can reformulate the definition of $\omega$ in the following way:

$$\begin{aligned}\omega_{\chi,\psi}(t) &= \sum_{i\in\mathbb{Z}} (-1)^i \dim H^i(\mathbb{D}\, R\operatorname{Hom}(\operatorname{IC}_\chi, \operatorname{IC}_\psi)) t^{i/2} \\ &= \sum_{i\in\mathbb{Z}} (-1)^i \dim \operatorname{Hom}^{-i}(\operatorname{IC}_\chi, \operatorname{IC}_\psi) t^{i/2}.\end{aligned} \tag{5.7}$$

REMARK 5.2. It is easy to see that our definitions of the polynomials $p, \tilde{p}, \lambda$, and $\omega$ can be extended to the case of simple perverse sheaves $S$ and $S'$ in different blocks. However, in this case, it is easy to see that $p_{S,S'} = \tilde{p}_{S,S'} = \omega_{S,S'} = 0$ by Theorem 3.5 and Corollary 3.7. Moreover, the main result of this section Theorem 5.6 becomes trivial in this case.

LEMMA 5.3. *For any simple perverse sheaves* $\operatorname{IC}_\chi$ *and* $\operatorname{IC}_\psi$ *in* $\mathcal{P}\mathrm{erv}_G(\mathcal{N})$, *we have that* $\omega_{\chi,\psi}(t) = \omega_{\psi,\chi}(t)$. *Furthermore, if* $\operatorname{IC}_\chi$ *and* $\operatorname{IC}_\psi$ *are in different blocks, then* $\omega_{\chi,\psi}(t) = \lambda_{\chi,\psi}(t) = 0$.

PROOF. If $\operatorname{IC}_\chi$ and $\operatorname{IC}_\psi$ are in different blocks, then $\omega_{\chi,\psi}(t) = \omega_{\psi,\chi}(t) = 0$ follows from Theorem 3.5 and Corollary 3.7 implies $\lambda_{\chi,\psi}(t) = \lambda_{\psi,\chi}(t) = 0$.

Now assume that $\mathrm{IC}_\chi$ and $\mathrm{IC}_\psi$ are in the same block. Proposition 4.7 implies

$$\mathrm{Hom}^i(\mathrm{IC}_\chi, \mathrm{IC}_\psi) = (V_\chi \otimes \mathrm{H}^i_L(\mathcal{O}) \otimes V^*_\psi)^{W(L)} \qquad \text{and}$$
$$\mathrm{Hom}^i(\mathrm{IC}_\psi, \mathrm{IC}_\chi) = (V^*_\chi \otimes \mathrm{H}^i_L(\mathcal{O}) \otimes V_\psi)^{W(L)}.$$

Since $W(L)$ is a Weyl group, we have $V \cong V^*$ for any $W(L)$-representation $V$. In particular, $\dim(V_\chi \otimes \mathrm{H}^i_L(\mathcal{O}) \otimes V^*_\psi)^{W(L)} = \dim(V^*_\chi \otimes \mathrm{H}^i_L(\mathcal{O}) \otimes V_\psi)^{W(L)}$. Hence, using equation (5.7) we obtain $\omega_{\chi,\psi}(t) = \omega_{\psi,\chi}(t)$. □

The following is a refinement of [**A**, Lemma 6.6] to include perverse sheaves that are not self dual.

LEMMA 5.4.

$$[\mathbb{D}\, R\mathrm{Hom}(j^*_{\mathcal{O}}\, \mathrm{IC}_\chi, j^!_{\mathcal{O}}\, \mathrm{IC}_\psi)] = \sum_{\{\phi,\phi' | \mathcal{O}_\phi = \mathcal{O}_{\phi'} = \mathcal{O}\}} p_{\chi,\phi}(t) \lambda_{\phi,\phi'}(t) p_{\psi^*,\phi'^*}(t) [\underline{\mathbb{C}}_{\mathrm{pt}}].$$

PROOF. First, we note that $\mathbb{D}(R\mathrm{Hom}(\mathcal{F}(n), \mathcal{G}(m))) = (\mathbb{D}\, R\mathrm{Hom}(\mathcal{F}, \mathcal{G}))(n-m)$. Thus, $\mathbb{D}\, R\mathrm{Hom}(-,-) : K(\mathcal{O}) \times K(\mathcal{O}) \to K(\mathcal{O})$ is $\mathbb{Z}[t^{1/2}, t^{-1/2}]$ linear in the first variable and antilinear in the second variable with respect to the involution $t^{1/2} \mapsto t^{-1/2}$. Hence,

$$[\mathbb{D}\, R\mathrm{Hom}(j^*_{\mathcal{O}}\, \mathrm{IC}_\chi, j^!_{\mathcal{O}}\, \mathrm{IC}_\psi)] = \sum_{\{\phi,\phi' | \mathcal{O}_\phi = \mathcal{O}_{\phi'} = \mathcal{O}\}} p_{\chi,\phi}(t) \tilde{p}_{\psi,\phi'}(t^{-1})\, \mathbb{D}[R\mathrm{Hom}(\mathcal{L}_\phi, \mathcal{L}_{\phi'})].$$

Now, we apply Lemma 5.1 to get

$$[\mathbb{D}\, R\mathrm{Hom}(j^*_{\mathcal{O}}\, \mathrm{IC}_\chi, j^!_{\mathcal{O}}\, \mathrm{IC}_\psi)] = t^{\dim \mathcal{O}} \sum_{\{\phi,\phi' | \mathcal{O}_\phi = \mathcal{O}_{\phi'} = \mathcal{O}\}} p_{\chi,\phi}(t) p_{\psi^*,\phi'^*}(t) [\mathbb{D}\, R\mathrm{Hom}(\mathcal{L}_\phi, \mathcal{L}_{\phi'})].$$

To finish the proof, it suffices to show $[\mathbb{D}\, R\mathrm{Hom}(\mathcal{L}_\phi, \mathcal{L}_{\phi'})] = t^{-\dim \mathcal{O}} \lambda_{\phi,\phi'}(t) [\underline{\mathbb{C}}_{\mathrm{pt}}]$. Let $a : \mathcal{O} \to \mathrm{pt}$. Then $R\mathrm{Hom}(\mathcal{L}_\phi, \mathcal{L}_{\phi'}) \cong R\Gamma(\mathcal{O}, \mathcal{L}^\vee_\phi \otimes \mathcal{L}_{\phi'}) \cong a_*(\mathcal{L}^\vee_\phi \otimes \mathcal{L}_{\phi'})$. Now, we apply Verdier duality to get $a_!(\mathcal{L}_\phi \otimes \mathcal{L}^\vee_{\phi'})[2\dim \mathcal{O}](\dim \mathcal{O})$. Hence, we have that $[\mathbb{D}\, R\mathrm{Hom}(\mathcal{L}_\phi, \mathcal{L}_{\phi'})] = t^{-\dim \mathcal{O}} [R\Gamma_c(\mathcal{O}, \mathcal{L}_\phi \otimes \mathcal{L}^\vee_{\phi'})]$. □

We need more notation for the proof of the following result. In particular, $\mathcal{O}_\phi \leq \mathcal{O}_{\phi'}$ means $\mathcal{O}_\phi \subset \overline{\mathcal{O}_{\phi'}}$ and $\mathcal{O}_\phi < \mathcal{O}_{\phi'}$ means $\mathcal{O}_\phi \subset \overline{\mathcal{O}_{\phi'}} \setminus \mathcal{O}_{\phi'}$. We also use the fact that $\mathcal{O}_\phi = \mathcal{O}_{\phi^*}$.

REMARK 5.5. The following theorem is proven in [**L5**, Theorem 24.8]. We include its proof here for completeness. The inductive steps of the proof illustrate a method for computing the unknown polynomials $p$ and $\lambda$ from $\omega$ which is known as the Lusztig–Shoji algorithm.

THEOREM 5.6. (1) *For all* $\chi, \psi$, $p_{\chi,\psi}(t) = p_{\chi^*,\psi^*}(t)$ *and* $\lambda_{\chi,\psi}(t) = \lambda_{\psi,\chi}(t)$.
(2) *The polynomials* $p$ *and* $\lambda$ *are the unique ones satisfying*

$$\omega_{\chi,\psi}(t) = \sum_{\phi,\phi'} p_{\chi,\phi}(t) \lambda_{\phi,\phi'}(t) p_{\psi,\phi'}(t)$$

*and conditions* (5.4) *and* (5.5).

PROOF. From Lemma 5.4 above and [**A**, Lemma 6.4], we have

$$\begin{aligned}[\mathbb{D}\, R\operatorname{Hom}(\mathrm{IC}_\chi, \mathrm{IC}_\psi)] &= \sum_{\mathcal{O}\subset N} [\mathbb{D}\, R\operatorname{Hom}(j_{\mathcal{O}}^* \mathrm{IC}_\chi, j_{\mathcal{O}}^! \mathrm{IC}_\psi)] \\ &= \sum_{\mathcal{O}\subset N} \sum_{\substack{\{\phi,\phi'\mid \\ \mathcal{O}_\phi=\mathcal{O}_{\phi'}=\mathcal{O}\}}} p_{\chi,\phi}(t)\lambda_{\phi,\phi'}(t)p_{\psi^*,\phi'^*}(t)[\underline{\mathbb{C}}_{\mathrm{pt}}].\end{aligned}$$

This can be improved because $\lambda_{\phi,\phi'}$ is zero when $\mathcal{O}_\phi \neq \mathcal{O}_{\phi'}$. Thus, we see that

$$\omega_{\chi,\psi}(t) = \sum_{\phi,\phi'} p_{\chi,\phi}(t)\lambda_{\phi,\phi'}(t)p_{\psi^*,\phi'^*}(t). \tag{5.8}$$

We will now use the above equation to prove (1), and thus, prove (2).

We prove both statements of (1) simultaneously by induction on $d = \dim \mathcal{O}_\psi$. If $d < 0$, we know the statement is trivially true. Let us suppose then that $p_{\chi,\psi}(t) = p_{\chi^*,\psi^*}(t)$ and $\lambda_{\chi,\psi}(t) = \lambda_{\psi,\chi}(t)$ for $d \leq k-1$. First, we will use (5.8) to show $\lambda_{\chi,\psi}(t) = \lambda_{\psi,\chi}(t)$ for $d = k$. Since we know $\lambda_{\chi,\psi}(t)$ vanishes otherwise, we can assume that $\mathcal{O}_\chi = \mathcal{O}_\psi$. Recall that $p_{\phi,\phi} = t^{-\dim \mathcal{O}_\phi/2}$ for any $\phi$. Now, with these conditions we have

$$\begin{aligned}\omega_{\chi,\psi}(t) &= p_{\chi,\chi}(t)p_{\psi^*,\psi^*}(t)\lambda_{\chi,\psi}(t) + \sum_{\substack{\mathcal{O}_\phi<\mathcal{O}_\chi \\ \mathcal{O}_{\phi'}<\mathcal{O}_\psi}} p_{\chi,\phi}(t)\lambda_{\phi,\phi'}(t)p_{\psi^*,\phi'^*}(t) \\ &= t^{-\dim \mathcal{O}_\chi}\lambda_{\chi,\psi}(t) + \sum_{\substack{\mathcal{O}_\phi<\mathcal{O}_\chi \\ \mathcal{O}_{\phi'}<\mathcal{O}_\psi}} p_{\chi,\phi}(t)\lambda_{\phi,\phi'}(t)p_{\psi,\phi'}(t).\end{aligned}$$

By the inductive hypothesis, $p_{\psi^*,\phi'^*}(t) = p_{\psi,\phi'}(t)$ since $d = \dim \mathcal{O}_{\phi'} \leq k-1$. After rearranging, we get

$$t^{-\dim \mathcal{O}_\chi}\lambda_{\chi,\psi}(t) = \omega_{\chi,\psi}(t) - \sum_{\substack{\mathcal{O}_\phi<\mathcal{O}_\chi \\ \mathcal{O}_{\phi'}<\mathcal{O}_\psi}} p_{\chi,\phi}(t)\lambda_{\phi,\phi'}(t)p_{\psi,\phi'}(t), \tag{5.9}$$

Since the right hand side of this equation is symmetric with respect to $\chi$ and $\psi$ by the inductive hypothesis and Lemma 5.3, we know the left hand side must be as well. Thus, $\lambda_{\chi,\psi}(t) = \lambda_{\psi,\chi}(t)$ for $d = k$.

Now, we need to show $p_{\chi,\psi}(t) = p_{\chi^*,\psi^*}(t)$ when $d = k$. From (5.4), we may assume $\mathcal{O}_\psi \leq \mathcal{O}_\chi$. Suppose $\alpha$ is such that $\mathcal{O}_\alpha = \mathcal{O}_\psi$, so that $\dim \mathcal{O}_\alpha = k$. Again, we use (5.8).

$$\begin{aligned}\omega_{\alpha,\chi}(t) &= \sum_{\mathcal{O}_\phi=\mathcal{O}_\alpha} p_{\alpha,\alpha}(t)p_{\chi^*,\phi^*}(t)\lambda_{\alpha,\phi}(t) + \sum_{\substack{\mathcal{O}_\phi<\mathcal{O}_\alpha \\ \mathcal{O}_\phi=\mathcal{O}_{\phi'}}} p_{\alpha,\phi}(t)\lambda_{\phi,\phi'}(t)p_{\chi^*,\phi'^*}(t) \\ &= t^{-(\dim \mathcal{O}_\alpha)/2} \sum_{\mathcal{O}_\phi=\mathcal{O}_\alpha} p_{\chi^*,\phi^*}(t)\lambda_{\alpha,\phi}(t) + \sum_{\substack{\mathcal{O}_\phi<\mathcal{O}_\alpha \\ \mathcal{O}_\phi=\mathcal{O}_{\phi'}}} p_{\alpha,\phi}(t)\lambda_{\phi,\phi'}(t)p_{\chi,\phi'}(t).\end{aligned}$$

Here, the induction hypothesis gives $p_{\chi^*,\phi'^*}(t) = p_{\chi,\phi'}(t)$ in the second sum since $d = \dim \mathcal{O}_{\phi'} < \dim \mathcal{O}_\alpha = k$. Similarly, we also have

$$\begin{aligned}\omega_{\chi,\alpha}(t) &= \sum_{\mathcal{O}_\phi = \mathcal{O}_\alpha} p_{\chi,\phi}(t) p_{\alpha^*,\alpha^*}(t) \lambda_{\phi,\alpha}(t) + \sum_{\substack{\mathcal{O}_\phi < \mathcal{O}_\alpha \\ \mathcal{O}_\phi = \mathcal{O}_{\phi'}}} p_{\chi,\phi}(t) \lambda_{\phi,\phi'}(t) p_{\alpha^*,\phi'^*}(t) \\ &= t^{-(\dim \mathcal{O}_\alpha)/2} \sum_{\mathcal{O}_\phi = \mathcal{O}_\alpha} p_{\chi,\phi}(t) \lambda_{\phi,\alpha}(t) + \sum_{\substack{\mathcal{O}_\phi < \mathcal{O}_\alpha \\ \mathcal{O}_\phi = \mathcal{O}_{\phi'}}} p_{\chi,\phi}(t) \lambda_{\phi,\phi'}(t) p_{\alpha,\phi'}(t).\end{aligned}$$

Keeping in mind the induction hypothesis and Lemma 5.3, we subtract these two formulas to get

$$\sum_{\mathcal{O}_\phi = \mathcal{O}_\alpha} \left( p_{\chi^*,\phi^*}(t) - p_{\chi,\phi}(t) \right) \lambda_{\alpha,\phi}(t) = 0. \tag{5.10}$$

The classes of irreducible local systems on $\mathcal{O}$ form a basis in $K(\mathcal{O})$, and therefore, the matrix $(\lambda_{\phi,\phi'})_{\mathcal{O}_\phi = \mathcal{O}_{\phi'}}$ must be nonsingular. Thus, (5.10) implies $p_{\chi^*,\phi^*}(t) = p_{\chi,\phi}(t)$. This completes the induction argument and the proof of (1). From (5.9) we get a recursive formula which implies the uniqueness of part (2) which completes the proof. □

## Appendix A. Central characters

Throughout this section we will assume that we have a connected algebraic group $G$ acting on a variety $X$ with finitely many orbits so that the center $Z$ acts trivially. The $G$-action induces an action of the quotient $G/Z$ on $X$, and the action map $G \times X \to X$ factors as

$$G \times X \overset{(q, id_X)}{\longrightarrow} G/Z \times X \overset{\alpha}{\to} X$$

where $q : G \to G/Z$ is the quotient.

PROPOSITION A.1. *The category of $G$-equivariant local systems on $G/Z$ (for the left translation action of $G$) is equivalent to the category of $Z/Z^\circ$-representations as a tensor category.*

PROOF. By induction equivalence, the category of $G$-equivariant sheaves on $G/Z$ is equivalent to the category of $Z$-equivariant sheaves on a point. Furthermore, the category of $Z$-equivariant sheaves on a point is equivalent to the category of $Z/Z^\circ$ representations. (The $Z$-equivariant fundamental group of a point is $Z/Z^0$.) □

For a character $\chi$ of $Z/Z^\circ$, we denote the corresponding local system under this equivalence by $\mathcal{L}_\chi$.

DEFINITION A.2. We say that a $G$-equivariant perverse (or constructible) sheaf $\mathcal{F}$ admits a central character $\chi$ if there is an isomorphism $\alpha^* \mathcal{F} \cong \mathcal{L}_\chi \boxtimes \mathcal{F}$.

LEMMA A.3. *A simple $G$-equivariant perverse sheaf admits a central character.*

PROOF. Let $P = \mathrm{IC}(\mathcal{O}, \mathcal{L})$ be a simple $G$-equivariant perverse sheaf with $\mathcal{O}$ a $G$-orbit in $X$. The $G$-equivariant fundamental group of $\mathcal{O}$ is

$$\pi_1^G(\mathcal{O}) = \mathrm{Stab}_G(x)/\mathrm{Stab}_G(x)^\circ,$$

where $x \in \mathcal{O}$. Since $Z$ acts trivially on $X$, we have a map $Z/Z^\circ \to \pi_1^G(\mathcal{O})$. Thus, if $\mathcal{L}$ is a local system on $\mathcal{O}$ corresponding to the $\pi_1^G(\mathcal{O})$ representation $V_{\mathcal{L}}$, the central character of $\mathcal{L}$ is the local system $\mathcal{L}_\chi$ corresponding to the restriction $V_{\mathcal{L}}|_{Z/Z^\circ}$. To see that $\mathcal{L}_\chi$ is the central character of $P$, use base change with the diagram

$$\begin{array}{ccc} G/Z \times \mathcal{O} & \xrightarrow{\alpha'} & \mathcal{O} \\ \big\downarrow{\scriptstyle j'} & & \big\downarrow{\scriptstyle j} \\ G/Z \times X & \xrightarrow{\alpha} & X \end{array}$$

Under base change, the natural map $\alpha^* j_! \mathcal{L} \to \alpha^* j_* \mathcal{L}$ goes to $j'_! \alpha'^* \mathcal{L} \cong j'_!(\mathcal{L}_\chi \boxtimes \mathcal{L}) \to j'_* \alpha'^* \mathcal{L} \cong j'_*(\mathcal{L}_\chi \boxtimes \mathcal{L})$. Hence, $\alpha^* P \cong \mathcal{L}_\chi \boxtimes P$. □

We record some facts about external tensor product for our convenience.

(1) $t$-exact
(2) $\mathbb{D}\mathcal{F} \boxtimes \mathbb{D}\mathcal{G} \cong \mathbb{D}(\mathcal{F} \boxtimes \mathcal{G})$
(3) $(f \times g)^*(\mathcal{F} \boxtimes \mathcal{G}) \cong f^*\mathcal{F} \boxtimes g^*\mathcal{G}$
(4) $(f \times g)_*(\mathcal{F} \boxtimes \mathcal{G}) \cong f_*\mathcal{F} \boxtimes g_*\mathcal{G}$
(5) $(f \times g)_!(\mathcal{F} \boxtimes \mathcal{G}) \cong f_!\mathcal{F} \boxtimes g_!\mathcal{G}$.

LEMMA A.4. *Suppose $\mathcal{F}$ has central character $\chi$ and $\mathcal{G}$ has central character $\psi$. Then $\mathcal{F} \otimes^L \mathcal{G}$ has central character $\chi \otimes \psi$*

PROOF.

$$\begin{aligned} \alpha^*(\mathcal{F} \otimes^L \mathcal{G}) &= \alpha^*\mathcal{F} \otimes^L \alpha^*\mathcal{G} \\ &= (\mathcal{L}_\chi \boxtimes \mathcal{F}) \otimes^L (\mathcal{L}_\psi \boxtimes \mathcal{G}) \\ &= p^*_{G/Z}\mathcal{L}_\chi \otimes p^*_X\mathcal{F} \otimes p^*_{G/Z}\mathcal{L}_\psi \otimes p^*_X\mathcal{G} \\ &= p^*_{G/Z}(\mathcal{L}_\chi \otimes \mathcal{L}_\psi) \otimes p^*_X(\mathcal{F} \otimes \mathcal{G}) \\ &= (\mathcal{L}_\chi \otimes \mathcal{L}_\psi) \boxtimes (\mathcal{F} \otimes \mathcal{G}). \end{aligned}$$

□

LEMMA A.5. *Suppose $\mathcal{F}$ has central character $\chi$. Then $\mathbb{D}\mathcal{F}$ has central character $\chi^{-1}$.*

PROOF. Since $\alpha$ is flat with $d = \dim G/Z$ dimensional fibers, we have $\alpha^! \cong \alpha^*[2d]$. Thus, we see that

$$\begin{aligned} \alpha^* \mathbb{D}\mathcal{F} &= \alpha^! \mathbb{D}\mathcal{F}[-2d] \\ &= \mathbb{D}\alpha^*\mathcal{F}[-2d] \\ &= \mathbb{D}(\mathcal{L}_\chi \boxtimes \mathcal{F})[-2d] \\ &= \mathbb{D}\mathcal{L}_\chi \boxtimes \mathbb{D}\mathcal{F}[-2d] \\ &= \mathcal{L}_\chi^\vee \boxtimes \mathbb{D}\mathcal{F}[2d - 2d] \\ &= \mathcal{L}_{\chi^{-1}} \boxtimes \mathbb{D}\mathcal{F}. \end{aligned}$$

□

LEMMA A.6. *Suppose that $f : X \to Y$ is a $G$-equivariant map between $G$-varieties on which $Z$ acts trivially. Then $f^!$, $f^*$, $f_*$, and $f_!$ preserve central character.*

PROOF. We consider the commutative diagram:

$$\begin{array}{ccc} G/Z \times X & \xrightarrow{id \times f} & G/Z \times Y \\ \Big\downarrow{\scriptstyle \alpha_X} & & \Big\downarrow{\scriptstyle \alpha_Y} \\ X & \xrightarrow{\quad f \quad} & Y \end{array}$$

First, we will show pull-backs preserve central character. Suppose $\mathcal{F}$ is a $G$-equivariant sheaf on $Y$ with central character $\chi$. Then we have

$$\begin{aligned} \alpha^* f^* \mathcal{F} &\cong (id \times f)^* \alpha^* \mathcal{F} \\ &\cong (id \times f)^* (\mathcal{L}_\chi \boxtimes \mathcal{F}) \\ &\cong \mathcal{L}_\chi \boxtimes f^* \mathcal{F}, \end{aligned}$$

$$\begin{aligned} \alpha^* f^! \mathcal{F} &\cong (id \times f)^! \alpha^* \mathcal{F} \\ &\cong (id \times f)^! (\mathcal{L}_\chi \boxtimes \mathcal{F}) \\ &\cong \mathcal{L}_\chi \boxtimes f^! \mathcal{F}. \end{aligned}$$

By base change and flatness of the action maps $\alpha$, we have that

$$\begin{aligned} \alpha^* f_* \mathcal{F} &\cong (id \times f)_* \alpha^* \mathcal{F} \\ &\cong (id \times f)_* (\mathcal{L}_\chi \boxtimes \mathcal{F}) \\ &\cong \mathcal{L}_\chi \boxtimes f_* \mathcal{F}. \end{aligned}$$

Similarly,

$$\begin{aligned} \alpha^* f_! \mathcal{F} &\cong (id \times f)_! \alpha^* \mathcal{F} \\ &\cong (id \times f)_! (\mathcal{L}_\chi \boxtimes \mathcal{F}) \\ &\cong \mathcal{L}_\chi \boxtimes f_! \mathcal{F}. \end{aligned}$$

□

LEMMA A.7. *Suppose $\mathcal{F}$ has central character $\chi$ and $\mathcal{G}$ has central character $\psi$. Then the sheaf $R\mathcal{H}\mathrm{om}(\mathcal{F}, \mathcal{G})$ has central character $\chi^{-1} \otimes \psi$.*

PROOF. We have an isomorphism

$$R\mathcal{H}\mathrm{om}(\mathcal{F}, \mathcal{G}) \cong \mathbb{D}(\mathcal{F} \otimes \mathbb{D}\mathcal{G}).$$

Of course, $\mathcal{F} \otimes \mathbb{D}\mathcal{G}$ has central character $\chi \otimes \psi^{-1}$ by Lemmas A.5 and A.4, and we apply Lemma A.5 again to get the result. □

PROPOSITION A.8. *Suppose that two perverse sheaves on $X$ have distinct central characters. Then $R\mathrm{Hom}(\mathcal{F}, \mathcal{F}') = 0$. In particular,*

$$\mathrm{Hom}^i(\mathcal{F}', \mathcal{F}) = \mathrm{Hom}^i(\mathcal{F}, \mathcal{F}') = 0$$

*for all $i$.*

PROOF. First, we note that on pt, the only possible central character is trivial. Let $\mathcal{F}$ and $\mathcal{F}'$ be perverse sheaves on $X$ with central characters $\chi$ and $\psi$ (and $\chi \neq \psi$). Let $a : X \to \mathrm{pt}$ be the constant map. Recall that $R\mathrm{Hom}(\mathcal{F}, \mathcal{F}') = a_* R\mathcal{H}\mathrm{om}(\mathcal{F}, \mathcal{F}')$. If $R\mathcal{H}\mathrm{om}(\mathcal{F}, \mathcal{F}') \neq 0$, then Lemma A.7 implies that $R\mathcal{H}\mathrm{om}(\mathcal{F}, \mathcal{F}')$

has nontrivial central character equal to $\chi^{-1} \otimes \psi$. Then, $a_* R\,\mathcal{H}om(\mathcal{F}, \mathcal{F}')$ also has central character $\chi^{-1} \otimes \psi$ by Lemma A.6 which is impossible. □

## Acknowledgements

We are very grateful to Pramod Achar for guidance in the early stages of this work and Jay Taylor for a number of useful comments and suggestions. We would also like to thank Syu Kato for explaining the action of Frobenius in Lemma 4.4.

## References

[A] P. N. Achar, *Green functions via hyperbolic localization*, Doc. Math. **16** (2011), 869–884. MR2861392 (2012j:20132)

[AHR] P. N. Achar, A. Henderson, and S. Riche, *Geometric Satake, Springer correspondence, and small representations II*, Represent. Theory **19** (2015), 94–166, DOI 10.1090/ert/465. MR3347990

[BBD] A. A. Beĭlinson, J. Bernstein, and P. Deligne, *Faisceaux pervers* (French), Analysis and topology on singular spaces, I (Luminy, 1981), Astérisque, vol. 100, Soc. Math. France, Paris, 1982, pp. 5–171. MR751966 (86g:32015)

[BL] J. Bernstein and V. Lunts, *Equivariant sheaves and functors*, Lecture Notes in Mathematics, vol. 1578, Springer-Verlag, Berlin, 1994. MR1299527 (95k:55012)

[BM1] W. Borho and R. MacPherson, *Représentations des groupes de Weyl et homologie d'intersection pour les variétés nilpotentes* (French, with English summary), C. R. Acad. Sci. Paris Sér. I Math. **292** (1981), no. 15, 707–710. MR618892 (82f:14002)

[BM2] W. Borho and R. MacPherson, *Partial resolutions of nilpotent varieties*, Analysis and topology on singular spaces, II, III (Luminy, 1981), Astérisque, vol. 101, Soc. Math. France, Paris, 1983, pp. 23–74. MR737927 (85j:14087)

[Br] T. Braden, *Hyperbolic localization of intersection cohomology*, Transform. Groups **8** (2003), no. 3, 209–216, DOI 10.1007/s00031-003-0606-4. MR1996415 (2004f:14037)

[CG] N. Chriss and V. Ginzburg, *Representation theory and complex geometry*, Birkhäuser Boston, Inc., Boston, MA, 1997. MR1433132 (98i:22021)

[De] P. Deligne, *La conjecture de Weil. II* (French), Inst. Hautes Études Sci. Publ. Math. **52** (1980), 137–252. MR601520 (83c:14017)

[L0] G. Lusztig, *Coxeter orbits and eigenspaces of Frobenius*, Invent. Math. **38** (1976/77), no. 2, 101–159. MR0453885 (56 #12138)

[L1] G. Lusztig, *Intersection cohomology complexes on a reductive group*, Invent. Math. **75** (1984), no. 2, 205–272, DOI 10.1007/BF01388564. MR732546 (86d:20050)

[L2] G. Lusztig, *Character sheaves. I*, Adv. in Math. **56** (1985), no. 3, 193–237, DOI 10.1016/0001-8708(85)90034-9. MR792706 (87b:20055)

[L3] G. Lusztig, *Character sheaves. II, III*, Adv. in Math. **57** (1985), no. 3, 226–265, 266–315, DOI 10.1016/0001-8708(85)90064-7. MR806210 (87m:20118a)

[L4] G. Lusztig, *Character sheaves. IV*, Adv. in Math. **59** (1986), no. 1, 1–63, DOI 10.1016/0001-8708(86)90036-8. MR825086 (87m:20118b)

[L5] G. Lusztig, *Character sheaves. V*, Adv. in Math. **61** (1986), no. 2, 103–155, DOI 10.1016/0001-8708(86)90071-X. MR849848 (87m:20118c)

[L6] G. Lusztig, *Cuspidal local systems and graded Hecke algebras. I*, Inst. Hautes Études Sci. Publ. Math. **67** (1988), 145–202. MR972345 (90e:22029)

[L7] G. Lusztig, *On the cleanness of cuspidal character sheaves* (English, with English and Russian summaries), Mosc. Math. J. **12** (2012), no. 3, 621–631, 669. MR3024826

[O] V. Ostrik, *A remark on cuspidal local systems*, Adv. Math. **192** (2005), no. 1, 218–224, DOI 10.1016/j.aim.2004.04.008. MR2122285 (2006j:20066)

[S] T. Shoji, *On the Green polynomials of classical groups*, Invent. Math. **74** (1983), no. 2, 239–267, DOI 10.1007/BF01394315. MR723216 (85f:20032)

[R] L. Rider, *Formality for the nilpotent cone and a derived Springer correspondence*, Adv. Math. **235** (2013), 208–236, DOI 10.1016/j.aim.2012.12.001. MR3010057

Department of Mathematics, Massachusetts Institute of Technology, Cambridge, Massachusetts
*E-mail address*: laurajoy@mit.edu

Department of Mathematics, University of Georgia, Athens, Georgia
*E-mail address*: arussell@math.uga.edu

Proceedings of Symposia in Pure Mathematics
Volume **92**, 2016
http://dx.doi.org/10.1090/pspum/092/01582

# Categorifying the tensor product of a level 1 highest weight and perfect crystal in type $A$

Monica Vazirani

Abstract. We use KLR algebras to categorify a crystal isomorphism between a highest weight crystal and the tensor product of a perfect crystal and another highest weight crystal, all in level 1 type $A$ affine. The nodes of the perfect crystal correspond to a family of trivial modules and the nodes of the highest weight crystal correspond to simple modules, which we may also parameterize by $\ell$-restricted partitions. In the case $\ell$ is a prime, one can reinterpret all the results for the symmetric group in characteristic $\ell$. The crystal operators correspond to socle of restriction and behave compatibly with the rule for tensor product of crystal graphs.

Contents



## 1. Introduction

Kang-Kashiwara [**10**] and Webster [**28**] show the cyclotomic Khovanov-Lauda-Rouquier (KLR) algebra $R^\Lambda$ categorifies the highest weight representation $V(\Lambda)$ in arbitrary symmetrizable type. (KLR algebras are also known as quiver Hecke algebras.) We will say the *combinatorial* version of this statement is that $R^\Lambda$ categorifies the crystal $B(\Lambda)$, where simple modules correspond to nodes, and functors that take socle of restriction correspond to arrows, i.e. the Kashiwara crystal operators. Webster [**28**] and Losev-Webster [**22**] categorify the tensor product of highest weight modules, and hence the tensor product of highest weight crystals. However,

2010 *Mathematics Subject Classification.* Primary 05E10; Secondary 20C08.

This work was partially supported by NSF-MSPRF, NSA grant H98230-12-1-0232, and the Simons Foundation.

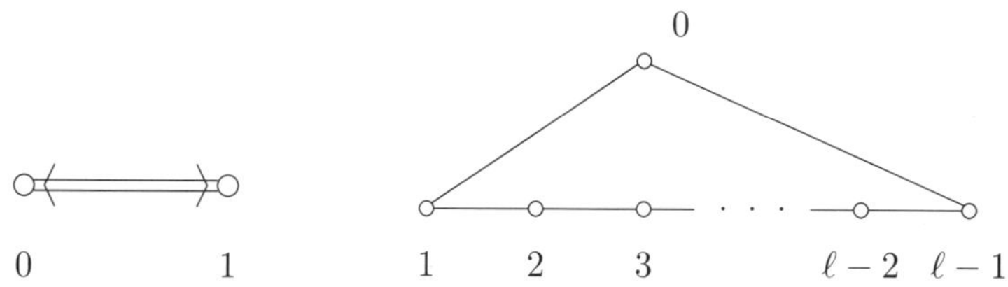


FIGURE 1. The Dynkin diagram for $A_1^{(1)}$ is on the left and and the Dynkin diagram for $A_{\ell-1}^{(1)}$ with $\ell > 2$ is on the right.

one can consider a tensor product of crystals

$$\mathcal{B} \otimes B(\Lambda) \simeq B(\Lambda') \tag{1.1}$$

where $\Lambda, \Lambda' \in P^+$ are of level $k$ and $\mathcal{B}$ is a perfect crystal of level $k$. In this paper, we (combinatorially) categorify the crystal isomorphism (1.1) in the case the level $k = 1$ for type $A_{\ell-1}^{(1)}$ and $\mathcal{B} = B^{1,1}$ which is drawn in Figure 2. Each node of $\mathcal{B}$ corresponds to a family of trivial modules, but note this does *not* give a categorification of $\mathcal{B}$. (By symmetry we have similar results for $\mathcal{B} = B^{\ell-1,1}$ of Figure 5 whose nodes correspond to sign modules.)

We note that this gives a construction of simple modules that is somewhat intermediate between the crystal operator construction and the Specht module construction. Combinatorially, the former corresponds to building an $\ell$-restricted partition one (good) box at a time. Our construction builds a partition one row at a time, or dually one column at a time. The Specht module construction (at least for $\mathbb{F}_\ell \mathcal{S}_n$ or the Hecke algebra of type $A$) builds the simple from the whole partition, constructing the simple as a subquotient of an induced trivial module from a parabolic subalgebra that corresponds to the partition. However, this paper also describes how socle of restriction interacts with the construction. One can also recover this construction for finite type $A_{\ell-1}$ as its Dynkin diagram is a subdiagram of that of type $A_{\ell-1}^{(1)}$, or recovers characteristic 0 constructions taking $\ell \to \infty$. For a construction of simple modules related to the crystal $B(\infty)$ for finite type KLR algebras see [**2**].

This paper is based on unpublished work of the author [**25, 26**] which was done for the affine Hecke algebra of type $A$ at an $\ell$th root of unity. We chose to rewrite this in the language of KLR algebras to appeal to the modern reader and also make it easier to then generalize the theorem to other affine types in [**20**].

I wish to thank Henry Kvinge for his help with the figures and whose feedback greatly improved the exposition.

## 2. Type $A$ Cartan datum and crystals

**2.1. Cartan datum for type $A_{\ell-1}^{(1)}$.** Fix an integer $\ell \geq 2$. In this paper we will work solely in type $A_{\ell-1}^{(1)}$. Let $I$ be the indexing set

$$I = \{0, 1, \ldots, \ell - 1\}. \tag{2.1}$$

Let $[a_{ij}]_{i,j\in I}$ denote the associated Cartan matrix. For $\ell > 2$ the type $A$ Cartan matrix is the $\ell \times \ell$ matrix

$$\begin{bmatrix} 2 & -1 & 0 & & \cdots & 0 & 0 \\ -1 & 2 & -1 & & \cdots & 0 & 0 \\ \vdots & & & \ddots & & & \vdots \\ 0 & 0 & & \cdots & -1 & 2 & -1 \\ 0 & 0 & & \cdots & 0 & -1 & 2 \end{bmatrix}.$$

When $\ell = 2$ it is

$$\begin{bmatrix} 2 & -2 \\ -2 & 2 \end{bmatrix}.$$

Following [**9**] we let $\mathfrak{h}$ be a Cartan subalgebra, $\prod = \{\alpha_0, \dots, \alpha_{\ell-1}\}$ its system of simple roots, $\prod^\vee = \{h_0, \dots, h_{\ell-1}\}$ its simple coroots, and $Q$ and $Q^\vee$ the root and coroot lattices respectively. Then

$$Q^+ = \bigoplus_{i\in I} \mathbb{Z}_{\geq 0}\alpha_i.$$

For an element $\nu \in Q^+$, we define its *height*, $|\nu|$, to be the sum of the coefficients, i.e. if $\nu = \sum_{i\in I} \nu_i\alpha_i$ then

$$|\nu| = \sum_{i\in I} \nu_i. \tag{2.2}$$

We also have a symmetric bilinear form

$$(\ ,\ ) : \mathfrak{h}^* \times \mathfrak{h}^* \to \mathbb{C}$$

which satisfies

$$a_{ij} = \langle h_i, \alpha_j \rangle = \frac{2(\alpha_i, \alpha_j)}{(\alpha_i, \alpha_i)}$$

where $\langle\ ,\ \rangle : \mathfrak{h} \times \mathfrak{h}^* \to \mathbb{C}$ is the canonical pairing. Using this pairing we define the fundamental weights $\{\Lambda_i \mid i \in I\}$ via

$$\langle h_j, \Lambda_i \rangle = \delta_{ji}.$$

The weight lattice is $\bigoplus_{i\in I} \mathbb{Z}\Lambda_i$ and the integral dominant weights are

$$P^+ = \bigoplus_{i\in I} \mathbb{Z}_{\geq 0}\Lambda_i.$$

REMARK 2.1. Because in this paper we work exclusively with Cartan datum associated with $A^{(1)}_{\ell-1}$, it is often convenient to identify elements of $I$ with $\mathbb{Z}/\ell\mathbb{Z}$, so when stating that $k \in I$, we will usually think of $k \in \mathbb{Z}/\ell\mathbb{Z}$ even if we neglect to write $\overline{k}$ or $k \bmod \ell$. We will often be considering a sequence of $k$ operators, $k \in \mathbb{N} := \mathbb{Z}_{\geq 0}$, but the $k$th operator may be indexed by $(k-1) \bmod \ell$, for which it is convenient to relax notation.

**2.2. Review of crystals.** We recall the tensor category of crystals following Kashiwara [**14**], see also [**12, 13, 15**].

A *crystal* is a set $B$ together with maps

- $\mathrm{wt}\colon B \longrightarrow P$,
- $\varepsilon_i, \varphi_i : B \longrightarrow \mathbb{Z} \sqcup \{-\infty\}$ for $i \in I$,
- $\widetilde{e}_i, \widetilde{f}_i\colon B \longrightarrow B \sqcup \{0\}$ for $i \in I$,

such that

C1. $\varphi_i(b) = \varepsilon_i(b) + \langle h_i, \mathrm{wt}(b)\rangle$ for any $i$.
C2. If $b \in B$ satisfies $\widetilde{e}_i b \neq 0$, then

$$\varepsilon_i(\widetilde{e}_i b) = \varepsilon_i(b) - 1, \qquad \varphi_i(\widetilde{e}_i b) = \varphi_i(b) + 1, \qquad \mathrm{wt}(\widetilde{e}_i b) = \mathrm{wt}(b) + \alpha_i.$$

C3. If $b \in B$ satisfies $\widetilde{f}_i b \neq 0$, then

$$\varepsilon_i(\widetilde{f}_i b) = \varepsilon_i(b) + 1, \qquad \varphi_i(\widetilde{f}_i b) = \varphi_i(b) - 1, \qquad \mathrm{wt}(\widetilde{f}_i b) = \mathrm{wt}(b) - \alpha_i.$$

C4. For $b_1, b_2 \in B$, $b_2 = \widetilde{f}_i b_1$ if and only if $\widetilde{e}_i b_2 = b_1$.
C5. If $\varphi_i(b) = -\infty$, then $\widetilde{e}_i b = \widetilde{f}_i b = 0$.

If $B_1$ and $B_2$ are two crystals, then a *morphism* $\psi\colon B_1 \to B_2$ of crystals is a map

$$\psi\colon B_1 \sqcup \{0\} \to B_2 \sqcup \{0\}$$

satisfying the following properties:

M1. $\psi(0) = 0$.
M2. If $\psi(b) \neq 0$ for $b \in B_1$, then

$$\mathrm{wt}(\psi(b)) = \mathrm{wt}(b), \qquad \varepsilon_i(\psi(b)) = \varepsilon_i(b), \qquad \varphi_i(\psi(b)) = \varphi_i(b).$$

M3. For $b \in B_1$ such that $\psi(b) \neq 0$ and $\psi(\widetilde{e}_i b) \neq 0$, we have $\psi(\widetilde{e}_i b) = \widetilde{e}_i(\psi(b))$.
M4. For $b \in B_1$ such that $\psi(b) \neq 0$ and $\psi(\widetilde{f}_i b) \neq 0$, we have $\psi(\widetilde{f}_i b) = \widetilde{f}_i(\psi(b))$.

A morphism $\psi$ of crystals is called *strict* if

$$\psi \circ \widetilde{e}_i = \widetilde{e}_i \circ \psi, \qquad \psi \circ \widetilde{f}_i = \widetilde{f}_i \circ \psi,$$

and an *embedding* if $\psi$ is injective.

Given two crystals $B_1$ and $B_2$ their tensor product $B_1 \otimes B_2$ (using the reverse Kashiwara convention) has underlying set $\{b_1 \otimes b_2; b_1 \in B_1, \text{ and } b_2 \in B_2\}$ where we identify $b_1 \otimes 0 = 0 \otimes b_2 = 0$. The crystal structure is given as follows:

$$\mathrm{wt}(b_1 \otimes b_2) = \mathrm{wt}(b_1) + \mathrm{wt}(b_2), \tag{2.3}$$

$$\varepsilon_i(b_1 \otimes b_2) = \max\{\varepsilon_i(b_2), \varepsilon_i(b_1) - \langle h_i, \mathrm{wt}(b_2)\rangle\}, \tag{2.4}$$

$$\varphi_i(b_1 \otimes b_2) = \max\{\varphi_i(b_2) + \langle h_i, \mathrm{wt}(b_1)\rangle, \varphi_i(b_1)\}, \tag{2.5}$$

$$\widetilde{e}_i(b_1 \otimes b_2) = \begin{cases} \widetilde{e}_i b_1 \otimes b_2 & \text{if } \varepsilon_i(b_1) > \varphi_i(b_2) \\ b_1 \otimes \widetilde{e}_i b_2 & \text{if } \varepsilon_i(b_1) \leq \varphi_i(b_2), \end{cases} \tag{2.6}$$

$$\widetilde{f}_i(b_1 \otimes b_2) = \begin{cases} \widetilde{f}_i b_1 \otimes b_2 & \text{if } \varepsilon_i(b_1) \geq \varphi_i(b_2) \\ b_1 \otimes \widetilde{f}_i b_2 & \text{if } \varepsilon_i(b_1) < \varphi_i(b_2). \end{cases} \tag{2.7}$$

Given a crystal $B$, we can draw its associated crystal graph with nodes (or vertices) $B$ and $I$-colored arrows (directed edges) as follows. When $\widetilde{e}_i b = a$ (so $b = \widetilde{f}_i a$) we draw an $i$-colored arrow $a \xrightarrow{i} b$. We also say $b$ has an incoming $i$-arrow and $a$ has an outgoing $i$-arrow.

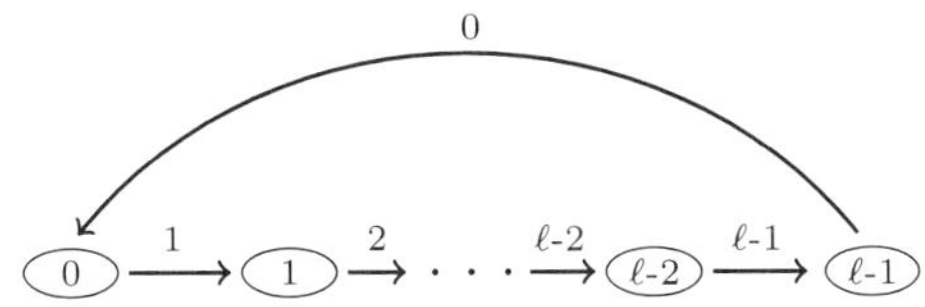

FIGURE 2. The level 1 perfect crystal $\mathcal{B}$, which is also denoted $B^{1,1}$.

## 3. Level 1 crystals in type $A_{\ell-1}^{(1)}$

The level 1 highest weight crystal, or fundamental crystal, $B(\Lambda_i)$ has a model (see Figure 3) with nodes *$\ell$-restricted* partitions, i.e. $\lambda = (\lambda_1, \ldots, \lambda_t)$ such that $\lambda_r \in \mathbb{Z}_{\geq 0}$, $0 \leq \lambda_r - \lambda_{r+1} < \ell$ for all $r$. Observe that for fixed $\ell$, as directed graphs $B(\Lambda_0)$ and $B(\Lambda_i)$ are identical. The edge labels or "colors" for $B(\Lambda_i)$ are obtained from those of $B(\Lambda_0)$ by adding $i$ mod $\ell$.

Let $\mathcal{B}$ be the crystal graph in Figure 2. $\mathcal{B}$ is an example of a level 1 *perfect* crystal. See [**11**] for the definition of a perfect crystal and for many of its important properties. $\mathcal{B}$, often denoted $B^{1,1}$ in the literature is also an example of a Kirillov-Reshetikhin crystal. Observe that we have parameterized the nodes of $\mathcal{B}$ so that

$$\varepsilon_i(\textcircled{k}) = \delta_{i,k}.$$

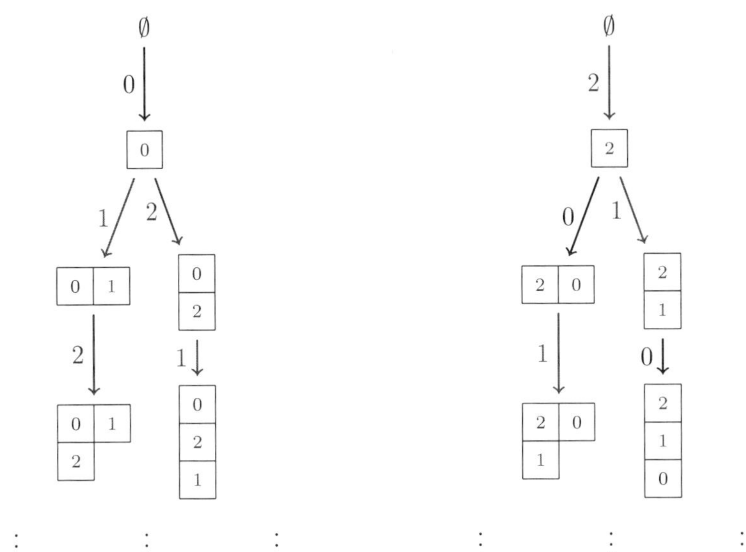

FIGURE 3. $B(\Lambda_0)$ and $B(\Lambda_2)$ for $\ell = 3$.

Then $\mathcal{T} : B(\Lambda_i) \xrightarrow{\simeq} \mathcal{B} \otimes B(\Lambda_{i-1})$ is an isomorphism of crystals. The isomorphism is pictured in Figure 4 for $i = 0$ and $\ell = 3$. Combinatorially, $\mathcal{T}(\lambda) = \textcircled{k} \otimes \mu$ where $k \equiv \lambda_1 + i - 1 \bmod \ell$ and $\mu = (\lambda_2, \ldots, \lambda_t)$ if $\lambda = (\lambda_1, \lambda_2, \cdots, \lambda_t)$.

So we obtain $\mu$ from $\lambda$ by removing its top row. In Figure 4, we draw

$$\mathcal{T}(\lambda) = \begin{matrix} \textcircled{k} \\ \otimes \\ \mu \end{matrix} \tag{3.1}$$

so the visual of the top row removal stands out. Note $\lambda_1 - \mu_1 = \lambda_1 - \lambda_2 < \ell$ means that $\mathcal{T}$ has a well-defined inverse.

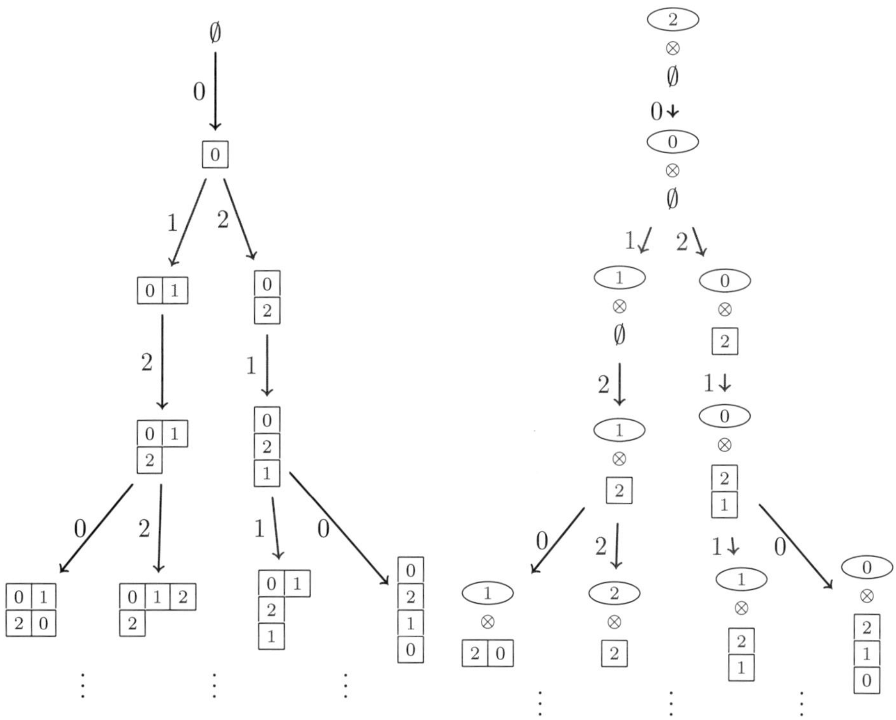

FIGURE 4. The isomorphism $B(\Lambda_0) \simeq \mathcal{B} \otimes B(\Lambda_2)$ for $\ell = 3$.

When drawing our model of $B(\Lambda_i)$, we label each box of an $\ell$-restricted partition with $k \in I$, such that the main diagonal gets label $i$, and labels increase by 1 mod $\ell$ as one increases diagonals (moving right). In this manner, the last box in the top row of $\lambda$ is labeled $k$ when $\mathcal{T}(\lambda) = \textcircled{k} \otimes \mu$. Note further that if we have a $k$-arrow $\gamma \xrightarrow{k} \lambda$ then the box $\lambda/\gamma$ is labeled $k$ (though not necessarily conversely). In fact, once one knows the structure of $\mathcal{B}$ and the tensor product rule for crystals, one can obtain the rule for which $k$-box $\widetilde{e}_k$ removes by iterating $\mathcal{T}$.

$\mathcal{B}^{\text{opp}}$, often denoted $B^{\ell-1,1}$ in the literature, is another level 1 perfect crystal and is also an example of a Kirillov-Reshetikhin crystal. $\mathcal{B}^{\text{opp}}$ is pictured in Figure 5; note it can be obtained from $\mathcal{B}$ from reversing orientation of all arrows, and we chose to relabel nodes so that still

$$\varepsilon_i(\textcircled{k}) = \delta_{i,k}.$$

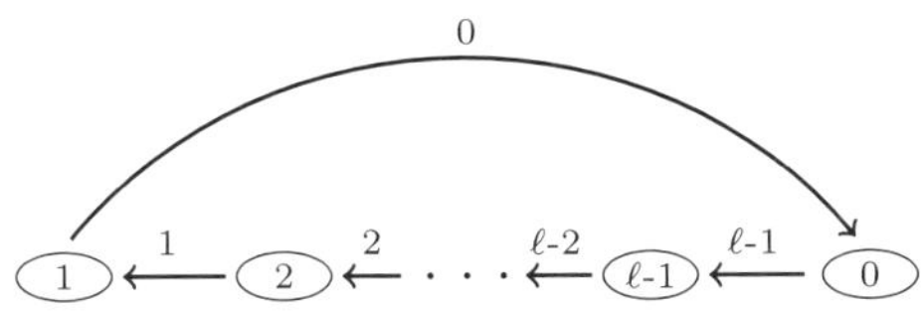


FIGURE 5. The level 1 perfect crystal $\mathcal{B}^{\text{opp}}$, which is also denoted $B^{\ell-1,1}$.

We have another crystal isomorphism $\mathcal{T}^{\text{opp}} : B(\Lambda_i) \xrightarrow{\simeq} \mathcal{B}^{\text{opp}} \otimes B(\Lambda_{i+1})$. See Figure 6. This isomorphism is compatible with the model of $B(\Lambda_i)$ that labels nodes with *$\ell$-regular* partitions, that is, those partitions $\mu$ such that the transposed diagram $\mu^T$ is $\ell$-restricted. Then the isomorphism $\mathcal{T}^{\text{opp}}$ corresponds to column removal, in the same way $\mathcal{T}$ corresponds to row removal.

While the underlying $I$-colored directed graphs $B(\Lambda_i)$ are identical, one does not obtain the $\ell$-regular model by merely transposing the partition indexing each node of the $\ell$-restricted model. See Section 5.0.1 for another model of $B(\Lambda_i)$ that comes from KLR algebras.

There are other level 1 perfect crystals besides $\mathcal{B}$ and $\mathcal{B}^{\text{opp}}$, but we do not consider them here.

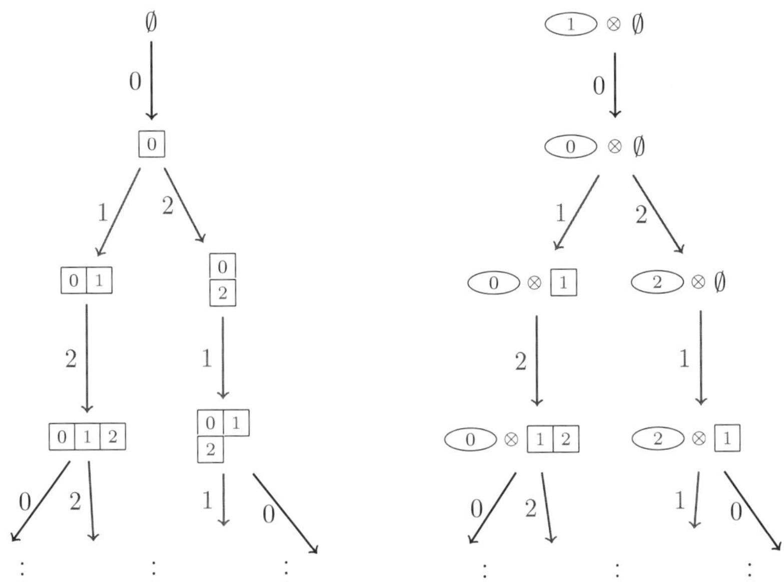

FIGURE 6. The isomorphism $B(\Lambda_0) \simeq \mathcal{B}^{\text{opp}} \otimes B(\Lambda_1)$ for $\ell = 3$.

## 4. Definition of the KLR algebra $R(\nu)$ and some functors

In what follows we let $[k]$ be the quantum integer in the indeterminant $q$,

$$[k] = q^{k-1} + q^{k-3} + \cdots + q^{1-k} \quad \text{and} \quad [k]! = [k][k-1]\ldots[1]. \tag{4.1}$$

For $\nu = \sum_{i \in I} \nu_i \alpha_i$ in $Q^+$ with $|\nu| = m$, we define $\mathrm{Seq}(\nu)$ to be all sequences

$$\underline{\mathbf{i}} = (i_1, i_2, \dots, i_m)$$

such that $i_k$ appears $\nu_k$ times. For $\underline{\mathbf{i}} \in \mathrm{Seq}(\nu)$ and $\underline{\mathbf{j}} \in \mathrm{Seq}(\mu)$, $\underline{\mathbf{ij}}$, will denote the concatenation of the two sequences unless otherwise specified. It follows that $\underline{\mathbf{ij}} \in \mathrm{Seq}(\nu + \mu)$. We write

$$i^n = \underbrace{(i, i, \dots, i)}_{n}. \tag{4.2}$$

There is a left action of the symmetric group, $\mathcal{S}_m$, on $\mathrm{Seq}(\nu)$ defined by,

$$s_k(\underline{\mathbf{i}}) = s_k\Big(i_1, i_2, \dots, i_k, i_{k+1}, \dots, i_m\Big) = (i_1, i_2, \dots, i_{k+1}, i_k, \dots, i_m) \tag{4.3}$$

where $s_k$ is the adjacent transposition in $\mathcal{S}_m$ that interchanges $k$ and $k+1$.

Since this paper only considers KLR algebras of type $A^{(1)}_{\ell-1}$, we simplify the definition below from that for general type. The definition for arbitrary symmetrizable types can be found in [**16**], [**17**], and [**24**]. Using the more general definition with Rouquier's parameters $Q_{i,j}(u, v)$ will not change the results or proofs in this paper, as they concern crystal-theoretic phenomena. There is also a diagrammatic presentation of KLR algebras which can be found in [**16**], [**17**]. By results of Brundan-Kleshchev [**4**], [**5**], there is an isomorphism between $R^\Lambda(\nu)$ and $H^\Lambda_\nu$ where $H^\Lambda_\nu$ is a block of the cyclotomic Hecke algebra $H^\Lambda_m$ as defined in [**1, 3, 6**]. Hence readers unfamiliar with KLR algebras can translate all statements and proofs in terms of Hecke algebras throughout the paper. We remark that historically, this is the original setting in which the theorems from this paper were proved [**25, 26**]. In fact the reader can think of all results as being stated for $\mathbb{F}_\ell \mathcal{S}_m$ in the case that $\ell$ is prime if the other algebras are not familiar.

For $\nu \in Q^+$ with $|\nu| = m$, the *KLR algebra* $R(\nu)$ is the associative, graded, unital $\mathbb{C}$-algebra generated by

$$1_{\underline{\mathbf{i}}} \text{ for } \underline{\mathbf{i}} \in \mathrm{Seq}(\nu), \quad x_r \text{ for } 1 \le r \le m, \quad \psi_r \text{ for } 1 \le r \le m-1, \tag{4.4}$$

subject to the following relations, where $\underline{\mathbf{i}}, \underline{\mathbf{j}} \in \mathrm{Seq}(\nu)$ and equality between $i_r$ and $i_t$ is taken to mean equality in $\mathbb{Z}/\ell\mathbb{Z}$.

$$1_{\underline{\mathbf{i}}} 1_{\underline{\mathbf{j}}} = \delta_{\underline{\mathbf{i}},\underline{\mathbf{j}}} 1_{\underline{\mathbf{i}}}, \qquad x_r 1_{\underline{\mathbf{i}}} = 1_{\underline{\mathbf{i}}} x_r, \qquad \psi_r 1_{\underline{\mathbf{i}}} = 1_{s_r(\underline{\mathbf{i}})} \psi_r, \qquad x_r x_t = x_t x_r, \tag{4.5}$$

$$\psi_r \psi_t = \psi_t \psi_r \qquad \text{if } |r - t| > 1, \tag{4.6}$$

$$\psi_r \psi_r 1_{\underline{\mathbf{i}}} = \begin{cases} 0 & \text{if } i_r = i_{r+1} \\ (x_r^{-a_{i_r i_{r+1}}} + x_{r+1}^{-a_{i_{r+1} i_r}}) 1_{\underline{\mathbf{i}}} & \text{if } i_r = i_{r+1} \pm 1 \\ 1_{\underline{\mathbf{i}}} & \text{otherwise,} \end{cases} \tag{4.7}$$

(4.8)

$$(\psi_r \psi_{r+1} \psi_r - \psi_{r+1} \psi_r \psi_{r+1}) 1_{\underline{\mathbf{i}}} = \begin{cases} 1_{\underline{\mathbf{i}}} & \text{if } \ell > 2 \text{ and } i_r = i_{r+2} = i_{r+1} \pm 1 \\ (x_r + x_{r+2}) 1_{\underline{\mathbf{i}}} & \text{if } \ell = 2 \text{ and } i_r = i_{r+2} = i_{r+1} \pm 1 \\ 0 & \text{otherwise,} \end{cases}$$

$$(\psi_r x_t - x_{s_r(t)} \psi_r) 1_{\underline{\mathbf{i}}} = \begin{cases} 1_{\underline{\mathbf{i}}} & \text{if } t = r \text{ and } i_r = i_{r+1} \\ -1_{\underline{\mathbf{i}}} & \text{if } t = r+1 \text{ and } i_r = i_{r+1} \\ 0 & \text{otherwise.} \end{cases} \tag{4.9}$$

The elements $1_{\underline{\mathbf{i}}}$ are idempotents in $R(\nu)$ by (4.5) and the identity element is given by

$$1_\nu = \sum_{\underline{\mathbf{i}} \in \mathrm{Seq}(\nu)} 1_{\underline{\mathbf{i}}}. \tag{4.10}$$

Thus, as a vector space $R(\nu)$ decomposes as,

$$R(\nu) = \bigoplus_{\underline{\mathbf{i}}, \underline{\mathbf{j}} \in \mathrm{Seq}(\nu)} 1_{\underline{\mathbf{i}}} R(\nu) 1_{\underline{\mathbf{j}}}.$$

The generators of $R(\nu)$ are graded as,

$$\deg(1_{\underline{\mathbf{i}}}) = 0, \quad \deg(x_r 1_{\underline{\mathbf{i}}}) = 2, \quad \deg(\psi_r 1_{\underline{\mathbf{i}}}) = -(\alpha_{i_r}, \alpha_{i_{r+1}}). \tag{4.11}$$

We define

$$R = \bigoplus_{\nu \in Q^+} R(\nu). \tag{4.12}$$

Notice that while $R(\nu)$ is unital, $R$ is not.

For each $w \in \mathcal{S}_m$ we fix once and for all a reduced expression

$$\widehat{w} = s_{i_1} s_{i_2} \dots s_{i_t}. \tag{4.13}$$

Observe that the $s_k$ are Coxeter generators of $\mathcal{S}_m$, and $t$ is the Coxeter length of $w$. Let $\psi_{\widehat{w}} = \psi_{i_1}\psi_{i_2}\dots\psi_{i_t}$ correspond to the chosen reduced expression $\widehat{w}$. For $\underline{\mathbf{i}}, \underline{\mathbf{j}} \in \mathrm{Seq}(\nu)$, let ${}_{\underline{\mathbf{j}}}\mathcal{S}_{\underline{\mathbf{i}}}$ be the permutations in $\mathcal{S}_m$ that take $\underline{\mathbf{i}}$ to $\underline{\mathbf{j}}$.

THEOREM 4.1. [**16**, Theorem 2.5] *As a $\mathbb{C}$-vector space $1_{\underline{\mathbf{j}}} R(\nu) 1_{\underline{\mathbf{i}}}$ has basis,*

$$\{\psi_{\widehat{w}} x_1^{b_1} \dots x_m^{b_m} 1_{\underline{\mathbf{i}}} \mid w \in {}_{\underline{\mathbf{j}}}\mathcal{S}_{\underline{\mathbf{i}}},\ b_r \in \mathbb{Z}_{\geq 0}\}.$$

It is known that all simple $R(\nu)$-modules are finite dimensional [**16**]. For this reason, in this paper we only consider the category of finite-dimensional KLR-modules $R(\nu)$-mod and $R$-mod.

We often refer to $1_{\underline{\mathbf{i}}} M$ as the $\underline{\mathbf{i}}$-*weight* space of $M$ and any $0 \neq v \in 1_{\underline{\mathbf{i}}} M$ as a *weight vector*. A *weight basis* is a basis consisting of weight vectors.

We define the graded character of an $R(\nu)$-module to be

$$\mathrm{Char}(M) = \sum_{\underline{\mathbf{i}} \in \mathrm{Seq}(\nu)} \mathrm{gdim}(1_{\underline{\mathbf{i}}} M) \cdot [\underline{\mathbf{i}}]. \tag{4.14}$$

Here $\mathrm{gdim}(1_{\underline{\mathbf{i}}} M)$ is an element of $\mathbb{Z}[q, q^{-1}]$, and hence $\mathrm{Char}(M)$ is an element of the free $\mathbb{Z}[q, q^{-1}]$-module generated by all $[\underline{\mathbf{i}}]$ for $\underline{\mathbf{i}} \in \mathrm{Seq}(\nu)$. We will let $\mathrm{supp}(M)$ denote the multiset that is the support of $\mathrm{Char}(M)$ so that

$$\mathrm{Char}(M)|_{q=1} = \sum_{[\underline{\mathbf{i}}] \in \mathrm{supp}(M)} [\underline{\mathbf{i}}]. \tag{4.15}$$

Our notational convention is to write $\underline{\mathbf{i}} \in \mathrm{Seq}(\nu)$ but write $[\underline{\mathbf{i}}] \in \mathrm{supp}(M)$. Since characters are an important combinatorial tool, it is worthwhile to set a special notation for them.

Because $R$ is a graded algebra, we will only work with homomorphisms between $R$-modules that are either degree preserving or degree homogeneous. We denote the $\mathbb{C}$-vector space of degree preserving homomorphisms between $R(\nu)$-modules $M$ and $N$ by $\mathrm{Hom}(M, N)$. Since any homogeneous homomorphism can be interpreted as degree preserving by shifting the grading on our target or source module, then we can write the $\mathbb{C}$-vector space of homogeneous homomorphisms between $M$ and $N$, $\mathrm{HOM}(M,N)$, by

$$\mathrm{HOM}(M,N) = \bigoplus_{k\in\mathbb{Z}} \mathrm{Hom}(M, N\{k\}). \tag{4.16}$$

While the grading is important, it was shown in [**16**] that there is a unique grading on a simple $R$-module up to overall grading shift. Since this paper concerns simple modules, we will rarely use or discuss the grading. All isomorphisms between modules will be taken up to overall grading shift.

REMARK 4.2. Because $x_r 1_{\underline{\mathbf{i}}} \in R(\nu)$ is always positively graded for $1 \le r \le |\nu|$ and $\underline{\mathbf{i}} \in \mathrm{Seq}(\nu)$, then on a finite dimensional $R(\nu)$-module, $M$, $x_r 1_{\underline{\mathbf{i}}}$ will always act nilpotently.

**4.1. Trivial and sign modules.** For $k \in \mathbb{N}, i \in I$, define positive roots

$$\gamma^+_{i;k} := \alpha_i + \alpha_{i+1} + \cdots + \alpha_{i+k-1} \quad \text{and} \quad \gamma^-_{i;k} := \alpha_i + \alpha_{i-1} + \cdots + \alpha_{i-k+1} \tag{4.17}$$

of height $k$. As in Remark 2.1 we interpret subscripts to be in $I$. Because 1-dimensional modules will play a key role in our main theorems below, we give the following classification.

PROPOSITION 4.3. If $M$ is a 1-dimensional $R(\nu)$-module with $|\nu| = m$, then $M$ has character

$$\mathrm{Char}(M) = [i, i+1, \ldots, i+m-2, i+m-1] \tag{4.18}$$

or

$$\mathrm{Char}(M) = [i, i-1, \ldots, i-m+2, i-m+1] \tag{4.19}$$

and $\nu = \gamma^+_{i;m}$ or $\nu = \gamma^-_{i;m}$ respectively. The entries in $\mathrm{Char}(M)$ should be taken modulo $\ell$.

PROOF. Our proof uses similar techniques to those used in [**19**] for the study of calibrated (or homogeneous) modules, of which the 1-dimensional $R(\nu)$-modules form a subset. Let $M$ be spanned by the vector $v$, and let $\underline{\mathbf{i}}$ be the unique element of $\mathrm{Seq}(\nu)$ such that $1_{\underline{\mathbf{i}}} M \neq \mathbf{0}$. We write

$$\underline{\mathbf{i}} = (i_1, i_2, \ldots, i_m).$$

Recall $x_r 1_{\underline{\mathbf{i}}}$, for $1 \le r \le m$ acts nilpotently on $M$ by Remark 4.2. Since $M$ is 1-dimensional, then $x_r 1_{\underline{\mathbf{i}}} v = 0$.

Suppose $i_r = i_{r+1}$ (recall that we interpret $i_r, i_{r+1} \in \mathbb{Z}/\ell\mathbb{Z}$). Then from relation (4.9),

$$(\psi_r x_r - x_{r+1}\psi_r) 1_{\underline{\mathbf{i}}} v = 1_{\underline{\mathbf{i}}} v = v, \tag{4.20}$$

but this is impossible as $x_r$ and $x_{r+1}$ both act as zero. Thus $i_r \neq i_{r+1}$.

Let $\psi_r 1_{\underline{\mathbf{i}}} v = a_r v$ for some constant $a_r$. Given we showed $i_r \neq i_{r+1}$, if additionally we suppose $i_{r+1} \neq \pm 1 + i_r$, then by relation (4.7),

$$a_r^2 v = \psi_r^2 1_{\underline{\mathbf{i}}} v = v, \tag{4.21}$$

so $a_r \neq 0$. Then, $0 \neq \psi_r 1_{\underline{\mathbf{i}}} v = 1_{s_r(\underline{\mathbf{i}})} \psi_r v \in 1_{s_r(\underline{\mathbf{i}})} M$. But $s_r(\underline{\mathbf{i}}) \neq \underline{\mathbf{i}}$ given $i_r \neq i_{r+1}$ which contradicts the fact that $M$ is 1-dimensional. Hence $i_{r+1} = \pm 1 + i_r$ and

$$a_r^2 v = \psi_r^2 1_{\underline{\mathbf{i}}} v = (x_r + x_{r+1}) 1_{\underline{\mathbf{i}}} v = 0 \tag{4.22}$$

showing $a_r = 0$. Thus $\psi_r 1_{\underline{\mathbf{i}}} v = 0$ for all $r$.

In the case when $\ell = 2$, $i_{r+1} = \pm 1 + i_r$ fully determines $\underline{\mathbf{i}}$ and agrees with the conclusions of the proposition, so for the rest of the proof we assume $\ell > 2$. Suppose $i_r = i_{r+2}$ for some $1 \leq r \leq m-2$. Since $i_{r+1} = \pm 1 + i_r = \pm 1 + i_{r+2}$, relation (4.8) gives

$$0 = (a_r^2 a_{r+1} - a_r a_{r+1}^2) 1_{\underline{\mathbf{i}}} v = (\psi_r \psi_{r+1} \psi_r - \psi_{r+1} \psi_r \psi_{r+1}) 1_{\underline{\mathbf{i}}} v = 1_{\underline{\mathbf{i}}} v = v \tag{4.23}$$

which is a contradiction as $a_r = 0$ but $v \neq 0$. So $i_r \neq i_{r+2}$ and $\underline{\mathbf{i}}$ has form (4.18) or (4.19). In particular $\nu = \gamma_{i;m}^{\pm}$. To show that such $M$ actually exist one need only check that setting $1_{\underline{\mathbf{i}}} v = v$ for $\underline{\mathbf{i}}$ as in (4.18) or (4.19), and $x_r v = \psi_r v = 1_{\underline{\mathbf{j}}} v = 0$ for $\underline{\mathbf{j}} \neq \underline{\mathbf{i}}$, satisfies all of the relations on the generators of $R(\gamma_{i;m}^{\pm})$. □

When $k > 0$, we denote the 1-dimensional $R(\gamma_{i;k}^{+})$-module $T$ with ascending character,

$$\operatorname{Char} T = [i, i+1, \ldots, i+k-1] \quad \text{as} \quad T = T_{i;k} = T(i, i+1, \ldots, i+k-1), \tag{4.24}$$

and the 1-dimensional $R(\gamma_{i;k}^{-})$-module $S$ with descending character,

$$\operatorname{Char} S = [i, i-1, \ldots, i-k+1] \quad \text{as} \quad S = S_{i;k} = S(i, i-1, \ldots, i-k+1). \tag{4.25}$$

We refer to $k$ as the *height* of $T_{i;k}$ and $S_{i;k}$ respectively. When $k = 0$, then $\gamma_{i;k}^{\pm} = 0$ and $T_{i;k} = \mathbb{1}$, the unique simple $R(0)$-module, which we will refer to as the *unit* module. In this paper we choose to work with 1-dimensional modules with ascending character. (These are analogous to trivial modules for the affine Hecke algebra or symmetric group.) One could also have chosen to use the 1-dimensional modules with descending character (analogous to sign modules) with obvious modifications. Hence we will informally refer to each type of module as a trivial or sign module, respectively.

EXAMPLE 4.4. In type $A_3^{(1)}$, with

$$\nu = 2\alpha_0 + 2\alpha_1 + \alpha_2 + \alpha_3,$$

of height 6, $R(\nu)$ has a simple 1-dimensional module $T = T_{0;6}$ with

$$\operatorname{Char}(T) = [0, 1, 2, 3, 0, 1],$$

but it is convenient to also write this as

$$\operatorname{Char}(T) = [0, 1, 2, 3, 4, 5].$$

**4.2. Induction and restriction.** It was shown in [**16**] and [**17**] that for $\nu, \mu \in Q^+$ there is a non-unital embedding

$$R(\nu) \otimes R(\mu) \hookrightarrow R(\nu + \mu). \tag{4.26}$$

This map sends the idempotent $1_{\underline{\mathbf{i}}} \otimes 1_{\underline{\mathbf{j}}}$ to $1_{\underline{\mathbf{ij}}}$. The identity $1_\nu \otimes 1_\mu$ of $R(\nu) \otimes R(\mu)$ has as its image

$$\sum_{\underline{\mathbf{i}} \in \operatorname{Seq}(\nu)} \sum_{\underline{\mathbf{j}} \in \operatorname{Seq}(\mu)} 1_{\underline{\mathbf{ij}}}. \tag{4.27}$$

Using this embedding one can define induction and restriction functors,

$$\begin{aligned} \operatorname{Ind}_{\nu,\mu}^{\nu+\mu} : (R(\nu) \otimes R(\mu))\text{-mod} &\to R(\nu+\mu)\text{-mod} \\ M &\mapsto R(\nu+\mu) \otimes_{R(\nu) \otimes R(\mu)} M \end{aligned} \tag{4.28}$$

and

$$\operatorname{Res}^{\nu+\mu}_{\nu,\mu} : R(\nu+\mu)\text{-mod} \to (R(\nu)\otimes R(\mu))\text{-mod}. \tag{4.29}$$

In the future we will write $\operatorname{Ind}^{\nu+\mu}_{\nu,\mu} = \operatorname{Ind}$ and $\operatorname{Res}^{\nu+\mu}_{\nu,\mu} = \operatorname{Res}$ when the algebras are understood from the context. More generally we can extend this embedding to finite tensor products

$$R(\nu^{(1)})\otimes R(\nu^{(2)})\otimes\cdots\otimes R(\nu^{(k)}) \hookrightarrow R(\nu^{(1)}+\nu^{(2)}+\cdots+\nu^{(k)}). \tag{4.30}$$

We refer to the image of this embedding as a *parabolic subalgebra* and denote it by $R(\underline{\nu}) \subset R(\nu^{(1)}+\cdots+\nu^{(k)})$. We denote the image of the identity under this embedding as $1_{\underline{\nu}}$. It follows from Theorem 4.1 that $R(\nu^{(1)}+\nu^{(2)}+\cdots+\nu^{(k)})1_{\underline{\nu}}$ is a free right $R(\underline{\nu})$-module and $1_{\underline{\nu}}R(\nu^{(1)}+\nu^{(2)}+\cdots+\nu^{(k)})$ is a free left $R(\underline{\nu})$-module. Let $m_i = |\nu^{(i)}|$ and set

$$P=(m_1,\dots,m_k) \quad\text{and}\quad \mathcal{S}_P = \mathcal{S}_{m_1}\times\mathcal{S}_{m_1}\times\cdots\times\mathcal{S}_{m_k}. \tag{4.31}$$

Let $\mathcal{S}_{m_1+\cdots+m_k}/\mathcal{S}_P$ be the collection of minimal length left coset representatives of $\mathcal{S}_P$ in $\mathcal{S}_{m_1+\cdots+m_k}$ and $\mathcal{S}_P\backslash\mathcal{S}_{m_1+\cdots+m_k}$ be the collection of minimal length right coset representatives. We construct a weight basis for an induced module as follows. If $M$ is an $R(\underline{\nu})$-module with weight basis $U$ then $\operatorname{Ind}^{\nu^{(1)}+\cdots+\nu^{(k)}}_{\underline{\nu}} M$ has weight basis

$$\{\psi_{\widehat{w}}\otimes u \mid w\in\mathcal{S}_{m_1+\cdots+m_k}/\mathcal{S}_P, u\in U\}. \tag{4.32}$$

Induction is left adjoint to restriction (a property known as Frobenius reciprocity),

$$\operatorname{HOM}_{R(\nu^{(1)}+\cdots+\nu^{(k)})}(\operatorname{Ind}^{\nu^{(1)}+\cdots+\nu^{(k)}}_{\underline{\nu}} M, N) \cong \operatorname{HOM}_{R(\underline{\nu})}(M, \operatorname{Res}^{\nu^{(1)}+\cdots+\nu^{(k)}}_{\underline{\nu}} N). \tag{4.33}$$

Given $\underline{\mathbf{i}}\in\operatorname{Seq}(\nu)$ and $\underline{\mathbf{j}}\in\operatorname{Seq}(\mu)$, a shuffle of $\underline{\mathbf{i}}$ and $\underline{\mathbf{j}}$ is an element $\underline{\mathbf{k}}$ of $\operatorname{Seq}(\nu+\mu)$ such that $\underline{\mathbf{k}}$ has $\underline{\mathbf{i}}$ as a subsequence and $\underline{\mathbf{j}}$ as the complementary subsequence. We denote by $\underline{\mathbf{i}}\sqcup\underline{\mathbf{j}}$ the formal sum of all shuffles of $\underline{\mathbf{i}}$ and $\underline{\mathbf{j}}$. The multi-set of all shuffles of $\underline{\mathbf{i}}$ and $\underline{\mathbf{j}}$ are in bijection with the minimal length left coset representatives $\mathcal{S}_{|\nu|+|\mu|}/\mathcal{S}_{|\nu|}\times\mathcal{S}_{|\mu|}$. Using the definition of degree from KLR algebras, we can associate to any shuffle a degree which we denote as $\deg(\underline{\mathbf{i}},\underline{\mathbf{j}},\underline{\mathbf{k}})$. Then the quantum shuffle of $\underline{\mathbf{i}}$ and $\underline{\mathbf{j}}$ is

$$\underline{\mathbf{i}}\sqcup\!\sqcup\underline{\mathbf{j}} = \sum_{\sigma\in\mathcal{S}_{|\nu|+|\mu|}/\mathcal{S}_{|\nu|}\times\mathcal{S}_{|\mu|}} q^{\deg(\underline{\mathbf{i}},\underline{\mathbf{j}},\sigma(\underline{\mathbf{ij}}))}\sigma(\underline{\mathbf{ij}}), \tag{4.34}$$

so that $\underline{\mathbf{i}}\sqcup\underline{\mathbf{j}} = (\underline{\mathbf{i}}\sqcup\!\sqcup\underline{\mathbf{j}})|_{q=1}$. Note that we will usually shuffle characters, hence we also write $[\underline{\mathbf{i}}]\sqcup\!\sqcup[\underline{\mathbf{j}}]$. For an $R(\mu)$-module $M$ and $R(\nu)$-module $N$ it was shown in [**16**] that

$$\operatorname{Char}(\operatorname{Ind}^{\mu+\nu}_{\mu,\nu} M\boxtimes N) = \operatorname{Char}(M)\sqcup\!\sqcup\operatorname{Char}(N). \tag{4.35}$$

This identity is referred to as the **Shuffle Lemma**.

**4.3. Simple modules of $R(n\alpha_i)$.** For $\nu = n\alpha_i$, induction allows for a particularly easy description of all simple $R(n\alpha_i)$-modules. Let $L(i)$ be the 1-dimensional $R(\alpha_i)$-module. (Note $x_1 1_i$ acts as zero.) Then the unique simple $R(n\alpha_i)$ module is

$$L(i^n) := \operatorname{Ind}^{n\alpha_i}_{\alpha_i,\alpha_i,\dots,\alpha_i} L(i)\boxtimes\cdots\boxtimes L(i) \tag{4.36}$$

up to overall grading shift, which we may shift to have character

$$\operatorname{Char}(L(i^n)) = [n]![i,i,\dots,i]. \tag{4.37}$$

**4.4. Crystal operators on the category $R$-mod.** In the previous section we defined induction and restriction for KLR algebras. Following the work of Grojnowski [**8**] where crystal operators were developed as functors on the category of modules over affine Hecke algebras of type $A$ (or [**18**] for $\mathbb{F}_\ell \mathcal{S}_m$), the KLR analogues of crystal operators were introduced in [**16**], and further developed in [**21**], [**10**]. For each $i \in I$, if $M \in R(\nu)$-mod and $\nu - \alpha_i \in Q^+$, define the functor $\Delta_i : R(\nu)\text{-mod} \to R(\nu - \alpha_i) \otimes R(\alpha_i)\text{-mod}$ as the restriction

$$\Delta_i M := \mathrm{Res}^{\nu}_{\nu-\alpha_i,\alpha_i} M. \tag{4.38}$$

Note that this is equivalent to multiplying $M$ by $1_{\nu-\alpha_i} \otimes 1_{\alpha_i}$. It is also sometimes useful to think of this functor as killing all weight spaces corresponding to elements of $\mathrm{Seq}(\nu)$ that do not end in $i$. If $\nu - \alpha_i \notin Q^+$ then $\Delta_i M = \mathbf{0}$. We similarly define

$$\Delta_{i^n} M := \mathrm{Res}^{\nu}_{\nu-n\alpha_i,n\alpha_i} M. \tag{4.39}$$

Next define the functor $e_i : R(\nu)\text{-mod} \to R(\nu - \alpha_i)\text{-mod}$ as the restriction,

$$e_i M := \mathrm{Res}^{R(\nu-\alpha_i)\otimes R(\alpha_i)}_{R(\nu-\alpha_i)} \Delta_i M \tag{4.40}$$

When $M$ is simple, we can further refine this functor by setting

$$\widetilde{e}_i M := \mathrm{soc}\, e_i M. \tag{4.41}$$

We measure how many times we can apply $\widetilde{e}_i$ to a simple module $M$ by

$$\varepsilon_i(M) := \max\{n \geq 0 \mid (\widetilde{e}_i)^n M \neq \mathbf{0}\}. \tag{4.42}$$

Let $\widetilde{f}_i : R(\nu)\text{-mod} \to R(\nu + \alpha_i)\text{-mod}$ be defined by

$$\widetilde{f}_i M := \mathrm{cosoc}\,\mathrm{Ind}\, M \boxtimes L(i). \tag{4.43}$$

We also set $\mathrm{wt}(M) = -\nu$ if $M \in R(\nu)$-mod, and $\varphi_i(M) = \varepsilon_i(M) - \langle h_i, \nu\rangle$. This data is all part of a crystal datum that defines the structure of the crystal graph $B(\infty)$ on the simple $R$-modules. See Section 5.0.1.

Some of the most important facts about $e_i, \widetilde{e}, \widetilde{f}_i$ stated in [**16**] are given in the following proposition.

PROPOSITION 4.5. Let $i \in I$, $\nu \in Q^+$, $n \in \mathbb{Z}_{>0}$.

(1) Let $M \in R(\nu)$-mod. Then

$$\mathrm{Char}(\Delta_{i^n} M) = \sum_{\underline{\mathbf{j}} \in \mathrm{Seq}(\nu - n\alpha_i)} \mathrm{gdim}(1_{\underline{\mathbf{j}} i^n} M) \cdot \underline{\mathbf{j}} i^n,$$

(2) Let $N \in R(\nu)$-mod be irreducible and $M = \mathrm{Ind}^{\nu+n\alpha_i}_{\nu,n\alpha_i} N \boxtimes L(i^n)$. Let $\varepsilon = \varepsilon_i(N)$. Then
 (a) $\Delta_{i^{\varepsilon+n}} M \cong (\widetilde{e}_i)^\varepsilon N \boxtimes L(i^{\varepsilon+n})$.
 (b) $\mathrm{cosoc}\, M$ is irreducible; further $\mathrm{cosoc}\, M \cong (\widetilde{f}_i)^n N$, $\Delta_{i^{\varepsilon+n}} (\widetilde{f}_i)^n N \cong (\widetilde{e}_i)^\varepsilon N \boxtimes L(i^{\varepsilon+n})$, and $\varepsilon_i((\widetilde{f}_i)^n N) = \varepsilon + n$.
 (c) $(\widetilde{f}_i)^n N$ occurs with multiplicity one as a composition factor of $M$.
 (d) All other composition factors $K$ of $M$ have $\varepsilon_i(K) < \varepsilon + n$.

(3) Let $\underline{\mu} = (\mu_1\alpha_i, \ldots, \mu_r\alpha_i)$ with $\sum_{k=1}^r \mu_k = n$.
 (a) All composition factors of $\mathrm{Res}^{n\alpha_i}_{\underline{\mu}} L(i^n)$ are isomorphic to $L(i^{\mu_1}) \boxtimes \cdots \boxtimes L(i^{\mu_r})$, and $\mathrm{soc}(\mathrm{Res}^{n\alpha_i}_{\underline{\mu}} L(i^n))$ is irreducible.
 (b) $\widetilde{e}_i L(i^n) \cong L(i^{n-1})$.

(4) Let $M \in R(\nu)$-mod be irreducible with $\varepsilon_i(M) > 0$. Then $\widetilde{e}_i M = \operatorname{soc}(e_i M)$ is irreducible and $\varepsilon_i(\widetilde{e}_i M) = \varepsilon_i(M) - 1$. Furthermore if $K$ is a composition factor of $e_i M$ and $K \not\cong \widetilde{e}_i M$, then $\varepsilon_i(K) < \varepsilon_i(M) - 1$.
(5) For irreducible $M \in R(\nu)$-mod let $m = \varepsilon_i(M)$. Then $e_i^m M$ is isomorphic to $(\widetilde{e}_i)^m M^{\oplus [m]!}$. In particular, if $m = 1$ then $e_i M = \widetilde{e}_i M$.
(6) For irreducible modules $N \in R(\nu)$-mod and $M \in R(\nu + \alpha_i)$-mod we have $\widetilde{f}_i N \cong M$ if and only if $N \cong \widetilde{e}_i M$.
(7) Let $M, N \in R(\nu)$-mod be irreducible. Then $\widetilde{f}_i M \cong \widetilde{f}_i N$ if and only if $M \cong N$. Assuming $\varepsilon_i(M), \varepsilon_i(N) > 0$, $\widetilde{e}_i M \cong \widetilde{e}_i N$ if and only if $M \cong N$.

On the level of characters, $e_i$ roughly removes an $i$ from the rightmost entry of a module's character. We can construct analogous functors for removal of $i$ from the left side of a module's character, as well as an analogue to $\widetilde{f}_i$. These are denoted by $e_i^\vee$, $\widetilde{e}_i^\vee$, $\widetilde{f}_i^\vee$ and we will use them extensively in this paper. We use the involution $\sigma$ introduced below to define them. Let $w_0$ be the longest element of $\mathcal{S}_{|\nu|}$. Then $\sigma : R(\nu) \to R(\nu)$ is defined as follows:

$$1_{\underline{\mathbf{i}}} \mapsto 1_{w_0(\underline{\mathbf{i}})} \tag{4.44}$$

$$x_r \mapsto x_{|\nu|+1-r} \tag{4.45}$$

$$\psi_r 1_{\underline{\mathbf{i}}} \mapsto (-1)^{\delta_{i_r, i_{r+1}}} \psi_{|\nu|-r} 1_{w_0(\underline{\mathbf{i}})}. \tag{4.46}$$

For an $R(\nu)$-module $M$, let $\sigma^* M$ be the $R(\nu)$-module $M$ but with the action of $R(\nu)$ twisted by $\sigma$,

$$r \cdot u = \sigma(r) u.$$

Now let $e_i^\vee : R(\nu)\text{-mod} \to R(\nu - \alpha_i)\text{-mod}$ be the restriction functor defined as

$$e_i^\vee := \sigma^* e_i \sigma = \operatorname{Res}^{R(\alpha_i) \otimes R(\nu - \alpha_i)}_{R(\nu - \alpha_i)} \circ \operatorname{Res}^{\nu}_{\alpha_i, \nu - \alpha_i}, \tag{4.47}$$

and similarly,

$$\widetilde{e}_i^\vee M := \sigma^*(\widetilde{e}_i(\sigma^* M)) = \operatorname{soc} e_i^\vee M, \tag{4.48}$$

$$\widetilde{f}_i^\vee M := \sigma^*(\widetilde{f}_i(\sigma^* M)) = \operatorname{cosoc} \operatorname{Ind}^{\nu + \alpha_i}_{\alpha_i, \nu} L(i) \boxtimes M, \tag{4.49}$$

$$\varepsilon_i^\vee(M) := \varepsilon_i(\sigma^* M) = \max\{n \geq 0 \mid (\widetilde{e}_i^\vee)^n M \neq \mathbf{0}\}. \tag{4.50}$$

Note that by the exactness of restriction, $e_i, e_i^\vee$ are exact functors, while $\widetilde{e}_i$ and $\widetilde{e}_i^\vee$ are only left exact, and $\widetilde{f}_i$ and $\widetilde{f}_i^\vee$ are only right exact. When $k \in \mathbb{N}$, the indices on $e_k$, $\widetilde{e}_k$, $e_k^\vee$, $\widetilde{e}_k^\vee$, $\widetilde{f}_k$, $\widetilde{f}_k^\vee$ should always be interpreted modulo $\ell$, i.e. we also identify $k \in I$.

Example 4.6. The module of Example 4.4 can be constructed as $T_{0;6} = \widetilde{f}_5 \widetilde{f}_4 \widetilde{f}_3 \widetilde{f}_2 \widetilde{f}_1 \widetilde{f}_0 \,\mathbb{1}$ or as $\widetilde{f}_5 \widetilde{f}_4 \widetilde{f}_3 \widetilde{f}_2 \widetilde{f}_1 L(0)$. Since $\ell = 4$ this is also $\widetilde{f}_1 \widetilde{f}_0 \widetilde{f}_3 \widetilde{f}_2 \widetilde{f}_1 \widetilde{f}_0 \,\mathbb{1}$, but for the purposes of this paper, we prefer the first expression.

Remark 4.7. There is a nice character-theoretic interpretation of $\varepsilon_i$ and $\varepsilon_i^\vee$. Let $M$ be a simple $R(\nu)$-module with $|\nu| = m$. Then

a.) $\varepsilon_i(M) = c$ implies that there exists

$$\underline{\mathbf{i}} = (i_1, \ldots, i_{m-c}, \underbrace{i, i, \ldots, i}_{c})$$

such that $1_{\underline{\mathbf{i}}} A \neq \mathbf{0}$. In other words $[\underline{\mathbf{i}}]$ is in the support of $M$; however no $[\underline{\mathbf{j}}]$ such that

$$\underline{\mathbf{j}} = (i_1, \ldots, i_{m-c-1}, \underbrace{i, i, \ldots, i}_{c+1}).$$

is in $\mathrm{supp}(M)$.

b.) $\varepsilon_i^\vee(M) = c$ implies that there exists $[\underline{\mathbf{i}}]$ in the support of $M$ of the form

$$\underline{\mathbf{i}} = (\underbrace{i, i, \ldots, i}_{c}, i_{c+1}, \ldots, i_m)$$

but no $[\underline{\mathbf{j}}]$ of the form

$$\underline{\mathbf{j}} = (\underbrace{i, i, \ldots, i}_{c+1}, i_{c+2}, \ldots, i_m).$$

**4.5. Serre relations.** Because the functors $e_i$, $i \in I$, are exact, they descend to well-defined linear operators on the Grothendieck group of $R$, $G_0(R)$. It is shown in [**16, 17**] that these operators satisfy the quantum Serre relations, and that these relations are in fact minimal. We have

$$\sum_{r=0}^{-a_{ij}+1} (-1)^r e_i^{(-a_{ij}+1-r)} e_j e_i^{(r)} [M] = \mathbf{0}. \tag{4.51}$$

for all $i \neq j \in I$ and $M \in R\text{-mod}$, where $e_i^{(r)} = \frac{1}{[r]!} e_i^r$ is the divided power. (Recall $a_{ij} = \langle h_i, \alpha_j \rangle$.) The minimality of these relations imply that, for $0 \leq c < -a_{ij} + 1$,

$$\sum_{r=0}^{c} (-1)^r e_i^{(c-r)} e_j e_i^{(r)} \tag{4.52}$$

is never the zero operator on $G_0(R)$ by the quantum Gabber-Kac Theorem [**23**] and the work of [**16, 17**], which essentially computes the kernel of the map from the free algebra on generators $e_i$ to $G_0(R)$.

**4.6. Jump.** When we apply $\tilde{f}_i$ to irreducible $R(\nu)$-module $M$ for $i \in I$, then Proposition 4.5.2 tells us that $\tilde{f}_i M$ is an irreducible $R(\nu + \alpha_i)$-module with

$$\varepsilon_i(\tilde{f}_i M) = \varepsilon_i(M) + 1. \tag{4.53}$$

We could also ask whether $\varepsilon_i^\vee(\tilde{f}_i M)$ and $\varepsilon_i^\vee(M)$ differ. Questions like this motivate the introduction of the function $\mathrm{jump}_i$, which is based on a concept for Hecke algebras in [**8**], and was introduced for KLR algebras and studied extensively in [**21**].

DEFINITION 4.8. Let $M$ be a simple $R(\nu)$-module, and let $i \in I$. Then

$$\mathrm{jump}_i(M) := \max\{J \geq 0 \mid \varepsilon_i^\vee(M) = \varepsilon_i^\vee(\tilde{f}_i^J M)\}.$$

LEMMA 4.9. [**21**] *Let $M$ be a simple $R(\nu)$-module. The following are equivalent:*

*(1)* $\mathrm{jump}_i(M) = 0$
*(2)* $\tilde{f}_i M \cong \tilde{f}_i^\vee M$
*(3)* $\mathrm{Ind}\, M \boxtimes L(i^m)$ *is irreducible for all* $m \geq 1$
*(4)* $\mathrm{Ind}\, M \boxtimes L(i^m) = \mathrm{Ind}\, L(i^m) \boxtimes M$ *for all* $m \geq 1$
*(5)* $wt_i(M) + \varepsilon_i(M) + \varepsilon_i^\vee(M) = 0$, *where* $wt_i(M) = -\langle h_i, \nu \rangle$.
*(6)* $\varepsilon_i(\tilde{f}_i^\vee M) = \varepsilon_i(M) + 1$
*(7)* $\varepsilon_i^\vee(\tilde{f}_i M) = \varepsilon_i^\vee(M) + 1$.

PROOF. See [**21**]. □

It is shown in [**21**] that

$$\operatorname{jump}_i(\widetilde{f}_i M) = \max\{0, \operatorname{jump}_i(M) - 1\} = \operatorname{jump}_i(\widetilde{f}_i^\vee M). \tag{4.54}$$

It is also shown in [**21**] that

$$\operatorname{jump}_i(M) = \operatorname{wt}_i(M) + \varepsilon_i(M) + \varepsilon_i^\vee(M). \tag{4.55}$$

Using information from $\operatorname{jump}_i$ we can also determine when the crystal operators commute with their $\sigma$-symmetric versions.

EXAMPLE 4.10. Suppose $\ell > 2$. Observe $\operatorname{jump}_1(L(0)) = 1$ and

$$\widetilde{f}_1^\vee \widetilde{f}_1 L(0) \cong \operatorname{Ind} L(1) \boxtimes T(0,1)$$

whose character has support $\{[1,0,1],[0,1,1],[0,1,1]\}$. However

$$\widetilde{f}_1 \widetilde{f}_1^\vee L(0) \cong \operatorname{Ind} S(1,0) \boxtimes L(1)$$

whose character has support $\{[1,0,1],[1,1,0],[1,1,0]\}$.

In the case $\ell = 2$, note $\operatorname{jump}_1(T(0,1)) = 1$ and we can similarly calculate $\widetilde{f}_1^\vee \widetilde{f}_1 T(0,1) \not\cong \widetilde{f}_1 \widetilde{f}_1^\vee T(0,1)$ (in fact the former is 8-dimensional while the latter is 4-dimensional).

We shall see below that this phenomenon is special to $\operatorname{jump}_i(M) = 1$.

LEMMA 4.11. *Let $M$ be a simple $R(\nu)$-module.*

*(1)* [**21**] *If $i \neq j$, then*
   *(a)* $\widetilde{f}_i \widetilde{f}_j^\vee M \cong \widetilde{f}_j^\vee \widetilde{f}_i M$.
   *(b) If* $\widetilde{e}_j^\vee M \neq \mathbf{0}$ *then* $\widetilde{f}_i \widetilde{e}_j^\vee M \cong \widetilde{e}_j^\vee \widetilde{f}_i M$.
   *(c) If* $\widetilde{e}_j M \neq \mathbf{0}$ *then* $\widetilde{f}_i^\vee \widetilde{e}_j M \cong \widetilde{e}_j \widetilde{f}_i^\vee M$.
   *(d) If further* $\widetilde{e}_i M \neq \mathbf{0}$ *then,* $\widetilde{e}_i \widetilde{e}_j^\vee(M) \cong \widetilde{e}_j^\vee \widetilde{e}_i(M)$.

*(2) (a)* $\operatorname{jump}_i(M) \neq 1$ *if and only if* $\widetilde{f}_i^\vee \widetilde{f}_i M \cong \widetilde{f}_i \widetilde{f}_i^\vee M$.
   *(b) If* $\widetilde{e}_i^\vee M \neq \mathbf{0}$*, then* $\operatorname{jump}_i(\widetilde{e}_i^\vee M) \neq 1$ *if and only if* $\widetilde{e}_i^\vee \widetilde{f}_i M \cong \widetilde{f}_i \widetilde{e}_i^\vee M$.
   *(c) If* $\widetilde{e}_i M \neq \mathbf{0}$*, then* $\operatorname{jump}_i(\widetilde{e}_i M) \neq 1$ *if and only if* $\widetilde{e}_i \widetilde{f}_i^\vee M \cong \widetilde{f}_i^\vee \widetilde{e}_i M$.

PROOF.

(1) Consider the short exact sequence,

$$\mathbf{0} \to K \to \operatorname{Ind} M \boxtimes L(i) \to \widetilde{f}_i M \to \mathbf{0} \tag{4.56}$$

and recall $\widetilde{f}_i M$ is the unique composition factor of $\operatorname{Ind} M \boxtimes L(i)$ such that $\varepsilon_i(\widetilde{f}_i M) = \varepsilon_i(M) + 1$, and that for all composition factors $N$ of $K$, $\varepsilon_i(N) \leq \varepsilon_i(M)$. By the exactness of induction there is a second short exact sequence

$$\mathbf{0} \to \operatorname{Ind} L(j) \boxtimes K \to \operatorname{Ind} L(j) \boxtimes M \boxtimes L(i) \to \operatorname{Ind} L(j) \boxtimes \widetilde{f}_i M \to \mathbf{0}, \tag{4.57}$$

and since $i \neq j$ the Shuffle Lemma tells us that for all composition factors $N'$ of $\operatorname{Ind} L(j) \boxtimes K$, $\varepsilon_i(N') \leq \varepsilon_i(M)$. By the Shuffle Lemma and Frobenius reciprocity

$$\varepsilon_i(\widetilde{f}_j^\vee \widetilde{f}_i M) = \varepsilon_i(\widetilde{f}_i \widetilde{f}_j^\vee M) = \varepsilon_i(M) + 1. \tag{4.58}$$

Hence there can be no nonzero map

$$\operatorname{Ind} L(j) \boxtimes K \to \widetilde{f}_i \widetilde{f}_j^\vee M, \tag{4.59}$$

so that the submodule $\operatorname{Ind} L(j) \boxtimes K$ is contained in the kernel of $\beta$, as pictured in (4.60).

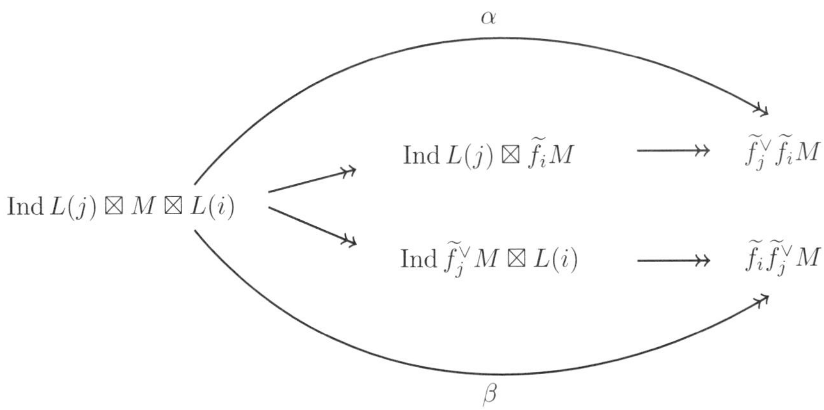


(4.60)

Hence $\beta$ induces a nonzero map (necessarily surjective)

$$\operatorname{Ind} L(j) \boxtimes \widetilde{f}_i M \twoheadrightarrow \widetilde{f}_i \widetilde{f}_j^\vee M. \tag{4.61}$$

Because $\operatorname{Ind} L(j) \boxtimes \widetilde{f}_i M$ has unique simple quotient $\widetilde{f}_j^\vee \widetilde{f}_i M$, then $\widetilde{f}_j^\vee \widetilde{f}_i M \cong \widetilde{f}_i \widetilde{f}_j^\vee M$. This proves 1a.

The three isomorphisms in 1b, 1c, and 1d all follow from 1a. For example, if $\widetilde{e}_j^\vee M$ is nonzero, then

$$\widetilde{f}_i M \cong \widetilde{f}_i \widetilde{f}_j^\vee \widetilde{e}_j^\vee M \cong \widetilde{f}_j^\vee \widetilde{f}_i \widetilde{e}_j^\vee M. \tag{4.62}$$

Applying $\widetilde{e}_j^\vee$ to both sides we get 1b. 1c and 1d follow similarly.

(2) We prove 2a. Let $c = \varepsilon_i^\vee(M), m = \varepsilon_i(M)$.

- Suppose $\operatorname{jump}_i(M) = 0$. Then also $\operatorname{jump}_i(\widetilde{f}_i M) = \operatorname{jump}_i(\widetilde{f}_i^\vee M) = 0$ by (4.54). Thus by Lemma 4.9

$$\widetilde{f}_i^\vee \widetilde{f}_i M \cong \widetilde{f}_i \widetilde{f}_i M \cong \widetilde{f}_i \widetilde{f}_i^\vee M. \tag{4.63}$$

- Suppose $\operatorname{jump}_i(M) = 1$. By Lemma 4.9 and Proposition 4.5, we have $\varepsilon_i(\widetilde{f}_i M) = m+1$ but $\varepsilon_i^\vee(\widetilde{f}_i M) = c$. While $\varepsilon_i(\widetilde{f}_i^\vee M) = m$ but $\varepsilon_i^\vee(\widetilde{f}_i^\vee M) = c+1$. Further by (4.54) $\operatorname{jump}_i(\widetilde{f}_i M) = \operatorname{jump}_i(\widetilde{f}_i^\vee M) = 0$. Hence $\varepsilon_i(\widetilde{f}_i^\vee \widetilde{f}_i M) = m+2$, $\varepsilon_i^\vee(\widetilde{f}_i^\vee \widetilde{f}_i M) = c+1$ whereas $\varepsilon_i(\widetilde{f}_i \widetilde{f}_i^\vee M) = m+1$, $\varepsilon_i^\vee(\widetilde{f}_i \widetilde{f}_i^\vee M) = c+2$. Thus the two modules cannot be isomorphic.
- Suppose $\operatorname{jump}_i(M) \geq 2$. Then $\operatorname{jump}_i(\widetilde{f}_i M) = \operatorname{jump}_i(\widetilde{f}_i^\vee M) \geq 1$. We calculate

$$\varepsilon_i(\widetilde{f}_i \widetilde{f}_i^\vee M) = m+1 = \varepsilon_i(\widetilde{f}_i^\vee \widetilde{f}_i M) \tag{4.64}$$

$$\varepsilon_i^\vee(\widetilde{f}_i \widetilde{f}_i^\vee M) = c+1 = \varepsilon_i^\vee(\widetilde{f}_i^\vee \widetilde{f}_i M). \tag{4.65}$$

We will show there is no nonzero map

$$\operatorname{Ind} L(i) \boxtimes K \to \widetilde{f}_i \widetilde{f}_i^\vee M \tag{4.66}$$

for any proper submodule $K \subseteq \operatorname{Ind} M \boxtimes L(i)$. Given we have a surjection

$$\operatorname{Ind} L(i) \boxtimes M \boxtimes L(i) \twoheadrightarrow \widetilde{f}_i \widetilde{f}_i^\vee M \tag{4.67}$$

this means we must have a nonzero map

$$\operatorname{Ind} L(i) \boxtimes \widetilde{f}_i M \to \widetilde{f}_i \widetilde{f}_i^\vee M, \tag{4.68}$$

which will prove the lemma as

$$\widetilde{f}_i^\vee \widetilde{f}_i M = \operatorname{cosoc} \operatorname{Ind} L(i) \boxtimes \widetilde{f}_i M. \tag{4.69}$$

First note there is no nonzero map

$$\operatorname{Ind} L(i) \boxtimes \widetilde{f}_i^\vee M \to \widetilde{f}_i \widetilde{f}_i^\vee M \tag{4.70}$$

as $\operatorname{cosoc}(\operatorname{Ind} L(i) \boxtimes \widetilde{f}_i^\vee M) = (\widetilde{f}_i^\vee)^2 M$ and $\varepsilon_i^\vee((\widetilde{f}_i^\vee)^2 M) = c+2 \neq c+1 = \varepsilon_i^\vee(\widetilde{f}_i \widetilde{f}_i^\vee M)$. Let $D$ be any other composition factor of $\operatorname{Ind} M \boxtimes L(i)$ apart from $\widetilde{f}_i M$ or $\widetilde{f}_i^\vee M$ (recall the latter occur with multiplicity one as composition factors). Then by Proposition 4.5, $\varepsilon_i(D) \leq m$, $\varepsilon_i^\vee(D) \leq c$. If there were a nonzero map $\operatorname{Ind} L(i) \boxtimes D \to \widetilde{f}_i \widetilde{f}_i^\vee M$, it would imply $\widetilde{f}_i^\vee D \cong \widetilde{f}_i \widetilde{f}_i^\vee M$ and so $\varepsilon_i^\vee(\widetilde{f}_i^\vee D) = c+1$ meaning $\varepsilon_i^\vee(D) = c$. Also $m+1 = \varepsilon_i(\widetilde{f}_i^\vee D) \leq \varepsilon_i(D)+1$ by the Shuffle Lemma, forcing $\varepsilon_i(D) = m$. By Lemma 4.9 this forces $0 = \operatorname{jump}_i(D)$ and $\widetilde{f}_i D \cong \widetilde{f}_i^\vee D \cong \widetilde{f}_i \widetilde{f}_i^\vee M$ from above, forcing $D \cong \widetilde{f}_i^\vee M$, which we already ruled out. Hence there must be a nonzero map

$$\operatorname{Ind} L(i) \boxtimes \widetilde{f}_i M \to \widetilde{f}_i \widetilde{f}_i^\vee M. \tag{4.71}$$

Now that we have established that $\widetilde{f}_i^\vee \widetilde{f}_i M \cong \widetilde{f}_i \widetilde{f}_i^\vee M$ if and only if $\operatorname{jump}_i(M) \neq 1$, statements 2b and 2c follow directly from Proposition 4.5.6.

□

REMARK 4.12. Because $\widetilde{e}_i^\vee$ and $\widetilde{f}_j$ commute for $i \neq j$, then $\varepsilon_i^\vee(\widetilde{f}_j M) = \varepsilon_i^\vee(M)$. An equivalent statement holds for $\widetilde{e}_i$, $\widetilde{f}_j^\vee$, and $\varepsilon_i$. When $\operatorname{jump}_i(M) \neq 0$, $\varepsilon_i^\vee(\widetilde{f}_i M) = \varepsilon_i^\vee(M)$.

## 5. The functor $\mathrm{pr}_\Lambda$

For $\Lambda = \sum_{i \in I} \lambda_i \Lambda_i \in P^+$ define $\mathcal{I}_\nu^\Lambda$ to be the two-sided ideal of $R(\nu)$ generated by the elements $x_1^{\lambda_{i_1}} 1_{\underline{\mathbf{i}}}$ for all $\underline{\mathbf{i}} \in \operatorname{Seq}(\nu)$. When $\nu$ is clear from the context we write, $\mathcal{I}_\nu^\Lambda = \mathcal{I}^\Lambda$. The *cyclotomic KLR algebra of weight* $\Lambda$ is then defined as

$$R^\Lambda = \bigoplus_{\nu \in Q^+} R^\Lambda(\nu) \quad \text{where} \quad R^\Lambda(\nu) := R(\nu)/\mathcal{I}_\nu^\Lambda. \tag{5.1}$$

The algebra $R^\Lambda(\nu)$ is finite dimensional, [**4, 21**]. The category of finite dimensional $R^\Lambda(\nu)$-modules is denoted $R^\Lambda(\nu)\text{-mod}$ and the category of finite dimensional $R^\Lambda$-modules is denoted $R^\Lambda\text{-mod}$. The category of finite dimensional $R$-modules on which $\mathcal{I}^\Lambda$ vanishes is denoted

$$\operatorname{Rep}^\Lambda.$$

While we can identify $R^\Lambda$ - mod with $\mathrm{Rep}^\Lambda$, we choose to work with $\mathrm{Rep}^\Lambda$. We construct a right-exact functor, $\mathrm{pr}_\Lambda : R(\nu)\text{ - mod} \to R(\nu)\text{ - mod}$, via

$$\mathrm{pr}_\Lambda M := M/\mathcal{I}^\Lambda M. \tag{5.2}$$

It is customary in the literature to interpret $\mathrm{pr}_\Lambda$ as being a functor from $R(\nu)$ - mod to $R^\Lambda(\nu)$ - mod, but in this paper it will be more convenient to consider it as a functor $R(\nu)\text{ - mod} \to \mathrm{Rep}^\Lambda$. The reader may keep in mind that the image of $\mathrm{pr}_\Lambda$ consists of $R(\nu)$-modules which descend to $R^\Lambda(\nu)$-modules. Observe that in the opposite direction there is an exact functor $\mathrm{infl}_\Lambda : R^\Lambda(\nu)\text{ - mod} \to R(\nu)\text{ - mod}$, where $R(\nu)$ acts on $R^\Lambda(\nu)$-module $M$ through the projection map $R(\nu) \twoheadrightarrow R^\Lambda(\nu)$.

REMARK 5.1. If $M$ is a $R(\nu)$-module and $A$ is a simple module in $\mathrm{Rep}^\Lambda$ for $\Lambda \in P^+$, then since $\mathrm{pr}_\Lambda A \cong A$, the right exactness of $\mathrm{pr}_\Lambda$ implies that any surjection $M \twoheadrightarrow A$ gives a surjection $\mathrm{pr}_\Lambda M \twoheadrightarrow A$. Similarly, since there always exists a surjection $M \twoheadrightarrow \mathrm{pr}_\Lambda M$, given a surjection $\mathrm{pr}_\Lambda M \twoheadrightarrow A$ we immediately get a surjection $M \twoheadrightarrow A$. In such situations there is an equivalence between the two surjections $M \twoheadrightarrow A$ and $\mathrm{pr}_\Lambda M \twoheadrightarrow A$ which we will henceforth use freely.

If $M$ is simple then either $\mathrm{pr}_\Lambda M = \mathbf{0}$ or $\mathrm{pr}_\Lambda M = M$. There is a useful criterion for determining the action of $\mathrm{pr}_\Lambda$ on simple $R(\nu)$-modules given by the following proposition.

PROPOSITION 5.2. [**21**] Let $\Lambda = \sum_{i\in I} \lambda_i \Lambda_i \in P^+$, $\nu \in Q^+$, and let $M$ be a simple $R(\nu)$-module. Then $\mathcal{I}^\Lambda M = \mathbf{0}$ if and only if $\mathrm{pr}_\Lambda M = M$ if and only if $\mathrm{pr}_\Lambda M \neq \mathbf{0}$ if and only if

$$\varepsilon_i^\vee(M) \leq \lambda_i$$

for all $i \in I$. When these conditions hold $M \in \mathrm{Rep}^\Lambda$. Hence we may identify $M$ with $\mathrm{pr}_\Lambda M$ (or as an $R^\Lambda(\nu)$-module).

In this paper we will primarily consider $\Lambda = \Lambda_i$ in which case $\mathcal{I}_\nu^{\Lambda_i}$ is generated by $x_1 1_{ii_2\dots i_m}$ and $1_{ji_2\dots i_m}, j \neq i$ ranging over $\underline{\mathbf{i}} \in \mathrm{Seq}(\nu)$.

Notice that Proposition 5.2 immediately tells us that the 1-dimensional modules $T_{i;k} \in \mathrm{Rep}^{\Lambda_i}$ for any $k \geq 0$. For $\Lambda = \sum_{i\in I} \lambda_i \Lambda_i \in P^+$ and $M$ an irreducible $R(\nu)$-module set

$$\varphi_i^\Lambda(M) = \lambda_i + \varepsilon_i(M) + \mathrm{wt}_i(M). \tag{5.3}$$

Notice that when $\Lambda = \Lambda_j$ this gives

$$\varphi_i^{\Lambda_j}(M) = \delta_{ij} + \varepsilon_i(M) + \mathrm{wt}_i(M). \tag{5.4}$$

REMARK 5.3. By formula (5.4) if $M$ is a simple module in $\mathrm{Rep}^{\Lambda_j}$ it follows that

$$\varphi_i^{\Lambda_j}(M) = \begin{cases} \delta_{ij} & \text{if } M = \mathbb{1}, \\ \mathrm{jump}_i(M) & \text{otherwise.} \end{cases} \tag{5.5}$$

PROPOSITION 5.4. Let $M$ be a simple $R(\nu)$-module with $\mathrm{pr}_\Lambda M \neq \mathbf{0}$. Then

$$\varphi_i^\Lambda(M) = \max\{n \in \mathbb{Z} \mid \mathrm{pr}_\Lambda \widetilde{f}_i^n M \neq \mathbf{0}\}. \tag{5.6}$$

From property (4.54) of $\mathrm{jump}_i$ it is clear that if we apply $\widetilde{f}_i$ sufficiently many times to any module $M \in R^\Lambda(\nu)$ - mod, then eventually we will eventually reach an $n$ for which

$$\varepsilon_i^\vee(\widetilde{f}_i^n M) > \lambda_i \tag{5.7}$$

and so $\mathrm{pr}_\Lambda \widetilde{f}_i^n M = \mathbf{0}$. Proposition 5.4 shows that $\varphi_i^\Lambda$ measures this for simple modules in $\mathrm{Rep}^\Lambda$. In fact it is true that $\mathrm{pr}_\Lambda M \neq \mathbf{0}$ if and only if $\varphi_i^\Lambda(M) \geq 0$ for all $i \in I$. We remark below that the function $\varphi_i^\Lambda$ is part of a crystal datum.

5.0.1. *Module-theoretic model of* $B(\Lambda)$. Let $M$ be a simple $R(\nu)$-module. Set

$$\mathrm{wt}(M) = -\nu \quad \text{and} \quad \mathrm{wt}_i(M) = -\langle h_i, \nu\rangle. \tag{5.8}$$

Let $\mathrm{Irr}\, R$ be the set of isomorphism classes of simple $R$-modules and $\mathrm{Irr}\, R^\Lambda$ be the set of isomorphism classes of simple modules in $\mathrm{Rep}^\Lambda$. In [**21**] it was shown that the tuple $(\mathrm{Irr}\, R, \varepsilon_i, \varphi_i, \widetilde{e}_i, \widetilde{f}_i, \mathrm{wt})$ defines a crystal isomorphic to $B(\infty)$ and $(\mathrm{Irr}\, R^\Lambda, \varepsilon_i, \varphi_i^\Lambda, \widetilde{e}_i, \widetilde{f}_i, \mathrm{wt})$ defines a crystal isomorphic to the highest weight crystal $B(\Lambda)$.

### 5.1. Interaction of $\mathrm{pr}_\Lambda$ and induction.

The following is a list of useful facts about the way that the functor $\mathrm{pr}_\Lambda$ interacts with the functor of induction.

PROPOSITION 5.5. Fix $\Lambda \in P^+$, let $\mu, \nu \in Q^+$, $M$ be a simple $R(\mu)$-module and $N$ a simple $R(\nu)$-module.

(a) If $\mathrm{pr}_\Lambda M = \mathbf{0}$ then $\mathrm{pr}_\Lambda \mathrm{Ind}\, M \boxtimes N = \mathbf{0}$.
(b) If $\mathrm{pr}_\Lambda \mathrm{Ind}\, M \boxtimes L(i^c) = \mathbf{0}$ and $\varepsilon_i^\vee(N) \geq c$ then $\mathrm{pr}_\Lambda \mathrm{Ind}\, M \boxtimes N = \mathbf{0}$.
(c) If $c > \varphi_i^\Lambda(M)$ then $\mathrm{pr}_\Lambda \mathrm{Ind}\, M \boxtimes L(i^c) = \mathbf{0}$.
(d) Let $\varphi = \varphi_i^\Lambda(M)$, then $\mathrm{pr}_\Lambda \mathrm{Ind}\, M \boxtimes L(i^\varphi) \cong \widetilde{f}_i^\varphi M$.
(e) If $\mathrm{pr}_\Lambda C = M$ then $\mathrm{pr}_\Lambda \mathrm{Ind}\, C \boxtimes N \cong \mathrm{pr}_\Lambda \mathrm{Ind}\, M \boxtimes N$.

PROOF. We let

$$\Lambda = \sum_{i \in I} \lambda_i \Lambda_i. \tag{5.9}$$

(a) If $\mathrm{pr}_\Lambda M = \mathbf{0}$, then by Proposition 5.2 there is some $i \in I$ such that $\varepsilon_i^\vee(M) > \lambda_i$. Suppose that $\mathrm{pr}_\Lambda \mathrm{Ind}\, M \boxtimes N \neq \mathbf{0}$. Then it has some simple quotient $Q$, and there are surjections

$$\mathrm{Ind}\, M \boxtimes N \twoheadrightarrow \mathrm{pr}_\Lambda \mathrm{Ind}\, M \boxtimes N \twoheadrightarrow Q. \tag{5.10}$$

Frobenius reciprocity and Proposition 5.2 imply that $\mathrm{pr}_\Lambda Q = Q$. By Frobenius reciprocity $\mathrm{Res}_{\mu,\nu}^{\mu+\nu} Q$ has $M \boxtimes N$ as a $(R(\mu) \otimes R(\nu))$-submodule. But Remark 4.7 then implies $\varepsilon_i^\vee(Q) > \lambda_i$ so that $\mathrm{pr}_\Lambda Q = \mathbf{0}$, a contradiction.

(b) If $\varepsilon_i^\vee(N) \geq c$ then there is a surjection,

$$\mathrm{Ind}\, L(i^c) \boxtimes (\widetilde{e}_i^\vee)^c N \twoheadrightarrow N \tag{5.11}$$

and by the exactness of induction a surjection

$$\mathrm{Ind}\, M \boxtimes L(i^c) \boxtimes (\widetilde{e}_i^\vee)^c N \twoheadrightarrow \mathrm{Ind}\, M \boxtimes N. \tag{5.12}$$

If $\mathrm{pr}_\Lambda \mathrm{Ind}\, M \boxtimes L(i^c) = \mathbf{0}$, then by part (a) above and the right exactness of $\mathrm{pr}_\Lambda$, $\mathrm{pr}_\Lambda \mathrm{Ind}\, M \boxtimes N = \mathbf{0}$.

(c) This follows from Proposition 5.4 and the fact that the induced module has unique simple quotient $\widetilde{f}_i^c M$; or see [**21**].

(d) Consider the exact sequence,

$$\mathbf{0}\to K\to \operatorname{Ind} M\boxtimes L(i^{\varphi})\to \widetilde{f}_i^{\varphi}M\to \mathbf{0}. \tag{5.13}$$

$\widetilde{f}_i^{\varphi}M$ is the unique composition factor of $\operatorname{Ind} M\boxtimes L(i^{\varphi})$ such that $\varepsilon_i(\widetilde{f}_i^{\varphi}M) = \varphi+\varepsilon_i(M)$, so $\varepsilon_i(D)<\varphi+\varepsilon_i(M)$ for all composition factors $D$ of $K$ by Proposition 4.5. All composition factors $D$ of $K$ have the same weight as $\widetilde{f}_i^{\varphi}M$. By (5.3) and Proposition 4.5, $\varphi_i^{\Lambda}(D)=\lambda_i+\varepsilon_i(D)+\mathrm{wt}_i(D)<\lambda_i+\varepsilon_i(\widetilde{f}_i^{\varphi}M)+\mathrm{wt}_i(\widetilde{f}_i^{\varphi}M)=\varphi_i^{\Lambda}(\widetilde{f}_i^{\varphi}M)=0$. In particular this shows $\mathrm{pr}_\Lambda K=\mathbf{0}$ so by the right exactness of $\mathrm{pr}_\Lambda$ we get (d).

(e) Consider the diagram in (5.14),

(5.14)

$$\begin{array}{ccccccc}
 & & & \mathbf{0} & & & \\
 & & & \downarrow & & & \\
 & & & \operatorname{Ind}\mathcal{I}^{\Lambda}C\boxtimes N & \overset{\beta\circ\alpha}{\searrow} & & \\
 & & & \alpha\downarrow & & & \\
\mathbf{0} & \longrightarrow & \mathcal{I}^{\Lambda}(\operatorname{Ind}C\boxtimes N) & \longrightarrow \operatorname{Ind}C\boxtimes N & \xrightarrow{\beta} & \mathrm{pr}_{\Lambda}(\operatorname{Ind}C\boxtimes N) & \longrightarrow \mathbf{0}\\
 & & & \gamma\downarrow & & \overset{g}{\nearrow}\ (\text{dotted}) & \\
 & & & \operatorname{Ind}M\boxtimes N & & & \\
 & & & \downarrow & & & \\
 & & & \mathbf{0} & & &
\end{array}$$

where the horizontal and vertical sequences are exact. Recall that $\mathcal{I}_{\mu}^{\Lambda}$ in $R(\mu)$ is generated by the set $\{x_1^{\lambda_{i_1}}1_{\underline{\mathbf{i}}}\}_{\underline{\mathbf{i}}\in\mathrm{Seq}(\mu)}$ where $\underline{\mathbf{i}}=i_1i_2\dots i_m$ and $|\mu|=m$. Under the embedding

$$R(\mu)\hookrightarrow R(\mu)\otimes R(\nu)\hookrightarrow R(\mu+\nu),$$

this set maps to the set

$$\Big\{\sum_{\underline{\mathbf{j}}\in\mathrm{Seq}(\nu)} x_1^{\lambda_{i_1}}1_{\underline{\mathbf{ij}}}\Big\}_{\underline{\mathbf{i}}\in\mathrm{Seq}(\mu)}$$

in $R(\mu+\nu)$. This set is contained in the ideal generated by $\{x_1^{\lambda_{i_1}}1_{\underline{\mathbf{k}}}\}_{\underline{\mathbf{k}}\in\mathrm{Seq}(\mu+\nu)}$ which generates $\mathcal{I}_{\mu+\nu}^{\Lambda}$. It follows that

$$R(\mu+\nu)\mathcal{I}_{\mu}^{\Lambda}\subseteq\mathcal{I}_{\mu+\nu}^{\Lambda}, \tag{5.15}$$

and hence

$$\operatorname{Ind}\mathcal{I}_{\mu}^{\Lambda}C\boxtimes N\subseteq\mathcal{I}_{\mu+\nu}^{\Lambda}(\operatorname{Ind}C\boxtimes N). \tag{5.16}$$

This tells us that the composition $\beta\circ\alpha$ from the diagram in (5.14) is zero, so there exists a surjective homomorphism $g:\operatorname{Ind}M\boxtimes N\to\mathrm{pr}_\Lambda\operatorname{Ind}C\boxtimes N$. Applying $\mathrm{pr}_\Lambda$ to the diagram in (5.14), and denoting the resulting maps from $\gamma$, $\beta$, and $g$ as $\tilde{\gamma}$, $\tilde{\beta}$, and $\tilde{g}$ respectively, right exactness yields $\tilde{\gamma}$, $\tilde{\beta}$, and $\tilde{g}$ are surjections as shown in (5.17). It follows from considerations of dimension and that $\mathrm{pr}_\Lambda C = M$ that $\tilde{g}$ must be an isomorphism.

$$\begin{array}{ccccc} \mathrm{pr}_\Lambda(\mathrm{Ind}\, C \boxtimes N) & \xrightarrow{\tilde{\beta}} & \mathrm{pr}_\Lambda(\mathrm{Ind}\, C \boxtimes N) & \longrightarrow & \mathbf{0} \\ \tilde{\gamma}\downarrow & \nearrow \tilde{g} & & & \\ \mathrm{pr}_\Lambda(\mathrm{Ind}\, M \boxtimes N) & & & & \\ \downarrow & & & & \\ \mathbf{0} & & & & \end{array} \tag{5.17}$$

□

**5.2. Applying Proposition 5.5 to $T_{i;k}$.** We will frequently need to compute $\mathrm{jump}_j$ for the 1-dimensional "trivial" $R(\gamma_{i;k}^+)$-module $T_{i;k}$. When $k = 0$, we compute for the unit module that $\mathrm{jump}_j(\mathbb{1}) = 0$ but $\varphi_j^{\Lambda_i}(\mathbb{1}) = \delta_{ij}$. When $k \geq 1$,

$$\mathrm{wt}_j(T_{i;k}) = -\langle h_j, \gamma_{k;i}^+ \rangle = \delta_{j,i-1} - \delta_{j,i} + \delta_{j,i+k} - \delta_{j,i+k-1}. \tag{5.18}$$

Note that here as elsewhere, the indices $p, q$ in $\delta_{p,q}$ should be taken modulo $\ell$. Then,

$$\mathrm{jump}_j(T(i, i+1, \ldots, i+k-1)) = \delta_{j,i-1} + \delta_{j,i+k}. \tag{5.19}$$

Similarly,

$$\varphi_j^{\Lambda_i}(T(i, i+1, \ldots, i+k-1)) = \delta_{j,i-1} + \delta_{j,i+k}. \tag{5.20}$$

Here we record some useful facts concerning the way that the modules $T_{0;k}$ interact with the functors induction and $\mathrm{pr}_{\Lambda_0}$. Notice that all these facts hold for $T_{i;k}$ and $\mathrm{pr}_{\Lambda_i}$ after making obvious modifications.

PROPOSITION 5.6. Fix $k \in \mathbb{N}$, $k > 0$.

1. If $j \not\equiv -1, k$ then $\mathrm{pr}_{\Lambda_0} \mathrm{Ind}\, T(0, 1, \ldots, k-1) \boxtimes L(j) = \mathbf{0}$.
2. If $k \not\equiv -1$ then $\mathrm{pr}_{\Lambda_0} \mathrm{Ind}\, T(0, 1, \ldots, k-1) \boxtimes L(k) \cong T(0, 1, \ldots, k)$.
3. If $k \not\equiv -1$ then $\mathrm{pr}_{\Lambda_0} \mathrm{Ind}\, T(0, 1, \ldots, k-1) \boxtimes L(k) \boxtimes L(k) = \mathbf{0}$.
4. $\mathrm{pr}_{\Lambda_0} \mathrm{Ind}\, T_{0;k} \boxtimes L(-1) \cong \left(\mathrm{Ind}\, T_{0;k} \boxtimes L(-1)\right) \big/ T(-1, 0, \ldots, k-1)$ and further
   (a) If $k \not\equiv -1$ then $\mathrm{pr}_{\Lambda_0} \mathrm{Ind}\, T_{0;k} \boxtimes L(-1)$ is irreducible and
   $$\mathrm{pr}_{\Lambda_0} \mathrm{Ind}\, T_{0;k} \boxtimes L(-1) \boxtimes L(-1) = \mathbf{0}.$$
   (b) If $k \equiv -1$, then
   $$\mathrm{pr}_{\Lambda_0} \mathrm{Ind}\, T_{0;k} \boxtimes L(-1) \boxtimes L(-1) \cong \widetilde{f}_{-1}^2 T_{0;k}$$
   $$\mathrm{pr}_{\Lambda_0} \mathrm{Ind}\, T_{0;k} \boxtimes L(-1) \boxtimes L(-1) \boxtimes L(-1) = \mathbf{0}.$$
   Further, if $k > 1$ then $\mathrm{pr}_{\Lambda_0} \mathrm{Ind}\, T_{0;k} \boxtimes L(-1)$ has two composition factors. But if $k = 1$ (so $\ell = 2$) then $\mathrm{pr}_{\Lambda_0} \mathrm{Ind}\, T_{0;k} \boxtimes L(-1) = T(0,1)$ is irreducible.

PROOF. Note the hypothesis $k > 0$ implies $T_{0;k} \neq \mathbb{1}$, i.e. $\nu \neq 0$.
Below we write $=$ for $\equiv \bmod \ell$ or equality in $I$.

1. When $j \neq -1, k$, formula (5.20) gives
$$\varphi_j^{\Lambda_0}(T(0,1,\dots,k-1)) = 0.$$
Hence by Proposition 5.5.c
$$\mathrm{pr}_{\Lambda_0} \operatorname{Ind} T(0,1,\dots,k-1) \boxtimes L(j) = \mathbf{0}. \tag{5.21}$$
2. When $k \neq -1$, formula (5.20) gives
$$\varphi_k^{\Lambda_0}(T(0,1,\dots,k-1)) = 1.$$
Hence by Proposition 5.5.d
$$\mathrm{pr}_{\Lambda_0} \operatorname{Ind} T(0,1,\dots,k-1) \boxtimes L(k) \cong \widetilde{f}_k T(0,1,\dots,k-1) \tag{5.22}$$
$$\cong T(0,1,\dots,k-1,k), \tag{5.23}$$
where the second isomorphism holds by Frobenius reciprocity and the irreducibility of $\widetilde{f}_k T_{0;k}$.
3. As noted above
$$\varphi_k^{\Lambda_0}(T(0,\dots,k-1)) = 1.$$
Proposition 5.5.c then implies,
$$\mathrm{pr}_{\Lambda_0} \operatorname{Ind} T(0,\dots,k-1) \boxtimes L(k) \boxtimes L(k) = \mathbf{0}. \tag{5.24}$$
4. Let $v \otimes u$ span the 1-dimensional module $T_{0;k} \boxtimes L(-1)$. Then as in (4.32), $M := \operatorname{Ind} T_{0;k} \boxtimes L(-1)$ has basis
$$\{\ 1_{\underline{\mathbf{i}}} \otimes (v \otimes u),\ \psi_k 1_{\underline{\mathbf{i}}} \otimes (v \otimes u),\ \dots,\ \psi_1 \cdots \psi_{k-1}\psi_k 1_{\underline{\mathbf{i}}} \otimes (v \otimes u)\ \}$$
where $\underline{\mathbf{i}} = (0,1,\dots,k-1,-1)$.

Recall that $\mathcal{I}^{\Lambda_0}$ is generated by $x_1 1_{\underline{\mathbf{j}}}$ where $j_1 = 0$ and by $1_{\underline{\mathbf{p}}}$ where $p_1 \neq 0$. In particular, for $\underline{\mathbf{p}} = (-1,0,\dots,k-1)$ we see
$$1_{\underline{\mathbf{p}}}\Big(\psi_1 \dots \psi_k 1_{\underline{\mathbf{i}}} \otimes (v \otimes u)\Big) = \psi_1 \dots \psi_k 1_{\underline{\mathbf{i}}} \otimes (v \otimes u) \in \mathcal{I}^{\Lambda_0} M,$$
but $1_{\underline{\mathbf{p}}}(\psi_r \dots \psi_{k-1}\psi_k 1_{\underline{\mathbf{i}}} \otimes (v \otimes u)) = 0$ for $r > 1$. Also note $1_{\underline{\mathbf{j}}} 1_{\underline{\mathbf{p}}} = 0$. We further calculate
$$x_1 1_{\underline{\mathbf{j}}}\Big(\psi_r \dots \psi_r 1_{\underline{\mathbf{i}}} \otimes (v \otimes u)\Big) = 1_{\underline{\mathbf{j}}} \psi_r \dots \psi_k 1_{\underline{\mathbf{i}}} \otimes (x_1 v \otimes u) = 0$$
whenever $r > 1$. Hence $\mathcal{I}^{\Lambda_0} M$ is spanned by $\psi_1 \dots \psi_k 1_{\underline{\mathbf{i}}} \otimes (v \otimes u)$ and so we conclude that $\mathrm{pr}_{\Lambda_0} M = M/T(-1,0,\dots,k-1)$ as stated.

4(a) Suppose $k \neq -1$. As $\varphi_{-1}^{\Lambda_0}(T_{0;k}) = 1$ by (5.20), Proposition 5.5.d tells us that $\mathrm{pr}_{\Lambda_0} M = \widetilde{f}_{-1} T_{0;k}$ is irreducible. In particular it has dimension $k$. Using the Shuffle Lemma (along with the calculation of $\mathcal{I}^{\Lambda_0} M$ above), one could easily compute its character. By Proposition 5.5.c we see $\mathrm{pr}_{\Lambda_0} \operatorname{Ind} T_{0;k} \boxtimes L(-1) \boxtimes L(-1) = \mathbf{0}$.

4(b) Next suppose $k = -1$ (i.e. $k \equiv -1 \bmod \ell$) and $k > 1$ (when considering $k \in \mathbb{N}$). Then (5.20) yields $\varphi_{-1}^{\Lambda_0}(T_{0;k}) = 2$ so Proposition 5.5.d immediately gives (4b). Further, it is easy to see $\widetilde{f}_{-1} T_{0;k} = \widetilde{f}_k T_{0;k} = T_{0;k+1}$ which is 1-dimensional so from above $\mathrm{pr}_{\Lambda_0} M$ has at least 2 composition factors when $k > 1$. Next, the Shuffle Lemma and Serre relations (4.52) tell us the $(k-1)$-dimensional subquotient, corresponding to the span of
$$\{\psi_k \otimes (v \otimes u), \dots, \psi_2 \cdots \psi_k \otimes (v \otimes u)\},$$
is irreducible.

The remaining case is $k \equiv -1 \bmod \ell$, $k = 1$, forcing $\ell = 2$, and also $T_{0;k} = L(0)$. As above $\varphi_{-1}^{\Lambda_0}(T_{0;1}) = 2$, yielding (4b). The only difference is that $\mathrm{pr}_{\Lambda_0} M$ has only 1 composition factor (namely $T_{0;k+1}$) as it is only 1-dimensional. □

## 6. Main theorems

As remarked in Section 5.0.1, the graph with nodes corresponding to isomorphism classes $[M]$ for $M$ a simple $R^{\Lambda_i}(\nu)$-module and arrows $[\widetilde{e}_j M] \xrightarrow{j} [M]$ is the crystal graph $B(\Lambda_i)$. We can also use $\ell$-restricted partitions $\lambda$ to label the nodes of $B(\Lambda_i)$ as $[M^\lambda]$. The main theorems show for the isomorphism $\mathcal{T} : B(\Lambda_i) \xrightarrow{\cong} \mathcal{B} \otimes B(\Lambda_{i-1})$ that $\textcircled{k-1+i} \otimes \mu = \mathcal{T}(\lambda)$ corresponds to

$$\operatorname{Ind} T_{i;k} \boxtimes [M^\mu] \twoheadrightarrow [M^\lambda]$$

for $k = r(M^\lambda)$ (defined below), and that the crystal operators commute with this surjection in the appropriate manner.

Another way to view the theorems is that they give a module-theoretic construction of $\mathcal{T}$ and justify it is an isomorphism of crystals.

THEOREM 6.1. *Let $A$ be a simple $R(\nu)$-module in $\mathrm{Rep}^{\Lambda_i}$ with $|\nu| \geq 1$.*

*(1) There exists $k \in \mathbb{N}, k \geq 1$ such that $\widetilde{e}^\vee_{i+k-1} \cdots \widetilde{e}^\vee_{i+1} \widetilde{e}^\vee_i A$ is a simple $R(\nu - \gamma^+_{i;k})$-module in $\mathrm{Rep}^{\Lambda_{i-1}}$.*

*(2) Let*

$$r(A) = k$$

*be the minimal $k$ such that statement (1) holds and let*

$$\mathcal{R}(A) = \widetilde{e}^\vee_{i+k-1} \cdots \widetilde{e}^\vee_{i+1} \widetilde{e}^\vee_i A.$$

*Then there exists a surjection*

$$\mathrm{pr}_{\Lambda_i} \operatorname{Ind} T(i, i+1, \ldots, i+k-1) \boxtimes \mathcal{R}(A) \twoheadrightarrow A. \tag{6.1}$$

PROOF. For ease of exposition, we set $i = 0$ in the proof. For $t \in \mathbb{N}$ set $\mathcal{R}_0(A) = A$ and let

$$\mathcal{R}_t(A) = \widetilde{e}^\vee_{t-1} \cdots \widetilde{e}^\vee_1 \widetilde{e}^\vee_0 A. \tag{6.2}$$

We show by induction on $t \leq r(A)$ that $\mathcal{R}_t(A) \in \mathrm{Rep}^{\Lambda_t + \Lambda_{-1}}$ and there exists a surjection

$$\operatorname{Ind} T(0, 1, \ldots, t-1) \boxtimes \mathcal{R}_t(A) \twoheadrightarrow A. \tag{6.3}$$

In the base case $t = 0$, $\mathcal{R}_0(A) = A$. If $|\nu| = 1$, then $A = L(0)$ and $\widetilde{e}^\vee_0 L(0) \cong \mathbb{1} \in \mathrm{Rep}^{\Lambda_{-1}}$, so $r(A) = 1$. The existence of the surjection in this case is vacuous. Assume that $|\nu| > 1$. Then $\mathcal{R}_1(A) = \widetilde{e}^\vee_0 A \neq \mathbf{0}, \mathbb{1}$. By Proposition 4.5.2 there is a surjection

$$\operatorname{Ind} L(0) \boxtimes \mathcal{R}_1(A) \twoheadrightarrow A. \tag{6.4}$$

It follows directly from Proposition 5.6 and Proposition 5.5.b that for $t \geq 1$ if $\mathrm{pr}_{\Lambda_0} \operatorname{Ind} T_{0;t} \boxtimes D \twoheadrightarrow A$ then $D \in \mathrm{Rep}^{\Lambda_t + \Lambda_{-1}}$.

In more detail, Proposition 5.2 implies $D \in \mathrm{Rep}^\Lambda$ where $\Lambda = \sum \varepsilon^\vee_i(D)\Lambda_i \in P^+$. Proposition 5.5.b tells us $\mathrm{pr}_{\Lambda_0} \operatorname{Ind} T_{0;t} \boxtimes D \neq 0$ implies $\mathrm{pr}_{\Lambda_0} \operatorname{Ind} T_{0;t} \boxtimes L(i^{\varepsilon^\vee_i(D)}) \neq 0$. Thus for $i \neq -1, t$ we have $\varepsilon^\vee_i(D) = 0$ by Proposition 5.6.1.

If $t \neq -1$, Proposition 5.6.4a implies $\varepsilon_{-1}^{\vee}(D) \leq 1$ and Proposition 5.6.3 implies $\varepsilon_t^{\vee}(D) \leq 1$ so $D \in \mathrm{Rep}^{\Lambda_t+\Lambda_{-1}}$. If $t = -1$ then Proposition 5.6.4b implies $\varepsilon_{-1}^{\vee}(D) \leq 2$ so $D \in \mathrm{Rep}^{2\Lambda_{-1}} = \mathrm{Rep}^{\Lambda_t+\Lambda_{-1}}$.

Since $\mathrm{pr}_{\Lambda_0} A = A$, observe any surjection $M \twoheadrightarrow A$ factors through $M \twoheadrightarrow \mathrm{pr}_{\Lambda_0} M \twoheadrightarrow A$. Assume our inductive hypothesis (6.3) holds. Then from above, $\mathcal{R}_t(A) \in \mathrm{Rep}^{\Lambda_t+\Lambda_{-1}}$. If in fact $\mathcal{R}_t(A) \in \mathrm{Rep}^{\Lambda_{-1}}$then we are done (and $t \geq r(A)$). If not, then $\mathcal{R}_{t+1}(A) = \widetilde{e}_t^{\vee}\mathcal{R}_t(A) \neq 0$.

Transitivity and exactness of induction give us a surjection

$$\mathrm{Ind}\, T(0, \ldots, t-1) \boxtimes L(t) \boxtimes \mathcal{R}_{t+1}(A) \twoheadrightarrow A. \tag{6.5}$$

In the first case, suppose $t \neq -1$. Then Proposition 5.5.e and Proposition 5.6.2 imply

$$\mathrm{pr}_{\Lambda_0} \mathrm{Ind}\, T_{0;t} \boxtimes L(t) \boxtimes \mathcal{R}_{t+1}(A) \cong \mathrm{pr}_{\Lambda_0} \mathrm{Ind}\, T_{0;t+1} \boxtimes \mathcal{R}_{t+1}(A) \tag{6.6}$$

and we get

$$\mathrm{pr}_{\Lambda_0} \mathrm{Ind}\, T(0, \ldots, t-1, t) \boxtimes \mathcal{R}_{t+1}(A) \twoheadrightarrow A. \tag{6.7}$$

In the case $t = -1$ then by the inductive hypothesis $\mathcal{R}_t(A) \in \mathrm{Rep}^{2\Lambda_{-1}}$ and $\mathcal{R}_t(A) \notin \mathrm{Rep}^{\Lambda_{-1}}$ as we are assuming $t < r(A)$. Thus $\varepsilon_{-1}^{\vee}(\widetilde{e}_{-1}^{\vee}\mathcal{R}_t(A)) = \varepsilon_{-1}^{\vee}(\mathcal{R}_{t+1}(A)) = 1$. If $K$ is any composition factor of $\mathrm{Ind}\, T_{0;t} \boxtimes L(-1)$ other than $T_{0;t+1}$ then $\varphi_{-1}^{\Lambda_0}(K) \leq 0$ by (5.3) so $\mathrm{pr}_{\Lambda_0} \mathrm{Ind}\, K \boxtimes L(-1) = \mathbf{0}$ which implies $\mathrm{pr}_{\Lambda_0} \mathrm{Ind}\, K \boxtimes \mathcal{R}_{t+1}(A) = \mathbf{0}$. So (6.5) must factor through

$$\mathrm{Ind}\, T(0, \ldots, -2, -1) \boxtimes \mathcal{R}_{t+1}(A) \twoheadrightarrow A. \tag{6.8}$$

This completes the induction.

We take $r(A)$ to be the smallest $k$ such that $\mathcal{R}_k(A) \in \mathrm{Rep}^{\Lambda_{-1}}$. Note that the process above must terminate as $r(A) \leq |\nu|$. In fact, in the case $r(A) = |\nu|$ we must have $A = T_{0;|\nu|}$ and $\mathcal{R}_{|\nu|}(A) = \mathcal{R}(A) = \mathbb{1} \in \mathrm{Rep}^{\Lambda_{-1}}$. □

By considering sign in place of trivial modules, a very similar proof yields the following theorem.

THEOREM 6.2. *Let $A$ be a simple $R(\nu)$-module in $\mathrm{Rep}^{\Lambda_i}$ with $|\nu| \geq 1$.*

*(1) There exists $k \in \mathbb{N}, k \geq 1$ such that $\mathbf{0} \neq \widetilde{e}_{i-k+1}^{\vee} \cdots \widetilde{e}_{i-1}^{\vee}\widetilde{e}_i^{\vee} A$ is a simple $R(\nu - \gamma_{i;k}^{-})$-module in $\mathrm{Rep}^{\Lambda_{i+1}}$.*

*(2) Let $c(A) = k$ be the minimal $k$ such that* (1) *holds and let $\mathcal{C}(A) = \widetilde{e}_{i-k+1}^{\vee} \cdots \widetilde{e}_{i-1}^{\vee}\widetilde{e}_i^{\vee} A$. Then there exists a surjection*

$$\mathrm{pr}_{\Lambda_i} \mathrm{Ind}\, S(i, i-1, \ldots, i-k+1) \boxtimes \mathcal{C}(A) \twoheadrightarrow A. \tag{6.9}$$

CONJECTURE 6.3. With hypotheses as above,

$$A = \mathrm{cosoc}\, \mathrm{pr}_{\Lambda_i} \mathrm{Ind}\, T_{i;r(A)} \boxtimes \mathcal{R}(A),$$
$$A = \mathrm{cosoc}\, \mathrm{pr}_{\Lambda_i} \mathrm{Ind}\, S_{i;c(A)} \boxtimes \mathcal{C}(A).$$

**6.1. Relation to Specht modules.** A Specht module for $\mathcal{S}_n$ is constructed as a submodule of the induction of a trivial module from a Young subgroup $\mathcal{S}_\lambda$ (this is one of our $\mathcal{S}_P$ as in (4.31)). Specht modules can also be constructed for the Hecke algebra of type $A$ as in [**7**]. They are equipped with an integral form that allows one to specialize the Specht modules over $\mathbb{F}_\ell$ in the former case, to an $\ell$-th root of unity in the latter.

When $\lambda$ is $\ell$-regular, the specialization of the Specht module $\bar{S}^\lambda$ has unique simple quotient $D^\lambda$. In other words, $D^\lambda$ is a subquotient of a module induced from a 1-dimensional module. Further $\{D^\lambda \mid \lambda \vdash n, \lambda \text{ is } \ell\text{-regular}\}$ is a complete set of simple modules of $\mathbb{F}_\ell \mathcal{S}_n$ or the finite Hecke algebra at an $\ell$-th root of unity. The crystal structure on these simples by taking socle of restriction agrees with the model of $B(\Lambda_0)$ taking nodes to be $\ell$-regular partitions [**18**]. This is the model compatible with tensoring by $\mathcal{B}^{\mathrm{opp}}$. (See Section 3.)

Repeating the construction of Theorem 6.2 yields $D^\lambda$ as the quotient of a module induced from a (possibly conjugate) parabolic subalgebra of shape $\lambda^T$, where the module being induced is a (parabolic) sign module. In other words, the restriction of $D^\lambda$ to that parabolic subalgebra contains a $\boxtimes$ of sign modules $S_{i;k}$. On the other hand, in the construction of Specht modules for the finite Hecke algebra of type $A$ given in [**7**], the Specht module contains a special vector that is anti-symmetrized according to a parabolic subalgebra of shape $\lambda^T$. In other words, the same induced module that has $D^\lambda$ as a quotient also has a nonzero map to $S^\lambda$.

In fact $\mathbb{Q} \otimes_{\mathbb{Z}} S^\lambda$ can be characterized as the unique irreducible $\mathbb{Q}\mathcal{S}_n$-module such that $\mathrm{Res}_{\mathcal{S}_\lambda}$ contains a trivial module and $\mathrm{Res}_{\mathcal{S}_{\lambda^T}}$ contains a sign module.

When $\ell$ is a root of unity (or we work over $\mathbb{F}_\ell$) the difficulty is in specializing quotients of (induced) modules. The existence of a map from and induced sign module to $D^\lambda$ is not a surprise, but the result on how the crystal operators act is nontrivial.

**6.2. The action of crystal operators $\widetilde{f}_j$ and $\widetilde{e}_j$.** Next we study the action of the crystal operators $\widetilde{e}_j$ and $\widetilde{f}_j$ to show (6.1) categorifies our crystal isomorphism $\mathcal{T}$. We refer the reader back to Section 3.

Compare the theorems below with (2.6) and (2.7). As in [**21**] simple modules correspond to nodes in $B(\Lambda_i)$. Each node of the perfect crystal $\mathcal{B}$ (respectively $\mathcal{B}^{\mathrm{opp}}$) corresponds to a family of trivial (respectively sign) modules $T_{i;k+t\ell}, t \in \mathbb{N}$. (However this does not give a categorification of $\mathcal{B}$ itself.) It is in this manner that the main theorems of this paper give a categorification of the crystal isomorphism $\mathcal{T}$ (resp. $\mathcal{T}^{\mathrm{opp}}$).

THEOREM 6.4. *Let $A \in \mathrm{Rep}^{\Lambda_i}$ be simple. Let $j \in I$ be such that $\widetilde{e}_j A \neq 0$, and let $k = r(A)$. Then there exists a surjection*

$$\begin{aligned} &\mathrm{Ind}\left(\widetilde{e}_j T_{i;k} \boxtimes \mathcal{R}(A)\right) \twoheadrightarrow \widetilde{e}_j A \qquad \textit{if} \quad \varepsilon_j(T_{i;k}) > \varphi_j^{\Lambda_{i-1}}(\mathcal{R}(A)) \\ &\mathrm{Ind}\left(T_{i;k} \boxtimes \widetilde{e}_j \mathcal{R}(A)\right) \twoheadrightarrow \widetilde{e}_j A \qquad \textit{if} \quad \varepsilon_j(T_{i;k}) \leq \varphi_j^{\Lambda_{i-1}}(\mathcal{R}(A)). \end{aligned} \tag{6.10}$$

THEOREM 6.5. *Let $A \in \mathrm{Rep}^{\Lambda_i}$ be simple. Let $j \in I$ be such that $\mathrm{pr}_{\Lambda_i} \widetilde{f}_j A \neq 0$, and let $k = r(A)$. Then there exists a surjection*

$$\begin{aligned} &\mathrm{Ind}\left(\widetilde{f}_j T_{i;k} \boxtimes \mathcal{R}(A)\right) \twoheadrightarrow \widetilde{f}_j A \qquad \textit{if} \quad \varepsilon_j(T_{i;k}) \geq \varphi_j^{\Lambda_{i-1}}(\mathcal{R}(A)) \\ &\mathrm{Ind}\left(T_{i;k} \boxtimes \widetilde{f}_j \mathcal{R}(A)\right) \twoheadrightarrow \widetilde{f}_j A \qquad \textit{if} \quad \varepsilon_j(T_{i;k}) < \varphi_j^{\Lambda_{i-1}}(\mathcal{R}(A)). \end{aligned} \tag{6.11}$$

Theorem 6.4 follows directly from Theorem 6.5, therefore will only prove the latter. Similar theorems hold using sign modules and $\mathcal{C}(A)$.

Before doing this, we need to establish several lemmas.

PROPOSITION 6.6. [**27**] Let $m = |\nu|$. Let $M$ be a simple $R(\nu)$-module. If $M \in \mathrm{Rep}^{\Lambda_i}$ and $\mathbf{0} \neq \widetilde{e}_i^\vee(M) \in \mathrm{Rep}^{\Lambda_{i+1}}$ then $M = T_{i;m}$.

PROOF. This can be directly adapted from Theorem 3.7 of [**27**], replacing $\widetilde{e}_i$ with $\widetilde{e}_i^\vee$ and noting $\varepsilon_j^\vee(M) = \delta_{i,j}$, $\varepsilon_j^\vee(\widetilde{e}_i^\vee M) = \delta_{i+1,j}$ (assuming $m \geq 2$). It was proved in the context of $B(\Lambda_i)$, hence holds for $\mathrm{Rep}^{\Lambda_i}$ by [**21**]. □

PROPOSITION 6.7. Let $A$, $\mathcal{R}_t(A)$ be as in (6.2) and $m = |\nu|$. If there exists $1 \leq t < r(A)$ with $\mathcal{R}_t(A) \in \mathrm{Rep}^{\Lambda_{i+t}}$, then in fact $\mathcal{R}_j(A) \in \mathrm{Rep}^{\Lambda_{i+j}}$ for all $1 \leq j \leq r(A)$, $r(A) = \min\{\ell - 1, m\}$, and $A = T_{i;m}$.

PROOF. As usual, we set $i = 0$ for ease of exposition. We have already shown $\mathcal{R}_t(A) \in \mathrm{Rep}^{\Lambda_t + \Lambda_{-1}}$. Given $A = \mathcal{R}_0(A) \in \mathrm{Rep}^{\Lambda_0}$, suppose $e_0^\vee A = \widetilde{e}_0^\vee A = \mathcal{R}_1(A) \in \mathrm{Rep}^{\Lambda_1}$ then by Proposition 6.6, $A = T_{0;m}$ and we are done. Further in the case $\ell = 2$ this means $r(A) = 1$. If $\ell > 2$ then $r(A) \leq \ell - 1$. Assume otherwise.

*Case 1*: $\ell \neq 2$. Then $\varepsilon_{-1}^\vee(\mathcal{R}_1(A)) = 1$. This means there is $[\underline{\mathrm{i}}] = [i_1, i_2, \ldots, i_m] \in \mathrm{supp}(A)$ with $i_1 = 0$, $i_2 = -1$. Recall we have

$$\mathrm{Ind}\, T_{0;t-1} \boxtimes \mathcal{R}_t(A) \twoheadrightarrow A. \tag{6.12}$$

By the Shuffle Lemma, the only way to have $[\underline{\mathrm{i}}] \in \mathrm{supp}(A)$ is if $\varepsilon_{-1}^\vee(\mathcal{R}_t(A)) \geq 1$.
*Case 2*: $\ell = 2$. Then $\varepsilon_{-1}^\vee(\mathcal{R}_1(A)) = 2$, as we assumed $r(A) > 1$. Then there is $[\underline{\mathrm{i}}] \in \mathrm{supp}(A)$ with $i_1 = 0$, $i_2 = -1$, $i_3 = -1$. Again, by the Shuffle Lemma, this is only possible if $\varepsilon_{-1}^\vee(\mathcal{R}_t(A)) \geq 1$.

Furthermore, when $t \equiv -1 \bmod \ell$ for $0 < t < r(A)$ we have $\varepsilon_{-1}^\vee(\mathcal{R}_t(A)) = 2$ by the minimality of $r(A)$. □

LEMMA 6.8. *Let $A$ be a simple $R^{\Lambda_0}(\nu)$-module with $k = r(A)$, $\mathbb{1} \neq \mathcal{R}(A) \in \mathrm{Rep}^{\Lambda_{-1}}$. Fix $j \in I$. Let $J = jump_j(\mathcal{R}(A))$. If $k \neq j+1$ then for $t \in \mathbb{N}$, $0 \leq t \leq k$,*

$$jump_j(\mathcal{R}_t(A)) = \begin{cases} J & t \not\equiv j+1 \bmod \ell \\ J+1 & t \equiv j+1 \bmod \ell. \end{cases} \tag{6.13}$$

*If $k = j+1$ and $J \neq 0$, then for $0 \leq t \leq k$,*

$$jump_j(\mathcal{R}_t(A)) = \begin{cases} J-1 & t \not\equiv j+1 \bmod \ell \\ J & t \equiv j+1 \bmod \ell. \end{cases} \tag{6.14}$$

*If $k = j+1$ and $J = 0$, then for $0 < t < k$*

$$jump_j(\mathcal{R}_t(A)) = \begin{cases} 0 & t \not\equiv j+1 \bmod \ell \\ 1 & t \equiv j+1 \bmod \ell. \end{cases} \tag{6.15}$$

*and $jump_j(A) = 0$.*

PROOF. We will first prove the lemma in the case $A = T_{0;m}$, where $m = |\nu|$. Then $k \leq \ell - 1$. For $0 \leq t \leq k$ we have $\mathcal{R}_t(A) = T(t, t+1, \ldots, m-1)$. From (5.19), $\mathrm{jump}_j(\mathcal{R}_t(A)) = \delta_{j,t-1} + \delta_{j,m}$ (recalling none of these modules are $\mathbb{1}$ by hypothesis). One can easily check the Lemma holds.

From now on, we assume $A$ is not a trivial module.

We now continue with the third case. Suppose $k = j+1$ and $J = 0$. Then $\mathcal{R}(A) = \mathcal{R}_k(A) \in \mathrm{Rep}^{\Lambda_{-1}}$, so $\varepsilon_j^\vee(\mathcal{R}(A)) = \delta_{j,-1}$. $\mathcal{R}_{k-1}(A) = \widetilde{f}_j^\vee \mathcal{R}(A) = \widetilde{f}_j \mathcal{R}(A)$ so $\varepsilon_j^\vee(\mathcal{R}_{k-1}(A)) = \varepsilon_j^\vee(\mathcal{R}(A)) + 1$ and $\varepsilon_j(\mathcal{R}_{k-1}(A)) = \varepsilon_j(\mathcal{R}(A)) + 1$. In particular $\mathcal{R}_t(A) \in \mathrm{Rep}^{\Lambda_t + \Lambda_{-1}}$ but we may assume $\mathcal{R}_t(A) \notin \mathrm{Rep}^{\Lambda_t}$ or else by [**27**] this would force $\mathcal{R}(A)$ and $A$ itself to be trivial. Further $\mathrm{wt}_j(\mathcal{R}_{k-1}(A)) = \mathrm{wt}_j(\widetilde{f}_j^\vee \mathcal{R}(A)) = \mathrm{wt}_j(\mathcal{R}(A)) - 2$. Hence

$$\begin{aligned}\mathrm{jump}_j(\mathcal{R}_{k-1}(A)) &= \varepsilon_j^\vee(\mathcal{R}_{k-1}(A)) + \varepsilon_j(\mathcal{R}_{k-1}(A)) + \mathrm{wt}_j(\mathcal{R}_{k-1}(A)) \\ &= \varepsilon_j^\vee(\mathcal{R}(A)) + 1 + \varepsilon_j(\mathcal{R}(A)) + 1 + \mathrm{wt}_j(\mathcal{R}(A)) - 2 \\ &= 0.\end{aligned} \tag{6.16}$$

For $\mathcal{R}_{k-2}(A)$,

$$\varepsilon_j^\vee(\mathcal{R}_{k-2}(A)) = \delta_{j,k-2} + \delta_{j,-1} = \delta_{j,-1} = \varepsilon_j^\vee(\mathcal{R}(A)). \tag{6.17}$$

Also

$$\varepsilon_j(\mathcal{R}_{k-2}(A)) = \varepsilon_j(\widetilde{f}_{j-1}^\vee \mathcal{R}_{k-1}(A)) = \varepsilon_j(\mathcal{R}_{k-1}(A)) \tag{6.18}$$

by Remark 4.12.

If $k-2 \neq j+1$ (i.e. $\ell \neq 2$) then

$$\begin{aligned}\mathrm{jump}_j(\mathcal{R}_{k-2}(A)) &= \varepsilon_j^\vee(\mathcal{R}(A)) + \varepsilon_j(\mathcal{R}(A)) + 1 + \mathrm{wt}_j(\mathcal{R}(A)) - 2 + 1 \\ &= 0.\end{aligned} \tag{6.19}$$

Since $\varepsilon_j(\widetilde{f}_i^\vee B) = \varepsilon_j(B)$, $\varepsilon_j^\vee(\widetilde{f}_i^\vee B) = \varepsilon_j^\vee(B)$, and $\mathrm{wt}_j(\widetilde{f}_i^\vee B) = \mathrm{wt}_j(B)$ when $i \notin \{j-1, j, j+1\}$, similar computations show $\mathrm{jump}_j(\mathcal{R}_t(A)) = 0$ for $k - \ell < t \leq k$.

If $k - \ell > 0$, we check

(6.20)

$$\begin{aligned}\mathrm{jump}_j(\mathcal{R}_{k-\ell}(A)) &= \delta_{j,j+1} + \delta_{j,-1} + \varepsilon_j(\widetilde{f}_{j+1}^\vee \mathcal{R}_{k-\ell+1}(A)) + \mathrm{wt}_j(\widetilde{f}_{j+1}^\vee \mathcal{R}_{k-\ell+1}(A)) \\ &= \mathrm{jump}_j(\mathcal{R}_{k-\ell+1}(A)) + 1 = 1.\end{aligned}$$

(Note that when $\ell = 2$, $\varepsilon_j^\vee(\mathcal{R}_{k-\ell}(A)) = \varepsilon_j^\vee(\mathcal{R}_{k-\ell+1}(A)) - 1$ but the equality still holds as $\mathrm{wt}_j(\widetilde{f}_{j+1}^\vee \mathcal{R}_{k-\ell+1}(A)) = \mathrm{wt}_j(\mathcal{R}_{k-\ell+1}(A)) + 2$.)

Next $\mathrm{jump}_j(\mathcal{R}_{k-\ell-1}(A)) = \mathrm{jump}_j(\widetilde{f}_j^\vee \mathcal{R}_{k-\ell}(A)) = 1 - 1 = 0$. Now all other inductive computations for $\mathrm{jump}_j(\mathcal{R}_t(A))$ are identical to the above computations, down to $t = 0$, for which $\mathcal{R}_0(A) = A$. Here if $0 = t \equiv j+1 \bmod \ell$ then because $\varepsilon_j^\vee(A) = 0 \neq \delta_{j,-1} + \delta_{j,0}$ we instead get $\mathrm{jump}_j(A) = 0$.

The second case, $k = j+1$ but $J > 0$, is also very similar to the above. The only difference is that $\mathrm{jump}_j(\mathcal{R}(A)) \neq 0$ hence $\varepsilon_j(\mathcal{R}_{k-1}(A)) = \varepsilon_j(\widetilde{f}_j^\vee \mathcal{R}(A)) = \varepsilon_j(\mathcal{R}(A))$. Regardless $\mathrm{jump}_j(\mathcal{R}_{k-1}(A)) = \mathrm{jump}_j(\widetilde{f}_j^\vee \mathcal{R}(A)) = J - 1$. We check

$$\begin{aligned}\mathrm{jump}_j(\mathcal{R}_{k-2}(A)) &= \delta_{j,k-2} + \delta_{j,-1} + \varepsilon_j(\widetilde{f}_{j-1}^\vee \mathcal{R}_{k-1}(A)) + \mathrm{wt}_j(\widetilde{f}_{j-1}^\vee \widetilde{f}_j^\vee \mathcal{R}(A)) \\ &= 0 + \varepsilon_j^\vee(\mathcal{R}(A)) + \varepsilon_j(\mathcal{R}(A)) + \mathrm{wt}_j(\mathcal{R}(A)) - 2 - \langle h_j, \alpha_{j-1}\rangle \\ &= \begin{cases} J-1 & \text{if } \ell \neq 2 \\ J & \text{if } \ell = 2. \end{cases}\end{aligned}$$

Note in the case $\ell = 2$ that $k-2 \equiv j+1$, so this is consistent with the statement of the lemma.

The rest of the proof is identical to that in Case 1.

Finally we consider $k \not\equiv j+1 \bmod \ell$. Letting $j_0 \in \mathbb{Z}$, $j_0 < k$ be maximal such that $j_0 \equiv j \bmod \ell$, it is clear that $\mathrm{jump}_j(\mathcal{R}_t(A)) = \mathrm{jump}_j(\mathcal{R}(A))$ for all $t > j_0 + 1$.

Then

$$\begin{aligned}\mathrm{jump}_j(\mathcal{R}_{j_0+1}(A)) &= \delta_{j,j_0+1}+\delta_{j,-1}+\varepsilon_j(\widetilde{f}^\vee_{j+1}\mathcal{R}_{j_0+2}(A))+\mathrm{wt}_j(\widetilde{f}^\vee_{j+1}\mathcal{R}_{j_0+2}(A))\\ &= 0+\delta_{j,-1}+\varepsilon_j(\mathcal{R}_{j_0+2}(A))+\mathrm{wt}_j(\mathcal{R}_{j_0+2}(A))-\langle h_j,\alpha_{j+1}\rangle\\ &= \mathrm{jump}_j(\mathcal{R}_{j_0+2}(A))+1=J+1.\end{aligned}$$

In the case $\ell\neq 2$ this follows as $\delta_{j,-1}=\varepsilon^\vee_j(\mathcal{R}_{j_0+2}(A))$ and $\langle h_j,\alpha_{j+1}\rangle=-1$. In the case $\ell=2$, we have $\varepsilon^\vee_j(\mathcal{R}_{j_0+2}(A))=1+\delta_{j,-1}$ but $\langle h_j,\alpha_{i+1}\rangle=-2$.

Next $\mathrm{jump}_j(\mathcal{R}_{j_0}(A))=\mathrm{jump}_j(\widetilde{f}^\vee_j\mathcal{R}_{j_0+1}(A))=\mathrm{jump}_j(\mathcal{R}_{j_0+1}(A))-1=J$.

Also for $\ell\neq 2$,

$$\begin{aligned}\mathrm{jump}_j(\mathcal{R}_{j_0-1}(A)) &= \delta_{j,j-1}+\delta_{j,-1}+\varepsilon_j(\widetilde{f}^\vee_{j-1}\mathcal{R}_{j_0}(A))+\mathrm{wt}_j(\widetilde{f}^\vee_{j-1}\mathcal{R}_{j_0}(A))\\ &= \mathrm{jump}_j(\mathcal{R}_{j_0}(A))=J\end{aligned}$$

as $\varepsilon^\vee_j(\mathcal{R}_{j_0}(A))=1+\delta_{j,-1}$ and $\langle h_j,\alpha_{j-1}\rangle=-1$. We don't consider $\ell=2$ as $j_0-1\equiv j_0+1$ and we have already considered that case. We note that the calculations of $\mathrm{jump}_j(\mathcal{R}_t(A))$ only depend on $t \bmod \ell$ and so we are done. □

PROOF OF THEOREM 6.5. For ease of exposition we set $i=0$ in the proof. In fact we will prove a slightly stronger statement, that when $\varepsilon_j(T_{0;r(A)})<\varphi_j^{\Lambda_{-1}}(\mathcal{R}(A))$ and $\mathrm{pr}_{\Lambda_0}\widetilde{f}_jA\neq\mathbf{0}$ then $r(\widetilde{f}_jA)=r(A)$ and for $0<t\leq r(A)$, $\mathcal{R}_t(\widetilde{f}_jA)=\widetilde{f}_j\mathcal{R}_t(A)$.

First note that in the case $\mathcal{R}(A)=\mathbb{1}$ the theorem is obvious as $\widetilde{f}_j\mathcal{R}(A)=L(j)$ and $\widetilde{f}_jT_{0;k}=\operatorname{cosoc}\operatorname{Ind}T_{0;k}\boxtimes L(j)$. So from now on assume $\mathcal{R}(A)\neq\mathbb{1}$.

*Case 1*: Suppose $\varepsilon_j(T_{0;k})=0$. In particular, $j\neq k-1$.

- *Case 1a*: Suppose $\mathrm{jump}_j(\mathcal{R}(A))=0$. By Lemma 6.8,

$$\mathrm{jump}_j(A)=\begin{cases}0 & 0\not\equiv j+1 \bmod \ell,\\ 1 & 0\equiv j+1 \bmod \ell.\end{cases}\tag{6.21}$$

So when $j\neq -1$ as $\mathrm{jump}_j(A)=0$, $\mathrm{pr}_{\Lambda_0}\widetilde{f}_jA=\mathbf{0}$ and so we need not consider this case. Hence we may assume $j=-1$. Further, as $\mathrm{jump}_{-1}(\mathcal{R}(A))=0$ and $\varepsilon^\vee_{-1}(\mathcal{R}(A))=1$ we have

$$\begin{aligned}\operatorname{Ind}T_{0;k}\boxtimes L(-1)\boxtimes\mathcal{R}(A) &\xrightarrow{\cong}\operatorname{Ind}T_{0;k}\boxtimes\mathcal{R}(A)\boxtimes L(-1)\\ &\twoheadrightarrow\operatorname{Ind}A\boxtimes L(-1)\twoheadrightarrow\widetilde{f}_{-1}A.\end{aligned}$$

This implies $\mathrm{pr}_{\Lambda_0}\operatorname{Ind}T_{0;k}\boxtimes L(-1)\boxtimes L(-1)\neq\mathbf{0}$, forcing $k\equiv -1 \bmod \ell$ by Propositions 5.4 and 5.5. But then we have

$$\operatorname{Ind}T(0,\ldots,-2)\boxtimes L(-1)\boxtimes\mathcal{R}(A)\twoheadrightarrow\widetilde{f}_{-1}A.$$

Because

$$\operatorname{Ind}T(0,\ldots,-2)\boxtimes\mathcal{R}(A)\twoheadrightarrow A$$

we see $\varepsilon_{-1}(\mathcal{R}(A))=\varepsilon_{-1}(A)=\varepsilon_{-1}(\widetilde{f}_{-1}A)-1$. If we had

$$\operatorname{Ind}N\boxtimes\mathcal{R}(A)\twoheadrightarrow\widetilde{f}_{-1}A$$

for any composition factor $N$ of $\operatorname{Ind}T_{0;k}\boxtimes L(-1)$ other than $\widetilde{f}_{-1}T_{0;k}$, the Shuffle Lemma would yield $\varepsilon_{-1}(\widetilde{f}_{-1}A)=\varepsilon_{-1}(\mathcal{R}(A))$, a contradiction. Hence we have

$$\operatorname{Ind}T_{0;k+1}\boxtimes\mathcal{R}(A)=\operatorname{Ind}\widetilde{f}_{-1}T_{0;k}\boxtimes\mathcal{R}(A)\twoheadrightarrow\widetilde{f}_{-1}A.\tag{6.22}$$

- *Case 1b*: Suppose that $\mathrm{jump}_j(\mathcal{R}(A)) = J > 0$. Again by Lemma 6.8

$$\mathrm{jump}_j(A) = \begin{cases} J & 0 \not\equiv j+1 \bmod \ell, \\ J+1 & 0 \equiv j+1 \not\equiv k \bmod \ell. \end{cases} \tag{6.23}$$

Note $\widetilde{f}_j\mathcal{R}(A) \in \mathrm{Rep}^{\Lambda_{-1}}$ as $\mathrm{jump}_j(\mathcal{R}(A)) > 0$.

We compute

$$\widetilde{f}_j\mathcal{R}_{k-1}(A) = \widetilde{f}_j\widetilde{f}^\vee_{k-1}\mathcal{R}(A) = \widetilde{f}^\vee_{k-1}\widetilde{f}_j\mathcal{R}(A) \tag{6.24}$$

as $j \neq k-1$ as in Case 1. Clearly $\widetilde{f}^\vee_{k-1}\widetilde{f}_j\mathcal{R}(A) \in \mathrm{Rep}^{\Lambda_{k-1}+\Lambda_{-1}}$ but $\widetilde{f}^\vee_{k-1}\widetilde{f}_j\mathcal{R}(A) \notin \mathrm{Rep}^{\Lambda_{-1}}$. Assume we have shown

$$\widetilde{f}_j\mathcal{R}_t(A) = \widetilde{f}^\vee_t \cdots \widetilde{f}^\vee_{k-1}\widetilde{f}_j\mathcal{R}(A) \in \mathrm{Rep}^{\Lambda_t+\Lambda_{-1}} \setminus \mathrm{Rep}^{\Lambda_{-1}}. \tag{6.25}$$

Then we compute $\widetilde{f}_j\mathcal{R}_{t-1}(A) = \widetilde{f}_j\widetilde{f}^\vee_{t-1}\mathcal{R}_t(A)$. If $t-1 \not\equiv j \bmod \ell$ this is equal to $\widetilde{f}^\vee_{t-1}\widetilde{f}_j\mathcal{R}_t(A) = \widetilde{f}^\vee_{t-1}\cdots\widetilde{f}^\vee_{k-1}\widetilde{f}_j\mathcal{R}(A)$ by the inductive hypothesis. If instead $t \equiv j+1 \bmod \ell$ then

$$\mathrm{jump}_j(\mathcal{R}_t(A)) = J+1 > 1. \tag{6.26}$$

So by Lemma 4.11.2

$$\widetilde{f}_j\mathcal{R}_{t-1}(A) = \widetilde{f}_j\widetilde{f}^\vee_j\mathcal{R}_t(A) = \widetilde{f}^\vee_j\widetilde{f}_j\mathcal{R}_t(A) = \widetilde{f}^\vee_j\widetilde{f}^\vee_t\cdots\widetilde{f}^\vee_{k-1}\widetilde{f}_j\mathcal{R}(A) \tag{6.27}$$

$$= \widetilde{f}^\vee_{t-1}\cdots\widetilde{f}^\vee_{k-1}\widetilde{f}_j\mathcal{R}(A). \tag{6.28}$$

By downwards induction $\widetilde{f}_jA = \widetilde{f}_j\mathcal{R}_0(A) = \widetilde{f}^\vee_0\cdots\widetilde{f}^\vee_{k-1}\widetilde{f}_j\mathcal{R}(A)$. Certainly, as $\mathrm{jump}_j(\mathcal{R}_{t-1}(A)) > 0$ and $\mathcal{R}_{t-1}(A) \in \mathrm{Rep}^{\Lambda_{t-1}+\Lambda_{-1}} \setminus \mathrm{Rep}^{\Lambda_{-1}}$ when $t-1 > 0$ we have $\widetilde{f}_j\mathcal{R}_{t-1}(A) \in \mathrm{Rep}^{\Lambda_{t-1}+\Lambda_{-1}} \setminus \mathrm{Rep}^{\Lambda_{-1}}$. And we already know $\widetilde{f}_jA \in \mathrm{Rep}^{\Lambda_0}$. This completes the induction and furthermore shows

$$r(\widetilde{f}_jA) = r(A), \qquad \mathcal{R}(\widetilde{f}_jA) = \widetilde{f}_j\mathcal{R}(A) \tag{6.29}$$

as well as the stronger statement $\mathcal{R}_t(\widetilde{f}_jA) = \widetilde{f}_j\mathcal{R}_t(A)$ for $0 \le t \le k$. In other words we have

$$\mathrm{Ind}\, T_{0;k} \boxtimes \widetilde{f}_j\mathcal{R}(A) \to \widetilde{f}_jA. \tag{6.30}$$

*Case 2*: Suppose that $\varepsilon_j(T_{0;k}) = 1$. This is the only other possibility as $\varepsilon_j(T_{0;k}) \le 1$ for all $j \in I$. Note $k \equiv j+1 \bmod \ell$.

- *Case 2a*: If $\varphi_j^{\Lambda_{-1}}(\mathcal{R}(A)) = 0$ then $\mathrm{jump}_j(A) = 0$ by Lemma 6.8 so $\mathrm{pr}_{\Lambda_0}\widetilde{f}_jA = 0$ and we need not consider this case.
- *Case 2b*: If $\varphi_j^{\Lambda_{-1}}(\mathcal{R}(A)) = 1$ then again by Lemma 6.8 as $\mathcal{R}(A) = \mathcal{R}_k(A)$ and $k \equiv j+1 \bmod \ell$

$$\mathrm{jump}_j(A) = \mathrm{jump}_j(\mathcal{R}_0(A)) = \begin{cases} 0 & 0 \not\equiv j+1 \bmod \ell \\ 1 & 0 \equiv j+1 \bmod \ell. \end{cases} \tag{6.31}$$

So we need not consider this case unless $j = -1$. However we will show, this case cannot arise as we assumed $\mathcal{R}(A) \neq \mathbb{1}$. Note that $\mathrm{jump}_j(\mathcal{R}_{k-1}(A)) = \mathrm{jump}_{-1}(\widetilde{f}^\vee_{-1}\mathcal{R}(A)) = 0$ and $\mathcal{R}_{k-1}(A) \in \mathrm{Rep}^{2\Lambda_{-1}}$. Thus we have

$$\mathrm{Ind}\, T_{0;k-1} \boxtimes L(-1) \boxtimes \mathcal{R}_{k-1}(A) \xrightarrow{\cong} \mathrm{Ind}\, T_{0;k-1} \boxtimes \mathcal{R}_{k-1}(A) \boxtimes L(-1)$$
$$\twoheadrightarrow \mathrm{Ind}\, A \boxtimes L(-1) \twoheadrightarrow \widetilde{f}_{-1}A.$$

However $\mathrm{pr}_{\Lambda_0} \operatorname{Ind} T(0, \dots, k-1) \boxtimes L(-1) \boxtimes L(-1) \boxtimes L(-1) = \mathbf{0}$ by Proposition 5.6.4b, which is a contradiction to $\mathrm{pr}_{\Lambda_0} \widetilde{f}_{-1} A \neq \mathbf{0}$.

- *Case 2c*: When $\varphi_j^{\Lambda_{-1}}(\mathcal{R}(A)) = J > 1$ the argument is similar to Case 1*b*. □

**6.3. Other types.** We can repeat the arguments above to prove similar theorems in affine type $B, C$, and $D$, using the Kirillov-Reshetikhin crystal $B^{1,1}$ of appropriate type, and the modules corresponding to the nodes studied in [**27**] in place of $T_{i;k}$. This is done in [**20**].

## References

[1] Susumu Ariki and Kazuhiko Koike, *A Hecke algebra of* $(\mathbf{Z}/r\mathbf{Z}) \wr \mathfrak{S}_n$ *and construction of its irreducible representations*, Adv. Math. **106** (1994), no. 2, 216–243, DOI 10.1006/aima.1994.1057. MR1279219 (95h:20006)

[2] Georgia Benkart, Seok-Jin Kang, Se-jin Oh, and Euiyong Park, *Construction of irreducible representations over Khovanov-Lauda-Rouquier algebras of finite classical type*, Int. Math. Res. Not. IMRN **5** (2014), 1312–1366. MR3178600

[3] Michel Broué and Gunter Malle, *Zyklotomische Heckealgebren* (German), Astérisque **212** (1993), 119–189. Représentations unipotentes génériques et blocs des groupes réductifs finis. MR1235834 (94m:20095)

[4] Jonathan Brundan and Alexander Kleshchev, *Blocks of cyclotomic Hecke algebras and Khovanov-Lauda algebras*, Invent. Math. **178** (2009), no. 3, 451–484, DOI 10.1007/s00222-009-0204-8. MR2551762 (2010k:20010)

[5] Jonathan Brundan and Alexander Kleshchev, *Graded decomposition numbers for cyclotomic Hecke algebras*, Adv. Math. **222** (2009), no. 6, 1883–1942, DOI 10.1016/j.aim.2009.06.018. MR2562768 (2011i:20002)

[6] I. V. Cherednik, *A new interpretation of Gel′fand-Tzetlin bases*, Duke Math. J. **54** (1987), no. 2, 563–577, DOI 10.1215/S0012-7094-87-05423-8. MR899405 (88k:17005)

[7] Richard Dipper and Gordon James, *Representations of Hecke algebras of general linear groups*, Proc. London Math. Soc. (3) **52** (1986), no. 1, 20–52, DOI 10.1112/plms/s3-52.1.20. MR812444 (88b:20065)

[8] I. Grojnowski, *Affine sl_p controls the representation theory of the symmetric group and related Hecke algebras*, arXiv:9907129 (1999).

[9] Victor G. Kac, *Infinite-dimensional Lie algebras*, 2nd ed., Cambridge University Press, Cambridge, 1985. MR823672 (87c:17023)

[10] Seok-Jin Kang and Masaki Kashiwara, *Categorification of highest weight modules via Khovanov-Lauda-Rouquier algebras*, Invent. Math. **190** (2012), no. 3, 699–742, DOI 10.1007/s00222-012-0388-1. MR2995184

[11] Seok-Jin Kang, Masaki Kashiwara, Kailash C. Misra, Tetsuji Miwa, Toshiki Nakashima, and Atsushi Nakayashiki, *Perfect crystals of quantum affine Lie algebras*, Duke Math. J. **68** (1992), no. 3, 499–607, DOI 10.1215/S0012-7094-92-06821-9. MR1194953 (94j:17013)

[12] M. Kashiwara, *On crystal bases of the Q-analogue of universal enveloping algebras*, Duke Math. J. **63** (1991), no. 2, 465–516, DOI 10.1215/S0012-7094-91-06321-0. MR1115118 (93b:17045)

[13] Masaki Kashiwara, *Bases cristallines* (French, with English summary), C. R. Acad. Sci. Paris Sér. I Math. **311** (1990), no. 6, 277–280. MR1071626 (91g:17017)

[14] Masaki Kashiwara, *On crystal bases*, Representations of groups (Banff, AB, 1994), CMS Conf. Proc., vol. 16, Amer. Math. Soc., Providence, RI, 1995, pp. 155–197. MR1357199 (97a:17016)

[15] Masaki Kashiwara and Yoshihisa Saito, *Geometric construction of crystal bases*, Duke Math. J. **89** (1997), no. 1, 9–36, DOI 10.1215/S0012-7094-97-08902-X. MR1458969 (99e:17025)

[16] Mikhail Khovanov and Aaron D. Lauda, *A diagrammatic approach to categorification of quantum groups. I*, Represent. Theory **13** (2009), 309–347, DOI 10.1090/S1088-4165-09-00346-X. MR2525917 (2010i:17023)

[17] Mikhail Khovanov and Aaron D. Lauda, *A diagrammatic approach to categorification of quantum groups II*, Trans. Amer. Math. Soc. **363** (2011), no. 5, 2685–2700, DOI 10.1090/S0002-9947-2010-05210-9. MR2763732 (2012a:17021)

[18] A. S. Kleshchev, *Branching rules for modular representations of symmetric groups. I*, J. Algebra **178** (1995), no. 2, 493–511, DOI 10.1006/jabr.1995.1362. MR1359899 (96m:20019a)

[19] Alexander Kleshchev and Arun Ram, *Homogeneous representations of Khovanov-Lauda algebras*, J. Eur. Math. Soc. (JEMS) **12** (2010), no. 5, 1293–1306, DOI 10.4171/JEMS/230. MR2677617 (2011m:20010)

[20] Henry Kvinge and Monica Vazirani, *Categorifying the tensor product of the Kirillov-Reshetkhin crystal* $B^{1,1}$ *and a fundamental crystal*, (2015), arXiv:1508.04182.

[21] Aaron D. Lauda and Monica Vazirani, *Crystals from categorified quantum groups*, Adv. Math. **228** (2011), no. 2, 803–861, DOI 10.1016/j.aim.2011.06.009. MR2822211 (2012i:17013)

[22] Ivan Losev and Ben Webster, *On uniqueness of tensor products of irreducible categorifications*, Selecta Math. (N.S.) **21** (2015), no. 2, 345–377, DOI 10.1007/s00029-014-0172-8. MR3338680

[23] George Lusztig, *Introduction to quantum groups*, Modern Birkhäuser Classics, Birkhäuser/Springer, New York, 2010. Reprint of the 1994 edition. MR2759715 (2011j:17028)

[24] R. Rouquier, *2-Kac-Moody algebras*, arXiv:0812.5023 (2008).

[25] Monica Vazirani, *A Hecke theoretic shadow of tensoring the crystal of the basic representation with a level 1 perfect crystal*, (1999), Unpublished notes.

[26] ———, *A Hecke theoretic shadow of tensoring the crystal of the basic representation with a level 1 perfect crystal*, Mathematisches Forschungsinstitut Oberwolfach Report **Report No. 14/2003** (2003).

[27] Monica Vazirani, *An observation on highest weight crystals*, J. Algebra **315** (2007), no. 2, 483–501, DOI 10.1016/j.jalgebra.2007.05.021. MR2351874 (2009a:17025)

[28] B. Webster, *Knot invariants and higher representation theory I: diagrammatic and geometric categorification of tensor products*, arXiv:1001.2020 (2010).

Mathematics Department, One Shields Ave, Davis, California 95616

Proceedings of Symposia in Pure Mathematics
Volume **92**, 2016
http://dx.doi.org/10.1090/pspum/092/01584

# On the unitary representations of the affine $ax+b$-group, $\widehat{sl(2,\mathbb{R})}$ and their relatives

Anton M. Zeitlin

ABSTRACT. This article focuses on two related topics: unitary representations of the loop $ax+b$-group and their relation to a loop version of the $\Gamma$-function and the construction of continuous series for the $\widehat{sl(2,\mathbb{R})}$-algebra. Mainly this is a survey of some results obtained by the author and in collaboration with I.B. Frenkel, along with the motivation for them from the physical and mathematical points of view.

## 1. Introduction

Recently, the continuous series of unitary representations of $U_q(sl(2,\mathbb{R}))$ were constructed [**28**], [**6**], so that this set of representations is closed under the tensor product [**25**], generating a "continuous" tensor category. These representations are constructed via the representations of quantum plane and do not have the classical limit. The representation theory of the quantum plane is very subtle (see e.g. [**18**], [**10**]) and is related to the properties of quantum dilogarithm. In [**18**] it was shown that the representations of quantum plane and furthermore, the intertwining operators of the tensor product representation have a classical limit, where they correspond to appropriate objects in the representation theory of the $ax+b$-group, the affine group of the line (below, it is denoted as $G$). In the work of I.B. Frenkel and H.-K. Kim [**10**] it was shown that basic structures from the quantum Teichmueller theory [**1**] can be derived from the tensor products of the representations of the quantum plane, which is very important for the proper understanding of the Chern-Simons theory associated with $sl(2,\mathbb{R})$ (see below for more details).

One important problem to think about is the construction of the affine analogue of the above formalism. Namely, one of the main problems for investigation would be to generalize the Kazhdan-Lusztig equivalence of categories [**21**], [**5**] to the case of the quantum plane and $U_q(sl(2,\mathbb{R}))$. It was conjectured in [**11**] by I.B. Frenkel and the author, that there exists a braided tensor category of representations of the affine algebra $\widehat{sl}_k(2,\mathbb{R})$, equivalent to the braided tensor category of the mentioned above continuous series of $U_q(sl(2,\mathbb{R}))$.

One fact supporting this conjecture is that there exists a category of representations of the Virasoro algebra related to Liouville theory [**26**],[**29**], which is equivalent to the mentioned category of representations of the Virasoro algebra. Therefore, the corresponding category of representations of the affine algebra $\widehat{sl}_k(2,\mathbb{R})$ is a missing piece in this equivalence of categories.

However, a first natural problem to solve is to find unitary representations for the $\widehat{ax+b}_k$-group, the loop version of the $ax+b$-group with central extension (with central charge $k$) for the $a$-subgroup, and compare them and their tensor structure with the representation theory of quantum plane.

In [**33**] the representation theory of the loop counterpart of the $ax+b$-group ($\Omega G$) and its central extension ($\hat{G}$) were studied. There the unitary representations of $\hat{G}$ are constructed, which naturally generalize the irreducible representations of $G$.

It is known that the irreducible unitary representations of $G$ divide into three classes (up to equivalence). The first two classes contain infinite-dimensional representations and there is only one representation in each class, however the third class consists of 1-dimensional representations, labeled by the real parameter. The infinite dimensional representations of $G$ can be realized in the Hilbert space $L^2(\mathbb{R}_+, \frac{dx}{x})$. These representations are produced by the canonical construction of induced representations. This classical material is surveyed in Section 2.1.

However, in the case of the loop $ax+b$-group the situation is more subtle. In [**33**] the $L^2$ space with respect to the Wiener measure is considered as a representation space of $\hat{G}$. The constructed unitary representations of $\hat{G}$ are labeled by a certain function. It is proven in [**33**] that certain representations in this class (e.g. when this function is constant) are irreducible. We review these results in Section 3.

An interesting issue studied in [**33**] is based on the mentioned relations of the representations of group $G$ and the theory of special functions. It is known that the representations of $G$ are related to the $\Gamma$- and $B$-functions (see e.g. [**31**]). Therefore, one may hope to construct their loop counterparts while studying the associated loop group. In [**33**] the first step in this direction is made. It is known, that one can relate (via the bilateral Laplace transform) the action of the group element of representation of $G$ in the representation space with the integral operator, so that its kernel is expressed via the $\Gamma$-function. In [**33**] the loop analogue of the $\Gamma$-function was defined, considering similar arguments applied to $\hat{G}$. It is a functional on the space of paths, which we denote as $\hat{\Gamma}_\mu$, depending on some continuous function $\mu$, such that $\mu(u) > 0$ on $[0, 2\pi]$, and has a property which generalizes the famous property of the $\Gamma$-function, $\Gamma(x+1) = x\Gamma(x)$:

$$\int_0^{2\pi} g(v)\mu(v)\hat{\Gamma}_\mu(z+\delta_v)dv = \int_0^{2\pi} g(v)z(v)dv\hat{\Gamma}_\mu(z) + \frac{1}{t}\int_0^{2\pi} g''(v)\frac{\delta}{\delta z(v)}\hat{\Gamma}_\mu(z)dv, \tag{1}$$

where $g(v)$ is any twice differentiable function on $[0, 2\pi]$, such that $g(0) = g(2\pi) = 0$, $\delta_v = \delta(u-v)$ is a delta-function on the interval $[0, 2\pi]$ and $t$ is the parameter of the Wiener measure. We describe this construction of the loop $\Gamma$-function from [**33**] in Section 4.

It is known (see e.g. [**18**]) that the unitary representations of $G$ are closed under the tensor product and there are three types of "simple" objects in the category of unitary representations of $G$. It appears that their tensor products decompose as direct integrals of these "simple" objects. Using the principles discussed in the beginning of this section, it is expected to have the braided tensor category for

$\hat{G}$, where the braiding is related to the value of the central charge. One can also hope to obtain a differential equation governing the intertwining operators, i.e. the analogue of the Knizhnik-Zamolodchikov equation.

However, the results related to the $\widehat{ax+b_k}$-group are only part of the main task, namely the construction of the continuous series of unitary representations of $\widehat{sl}_k(2,\mathbb{R})$.

Unfortunately, the standard approach of inducing representation of $\widehat{sl}_k(2,\mathbb{R})$ wouldn't fit the construction, since the resulting modules appear to be nonunitary. It turns out one can use the results obtained in [**33**] to construct new modules for $\widehat{sl}_k(2,\mathbb{R})$ by means of the Wakimoto-type formalism, using the "currents" corresponding to the Lie algebra elements of $(ax+b)_k$ and infinite dimensional Heisenberg algebra free fields. This was done in [**11**], inspired by the formalism of [**12**]. It turned out that the correlators of the resulting $\widehat{sl}_k(2,\mathbb{R})$-currents, defining the pairing in the representation, diverge, and therefore we had to describe the scheme of eliminating those divergencies. This leads to a very interesting graphical formalism, similar to the Feynman diagram technique, where the divergences corresponded to 1-loop graphs. The regularization scheme involves dependence on the infinite family of parameters: one parameter for each loop with a given number of vertices. These modules were called *continuous series* for $\widehat{sl}_k(2,\mathbb{R})$ in [**11**], since the described above Wakimoto-type realization is a natural affinization of certain realization of standard continuous series of $sl(2,\mathbb{R})$. These representations are labeled by the continuous parameter arizing from the highest weight of the Fock module of the infinite-dimensional Heisenberg algebra. One can generalize this construction of continuous series to the higher rank case, using similar procedure, that works for quantum groups, introduced in [**9**]. We describe the construction of the above $\widehat{sl}_k(2,\mathbb{R})$ representations from [**11**] in Sections 5 and 6 of this paper. Section 5 is devoted to the construction of Wakimoto-like modules, inspired by the construction of the standard continuous series of $sl(2,\mathbb{R})$, sketched in Sections 2.2 and 2.3. Section 6 is devoted directly to the construction of the $\widehat{sl}_k(2,\mathbb{R})$-modules of [**11**] using regularization and graphical calculus.

An important question one can immediately ask is for which values of regularization parameters the resulting representations are unitary. Answering this question will involve studying the resulting bilinear form and analyzing it using the graph formalism that was a cornerstone of the definition of those representations in [**11**]. However, we do believe that finding the representations forming a braided tensor category will single out the necessary family of unitary representations. Therefore the next question to ask is what are the intertwining operators for the proposed tensor category. One of the ways to do that is to construct them in a similar fashion as in the Virasoro case [**26**], [**29**].

There is an interesting physical perspective on the proposed tensor categories discussed, since currently the study of Chern-Simons theories with noncompact gauge groups has become very important from the point of view of both mathematics and physics, see e.g. [**4**] for a review. In particular, it was argued since the late 80s [**30**] that the canonical quantization of $SL(2,\mathbb{R})$ Chern-Simons theory is connected with both the quantum Teichmueller theory [**1**], which is related to the representation theory of the quantum plane and the Liouville theory, related to representation theory of $U_q(sl(2,\mathbb{R}))$. It is known that the Liouville theory can be obtained from the $SL(2,\mathbb{R})$ WZW theory by means of the Drinfeld-Sokolov reduction. At the same time, the study of Chern-Simons theory with the compact gauge group $G$ showed, that its space of states in the presence of Wilson lines is isomorphic to the space of conformal blocks of the WZW model associated with $G$ [**32**]. Therefore, we may expect that this works for noncompact Lie groups too, and the construction of the continuous series of unitary representations of $\widehat{sl}_k(2,\mathbb{R})$ will provide an important link between the quantum Teichmueller theory, the $SL(2,\mathbb{R})$ Chern-Simons, the $SL(2,\mathbb{R})$ WZW model and its reduction, the Liouville theory.

In the end of this introduction let us summarize shortly for convenience the contents of this paper. In Section 2 we gather some standard facts about the unitary representations of $ax+b$-group and $sl(2,\mathbb{R})$ Lie algebra. In Section 3, we describe some unitary representations of the loop $ax+b$-group, which are generalizations of the ones considered in Section 2. Section 4 is devoted to the construction of a loop analogue $\Gamma$-function, which appears naturally from the representations of loop $ax+b$-group. Section 5 is devoted to the study of auxiliary current algebras, which we will use in Section 6 to construct representations of $\widehat{sl}(2,\mathbb{R})$.

**Acknowledgements.** I am indebted to Igor Frenkel for proposing the areas described in this review and for our fruitful and pleasant collaboration. I am very grateful to the organizers of the Southeastern Lie Theory Workshop for inviting me to a talk and for stimulating discussions. I am also indebted to the referee for her/his comments and to Antonina N. Fedorova for a careful reading of the manuscript.

## 2. Unitary representations of the affine group of a line and the related realizations of $sl(2,\mathbb{R})$

**2.1. $ax+b$-group and $\Gamma$-function.** In this subsection we remind the reader of all the necessary facts (for more information and references one can consult [**31**], chapter 5) about the unitary representations of $ax+b$-group, which is the affine group of the real line. In other words, each group element $g=g(a,b)$ is determined by the pair of real numbers $a,b$ such that $a>0$. The composition law is defined as follows: $g(a_1,b_1)g(a_2,b_2)=g(a_1a_2,a_1b_2+b_1)$. In the following we will call this group $G$.

The unitary representations of this group are constructed by means of the method of induced representations. Let us consider the representation $r_\lambda$ of the subgroup $B$ of $G$ generated by $g(1,b)$-elements, such that $r_\lambda(g(1,b))=e^{\lambda b}$ (here $\lambda$ is complex). Then according to the construction of induced representations we have to consider the space of complex valued functions on $G$, i.e. $f(g(a,b))\equiv f(a,b)$ such that

$$f(a,b+b_0)=e^{\lambda b_0}f(a,b). \tag{2}$$

Therefore $f(a,b)=e^{\lambda b}f(a,0)$ i.e. the function $f$ can be expressed in terms of the function $\phi(a)=f(a,0)$ on the subgroup $A$ generated by $g(a,0)$-elements. Then the operator of induced representation $R_\lambda$ acts on $f(g)$ as $R_\lambda(g_0)f(g)=f(gg_0)$, and the resulting formula for representation on the functions $\phi(a)$ is:

$$R_\lambda(g_0)\phi(a)=e^{\lambda ab_0}\phi(a_0a), \tag{3}$$

or, in other words, since the space of functions $\phi$ is just a space of functions on the ray $0<x<\infty$, we have $R_\lambda(g)\phi(x)=e^{\lambda bx}\phi(ax)$. One can consider the invariant measure on $A$, it is just $\frac{dx}{x}$. Then the following theorem is valid [**31**].

**Theorem 2.1.** *i) Representation $R_\lambda$ of $G$ is unitary on $L^2(\mathbb{R}_+, \frac{dx}{x})$ if $\lambda \in i\mathbb{R}$.*
*ii) Representations $R_\lambda$ and $R_{e^\xi \lambda}$ are equivalent if $\xi \in \mathbb{R}$.*
*iii) Consider a semigroup $G_+$(resp. $G_-$), consisting of such $g(a,b)$, that $b > 0$ (resp. $b < 0$). Then $R_\lambda$ with $\lambda \in \mathbb{C}$, such that $Re\lambda < 0$ (resp. $Re\lambda > 0$) is a representation for $G_+$(resp. $G_-$) on $L^2(\mathbb{R}_+, \frac{dx}{x})$.*

As a consequence, one can see that there are only three classes of unequivalent unitary representations $R_\lambda$: $R_{\pm i}$ and $R_0$. The representations $R_{\pm i}$ are irreducible, while $R_0$ decomposes into the direct integral of 1-dimensional representations $T_\rho$, such that $T_\rho(g(a,b)) = a^{i\rho}$. One can show that any other irreducible representation of $G$ falls into one of the classes $R_{\pm i}$, $T_\rho$.

In order to generalize representations $R_\lambda$ to the loop case, it is useful to consider another form of the representations $R_\lambda$. Namely, instead of the space $L^2(\mathbb{R}_+, \frac{dx}{x})$ one can consider $L^2(\mathbb{R}, dt) \equiv L^2(\mathbb{R})$, by substitution $x = e^t$. Therefore, the formula for the representation $R_\lambda$ can be rewritten as follows:

$$R_\lambda(g(e^\alpha, b))f(t) = e^{i\lambda b e^t} f(t+\alpha), \tag{4}$$

where $f \in L^2(\mathbb{R}, dt)$ and we represented $a$ as $e^\alpha$.

Now we discuss the relation of the representation $R_\lambda$ and $\Gamma$-function. It is well known, that the bilateral Laplace transform (or the Fourier transform in the complex domain)

$$\mathcal{L}f(p) = \frac{1}{\sqrt{2\pi}} \int_{\mathbb{R}} e^{ipt} f(t)dt, \tag{5}$$

where $p$ is a complex number, has the inverse

$$\mathcal{L}^{-1}g(t) = \frac{1}{\sqrt{2\pi}} \int_{\mathbb{R}+iT} e^{-ipt} g(p)dp, \tag{6}$$

where $T$ is a real number, so that the contour of integration is in the region of convergence of $g(p)$. Therefore, if one can make sense of $\mathcal{L}R_\lambda\mathcal{L}^{-1}$, it gives us a representation, equivalent to $R_\lambda$. Let $\mathcal{D} \subset L^2(\mathbb{R})$ be the space of $C^\infty$ functions with finite support.

**Proposition 2.1.** [**31**] *The bilateral Laplace transform of an element of $\mathcal{D}$ is an analytic function in the entire complex plane of the exponential type, i.e. $|\mathcal{L}f(x+iy)| < Ce^{b|y|}$, where $C > 0$, $b > 0$. At the same time, the inverse Laplace transform of such an analytic function belongs to $\mathcal{D}$. Moreover, we have the following property:*

$$\int_{\mathbb{R}} |\mathcal{L}f(x+iy)|^2 dx < \infty, \tag{7}$$

*where $f \in \mathcal{D}$.*

We notice that $\mathcal{D}$ is invariant under the action of the operators $R_\lambda(g)$. Let $\Gamma(z)$ denote the $\Gamma$-function:

$$\Gamma(z) = \int_0^\infty e^{-x} x^{z-1} dx. \tag{8}$$

Then we have a Proposition.

**Proposition 2.2.** [**31**] *i) Consider the action of the* $\mathcal{L}R_\lambda(g)\mathcal{L}^{-1}$ *on* $\mathcal{LD}$, *when* $g = g(a,b) \in G_+$ *and* $\lambda < 0$. *Then*

$$\mathcal{L}R_\lambda(g)\mathcal{L}^{-1}f(t_1) = \frac{1}{2\pi}\int_{\mathbb{R}+i0}\Gamma(it_1 - it_2)a^{-it_1}\left(\frac{-\lambda b}{a}\right)^{it_2 - it_1} f(t_2)dt_2, \tag{9}$$

*where* $f$ *is an analytic function on* $\mathbb{C}$ *of the exponential type.*

*ii) One can analytically continue the expression above with respect to* $-\lambda b$ *to all complex plane except for the negative real axis.*

**2.2.** $\mathcal{A}$ **and** $\mathcal{K}$ **algebras and their unitary representations.** In the rest of this section we present a construction of continuous series of $sl(2,\mathbb{R})$ in a special way, which will be convenient for generalizations to the loop case. First, we construct certain modifications of the Lie algebra of $ax+b$-group, which we call $\mathcal{A}$-algebra and $\mathcal{K}$-algebra: they have a similar algebraic structure, but different $*$-structure. The $\mathcal{A}$-algebra is an algebra with three generators: $h, e^\pm$, so that $ih, ie^+$ generate the Lie algebra of ax+b-group, however we make the b-subgroup generator $e^+$ to be invertible, so that the inverse element is $e^-$. The commutation relations and the star structure are as follows:

$$[h, e^\pm] = \pm ie^\pm, \quad e^\pm e^\mp = 1, \quad h^* = h, \quad e^{\pm^*} = e^\pm. \tag{10}$$

The generators $h, \alpha^\pm$ of $\mathcal{K}$-algebra have similar commutation relations, but the star-structure is different:

$$[h, \alpha^\pm] = \mp\alpha^\pm, \quad \alpha^\pm\alpha^\mp = 1, \quad h^* = h, \quad \alpha^{\pm^*} = \alpha^\mp. \tag{11}$$

To construct the unitary representations of the above $*$-algebras, one can realize the generators $h, e^\pm$ as unbounded self-adjoint operators $i\frac{d}{dx}, e^{\pm x}$ in the Hilbert space of square-integrable functions on the real line $L^2(\mathbb{R})$ as well as generators $h, \alpha^\pm$ as the operators $i\frac{d}{d\phi}, e^{\pm i\phi}$ in the Hilbert space of square integrable functions on the circle $L^2(S^1)$. These generators acting on a specific vector in the appropriate Hilbert space generate a dense set. More explicitly one has the following proposition (see e.g. [**11**], Proposition 2.1).

**Proposition 2.3.** *i) Let* $D_A$ *be the space spanned by the vectors* $a \cdot v_0$, *where* $v_0 = e^{-tx^2} \in L^2(\mathbb{R})$, $t > 0$ *and* $a$ *belongs to the universal enveloping algebra of* $\mathcal{A}$-*algebra, such that the action of generators is realized as* $h = i\frac{d}{dx}, e^\pm = e^{\pm x}$. *Then* $D_A$ *is a dense set in* $L^2(\mathbb{R})$.

*ii) Let* $D_K$ *be the space spanned by the vectors* $a \cdot v_0$, *where* $v_0 = 1 \in L^2(S^1)$ *and* $a$ *belongs to the universal enveloping algebra of* $\mathcal{K}$-*algebra, such that the action of generators is realized as* $h = i\frac{d}{d\phi}, \alpha^\pm = e^{\pm i\phi}$, *where* $\phi \in [0, 2\pi]$ *is the angle coordinate on* $S^1$. *Then* $D_K$ *is a dense set in* $L^2(S^1, d\phi) \equiv L^2(S^1)$.

**2.3. Continuous series of** $sl(2,\mathbb{R})$ **via** $\mathcal{A}$ **and** $\mathcal{K}$ **algebras.** First of all, let us introduce some notations. We are interested in unitary representations of $sl(2,\mathbb{R})$

algebra, i.e. Lie algebra with generators $E, F, H$ such that

$$
\begin{aligned}
&[E,F]=H, \quad [H,E]=2E, \quad [H,F]=-2F \\
&E=-E^*, \quad F=-F^*, \quad H=-H^*.
\end{aligned} \tag{12}
$$

We note that there are two standard realizations of continuous series of $sl(2,\mathbb{R})$, one is related to inducing the representations from the diagonal subgroup of $sl(2,\mathbb{R})$, corresponding to the generator $H$, the other one is related to inducing representations from maximally compact subgroup generated by $J^3 = E - F$. It is convenient to introduce the following change of generators (which provides the correspondence between $sl(2,\mathbb{R})$ and $su(1,1)$):

$$
J^{\pm} = E + F \mp iH, \tag{13}
$$

so that the relations (12) can be rewritten in the following way:

$$
\begin{aligned}
&[J^3, J^{\pm}] = \pm 2iJ^{\pm}, \quad [J^+, J^-] = -iJ^3, \\
&{J^+}^* = -J^-, \quad {J^3}^* = -J^3.
\end{aligned} \tag{14}
$$

Now we will give two realizations of $sl(2,\mathbb{R})$ via generators of $\mathcal{A}$- and $\mathcal{K}$-algebras. Using $\mathcal{A}$-algebra, one can write down the expressions for $E, F, H$:

$$
\begin{aligned}
E &= \frac{i}{2}(e^+h + he^+) + i\lambda e^+, \\
F &= -\frac{i}{2}(e^-h + he^-) + i\lambda e^-, \\
H &= -2ih,
\end{aligned} \tag{15}
$$

where $\lambda$ is a real parameter. Similarly, using $\mathcal{K}$-algebra, one can represent $J^3, J^{\pm}$ in a similar way:

$$
\begin{aligned}
J^+ &= \frac{i}{2}(\alpha^+h + h\alpha^+) - \lambda\alpha^+, \\
J^- &= \frac{i}{2}(\alpha^-h + h\alpha^-) + \lambda\alpha^-, \\
J^3 &= 2ih.
\end{aligned} \tag{16}
$$

Therefore the following theorem is valid (see e.g. [**11**], Theorem 2.1).

**Theorem 2.2.** *Let $\mathcal{D}_A$ be the space spanned by the vectors $a \cdot v_0$, where $v_0 = e^{-tx^2} \in L^2(\mathbb{R})$ and $a$ belongs to the universal enveloping algebra of $sl(2,\mathbb{R})$-algebra, such that the action of generators is realized as in (15), so that $h = i\frac{d}{dx}, e^{\pm} = e^{\pm x}$. Then $\mathcal{D}_A$ is a dense set in $L^2(\mathbb{R})$ and it is a representation space for $sl(2,\mathbb{R})$.*
*ii) Let $\mathcal{D}_K$ be the space spanned by the vectors $a \cdot v_0$, where $v_0 = 1 \in L^2(S^1)$ and $a$ belongs to the universal enveloping algebra of $sl(2,\mathbb{R})$-algebra, such that the action of generators is realized as in (16), so that $h = i\frac{d}{d\phi}, e^{\pm} = e^{\pm i\phi}$, where $\phi \in [0, 2\pi] \cong S^1$. Then $\mathcal{D}_K$ is a dense set in $L^2(S^1)$ and it is representation space for $sl(2,\mathbb{R})$.*

## 3. Unitary representations of the path and loop versions of $ax+b$- group

**Notation.** In the following we will use several functional spaces, so let us fix notations. Let $\mu$ be the $\sigma$-additive measure on some space $M$. Then we denote as

$L^2(M, d\mu; k)$ the Hilbert space of square-integrable complex-valued functions (here $k$ stands for $\mathbb{C}$ or $\mathbb{R}$). Since in the most of cases we will deal with $L^2(M, d\mu; \mathbb{C})$, we drop $\mathbb{C}$, i.e. $L^2(M, d\mu) \equiv L^2(M, d\mu; \mathbb{C})$. The space of real-valued continuous functions on the interval $[a, b]$ will be referred to as $C[a, b]$. The space of real-valued absolutely continuous functions on $[a, b]$, i.e. differentiable functions from $C[a, b]$, whose derivative belong to $L^2([a, b], dx; \mathbb{R})$, where $dx$ is the Lebesgue measure on $[a, b]$ will be denoted as $C'[a, b]$.

We introduce $C'_0[0, 2\pi]$, the Hilbert space of real-valued absolutely continuous functions, such that $x(0) = 0$ for any $x \in C'_0[0, 2\pi]$. The completion of $C'_0[0, 2\pi]$ is the Banach space $C_0[0, 2\pi]$ of real-valued continuous functions such that $x(0) = 0$ for any $f \in C_0[0, 2\pi]$ (see Appendix). Finally, let $C_{0,X}[0, 2\pi]$ denote the closed subspace of $C_0[0, 2\pi]$, consisting of real-valued continuous functions such that $x(2\pi) = X$ for some $X \in \mathbb{R}$.

**3.1. Path groups and loop groups associated to $G$.** In this subsection, we define all the path and loop groups, which we study in this article. All the necessary facts concerning the Wiener measure can be found in the Appendix.

At first we define the path group $PG$. Let us consider a set of elements of the form $g(e^\alpha, b)$, where $\alpha$ and $b$ are real valued absolutely continuous functions on $[0, 2\pi]$. We can define the bilinear operation as follows:

$$g(e^{\alpha_1}, b_1) \cdot g(e^{\alpha_2}, b_2) = g(e^{\alpha_1+\alpha_2}, e^{\alpha_1} b_2 + b_1), \tag{17}$$

where $e^{\alpha_1} b_2$ stands for pointwise multiplication of the absolutely continuous functions $e^{\alpha_1}$ and $b_2$. It is clear that the operation is well defined and satisfies the group laws. Let us denote the group which is a set of all such elements $\mathrm{g}(e^\alpha, b)$ as $PG$.

In the following we will sometimes need $\alpha$ to be in the Cameron-Martin space for the Wiener measure, so it is useful to introduce the based path group $P_0G \subset PG$, generated by $g(e^\alpha, b)$, where $\alpha(0) = 0$.

Similarly, one can define loop groups $\Omega G$ and $\Omega_0 G$. The loop group $\Omega G \subset PG$ is generated by $(e^\alpha, b)$, where $\alpha, b$ are such that $\alpha(0) = \alpha(2\pi)$, $b(0) = b(2\pi)$, while the based loop group $\Omega_0 G \equiv P_0 G \cap \Omega G$.

Finally, one can define the central extended versions of $\Omega G$ (resp. $\Omega_0 G$). Let us consider the elements $g(e^\alpha, b, s)$, where $s \in \mathbb{R}$ and $(e^\alpha, b) \in \Omega G$ (resp. $\Omega_0 G$). Then the multiplication law can be modified in the following way:

$$\begin{aligned} &g(e^{\alpha_1}, b_1, s) \cdot g(e^{\alpha_2}, b_2, t) = \\ &g(e^{\alpha_1+\alpha_2}, e^{\alpha_1} b_2 + b_1, t + s + k \int_0^{2\pi} \alpha_1(u)\alpha'_2(u) \mathrm{d}u), \end{aligned} \tag{18}$$

where $k \in \mathbb{R}$. We denote the corresponding group $\hat{G}$ (resp. $\hat{G}_0$) and we will refer to $k$ as central charge.

**3.2. Unitary representations.** In this subsection, we will describe the unitary representations of both $PG$ and $\Omega G$, as well as their subgroups $P_0 G$ and $\Omega_0 G$, which are the appropriate generalizations of the unitary representations of $G$, considered in Section 2.

We will start from $P_0 G$ and use the same approach as before, i.e. we will use the method of induced representations. Let us take one-dimensional representations of the $B$-subgroup (i.e. subgroup of elements of the form $g(1, b)$) of the form

$\tilde{r}_\lambda(g(1,b)) = e^{\int_0^{2\pi}\lambda(u)b(u)du}$, where $\lambda \in L^2[0,2\pi]$. Following the method of induced representations, like we did in the case of group $G$, we arrive to the following formula for the representations on the space of functionals on the $A$-subgroup:

$$\tilde{R}_\lambda(g(e^{\alpha_0}, b_0))f(\alpha) = e^{\int_0^{2\pi}\lambda(u)b_0(u)e^{\alpha(u)}du} f(\alpha+\alpha_0). \tag{19}$$

However, in order to make these representations unitary one needs to define the proper inner product on the space of functionals. To do that, one has to consider the Wiener measure (see Appendix). It is not invariant under translations, so one should improve the formula for the representations in order to make them unitary. Let us consider the Hilbert space $H_p^0 = L^2(C_0[0,2\pi], dw^t; \mathbb{C})$. The following statement is true.

**Theorem 3.1.** *The following action of $P_0G$ on the space of functionals of continuous functions*

$$\begin{aligned}&\rho_\lambda(g(e^\alpha, b))(f)(x) = \\ &e^{-\frac{1}{4t}\int_0^{2\pi}\alpha'(u)\alpha'(u)du - \frac{1}{2t}\int_0^{2\pi}\alpha'(u)dx(u)} e^{\int_0^{2\pi}\lambda(u)b(u)e^{x(u)}du} f(x+\alpha)\end{aligned} \tag{20}$$

*defines the unitary representation of $P_0G$ on $H_p^0$ iff $i\lambda(u) \in L^2([0,2\pi];\mathbb{R})$.*

**Proof.** First we prove that this action is actually an action of a group, i.e. $\rho_\lambda(g(e^{\alpha_1}, b_1))\rho_\lambda(g(e^{\alpha_2}, b_2)) = \rho_\lambda(g(e^{\alpha_1+\alpha_2}, e^{\alpha_1}b_2 + b_1))$. We check it, writing the explicit expressions:

$$\begin{aligned}&\rho_\lambda(g(e^{\alpha_1}, b_1))\rho_\lambda(g(e^{\alpha_2}, b_2))f(x) = \\ &\rho_\lambda(g(e^{\alpha_1}, b_1))(e^{-\frac{1}{4t}\int_0^{2\pi}\alpha_2'(u)\alpha_2'(u)du - \frac{1}{2t}\int_0^{2\pi}\alpha_2'(u)dx(u)} \\ &e^{\int_0^{2\pi}\lambda(u)b_2(u)e^{x(u)}du} f(x+\alpha_2)) = \\ &e^{-\frac{1}{4t}\int_0^{2\pi}\alpha_2'(u)\alpha_2'(u)du - \frac{1}{4t}\int_0^{2\pi}\alpha_1'(u)\alpha_1'(u)du - \frac{1}{2t}\int_0^{2\pi}\alpha_1'(u)\alpha_1'(u)du} \\ &e^{-\frac{1}{2t}\int_0^{2\pi}(\alpha_1'(u)+\alpha_2'(u))dx(u)} e^{\int_0^{2\pi}\lambda(u)(b_2(u)e^{\alpha_1(u)}+b_1(u))e^{x(u)}du} f(x+\alpha_1+\alpha_2)) = \\ &\rho_\lambda(e^{\alpha_1+\alpha_2}, e^{\alpha_1}b_2 + b_1)f(x).\end{aligned} \tag{21}$$

Now we prove that the action $\rho_\lambda$ defines a unitary representation for $P_0G$. In order to do that we just need to use the definition of unitarity and the translation invariance property of the Wiener measure:

$$\begin{aligned}&(\rho_\lambda(g(e^\alpha, b))(f), \rho_\lambda(g(e^\alpha, b))(h))_{H_p} = \\ &\int_{C_0[0,2\pi]} \overline{\rho_\lambda(g(e^\alpha, b))(f)(x)}\rho_\lambda(g(e^\alpha, b))(h)(x)dw^t(x) = \\ &\int_{C_0[0,2\pi]} e^{-\frac{1}{2t}\int_0^{2\pi}\alpha'(u)\alpha'(u)du - \frac{1}{t}\int_0^{2\pi}\alpha'(u)dx(u)}\overline{f(x+\alpha)}g(x+\alpha)dw^t(x) = \\ &\int \overline{f(x)}g(x)dw(x) = (f,g)_{H_p}.\end{aligned} \tag{22}$$

Thus the theorem is proven. ■

In the following we will denote representations $\rho_\lambda(g(e^\alpha, b)) \equiv \rho_\lambda(e^\alpha, b)$.

Using the same steps as in the proof of the theorem above, one can show that analogous fact for the Hilbert space $H_p^{0,X} = L^2(C_{0,X}[0,2\pi], dw_X^t; \mathbb{C})$ and group

$\Omega_0 G$ is true.

**Proposition 3.1.** *i) Formula (20) defines a unitary representation* $\rho_\lambda^X$ *of* $\Omega_0 G$ *on* $H_p^{0,X}$.

*ii) Representation* $\rho_\lambda^{X_1}$ *on* $H_p^{0,X_1}$ *is equivalent to the representation of* $\rho_{e^\eta\lambda}^{X_2}$ *on* $H_p^{0,X_2}$, *where* $\eta$ *is absolutely continuous,* $\eta(0) = 0$, $\eta(2\pi) = X_1 - X_2$.

Therefore, all representations $\rho_\lambda^X$ are equivalent to the representations $\rho_{e^\xi\lambda}^0$ on $H_l^0 \equiv L^2(C_{0,0}[0, 2\pi], dw^t; \mathbb{C})$, where $\xi$ is absolutely continuous, $\xi(0) = 0, \xi(2\pi) = X$. Moreover, one can show that this equivalence does not depend on the choice of such $\xi$.

However, we note here, that since $\Omega_0 G \subset P_0 G$, the Hilbert space $H_p^0$ is a representation space for $\Omega_0 G$ too. Therefore we have a Proposition, arising from the direct integral decomposition of $H_p^0$.

**Proposition 3.2.** *The representation* $\rho_\lambda$ *of* $\Omega_0 G$ *on* $H_p^0$ *is the direct integral of the representations of* $\Omega_0 G$*:*

$$\rho_\lambda = \int_{\mathbb{R}}^{\oplus} \rho_\lambda^X dX, \tag{23}$$

*where* $\rho_\lambda^X$ *is equivalent to* $\rho_{\lambda_X}^0$ *such that* $\lambda_X = e^\xi \lambda$, $\xi \in C'[0, 2\pi]$ *and* $\xi(0) = 0$, $\xi(2\pi) = X$.

In order to define the representation of $PG$ and $\Omega G$, one needs to extend the Hilbert space, i.e. one has to consider the Hilbert spaces $H_p = L^2(C_0[0, 2\pi] \oplus \mathbb{R}, dw^t \times dx_0; \mathbb{C})$ and $H_l = L^2(C_{0,0}[0, 2\pi] \oplus \mathbb{R}, dw^t \times dx_0; \mathbb{C})$, where $dx_0$ is the Lebesgue measure on $\mathbb{R}$. We also note that every element $(e^\alpha, b)$ of $PG$ (resp. $\Omega G$) can be represented as $(e^\alpha, b) = (e^{\alpha_0} e^{\tilde{\alpha}}, b)$, where $\alpha_0 = \alpha(0)$ and $\tilde{\alpha}(0) = 0$. Then the following theorem is true.

**Theorem 3.2.** *i) The following action of the group element* $g(e^\alpha, b)$ *on the space of functionals of* $x(u), x_0$*:*

$$\begin{aligned} &\tilde{\rho^p}_\lambda(e^\alpha, b)(f)(x, x_0) = \\ &\qquad e^{-\frac{1}{4t}\int_0^{2\pi} \alpha'(u)\alpha'(u)du - \frac{1}{2t}\int_0^{2\pi} \alpha'(u)dx(u)} e^{\int_0^{2\pi} \lambda(u)b(u)e^{x(u)+x_0}du} f(x + \tilde{\alpha}, x_0 + \alpha_0) \end{aligned} \tag{24}$$

*defines a unitary representation of* $PG$ *on the Hilbert space* $H_p$ *iff* $\lambda \in iL^2([0, 2\pi]$.
*ii) The formula (24) defines the unitary representation* $\tilde{\rho}_\lambda^l$ *of* $\Omega G$ *on the Hilbert space* $H_l$ *iff* $\lambda \in iL^2([0, 2\pi]; \mathbb{R})$.

Note, that the representations $\rho_\lambda^p$, $\rho_\lambda^l$ are equivalent to $\rho_{e^t\lambda}^p$, $\rho_{e^t\lambda}^l$ correspondingly, for any $t \in \mathbb{R}$. In particular, in the case when $\lambda = const$ there exist only three inequivalent classes of unitary representations of $PG$ ($\Omega G$), corresponding to $\tilde{\rho}_0$, $\tilde{\rho}_i$, $\tilde{\rho}_{-i}$ respectively, like it was in the finite dimensional case.

**3.3. Central extension and Lie algebra generators.** In this subsection, we will construct the unitary representations of groups $\hat{G}$ on $H_l$. Let $(e^\alpha, b, s)$ be an element of $\hat{G}$. Let us consider the following action of this element on the function

$H_l$:

$$\rho^l_{\lambda,k}((e^\alpha,b,s))(f)(x,x_0)=e^{is}e^{-\frac{1}{4t}\int_0^{2\pi}\alpha'(u)\alpha'(u)du-\frac{1}{2t}\int_0^{2\pi}\alpha'(u)dx(u)}$$

$$e^{ik\int_0^{2\pi}\alpha(u)dx(u)}e^{\int_0^{2\pi}\lambda(u)b(u)e^{x(u)+x_0}du}f(x+\tilde{\alpha},x_0+\alpha_0). \tag{25}$$

**Theorem 3.3.** *Formula (25) defines a unitary representation of $\hat{G}$ for the central charge $k\in\mathbb{R}$, and $\lambda\in iL^2([0,2\pi];\mathbb{R})$. The representation ${\rho^l}_{e^\xi\lambda,k}$ is equivalent to ${\rho^l}_{\lambda,k}$ if $\xi$ is any absolutely continuous function on $[0,2\pi]$, such that $\xi(0)=\xi(2\pi)$.*

Ignoring $x_0$ dependence in (25), one can define the unitary representation of $\hat{G}_0$ on $H_l^0$, for which we will keep the same notation $\rho^l_{\lambda,k}$. We also mention that the operators $\rho^l_{\lambda,k}(e^{\alpha_1},0,s_1)$ and $\rho^l_{\lambda,-k}(e^{\alpha_2},0,s_2)$ corresponding to the action of the $A$ subgroup of $\hat{G}$ commute.

One can write down the expressions for the Lie algebra generators. As we can see, the generator which is responsible for the action of the "$B$-subgroup" has the form:

$$T_bf(x,x_0)=\frac{d}{d\epsilon}_{|_{\epsilon=0}}\rho^l_{\lambda,k}(1,\epsilon b,0)=\int_0^{2\pi}\lambda(u)b(u)e^{x(u)+x_0}du. \tag{26}$$

It is well-defined for all the functionals $f(x,x_0)\in H_l$ such that $e^{x_0}f(x,x_0)\in H_l$. However, the $a$-subgroup generators are more peculiar:

$$D_{k,\alpha}f(x)=\frac{d}{ds}_{|_{\epsilon=0}}\rho^l_{\lambda,k}(e^{\epsilon\alpha},0,0)= \tag{27}$$

$$(ik\int_0^{2\pi}\alpha(u)dx(u)+\frac{1}{2t}\int_0^{2\pi}\alpha'(u)dx(u))f(x)+\frac{d}{d\epsilon}_{|_{\epsilon=0}}f(x+\epsilon\tilde{\alpha},x_0+\epsilon\alpha_0).$$

Therefore the action of the Lie algebra element $T_\alpha$ is defined only on the set of weakly differentiable functionals, which form a dense subset in $H_l$. In the notations of variational calculus, one can introduce the operators:

$$D_{\alpha,k}(\alpha)=\alpha_0\frac{\partial}{\partial x_0}+\int_0^{2\pi}\tilde{\alpha}(u)\frac{\delta}{\delta x(u)}du+\frac{1}{2t}\int_0^{2\pi}\tilde{\alpha}'(u)x'(u)du+$$

$$ik\int_0^{2\pi}du\tilde{\alpha}(u)x'(u)du,$$

$$T_b=\int_0^{2\pi}\lambda(u)b(u)e^{x_0+x(u)}du. \tag{28}$$

One can see that (28) satisfies the needed relation:

$$[D_{\alpha_1,k},D_{\alpha_2,k}]=-2ik\int_0^{2\pi}\alpha_1'(u)\alpha_2(u)du,\quad [D_{\alpha,k},T_b]=T_{\alpha b}. \tag{29}$$

Finally, let us consider the following two semigroups of $\hat{G}_+,\hat{G}_-\in\hat{G}$, such that they consist of group elements $g(e^\alpha,b)$, where $b(u)>0$ or $b(u)<0$ for all $u$ correspondingly. The following statement will be important will be important for the construction of loop $\Gamma$-function in the next section.

**Proposition 3.3.** *Let $Im\lambda, Re\lambda\in L^2([0,2\pi];\mathbb{R})$. Then (25) defines a representation of the semigroup $\hat{G}_+$ (resp. $\hat{G}_-$) if $Re\lambda(u)<0$ (resp. $\lambda>0$) on $[0,2\pi]$.*

In the following we will denote the representations of $\hat{G}_+, \hat{G}_-$ the same way, namely $\rho^l_{\lambda,k}$.

**3.4. Irreducibility and classification of unitary representations.** In Section 2, we learned that the irreducible unitary representations of $G$ turn out to be equivalent to either $R_{\pm i}$ or $T_\rho$. In the loop case the following weaker result holds [**33**].

**Theorem 3.4.** *Representations $\rho^l_{\lambda,k}$, such that $\lambda(u) \equiv \lambda \in \mathbb{R}\backslash 0$ of $\hat{G}$ $(\hat{G}_0)$ on $H_l$ $(H^0_l)$, are irreducible.*

One can prove that, if $i\lambda$ is either a positive or a negative function from $L^2([0,2\pi],\mathbb{R})$, the representation $\rho^l_{\lambda,k}$ is also irreducible. However, if $\lambda(u)=0$ $\rho^l_{\lambda,k}$ reduces to the representation of the $A$-subgroup, i.e. the representations of the loop Heisenberg group. It is not irreducible since $[\rho^l_{\lambda,k}(e^{\alpha_1},0,s),\rho^l_{\lambda,-k}(e^{\alpha_2},0,t)]=0$. To author's knowledge the classification of the unitary representations of the loop Heisenberg group is not yet known (see e.g. [**7**] for review of the subject). The same argument as in Theorem 3.4, shows that if $\lambda(u)=0$ for $u\in[a,b]$ the representation $\rho^l_{\lambda,k}$ is not irreducible. Therefore, we make the following conjecture about the classification of the irreducible representations of $\hat{G}$. Hence the irreducible unitary representations of $\hat{G}$ which we considered, are either equivalent to the representations of the $A$-subgroup, or equivalent to the representations $\rho^l_{\lambda,k}$, where $i\lambda\in L^2([0,2\pi],\mathbb{R})$ is strictly positive or negative function on $[0,2\pi]$. Moreover, we know that $\rho^l_{\lambda,k}$ and $\rho^l_{e^\xi\lambda,k}$ are equivalent to each other if $\xi\in C'_{0,0}[0,2\pi]$. Therefore, an interesting problem to study is the classification of all finite-dimensional representations of $\hat{G}$. A reasonable conjecture (by analogy with the finite-dimensional case) would be that the three discussed classes of representations, namely $\rho^l_{\lambda,k}$ for positive and negative $\lambda$ and irreducible unitary representations of the loop Heisenberg group exhaust all irreducible unitary representations of $\hat{G}$.

## 4. (Loop) Γ-function and the action of the affine loop group

**4.1. Fourier transform for the classical Wiener measure.** In this section, we consider the generalizations of the formula relating the action of the group $G$ and the Γ-function (see Section 1). In particular, we introduce a new object, which we will refer to as the *loop* $\Gamma$-*function*. In order to do that one needs to construct the generalization of the Fourier/Laplace transform in the case of the Wiener measure. We have already seen a unitary transformation on the $L^2$ space for the abstract Wiener measure, called the Fourier-Wiener transform, but we will choose another transformation here.

Let us consider the following transformation on the Hilbert space $H^0_l$:

$$\mathcal{F}f(p)=\int_{C_{0,0}[0,2\pi]} e^{i\int_0^{2\pi}p(u)x(u)du}f(x)dw^t_0(x), \tag{30}$$

where $p(u)\in C[0,2\pi]$ and $p(0)=p(2\pi)=0$. Unlike the usual Fourier transform for the Lebesgue measure on the real line, the transformation $\mathcal{F}$ is not a unitary operator.

**Proposition 4.1.** *The operator $\mathcal{F}$ is a compact normal operator on $H_l^0$ with no zero eigenvalues.*

**Proof.** The general condition [2] for the general integral operator $K : L^2(X, d\mu(x)) \to L^2(Y, d\nu(y))$, such that

$$Kf(y) = \int_X K(y,x)d\mu(x) \tag{31}$$

to be compact, is that $\int\int |K(x,y)|^2 d\mu(x)d\nu(y) < \infty$. In order to prove that it is normal, i.e. $\mathcal{F}\mathcal{F}^* = \mathcal{F}^*\mathcal{F}$, one just needs to write explicitly the resulting expressions and then use the Fubini theorem. To show that the operator $\mathcal{F}$ has no nonzero eigenvalues, one needs to use the fact that the exponentials of the form $e^{\int_0^{2\pi}\alpha'(u)dx(u)}$, where $\alpha(u)$ is absolutely continuous, form a dense subset in $H_l^0$ [17]. ∎

Taking into account that $\mathcal{F}$ is a normal operator and using the polar decomposition theorem, one can decompose it as $\mathcal{F} = U_{\mathcal{F}}K$, where $U_{\mathcal{F}}$ is a unitary operator on $H_l^0$ and $K$ is a compact self-adjoint operator, such that $K = \sqrt{\mathcal{F}\mathcal{F}^*}$.

It is obvious that the operators $\mathcal{F}$ and $U_{\mathcal{F}}$ can be continued to the space $H_l$. At the same time, we define another unitary operator $F$, which is a standard Fourier transform with respect to measure $dx_0$. One can show that $\mathcal{F}$ and $F$ commute.

**4.2. Loop Γ-function.** Let us consider the following expression:

$$\rho_{\lambda,k}^{l,\mathcal{F}}(e^\alpha, b, s) = U_{\mathcal{F}}\rho_{\lambda,k}^{l}(e^\alpha, b, s)U_{\mathcal{F}}^*. \tag{32}$$

Since $U_{\mathcal{F}}$ is unitary, $\rho_{\lambda,k}^{l,\mathcal{F}}$ defines an equivalent representation of $\hat{G}$ on $H_l$. Because of the results of subsection 4.1, on the image of $K$, one can rewrite it as follows:

$$\rho_{\lambda,k}^{l,\mathcal{F}}(e^\alpha, b, s) = K^{-1}\mathcal{F}\rho_{\lambda,k}^{l}(e^\alpha, b, s)\mathcal{F}^*K^{-1}. \tag{33}$$

Here, $K$ is a fixed self-adjoint operator, thus we are interested in the object $\mathcal{F}\rho_{\lambda,k}^{l}(e^\alpha, b)\mathcal{F}^*$. We consider the case when $(e^\alpha, b)$ in $\hat{G}_+$ and $Im\lambda = 0$, $Re\lambda(u) > 0$. Let us write it down explicitly:

$$\begin{aligned}
&\mathcal{F}\rho_{\lambda,k}^{l}(e^\alpha, b, s)\mathcal{F}^*f(x,x_0) = \\
&e^{is}e^{-\frac{1}{4t}\int_0^{2\pi}\alpha'(u)\alpha'(u)du}\int_{C_{0,0}[0,2\pi]} e^{-i\int_0^{2\pi}(p(u)x(u))du}e^{-\frac{1}{2t}\int_0^{2\pi}\alpha'(u)dp(u)} \\
&e^{ik\int_0^{2\pi}\alpha(u)dp(u)}e^{\int_0^{2\pi}\lambda(u)b(u)e^{p(u)+x_0}du} \\
&\int_{C_{0,0}[0,2\pi]} e^{i\int_0^{2\pi}(p(u)+\tilde{\alpha}(u)y(u)du}f(y,x_0+\alpha_0)dw_0^t(y)dw_0^t(p).
\end{aligned} \tag{34}$$

Using the Fubini theorem, one can rewrite it as follows:

$$\begin{aligned}
&\mathcal{F}\rho_{\lambda,k}^{l}(e^\alpha, b, s)\mathcal{F}^*f(x,x_0) = \\
&\int_{C_{0,0}[0,2\pi]} \mathbb{K}_{s,\alpha,b}^{\lambda,k}(x-y,x_0)e^{i\int_0^{2\pi}\tilde{\alpha}(u)y(u)du}f(y,x+\alpha_0)dw_0^t(y),
\end{aligned} \tag{35}$$

where

$$\mathbb{K}^{\lambda,k}_{s,\alpha,b}(x-y,x_0)=e^{is}e^{-\frac{1}{4t}\int_0^{2\pi}\alpha'(u)\alpha'(u)}$$
$$\int_{C_{0,0}[0,2\pi]} e^{i\int_0^{2\pi}p(u)(x(u)-y(u))du}e^{\int_0^{2\pi}\lambda(u)b(u)e^{p(u)+x_0}du}$$
$$e^{ik\int_0^{2\pi}\alpha(u)dp(u)}e^{-\frac{1}{2t}\int_0^{2\pi}\alpha'(u)dp(u)}dw_0^t(p). \tag{36}$$

In the case when $\alpha$ is twice differentiable, one can see that the object $\mathbb{K}^{\lambda,k}_{s,\alpha,b}(z,x_0)$ up to factors independent of $x, y$ is a particular case of the following functional:

$$\hat{\Gamma}_\mu(z)=\int_{C_{0,0}[0,2\pi]} e^{\int_0^{2\pi}p(u)z(u)du}e^{-\int_0^{2\pi}\mu(u)e^{p(u)}du}dw_0^t(p), \tag{37}$$

where $Rez, Imz\in L^2([0,2\pi];\mathbb{R})$, $\mu\in L^2([0,2\pi];\mathbb{R})$. We will call $\hat{\Gamma}_\mu(z)$ the *loop Gamma function* or simply $\hat{\Gamma}$-*functional*. It has the following properties [**33**].

**Theorem 4.1.**
*i)* $\hat{\Gamma}_\mu(z)$ *is well defined for any* $Rez, Imz\in L^2([0,2\pi];\mathbb{R})$, $\mu\in L^2([0,2\pi];\mathbb{R})$ *and* $\mu(u)\geq 0$ *on* $[0,2\pi]$.
*ii) The following relation is valid:*

$$\int_0^{2\pi} g(v)\mu(v)\hat{\Gamma}_\mu(z+\delta_v)dv=\int_0^{2\pi} g(v)z(v)dv\hat{\Gamma}_\mu(z)+$$
$$\frac{1}{t}\int_0^{2\pi} g''(v)\frac{\delta}{\delta z(v)}\hat{\Gamma}_\mu(z)dv, \tag{38}$$

*where* $g(v)$ *is any twice differentiable function on* $[0,2\pi]$, *such that* $g(0)=g(2\pi)=0$, $\delta_v=\delta(u-v)$ *is a delta-function on the interval* $[0,2\pi]$ *and* $\int_0^{2\pi}\xi(v)\frac{\delta}{\delta z(v)}\hat{\Gamma}_\mu(z)=\frac{d}{d\epsilon}_{|\epsilon=0}\hat{\Gamma}_\mu(z+\epsilon\xi)$ *for any continuous function* $\xi$.

**Proof.** An important step in the proof is the consideration of the infinitesimal form of the translation invariance. Let $f\in H_l^0$ such that it is weakly differentiable. Then we have the following property:

$$\int_{C_{0,0}[0,2\pi]} e^{-\frac{1}{2t}\int_0^{2\pi}\epsilon^2 g'(u)g'(u)du+\frac{1}{t}\int_0^{2\pi}\epsilon g''(u)x(u)du}f(x+\epsilon g)dw_0^t(x)=$$
$$\int f(x)dw^t(x), \tag{39}$$

where $g\in C^2_{0,0}[0,2\pi]$. This is just the translation property (the translation is with respect to function $\epsilon g$), where $\epsilon$ is some real parameter. Then, if we differentiate with respect to $\epsilon$ at zero, we obtain the following formula:

$$\int_{C_{0,0}[0,2\pi]}\Big(\frac{1}{t}\int_0^{2\pi} g''(u)x(u)du\Big)f(x)dw_0^t(x)=$$
$$-\int_{C_{0,0}[0,2\pi]}\frac{d}{d\epsilon}_{|\epsilon=0}f(x+\epsilon g)dw_0^t(x). \tag{40}$$

Let us apply this property to the integrand of the $\hat{\Gamma}$-functional, i.e. let $f(x) = F(x,z) \equiv e^{\int_0^{2\pi} x(u)z(u)du} e^{-\int_0^{2\pi} \mu(u)e^{x(u)}du}$. Then we have:

$$\int_{C_{0,0}[0,2\pi]} \Big(\frac{1}{t}\int_0^{2\pi} g''(u)x(u)\mathrm{d}u\Big)F(x,z)dw_0^t(x) =$$

$$\frac{1}{t}\int_0^{2\pi} g''(v)\frac{\delta}{\delta z(v)}\hat{\Gamma}_\mu(z) \tag{41}$$

and

$$-\int_{C_{0,0}[0,2\pi]} \frac{d}{d\epsilon}_{|\epsilon=0} F(x+\epsilon g, z)dw_0^t(x) =$$

$$-\int_{C_{0,0}[0,2\pi]} \Big(\int_0^{2\pi} g(v)z(v)dv - \int_0^{2\pi} \mu(v)g(v)e^{x(v)}dv\Big)F(x,z)dw_0^t(x) =$$

$$-\int_0^{2\pi} g(v)z(v)dv\hat{\Gamma}_\mu(z) + \int_0^{2\pi} g(v)\mu(v)\hat{\Gamma}(z+\delta_v)dv. \tag{42}$$

Therefore, combining (40), (41) and (42) we obtain (38). ■

The property ii) from the theorem above is a natural generalization of the property of the ordinary $\Gamma$-function: $\Gamma(z+1) = z\Gamma(z)$. We notice, however, that there is also an extra term, depending on the $t$-parameter, related to the choice of the measure. So, the proper analogue of the functional $\hat{\Gamma}$ would be the "regularized" $\Gamma$-function:

$$\Gamma_{\mu,t}(z) = \int_{\mathbb{R}} e^{-\mu e^x} e^{zx} e^{-\frac{x^2}{2t}} dx, \tag{43}$$

where $\mu, t \in R_+$. It satisfies the equation

$$\mu\Gamma_{\mu,t}(z+1) = z\Gamma_{\mu,t}(z) - \frac{1}{t}\frac{d\Gamma(z)}{dz}. \tag{44}$$

This function is well defined for all complex values of $z$. Its relation to the original $\Gamma$-function can be obtained by considering $t \to \infty$ limit:

$$\lim_{t\to\infty} \Gamma_{\mu,t}(z+1) = \mu^{-z}\Gamma(z). \tag{45}$$

Also, we note here that this function can be related to the matrix elements of the representations of $G$ (see Section 2) if we would consider the Gaussian measure instead of the Lebesgue measure on the real line.

## 5. Representations of $\widehat{\mathcal{A}}$ and $\widehat{\mathcal{K}}$

**5.1. Definitions.** In this section we consider the following *-algebras, which we denote $\widehat{\mathcal{A}}$ and $\widehat{\mathcal{K}}$ that are the affine analogues of $\mathcal{A}, \mathcal{K}$, considered in Section 2. The first algebra is close to the Lie algebra of a loop $ax+b$-group (see Section 3), which is generated by $h_n$, $e_n^{\pm}$, $n \in \mathbb{Z}$, so that the generating "currents" are:

$$h(u) = \sum_{n\in\mathbb{Z}} h_{-n}e^{inu}, \quad e^{\pm}(u) = \sum_{n\in\mathbb{Z}} e^{\pm}_{-n}e^{inu} \tag{46}$$

and obey the following commutation relations, expressed via the generating functions:

$$
\begin{aligned}
&[h(u),h(v)]=0,\quad [e^{\pm}(u),e^{\pm}(v)]=0,\quad [e^{\pm}(u),e^{\mp}(v)]=0,\\
&[h(u),e^{\pm}(v)]=\pm i\delta(u-v)e^{\pm}(v),\quad e^{\pm}(u)\cdot e^{\mp}(u)=1.
\end{aligned} \tag{47}
$$

The *-structure is such that $h(u)$, $e^{\pm}(u)$ are Hermitian, i.e.

$$
h(u)^*=h(u),\quad e^{\pm}(u)^*=e^{\pm}(u). \tag{48}
$$

The algebra $\widehat{\mathcal{K}}$ is generated by $h_n$, $\alpha_n^{\pm}$, $n\in\mathbb{Z}$, it has similar commutation relations, but a different *-structure:

$$
\begin{aligned}
&h(u)=\sum_n h_{-n}e^{inu},\quad \alpha^{\pm}(u)=\sum_n \alpha^{\pm}_{-n}e^{inu},\\
&[h(u),h(v)]=0,\quad [\alpha^{\pm}(u),\alpha^{\pm}(v)]=0,\quad [\alpha^{\pm}(u),\alpha^{\mp}(v)]=0,\\
&[h(u),\alpha^{\pm}(v)]=\mp\delta(u-v)\alpha^{\pm}(v),\quad \alpha^{\pm}(u)\alpha^{\mp}(u)=1,\\
&h(u)^*=h(u),\quad \alpha^{\pm}(u)^*=\alpha^{\mp}(u).
\end{aligned} \tag{49}
$$

Using Gaussian integration on Hilbert spaces, in the next subsection we will construct some representations of these algebras.

**5.2. Construction of representations.** In order to construct the representations of $\widehat{\mathcal{A}}$ and $\widehat{\mathcal{K}}$ one can use the Gaussian measure on a Hilbert space (see Appendix). Let us consider the Fourier series of a function from $L^2(S^1,\mathbb{R})$:

$$
x(u)=\sum_{n\in\mathbb{Z}} x_{-n}e^{inu},\quad x_0\in\mathbb{R},\quad x_n^*=x_{-n}. \tag{50}
$$

Let us introduce two quadratic forms defining two types of trace-class operators on $L^2(S^1,\mathbb{R})$, which will determine the appropriate Gaussian measures:

$$
\begin{aligned}
&B_A(x,x)=\frac{1}{2}\sum_{n\geq 1}\xi_n^{-1}x_nx_{-n}+\xi_0^{-1}x_0^2,\\
&B_K(x,x)=\frac{1}{2}\sum_{n\geq 1}\xi_n^{-1}x_nx_{-n},
\end{aligned} \tag{51}
$$

where $\xi_n>0$ for all $n$ and $\sum_n\xi_n<\infty$. The Gaussian measures we are interested in, heuristically can be expressed as follows:

$$
\begin{aligned}
&dw_A=(\sqrt{det(2\pi N_A)})^{-1}e^{-B_A(x,x)}dx_0\prod_{n=1}^{\infty}[\frac{i}{2}dx_n\wedge dx_{-n}],\\
&dw_K=(\sqrt{det(2\pi N_K)})^{-1}e^{-B_K(x,x)}d\phi\prod_{n=1}^{\infty}[\frac{i}{2}dx_n\wedge dx_{-n}],
\end{aligned} \tag{52}
$$

where $N_A,N_K$ are trace-class diagonal operators determined by the quadratics forms (51). Here as before the range for $\phi$ is $[0,2\pi]$. Literally the difference between two measures is that in the second one we compactified the zero mode $x_0$ on a circle with parameter $\phi$ for $dw_K$. Hilbert spaces of square-integrable functions with respect to these measures are denoted in the following as $\mathcal{H}_A$ and $\mathcal{H}_K$.

Let us construct the unitary representations of $\widehat{\mathcal{A}}$ and $\widehat{\mathcal{K}}$-algebras in the Hilbert spaces $\mathcal{H}_A$ and $\mathcal{H}_K$, correspondingly.

We will explicitly define the operators, which will represent the generators. Let us start from $\widehat{\mathcal{A}}$-algebra. First of all, let us extend the index of $\xi_n$ to all integers, so that $\xi_{-n}=\xi_n$, $n\in\mathbb{Z}$. We need the following differential operators:

$$b_{-n}=i(\partial_n-\xi_n^{-1}x_{-n}),\quad a_{-n}=i\partial_n, \tag{53}$$

where $\partial_n=\frac{\partial}{\partial x_n}$. These operators are formally conjugate to each other,

$$a_n^*=b_{-n}, \tag{54}$$

being considered on a certain dense set $D_A^l$, i.e. functions which are the sums of monomials

$$\prod_{k=1}^{n}\langle\mu_k,x\rangle\prod_{s=1}^{m}\langle\lambda_s,e^{\pm x}\rangle, \tag{55}$$

where $\langle,\rangle$ is the standard $L^2(S^1,\mathbb{R})$ pairing and $\mu_k,\lambda_k$ are trigonometric polynomials. We define the operators $h_n$ as follows:

$$h_{-n}=\frac{1}{2}(a_{-n}+b_{-n})=i(\partial_n-\frac{1}{2}\xi_n^{-1}x_{-n}). \tag{56}$$

Hence, on a dense set $h_n^*=h_{-n}$, so that the current $h(u)=\sum_{n\in\mathbb{Z}}h_{-n}e^{inu}$ is Hermitian. Also, note that Hermitian currents $h(u),x(v)$ generate the infinite-dimensional Heisenberg algebra:

$$[h(u),x(v)]=i\delta(u-v). \tag{57}$$

We also define the currents $e^{\pm}(u)$:

$$e^{\pm}(u)=e^{\pm x(u)}. \tag{58}$$

It also follows that they satisfy the commutation relations (47). One can show that $e^{\pm x(u)}$ for any $u$ are Hermitian operators, considered on the dense set $D_A^l$. Therefore, this gives a unitary representation of $\widehat{\mathcal{A}}$-algebra.

Similarly one can construct the unitary representations of $\widehat{\mathcal{K}}$-algebra. Let us consider the dense set $D_A^l$ in $\mathcal{H}_K$ of the following form:

$$\prod_{k=1}^{n}\langle\mu_k,x\rangle\prod_{s=1}^{m}\langle\lambda_s,e^{\pm ix^c}\rangle, \tag{59}$$

where $\lambda_i$ are trigonometric polynomials without the constant term and $x^c(u)=\phi+\sum_{n\neq 0}x_ne^{-inu}$. The currents $h(u),\alpha^{\pm}(u)$ are defined by the following formulas:

$$\begin{aligned} h(u)&=\sum_{n\neq 0}i(\partial_n-\frac{1}{2}\xi_n^{-1}x_{-n})e^{inu}+i\partial_\phi,\\ \alpha^{\pm}(u)&=e^{\pm ix^c(u)}. \end{aligned} \tag{60}$$

It is possible to introduce operators $a_n,b_n$ and define them by the same formulas as in (53) for all $n\neq 0$. For $n=0$ we put $a_0=b_0=i\partial_\phi$. Therefore, the elements of the form $a\cdot v_0$, where $v_0\equiv 1\in\mathcal{H}_K$ and $a$ belongs to universal enveloping algebra of $\widehat{\mathcal{K}}$ and the action of the generators is given by the formulas (60). Such elements generate a dense set $D_K^l$ in $\mathcal{H}_K$, which is a unitary representation of $\widehat{\mathcal{K}}$-algebra.

An important notion which is necessary for our construction is the correlator associated with the representation.

By the correlator of generators $T_1, \ldots, T_n$ of $\widehat{\mathcal{A}}$-algebra (resp. $\widehat{\mathcal{K}}$-algebra) we mean the following expression:

$$< T_1 \ldots T_n > \equiv \langle v_0, T_1 \ldots T_n v_0 \rangle, \tag{61}$$

where the pairing $\langle , \rangle$ is of the Hilbert space $\mathcal{H}_A$ (resp. $\mathcal{H}_K$), $v_0$ is the vector corresponding to the constant function 1 and $T_1, \ldots, T_n$ are the generators $h_n, e^{\pm}_m$ (resp. $h_n, \alpha^{\pm}_m$).

We remind that $h_n = \frac{1}{2}(a_n + b_n)$. It is crucial for the following, that the correlators

$$< T_1 \ldots T_n a_k > \quad , \quad < b_k T_1 \ldots T_n > \tag{62}$$

vanish for any generators $T_1, \ldots, T_n$, as a simple consequence of properties of the Gaussian integration.

It is natural to call operators $a_n$ *annihilation* operators and $b_n$ *creation* operators. This allows us to define the normal ordering. Namely, when we write down the expression $: T_1 \ldots T_n :$ for the product of $n$ generators, we reorder them in such a way that the creation operators will be to the left and the annihilation ones to the right.

This procedure together with the vanishing of the correlators (62) also gives an easy method to compute the correlators: by means of commutation relations of generators, one can reduce the products $T_1, \ldots, T_n$ to the normally ordered expressions. Therefore, the result will reduce to the correlators of the generators $e^{\pm}_n$ or $\alpha^{\pm}_n$. We also note that in the case of $\widehat{\mathcal{K}}$-algebra the corresponding correlators are nonzero only if they have an equal number of $\alpha^+$ and $\alpha^-$ generators.
In order to compute the correlators of these generators it is easier to consider the appropriate currents instead of modes and use the Gaussian integration.

There are the following expressions for correlation functions:

$$\begin{aligned}
&\langle e_+(u_1) \ldots e_+(u_n) e_-(v_1) \ldots e_-(v_m) \rangle = \\
&\exp\Big( \sum_{i<j; i,j=1}^{n} N_A(u_i, u_j) + \sum_{r<s; r,s=1}^{m} N_A(v_r, v_s) - \sum_{k=1}^{n} \sum_{l=1}^{m} N_A(u_k, v_l) + \\
&\frac{n+m}{2} N_A(0,0)\Big),
\end{aligned} \tag{63}$$

$$\begin{aligned}
&\langle \alpha_+(u_1) \ldots \alpha_+(u_n) \alpha_-(v_1) \ldots \alpha_-(v_m) \rangle = \\
&\delta_{n,m} \exp\Big(- \sum_{i<j; i,j=1}^{n} N_K(u_i, u_j) - \sum_{i<j; i,j=1}^{n} N_K(v_i, v_j) + \sum_{k,l=1}^{n} N_K(u_k, v_l) \\
&+ n N_K(0,0)\Big),
\end{aligned} \tag{64}$$

where

$$\begin{aligned}
N_A(u, v) &= 2 \sum_{n \geq 0} cos(n(u - v)) \xi_n, \\
N_K(u, v) &= 2 \sum_{n > 0} cos(n(u - v)) \xi_n.
\end{aligned} \tag{65}$$

We note here that in the case of correlators of the currents involving not only $e^{\pm}, \alpha^{\pm}$, but also $h(u)$, the resulting expression will consist of monomials (64) multiplied on a certain product of delta-functions, coming from commutation relations of $\widehat{\mathcal{K}}$, $\widehat{\mathcal{A}}$-algebras.

We have shown above, that using the Gaussian measure, one can construct unitary representations of $\widehat{\mathcal{A}}$, $\widehat{\mathcal{K}}$-algebras. However, it is possible to simplify those representations by making $a_k$ commute with $b_s$ for any $k$ and $s$. However, the resulting module will not be unitary, i.e. the pairing though nondegenerate will lose its positivity.

At first, let us give the explicit description of such module for $\widehat{\mathcal{K}}$-algebra. For this we consider the vacuum vector $v_0$, such that $v_0$ is annihilated by $a_k$:

$$a_k v_0 = 0. \tag{66}$$

Then the module which we will refer to as $\mathcal{V}_A$ is spanned by the following vectors:

$$b_{m_1}\dots b_{m_s} e^{\pm}_{n_1}\dots e^{\pm}_{n_r} v_0, \tag{67}$$

where $n_1,\dots,n_r,m_1,\dots,m_s \in \mathbb{Z}$. Let us begin to define the pairing with the postulation of the points: $e(u)=\sum_n e_n e^{inu}$ is a Hermitian current, $b(u)^* = a(u)$, so that $b(u)=\sum_n b_n e^{-inu}$, $a(u)=\sum_n a_n e^{-inu}$ and

$$[b(u),a(v)] = 0. \tag{68}$$

The pairing is uniquely defined by the correlator of $e^{\pm}$ currents is given by (63). Similarly one can define the module $\mathcal{V}_K$ for $\widehat{\mathcal{K}}$ algebra with the same conditions, just replacing $e^{\pm}$ with $\alpha^{\pm}$, as a result $\mathcal{V}_K$ is spanned by

$$b_{m_1}\dots b_{m_s}\alpha^{\pm}_{n_1}\dots\alpha^{\pm}_{n_r} v_0. \tag{69}$$

One can define the pairing on $\mathcal{V}_K$ which is uniquely determined by the correlators of $\alpha^{\pm}$ currents (64). This pairing is Hermitean and nondegenerate. It gives a structure of nonunitary representations of the *-algebras $\widehat{\mathcal{A}}$, $\widehat{\mathcal{K}}$ on the spaces $\mathcal{V}_A$, $\mathcal{V}_K$ correspondingly.

We note here that clearly, these representations are nonunitary, because you can easily find vectors $v\in\mathcal{V}_A,\mathcal{V}_K$, such that $\langle v,v\rangle = 0$.

## 6. Construction of representations for $\widehat{sl(2,\mathbb{R})}$

**6.1. Regularization and commutator.** In order to construct the $\widehat{sl(2,\mathbb{R})}$ representations via $\widehat{\mathcal{A}}$, $\widehat{\mathcal{K}}$, one has to find an affine analogue of the formulas for $E,F,H$ and $J^3,J^{\pm}$. In the case of nontrivial central extension, the $\widehat{\mathcal{A}}$, $\widehat{\mathcal{K}}$ representations appear to be insufficient, one has to introduce the representation of the infinite-dimensional Heisenberg algebra, so that the generating current is $\rho(u)=\sum_{n\in\mathbb{Z}}\rho_n e^{-inu}$ and the commutation relations are:

$$[\rho_n,\rho_m] = 2\kappa n\delta_{n,-m}, \tag{70}$$

where $\kappa\in\mathbb{R}_{>0}$. The irreducible module, the so-called Fock module $F_{\kappa,p}$ of this algebra is defined as follows. We introduce a vector $vac_p$ with the property $\rho_n vac_p = 0$, $p\in\mathbb{R}$, $n>0$ so that

$$F_{\kappa,p} = \{\rho_{-n_1}\dots\rho_{-n_k} vac_p;\quad n_1,\dots,n_k>0,\quad \rho_0 vac_p = p\cdot vac_p\}. \tag{71}$$

The Hermitian pairing is defined so that

$$\langle vac_p, vac_p\rangle = 1,\quad \rho_n^* = \rho_{-n}. \tag{72}$$

Another object required in this section is the regularized version of the $\rho$, $h$, $e^{\pm}$, $\alpha^{\pm}$ currents. Namely, for any $\varphi$, which stands for any of $\rho$, $h$, $x(u)$ or $x^c(u)$ we consider

$$\varphi(z,\bar{z}) = \sum_{n\geq 0}\varphi_n\bar{z}^n + \sum_{n>0}\varphi_{-n}z^n, \tag{73}$$

where $z = re^{iu}$, so that $0 < r \leq 1$. We denote $e^{\pm}(z,\bar{z}) \equiv e^{\pm x(z,\bar{z})}$ and $\alpha^{\pm}(z,\bar{z}) \equiv e^{\pm ix^c(z,\bar{z})}$. The Wick theorem implies that the correlators of the regularized Heisenberg currents $\rho(z_1,\bar{z}_1),\dots,\rho(z_n,\bar{z}_n)$ are finite as long as $0 < |z_i| < 1$, i.e. the expressions

$$\langle vac_p, \rho(z_1,\bar{z}_1)\dots\rho(z_n,\bar{z}_n)vac_p\rangle \tag{74}$$

are finite. One can consider the limit $|z_i| \to 1$ in the sense of distributions, so that (74) is the sum of products of distributions, i.e.

$$\langle vac_p, \rho(u_1)\rho(u_2)vac_p\rangle = \frac{2\kappa}{(1-e^{i(u_2-u_1-i0)})^2} + p^2. \tag{75}$$

Next we consider the following sets of composite regularized currents

$$\begin{aligned} &J^{\pm}(z,\bar{z}) = \\ &\frac{i}{2}(b(z,\bar{z})\alpha^{\pm}(z,\bar{z}) + \alpha^{\pm}(z,\bar{z})a(z,\bar{z})) \pm \kappa\partial_u\alpha^{\pm}(z,\bar{z}) \pm \rho(z,\bar{z})\alpha^{\pm}(z,\bar{z}),\\ &J^3(z,\bar{z}) = 2ih(z,\bar{z}) - 2\kappa\alpha^-(z,\bar{z})\partial_u\alpha^+(z,\bar{z}) \end{aligned} \tag{76}$$

and

$$\begin{aligned} &E(z,\bar{z}) = \\ &\frac{i}{2}(b(z,\bar{z})e^+(z,\bar{z}) + e^+(z,\bar{z})a(z,\bar{z})) + i\kappa\partial_u e^+(z,\bar{z}) + i\rho(z,\bar{z})e^+(z,\bar{z}),\\ &F(z,\bar{z}) = \\ &-\frac{i}{2}(b(z,\bar{z})e^-(z,\bar{z}) + e^-(z,\bar{z})a(z,\bar{z})) + i\kappa\partial_u e^-(z,\bar{z}) + i\rho(z,\bar{z})e^-(z,\bar{z}),\\ &H(z,\bar{z}) = -2ih(z,\bar{z}) + 2i\kappa e^-(z,\bar{z})\partial_u e^+(z,\bar{z}). \end{aligned} \tag{77}$$

satisfying the Hermicity conditions:

$$\begin{aligned} &E(z,\bar{z})^* = -E(z,\bar{z}), \quad F(z,\bar{z})^* = -F(z,\bar{z}), \quad H(z,\bar{z})^* = -H(z,\bar{z}),\\ &J^3(z,\bar{z})^* = -J^3(z,\bar{z}), \quad J^{\pm}(z,\bar{z})^* = -J^{\mp}(z,\bar{z}). \end{aligned} \tag{78}$$

Then if $\phi_k$ denote $E, F, H$ or $J^3, J^{\pm}$ then the correlators

$$\langle\phi_1(z_1,\bar{z}_1)\dots\phi_n(z_n,\bar{z}_n)\rangle_p \equiv \langle v_0\otimes vac_p, \phi_1(z_1,\bar{z}_1)\dots\phi_n(z_n,\bar{z}_n)v_0\otimes vac_p\rangle, \tag{79}$$

are well-defined for $0 < |z_i| < 1$. Actually, after the normal ordering procedure $\phi_1(z_1,\bar{z}_1)\dots\phi_n(z_n,\bar{z}_n)$ will be represented as normally ordered products with coefficients which are continuous functions of $u_1,\dots,u_n$, where $z_i = r_ie^{u_i}$. Therefore, the expression $\langle\phi_1(z_1,\bar{z}_1)\dots\phi_n(z_n,\bar{z}_n)\rangle_p$ will be given by the sum of correlators of exponentials $\alpha^{\pm}$ or $e^{\pm}$ which are well-defined even on the unit circle as we know from the formulas (63). Notice, that if only one of the currents $\phi_k(z_k,\bar{z}_k)$ is such that $|z_k| = e^{iu_k}$, i.e. it lies on the unit circle, the situation will not change. This is because during the normal ordering procedure, the commutators of $a$- and $b$- parts of this generator with other ones again produce continuous functions of $u_1,\dots,u_n$, since all other $z_i$ are inside the unit circle and the normal ordering wouldn't produce any distributions.

This allows us to make sense of the commutator of two currents on the unit circle. We consider the following limit of the difference of two correlators

$$\begin{aligned}(80)\qquad &\langle\phi_1(z_1,\bar z_1)\dots[\xi(u_1),\eta(u_2)]\dots\phi_n(z_n,\bar z_n)\rangle_p\equiv\\ &\lim_{r_1,r_2\to 1}\big(\langle\phi_1(z_1,\bar z_1)\dots\xi(w_1,\bar w_1)\eta(w_2,\bar w_2)\dots\phi_n(z_n,\bar z_n)\rangle_p-\\ &\langle\phi_1(z_1,\bar z_1)\dots\eta(w_2,\bar w_2)\xi(w_1,\bar w_1)\dots\phi_n(z_n,\bar z_n)\rangle_p\big),\end{aligned}$$

Then the following result holds (see [**11**] Proposition 4.2, Theorem 4.1).

**Theorem 6.1.** *i) The expression (80) exists in the sense of distributions, more specifically the answer will contain a linear combination of delta-functions $\delta(u_1-u_2)$ and their derivatives. Here $\phi_k$ stands for $E,F,H$ or $J^3,J^\pm$, $w_i=r_ie^{iu_i}$, $\xi(u_1)\equiv\xi(e^{iu_1},e^{-iu_1})$, $\eta(u_2)\equiv\eta(e^{iu_2},e^{-iu_2})$,*
*ii) The commutator (80) of the currents $\xi,\eta=E,F,H$ or $\xi,\eta=J^3,J^\pm$ exists and satisfies the commutation relations for $\widehat{sl(2,\mathbb{R})}$ algebra with the central charge $\kappa$:*

$$\begin{aligned}&[E(u),F(v)]=H(v)\delta(u-v)-4i\kappa\delta'(u-v),\quad [H(u),H(v)]=8i\kappa\delta'(u-v),\\ (81)\quad &[H(u),E(v)]=2E(v)\delta(u-v),\quad [H(u),F(v)]=-2F(v)\delta(u-v)\end{aligned}$$

*and*

$$\begin{aligned}&[J^+(u),J^-(v)]=iJ^3(v)\delta(u-v)+4i\kappa\delta'(u-v),\\ &[J^3(u),J^3(v)]=-8i\kappa\delta'(u-v),\\ (82)\qquad &[J^3(u),J^\pm(v)]=\pm 2iJ^\pm(v)\delta(u-v).\end{aligned}$$

It is important to mention that though we managed to define the commutator, based on the regularized commutators, this definition doesn't provide a representation, since the correlation functions of $E,F,H$ or $J^3,J^\pm$ do not exist, if more than one of the arguments lies on a circle. Then it is clear that the spaces from which we started, i.e. $\mathcal{H}_K\otimes F_p$ or $\mathcal{H}_A\otimes F_p$ are not suitable to be spaces for $\widehat{sl(2,\mathbb{R})}$-module. However, we still have the regularized correlators which obey commutation relations. If we manage to eliminate divergencies in such a way that commutation relations and Hermicity conditions would be preserved, then the correlators will determine representation with the Hermitian bilinear form. In the next subsection we show a method, how to get rid of the divergencies and redefine the correlator, so that it is well-defined when all the arguments are on a circle.

**6.2. Renormalization of correlators and construction of representations of $\widehat{sl}(2,\mathbb{R})$.** In this subsection, we will renormalize the correlators with currents on a circle for the cases of $\mathcal{V}_A$ and $\mathcal{V}_K$ representations. In both cases the description is very similar, so we focus on $\mathcal{V}_K$ case and the correlators of $J^3,J^\pm$ currents. Generalization to $\mathcal{V}_A$ and $E,F,H$ goes along the same path.

In order to renormalize, at first we have to understand what kind of divergencies we are dealing with. For that purpose it is convenient to write down the expression for the correlator in the graphic form using the Feynman-like diagrams.

Recall, that the easiest way to compute the correlator for $\widehat{\mathcal{K}}$ is to reduce the whole expression to the normal ordered form, i.e. all the creation operators $b(z,\bar z)$ are moved to the left and all the annihilation operators $a(z,\bar z)$ to the right.

Once we move the creation operator $b(z,\bar{z})$, which is part of a certain generator at the point to the left (or the annihilation operator $a(z,\bar{z})$ to the right), it may produce the following terms arising from the commutation with the $\alpha^{\pm}$-generator of $J^{\pm}$ currents on the right (on the left in the case of $a(z,\bar{z})$) of the given generator:

$$\begin{aligned}[a(z,\bar{z}),\alpha^{\pm}(w,\bar{w})] &= \mp\alpha^{\pm}(w,\bar{w})\delta(z,w),\\ [\alpha^{\pm}(w,\bar{w}),b(z,\bar{z})] &= \pm\alpha^{\pm}(z,\bar{z})\delta(z,w),\end{aligned} \tag{83}$$

where $\delta(z,w) = \sum_{n\geq 0}(z\bar{w})^n + \sum_{n>0}(\bar{z}w)^n$. If $z,w$ are on the circle, i.e. $z = e^{iu}, w = e^{iv}$, we get $\delta(z,w) = \delta(u-v)$. We will depict every term of the form (83), which we obtain during the normal ordering procedure, as a line from one vertex to another:

(84)

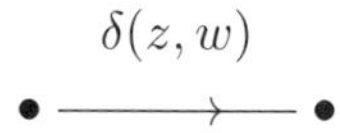


Here the initial vertex corresponds to the term containing creation/annihilation operator and the terminal vertex correspond to the term containing $\alpha^{\pm}$. The direction of the arrow (to the right or to the left) indicates whether it was annihilation or creation operator.

At the same time, each generator $J^{\pm}$ contributes terms like that

$$\alpha^{\pm}(z,\bar{z})a(z,\bar{z}), \quad b(z,\bar{z})\alpha^{\pm}(z,\bar{z}). \tag{85}$$

They enter the diagram as vertices with one outgoing line to the right or to the left and any amount of incoming lines:

(86)

Signs $\pm$ over the vertices correspond to $\alpha^{\pm}$, while we neglect for simplicity the dependence on $z,\bar{z}$ variables. The incoming line in the vertex forms when during the normal ordering procedure the creation/annihilation operators from other vertices leave the commutator term (83) with $\alpha^{\pm}$ at a given vertex. The outgoing line forms when the creation/annihilation operator of a given vertex leaves a commutator term $\alpha^{\pm}$ from another vertex.

On the other hand, $J^{\pm}$ also have terms of the form

$$\kappa\partial_u\alpha^{\pm}(z,\bar{z}), \quad \rho(z,\bar{z})\alpha^{\pm}(z,\bar{z}). \tag{87}$$

According to the strategy formulated, we denote the first term from (87) as a terminal vertex for graphs, because it contains only $\alpha^{\pm}$:

(88)

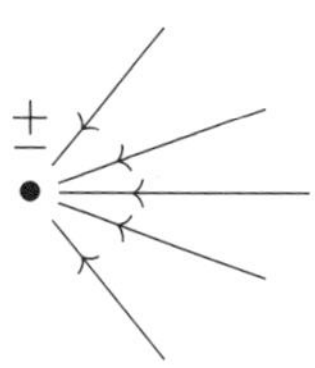

However the second term from (87) is composite: it has contribution from $\alpha^{\pm}(z,\bar{z})$ and $\rho(z,\bar{z})$. We consider this term as a terminal vertex for the graphs coming from the normal ordering of $a, b$-operators, however, we should add one outgoing "wavy" line, corresponding to the normal ordering in the Fock space:

(89) 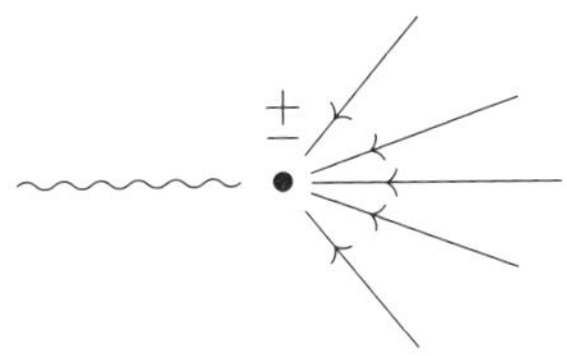

Each wavy line produces the term from 2-point correlator (75) and may appear only once in the connected graph: it is so, because of the combinatorial formula, expressing the Fock space n-point correlator via 2-point correlators.

Finally, there are terms coming from $J^3$-generator, namely

$$\frac{1}{2}a(z,\bar{z}), \quad \frac{1}{2}b(z,\bar{z}), \tag{90}$$

$$2\kappa\alpha^{-}(z,\bar{z})\partial_u\alpha^{+}(z,\bar{z}). \tag{91}$$

According to our conventions, we denote first two elements (90) as initial vertices with one outgoing line to the right and to the left correspondingly:

(92) 

Zeros over the vertices represent the absence of $\alpha^{\pm}$ contribution from these terms. Finally, (91) will correspond to the terminal vertex, because it has neither creation or annihilation operators:

(93) 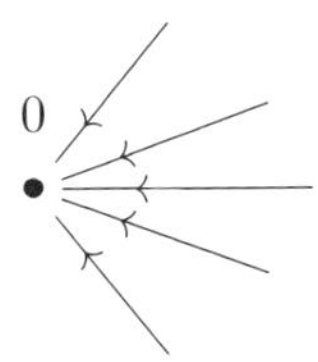

However, in this exceptional case we will consider every line as a sum of two terms contributing $\delta(z,w)$, because there are two $\alpha$'s in this vertex corresponding to the term $\alpha^{-}(z,\bar{z})\partial_u\alpha^{+}(z,\bar{z})$, so we consider contribution of the commutator with each of them. Moreover, formally on the circle (if such a limit exists), there can be only one incoming line into this vertex, because of the commutation relations:

$$[a(u), \alpha^{+}(v)\partial_v\alpha^{-}(v)] = [b(u), \alpha^{+}(v)\partial_v\alpha^{-}(v)] = -\delta'(u-v). \tag{94}$$

Here are two sample diagrams

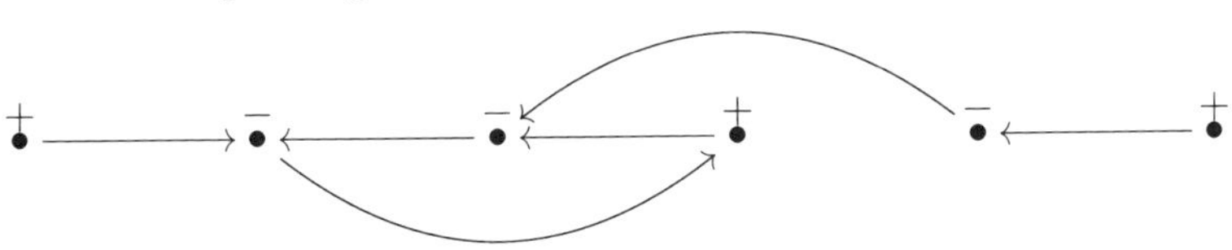

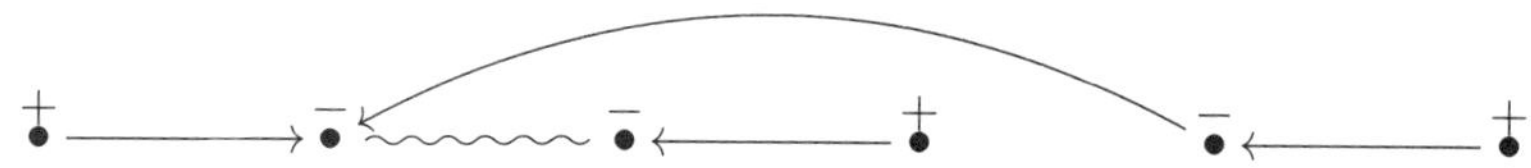

which belong to the expansion of the 6-point correlator

$$\langle J^+(z_1,\bar z_1)J^-(z_2,\bar z_2)J^-(z_3,\bar z_3)J^+(z_4,\bar z_4)J^-(z_5,\bar z_5)J^+(z_6,\bar z_6)\rangle. \tag{95}$$

Notice that every connected graph contributing to the correlator

$$\langle \phi_1(z_1,\bar z_1)\ldots.\phi_n(z_n,\bar z_n)\rangle_p, \tag{96}$$

where $\phi_i = J^3, J^{\pm}$, has at most one loop. Actually, each vertex has possibly many incoming lines, but at most one outgoing line. Let us consider the graph with one loop. It means that all outgoing lines of the vertices in this loop are included in the loop and all other lines are incoming. Therefore none of these vertices can participate in a different loop, since all outgoing lines are included in the first loop. Let us assume that there is another loop with different set of vertices in the same graph. However, since for all vertices participating all outgoing lines are "circulating" inside the loop there is no possibility to make a connected graph out of two loops.

However, we see that each loop diagram, though producing a well-defined expression for currents inside the circle, in the limit $|z_i|\to 1$ produce a divergence of the type $\delta(0)$, since every line in the loop produces delta-function. It is easy to see, say in the case of the correlator that

$$\langle J^+(z_1,\bar z_1)J^-(z_2,\bar z_2)\rangle_p. \tag{97}$$

When the creation operator from $J^-$ contributes the commutator term (83) with the exponent in $J^+$ and at the same time, the annihilation operator from $J^-$ contributes the commutator term (83) with the exponent in $J^+$, the following diagram is produced:

(98)

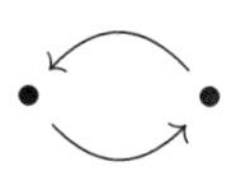

and leads to the divergent term $\delta(u_1-u_2)\cdot\delta(u_1-u_2)$ when considered on the circle.

The simplest way to regularize correlators to preserve commutation relations is to throw away all the loop diagrams. This allows to make sense of the correlators on a circle and leads to the following Theorem [**11**].

**Theorem 6.2.a.** *Let us consider the renormalized correlators*

$$\langle \phi_1(z_1,\bar z_1)\ldots.\phi_n(z_n,\bar z_n)\rangle_p^R, \tag{99}$$

*where $\phi_i = J^3, J^{\pm}$, which contain only the tree graph contributions to (96). Then the limit $r_k\to 1$ (where $z_k = r_ke^{iu_k}$) of (99) exists as a distribution. Moreover, the commutation relation between $J^3, J^{\pm}$ under the renormalized correlator reproduces $\widehat{sl(2,\mathbb{R})}$. Therefore, the correlators (99) considered on a circle define a module for $\widehat{sl(2,\mathbb{R})}$ with the Hermitian bilinear form.*

**Proof.** To prove this theorem, it is enough to look at relevant contibutions for the commutators. It appears that they all come from initial/terminal vertices and their closest neighbors. For example, let us consider the diagram of this sort:

(100) 

corresponding to the correlator

$$\langle \dots J^+(z_1,\bar{z}_1)J^-(z_2,\bar{z}_2)\dots\rangle_p. \tag{101}$$

There is a diagram with equal contribution

(102) 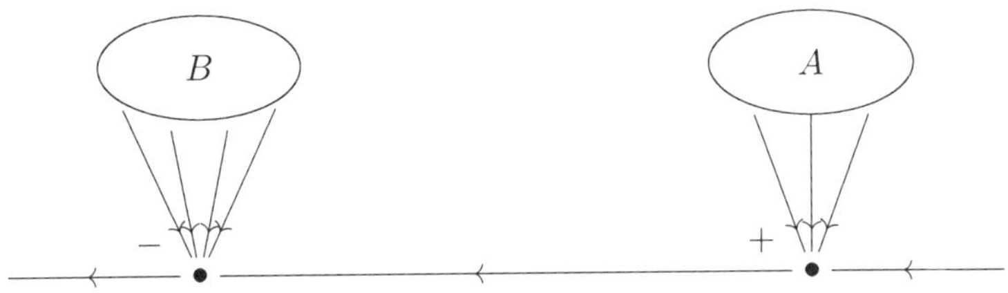


from the correlator

$$\langle \dots J^-(z_2,\bar{z}_2)J^+(z_1,\bar{z}_1)\dots\rangle_p. \tag{103}$$

Therefore, they cancel each other when we take a commutator of $J^+(u_1), J^-(u_2)$, since the middle line produces $\delta(u_1-u_2)$. One can understand it in the following way: the commutator $[J^+(u_1), J^-(u_2)]$ involves $J^3$-term and $\delta'$-term, $J^3$-term contains only initial and terminal vertices, therefore the diagrams, which contribute to this commutator should have also $J^+$ or $J^-$ as initial or terminal vertices. As for the commutators of $J^3$ with $J^\pm$, the picture above indicates that only terminal/initial vertices and their closest neighbors contribute, because again, all terms in $J^3$ are depicted as initial/terminal vertices. Therefore, it is clear that one can cancel all loop contributions to the correlators and still the commutation relations of Theorem 6.1 will be valid. However, if all the loops are eliminated, one can consider the limit $|z_i|\to 1$, putting all currents on a circle, since all divergent graphs are gone. Therefore, we have the well-defined correlators of the generators of $\widehat{sl(2,\mathbb{R})}$ Lie algebra. One can see that these correlators define a Hermitian bilinear form, because after the conjugation one can obtain one-to-one correspondence between tree graphs in conjugated correlators. ■

It appears that the theorem above can be generalized: instead of eliminating loops completely, one can renormalize them, i.e. eliminate the divergence of the type $\delta(0)$. Namely, one can associate a real number $\mu_k$ with every loop which is the contribution of $k$ $J^\pm$-currents at the points $z_1,\dots,z_k$. This number $\mu_k$ enters the loop in the following way. Suppose currents $J^\pm(z_i)$ appear in the correlator in the given order from left to right. Our loop contains the following term

$$\delta(z_1,z_2)\cdot\delta(z_2,z_3)\dots\delta(z_{k-1},z_k)\cdot\delta(z_k,z_1), \tag{104}$$

where $\delta(z,w)=\sum_{n\geq 0}(z\bar{w})^n+\sum_{n<0}(w\bar{z})^n$. In the limit $|z_k|\to 1$ we will substitute it with the expression

$$\mu_k\cdot\delta(u_1-u_2)\cdot\delta(u_2-u_3)\ldots\delta(u_{k-1}-u_k), \tag{105}$$

where we remind that $z_k=r_ke^{iu_k}$. This procedure again does affect neither commutation relations inside the correlator nor Hermicity condition for the resulting bilinear form, because loops do not participate in commutation relations. Therefore, we have a new version of the theorem above [**11**].

**Theorem 6.2.b.** *Let us consider the renormalized correlators*

$$\langle\phi_1(z_1,\bar{z}_1)\ldots.\phi_n(z_n,\bar{z}_n)\rangle_p^{R,\{\mu_n\}}, \tag{106}$$

*where* $\phi_i=J^3,J^{\pm}$, *such that the loop contributions to* (96) *are replaced by their renormalized analogues with the family of arbitrary real parameters* $\{\mu_n\}$. *Then the limit* $r_k\to 1$ *(where* $z_k=r_ke^{iu_k}$*) of* (106) *exists and the commutation relations between* $J^3,J^{\pm}$ *under the renormalized correlator reproduce Lie algebra* $\widehat{sl(2,\mathbb{R})}$, *and correlators* (106) *considered on a circle define a module for* $\widehat{sl(2,\mathbb{R})}$ *with the Hermitian bilinear form.*

**Proof.** The proof goes along the same lines as in the part $a$ of the theorem. When we consider the loop contribution to the commutator, there will be two loop diagrams, each emerging from one of the terms precisely as in the pictures (100), (101) which again will cancel each other, because of the delta function line between two vertices and because there always is one incoming line and outgoing line for each vertex associated with the currents, participating in the commutator in these two terms. Therefore, the commutation relations are unaffected in the presence of loops. It is easy to see that the same applies to the Hermicity condition. ∎

We mention that in this article the construction of $\widehat{sl}(2,\mathbb{R})$-modules based on unitary modules of $\mathcal{K}$-algebra is not considered. The renormalization procedure in that case is very similar, the difference is that there are several tree diagrams in teh expansion of correlators, which also require renormalization procedure. For more details on the treatment of unitary representations of unitary modules of $\mathcal{K}$-algebra one can consult [**11**].

Finally, it is important to remark that the realization of $\widehat{sl(2,\mathbb{R})}$ from Section 6.1 can be generalized in the higher rank case via multiple $\mathcal{K}$-algebras leading to the formulas similar to (76), (77) (in the case of quantum group there is a close construction using algebras of quantum planes [**19**], [**20**]). It will be a pure technical problem to modify Theorem 6.2.a and 6.2.b in the case of a higher rank.

## Appendix: Classical Wiener Measure and Gaussian Integration on Hilbert spaces

**A.1. A brief review of the abstract approach to Wiener measure.** In this section we collect all the necessary facts about the Wiener measure. For a more detailed exposition of this subject one can consult [**23**], [**17**], [**14**], [**24**], [**8**].

The abstract Wiener measure is the Gaussian measure on a Banach space with certain properties. The construction is as follows. We start from the Hilbert space

$\mathcal{H}$ and consider a Gaussian measure associated with the unital operator and the norm $||\cdot||$ on the Hilbert space, such that heuristically measure can be represented as follows:

$$d\tilde{w}^t \sim e^{-\frac{||x||^2}{2t}}[dx]. \tag{107}$$

However, this Gaussian measure is not $\sigma$-additive. In order to make it $\sigma$-additive, one has to consider a weaker norm $|\cdot|$ with certain conditions on it with respect to measure $d\tilde{w}^t$. Then one can consider a completion $\mathcal{H}$ with respect to the norm $|\cdot|$. This will give a Banach space $\mathcal{B}$. Now the Gaussian measure $d\tilde{w}^t$ can be extended to the Banach space, where it becomes $\sigma$-additive.

The important example, which we will consider in the next subsection, is constructed as follows. Let us take the Hilbert space $C_0'[0,2\pi]$ of real-valued absolutely continuous functions, such that $x(0)=0$ for any $x\in C_0'[0,2\pi]$ and $\int_0^{2\pi}(x'(u))^2du<\infty$. The inner product is given by: $\langle x_1,x_2\rangle=\int_0^{2\pi}x_1'(u)x_2'(u)du$. Then one can consider the weaker norm on this space: $|x|=\sup_{u\in[0,2\pi]}x(u)$. The completion of $C_0'[0,2\pi]$ with respect to $|\cdot|$ is the Banach space $C_0[0,2\pi]$ of real-valued continuous functions such that $x(0)=0$ for any $f\in C_0[0,2\pi]$.

It appears that the norm $|\cdot|$ satisfies all necessary properties and therefore there exists a $\sigma$-additive Gaussian measure $dw^t$ on $C_0[0,2\pi]$. This measure is called the classical Wiener measure. In the next subsection we will give its direct construction, using another approach.

**A.2. The construction of the classical Wiener measure and its basic properties.** Consider the space of continuous functions $C_0[0,2\pi]$. Let $C_{0,X}[0,2\pi]$ denote the closed subspace of $C_0[0,2\pi]$, consisting of real-valued continuous functions such that $x(2\pi)=X$ for some $X\in\mathbb{R}$.

The following subsets of $C_0[0,2\pi]$ are called cylinder sets:

$$\{x\in C_0[0,2\pi]: x(\tau_1)\in A_1,\dots,x(\tau_n)\in A_n, 0<\tau_1<\dots\tau_n\leq 2\pi\}, \tag{108}$$

where $A_1,A_2,\dots A_n$ are Borel subsets of $\mathbb{R}$.
The Wiener measure with variance $t\geq 0$ is defined on the cylindrical sets of $C_0[0,2\pi]$ as follows:

$$\begin{aligned}&w^t(x(\tau_1)\in A_1,\dots,x(\tau_n)\in A_n)=\\&\int_{A_1}\dots\int_{A_n}u_t(\Delta x_1,\Delta\tau_1)\dots u_t(\Delta x_n,\Delta\tau_n)dx_1\dots dx_n.\end{aligned} \tag{109}$$

The conditional Wiener measure of variance $t>0$ is defined on the cylinder sets of $C_{0,X}[0,2\pi]$ as follows:

$$\begin{aligned}&w_X^t(x(\tau_1)\in A_1,\dots,x(\tau_n)\in A_n)=\\&\int_{A_1}\dots\int_{A_{n-1}}u_t(\Delta x_1,\Delta\tau_1)\dots u_t(\Delta x_n,\Delta\tau_n)dx_1\dots dx_{n-1},\end{aligned} \tag{110}$$

where $u_t(x,s)=\frac{1}{2\pi\sqrt{ts}}e^{-\frac{x^2}{4\pi st}}$, $dx$ stands for the Lebesgue measure on $\mathbb{R}$, $\Delta x_k=x_k-x_{k-1}$, $\Delta\tau_k=\tau_k-\tau_{k-1}$, $x_0=0$, $\tau_0=0$. In the case ii) $x_n=X$, $\tau_n=2\pi$.

It appears that the resulting measures are $\sigma$-additive and moreover $w^t(C_0[0,2\pi])=1$, $w_X^t(C_{0,X}[0,2\pi])=u_t(X,2\pi)$.

One can show that the description of the Wiener measure that we gave in subsection A.1. and the one introduced in this subsection agree (see Section I.5. of

[**23**]). Now we want to relate $L^2$ spaces with respect to the Wiener measure and the conditional Wiener measure. In order to do that, notice the following property. If $f$ is an integrable function on $C_{0,X}([0,2\pi])$ for almost all $X$, then by the definition of the Wiener measure one has:

$$\int_{C_0[0,2\pi]} f(x)dw^t(x) = \int_{\mathbb{R}} \Big( \int_{C_{0,X}[0,2\pi]} f(x)dw^t_X(x) \Big) dX. \tag{111}$$

Therefore, the Hilbert space of square-integrable functions with respect to $dw^t$ decomposes as a direct integral:

$$L^2(C_0[0,2\pi], dw^t) = \int_{\mathbb{R}}^{\oplus} L^2(C_{0,X}([0,2\pi]), dw^t_X)dX. \tag{112}$$

The Wiener measure and the conditional Wiener measure have translation property. Let us describe this property in detail.

Let $f$ be an integrable function on $C_0([0,2\pi])$ and $y \in C_0'[0,2\pi]$, then

$$\begin{aligned} &\int_{C_0[0,2\pi]} f(x)dw^t(x) = \\ &\int_{C_0[0,2\pi]} f(x+y)e^{-\frac{1}{t}\int_0^{2\pi} y'(u)dx(u) - \frac{1}{2t}\int_0^{2\pi} y'(u)y'(u)du} dw^t(x), \end{aligned} \tag{113}$$

where $\int_0^{2\pi} y'(u)dx(u)$ is the Stieltjes integral.

If $f$ is an integrable function in $C_{0,X}([0,2\pi])$ and $y$ as above, such that $y(2\pi) = Y$ then,

$$\begin{aligned} &\int_{C_{0,X+Y}[0,2\pi]} f(x)dw^t_{X+Y}(x) = \\ &\int_{C_{0,X}[0,2\pi]} f(x+y)e^{-\frac{1}{t}\int_0^{2\pi} y'(u)dx(u) - \frac{1}{2t}\int_0^{2\pi} y'(u)y'(u)du} dw^t_X(x). \end{aligned} \tag{114}$$

The space of Wiener measure translations, i.e. $C_0'[0,2\pi]$ is usually called the Cameron-Martin space.

One can define a unitary operator on $L^2$ space for any abstract Wiener measure, which is similar to the Fourier transform. We will need it only in the case of measure $dw^t_0$ on $C_{0,0}[2\pi]$. The formula is as follows:

$$Ff(y) = \int_{C_{0,0}[0,2\pi]} f(x+iy)dw^{2t}_0(x). \tag{115}$$

Here $f \in L^2(C_{0,0}[0,2\pi], dw^t_0)$. The inverse transformation is given by the changing sign of $y$ in the formula above.

**A.3. Gaussian integration on Hilbert spaces.** In this subsection we recall a few basic facts and formulas, for more information, see e.g. [**3**], [**23**]. Suppose we have a real separable Hilbert space $H$ with the orthonormal basis $\{e^i\}$, $i \in \mathbb{N}$ and the pairing $\langle \cdot, \cdot \rangle$. Every element $x$ of this Hilbert space can be expressed as $x = \sum_i x_i e^i$. Let us introduce positive real numbers $\lambda_i$, $i \in \mathbb{N}$, so that $\sum_i \lambda_i < \infty$. Then one can say that the numbers $\lambda_i$ define a diagonal trace class operator $A$ on our Hilbert space. Than it appears possible to define the sigma-additive measure

$d\mu_A$ on $H$ and heuristically express it as follows:

$$d\mu_A(x) = (\sqrt{\det 2\pi A})^{-1} \cdot e^{-\frac{1}{2}\langle x, A^{-1}x\rangle}[dx], \tag{116}$$

which can be thought of as the infinite product of 1-dimensional Gaussian measures for each $i$: $d\mu_i = \sqrt{2\pi\lambda_i}^{-1} e^{-\lambda_i^{-1}x_i^2}$.

Since it is a sigma-additive measure, one can define the space of square-integrable functions $L^2(H, d\mu_A)$ with respect to it. One of the basic formulas is the translational shift in the measure. Namely, if $b \in ImA$, then

$$\int f(x)d\mu_A(x) = \int f(x+b)e^{-\frac{1}{2}\langle b, A^{-1}b\rangle - \langle x, A^{-1}b\rangle} d\mu_A(x). \tag{117}$$

One can consider also an infinitesimal version of this formula. Making $b$ infinitesimal and parallel to $e_i$, we obtain that

$$\int D_i f(x) d\mu_A(x) = 0, \quad D_i = \partial_{x_i} - \lambda_i^{-1}x_i, \tag{118}$$

if $\partial_{x_i} f(x) \in L^2(H, d\mu_A)$. It should be noted that the following monomials

$$(\prod_{i=1}^k \langle \alpha_i, x\rangle)e^{\langle \beta, x\rangle}, \tag{119}$$

where $\alpha_i, \beta$ are the elements of the complexified Hilbert space, are always integrable with respect to $d\mu_A$, moreover, they belong to $L^2(H, d\mu_A)$. There is an explicit formula for the integral of the function (119), which can be derived from the simple result:

$$\int e^{\langle \beta, x\rangle} d\mu_A(x) = e^{\frac{1}{2}\langle \beta, A\beta\rangle}. \tag{120}$$

## References


[1] V. V. Fok and L. O. Chekhov, *Quantum Teichmüller spaces* (Russian, with Russian summary), Teoret. Mat. Fiz. **120** (1999), no. 3, 511–528, DOI 10.1007/BF02557246; English transl., Theoret. and Math. Phys. **120** (1999), no. 3, 1245–1259. MR1737362 (2001g:32034)

[2] J. B. Conway, *A course in functional analysis*, Graduate Texts in Mathematics, vol. 96, Springer-Verlag, New York, 1985. MR768926 (86h:46001)

[3] G. Da Prato, *An introduction to infinite-dimensional analysis*, Universitext, Springer-Verlag, Berlin, 2006. Revised and extended from the 2001 original by Da Prato. MR2244975 (2009a:46001)

[4] T. Dimofte, *Quantum Riemann surfaces in Chern-Simons theory*, Adv. Theor. Math. Phys. **17** (2013), no. 3, 479–599. MR3250765

[5] P. I. Etingof, I. B. Frenkel, and A. A. Kirillov Jr., *Lectures on representation theory and Knizhnik-Zamolodchikov equations*, Mathematical Surveys and Monographs, vol. 58, American Mathematical Society, Providence, RI, 1998. MR1629472 (2001b:32028)

[6] L. Faddeev, *Modular double of a quantum group*, Conférence Moshé Flato 1999, Vol. I (Dijon), Math. Phys. Stud., vol. 21, Kluwer Acad. Publ., Dordrecht, 2000, pp. 149–156. MR1805888 (2002a:81126)

[7] M. Florig and S. J. Summers, *Further representations of the canonical commutation relations*, Proc. London Math. Soc. (3) **80** (2000), no. 2, 451–490, DOI 10.1112/S0024611500012259. MR1734324 (2001g:81132)

[8] I. B. Frenkel, *Orbital theory for affine Lie algebras*, Invent. Math. **77** (1984), no. 2, 301–352, DOI 10.1007/BF01388449. MR752823 (86d:17014)

[9] I. B. Frenkel and I. C. H. Ip, *Positive representations of split real quantum groups and future perspectives*, Int. Math. Res. Not. IMRN **8** (2014), 2126–2164. MR3194015

[10] I. B. Frenkel and H. K. Kim, *Quantum Teichmüller space from the quantum plane*, Duke Math. J. **161** (2012), no. 2, 305–366, DOI 10.1215/00127094-1507390. MR2876932

[11] I. B. Frenkel and A. M. Zeitlin, *On the continuous series for* $\widehat{sl(2,\mathbb{R})}$, Comm. Math. Phys. **326** (2014), no. 1, 145–165, DOI 10.1007/s00220-013-1832-9. MR3162487

[12] I. B. Frenkel and Y. Zhu, *Vertex operator algebras associated to representations of affine and Virasoro algebras*, Duke Math. J. **66** (1992), no. 1, 123–168, DOI 10.1215/S0012-7094-92-06604-X. MR1159433 (93g:17045)

[13] A. M. Veršik, I. M. Gel′fand, and M. I. Graev, *Representations of the group* SL(2, **R**), *where* **R** *is a ring of functions* (Russian), Uspehi Mat. Nauk **28** (1973), no. 5(173), 83–128. MR0393337 (52 #14147)

[14] I.M. Gelfand, A.M. Yaglom, *Integration in functional spaces with applications to quantum physics*, J. Math. Physics **1** (1969) 48-96.

[15] A. Gerasimov, S. Kharchev, and D. Lebedev, *Representation theory and quantum integrability*, Infinite dimensional algebras and quantum integrable systems, Progr. Math., vol. 237, Birkhäuser, Basel, 2005, pp. 133–156, DOI 10.1007/3-7643-7341-5_4. MR2160844 (2006j:17016)

[16] A. Gerasimov, S. Kharchev, D. Lebedev, S. Oblezin, *On a Class of Representations of Quantum Groups*, arXiv:math/0501473.

[17] T. Hida, *Brownian motion*, Applications of Mathematics, vol. 11, Springer-Verlag, New York-Berlin, 1980. Translated from the Japanese by the author and T. P. Speed. MR562914 (81a:60089)

[18] I. C. H. Ip, *The classical limit of representation theory of the quantum plane*, Internat. J. Math. **24** (2013), no. 4, 1350031, 40, DOI 10.1142/S0129167X13500316. MR3062971

[19] I.C.H. Ip, *Positive Representations of Split Real Simply-laced Quantum Groups*, arXiv:1203.2018.

[20] I. C.-H. Ip, *Positive representations of split real quantum groups*, Symmetries and groups in contemporary physics, Nankai Ser. Pure Appl. Math. Theoret. Phys., vol. 11, World Sci. Publ., Hackensack, NJ, 2013, pp. 175–180, DOI 10.1142/9789814518550_0020. MR3221459

[21] D. Kazhdan and G. Lusztig, *Tensor structures arising from affine Lie algebras. I, II*, J. Amer. Math. Soc. **6** (1993), no. 4, 905–947, 949–1011, DOI 10.2307/2152745. MR1186962 (93m:17014)

[22] S. Kharchev, D. Lebedev, and M. Semenov-Tian-Shansky, *Unitary representations of* $U_q(\mathfrak{sl}(2,\mathbb{R}))$, *the modular double and the multiparticle q-deformed Toda chains*, Comm. Math. Phys. **225** (2002), no. 3, 573–609, DOI 10.1007/s002200100592. MR1888874 (2003b:17019)

[23] H. H. Kuo, *Gaussian measures in Banach spaces*, Lecture Notes in Mathematics, Vol. 463, Springer-Verlag, Berlin-New York, 1975. MR0461643 (57 #1628)

[24] Y. J. Lee, *Applications of the Fourier-Wiener transform to differential equations on infinite-dimensional spaces. I*, Trans. Amer. Math. Soc. **262** (1980), no. 1, 259–283, DOI 10.2307/1999982. MR583855 (82k:35098)

[25] B. Ponsot and J. Teschner, *Clebsch-Gordan and Racah-Wigner coefficients for a continuous series of representations of* $\mathcal{U}_q(\mathfrak{sl}(2,\mathbb{R}))$, Comm. Math. Phys. **224** (2001), no. 3, 613–655, DOI 10.1007/PL00005590. MR1871903 (2005a:17011)

[26] B. Ponsot, J. Teschner, *Liouville bootstrap via harmonic analysis on a noncompact quantum group*, hep-th/9911110.

[27] G. Moore and N. Seiberg, *Classical and quantum conformal field theory*, Comm. Math. Phys. **123** (1989), no. 2, 177–254. MR1002038 (90e:81216)

[28] K. Schmüdgen, *Operator representations of* $\mathcal{U}_q(\mathrm{sl}_2(\mathbf{R}))$, Lett. Math. Phys. **37** (1996), no. 2, 211–222, DOI 10.1007/BF00416024. MR1391203 (97d:17010)

[29] J. Teschner, *A lecture on the Liouville vertex operators*, Proceedings of 6th International Workshop on Conformal Field Theory and Integrable Models, Internat. J. Modern Phys. A **19** (2004), no. May, suppl., 436–458, DOI 10.1142/S0217751X04020567. MR2087125 (2005g:81281)

[30] H. Verlinde, *Conformal field theory, two-dimensional quantum gravity and quantization of Teichmüller space*, Nuclear Phys. B **337** (1990), no. 3, 652–680, DOI 10.1016/0550-3213(90)90510-K. MR1057726 (91c:81149)

[31] N. Ja. Vilenkin, *Special functions and the theory of group representations*, Translated from the Russian by V. N. Singh. Translations of Mathematical Monographs, Vol. 22, American Mathematical Society, Providence, R. I., 1968. MR0229863 (37 #5429)

[32] E. Witten, *Quantum field theory and the Jones polynomial*, Comm. Math. Phys. **121** (1989), no. 3, 351–399. MR990772 (90h:57009)

[33] A. M. Zeitlin, *Unitary representations of a loop* $ax+b$ *group, Wiener measure and* $\Gamma$*-function*, J. Funct. Anal. **263** (2012), no. 3, 529–548, DOI 10.1016/j.jfa.2012.05.004. MR2925913

Department of Mathematics, Columbia University, 2990 Broadway, New York, New York 10027 – and – IPME RAS, V.O. Bolshoj pr., 61, 199178, St. Petersburg

*E-mail address*: zeitlin@math.columbia.edu

*URL*: http://math.columbia.edu/~zeitlin; http://www.ipme.ru/zam.html

PUBLISHED TITLES IN THIS SERIES




ISBN 978-1-4704-1844-1

9 781470 418441

PSPUM/92